AF615201

ADVANCES IN LASER SCIENCE–I

AMERICAN INSTITUTE OF PHYSICS
CONFERENCE PROCEEDINGS NO. 146
NEW YORK 1986

OPTICAL SCIENCE AND ENGINEERING SERIES 6

SERIES EDITOR: RITA G. LERNER

ADVANCES IN LASER SCIENCE–I

PROCEEDINGS OF THE FIRST INTERNATIONAL
LASER SCIENCE CONFERENCE

DALLAS, TX 1985

EDITORS:
WILLIAM C. STWALLEY
UNIVERSITY OF IOWA

MARSHALL LAPP
SANDIA NATIONAL LABORATORIES

L.C. Catalog Card No. 86-71536
ISBN 0-88318-345-5
DOE CONF-8511166

Printed in the United States of America

Preface

Laser Science is an emerging technical area with a strong interdisciplinary flavor. It is based on a wide range of traditional interest areas, including (but undoubtedly not limited to) atomic and molecular physics, chemical physics, condensed matter physics, optical physics and engineering, plasma physics, physical chemistry, photochemistry, materials science and engineering, electrical engineering, gaseous electronics, quantum electronics, and electro-optics. At the core of laser science are the mechanisms of the lasers themselves and the interaction of the laser photons with matter (spectroscopy and photoprocesses). Surrounding this core is the wide spectrum of scientific applications of lasers, not only in the disciplines mentioned above, but also in virtually every other area of science and technology. The primary purpose of the International Laser Science Conference (ILS) is to survey annually both the laser and spectroscopy/photoprocesses core areas and a wide variety of selected scientific applications of lasers. Secondary goals include improved cross fertilization among the areas listed above and improved international scientific communication.

The first such ILS Conference (ILS—I) was held at the University of Texas at Dallas Conference Center, November 18-22, 1985. The conference was a Topical Conference of the American Physical Society, and was designated the Annual Meeting of the newly formed Topical Group on Laser Science. Carl B. Collins (University of Texas at Dallas) was Conference Chair, and also headed the local organizing efforts, with exemplary secretarial assistance by Lynda Horne and her colleagues. The Center itself provided an outstanding environment for the conference with both ready access to the meeting rooms and abundant space for informal discussion. Richard C. Powell (Oklahoma State University) was Conference Co-Chair, contributing administrative expertise, especially with financial arrangements. Rolf Gross (Aerospace Corporation) was International Co-Chair, a post designed to aid international participation in the conference. His diligent efforts provided a solid international base for this first meeting. The program was assembled by us, with generous advice from the Program Committee and the tireless efforts of those members of the Program Committee who agreed to organize sessions. Receipt, compilation, correction, and acknowledgement of abstracts and assembly of the Program (printed in the Bulletin of the American Physical Society) was done by the expert and dedicated secretarial staff at the University of Iowa Laser Facility, headed by Lynn Borders, the ILS—II Administrative Assistant. Generous support for ILS—I was provided by the Air Force Office of Scientific Research, the Army Research Office, the National Science Foundation, the University of Texas at Dallas, and the University of Iowa.

The conference consisted of five parallel sessions over five days and included four outstanding Plenary Talks. Poster sessions (including many postdeadline papers), which allowed for greater individual discussion, were presented late Tuesday. Session organizers were encouraged to make thoughtful development of session topics a prime consideration. Contributed talks were included only when they meshed well with invited and overview talks.

The speakers at the ILS Conference were given instructions for preparation of the brief camera-ready manuscripts at the meeting. They did, with few exceptions, an outstanding job (as they had in their oral and poster presentations), thereby reducing the editorial burden and also the reviewing burden, borne for the most part by the Session Organizers (listed in the Table of Contents) and the International Co-Chair. The papers here are organized by subject, rather than chronologically, with the poster papers in some cases being rather arbitrarily assigned. Final responsibility for the physical assembly of the 258 papers in this volume went to Lynn Borders, whom we thank most sincerely for an exceedingly impressive and heroic job.

William C. Stwalley
University of Iowa,
Iowa City
Program Chair
International Laser Science Conference—I

Marshall Lapp
Sandia National Laboratories,
Livermore
Program Co-Chair

CONTENTS

I. PANEL: HAVE ALL THE BEST LASERS BEEN DISCOVERED?

II. GAMMA RAY LASERS

III. X-Ray Lasers

Session organized by P. Hagelstein, Lawrence Livermore National Laboratory

IV. Visible and Near Visible Lasers and Novel Laser Concepts

Sessions organized by F. K. Tittel, Rice University; C. M. Tang and P. Sprangle, Naval Research Laboratory; and M. E. Koch, Vought Corporation

VII. Theory

Sessions organized by D. W. Noid, Institute for Defense Analyses; and L. Narducci and J. M. Yuan, Drexel University

VIII. Atomic Spectroscopy

Sessions organized by M. H. Nayfeh, University of Illinois, Urbana

IX. Molecular Spectroscopy

Sessions organized by W. S. Warren, Princeton University; and R. J. Saykally, University of California, Berkeley

X. Laser Chemistry, Ionization, Dissociation, and Collision Dynamics

Sessions organized by P. M. Dehmer, Argonne National Laboratory; J. L. Gole, Georgia Institute of Technology; P. D. Kleiber, University of Iowa; and D. R. Crosley, SRI International

XI. Ultrafast Spectroscopy

Sessions organized by G. A. Kenney-Wallace, University of Toronto

XII. Laser Diagnostics, Analysis, and Sensing

Sessions organized by R. K. Hanson, Stanford University; D. M. Lubman, University of Michigan; R. M. Measures, University of Toronto; A. C. Tam, IBM Research Laboratories; and D. K. Killinger, MIT Lincoln Laboratory

XIII. LASERS IN BIOLOGY AND MEDICINE

Sessions organized by J. LoCicero, Northwestern University; M. W. Berns, University of California, Irvine; R. M. Hochstrasser, University of Pennsylvania; and C. K. Johnson, University of Kansas

XIV. LASER INTERACTIONS WITH SURFACES AND PARTICULATES

Sessions organized by R. P. Van Duyne, Northwestern University; R. K. Chang, Yale University; W. Kiefer, Universitaet Graz, Austria; N. Winograd, Pennsylvania State University; and T. F. George, State University of New York at Buffalo

XV. Potpourri

Have All the Best Lasers Been Discovered?

Panel Discussion held at the International Laser Science Conference

Alexander J. Glass, Moderator

Four speakers comprised the panel which discussed the question, "Have all the best lasers been discovered?" They were Dr. George Baldwin of the Los Alamos National Laboratory (LANL), Dr. William Krupke of the Lawrence Livermore National Laboratory (LLNL), Dr. C. R. Jones also from LANL, and Professor Charles Rhodes of the University of Illinois, Chicago Circle Campus. The discussion leader was Dr. Alexander J. Glass of KMS Fusion, Inc.

Dr. Krupke examined the question posed to the panel. He pointed out that the "best" laser is the one which most effectively and economically satisfies the needs of a particular application. For several industrial, research, and military applications, lasers operating today fill the needs very well, and are unlikely to be replaced by newer devices. These include Nd:YAG, CO_2, GaAs, etc., in particular applications.

The full range of possible laser applications is far from realized, however. For these, the "best" laser clearly has not yet been discovered. These new lasers may require entirely new excitation mechanisms or new gain media, or may simply include already existing pumping techniques and gain media in a new configuration.

Dr. Krupke, and subsequent speakers, noted a trend away from the "discovery" of new lasers toward the "design" or "invention" of new lasers. It is now possible to design laser properties to satisfy a given application, rather than relying on serendipity in the random exploration implicit in the term "discovery".

Dr. Krupke supported this view with the following examples drawn from solid-state laser technology:

> "A conventional semiconductor laser acquires its characteristics directly from the electronic structure of the bulk lattice formed by the constituent elements. The recently acquired ability to fabricate semiconductor structures atomic-layer by atomic-layer, using novel molecular beam epitaxy and organometallic chemical vapor disposition techniques, has permitted the formation of <u>super-lattice</u> structures possessing quantum electronic properties differing markedly from those of the bulk semiconductor; as a result, it is expected that efficient visible wavelength lasers, comparable in performance to today's infrared lasers, can be tailored using these basic innovations in solid state physics.

These same advances in fabrication techniques now permit growth of semiconductor layers over wafer-scale areas with sufficient electrical and optical uniformity that 1-D (bar) and 2-D arrays of high power laser diodes can be built. By clever design, the array of diodes may be locked in phase giving rise to a revolutionary type of high power (~1000 W/cm^2 instantaneous, ~100 W/cm^2 average) high efficiency (> 30%) coherent source of exceptional compactness.

In the field of dielectric solid state lasers, a Renaissance of innovation not seen since the early 1960s began a few years ago. The reasons: (1) the recent introduction of new laser materials exhibiting significantly increased efficiency and power scalability compared to earlier materials (e.g. Nd:Cr:GSGG); (2) demonstration of a variety of broadly tunable solid state lasers (e.g., alexandrite, emerald, and transition-metal-ion lasers based on Ti, V, Co, Ni ions); (3) advances in the design and development of slab and flow-cooled disk laser devices capable of being scaled to extremely high average power with good beam quality; (4) the above-mentioned development of 1-D and 2-D high power semiconductor laser diode arrays suitable for resonantly pumping energy-storage solid state lasers.

By combining recent empirical and fundamental advances in solid state physics with innovations in excitation techniques, fabrication techniques, and novel device geometries, it should be possible, as never before, for valued applications to drive laser R&D toward those physics systems and techniques most likely to satisfy the laser requirements."

Dr. Jones took a slightly different point of view. He characterized the process of invention as being driven by "creativity and the challenge of invention, without regard to ultimate applications." He contrasted this step to development, which is carried out with particular applications in mind, and represents a long, dedicated, and often expensive process. Both invention and development are required to create a "good" (useful and affordable) laser technology.

Dr. Jones surveyed the field of laser technology, and identified several areas in which he anticipates significant growth and new developments. Two were major topics of discussion at the Conference: x-ray lasers and gamma-ray lasers. Both are in the invention state. Their potential applications are not even clearly identified at this time, and lie far in the future.

In the nearer term, Dr. Jones foresees the development of two existing types of lasers: free-electron lasers (FEL) and semiconductor diode lasers (SDL). He commented as follows:

"Free-electron laser development is driven primarily by military interest today, but the versatility of this laser is too great for civilian users to not play larger roles. Electron accelerator advancements, especially toward more compact structures, will be a major force in FEL development. Hence, in a single device, broad tuning ranges at moderate optical powers will be produced in a laboratory-size device. Many different users sharing this goal will spur on FEL development. Much progress will be made within the next 10 years, but closer to 20 will be required to meet all the goals of wide tunability, compactness, high efficiency, and simplicity of operation.

Despite major progress over the past decades, semiconductor lasers will continue to advance in power output, wavelength, versatility, and applications. The high efficiency and compactness of these devices is extremely appealing to many users. The military will be a major player in developing both SDL-pumped solid state lasers and high-power arrays of SDLs. Industrial and commercial applications at moderate power levels will conspire with military uses to rapidly accelerate development in these SDL configurations."

Dr. Jones also envisions the continuing development of excimer lasers, for a broad class of military, commercial and industrial uses, and of chemically-excited lasers (HF, iodine, and visible systems still undiscovered), primarily for military use. He stated that the simplicity of electrically-driven systems makes them preferable for industrial uses. He agreed with Dr. Krupke that optically-pumped, solid-state lasers would advance, with the introduction of new laser host materials.

In brief comments, Professor Rhodes proposed that advances in laser technology often have been paced by advances in excitation techniques. Thus, the development of CO_2 laser technology was spurred by the advance from glow discharge excitation, to the gas dynamic laser, and subsequently to electron-beam excitation. For solid-state lasers, the replacement of flashlamp pumping with diode pumping represents a similar advance. Professor Rhodes commented that new excitation technologies remain to be discovered, as do entirely new laser mechanisms. He further commented that new research applications, like the use of x-ray lasers in biological science, would provide a further impetus to development.

Dr. Baldwin approached the topic of discussion from a broader historical perspective. He examined the progress made since the mid-1800s in sources of coherent electromagnetic radiation. The accompanying figure, taken from Reference 1, shows a steady advance in generating higher frequencies, culminating with the development of the visible-light laser. However, progress to shorter wavelengths is seen to be slower. Dr. Baldwin asked, why is this so? He pointed out that our current lasers are still far from any fundamental limitation on short wavelength generation, but that past progress has resulted from radical innovation. He suggested that the progress of lasers to shorter wavelengths may depend on the introduction of new ideas from other, yet untapped, areas of physics and technology. Thus, the gamma-ray laser, if it is ever developed, may depend on the use of the ideas and techniques drawn from nuclear physics rather than from microwave or optical technology.

Dr. Baldwin identified several conceivable future developments, including the use of particle accelerators as excitation sources, muonic atoms, charge-annihilation, and other novel techniques to prepare population inversions. He discussed, at length, one possible x-ray laser scheme based on accelerator-pumping (Reference 2).

Dr. Baldwin closed with the statement that our progress in developing new lasers was limited by nontechnical factors rather than physical limitations. He commented as follows:

> "The main obstacles to progress are human, not physical: overspecialization, parochial attitudes, timidity, betting on sure things rather than taking a gamble, all act to inhibit progress. Moreover, the way we support R&D in this country does not encourage innovation. Projects that have a long term to payoff (such as my own graser field) must compete with short term, sure-thing programs. Frequent rejustification to management, which tends to think short-term, is an inhibiting factor. Another problem, ... is that authorities usually tend to be overconservative."

Dr. Baldwin pointed out that the history of the laser could have been different. He indicated that the laser could well have been invented by Ladenberg, in 1934, using technology available at the time. Furthermore, he introduced evidence that Soviet scientists contemplated the application of gamma-ray lasers to strategic defense before 1978. Dr. Baldwin commented that an overall, institutional objective like SDI could provide a much greater impetus to technology development than purely scientific needs.

In conclusion, all the panelists agreed that the "best" lasers have not all been invented. They indicated that as the technology matures, new lasers will be "designed" and "developed" with applications in mind, rather than found by the random discovery process. All of the panelists pointed to new technologies for excitation as a key part of laser development, and indicated that many of these technologies may come from other fields of science.

References

1. George C. Baldwin and Johndale C. Solem, "Is the Time Ripe? or Must We Wait So Long for Breakthroughs?", Laser Focus Magazine (June 1982).

2. George C. Baldwin, "Multistep Pumping Schemes for Short-Wave

Lasers", H. Hora and G. H. Miley, editors, Laser Interaction and Related Plasma Phenomena, Volume 6 (Plenum, New York, 1984), pages 107-123.

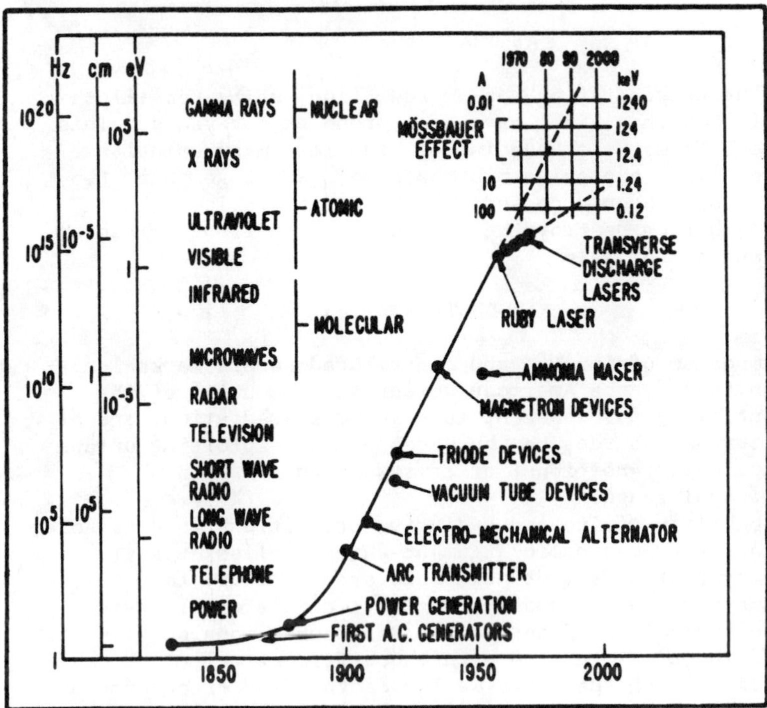

Fig 1 *Progress in the development of sources for coherent electromagnetic radiation.*

A CRITICAL REVIEW OF GAMMA-RAY LASER PROPOSALS*

George C. Baldwin
Los Alamos National Laboratory, Los Alamos NM 87545

ABSTRACT

Laser sources generating sub-nm radiation, using recoilless nuclear transitions in solids, have been proposed for years. This review examines, from the standpoint of kinetics, many solutions (viz., narrowed line, explosive neutron pump, two-stage pump, two-step pump) that have been proposed for the basic problem: that pumping can inhibit or destroy the Mossbauer and Borrmann effects, which are essential for gain.

INTRODUCTION

This symposium of invited and contributed papers marks long-overdue recognition by the American scientific community of the enormous potentiality for coherent sub-nanometer radiation, and of the possibility that sources can be developed by exploiting unique features of nuclear transitions, of crystals, and of this particular spectral range.[1]

The probability of induced radiative transitions is derivable by an elementary theromodynamic argument which applies at any energy to any type of system or multipole order. Thus, for Mossbauer gamma-ray lines (usually, magnetic dipole or electric quadrupole) in typical solid hosts, the resonance cross sections for nuclei are identical in form with those for optical transitions, except for the Debye-Waller factor, and often exceed the photoelectric absorption cross section by several orders of magnitude [Hanna].[2] In crystals, Bragg reflections create collimated radiation-field modes with greatly reduced absorption - the Borrmann effect [Post] - but enhanced interaction with resonant nuclei at lattice sites [Hannon]. Hence, an inverted nuclear population in a cool solid should amplify; mirrors are unnecessary. The problem is to establish a population inversion without destroying the cool crystal host essential to the Mossbauer and Borrmann effects. In this paper, we examine the major proposals for pumping grasers from the standpoint of their kinetic behavior.

To excite or stimulate deexcitation of any resonant system requires time, to define the frequency of the exciting vibration and the ratio of its bandwidth to the sharpness of resonance. Thus, the stimulation cross section is time-dependent. Atoms, emitting optical lines, respond rapidly to Doppler-broadened radiation. Resonant nuclei, on the other hand, respond slowly (on a time scale comparable with the upper-state lifetime, in the range 1 ns<T<10 us), because, for appreciable gain, the Mossbauer line must approach its homogeneous, natural linewidth.

UNPROMISING PROPOSALS

Early suggestions, to dope elongated crystals with radiochemically prepared long-lived nuclei ("isomers"), then cool or align the nuclear spins to increase the gain and thereby initiate lasing, were abandoned when it was realized that transitions of lifetime long enough to permit the preparation steps are inhomogenously broadened [Hanna, Hoy] to an extent that gain is impossible.

Subsequently, NMR techniques were suggested to reduce, average out or compensate three of the numerous interactions that broaden the nuclear line. They are complicated; none has been reduced to practice, despite obvious benefit to Mossbauer spectroscopy. We are not hopeful: the time needed to define the width of a narrowed line must always exceed the inverse linewidth to be achieved; even full initial inversion decays in only one half-life.

There are proposals to form highly excited compound nuclei by neutron capture; the ensuing radiative decay cascade might populate short-lived Mossbauer states. To maximize capture cross section and minimize damage to the host, the neutrons must be slow. Even then, enormous densities (~10^{22} cm^{-3}) and, therefore, explosive sources of neutrons are required. Neutron moderation times (and their fluctuations) exceed the lifetimes of typical Mossbauer states; thus, time-spread in the neutron pump is added to the resonant time-lag of the response of the decaying nuclei. Using the coupled Maxwell-Schroedinger equations with an appropriate time-dependence for the neutron capture rate in ^{82}Kr, we find that the 9.3-keV, 147-ns transition in ^{83}Kr has inappreciable gain unless the slow-neutron density exceeds 10^{22} cm^{-3} - but simple infinite-medium-case heat balance shows that one cannot moderate more than 5 x 10^{17} cm^{-3} fission neutrons (2.5 MeV) to the 40-eV ^{82}Kr capture resonance.

Another proposal would pump a three-level graser with Mossbauer radiation excited in a blanket by neutron capture. The additional kinetic delay and poor geometric efficiency (the graser filament can intercept only a small fraction of the pump radiation), dispose of this concept.

One might eliminate lower-state resonance absorption in a four-level scheme by polarizing the nuclei to be pumped. For example, in ^{161}Dy, line x-rays from a Ra target might invert a 75-keV m = -3/2 upper laser state with respect to substates of the 24-keV, I = -5/2 lower laser state via a 103-keV transition from an m = -7/2 ground state. High magnetic field and very low temperature are required; the necessary high pump power (not all the x-ray lines will be resonant) would inevitably overheat the Mossbauer medium.

Optically pumping the respective states into extreme hyperfine sublevels of opposite sign could also eliminate initial terminal-state resonance absorption. However, optical pumping is too slow to polarize the previously excited states before they have decayed.

TWO-STEP PUMPING

The best proposal for reducing pumping power is to incorporate a "storage isomer" into a host, then, with a small additional amount of energy, induce transfer to a level that can lase. Its feasibility depends upon:

1) Finding a nuclide with a short-lived level only slightly higher than the storage level. One known pair of nuclear states is separated by only 73 eV--if others exist, how do we find them [Martin, Strottman, Dietrich, Gove, Haight, Collins, Yaakobi]?

2) Producing and separating enough active isomer from other nuclear reaction products. This problem is soluble, at least in particular cases [Dyer].

3) Devising a rapid and efficient transfer process. Transfer by resonant radiation (the essence of Mossbauer spectroscopy!) is clearly feasible - but is it fast enough, and how do we generate it, or is there a better way [Solem, Rinker, Biedenharn, Reiss, Wender]? To enhance its rate of absorption by what will probably be a higher-multipole transition, the active nuclei in the graser host should be disposed in layers, so as to form a superlattice, spaced to enable Bragg reflections of the transfer radiation at the same angle as the radiation to be stimulated.

Assuming that nearly complete inversion can be established by interlevel transfer while preserving host integrity, Dicke superradiance[3] in Borrmann channels [Feld] provides an efficient, fast mechanism for creating a multi-beam coherent output pulse. Table I illustrates the kinetics for a hypothetical case, computed using relations derived by Feld[4] and explained in his paper. Note the high ratio of in-beam to off-beam radiation; also, that less than 10^{14} active nuclei need be created in the transfer step to generate a pulse carrying several megawatts of peak power. Whether it is possible to implant them into the host material, cubic boron nitride (an artificial crystal of exceptionally high Debye temperature) and then transfer from 40-h ^{133m}Ba to invert the 8-ns Mossbauer level is, of course, unknown.

TABLE I: SUPERRADIANCE OF ^{133}Ba IN BORAZON

Nuclide	Host
^{133}Ba	Cubic BN
Storage level, 40-h, 288 keV	Debye temp. 1700
Upper laser level, 8-ns, 12.3 keV	Bragg mode 222
Lower laser level, 10.7-y, 0 keV	

Performance			
Length, mm	5	10	20
Diameter, um	0.80	1.13	1.60
Active nuclei (times 10^{13})	2.4	5.0	10.4
Peak Power, Mw	17.2	35.9	75.1
Pulse Width, ns	0.68	0.67	0.65
Energy Ratio, on/off axis, times 10^9	3.9	7.7	15.0

CONCLUSIONS

To summarize, although solving the interdisciplinary problem of stimulating recoilless nuclear transitions will not be easy, the expected performance and many potential applications of a superradiant graser featuring anomalous emission of Mossbauer radiation into Borrmann modes from nuclei excited by gentle but fast transfer from a long-lived storage isomer, probably distributed to form a superlattice, justify a coordinated research program featuring identification of suitable nuclides, development of methods for preparing and incorporating pure storage isomer in appropriate single-crystal hosts of suitable geometry, and for rapid interlevel transfer that will not inhibit the Mossbauer and Borrmann effects.

REFERENCES

* Supported by the Division of Advanced Energy Projects, U. S. Department of Energy.

1. G. C. Baldwin, J. C. Solem, V. I. Gol'danskii, Rev.Mod. Phys. 53, 687 (1981).
2. Bracketed names refer, by first author, to other papers in this symposium.
3. R. H. Dicke, Phys. Rev. 93, 99 (1954); J. H. Terhune, G. C. Baldwin, Phys. Rev. Lett. 14, 589 (1965).
4. J. C. MacGillivray, M. S. Feld, Phys. Rev. A14, 1163 (1976).

ENSDF: THE EVALUATED NUCLEAR STRUCTURE DATA FILE

M. J. Martin
Oak Ridge National Laboratory, Oak Ridge, TN 37831*

ABSTRACT

The structure, organization, and contents of the Evaluated Nuclear Structure Data File, ENSDF, will be discussed. This file summarizes the state of experimental nuclear structure data for all nuclei as determined from consideration of measurements reported world wide. Special emphasis will be given to the data evaluation procedures and consistency checks utilized at the input stage and to the retrieval capabilities of the system at the output stage.

INTRODUCTION

The Nuclear Data Project, NDP, is a data evaluation center within the Physics Division at ORNL. We are a member of the international Nuclear Structure and Decay Data network, NSDD, which consists of 15 centers in 10 countries. Other US centers are located at Brookhaven National Laboratory, Lawrence Berkeley Laboratory, Idaho National Engineering Laboratory, and the University of Pennsylvania. Responsibility for coordination of the network resides with the National Nuclear Data Center at BNL.

The NSDD has responsibility for the evaluation of nuclear structure and decay data for all nuclides. The centers evaluate measurements of quantities such as excitation energies, spins and parities, half-lives, radiation energies and intensities, and static moments. Regular publication of the evaluations takes place in the journal Nuclear Data Sheets[1] for $A > 45$ and in the journal Nuclear Physics for $A \leq 45$.

Our responsibility within this network is two-fold. First, we are responsible for the evaluation of mass chains with $A > 195$ (excluding even $A = 238$ to 244). Second, the NDP houses the position of Editor-in-Chief of the Nuclear Data Sheets and so has the ultimate responsibility for the quality of the contents of ENSDF and, consequently, the quality of what is published in the Nuclear Data Sheets.

STRUCTURE OF ENSDF

In 1971, the NDP designed a formal structure for entering data into a computer file.[2] This structure has since been adopted by the international network.

The ENSDF file consists of "data sets", each of which summarizes the results from all relevant papers on one type of experiment. Each of these data sets is a collection of records for the levels

*Operated by Martin Marietta Energy Systems, Inc., for the US Department of Energy.

and radiations observed in the experiment. In addition to the "condensation" involved in the preparation of a data set, the properties of levels and radiations for each nucleus, based on all pertinent data sets, are further summarized in a data set labelled "Adopted Levels, Gammas". Examples of these data sets can be seen in any issue of the Nuclear Data Sheets.

Documentation for the quantities contained in the file consists of references to the published or informal scientific literature, as well as comments by the evaluator explaining the choice of "best" or "adopted" experimental numbers.

EVALUATION PROCEDURES

Work on a given mass chain begins with a reference list containing references for all available articles on nuclei with that mass. This reference list is prepared at BNL by scanning all the pertinent published literature and all the available secondary sources (theses, conference reports, laboratory reports, etc.). The coverage of the published literature is virtually complete.

For each reaction or decay, the evaluator then reads all the papers, evaluates the contents, extracts the pertinent data, intercompares such data from each source, and finally adopts a value, with an uncertainty, for each quantity to be included in the ENSDF file. To arrive at an adopted value, the evaluator may choose to take a weighted (or unweighted) average of all the experimental numbers or, at the other extreme, to take the result of one paper if the value from that paper is judged to be clearly superior to the others.

In the course of extracting data, the evaluator must check for the possible effect of more recent reliable energy standards, changes in the accepted values of physical "constants," or in reference values. For example, earlier $E\gamma$ data may need to be corrected for the presently accepted ^{198}Au 411 energy; extraction of a Q value from $\varepsilon_K/\varepsilon_L$ may have used a value for the fluorescent yield which has subsequently been revised; a g factor may have been deduced using a half-life for which a better value is presently available. In all cases where changes are made, such changes are carefully documented.

When all data sets yielding information on a given nucleus are completed, the evaluator then prepares the Adopted data sets. Bases for the adopted values are given, and arguments for spin and parity assignments based on established rules are given. In the preparation of all data sets, the evaluator makes use of an extensive set of analysis and checking programs. These range from straightforward calculational programs such as those for internal conversion coefficients, log ft's, and least-squares evaluation of level energies and β feedings in a radioactive decay data set, to programs that check adopted $J\pi$ values against L values in reactions and against gamma-ray multipole character. All analysis and checking programs operate directly on the data sets, and the output from, for example, the conversion coefficient and log ft programs, can if desired be automatically fed back into the data sets. At all stages following data entry, the hand copying of numbers is kept to a minimum.

isomers with $T_{1/2}$>1 s

Nucleus	E(level)	$T_{1/2}$	Jπ
^{171}Lu	71.3 2	79 s 2	1/2-
^{172}Lu	41.86	3.7 m 5	(1-)
^{174}Lu	170.86 4	142 d 2	(6-)
^{176}Yb	1050.7 7	11.4 s 5	(8)-
^{176}Lu	126.50	3.68 h 1	1-
^{177}Yb	331.5 3	6.41 s 2	1/2-
^{177}Lu	970.15 5	160.9 d 3	23/2-
^{177}Hf	1315.4 1	1.08 s 6	23/2+
	2740.0 3	51.4 m 5	37/2-
^{178}Lu	≈300	22.7 m 4	(7,8,9)
^{178}Hf	1147.44	4.0 s 2	8-
	≈2500	31 y 1	(16,17+)
^{179}Hf	374.8	18.68 s 6	(1/2)-
	1105.7	25.1 d 3	
^{179}W	221.9	6.7 m 3	(1/2-)
^{180}Hf	1141.62	5.5 h 1	8-
^{180}Ta	≈32	8.1 h 1	1+
^{181}Os	0.0+X	105 m 3	1/2-
^{182}Hf	1172.9	61.5 m 15	(8-)
^{182}Ta	519.7 4	15.84 m 10	10-
^{183}W	309.491 4	5.15 s 3	(11/2)+
^{183}Os	170.72 9	9.9 h 3	(1/2-)
^{184}Re	188.01 4	165 d 5	8+

gammas from isomers with $T_{1/2}$>1 s

Nucleus	E_γ	E(level)	I_γ	α	level $T_{1/2}$
^{171}Lu	71.1 2	71.3		484	79 s 2
^{174}Lu	59.08 2	170.86	4.4 2	3320	142 d 2
	126.2	170.86	6.4 25	266	142 d 2
^{176}Yb	96.1 3	1050.9	79 8	0.382	11.4 s 5
^{177}Yb	227.0 2	331.5		7.13	6.41 s 2
^{177}Lu	115.83 4	970.15	5.0 5	31.3	160.9 d 3
(^{177}Hf)	(14.16)	1315.4			1.08 s 6
	(55.15 2)	1315.4			1.08 s 6
^{177}Hf	214.0	2740.0	269 25	1.535	51.4 m 5
	228.44 6	1315.4	318 24	0.1874	1.08 s 6
^{178}Hf	88.88	1147.44		0.492	4.0 s 2
^{179}Hf	21.03	1105.7		1.196×10^4	25.1 d 3
	160.7	374.8		35.0	18.68 s 6
	257.32	1105.7		0.678	25.1 d 3
	374.8	374.8			18.68 s 6
^{179}W	101.6	221.5		1353	6.7 m 3
	221.5	221.5		10.47	6.7 m 3
^{180}Hf	57.549	1141.62		0.299	5.5 h 1
	500.714	1141.62		0.0618 5	5.5 h 1

Fig. 1. Isomeric levels and gammas with $T_{1/2}$ > 1 s.

When the mass chain is finished and the evaluator has submitted it for publication, it is subjected to a thorough review. The final version of the mass chain then appears in the Nuclear Data Sheets and is entered into the ENSDF file. Once in the file, the data in that mass chain become part of an extensive bank of data that can be retrieved in a variety of forms.

OUTPUT FROM ENSDF

As of June 1985, ENSDF contained 1948 Adopted Levels data sets, 2396 decay data sets, and 4257 reaction data sets. Information in these data sets is retrievable by nucleus (A, Z, and/or N), by class (odd-A, even-N = 200-214, etc.), by reaction or decay, or by reference. The output can be of the form of level-scheme drawings,[1] tables of nuclear properties,[1] tables of atomic and nuclear radiations,[3] or computer files of nuclear and radiation properties. See further references[4] for other examples of output from ENSDF.

Of particular interest to attendees at this talk might be a listing of nuclear isomers within a given range of half-lives. Figure 1 is a subset of a listing of isomers with $T_{1/2} > 1$ s and the properties of the γ's from these isomers.

This has been a brief overview of the Evaluated Nuclear Structure Data File, ENSDF, with a few examples of those types of information that can be extracted from that file. Those of you who may be interested in accessing the file or who would like more information on the system are invited to contact me or the National Nuclear Data Center at BNL.

REFERENCES

1. Nuclear Data Sheets, published by Academic Press, New York.
2. W. B. Ewbank and M. R. Schmorak, "Evaluated Nuclear Structure Data File — A Manual for Preparation of Data Sets," Oak Ridge National Laboratory Report ORNL-5054/R1 (February 1978).
3. M. J. Martin, "Nuclear-Decay Data for Selected Radionuclides," <u>A Handbook of Radioactivity Measurements Procedures</u>, Appendix A of NCRP Report No. 58, second edition (1985); M. J. Martin, "Radioisotopes," in <u>Encyclopedia of Chemical Technology</u>, Vol. 19 (John Wiley, New York, 1982), pp. 682-785.
4. W. B. Ewbank, "Versatile Output from a Simple Numeric Data File," in <u>Proceedings of the Sixth International CODATA Conference</u>, edited by B. Dreyfus (Pergamon Press, New York, 1979), pp. 359-369; W. B. Ewbank, "Systematics of Yrast Levels in Nuclei," in <u>Proceedings of the International Conference on Nuclear Physics</u>, LBL-11118 (1980), p. 322; W. B. Ewbank, "Status of Transactinium Nuclear Data in the Evaluated Nuclear Structure Data File," in <u>Proceedings, IAEA Advisory Group Meeting on Transactinium Isotope Nuclear Data</u>, IAEA-TECDOC-232 (1980), pp. 109-141; W. B. Ewbank, Y. A. Ellis, and M. R. Schmorak, Nucl. Data Sheets <u>26</u>, 1 (1970).

NUCLEAR STRUCTURE PROPERTIES FOR GAMMA-RAY LASERS

D. Strottman, E. D. Arthur, and D. G. Madland
Theoretical Division, Los Alamos National Laboratory
Los Alamos, N.M. 87545

ABSTRACT

We summarize some initial results in our investigation of the nuclear physics issues of gamma-ray lasers. We describe what is known thus far from existing experimental data and illustrate how theoretical models may be employed for systematic searches of candidate nuclei.

SURVEY OF EXPERIMENTAL DATA

We have earlier reported on simple nuclear physics considerations that indicate which mass regions may be most fruitful to search[1] for gamma-ray lasers and have cursorily examined the appropriate experimental data and theoretical tools.[2] Since then we have searched two computerized nuclear structure data libraries. The first, CDRL82,[3] is based on the particle-reaction and nuclear decay data contained in the 1978 Table of Isotopes[4] compilation. The second library used, ENSDF,[5] is from Nuclear Data Sheets[6] and is current to the 1980s for some mass chains.

The criterion used to perform the search was initial identification of isomeric states having a lifetime greater than 5 seconds. After identification of such a state, the spacings of nearby levels were examined to determine which ones (if any) fell within a specified excitation energy window, ΔE, of width 1 keV or 5 keV. If one relaxes the constraint that one member of the closely-spaced level pair be an isomeric level, then a substantially larger number of closely-spaced level pairs can be identified. Figure 1 illustrates, as a function of atomic mass, the number of level pairs having spacings $\leqq$ 1 keV lying below an excitation energy of 1.5 MeV, and which exclude ground-states as one member of the pair. A total of 242 such pairs were identified. However, imposition of the requirement that one of the levels be an isomer with a lifetime greater than five seconds results in the six candidate nuclei shown in Fig. 2. Of these, ^{179}Hf is of particular interest because it exhibits the closest level spacing (~ 200 eV) of any nucleus identified.

We have also searched both nuclear structure files for all known isomeric states. We have performed scans on each file, for isomeric state halflives, $T_{\frac{1}{2}}$, that are greater than 1 s, 1 min, 10 min, and 1 h. For each of these ranges we have calculated the average number of isomeric states as a function of the nuclear mass number A. Figure 3 illustrates the results for the range $T_{\frac{1}{2}} > 10$ min for the ENSDF file. Strong dependencies on mass number are evident. As we expected,[1,2] regions of shell closure in neutron number N and proton number Z, as well as deformed rare earth and actinide nuclei, contain the largest numbers of isomeric states.

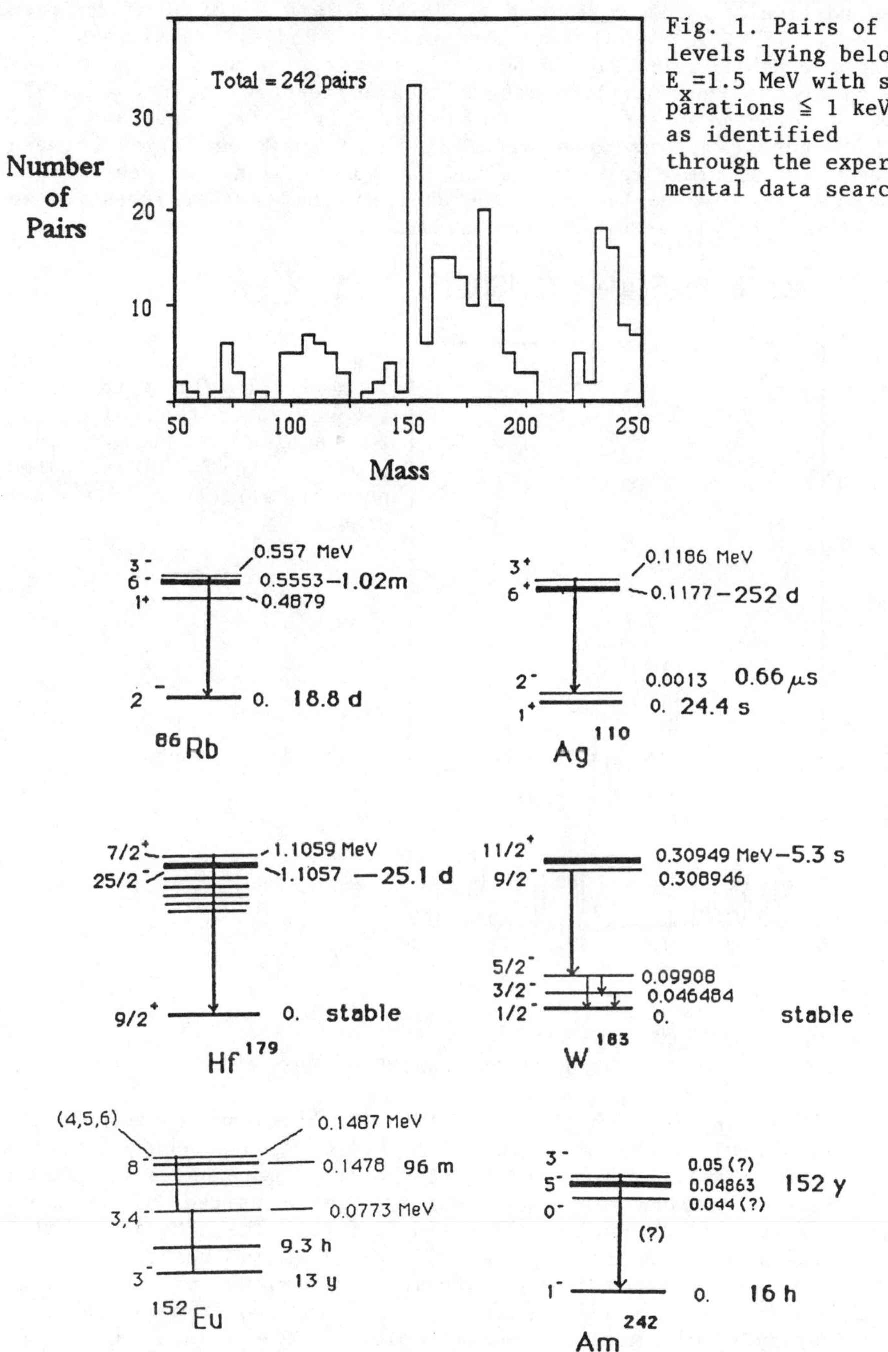

Fig. 1. Pairs of levels lying below E_x=1.5 MeV with separations ≦ 1 keV as identified through the experimental data search.

Fig. 2. Possible nuclear candidates fulfilling the level spacing and lifetime requirements discussed in the text.

In particular, the regions N = (28,50,82), Z = (28,50), the rare earths (150 ≦ A ≦ 190), and the actinides and transactinides (A > 220), exhibit well-defined peaks. However, the largest peak (for each of the four halflife ranges studied) occurs for A = 195-197, which corresponds to isotopes of $_{76}$Os, $_{77}$Ir, $_{78}$Pt, $_{79}$Au, and $_{80}$Hg. These nuclei are representative of the transition region between the heavy deformed rare earths and the spherical nuclei near doubly magic ^{208}Pb. Clearly, this region could be further investigated for possible gamma-ray laser considerations.

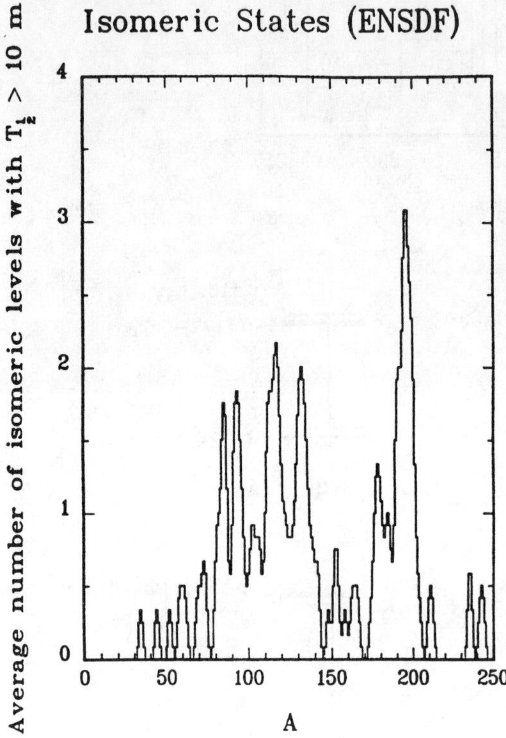

Fig. 3. Average number of isomeric levels with half-lives greater than 10 min as a function of nuclear mass number. The ENSDF evaluated nuclear structure file[5] was used.

THEORETICAL CONSIDERATIONS

It is apparent that none of the six above candidates meet the present requirements of a gamma-ray laser. Consequently, experimental work will be required to search for a more suitable nuclear candidate. Because of the amount of time required, this effort must be guided by theoretical considerations. The accuracy of theoretical models in the regions of interest is, at best, a few keV. This inaccuracy stems from an incomplete knowledge of the nucleon-nucleon interaction and mesonic corrections, and from the inherent inaccuracy of the nuclear models employed. The nucleus is a true many-body system with the interaction between any two nucleons depending upon all the other nucleons in the nucleus.

A very brief review of the mechanisms that may produce isomers was given in Ref. 2. Also discussed there were models that may be used to calculate the properties of levels in nuclei of interest. It was noted that odd-odd nuclei in the rare-earth region would be a likely place to begin a search for appropriate gamma-ray laser candidates.

We briefly illustrate the usefulness of nuclear models by discussing the structure of such a nucleus which has been proposed as a likely candidate. ^{186}Re has an isomer at an excitation of approximately 150 keV which has been tentatively identified as an 8^+. The lifetime of this level is 2×10^5 yr. In order for Re to meet the requirements of a laser, there must be a 7^- level within a keV of the 8^+ isomer. No such level is known, although it is easily conceivable that such a state would have been missed in the earlier experiments.

The structure of ^{186}Re is modeled by assuming the odd neutron and odd proton move in deformed orbits and interact via a phenomenological neutron-proton interaction. The energies of the odd nucleon may be checked by comparing with ^{185}Re and ^{185}W. The two lowest bands of ^{186}Re are a $K = 1^-$ (composed of a $K_n = 3/2^-$ neutron and a $K_p = 5/2^+$ proton) and a $K = 3^-$ ($K_n = 1/2^-$, $K_p = 5/2^+$). Neither band has a 7^- state below 500 keV. The lowest 7^- arises from a $K = 6^-$ band ($K_n = 7/2^-$, $K_p = 5/2^+$) and lies at approximately 400 keV. Although the error in the predicted energies may be as much as 30 keV in this initial calculation, nevertheless, the accuracy is sufficient to eliminate ^{186}Re as a viable candidate. As the calculations are refined, they will have an order of magnitude better accuracy, although the models are already very useful.

REFERENCES

1. E. D. Arthur, Comp., Los Alamos National Laboratory report LA-10288-PR (January 1985), p. 26.

2. D. Strottman, E. D. Arthur, and D. G. Madland, Los Alamos National Laboratory informal document LA-UR-85-2701 (May 1985).

3. R. J. Howerton, Lawrence Livermore National Laboratory report UCRL-50400 Vol. 23. Addendum (1983).

4. C. Michael Lederer and Virginia Shirley, Eds., Table of Isotopes, Seventh Edition (John Wiley and Sons, Inc., New York, 1978).

5. Evaluated Nuclear Structure Data File, available from the National Nuclear Data Center, Brookhaven National Laboratory, Upton, New York.

6. Atomic Data and Nuclear Data Tables, Section B: Nuclear Data Sheets, Academic Press, New York.

COHERENT AND INCOHERENT UPCONVERSION SCHEMES FOR PUMPING A GAMMA-RAY LASER*

C. B. Collins
Center for Quantum Electronics, University of Texas at Dallas
P.O. Box 830688, Richardson, TX 75083-0688

ABSTRACT

Since 1978 our research group has pursued an approach to the gamma-ray laser that is based upon the upconversion of long wavelength radiation incident upon isomeric nuclear populations. This approach can avoid many of the difficulties encountered with traditional concepts of single photon pumping. Recent experiments have confirmed the general feasibility indicating that a gamma-ray laser is feasible <u>if the right material exists</u>. The interaction energies and couplings are large enough so that the gamma-ray laser is probably more accessible to reasonably extrapolated technology than is the X-ray laser--simply because of the time scale for energy storage and more favorably narrow width of the transition. This paper reviews the background and context of our current efforts to prove the feasibility of a gamma-ray laser excited by the upconversion of incident radiation.

INTRODUCTION

By definition, a gamma-ray laser must draw its energy from nuclear excitation and so there exists the opportunity of exploiting the Mossbauer effect to insure that the maximum laser cross section is achieved. Provided the macroscopic host for the nuclei is sufficiently rigid and provided temperatures are sufficiently reduced, nuclear transitions at energies in the range from 1-100 keV can occur with widths corresponding to the natural lifetimes of the levels. Under these conditions the storage of excitation energies in the Mossbauer range can approach tera-Joules (10^{12} J) per liter for thousands of years. The stimulated release of this energy would occur at the rate at which resonant electromagnetic radiation passed through the laser medium and could lead to output powers as great as 3×10^{21} Watts/liter. This is an astronomical level of intensity representing 0.03% of the total power output from the sun, and implying that even a small gamma-ray laser would have very significant levels of output. Over the past seven years our research group has described[1-5] the several viable means through which such a device might be realized that are reviewed in this paper.

ENERGETICS

By involving two distinct steps the schemes we have proposed for pumping a gamma-ray laser avoid the severe relationships between storage times and spontaneous powers wasted at threshold that were imposed on the single-step processes.[6] Replacement power that is required, falls within a technically accessible range.

These two-step, upconversion processes for optically pumping nuclear reactions can be divided into two categories that correspond to the type of pumping employed: coherent and incoherent, as shown in Fig. 1. The critical concept here is that either transfers the stored population to a state at the head of a cascade leading to the upper laser level. To be effective the pumping processes cannot transfer too many quanta of angular momenta from the fields and the cascade provides a mechanism for further changes that may be necessary to reach the laser levels. Then the ultimate viability of these pump schemes will depend upon: (1) spectroscopic studies locating a suitable configuration of nuclear energy levels, and (2) "kinetic" studies providing an efficient path of cascading from the intermediate or dressed state to the upper laser level.

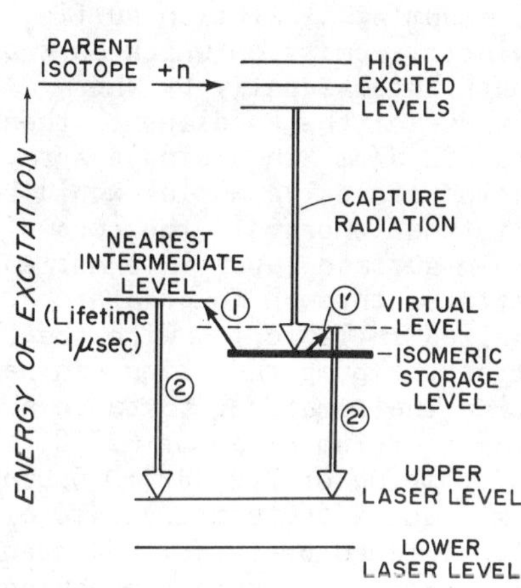

Figure 1

Schematic diagram showing the energetically excited levels of a typical nucleus of interest to the development of a gamma-ray laser. Lifetimes of the stored energies in the isomeric level produced by the initial capture can range from days to hundreds of years. The first phase of the two-step process for the stimulated release of the stored energy is shown in the figure by the solid arrows. Both correspond to the use of longer wavelength radiation to lift a nucleus from the storage level to a higher level of excitation that has a much shorter lifetime. The arrow marked (1) illustrates the incoherent pumping of the storage level through the absorption of a laser plasma X-ray that is resonant with the energy separation between the storage level and the next higher level of proper symmetry. The arrow marked (1′) represents the alternative process of coherent pumping through the non-resonant absorption of a photon from the radiation field in order to create a virtual or dressed state of excitation shown by the dashed level in the figure. In either case the gamma-ray output ultimately results from the upper laser level populated by a cascade occurring as a second step, as shown in the figure by either of the double arrows, (2) and (2′).

THRESHOLD ESTIMATES

Both the coherent and incoherent schemes for pumping a gamma-ray laser make stringent demands upon the arrangement of nuclear energy levels in the potential laser material. Both also depend upon the successful arrangement of an input source of radiation either to mix the properties of the storage level with those of some other state to release the metastability or to simply transfer the populations from the storage level to the other state. However, before focusing upon those problems in our following paper, it is reasonable to determine whether threshold could be conceivably achieved with extant devices *if a suitably ideal material existed in reality*.

In 1982 we published[2] the details of a modeling study performed upon idealized nuclei which showed that an input to transition 1 in Fig. 1 of the order of 10J at the source in the spectral width of a typical X-ray line at 10 keV from a filamentary plasma source 30 μm in diameter would be sufficent to pump a nearby slab to threshold. Since then, refinement of the Borrman-Post effect[7] has indicated that absorption losses at the output wavelength due to the photoelectric emission of inner shell electrons from the matter supporting the nuclei could be reduced by a factor of 50. This is the only major evolution of concepts since 1982 that would affect threshold estimates and applying it to those results gives a threshold fluence of 131 J/cm^2 for a 10 keV output transition.

As long as the Mossbauer width of the absorption line is preserved, the absorption length for the pump radiation can be very short[2] and the possibility of overheating or even vaporizing the pumped layer must be avoided by diluting the active element in a low-Z host[2] such as beryllium. For example, assuming a dilution sufficient to insure that the loss to photoelectric emission which must be balanced by the gain at threshold is contributed equally by the electrons of the active element and by those of the Be diluent, then the threshold fluence would be raised to 262 J/cm^2 in a single X-ray line and the minimum relative concentration of active nuclei would be of the order of 0.04%. The net result is that energy in the core of the X-ray line is still absorbed in a thin surface layer into nuclear excitation while that in the wings penetrates through to greater depths ejecting photoelectrons as summarized in Table I. With the proper geometry the scattered component of the pump radiation can be arranged to escape from the sample leaving the remainder to cause a negligable rise in temperature of the laser medium of about 62° C.

Scaling to a more vigorous level of pumping of 2.2 KJ/cm^2 would result in a small signal gain of 0.36 cm^{-1} and a temperature rise of 1050° C. The Be lattice is sufficiently rigid to preserve the essential Mossbauer component to a reasonable degree, but some broadening might actually be beneficial in order to increase the saturation intensity and hence, the final possible output.

Table I

Economy of the disposition of the 262 J/cm^2 pump fluence at threshold for this idealized material structure

Recipient	Energy Deposition
Nuclear Excitation	3.1 J/cm^3
Electrons	
Concentrate	156 J/cm^3
Diluent	101 J/cm^3
Scattered X-rays	55 J/cm^3

While the previous case is encouraging, there is much room for improvement. As the photon energy increases, there is much less absorption in the Be diluent and full escape of the scattered component occurs. Moreover, the medium can be layered so that escape of the primary photoelectrons generated by the wings of the pump line can occur before their energy is transferred into heating of the lattice. Once the best real nuclei are identified, a specific solution to the problem of the thermal economy can be tailored.

Accurate estimates of the threshold for the other pumping scheme designed to coherently mix the nuclear states with low (E $< 10^2$ eV) energy photons are not yet available, but it has been shown[2] that the situation is not worse than for incoherent pumping. Threshold fluences should range downward from a few KJ/cm^2 and thermal economies of the wasted radiation should be comparable. Moreover, it has been shown that the laser threshold can be reduced further by three to six orders of magnitude by manipulating the bulk ferromagnetic or ferroelectric properties of the material in which the nuclei are diluted. In the condensed state which is necessary for the Mossbauer and Borrmann effects the effect of the pump is intensified, first by the interactions between the nuclei and the atomic electrons and secondly by the cooperation of the electrons that gives rise to the ferromagnetic or ferroelectric properties as discussed in the subsequent companion paper.

CONCLUSIONS

The principal conclusion to be drawn from the present state-of-the-art is that a gamma-ray laser is feasible if a real nuclear material exists that has its properties sufficiently close to those of the ideal cases that have been modeled. This conclusion had been indicated by our studies published[2] in 1982 and the new results presented here dispel lingering concerns about material survival at the levels of input pump fluence that are sufficiently high to reach the threshold for gamma-ray laser output. While many enhancements to the configuration of a gamma-ray laser might be envisioned ultimately to result from studies of even more exotic effects such as superradiance, giant resonances, nuclear hole states and nuclear excimers, the absolutely critical factors determining the feasibility and means of realizing a gamma-ray laser have been identified as being: (1) the identity of the best candidate; (2) the threshold level for laser output; and (3) the upconversion driver for that material.

REFERENCES

*Supported in part by the Office of Naval Research and in part by the Strategic Defense Initiative Office (IST).

1. C. B. Collins, S. Olariu, M. Petrascu, and I. Popescu, Phys. Rev. Lett. 42, 1397 (1979).
2. C. B. Collins, F. W. Lee, D. M. Shemwell, B. D. DePaola, S. Olariu, and I. I. Popescu, J. Appl. Phys. 53, 4645 (1982); and references therein.
3. B. D. DePaola and C. B. Collins, J. Opt. Soc. Am. B 1, 812 (1984).
4. C. B. Collins and B. D. DePaola, Optics Lett. 10, 25 (1985).
5. B. D. DePaola, S. S. Wagal and C. B. Collins, J. Opt Soc. Am. B 2, 541 (1985).
6. G. C. Baldwin, J. C. Solem and V. I. Goldanskii, Rev. Mod. Phys. 53, 687 (1981).
7. B. Post, Paper TH1 in the subsequent session of this conference.

INTERLEVEL TRANSFER MECHANISMS AND THEIR APPLICATION TO GRASERS*

J. C. Solem
Los Alamos National Laboratory, Los Alamos N.M. 87545

ABSTRACT

Within the gamma-ray laser (GRASER) research community, much attention is being given to two-step schemes that store energy in a long-lived isomeric state and achieve lasing by transferring population to a short-lived state. Because the electron system exhibits large multipole moments and is in the near field of the nucelus, it can be used as an intermediate mechanism for transferring energy, angular momentum, and parity change. Two distinct electron-nucleus interaction mechanisms are discussed: (1) resonant electronic transitons and (2) collective outer-shell excitations.

INTRODUCTION

For nearly two decades, researchers[1] have been proposing GRASERS using interlevel transer. The long-lived isomer would be separated and concentrated by laser[2] or radio-chemical means[3] and implanted in a host crystal. If rapid interlevel transfer can be accomplished, superradiance will develop and produce an intense, highly directed burst of semicoherent gamma radiation. Calculations suggest that a GRASER using the Mössbauer and Borrmann effects in a perfect crystal can be superradiant with as few as 10^{13} active nuclei[10].

Direct interlevel transfer can clearly be accomplished: it is the basis of Mössbauer spectroscopy. However, it becomes difficult at long wave lengths because linewidths are very narrow and we cannot transfer radidly enough for superradiance. Assuming they obey the Weisskopf formula, low-energy transition rates go like $(\Delta E)^{2L+1}$, where L is the multipolarity. For instance, an M2 transition at ΔE = 100eV, would have a linewidth of about 10^{-11}Hz.

Electronic transitions, on the other hand, have much larger multipole moments and much broader resonances. Therefore, we propose to excite mixed nuclear-electronic transitions made possible by the near-field interaction.

RESONANT ELECTRONIC TRANSITIONS

This technique uses a laser to eject an electron selected so the most probably transition to fill its hole is resonant with the nuclear interlevel-transfer transition and has the correct multipolarity and parity. The process is identical to NEET[15]. The technique lacks energy efficiency because the energy to create the hole will generally exceed the interlevel transfer energy. A substantial fraction of the wasted energy will be deposited in the lattice and impair the Mössbauer and Borrman effects. Furthermore, Auger and Coster-Kronig transitions are also near-field effects and compete with the interlevel transfer.

To assess the feasibility of this mechanism, we obtained analytic solutions for interlevel transfer probability assuming hydrogenic electron wavefunctions, and also obtained numerical results using a Hartree-Fock algorithm for the electrons and a modified muonic atom code[6] for the electron-nucleus interaction.

Comparison of analytic and numerical results show that the order of magnitude of the interlevel transfer probability is

$$P \cong K_1 \frac{(K_2 A)^{2L/3}}{\Gamma_i(\Gamma_i+\Gamma_f)} B^{L+1} \tag{1}$$

where B is the binding energy of the electron hole, A is atomic weight, Γ_i and Γ_f are the initial and final electronic linewidths, and the constants K_1 = 1eV and $K_2 = 10^{-15}$ $eV^{-3/2}$. Some salient features of Eq. (1) are: (1) transitions with higher binding energy are favored, but such transitions of higher Z atoms are greatly broadened owing to the abundance of electrons in the near field; and (2) lower multipolarities are strongly favored owing to the magnitude of K_2. The first feature would favor light nuclei, which unfortunately would be unlikely to display isomerism or closely spaced levels. The second feature places some limitation on the ratio of lifetimes between the storage and the upper lasing state. For a dipole transition, the lifetime ratio does not exceed 10^6, except for the unlikely circumstance of a forbidden transition between the storage and lower lasing state. An alternative is to arrange for an intermediate transfer state between the storage and upper lasing level and thereby effect a quadrapole transition by two successive dipole transitions. The likelyhood of such a favorable constellation of states is also small.

A numerical search of electronic configurations revealed a few atoms that may support interlevel transfer if there were an appropriate coincidence with nuclear levels. One of these is ^{24}Mg, which has no known nuclear-level coincidence, but is highly deformed. We considered a $2S_{\frac{1}{2}}$ hole filled by the $2P_{\frac{1}{2}}$ electron with transition of the nucleus from 1^- to 0^+ at the same energy. The dominant broadening for both states in Auger effect, which gives 8.8mV for the $2S_{\frac{1}{2}}$ and 0.6mV for the $2P_{\frac{1}{2}}$. The transition energy is only 38eV. The matrix element for the transition from numerical calculation is 1.4mV for one Weisskopf unit, which leads to an interlevel transfer probability of about 3%. This may be sufficient for gamma-ray lasing if all other conditions were favorable.

COLLECTIVE ELECTRONIC EXCITION

Transition rates can be greatly enhanced if several electrons act collectively to drive a nuclear interlevel transfer. For example, if 10 electrons were involved, the matrix element would be 10 times larger, and the transition rate would be 100 times larger. Furthermore, if collective excitations are generated by multiple photon absorption, a low-quantum-energy laser can be used to drive a high-quantum-energy interlevel transfer[7,8]. The simultaneous absorption of as many as 99 photons have been reported for 6.4eV radiation in uranium[9]. Because reasonably low energy quadrapole transitions are known in nuclei and the quadrapole transition rates generally exceed those expected from the single particle model, it would be advantageous if the collective outershell excitations had a substantial quadrapole component. This should be straightforward to calculate by standard numerical techniques. Lacking such calculations, but judging from experience with the liquid-drop model, we would conjecture that as much as half the excitation energy could be quadrapole oscillations.

To estimate the matrix elements for coupling between the collective excitations and the nucleus, we simply sum over the N electrons that are acting coherently. The near-field interaction Hamiltonian is:

$$H' = -\sum_{e}^{n} \frac{Ze^2}{|\vec{r}_n - \vec{r}_e|} , \tag{2}$$

assuming the magnetic interaction is small.

If we knew the initial and final electron and nucleus wave-functions $\psi_{ei}, \psi_{ef}, \psi_{ni}, \psi_{nf}$, we could find the interaction matrix element by expanding the Hamiltonian in spherical harmonics:

$$E' = \langle\Psi_{ni}\Psi_{ei}|H'|\Psi_{nf}\Psi_{ef}\rangle$$

$$= 4\pi e^2 Z \sum_{e}^{N} \sum_{lm} \frac{1}{2L+1} \int_0^{\infty} \Psi^*_{ni}\, Y_{lm}(\theta_n,\phi_n)\, r_n^L\, \Psi_{nf}\, d\Omega_n^2 r_n dr_n \tag{3}$$

$$\times \int_0^{\infty} \Psi^*_{ei}\, Y^*_{lm}(\theta_e,\phi_e)\, r_e^{-L-1}\, \Psi_{ef}\, d\,\Omega_e r_e^2 dr_e ,$$

where we have ignored the contribution for the electron wave-functions inside the nucleus. A rough estimate can be obtained by guessing the radial matrix elements and assuming the N electrons act in perfect coherence.

$$E' = -\frac{4\pi NZe}{2L+1} \langle r_n^L \rangle \langle r_e^{-L-1} \rangle \qquad (4)$$

For a dipole transition in a heavy element, say uranium, I estimate the nuclear moment $Z\langle r_n \rangle \sim 10^{-13}$cm. A reasonable estimate of the collective electronic excitation might be 10 electrons participating coherently at a radius of about 5×10^{-12} cm. This gives a matrix element of about 3eV. For a quadrapole transition, $Z\langle r_n^2 \rangle \sim 10^{-23}\text{cm}^2$, which gives a matrix element of about ½eV. At this time we have no estimates of the corresponding linewidths, but it seems reasonable that interlevel transfer probabilities will be substantial. The most serious question is whether Mössbauer and Borrmann effect can be maintained with the laser intensities required for collective excitation. Certainly a traveling wave transfer scheme will be necessary.

The strong coupling suggests that nuclear transitions may provide a sensitive probe of collective electron excitation, especially for determining their total energy and multipolarity. The broad range of possible experiments may usher in a new era of whole-atom physics.

REFERENCES

1. J. W. Eerkens, U. S. Patent 3, 430,046 (1969). E. V. Baklanov and V. P. Chebotaev, Zh. Eksp. Teor. Fiz. Pis'ma Red. 21, 286 (1975); P. Kamenov and T. Bonchev, C. R. Acad. Bulg. Sci. 28, 1175 (1975); L. A. Rivlin, Sov. J. Quantum Electron. 8, 1412 (1977); L. A. Rivlin, Sov. J. Quantum Electron. 7, 380 (1977); B. S. Arad, S. Eliezer, Y. Paiss, Phys. Lett. A 74, 395 (1979); C. Collins, F. Lee, D. Shemwell, and B. DePaola, J. Appl. Phys. 53, 4645 (1982).
2. P. Dyer, G. Baldwin, C. Kittrel, D. Imre, and E. Abramson, Appl. Phys. Lett. 42, 311 (1983).
3. L. Szilard and T. Chalmers, Nature 134, 462 (1934).
4. M. S. Feld, Bull. Am. Phys. Soc. 30, 1815 (1985).
5. M. Morita, Progr. Theor. Phys. 49, 1575 (1974).
6. G. Rinker, Comput. Phys. Commun. 16, 221 (1979).
7. K. Boyer and C. K. Rhodes, Phys. Rev. Lett. 54, 1490 (1985).
8. C. K. Rhodes, Science, 30, 1345 (1985).
9. T. S. Luk, H. Pummer, K. Boyer, M. Shahidi, H. Egger, and C. K. Rhodes, Phys. Rev. Lett. 51, 110 (1983).

* Supported by the U. S. Department of Energy.

Kinetics of Nuclear Superradiance*

George C. Baldwin
Los Alamos National Laboratoary
Physics Division
Los Alamos, New Mexico 87545
and
Michael S. Feld+
G. R. Harrison Laboratory of Spectroscopy
Massachusetts Institute of Technology
Cambridge, Massachusetts 06730

Realization of gamma-ray lasers is still far in the future; still, much is now known about the probable form the system will take. Most likely, the device will be a single-shot, high-gain, single-pass system, comprising a long crystal without mirrors or reflecting surfaces. The pumping process that initiates lasing will rapidly transfer the active nuclei into the upper laser level. For reasons that will appear, the preferred gamma-ray emission mechanism is likely to be superradiance (SR), rather than the so-called amplified spontaneous emission, which occurs in high-gain, mirrorless lasers that operate on Doppler-broadened transitions.

Superradiance is produced in a sample of N two-level radiators by rapidly preparing a state of high gain. A typical SR output pulse, shown in Figure 1, consists of an intense burst of radiation, often observed with ringing, preceded by a significant delay T_D. Its peak intensity I_p increases as N^2 and its width T_W varies inversely with N. The entire emission process is completed in a time short in comparison with the radiative lifetime T_{rad} of the transition, leaving the system completely de-excited.

The buildup of radiation is triggered by spontaneous emission from one of the excited atoms; the subsequent evolution can be described semiclassically by coupled Maxwell-Schrödinger equations. The output of such a high-gain system can range from conventional pulsed laser emission (peak power proportional to the number of radiators, N) to strong SR (peak power proportional to N^2). In the limit of strong superradiance the principal features can be described analytically; the emitted pulse obeys certain scaling relationships that can be illustrated by the normalized curve of SR output power I vs time T (Figure 1). The behavior is characterized by a single parameter T_R, the cooperative spontaneous emission time

$$T_R = T_{rad}\ (8\pi/n\lambda^2 L), \tag{1}$$

with λ the wavelength corresponding to the transition energy E, n the inversion density of the SR sample and L its length. T_{rad} is the radiative lifetime of the transition. In the nuclear cases con-

---

* This paper is a condensed version of a larger article to be published in the Journal of Applied Physics in June 1986. References for this article are given therein. Supported by the Division of Basic Energy Sciences, U. S. Department of Energy.

\+ Consultant.

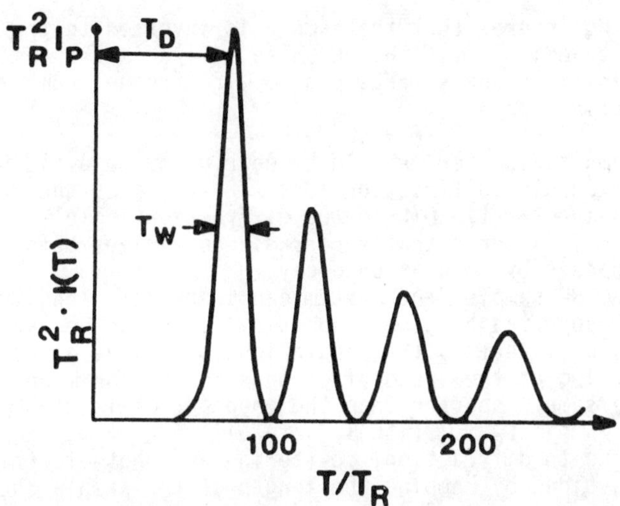

Figure 1. Normalized intensity vs time for strong SR output. In the ideal case shown here, all times scale as T_R and the output power as T_R^{-2}. In practice, the ringing is usually less pronounced.

sidered,

$$T_{rad} = T_u\ (1 + \alpha)/\beta, \qquad (2)$$

where T_u is the actual lifetime, α is the internal conversion coefficient and β is the branching ratio.

The parameters shown in Figure 1 are given by the following expressions:

$$T_D = (\phi^2/4)\ T_R, \qquad (3)$$

$$T_W = \phi T_R \qquad (4)$$

$$I_p = (4/\phi)\ N\ E/T_W, \qquad (5)$$

$$= [n\ \lambda^2 L/(2\pi\phi^2)][NE/T_{rad}], \qquad (6)$$

with

$$\phi = (1/2)\ \ln\ \{2\pi N\}, \qquad (7)$$

and N the total number of active nuclei in the sample.

The peak power therefore exceeds that emitted in ordinary spontaneous emission by the first factor on the right member of equation 6, $n\lambda^2 L/(2\pi\phi^2)$.

The basic requirement for superradiant emission is: rapid inversion of a system of two-level radiators to a state of high gain. The atom-field system will then evolve in a transient manner, rather than "ordinary" laser emission.

High gain insures that the cooperative spontaneous emission rate exceeds that of incoherent radiative decay. The electric field gain coefficient can be written in terms of n, the inversion density and σ, the stimulated emission cross section:

$$g = n\sigma/2 \qquad (8)$$

$$\sigma = f_T\ (2\pi\ \bar{\lambda}^2)/(1 + \alpha). \qquad (9)$$

Equation 8, using equations 1 and 9, gives

$$gL = 2(T_u/T_R)f_T. \qquad (10)$$

Rapid inversion insures that the sample is inverted to the necessary extent prior to emission of the SR pulse. T_{inv} is the time over which the transfer process takes place. The Fresnel number of the sample, F, defined as

$$F = \pi D^2/4\lambda L, \quad (11)$$

with D the sample diameter, should be near unity to avoid beam breakup and excessive diffraction loss. F expresses the ratio of the sample's aspect ratio D/L to the diffraction angle $\sim\lambda/D$. The condition $T_D < T_u$ insures that superradiance evolves before the inversion disappears by population decay.

In a long SR sample, the system can break into longitudinal sections that superradiate independently. This can be avoided using swept inversion (inverting the populations by a wave of transfer radiation advancing at the speed of photons in the Borrmann mode) or by making the sample shorter than the cooperation length L_c:

$$L_c = \phi\ [4\pi c\ T_{rad}/(n\lambda^2)]^{1/2}. \quad (12)$$

Losses due to diffraction, scattering and photoelectric absorption can be avoided by keeping the length of the sample short compared with

$$L < L_a = \phi/4\mu, \quad (13)$$

with μ the average electric-field loss per cm.

Assuming that the necessarily fast transfer rate can be achieved, we wish to calculate a) the smallest quantity of active nuclei required for gamma-ray SR, and b) dimensions of the smallest host crystal needed to contain them. Requirement a) minimizes both the pumping requirements and the demands on the activation and separation processes for preparing storage isomers. Requirement b) minimizes the difficulty of preparing and doping a single crystal host and, assuming that the transfer radiation can be concentrated, maximizes the transfer pump intensity for a given pump power.

For our minimum design parameters, we take the smallest values allowed by conditions of high gain, rapid inversion, moderate Fresnel number, rapid evolution of SR, swept inversion or short sample, and low losses:

i. $gL = \phi$ (smallest possible value of nL). (14)

ii. $T_{inv} = T_D$ (lowest acceptable transfer rate). (15)

iii. $F = 1$ (smallest sample volume - see below). (16)

iv. $T_D = T_u/\varepsilon$, (17)

where the reduction factor $\varepsilon > 1$, but is not too large, insuring that the initial inversion density does not decay excessively before the collective emission can develop.

v. If $L > L_c$, use swept inversion. (18)

vi. $L < \phi/4\mu = L_a$, (19)

insuring that loss will not cause deterioration of the SR output.

Smallest nL value: Using equations 10, 3 and 17 gives

$$gL = (\varepsilon/2)\ \phi^2\ f_T, \quad (20)$$

which is usually more restrictive than equation 14 and thus automatically satisfies it. This condition restricts the product nL. In addition, L must not exceed either L_a (equation 19) or (unless swept inversion is used) L_c (equation 18).

Smallest volume: In general, the sample volume may be written in terms of the Fresnel number, equation 11:

$$V = (\pi D^2/4)L = F \lambda L^2. \quad (21)$$

Thus, minimizing F also minimizes the sample volume for a given length L. In the following, we shall keep F arbitrary, although recognizing that F = 1 (equation 16) will minimize design parameters.

Derivation of minimum sample parameters: For determining the sample parameters we then have, from equation 20, using equations 8 and 9,

$$n = (1 + \alpha)\, \varepsilon\phi^2/(2\pi\lambda\!\!\bar{}^{\,2}L). \quad (22)$$

The total number of active nuclei nV is

$$N = (\pi/4)\, D^2\, L\, n, \quad (23a)$$

and, using equation 21, this is

$$N = F\, L^2 \lambda n. \quad (23b)$$

When L, F and ε have been specified, values of the three parameters n, N and ϕ can be determined by iterative solution of the three-equation set: equations 7, 22 and 23b, starting with an initial trail value for one parameter.

The SR pulse width, equation 4, is set by the choice of T_D, equation 17:

$$T_W = (4/\phi\, \varepsilon)\, T_u, \quad (24)$$

giving a peak power [equation 5]

$$I_p = (4/\phi)\, (N/e)\, (E/T_W), \quad (25a)$$

where E is the photon energy, or, combining equations 6 and 22,

$$I_p = \varepsilon(1 + \alpha)\, (N\, E/T_{rad}), \quad (25b)$$

which differs by a factor $\varepsilon(1 + \alpha)$ from the incoherent spontaneous emission power - a reasonable result, because the stored energy is released more rapidly by superradiance than by normal decay, so that fewer nuclei remain to decay via the conversion route.

However, the SR gamma-ray pulse is emitted during the short time T_W into a very small solid angle centered on the axis of the graser filament and thus will be far brighter than the isotropic incoherent spontaneous emission signal.

To illustrate the application of the above relations, we consider a hypothetical case, the well-known 24 keV (λ = 0.052 nm) Mössbauer transition of ^{119}Sn. We choose a diamond host lattice, because of its low removal cross section (6.6 barns) and because its high Debye temperature (2230 K) insures a high recoilless fraction.

The I = 3/2 upper level of the Mössbauer transition decays to the I = 1/2 ground state by an unbranched M1 transition. Hence g_u/g_l = 2 and β = 1. The internal conversion coefficient, 5.1, is relatively small. The upper-level lifetime T_u is 25.7 ns; a full initial inversion lasts for 10.4 ns. The gamma-ray removal cross section, primarily photoelectric absorption, is 2850 b. The free nuclear recoil energy is 0.532 x 10^{-4} x $(23.87)^2$/119, or 0.255 meV.

We assume that the ^{119}Sn Mössbauer linewidth in a single-crystal diamond host is determined essentially by uncertainty broadening of the upper level. The effective Debye temperature for the impurity dopant is 2232 x $(12/119)^{1/2}$ = 708 K, and the Debye-Waller factor (recoilless fraction) at low temperature is f = 0.94. We shall assume that the dopant is substitutional. The first three orders of Bragg reflections by the 220 planes occur at Bragg angles, respectively 19.1, 40.8 and 78.6 degrees. The corresponding coupling factors for E1 transitions (e.g., photoelectric) are 0.014, 0.055 and 0.123; for M1 (e.g., nuclear) transitions, they are 0.213, 0.854 and 1.921. Therefore, nonresonant attenuation will be reduced for all orders, the nuclear resonance reduced only slightly for the second order and enhanced for the third. We shall assume that the geometry of the graser body has been designed to take advantage of this feature, so as to select a 220 Borrmann mode using the third order Bragg reflection.

To begin, let us take a Fresnel number of unity, choose the length to be 5 mm, and assume that transfer to the upper level is complete and "instantaneous". An approximate trial value of N = 10^{13} gives ϕ = 15.9. Taking ε = 4.2 in equation 22 gives Ln = 1.52 x 10^{21} cm^{-2}. Multiplication by the area FLλ (equation 21) gives an improved value for N of 3.95 x 10^{12} and ϕ = 15.4, Ln = 1.50 x 10^{21} cm^{-2}; using this N, a second iteration gives ϕ = 15.39; further iteration is unnecessary, because the formula for ϕ is an approximation. The diameter of the filamentary region of the host that is to be doped is only D = 0.58 μm.

The output parameters are as follows: the delay time is 6.5 ns; the pulse width is 1.7 ns. The peak SR power (equation 25) is 2.34 MW, the pulse energy is 3.7 mJ, the emission solid angle is 6.5 x 10^{-8} sr, and the energy enhancement factor is 7.3 x 10^{8}.

Thus, with this length and Fresnel number, the host should be doped with 1.6% active nuclei. This dopant concentration shortens the nonresonant loss length to 0.11 cm, so the Borrmann effect, for which it is 0.87 cm, is essential. On the other hand, swept pumping is unnecessary, because the cooperation length is 1.4 cm.

Note that the value of the overall gain reduction factor never enters into the calculations. This is confirmed by more detailed and complete calculations that evaluate the gain, Borrmann parameters, etc.

The model has assumed that the stimulated emission process will be superradiance rather than ordinary laser amplification, because, lacking good mirrors at gamma-ray wavelengths, a single-pass system with very high gain will be required. With an emission line of nearly natural width, ordinary laser amplification requires a long buildup time; superradiant response in the transient regime, having a short T_R, is required.

These considerations suggest searching for candidates for the SR transition having small statistical-weight ratios or, better, comparable upper- and lower-state lifetimes.

Although the results presented here are encouraging, we must emphasize that: (1) we have assumed that there is a mechanism by which inversion can be created very rapidly, without inhibiting the Mössbauer effect, and (2) the analytical modeling presented here is approximate. Numerical analyses using the coupled Maxwell-Schrödinger equations should be used for more accuracy, and to simulate the details of the transfer process.

Nevertheless, analyses of the kinetics of superradiant nuclear gamma-ray emission for this ideal case encourages hope for the eventual identification and development of gamma-ray laser systems with realistic parameters.

ACTIVITIES AT LLNL RELEVANT TO GAMMA-RAY LASER CONCEPTS

F. S. Dietrich
Lawrence Livermore National Laboratory, Livermore, CA 94550

ABSTRACT

Current and planned gamma-ray laser activities at LLNL include both theoretical and experimental studies of spin and shape isomers, construction of a new accelerator laboratory that will include an optimal facility for gamma-ray laser candidate searches, evaluation of the prospects for developing ultrahigh resolution detectors, and the study of a positron-annihilation laser scheme.

Identifying gamma-ray laser candidates will require major advances in both theoretical and experimental techniques for nuclear spectroscopy. While the systematics of long-lived isomeric states that may serve as storage levels is reasonably well understood[1], transition matrix elements from these levels to neighboring levels and the properties of short-lived states that may serve as upper levels of a laser transition are not well predicted by theory.

We have recently completed a state-of-the art gamma spectrometer that includes a Compton polarimeter and an intrinsic germanium detector with an anticoincidence shield. As a first test of this system, we are studying the level schemes of isomeric nuclei in the mass region near A=90 formed by bombarding ^{89}Y targets with ^{7}Li beams from the LLNL cyclograaff accelerator. The results will be used to improve the predictions of the LLNL large-basis shell-model code[2] in this mass region. We are planning to study the properties of odd-odd deformed nuclei, which are easy to reach via (p,n) reactions on even-even targets, and (d,p) and (d,n) on odd-mass targets. These nuclei are interesting because they contain numerous isomers, and the rotational enhancement of the level density increases the probability of finding a nearby level that may serve as a transfer level[1,3]. The actinide nuclei are most likely to have such levels, and moreover have not been as thoroughly studied experimentally as the rare-earth nuclei.

Shape isomers[4], which have long lifetimes because they have much larger quadrupole deformations than the ground state, are imbedded in a sea of closely-spaced levels with normal deformation. The known shape isomers (in the actinides) decay by fission, which limits their half-lives to 15 msec or less. We have begun a program to search for shape isomers in lighter nuclei that may have longer lifetimes. Theoretical guidance will come from constrained Hartree-Fock calculations, with emphasis on the regions around ^{24}Mg (to test the techniques), Os, and Rn-Ra. We are presently performing experiments in the U-Np region to search for gamma decay from shape isomers.

An experimental facility tailored for gamma-ray laser studies requires beams suitable for producing nuclei near the valley of stability, a flexible beam-chopping system useful for studying lifetimes in the nanosecond to second range, state-of-the art gamma and conversion-electron spectrometry, and the ability to handle difficult targets that are radioactive or chemically unstable. These features will be incorporated in a new accelerator facility at LLNL, which will be completed in 1987. This laboratory will be based on an FN tandem accelerator (10 MV), and will provide beams of p, d, t, ^{3}He, ^{4}He, and "light" heavy ions (e.g., Li, O). The gamma spectrometers and Compton polarimeter will be moved from the cyclograaff to the new laboratory, and in addition the LLNL conversion electron spectrometer[5], presently located at the LANL tandem, will be moved to the new facility.

The complete determination of nuclear level schemes and the separation of closely-spaced levels are presently hindered by the resolution of solid-state detectors, which is typically in the 1 keV region. Recent progress in the development of superconducting junction detectors[6-8] shows the promise of an order-of-magnitude improvement in resolution; the Oxford group[7] has already inferred a resolution in the neighborhood of 150 eV in experiments with crossed indium films. We are proposing a system study to evaluate the practicality of developing a large-scale detector. This study would include selection of superconducting materials, evaluation of problems in developing detectors in the 1 cm^3 range, and preliminary design of ultralow-noise electronics and suitable ADC's.

We are also studying the scientific issues underlying a positron-annihilation laser scheme employing laser cooling of positronium[9]. Initial calculations indicate positive gain for a Doppler width less than 0.27 eV; a possible mechanism is rapid laser cooling of orthopositronium followed by r.f. spin-flipping to the much shorter-lived parapositronium. Theoretical activities will include investigation of optimal laser-cooling schemes and modeling of the r.f. transition between states; experimental activities will use the LLNL 100-MeV electron linac to study the optimal production of slow positrons and positronium.

This work was performed under the auspices of the U. S. Department of Energy by the Lawrence Livermore National Laboratory under contract number W-7405-ENG-48.

REFERENCES

1. D. Strottman, E. D. Arthur, and D. G. Madland, proceedings of IDA Gamma-Ray Laser Workshop, May, 1985.
2. S. D. Bloom, LLNL, private communication.
3. F. S. Dietrich, proceedings of IDA Gamma-Ray Laser Workshop, May, 1985.
4. V. Metag et al., Physics Reports 65,1(1980).
5. W. Stoffl and E. A. Henry, Nucl. Inst. and Meth. 227,77(1984).
6. M. Kurakado, Nucl. Inst. and Meth. 196,275(1982).
7. N. E. Booth et al., in proceedings of Conference on Solar Neutrinos and Neutrino Astronomy, Lead, South Dakota, Aug. 1984.
8. A. Barone et al., Nucl. Inst. and Meth. A234,61(1985).
9. E. P. Liang and R. H. Howell, LLNL, private communication.

NUCLEAR ISOMER SEPARATION*

P. Dyer
Physics Division, Los Alamos National Laboratory
Los Alamos, New Mexico 87545

ABSTRACT

We report experiments on selective photoionization of atoms containing isomeric nuclei of ^{197}Hg. Other isomer separation techniques and their limitations are discussed.

INTRODUCTION

To produce a nuclear population inversion for a gamma-ray laser, isomer separation is generally required, as nuclear reactions usually produce greater quantities of ground-state nuclei than excited-state nuclei. The laser isomer separation techniques discussed here are the same as those of isotope separation, but the technical problems are much more difficult, as the sample size is many orders of magnitude smaller and the nuclear states are short-lived. Here we discuss a resonance ionization demonstration experiment with ^{197}Hg and initial measurements of the optical piston with sodium. We conclude with a comparison of techniques.

RESONANCE IONIZATION OF ^{197m}Hg

We have demonstrated isomerically-selective photoionization of ^{197m}Hg (nuclear half-life 24 hours) via the atomic excitation sequence 6^1S_0 - 6^3P_1 - 8^1S_0 - Hg^+.[1] Three collinear pulsed dye laser beams were used: 254, 286, and 696 nm, selectively exciting the first two transitions and ionizing through an autoionization state in the continuum.

Gold target foils were bombarded by deuterons at the Los Alamos tandem Van de Graaff accelerator, to generate ^{197}Hg by the (d,2n) reaction. These target foils were then heated in vacuum to distill mercury onto a second gold "catcher" foil, which was then sealed in a shielded capsule for transportation to MIT.

Meanwhile, at the MIT Laser Center, optical excitation experiments were performed with two vapor cells, one containing natural mercury and the other, mercury enriched in ^{202}Hg, for adjustment and calibration of the apparatus. Upon arrival of the radioactive sample at the Laser Center (sixteen hours after the end of bombardment), the active catcher foils were introduced into a clean irradiation cell and heated to expel mercury. The Pyrex irradiation cells were 12-cm long and 15-mm in diameter, with fused-silica Brewster windows at each end. No materials that had been exposed to natural mercury were used in constructing the ^{197}Hg cell. Other materials to which mercury was exposed in the chamber were limited to Teflon, Viton O-rings, ceramic adhesive and clean iron; all had been previously found, using ^{197}Hg as a tracer, to have low tendency to adsorb mercury.

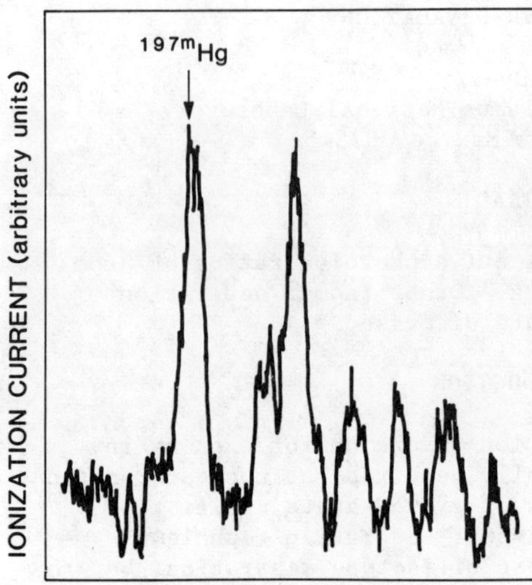

Fig 1. Ionization current, as a function of 254-nm scanning frequency, with 286-nm radiation fixed at the ^{197m}Hg-c hyperfine component.

The collecting electrodes were a pair of magnetically supported Fe wires on opposite sides of the laser beam. A collecting potential of 400 volts was applied to the electrodes. Currents from a phototube and from the ion-collector were amplified, passed to a boxcar integrator, and registered on a chart recorder. The quantities of ^{197m}Hg released into the cell were of the order of 2×10^{12} atoms. Ion-collection rates were of the order of 4×10^{6} s^{-1}.

Figure 1 shows the chart record of the ionization current when the 286-nm laser was fixed at the ^{197m}Hg-c peak, the 254-nm radiation was scanned, and the ions that were created in the final transition to the continuum were collected. A large peak in the ionization current is observed at the expected position. The combined excitation by both 254- and 286- nm radiations was sufficient to achieve clean separation of ^{197m}Hg.

However, we did not obtain an enriched sample outside the cell. An attempt to measure enrichment by counting gamma rays from the positive and negative electrodes failed. There was no significant difference in isomeric enrichment between the two electrodes. Moreover, the total number of radioactive atoms on the wire exceeded nearly 100-fold the number of ions collected, estimated from the ionization current and collection time. Presumably, the selectively ionized and collected portion was greatly diluted by nonselective adsorption of ^{197}Hg, despite precautions to use clean Fe electrodes.

The isomer ^{142m}Eu has been resonantly ionized by Alkhazov et al.[2] In this experiment, three laser beams intersected the atomic beam from a mass separator on-line to a proton synchrocyclotron.

THE OPTICAL PISTON

An alternative possible technique for laser isomer separation is light-induced drift.[3,4] Consider a laser beam incident along the axis of a capillary cell containing the atoms of interest, e.g. sodium, in a buffer gas. When the laser frequency is tuned just below the center of a Doppler-broadened absorption line, only those sodium atoms moving toward the laser beam will be excited. As the velocity-changing collision cross section for excited sodium atoms colliding with the buffer gas is greater than that for ground-state atoms, there will be a drift of sodium atoms along the direction of the laser beam, toward the downstream end of the capillary. If the sodium is optically dense, the "optical piston" is manifested as a front moving along the cell, as observed by Werij et al.[5]

We have begun measurements to repeat those of Werij et al. with sodium. Thus far a front movement of about 2 cm has been observed. We hope to extend these measurements to an element for which the piston effect is not limited by interaction of the active atoms with the cell walls. Enrichments for isotope separation, and subsequently isomer separation, will then be measured.

COMPARISON OF TECHNIQUES

A number of laser techniques are available for isomer separation: resonance ionization (where the ionization step may be performed by a laser or by electric fields or collisions acting on Rydberg states), the optical piston, photochemistry, radiation pressure, magnetic or electric deflection of an optically pumped atomic beam. The optimum choice will depend on atomic state energies, hyperfine structure, vapor pressure, chemistry (especially surface), nuclear state lifetime, isomer production rate, and initial enrichment factors. Various factors limit the efficiency of the separation, the enrichment achieved, and the time required to perform the separation. In general, separations in cells have high efficiency, but low resolution, whereas the opposite is true for separations in atomic beams.

Formation of an atomic beam is necessarily an inefficient process. In a cell there is the potential for a given atom to pass through the laser beam many times, but to achieve this, there must be little loss of the material to adsorption on the walls. A further factor limiting the efficiency for resonance ionization, particularly in cells, is space charge.

If the separation is performed in a cell, and the Doppler width is greater than the hyperfine splitting of the lines involved in discrete transitions, loss of enrichment results. Collisions in a cell also limit enrichment. In the case of resonance ionization, resonance charge exchange results in non-specific collection of ions. In the cases of the optical piston or of photochemistry, inelastic collisions at high sample densities dilute the enrichment. Multiphoton ionization is also a source of non-selective background in the case of resonance ionization by photons, whether in a cell or an atomic beam.

There are various considerations involved in the choice of pulsed versus CW lasers. CW lasers offer narrower bandwidth and high duty factors. Pulsed lasers are more suitable for producing UV wavelengths by frequency doubling. They are also better suited to multistep processes, such as resonance ionization with the ionization step performed by photons.

The most widely applicable technique thus far appears to be that of resonance ionization. In the case of isomer separation for a gamma-ray laser, this technique offers the advantage that implantation into a crystal may be performed by the field that collects the ions. If far UV wavelengths are not required, if ionization is performed by electric fields rather than photons, and if the time contraints are not too severe, CW lasers may be used. Otherwise pulsed laser excitation is necessary. The time to separate the required number of isomers for a gamma-ray laser, once the sample is in the laser beam, can be much shorter than a second. Thus the entire process of producing implanted isomers will be more limited by the time to produce the isomers in nuclear reactions and to transfer them from the reaction target to the laser beam, than it is by the laser ionization time.

REFERENCES

*Work supported in part by the Division of Advanced Energy Projects of the United States Department of Energy. Part of this work was performed while the author was a Visiting Scientist at the MIT Laser Research Center, which is a National Science Foundation Regional Instrumentation Facility.

1. P. Dyer, G. C. Baldwin, A. M. Sabbas, C. Kittrell, E. L. Schweitzer, E. Abramson, and D. G. Imre, J. Appl. Phys. 58, 2431 (1985).
2. G. D. Alkhazov, A. E. Barzakh, E. E. Berlovich, V. P. Denisov, A. G. Dernyatin, V. S. Ivanov, V. S. Letokhov, V. I. Mishin, and V. N. Fedoseev, JETP Lett. 40, 836 (1985).
3. F. Kh. Gel'mukhanov and A. M. Shalagin, JETP Lett. 29, 711 (1979).
4. Gerard Nienhuis, Phys. Rev. A31, 1636 (1985).
5. H. G. C. Werij, J. P. Woerdman, J. J. M. Beenakker, and I. Kuscer, Phys. Rev. Lett. 52, 2237 (1984).

THE ROCHESTER-STANFORD PROPOSAL FOR A SEARCH FOR NUCLEAR ISOMERIC STATES AS CANDIDATES FOR SHORT WAVELENGTH LASERS

H. E. Gove, D. Cline and T. M. Cormier
Nuclear Structure Research Laboratory, University of Rochester
Rochester, New York 14627

S. S. Hanna
Department of Physics, Stanford University
Stanford, California 94304

ABSTRACT

A research program is proposed to search for new isomers which may be suitable for the stimulated emission of nuclear gamma rays (grasers) and to measure their properties. The proposed research is a relatively straight forward extension of programs which have been underway at Stanford and Rochester for several years.

DESCRIPTION OF RESEARCH PROGRAM

At Rochester the work will be carried out using an MP tandem Van de Graaff accelerator which operates at a terminal potential of 12 to 13 MV. It is presently being upgraded to a terminal voltage of 18 MV. This upgrade which will be completed by the summer of 1986 will lead to the proposal of a further upgrade comprising the addition of a superconducting linear accelerator. The first upgrade will permit beams of mass 80 or less to be accelerated over the Coulomb barrier of a uranium target. The addition of the post accelerator will permit the acceleration of uranium beams to an energy exceeding the Coulomb barrier of a uranium target. The post accelerator configuration will permit an unrestricted isomer search throughout the whole periodic table both on and well off the line of beta stability. Even with the first upgrade, however, an enormous number of nuclides with potentially useful isomers can be explored.

Experimental programs at the University of Rochester's Nuclear Structure Research Laboratory (NSRL) and the Physics Department at Stanford University which have been underway for a number of years and which have direct relevance to the proposed graser isomer search are the following.

1. Study of High Spin States in Deformed Nuclei with Well Developed Collective Degrees of Freedom. This work is carried out by Cline and co-workers and involves a variety of nuclear reactions including (HI,xn) reactions[1,2], multiple heavy ion Coulomb excitation[3] and single and two neutron transfer in heavy ion collisions[4]. A major piece of apparatus used in the work and highly suitable for graser isomer searches is a so called "hedgehog" designed by Cline. It is a multiple gamma ray and particle detector device which permits measurements of multi particle/gamma, and gamma/gamma coincidences.

2. <u>Study of Nuclides with Masses Near the Combined Projectile and Target Masses</u>. This work is carried out by Cormier and co-workers and involves the use of a unique device called a recoil mass spectrometer (RMS) designed by Cormier et al.[5,6]. The properties of this RMS are really quite remarkable. It has a very high mass resolution, $M/\Delta M$ of 750, a large solid angle of 2msr and a wide energy range of 19%. The spectrometer operates at 0° to the beam direction and has a primary beam rejection efficiency of greater than 10^{13}. The device has been used in a number of (HI,xn) studies including measurements of nuclear shape changes at high spin involving (^{32}S,xn) and (^{34}S,xn) reactions on ^{126,128,130}Te[7] and the high spin evolution of quasicontinuum γ radiation in ^{156}Er using the reaction ^{128}Te(^{32}S,4n)^{156}Er[8]. More recently measurements[9] have been made on the gamma ray decay spectrum of ^{155}Dy a neutron deficient isotope of dysprosium using the ^{130}Te(^{30}Si,5n) reaction. The mass peaks 154, 155 and 156 were clearly resolved and gates on each of the three peaks in coincidence with the gamma detector produced the gamma decay spectra. It is also possible, because of the large solid angle of the RMS, to measure triple coincidences as well as double. A recent example[9] are studies of the mass 72 gamma decay spectrum of ^{72}Br, ^{72}Kr and ^{72}Se using the ^{28}Si on ^{50}Cr reaction and selecting the (αpn), (α2n) and (α2p) channels in a triple coincidence involving the mass peak, neutrons and gamma rays.

3. <u>Other Studies Related to the Properties of Possible Graser Candidates</u>. In order to asses the suitability of an isomeric level as a graser candidate it is necessary to have a nearly complete knowledge of the properties of the state. Techniques for measuring many of the important properties have been developed and applied at Stanford University over the course of many years as well as, in some cases, at the University of Rochester. Lifetimes of states of interest in the range 10^{-8} to 1 sec must be measured and considerable experience in such research reside both at Rochester and Stanford. Formation and decay modes must be determined as well as spins and parities. This is an active research area in both institutions. Magnetic dipole and electric quadrupole moments and line widths must be measured. NMR and other methods are used both at Stanford[10] and Rochester[11] - the latter involving RMS. Since recoilless emission will probably be essential to successful operation of a graser - such a level would also be a candidate for the Mössbauer effect. There has been extensive experience at Stanford in Mössbauer research over the past 20 years[12]. Finally, knowledge of the solid state environment of nuclei implanted in various host materials is crucial. Techniques for measuring various environmental parameters are available and in use[13].

Various regions in the periodic table are of particular interest for searches for potential grasers. Strottman et al.[14] have made a literature search to delineate these regions. The present proposed research will be guided by the information they provide.

The combination of the upgraded MP Tandem Van de Graaff accelerator and the experimental equipment described briefly above along with the experience of the experimental nuclear physicists at

Stanford and Rochester make a joint research program to search for potential grasers involving the two institutions particularly appropriate. If graser candidates exist they will be found and their relevant properties measured. The importance of such research in a host of applied fields needs no special emphasis here.

REFERENCES

1) S. W. Yates, I. Y. Lee, N. R. Johnson, E. Eichler, L. C. Riedinger, A. C. Kahler, D. Cline, R. S. Simon, P. A. Butler, P. Colombani, M. W. Guidry, F. S. Stephens, R. M. Diamond, R. M. Ronningen, R. D. Hickman, J. H. Hamilton and E. L. Robinson, Physical Review C21 (1980) 2366.
2) C. J. Lister, G. R. Young, D. Cline, J. Srebrny, D. Elmore, P. A. Butler, R. Ledoux and R. Frideau, Physical Review C20 (1979) 605.
3) D. Cline, Proceedings of the Niels Bohr Centennial Conference on "Nuclear Structure 1985", Copenhagen, Denmark, May 20-25, 1985 - Edited by: R. A. Broglia, G. B. Hagemann and B. Herskind, (Elsevier) 1985, p. 313, NSRL-292
4) M. W. Guidry, S. Juutinen, X. T. Liu, C. R. Binghan, A. Larabee, L. L. Riedinger, C. Baktash, I. Y. Lee, M. L. Halbert, D. Cline, B. Kotlinski, W. J. Kernan, D. Sarantites, T. M. Semkow, K. Honkanen and M. Rajagopalen, Submitted to Physics Letters, 1985, NSRL-290.
5) T. M. Cormier and P. M. Stwertka, Nuclear Instruments and Methods, 184 (1981) 423.
6) T. M. Cormier, M. G. Herman, B. S. Lin and P. M. Stwertka, Nuclear Instruments and Methods, 212 (1983) 185.
7) T. M. Cormier, P. M. Cormier, M. Herman, N. G. Nicolis and P. M. Stwertka, Phys. Rev. Letts. 51 (1983) 542.
8) P. M. Stwertka, T. M. Cormier, M. G. Herman and N. G. Nicolis, Phys. Rev. Letts. 54 (1985) 1635.
9) T. M. Cormier, Private Communications
10) T. Minamisono, J. W. Hugg, D. G. Mavis, T. K. Saylor, Lazarus, H. F. Glavish and S. S. Hanna, Phys. Rev. Lett. 34 (1975) 1465.
11) W. Rogers and D. L. Clark, (Private Communications) and H. G. Berry, L. J. Curtis, D. G. Ellis and R. M. Schectman, Phys. Rev. Lett. 32 (1974) 751.
12) P. B. Russell, G. L. Latshaw, G. Kaindl and S. S. Hanna, Nucl. Phys. A210 (1973) 133.
13) T. Minamisono, J. W. Hugg, D. G. Mavis, T. K. Saylor, S. M. Lazarus, H. F. Glavish and S. S. Hanna, Hyperfine Interactions 2 (1976) 315; J. W. Hugg, Ph.D. Thesis, Stanford University, 1979.
14) D. Strottman, E. D. Arthur and D. G. Madland, preprint, Proceedings of the IDA Gamma-Ray Laser Workshop, Alexandria, VA May 1985.

PROGRAM CONSIDERATIONS FOR THE DEMONSTRATION OF THE FEASIBILITY OF A GAMMA RAY LASER BASED UPON UPCONVERSION*

C. B. Collins
Center for Quantum Electronics, University of Texas at Dallas
P.O. Box 830688, Richardson, TX 75083-0688

ABSTRACT

Plans for a gamma-ray laser pose problems of a broad interdisciplinary nature requiring the fusion of concepts taken from previously unrelated fields of physics. Any actual attempt to construct a gamma-ray laser at the present time would be frustrated by the paucity of supportive data, procedures and understanding. A laser-grade database of nuclear properties does not yet exist, but the techniques for constructing one are currently being developed. Reviewed in this presentation will be the new techniques of Nuclear Raman Spectroscopy (NRS) and Modulated Nuclear Radiation (MNR) being implemented in order to characterize the laser-like properties of nuclear materials.

INTRODUCTION

The central problem to the realization of a gamma ray laser is the identification of a proper material from a slate of 29 candidates. The ultimate success of incoherent pump schemes involving flash X-rays will require investigation of the nuclear properties of those materials that are analogous to the "kinetics" of a conventional laser medium through use of the technique of Modulated Nuclear Radiation (MNR) that we recently introduced[1]. This methodology is the nuclear analog of the optical double resonance studies which yielded much of the laser-grade database upon which rest the newer visible and UV lasers. A database of comparable quality for nuclear "kinetics" will be required for the screening of the 29 candidates and this will require a great amount of input radiation into implementations of the MNR process. Essential to the success of this technique is the accessiblity of a source of pulses of X-rays of nanoseconds duration that can emit a total of one Joule per keV of linewidth in a reasonably brief working period. Either laser plasmas or large e-beam machines can do this in a single shot, each of which requires about an hour of laboratory time to prepare; but costs are very high. As a result, none of these traditional light sources for the subAngstrom region could be used to complete an evaluation of the 29 materials before the turn of the century. In a following post-deadline paper we report recent successes with a prototype flash X-ray device producing 100 mW of average power from 10 nsec pulses at energies near 8 keV, a level of performance approaching that of a large synchrotron light source.

In 1982 we modeled a coherent pumping scheme for a gamma-ray laser,[2] as well as the approach requiring the flash x-rays that was emphasized in the preceding communication.[1] As its name implies, coherent upconversion would allow nuclear materials to be pumped with an intense but conventional coherent source such as a laser. Threshold powers would be much lower than usually projected for a gamma ray laser and the pump sources could be drawn from a more mature technology. This scheme for pumping a gamma ray laser depends upon the development of states of nuclear excitation dressed by the photons of the pump field. Of particular importance is that the metastability of a dressed isomeric state would be released, and with it the stored energy. Unfortunately, despite the many applications of beautiful and involved techniques of nuclear spectroscopy in this aspect as well, the current data base does not provide laser-grade coverage and resolution. Emphasized in this paper will be the new technique of Nuclear Raman Spectroscopy (NRS) that we recently introduced.[3-5] The impact of both new technologies of NRS and MNR upon the search for a viable candidate material for a gamma-ray laser will be assessed.

COHERENT PUMPING SCHEMES

A graphical representation of the states of the total system of radiation fields plus matter that is particularly useful in describing the coherent pumping scheme for a gamma-ray laser is shown in Fig. 1. There it has been assumed that a nucleus in the absence of the radiation field has three levels $|0\rangle$, $|-\rangle$ and $|+\rangle$ corresponding to a lower laser level, a storage level and an upper laser level, respectively, and that the spontaneous transition $|-\rangle \rightarrow |0\rangle$ is forbidden. The splitting between levels $|-\rangle$ and $|+\rangle$ is $\hbar\omega_0$ and is assumed to be nucleonic in origin. Although shown as a variable to facilitate correlations, in a laser candidate it would have some fixed value. The nucleus is assumed to be illuminated by a monochromatic source such as a laser at a frequency ν_L.

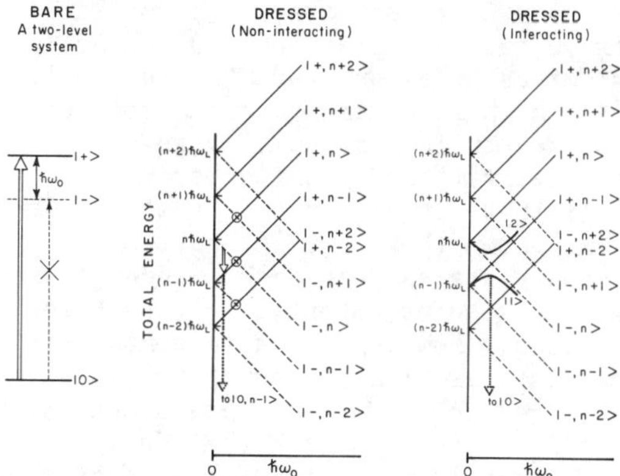

FIGURE 1: Energy level diagram for a three level quantum system immersed in electromagnetic radiation fields of various intensities.

The central panel of Fig. 1 depicts the quantization of nuclei plus fields in a low order of approximation in which accidental degeneracies have not been removed. For orientation the anti-Stokes Raman conversion of a laser photon to a gamma-ray with a negative detuning ΔE_{+-} has been shown by a sequence of arrows. The first step, consisting of the nonresonant excitation of $|+\rangle$, is represented by a downward arrow from $|-,n\rangle$ to $|+,(n-1)|\rangle$ since the total energy is being lost (because of the negative detuning).

The principal significance of the central panel is to identify by the circles some typical points at which accidental degeneracy occurs because of the resonance of the energies $E_{+}-E_{-}=h\nu_{L}$. At the next higher level of approximation when the interaction of the nuclear moments with the laser field is included, such degeneracies are removed. Then, the dressed states, such as $|1\rangle$ and $|2\rangle$, shown by the heavy curves in the right panel are obtained. The energies of these states are shifted but of most importance is that the properties of $|-\rangle$ and $|+\rangle$ are mixed in forming $|1\rangle$ and $|2\rangle$. As a result the transition from $|1\rangle$, the state evolving from $|-\rangle$ at higher energies, to the lower state $|0\rangle$ becomes allowed at a significant fraction of the amplitude of the original transition from $|+\rangle$ to $|0\rangle$. As a consequence, a transition from a storage state such as $|-\rangle$ can be "switched on" by using an intense laser field to dress the state. It is this concept which comprises the foundation of the scheme for pumping a gamma-ray laser through coherent upconversion.

NUCLEAR RAMAN SPECTROSCOPY

In 1984 the results[3] of a critical experiment were reported which showed that coherent superpositions, and even dressed nuclear states could be produced by immersing ferromagnetic media containing demonstration nuclei of ^{57}Fe in an oscillating radio-frequency field which served as the coherent pump. The amplitudes of the dressed states were found to agree with theory and in this way the estimates of the common matrix elements appearing in both the dressed state theory and in the threshold estimates were confirmed. Moreover, these experiments suggested a new technique of Nuclear Raman Spectroscopy that now promises to extend both the resolution and tuning range available for high resolution nuclear spectroscopy,[4,5] of the type needed in the search for potentially resonant intermediate states needed for applications of the dressing process.

The concept of these experiments is best appreciated by reference to Fig. 1. As shown there, an interaction potential is assumed to act upon stationary states such as $|-\rangle$ and $|+\rangle$ whose splittings are determined by a component potential of the $\hat{H}$ for which $|-\rangle$ and $|+\rangle$ are eigenfunctions. However, in a ferromagnetic medium in order to conform to the dictates of magnetohydrodynamics the effect of the interaction is arranged to add a time dependence which is periodic (but not sinusoidal) to the part of $\hat{H}$ that determines the splitting of $|-\rangle$ and $|+\rangle$. The result is that in the representation of Fig. 1, the horizontal axis must be considered to be undergoing cycles of stretching and shrinking with time. The physical effect is that new transition frequencies are generated at the sums and differences between the frequency of the gamma transition and those of an integral number of photons dressing the nuclear states.[3-5] By varying the frequency of the mixing radiation the energy of the sum frequency line can be swept, just as in the analogous processes implemented in the optical range and this provides the bases for the Nuclear Raman

Spectroscopy (NRS) technique.

While the addition of a tunable sideband to a nuclear source is more instinctively attractive, the complementary experiment with an absorber proves the same principles with less expense yielding the data shown in Fig. 2, taken from Ref. 5. The point of the experiment was not to characterize the energy levels of ^{57}Fe but to demonstrate the effectiveness of this new NRS technique for nuclear spectroscopy. In practice, the method would probably be most useful with the rf field being replaced by a microwave or infrared photon field. For example, the technique could be used to conduct Mossbauer spectroscopy with much higher resolution than previously achieved. This would be possible because no mechanical tuning needs to be used. In case microwave or infrared photons were used, the tuning range of Mossbauer spectroscopy would be dramatically increased. This NRS technique is now being fully implemented to facilitate the search for appropriate intermediate states for coherent upconversion. The most recent results are described in a postdeadline paper.

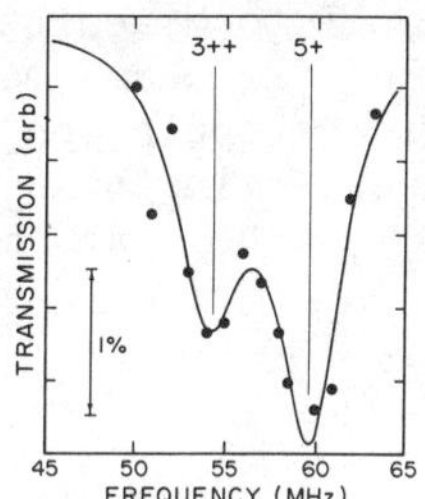

FIGURE 2: Gamma transmission intensity plotted as a function of rf photon frequency. The larger absorption line corresponds to the first order, sum-frequency sideband of the |1/2,-1/2>→|3/2,-1/2> nuclear transition of ^{57}Fe, while the smaller line corresponds to the second-order, sum-frequency sideband of the |1/2,1/2>→|3/2,-1/2> transition.

CONCLUSIONS

Preliminary experiments and analyses indicate that the new technologies of NRS and MNR can make significant contributions to the construction of a database of nuclear properties defined with laser-grade precision. Such a resource is essential for the identification of a suitable material for a gamma ray laser, if indeed one exists. While current technologies could not assess the suitability of the 29 best candidate isotopes by the turn of the century, the acceleration afforded by the new methodologies now being implemented should significantly affect the programmatic objectives of proving feasibility within a more reasonable time.

REFERENCES

*Supported in part by the Office of Naval Research and in part by the Strategic Defense Initiative Office (IST).

1. Our communciation TC4 in the preceding session.
2. C. B. Collins, F. W. Lee, D. M. Shemwell, B. D. DePaola, S. Olariu, and I. I. Popescu, J. Appl. Phys. 53 4645 (1982).
3. B. D. DePaola and C. B. Collins, J. Opt. Soc. Am. B 1, 812 (1984).
4. C. B. Collins and B. D. DePaola, Opt. Lett. 10, 25 (1985).
5. B. D. DePaola, S. S. Wagal and C. B. Collins, J. Opt. Soc. Am. B 2, 541 (1985).

CONDITIONS FOR γ-RAY LASING INTO MULTI-BEAM BORRMANN MODES OF MÖSSBAUER CRYSTALS

J. T. Hutton, G. T. Trammell, and J. P. Hannon
Rice University, Houston, Texas 77251

ABSTRACT

The advantages of utilizing the 2-beam Borrmann modes of a crystalline, rather than an amorphous, sample for the lasing modes of a γ-ray laser (reduced lasing threshold and higher gains above threshold, due to reduced photoabsorption and improved coupling between the lasing mode and the nuclei) are well known.[1,2] We discus here the further improvements which are obtained in multi-beam (3 or more) Borrmann modes. These modes would be fed automatically by emitters of multipole M1 or higher located within the crystal. As an example, the 6 beam mode which couples well to M1 emitters would have a lower threshold than a similar 2 beam mode. More importantly, the slope of the net gain vs. population inversion density curve would be three times greater in the 6 beam mode, allowing practical levels of gain to be reached at much lower inversion densities.

LASING IN A BORRMANN MODE

In our previous work,[2] we have shown that the critical population inversion requirement at threshold for a Mössbauer crystal, lasing in a 2-beam Borrmann mode, is substantially smaller than that required to induce lasing in an amorphous sample of the same material (or equivalently, off-Bragg in the crystal), so long as the emitters have multipolarity higher than E1. This is due to "anomalous emission,"[3] an effect whereby Mössbauer sources of M1 or higher multipolarity located deep within a crystal can emit into that crystal's Borrmann channels, but an E1 source cannot.

While it is likely that any graser will be a pulsed, rather than a steady state laser, the steady state equations are much simpler and serve to demonstrate the advantages of multi-beam modes. The Schawlow-Townes[4] steady state lasing condition is $K \geq 0$, where K, the net gain per unit length coefficient, is given by

$$K = K_o - \mu = \sigma_N g_N \Delta n - \sigma_A g_A n_o. \tag{1}$$

In this equation, σ_A is the nonresonant atomic absorption cross section, which for the Mössbauer energies will be almost entirely due to photoelectric absorption, while σ_N is the nuclear absorption-stimulated emission cross section at resonance.[5] Δn is the population inversion density, including spin degeneracy factors. g_N and g_A are the coupling factors of the nuclear resonators and the atomic electrons to the electromagnetic field of the lasing mode, respectively. For the usual plane wave modes (off Bragg or in an amorphous sample) $g_N=g_A=1$, but for the Borrmann modes, which consist of the coherent superposition of two or more plane waves, produced by Bragg reflections in perfect crystals, the coupling coefficients will not be one.[2] Note that any increases in g_N will be particularly significant, since this factor occurs in the term in K which is proportional to Δn (i.e. $\sigma_N g_N$ is the slope of the line K vs. Δn). Since any practical laser will require a net gain substantially above that near threshold, this increased slope will allow the required net gain per unit length to be obtained at a much lower population inversion density.

MULTI-BEAM BORRMANN MODES

Our recent results show that a substantial increase in the nuclear coupling coefficient, g_N, is obtained by going to multi-beam Borrmann modes. The number of available modes depends upon the length of $\lambda\!\!\!^{-}{}^{-1}$, which determines the size of the Ewald sphere. We find that _a crystalline needle, grown along an axis with either 4 or 6 fold symmetry, will lase into the mode with the largest number of beams at the largest available Bragg angle._ As an example, for ^{57}Fe grown along a (111) axis, this mode is a 12 beam mode at $\theta=76.6°$.

Even though there can be many beams in some of these modes (up to 12), they will be automatically excited by an internal source. An excited atom in the crystal will emit spherically, so that there is always radiation emitted into whatever is the proper set of directions to make the mode. There are no alignment problems, and the nuclei couple to the modes automatically.

Each multi-beam geometry (i.e. for a given value of m) has several Borrmann modes, and it is possible to find modes which couple well to any arbitrary multipole emitter (except E1). Of particular interest are modes which have symmetry properties so as to couple well to M1 emitters (a majority of Mössbauer transitions are M1). For these modes,

$$g_A = \frac{m}{2}\,\frac{\sin^2\theta}{\cos\theta}\,\frac{\langle x^2\rangle}{\lambda\!\!\!^{-}{}^2},\ \text{except } m=2 \qquad (2)$$

and

$$g_N = m\,\frac{\sin^2\theta}{\cos\theta}, \qquad (3)$$

where m is the number of beams in the mode, θ is the half-apex angle of the cone formed by the wavevectors, and $\langle x^2\rangle$ is the mean thermal plus zero point displacement of the atom from equilibrium. For the case m=2, there is an additional factor of 2 in g_A (Eq. 2). This occurs due to the symmetry change in going from 2 beams to 3 or more. In the 2 beam case, the wave field has nodal _planes_ at the atomic sites, while, in the case of 3 or more beams, there are nodal _lines._

Substituting (2) and (3) into (1) yields

$$K = \sigma_N m \frac{\sin^2\theta}{\cos\theta} [\Delta n - \Delta n_t] \tag{4}$$

where $\Delta n_t = [\langle x^2\rangle/2\lambda\!\!\!^{-2}][\sigma_A/\sigma_N]n_o$ is the threshold inversion density. (For the 2 beam case, $\Delta n_t = [\langle x^2\rangle/\lambda\!\!\!^{-2}][\sigma_A/\sigma_N]n_o$.) As an example, in Fig. 1 we plot three steady state K vs. Δn curves for an ^{57}Fe needle, grown along the (100) axis. Shown are the lines for off-Bragg lasing, and those for the 2 beam and 8 beam modes at the largest available cone angle. Note the improvement in slope obtained in the 8 beam mode.

These multi-beam Borrmann modes have interesting angular divergence properties as well. The angular divergence of the wave field emerging from an amorphous needle, or a crystalline needle lasing in an off-Bragg mode is determined by the "macro-geometry" of the sample; either by the physical collimation introduced by the size of the needle or by the diffraction limit, which depends on the needle diameter, whichever is greater.[1]

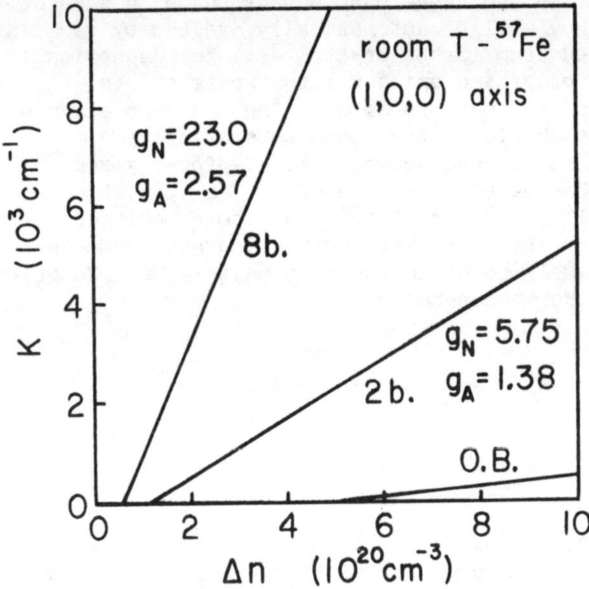

Fig. 1. Plots of K vs. Δn for an ^{57}Fe needle at room temperature, grown along the (100) axis. Lines are shown for off-Bragg, as well as for the M1 symmetry 2 and 8 beam modes with the largest available cone angle (here θ=71.7°).

Much better collimation may be obtained by utilizing the natural collimation which takes place during Bragg diffraction. In a 2 beam Borrmann mode, Bragg diffraction limits the angular divergence in the rocking curve direction to $\Delta\theta \approx W_D$, the Darwin width, typically 10^{-5} to 10^{-4} rad., regardless of the thickness of the crystal in that direction, (so long as the crystal is sufficiently thick to form the Borrmann modes, i.e. a few primary extinction lengths). Thus, in this case the angular divergence is determined by the "micro-geometry" of the sample, rather than the "macro-geometry."

In 2 beam modes, the angular divergence in the other direction is still determined by the size of the crystal, either geometry or the diffraction limit, as in the amorphous case. In multi-beam Borrmann modes, however, the radiation emerges in well defined beams, which are collimated in both directions to $\Delta\theta \approx W_D$.

SUPERRADIANCE

Finally, we find that when the inversion density has been increased sufficiently above the lasing threshold such as to lead to a practical level of gain[6,7] (say 10^6 times as many photons in the mode as would be expected for incoherent decay), the threshold for superradiance will also have been substantially exceeded. Since the time required to superradiate is much shorter than the time required to lase, the system will decay by superradiant, rather than stimulated emission. Nevertheless, the same benefits of multi-beam Borrmann modes, which are so important for lasing, also apply to the case of superradiance, so that once again the system will emit into the available multi-beam Borrmann mode with the largest cone angle.

REFERENCES

1. For an excellent review of all aspects of γ-ray laser research, see G. C. Baldwin, J. C. Solem, and V. I. Gol'danski, Rev. of Mod. Phys. 53, 687 (1981).
2. J. P. Hannon and G. T. Trammell, Optics Communications 15, 330 (1975).
3. J. P. Hannon, N. J. Carron, and G. T. Trammell, Phys. Rev. B9, 2810 (1974).
4. A. L. Schawlow and C. H. Townes, Phys. Rev. 112, 1940 (1958).
5. See for example Eq. A4 in Ref. 3.
6. G. T. Trammell and J. P. Hannon, Optics Communications 15, 325 (1975).
7. G. T. Trammell, J. T. Hutton, and J. P. Hannon, to appear in Proc. Wash. Workshop on X-Ray and γ-Ray Lasers, Wash., May 1985.

NUCLEAR EXCITATION THROUGH THE DYNAMIC HYPERFINE EFFECT*

George Rinker
T-1, Los Alamos National Laboratory, Los Alamos, NM 87545

ABSTRACT

Calculational methods developed for muonic atoms have been applied to the problem of nuclear interlevel transfer via resonant electronic states. Nuclear excitation probabilities as large as a few percent have been found. The most promising circumstances for the pumping of gamma-ray lasers will be discussed.

Figure 1 shows a prospective γ-ray laser scenario. A principal problem concerns pumping of the nuclear transition $I \to I'$. Direct photoabsorption is questionable because the cross section depends upon the nuclear radius as a dimensional parameter. This is too small to be useful for reasonable photon intensities. Photoabsorption by the electrons is much stronger because the relevant dimensional parameter is the atomic radius. This suggests a dual process in which the electron excitation energy is transferred to the nucleus by means of a resonant electron-nucleus coupling. The possibility of an enhanced result arises from the fact that the electron-nucleus interactions are obtained from multipole expansion of r^{-1} potentials, whereas the photoabsorption cross sections are obtained from plane waves. The phenomenon has been studied exhaustively in muonic atoms,[1] where for over 30 years it has been known as the dynamic hyperfine effect. Some implications in the present context have been discussed by Morita,[2] whose notation we adopt.

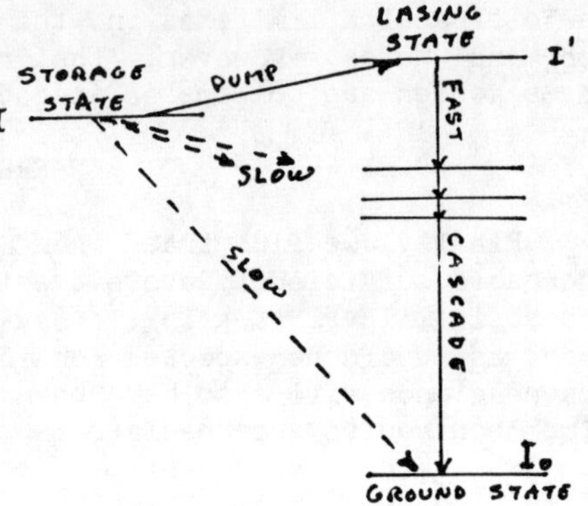

Fig. 1. γ-ray laser scenario. The metastable storage state I is populated by radiochemical means. Fast pumping initiates the lasing cascade $I' \to I_0$.

The quantity of interest is the steady-state population N' of nuclear state I': $N' \to 2\tan^2\theta \; T^2_{jj'}/T^2_{I'0}$ as $\theta \to 0$, where θ is the mixing angle in the coupled subspace $|Ij'\rangle$, $|I'j\rangle$, and j,j' are the resonant electron states. $T_{jj'}$ is the matrix element responsible for electron photoabsorption $\gamma j \to j'$, and $T_{I'0}$ is the nuclear decay matrix element $I' \to \gamma I_0$. The hamiltonian is trivial to diagonalize exactly in the coupled subspace once the mixing matrix element V_{23} is known. In the limit $V_{23} \to 0$, the perturbation result is obtained: $\theta \simeq V_{23}/(E_3 - E_2)$, where $E_3 = E_{I'} + E_j$, $E_2 = E_I + E_{j'}$. The effective coupling contains the energy denominator $E_3 - E_2$, which in principle can approach zero. In

*Supported by the U. S. Department of Energy

this limit, $\sin^2\theta \to 1/2$ and $N' \to T^2_{jj'}/T^2_{I'0}$. It should be pointed out that the finite lifetimes of the levels complicate the analysis. Various decay processes contribute. A time-dependent solution is required, and to my knowledge, the question is still open.

An adapted version of the computer code RURP[3] has been applied to the calculation of V_{23} for selected electric multipoles and single-particle electron excitations. Matrix elements for collective electron excitations from the same shell probably lie between $N^{1/2}$ and N times the single-particle results, where N is the number of participating electrons. Plotted in Figs.2-3 are the results of a general canvass of transitions for atomic numbers Z=12, 40, 68, and 92, in the energy range 100 to 10^5eV. Electron energies and wave functions are obtained from a self-consistent Dirac-Fock-Slater calculation for each neutral atom. Nuclear matrix elements used give transition strengths of $\simeq(4+L)^2$ single particle units, where L=1,2 is the multipolarity. These are probably optimistic. There is no striking correlation between transition energy and coupling matrix element. A similar search for L=3 transitions produced none with $V_{23}>0.5\times10^{-6}$eV. Depending upon the degree of energy degeneracy and the effects of competing decay processes, it is conceivable that the strongest dipole transitions could produce relative nuclear excitation probabilities of a few percent. Further details of these calculations are available from the author.

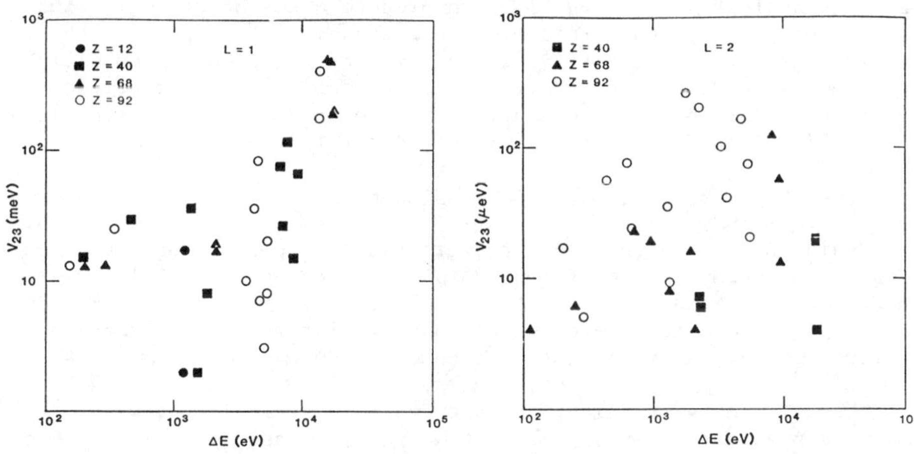

Fig. 2. L=1 matrix elements.

Fig. 3. L=2 matrix elements. Note vertical scale difference.

REFERENCES

1. See, e.g., E. Borie and G. A. Rinker, Rev. Mod. Phys. 54, 67 (1982), and references therein.

2. M. Morita, Prog. Theor. Phys. 49, 1574 (1973).

3. G. Rinker, Comput. Phys. Commun. 16, 221 (1979).

POSSIBILITY OF GRASING BY LASER-DRIVEN NUCLEAR EXCITATION*

L.C. Biedenharn, K. Boyer and Johndale C. Solem
Los Alamos National Laboratory
Theoretical Division, MS B210
Los Alamos, New Mexico 87545

The possibility of nuclear excitation by the sequential process--laser-driven coherent oscillations of outer atomic shells[1,2] which then couple via nuclear multipole moments to produce nuclear excitation[3]--raises the question as to whether grasing might be achieved through this mechanism. We examine here this possibility for laser driven excitation of the 73 ev first excited state of ^{235}U.

It is important to recognize that the electronic environment surrounding the ^{235}U nucleus will be substantially changed under the intense laser irradiation contemplated ($\sim 10^{17}$ W/cm^2). Multiple ionization along wich coherent oscillations of the outer electron shell (with excitation energy in the kilovolt range) is expected to exist. Under such circumstances it is reasonable to expect that there will be a sizeable density of available electronic configurations, of the right spin and parity, to couple resonantly with the nuclear 7/2- $\leftrightarrow$ 1/2+ transition. The result (by analogy with analog state mixing in nuclear physics) will be to produce a set of strongly mixed 'atomic-nuclear' states spread over an interval of the order of the coupling energy. The radiative widths of these coupled states will be dominated by the widths of their atomic components[4]. Moreover, in this highly perturbed environment, the large internal conversion width of the (1/2 +) isomeric state will be greatly reduced.

These circumstances may make grasing possible.

Ordinarily--assuming nuclear excitation and population inversion to exist--no stimulated emission could occur because of Doppler de-tuning. But in this highly perturbed nuclear environment, the mixed nuclear-atomic excited states should have radiative widths orders of magnitude larger, sufficient to overcome the de-tuning.

There is an interesting, and not especially difficult, way to test this possibility. Suppose that an initial laser irradiation has indeed produced a substantial density of 26-minute ^{235}U isomers. Then a second irradiation, creating coherent electron oscillations in this isomer population should produce transient mixed nuclear-atomic states, which with their substantially shortened lifetime, could then grase at 73 ev in a swept-gain superradiant way.

We propose that an experiment along these lines be attempted.

*Supported by U.S. Department of Energy

In a sense, our proposal is but a variant on the well-known two-step pumping scheme[5], with the (de-coupled, quiescent environment) 26-minute isomer as the storage state and the excited environment mixed atomic-nuclear states functioning as the transfer mechanism[6]. Energetically the (1/2+) ^{235}U isomer is nearly optimal for such a test, but the fact that the radiative transition is electric octupole is much less so. This is because the size of the electronshell-nuclear multipole coupling sets the scale for spreading of the mixed atomic-nuclear states. For octupole coupling this is small ($\sim$ mv). For a deformed heavy nucleus with a large collective quadrupole moment the situation is much more favorable ($\sim$ ev), especially at the level of coherent electron oscillations necessary to induce nuclear excitations in the kev range.

We wish to thank Dr. George Baldwin for discussions.

REFERENCES

1. K. Boyer and C.K. Rhodes, Phys. Rev. Lett. 54, 1490 (1985).
2. C.K. Rhodes, Science, 229, 1345 (1985).
3. L.C. Biedenharn, G.C. Baldwin, K. Boyer, and Johndale C. Solem, "Nuclear Excitation by Laser Driven Coherent Outer Shell Electron Oscillations", previous paper.
4. The anomolously large transition electric dipole moment for the ($\sim$ 200 ev) ground state doublet ($\pm 5/2$) of ^{229}Pr is attributed to mixing with coupled electronic states (Prof. W. Haxton, private communication).
5. G.C. Baldwin, "Two-step Pumping of a Superradiant Graser", Proceedings of the IDA Workshop on Gamma-Ray Lasers (21-22 May 1985, Alexandria, Va.) to be published.
6. J.C. Solem and G. Rinker, "Nuclear Interlevel Transfer Driven by Electronic Transitions", Proc. of the IDA Workshop on Gamma-Ray Lasers, loc. cit..

NUCLEAR EXCITATION BY LASER DRIVEN COHERENT OUTER SHELL ELECTRON OSCILLATIONS*

L.C. Biedenharn, G.C. Baldwin, K. Boyer
and Johndale C. Solem
Los Alamos National Laboratory, Theoretical Division, MS B210
Los Alamos, New Mexico 87545

Recent experiments[1,2] on multiphoton ionization of atoms by intense laser beams have shown unexpectedly large systematic effects. These experiments involved a high brightness ultraviolet laser operating at 193 nm (6.43 ev) with pulselength ~ 5 ps and intensities $10^{14} - 10^{17}$ W/cm^2. At this upper intensity the radiative field strength is greater than the characteristic atomic unit e/a_o^2.

In experiments of the type: $N\gamma + X \rightarrow X^{q+} + qe^-$, with X a heavy atom (Xe,...,U) excitations involving $N \sim 100$ photons and charges $q \sim 8$ are found. The characteristic features observed[3] are: (a) anomalously strong, non-linear effects; (b) pronounced shell structure effects; (c) maximum charge produced correlates with complete removal of outer shell; (d) coupling strength varies with the number of outer shell electrons; and (e) dramatic change as intensity exceeds 10^{15}W/cm^2.

These features have been interpreted as evidence for <u>laser driven coherent oscillations of the outer electron shell</u>[4]. A semi-quantitative analysis, based on a time-dependent Hartree-Fock approach, has been presented to support this view[5].

In the following we will accept the existence of coherent outer shell oscillations as validated, although this is by no means universally accepted[6]. Planned experiments in the intensity range $10^{17} - 10^{19}$ W/cm^2--where the radiative field strength is more than an order of magnitude larger than the atomic unit--should almost certainly achieve coherent oscillations.

We wish to emphasize that the existence of laser driven coherent multielectron oscillations will offer significant new possibilities for achieving <u>nuclear excitation</u>. In comparison to atomic excitation, nuclear excitation by direct radiative coupling is strongly disfavored on dimensional grounds, even neglecting shielding. By contrast, the indirect process--laser driven coherent outer shell oscillations which then couple electromagnetically to the nucleus to produce nuclear excitation--is strongly favored, since this latter process involves near-zone radiative coupling with irregular (r^{-L-1} as opposed to r^L) multipole moments. This feature is the operative principle behind the well known predominance of internal conversion over 'external' conversion and of Coulomb excitation vs radiative excitation of nuclei[7].

*Supported by U.S. Department of Energy

The nucleus ^{235}U affords a novel way to test this idea. This nucleus is exceptional in having an unusually low energy first excited state (1/2+) just 73 ev above the ground state (7/2-). Although the octupolar interaction energy induced by the coherent excitation of, say, the (4f) shell is very small compared to the nuclear excitation energy, the coupling should be strongly enhanced by near resonances. (There exists at least one close resonance: $5d_{3/2}$(-32.5 ev) ⟶ $6p_{3/2}$(-103.1 ev)[8,9], and the level density in uranium is unusually large.) It is highly probable that a significant fraction ($\sim$10%) of all coherently excited ^{235}U atoms results in excitation of the 73 ev isomeric level.

The signal for this excitation would be the emission of internal conversion electrons with $\sim$26 minute half-life following the laser pulse.

Let us note that excitation of the 73 ev level in ^{235}U has been observed in a laser heated plasma[9], but not so far in the collision-free regime characteristic of the presently proposed experiment.

We wish to thank Dr. Rinker for helpful discussions.

REFERENCES

1. T.S. Luk, H. Pummer, K. Boyer, M. Shahidi, H. Egger and C.K. Rhodes, Phys. Rev. Lett. 51, 110 (1983).
2. C.K. Rhodes, in Proceedings of the Third International Conference on Multiphoton Processes, P. Lambropoulos and S.J. Smith, Eds. (Springer-Verlag, Berlin, 1984).
3. C.K. Rhodes, Science, 229, 1345 (1985).
4. K. Boyer and C.K. Rhodes, Phys. Rev. Lett. 54, 1490 (1985).
5. A. Szöke and C.K. Rhodes, "A Theoretical Model of Inner-Shell Excitation by Outer-Shell Electrons", to be published.
6. P. Lambropoulos, Phys. Rev. Lett. 55, 2141 (1985).
7. L.C. Biedenharn and P.J. Brussaard, Coulomb Excitation, (Clarendon Press, Oxford, 1965.)
8. V.I. Goldanskii, R.N. Kuz'min, and V.A. Namiot, Sov. J. Nucl. Phys. 33, 319 (1981).
9. Y. Izawa and C. Yamanaka, Phys. Lett. 88B, 59 (1979).

NOVEL EXPERIMENTAL SCHEMES FOR OBSERVING THE MÖSSBAUER EFFECT IN LONG-LIVED NUCLEAR LEVELS

Gilbert R. Hoy
Old Dominion University, Norfolk, Virginia 23508

ABSTRACT

The development of gamma ray lasers (GRASER) will depend on the utilization of long-lived, recoilless, nuclear, gamma transitions, i.e., the Mössbauer effect. The lifetimes required from practical considerations must be on the order of seconds at least. It is not clear that the Mössbauer effect has ever been observed in such long-lived states. We propose some experimental techniques to unambiguously observe the effect in the ^{109}Ag first-excited nuclear states having a lifetime of about 40 seconds. The techniques proposed are: coincidence Mössbauer spectroscopy; conversion electron Mössbauer spectroscopy; and gravitational line sweeping.

INTRODUCTION

We proposed to apply novel Mossbauer spectroscopic methods to determine if recoilless, resonant, nuclear emission and absorption are occurring in long-lived (>>1μsec) nuclear transitions and the limits on the corresponding line widths. The methods proposed are (i) Coincidence Mössbauer Spectroscopy, (ii) Conversion Electron Mossbauer Spectroscopy, and (iii) Gravitational Line Sweeping based on the gravitational red shift.

The development of gamma ray lasers (GRASER) depends critically on a number of factors. It has been shown that the GRASER is only possible if recoilless gamma radiation is utilized. The need for long-lived nuclear states is crucial when one considers the practical problems associated with GRASER development.

PROPOSED EXPERIMENTS

We propose to study the long-lived ($\tau \approx 40$sec) first excited nuclear level ^{109}Ag. It appears that there are only three short papers[1-3] that deal with Mössbauer studies in ^{107}Ag and ^{109}Ag. Their results are not convincing. The major difficulty is due to the fact that the line width in the ^{109}Ag case is $\Gamma = 1.1 \times 10^{-17}$ eV and in ^{107}Ag, $\Gamma = 1.0 \times 10^{-17}$eV.

The idea used in the first experiment[1] was to observe the nuclear resonant effect <u>indirectly</u>. This experiment had a number of problems. They quote an effective line width $\Gamma_{eff} = 7.5 \times 10^5 \Gamma$. Even if true, the result is worthless from the GRASER point of view. The experiment by Wildner and Gonser[3] is far superior. In this experiment, the parent source (^{109}Cd) nuclei and the absorber ^{109}Ag ground state nuclei resided in the same silver single crystal. They quote an effective line width, $\Gamma_{eff} = 30\Gamma$ still much too large.

Our sample is planned to be a silver single crystal which will contain both the source and absorber nuclei, as in the Wildner-Gonser experiment.[3] Using such samples we propose to employ coincidence Mössbauer spectroscopic methods to measure the resonant self absorption. The t=0 pulse, in the present case, could be due to an x-ray produced following the electron capture in cadmium leading to the formation of the silver first excited nuclear level. The gamma ray resulting from the nuclear decay to the ground state is then counted, after passing through the rest of the silver sample, in delayed coincidence with the x-ray. The point here is that the measured decay curve will differ from the typical lifetime exponential shape. The calculated result using ^{109}Ag under ideal conditions is shown in Figure 1.

CALCULATED RESULT USING TIME FILTERING THEORY (^{109}Ag)

A promising technique is to look at the conversion electrons emitted after the resonantly absorbing nuclei are excited. The internal conversion coefficient, α, is approximately 20 for silver. Thus conversion electron Mössbauer spectroscopy (CEMS) is quite feasible. Consider Figure 2. In this case we are proposing to do CEMS, however the whole configuration is able to rotate about an axis perpendicular to the gravitational field. A simple calculation using ^{109}Ag shows that if the gamma ray photon falls ~ 1×10^{-4} cms in the gravitational field its energy is raised by about one line width. Rotating the crystal would "sweep" through the resonance.

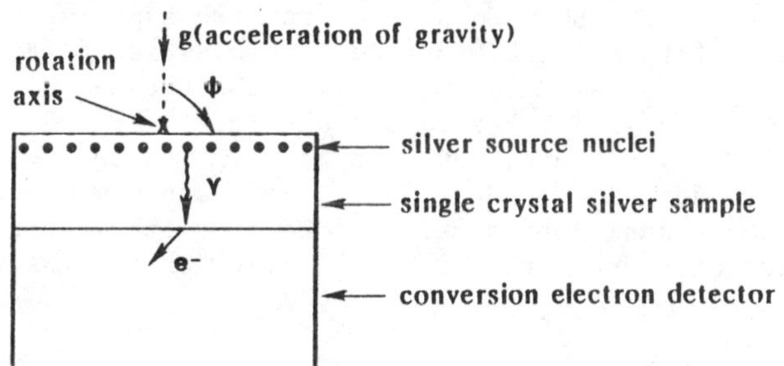

REFERENCES

1. G. E. Bizina, A. G. Beda, N. A. Burgov, and A. V. Davydov, Sov. Phys. JETP 18, 973 (1964).
2. V. G. Alpatov, A. G. Beda, G. E. Bizina, A. V. Davydov, and M. M. Korotokov, Proc. Int. Conf. on Mössbauer Spectroscopy, Bucharest, Romania (1977).
3. W. Wildner and U. Gonser, Journal de Physique 40, C2-47 (1979).

CYLINDRICAL MODEL FOR BAND SEQUENCES AND INTERBANDHEAD E2 TRANSITIONS IN EVEN-EVEN RARE EARTH NUCLEI*

F.X. Hartmann†, Y.Y. Sharon†† and R.A. Naumann
Princeton University, Princeton, NJ 08540

This work provides a cylindrical group-theoretical extension to the SU(3) limit of the nuclear interacting boson model (IBM). We use a six-dimensional cylindrical phonon basis [constructed according to the group chain $SU(3) \supset SU(2) \times U(1)$] to study the energies of, and the E2 transition matrix elements between, positive-parity intrinsic bandhead states. An interacting cylindrical boson (or phonon) Hamiltonian is constructed which leads to an improved description of the properties of low-lying rotational bands in deformed rare-earth nuclei. The deformations of intrinsic states are modeled by the breaking of the SU(3) spherical symmetry into a cylindrical symmetry.

In this report, we limit ourselves to assuming that the rotational and intrinsic states are exactly separable. We find that the SU(3) limit of the IBM does not provide the best description of rotational nuclei; additional terms of other IBM limits appear naturally in our more general cylindrically symmetric picture. We explain several observed features which cannot be described by the strict SU(3) limit of the IBM. We account for the experimentally observed ordering of rotational bandhead levels, which differs from nucleus to nucleus. For example, we explain the unusual 0^+ bandhead sequences in ^{156}Gd (see Table 1) and also of ^{168}Yb and ^{168}Er. In our broken SU(3) picture, interbandhead transition rates which are normally forbidden in the pure SU(3) limit can now occur [see Figure (1) for the transitions studied]. We have obtained the analytical dependences of these rates on the deformation and the boson number.[1]

*Supported in part by grants from the National Science Foundation.

†Present address: IDA, Alexandria, Virginia

††Permanent address: Stockton State College, Pomona, New Jersey

[1] F.X. Hartmann, Ph.D. Thesis, Princeton University, 1985.

Table I
Bandhead States in ^{156}Gd

Bandhead State Model	Expt.	Expt. Energy (keV)	Theor Energy (keV)
|g.s.>	0^+_1	0	0
|β>	0^+_2	1049	1049
|γ>	2^+_1	1154	1155
|2β>	0^+_3	1715	1717
|2γ>	2^+_2	1827	1830
|δ>	4^+_2	1861	1857
|α>	0^+_4	1168	1168

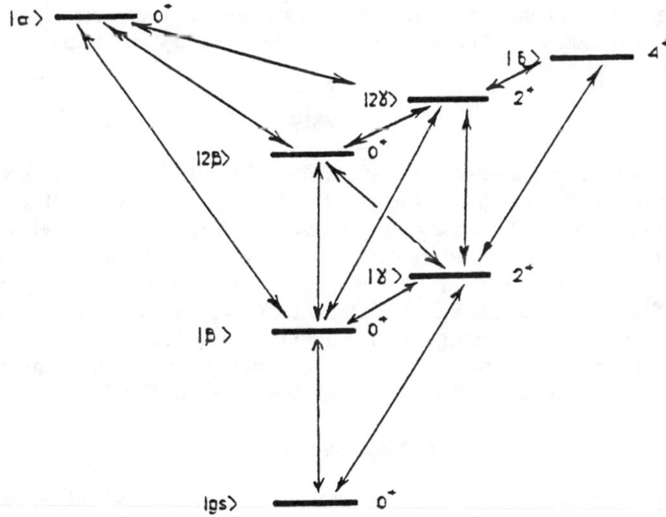

Figure 1 - The interbandhead E2 transitions studied, for the states listed in Table 1.

ASSESSMENT OF A METHOD PROPOSED FOR FINDING TRANSFER LEVELS FOR ISOMERIC DEEXCITATION*

R. C. Haight and G. C. Baldwin
Los Alamos National Laboratory, Los Alamos, NM 87545

ABSTRACT

The existence of closely spaced nuclear levels with disparate lifetimes is an essential feature of recent proposals for gamma-ray lasers. Resolving such pairs of levels is beyond the present capabilities of nuclear spectroscopy. In this work we consider the possibility that exposure of an isomer to a continuum of radiation, specifically a pulse of black body radiation produced by a laser, would reveal the second level. We compare the required couplings of the nucleus to the photon field with those observed. We conclude that this method probably is not appropriate for finding transfer levels.

INTRODUCTION

In two-stage pumping schemes for gamma-ray lasers, a required feature is the existence of a so-called transfer state (state 2 in Fig. 1) near an isomeric nuclear state (state i in Fig. 1). These two states should be very close together, separated by less than a few kilovolts, to reduce the energy imparted to the medium during excitation of the transfer level. Such closely-spaced doublets can be very difficult to resolve by conventional spectroscopy. We investigate here the possibility of finding transfer levels near isomers using broad-band radiation to excite the transfer level and thereby confirm the existence of such a level. Once this level has been found, spectroscopic tools of high resolution would be used to determine its energy, quantum numbers, and transition matrix elements.

ANALYSIS

For this assessment, we assume that black-body radiation, such as produced by laser-target interaction, illuminates the isomer. A very small part of this continuum photon spectrum (see Fig. 1) can convert the isomer to the transition level. We ask, what is the radiative width of the transition $i \rightarrow 2$ such that 0.1% of the isomer will be converted to the transfer level in 10^{-9} sec, the length of typical laser pulses. This minimum width depends on the number of photons incident on the isomer and therefore on the black-body temperature and photon (transition) energy. The results (solid curves) are compared with the widths of single particle transitions (one Weisskopf unit) for various multipolarities (dotted and dashed lines) in Fig. 2.

CONCLUSIONS

This method of searching for levels near isomeric states would be useful only under very restrictive conditions: The multipolarity of the transition could be only $E1$ or $M1$, the transition strengths should be close to the full single particle value, the transition energy should be neither too high ($E_\gamma < 10kT$) nor too low ($E_\gamma > 100eV$).

* Supported by Division of Basic Energy Sciences, U.S. Department of Energy

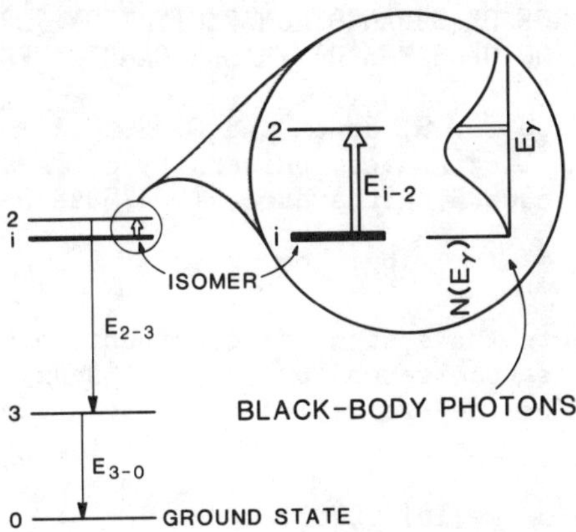

Fig. 1. Schematic approach of searching for levels (2) close to isomeric states (i) by irradiating the isomer with black-body photons. Signatures of successful transitions are depletion of the isomer, build-up of a residual state, or prompt radiation such as E_{2-3}, E_{3-0}, etc.

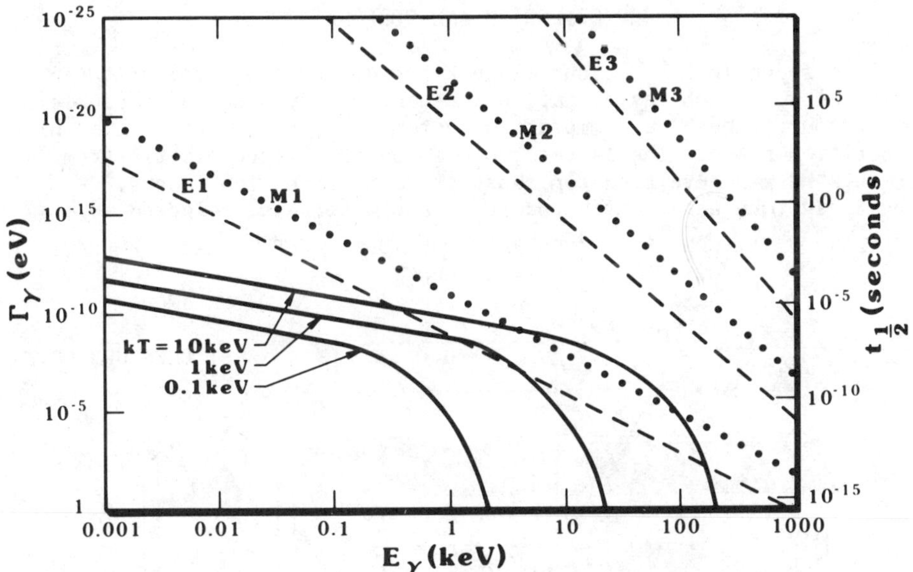

Fig. 2. Transition width required (solid curve) to convert 0.1% of the isomer to the transition level in 1 ns as a function of black-body temperature and transition energy. For comparison, single particle electric and magnetic transitions are shown (dashed and dotted lines respectively).

A FLASH SOURCE OF SUBANGSTROM EXCITATION FOR THE EVALUATION OF GAMMA RAY LASER CANDIDATES

F. Davanloo, T. S. Bowen and C. B. Collins
Center for Quantum Electronics, University of Texas at Dallas
P. O. Box 830688, Richardson, TX 75083-0688

ABSTRACT

This paper presents the design and construction of a flash X-ray device producing intense nanosecond pulses at 100 Hz rates to excite nuclear flourescence for the evaluation of candidate materials for a gamma ray laser.

INTRODUCTION

The central problem to the realization of a gamma-ray laser is the identification of a proper material from a slate of 29 candidates. The ultimate success of pump schemes involving flash X-rays will require investigation of the nuclear properties of those materials that are analogous to the "kinetics" of a conventional laser medium through use of the technique of Modulated Nuclear Radiation (MNR). This paper reports success in demonstrating a compact flash X-ray device producing 35mW and 45mW of average power isotropically from 10 nsec pulses at energies near 8 keV and 17 keV, respectively.

DEVICE DESIGN AND PERFORMANCE

As shown in Fig. 1, our flash X-ray device consisted of three critical subassemblies: 1) a low impedance X-ray tube, 2) a Blumlein power source and 3) a commutation system capable of operation at high repetition rates. The latter two components differed little from drivers we had developed for short pulse nitrogen ion lasers.[1] The X-ray tube was constructed from cast materials, selected to

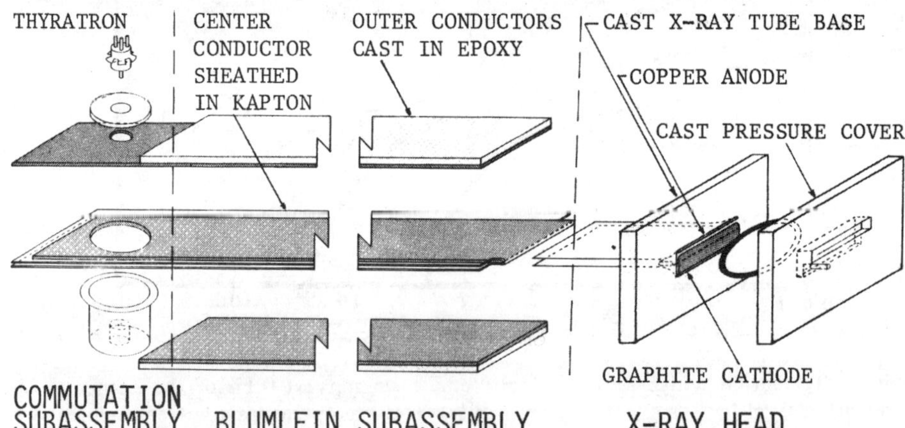

Figure 1: Schematic drawing of device characterized in this work.

minimize erosion and maximize heat tranfer. Copper foil strips 0.05 mm thick and 10 cm wide were fastened to the electrode mounts and then passed through the cast material forming the base of the X-ray head before it hardened. After emerging from the base, the foils were joined to the outermost copper plates of the Blumlein. The anode was simply a rod of Cu or Mo partially buried in the cast material forming the base while the cathode was demountable. It consisted of a strip of 0.381 mm thick graphite with a blade-like edge 10 cm wide and separated from the anode by a variable distance chosen to optimize the performance of this filamentary source of radiation.

Outputs were detected with a block of fast scintillator plastic and the resulting light output was measured with a faster photomultiplier and recorded with a Tektronix 7912AD transient digitizer.

Measurements of absolute intensities were made by comparing the time integrated flourescence from the plastic detector when illuminated with geometrically attenuated X-rays from the flash source directly with the level of excitation produced by a radioactive source of known characteristics. Figure 2 shows the resulting dependence of the average X-ray power emitted upon repetition rates for a charge voltage of 20 kV.

Attenuations measured with a combination of K-edge filters and known thicknesses of aluminum foil indicated that at least 25% of the total X-ray energy lay in the Cu and Mo K_{α} lines, in agreement with previous observations and expectations.[2,3] If it is assumed that these lines have their customary width of around 6eV, the best values of output emitted from the copper anode correspond to about 4.7×10^{13} keV/keV when expressed in terms customary for reporting laser plasma yields. Thus, in less than a minute of experimental time at 100 Hz repetition, over 2×10^{17} keV/keV could be emitted.

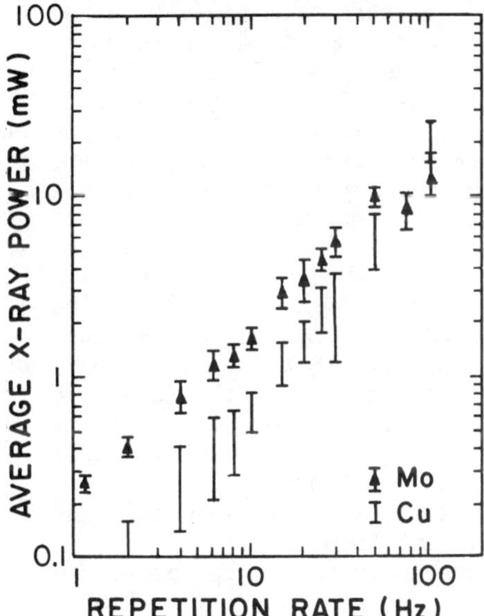

Figure 2: Average X-ray powers as function of repetition rates.

The authors acknowledge the support of this work in part by the Office of Naval Research and in part by the Innovative Science and Technology Program of the Stragetic Defense Initiative Office.

REFERENCES

1. C. B. Collins, IEEE J. Quantum Electron. QE-20, 47 (1984).
2. L. C. Bradley, A. C. Mitchell, Q. Johnson and I. D. Smith, Rev. Sci. Instrum. 55, 25 (1984).
3. Q. Johnson, A. C. Mitchell, and I. D. Smith, Rev. Sci. Instrum. 51, 741 (1980).

Observation of Dressed States of Nuclear Excitation in Fe-57*

S. S. Wagal and C. B. Collins
Center for Quantum Electronics University of Texas at Dallas
P.O. Box 830688, Richardson, TX 75083-0688

ABSTRACT

Mossbauer transitions excited in nuclei immersed in intense radiofrequency fields under some conditions show sidebands at energies displaced from intrinsic values by integral multiples of the energy of a photon of the fields. Reported here is a study of the details of this effect that support a dressed state model of the origins of the sideband transitions.

EXPERIMENT AND RESULTS

This paper reports the reexamination of an obscure Mossbauer effect first reported 17 years ago when it was observed that radiofrequency (rf.) sidebands could be produced on Mossbauer transitions.[1] Attributed to rather trivial origins involving magnetostriction, interest in the phenomenon did not persist. However, a recent repetition[2,3] of those classic experiments indicated that a multiphoton mechanism was responsible for the sidebands. Data were shown to be in agreement with a semiclassical model[3] in which the nuclear states were dressed, in the sense of Cohen-Tannoudji,[4,5] with photons of the rf. field. The sidebands were perceived as arising from gamma ray transitions occurring between the nuclear states whose energies had been shifted by integral multiples of the quanta of the dressing field. Considerable significance is attached to this interpretation in which relatively long wavelength radiation can serve for the efficient production of coherently mixed states of nuclear excitation.

In addition to suggesting many new experiments, the recent results evoked a controversial concern as to whether the multiphoton model was necessary or merely sufficient to explain the data. We report here a consideration of the necessity of the dressed state explanation. We addressed the question whether quantum structure could be found in the dependence of the sideband amplitude upon frequency and found an affirmative answer; a result seemingly incompatible with any magnetostrictive contribution.

Apparatus used in this experiment conformed to the pattern described in previous reports.[2,3,6] An unsplit source line was examined after transmission through a foil in which the absorbing nuclei were dressed with photons from an oscillating magnetic field driven by a coil surrounding the foil. Tuning of the source line resulted from controlled Doppler shifts produced by imparting to the source a velocity along the axis of propagation of the gamma radiation. As we described earlier,[3] the physical mechanism for the dressing of the states is a precession of the nuclear spins about a depolarizing field, H_d developed from time-varying magnetic poles induced on the surface of the magnetic foil by the concurrent precession of the internal saturation magnetization, M_s of the medium.

Thus, by forcing the precession of M_s with the modest fields from the external coil, very large surface polarizations (and H_d) are driven in response.

The spectra obtained in these experiments showed the prominent Mossbauer absorption lines normally displayed by the nuclei of Fe-57 together with the sidebands. The predictions of our previous model are shown in Fig. 1, together with new experimental data.

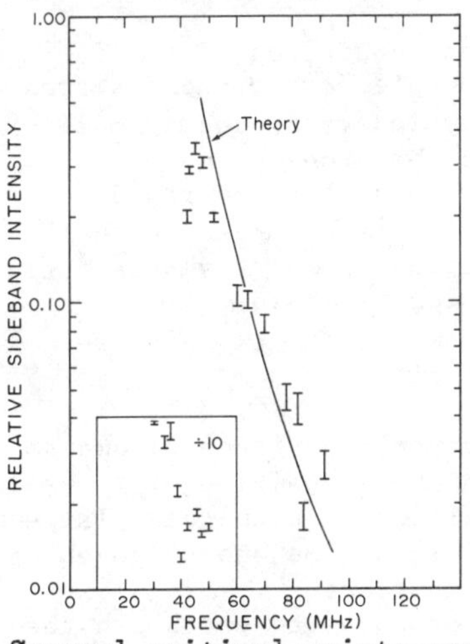

FIGURE 1. Plot of the relative intensity of the first order sideband to the parent intensity remaining when pumped by a product of radiofrequency power, P and quality factor Q of PQ= 600 W at the frequencies shown. Within the inset are results obtained with PQ=300 W and reduced by a factor of 10 for convenience in displaying. For comparison the predictions of a semiclassical, dressed state model of nuclear excitation are shown by the solid curve.

Several critical points emerge from consideration of Fig. 1. For the frequencies reasonably far from the resonance at 45 MHz, the semiclassical, dressed state model gives good agreement with measurements. Conversely, it is extremely significant that a strong resonance is observed in the conversion efficiency into the sidebands, just at the NMR frequency near 45 MHz for resonantly flipping the nuclei. Although this is a subtilty beyond the framework of the semiclassical model, it seems clearly a consequence of the mixing of states mechanism that is inherent in the dressed state model. Moreover, it seems a reasonable conclusion that the observed efficiency with which the nuclear states can be dressed tends to support an optimistic perspective on many of the other analogs of the processes of quantum electronics that may be realized at the nuclear level.

REFERENCES

*Supported in part by the Office of Naval Research and in part by the Strategic Defense Initiative Office (IST).

1. N. D. Heiman, L. Pfeiffer, J. C. Walker, Phys. Rev. Lett. 21, 93 (1968).
2. B. D. DePaola and C. B. Collins, J. Opt. Soc. Am. B 1, 812 (1984).
3. C. B. Collins and B. D. DePaola, Optics Lett. 10, 25 (1985).
4. C. Cohen-Tannoudji and S. Haroche, J. de Physique, 30, 125 (1969).
5. S. Haroche, Ann. Phys. 6, 189 (1971).
6. B. D. Depaola, S. S. Wagal, and C. B. Collins, J. Opt. Soc. Am. B 2, 541 (1985).

RELATIVISTIC ELECTRON - POSITRON GAMMA RAY LASER

F. Winterberg
Desert Research Institute, University of Nevada System,
Reno, Nevada 89506

ABSTRACT

A relativistic electron - positron superpinch, established by the coalescence of two counterstreaming electron - and positron beams is ideally suited for a gamma ray laser, producing multi-MeV photons. The concept ideally combines the property of a relativistic electron - positron superpinch, for the rapid attainment of high densities by magnetic pinch forces, with the greatly reduced annihilation cross section at relativistic energies, permitting to establish a high population inversion.

It was shown[(1)], that for an electron - positron plasma to serve as a gamma ray laser, particle number densities of $\simeq 4 \times 10^{20}$ cm^{-3} must be achieved. The obvious difficulty with these large densities is that they must be reached in a time short enough to prevent electron - positron pair annihilation.

To overcome this problem we have proposed to establish the needed dense electron - positron plasma configuration in a relativistic electron - positron superpinch[(2)], where due to the relativistic particle energies the annihilation cross section is greatly reduced. The establishment of a high particle density in a short time is ensured through relativistic synchrotron losses by axial electron and positron oscillations, by which the pinch is strongly cooled, rapidly reaching a high state of compression.

Relativistic electron - positron pinches have been carefully analyzed and found to be quite stable[(3)]. Forming a thin linear filament of high density with a high population inversion, they also happen to have the right geometric structure to serve as a laser.

The currents needed for the electron and positron beams are comparable in magnitude to those reached in intense relativistic electron beams. What therefore remains to be done before such a laser can be built, is the development of an efficient way to mass-produce positrons. A possible way this can be achieved is through the interaction of intense laser light with matter[(4)].

The idea is explained in Figure 1, showing the coalescence of an electron - with a positron beam, colliding head on. If the total current of both beams is large enough but less than the Alfven current, the beams will under emission of intense synchrotron radiation collapse into a narrow filament with a high population inversion of un-annihilated electron - positron pairs held together by magnetic pinch forces.

0094-243X/86/1460064-2$3.00

To ensure that the configuration can act as a gamma ray laser medium, it is required that the axial particle velocity component is small compared to the velocity of light, otherwise no photon avalanche can develop. This condition is met if the current is slightly less than the Alfven current, above which no plasma equilibrium exists.

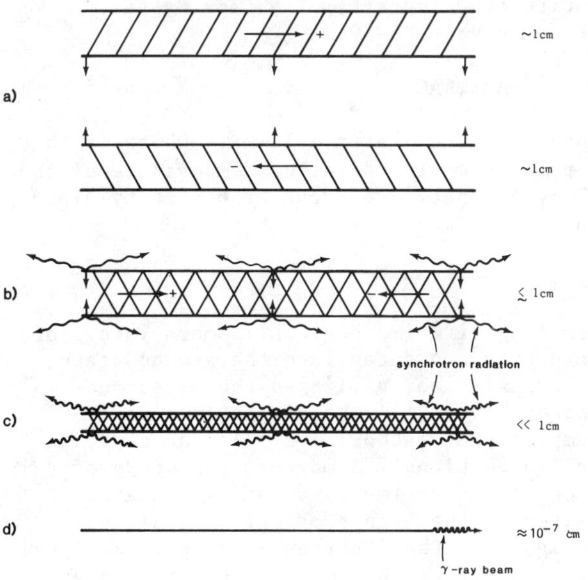

We find for the gain the following expression

$$G=\left[(r_o \ell \ln(2\gamma))/4r^2\gamma\right], \quad (1)$$

where ℓ and r are the length and radius of the pinch channel, r_o the classical electron radius and $\gamma = (1-v^2/c^2)^{-1/2}$ the usual relativistic factor. For the example ℓ = 100 cm, r ≃ 10^{-7} cm, γ = 20 (10MeV electrons and positrons), we find G = 100. The current turns out to be ~ 340000 Ampere, and there are ~ 7 × 10^{15} electron - positron pairs, each releasing two ~ 10MeV photons. The total laser energy released is ~ 20KJ.

Figure 1

During the pinch collapse the magnetic energy outside the pinch channel increases. To compensate for this imbalance in energy, the initial particle energies must be chosen larger. If the radius of the return current conductor is ~ 1 cm one finds in the quoted example that a ~ 7 times larger initial energy with two ~ 70MeV beams, would be needed.

REFERENCES

1. M. Bertolotti and C. Sibilia, Appl. Phys. 19 127 (1979).
2. F. Winterberg, Phys. Rev. A 19 1356 (1979).
3. B.E. Meierovich, Phys. Rev. A 26 1128 (1982), and J. Plasma Physics 29 361 (1983).
4. J.W. Shearer, J. Garrison, J. Wong and J.E. Swain, Phys. Rev. A 8 1582 (1973).

BEATING THE GRASER DILEMMA BY RAPID HEAT-REMOVAL UNDER HIGH PRESSURE

F. Winterberg
Desert Research Institute, University of Nevada System,
Reno, Nevada 89506

ABSTRACT

There is a good chance that the graser dilemma - which is that the heat produced during pumping would raise the temperature of the graser medium above the Debye temperature - can be beaten by electronic heat conduction under high pressure.

The idea to transform long lifetime nuclear isomers into short lifetime states, from where they can decay into the ground state through stimulated gamma ray emission, utilizing the Moessbauer effect, was first proposed by Eerkens[(1)] who suggested a rapid - or optical frequency pump. Other authors [(2,3,4,5)] have later come up with similar suggestions. A more recent study of this problem by C.B. Collins et al.[(6)] estimates an energy of about $\sim 10^{11}$ erg/cm^3 to be deposited in the graser material. This optimistic value was obtained ignoring the importance of spurious level broadening for life-times much larger than $\sim 10^{-6}$ sec, the upper time limit for the Moessbauer effect to work. But even an energy deposited of $\sim 10^{11}$ erg/cm^3, a typical value for high explosives, would make the concept unfeasible. We will propose a solution to this problem which employs high pressure techniques combined with rapid heat removal by electronic heat conduction.

If an energy e per unit volume is deposited in some material it will produce a pressure p and a temperature T:

$$p \sim e, \ T \sim e/fnk$$

where n is the atomic number density, k the Boltzmann constant and f a number rising with T and related to the internal degrees of freedom, with f ~ 5 a typical value. For the above given value of $e \sim 10^{11}$ erg/cm^3, $n \sim 5 \times 10^{22}$ cm^{-3}, $f \simeq 5$ one would have $p = 10^{11}$ dyn/cm$^2 \simeq 10^5$ atm and $T \sim 3 \times 10^4$ °K. For the more likely value $e \sim 10^{13}$ erg/cm^3, one would have $p \sim 10^{13}$ dyn/cm$^2 \simeq 10^7$ atm and $T \lesssim 10^6$ °K.

During the pumping time of $\sim 10^{-6}$ sec, dictated by the Moessbauer effect, we propose to put the graser medium under high pressure to reduce during this time the high temperature by electronic heat conduction below the Debye temperature.

The foregoing suggests the following graser - rod arrangement shown in Figure 1. The graser rod is positioned inside a cylindrical metallic tamp of good heat conductivity and at low temperature. Alongside the tamp the laser rod can be pumped by the optical radiation passing through a high pressure window.

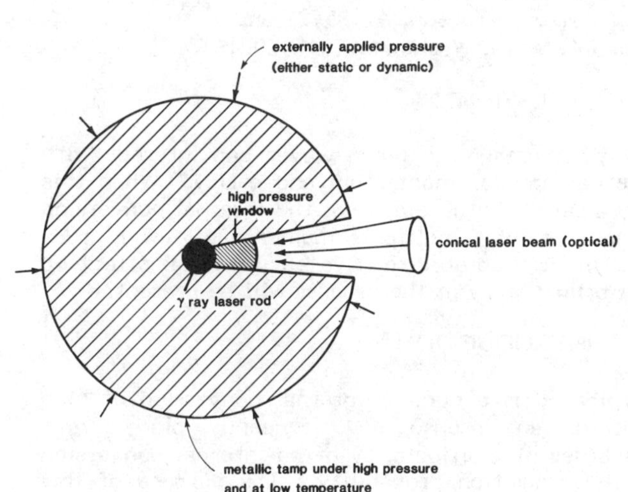

Figure 1

The rod, heated up to a high temperature by the pumping radiation, is momentarily transformed into a high temperature metal-vapor plasma possessing a large heat conductivity. The characteristic time to lose its high temperature by heat conduction is given by

$$\tau \sim \chi / r^2$$

where χ is the heat conduction number and r the rod radius. For $\chi \sim 1[\mathrm{cm}^2/\mathrm{sec}]$, and $r \sim 10^{-3}$ cm, follows $\tau \sim 10^{-6}$ sec, which is about the same as the pumping time.

To balance the high pressure, one can put the entire assembly into a static high press, which can produce pressures up to $\sim 10^5$ atm. For larger input energies and hence higher pressures, a convergent cylindrical pressure wave launched from the surface of the pressurizing tamp, by igniting a high explosive, can be used. For example a ~ 1 cm thick layer of high explosive detonated at a 10 cm tamp radius could produce a pressure of $\sim 10^7$ atm lasting $\sim 10^{-6}$ sec.

REFERENCES

1. J.W. Eerkens, Laser Focus, 10 (1974)
2. E.V. Baklanov and V.P. Chebatev, Sov. J. Quantum Electron. 6 345 (1976)
3. P. Kamenov and T. Bonchev C.R. Acad. Bulg. Sci. 28 1175 (1975)
4. L.A. Rivlin, Sov. J. Quantum Electron. 7, 380 (1977), 8 1412 (1977)
5. B. Arad, S. Eliezer and Y. Paiss, Phys. Lett. A 74, 395 (1979)
6. C.B. Collins, et al. J. Appl. Phys. 53 4645 (1982)

FIELD-ENHANCED INTERNAL CONVERSION AND ITS APPLICATION TO THE GAMMA RAY LASER

H. R. Reiss
University of Arizona, Tucson, AZ 85721 and
The American University, Washington, DC 20016

ABSTRACT

Intense low frequency electromagnetic plane waves can interact with nuclei so as to supply external angular momentum and parity. When this occurs in a state metastable against photon emission, the internal conversion component of the decay should exhibit multipole moments of lower order than normal for the decay. This is predicted to have a substantial effect on the lifetime of the state. A brief theory of the process will be presented.

INTRODUCTION

The process to be considered is a nuclear internal conversion of high multipole order, subjected to an intense low frequency plane wave electromagnetic field. The order of multipolarity of an internal conversion has a profound effect on the transition probability. The purpose of this investigation is to explore the extent to which the external electromagnetic field can transfer angular momentum to the process, thus reducing the effective multipolarity.

The primary mechanism is one in which the field induces angular momentum in the final state of the conversion electron. A classical model serves to exhibit the mechanism. Consider an intense circularly polarized field described in Coulomb gauge by the vector potential

$$\vec{A} = 2^{-1/2}\, a[\hat{x}\cos(\omega t-\vec{k}\cdot\vec{r}) + \hat{y}\sin(\omega t-\vec{k}\cdot\vec{r})];\quad \vec{k} = \hat{z}\omega/c \quad . \tag{1}$$

An exact classical relativistic solution in the simplest frame of reference shows that the electron moves in a circular orbit in a plane perpendicular to the propagation vector $\vec{k}$. This constitutes a field-induced angular momentum given in $\hbar$ units by

$$\vec{L} = \frac{\vec{r}\times\vec{p}}{\hbar} = \mp \frac{mc^2}{\hbar\omega}\, \frac{z_f}{(1+z_f)^{1/2}}\, \hat{z} \quad , \tag{2}$$

where z_f is the intensity parameter normally encountered in free-electron intense-field problems,

$$z_f = \frac{1}{2}(ea/mc^2)^2 \quad , \tag{3}$$

and where the ambiguous sign in Eq. (2) refers to right- or left-handed circular polarization. When a photon energy $\hbar\omega$ is much less than the electron rest energy mc^2, and when the field is very intense (z_f of order unity), then Eq. (2) expresses an extremely large field-caused angular momentum.

THEORY

A quantum theory of field-enhanced internal conversion starts with the covariant S matrix

$$S_{fi} = -i\int d^4x_n \int d^4x_e \, J_\mu^{(n)}(x_n)\, D^{\mu\nu}(x_n - x_e)\, J_\nu^{(e)}(x_e) \quad , \tag{4}$$

where indices n and e refer to nucleus and electron, the currents are given by

$$J^\mu = \overline{\Psi}_f(\pm e\gamma^\mu)\Psi_i \quad , \tag{5}$$

and the electromagnetic field propagator is

$$D^{\mu\nu} = -\frac{g^{\mu\nu}}{(2\pi)^4}\int d^4k \, \frac{e^{-ik\cdot(x_n - x_e)}}{k^2 + i\varepsilon} \quad . \tag{6}$$

Simple application of Eq. (4) gives the standard theory of internal conversion. The field-enhanced version of this theory comes about when the electron and nuclear states ψ in Eq. (5) are given by states in the presence of an externally applied field. For the electron states given by the standard Dirac Volkov solutions for linear polarization, the transition probability per unit time in the intense-field limit is

$$W = \int \frac{d^3p_e}{(2\pi)^3}\,\frac{1}{4}\, e^4 \int dV_n \int dV_e \, j_\mu^{(n)}\, j^{(e)\mu}\, \frac{e^{i\omega_0\rho}}{\rho} \int dV_n' \int dV_e'$$

$$\times\; j_\nu^{(n)\prime}\, j^{(e)\nu\prime}\, \frac{e^{-i\omega_0\rho'}}{\rho'}\, \frac{2^{3/2}}{m\beta\gamma z_f^{1/2}}\, e^{-i(\omega_0-\omega_e)\hat{q}\cdot(\vec{r}-\vec{r}')}$$

$$\times \left\{ \frac{e^{-i(m\gamma - \vec{p}_e\cdot\hat{\varepsilon})\hat{\varepsilon}\cdot(\vec{r}-\vec{r}')}}{[1-(\gamma-\vec{p}_e\cdot\hat{\varepsilon}/m)^2/2z_f]^{1/2}} + \frac{e^{i(m\gamma + \vec{p}_e\cdot\hat{\varepsilon})\hat{\varepsilon}\cdot(\vec{r}-\vec{r}')}}{[1-(\gamma+\vec{p}_e\cdot\hat{\varepsilon}/m)^2/2z_f]^{1/2}} \right\} . \tag{7}$$

In Eq. (7), the factors prior to the one containing $2^{3/2}$ are standard field-free results. The remaining factors relate to field-induced effects and they give rise to lower multipole orders than the first part when the transition is basically of high multipolarity. The quantities in Eq. (7) include the z_f of Eq. (3) and

$$\omega_0 = E_i^{(n)} - E_f^{(n)} \;,\quad \omega_e = E_f^{(e)} - E_i^{(e)} \;,\quad j^\mu = \overline{\phi}_f\,\gamma^\mu\,\phi_i \;,$$

$$\rho = |\vec{r}_n - \vec{r}_e| \;,\quad \beta = p\cdot q/m\omega \;,\quad \gamma = \left[\frac{\vec{p}_e\cdot\hat{\varepsilon}}{m^2} + 2\,\frac{(\omega_0-\omega_e)}{\beta m}\right]^{1/2} , \tag{8}$$

where the ϕ are the spatial parts of the field-free Ψ wave functions, and relativistic four-vector products are used with the convention $a\cdot b = a^0b^0 - \vec{a}\cdot\vec{b}$.

These results have possible application to direct field-induced transfer between storage (isomeric) and lasing (Mössbauer) levels in a gamma-ray laser.

PREDICTED CHANGES OF THE INTERNAL CONVERSION RATES IN ^{119}SN DUE TO ADMIXTURES OF LOWER MULTIPOLE ORDER

S. A. Wender, G. C. Baldwin and W. L. Talbert
Los Alamos National Laboratory, Los Alamos, NM 87545

H. R. Reiss
University of Arizona, Tuscon, AZ 87521

ABSTRACT

One conceivable basis for interlevel transfer from a long-lived isomer to a graser state might be to alter the multipolarity of the transfer step. As an example, in the case of ^{119}Sn any mechanism that introduces angular momentum to admix M3 radiation into the normally M4 transition from the 89.6 keV, 11/2$^-$ level to the 23.9 keV, 3/2$^+$ state could greatly shorten its normal 293 day halflife. We have calculated the K-shell internal conversion rate for the 89.6 keV level in ^{119}Sn for small admixtures of M3 radiation using Moszkowski formulas for the radiative transition rates and tabulated values of the internal conversion coefficients. We find that even very small admixtures of lower multipole order can produce very large changes in the internal conversion coefficient. This offers the basis for a simple and definitive test of the hypothesized mechanism for creating population inversion by introducing angular momentum.

INTRODUCTION

The internal conversion coefficient, α, is defined as the ratio of the number of decays by electron emission, T_e, to the number of decays by gamma emission, T_γ, from a given state. In general, when the change in angular momentum (multipole order) is large, gamma emission is hindered relative to electron decay and α is large. If the ejected electron could obtain additional angular momentum from some external field, the electron transition rate would increase and the lifetime of the state would decrease.

As an example, Fig. 1. shows the level diagram of ^{119}Sn. The K-shell internal conversion coefficient, α_k, for the 65.7 keV, M4 transition from the 89.6 Kev state is 1610.[1] If a photon from an externally applied field could supply one unit of angular momentum to the converting electron, the multipolarity of the transition would be M3. If this process happened X fraction of the time, α_k would be given by the following expressions:

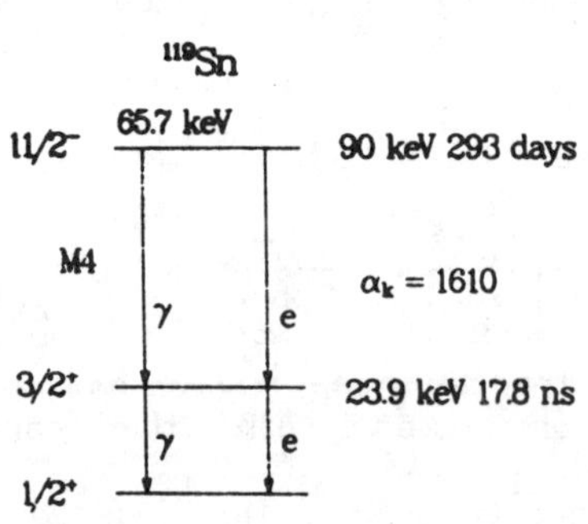

Fig. 1. Level diagram of ^{119}Sn

$$\alpha_k(X) = \frac{Te_k}{T_\gamma} = \frac{Te_k(M4) + X\, Te_k(M3)}{T_\gamma(M4)} = \frac{Te_k(M4)}{T_\gamma(M4)} + X\frac{Te_k(M3)}{T_\gamma(M4)}$$

$$\alpha_k(X) = \alpha_k(M4) + X\alpha_k(M3)\frac{T_\gamma(M3)}{T_\gamma(M4)}$$

Using $\alpha_k(M4) = 1610$[1], $\alpha_k(M3) = 211$[2], $T\gamma$[3] $(M3) = 2.6 \times 10^{-4}$ s^{-1}, and $T\gamma$[3] $(M4) = 1.5 \times 10^{-11}$, we obtain the following expression for α_k as a function of the parameter X.

$$\alpha_k(X) = 1610 + 3.67 \times 10^9\, X$$

This expression implies that if the X parameter is only 4×10^{-9}, then the change in α_k would be 1%, easily measureable.

The measurement of α_k can be performed in a very precise manner. The number of electron decays is proportional to the number of K x-rays divided by the fluorescence yield.[4] Because the photon energies are comparable, a single detector could then be used to simultaneously measure the K x-ray yield and the gamma ray yield. This ratio is proportional to α_k .

CONCLUSION

Measurement of the ratio of the K x-ray yield to the gamma ray yield provides a very sensitive test to determine whether an external field can introduce even an extremely (10^{-9}) small admixture of lower multipolarity.

REFERENCES

1. C. M. Lederer and V. S. Shirley, Table of Isotopes 7th ed., John Wiley & Sons, New York, NY (1978).

2. R. S. Hager and E. C. Seltzer, Nuclear Data Tables 4, (1968), 1.

3. Moszkowski transition rates given in ref. 1.

4. Walter Bambynek, Bernd Craseman, R. W. Fink, H. U. Freund, Hans Mark, C. D. Swift, R. E. Price, P. Venugopala Rao, Rev. Mod. Phys. 44, 716 (1972).

THE SANDIA X-RAY LASER PROGRAM

E. J. McGuire, K. Matzen, R. Spielman, M. A. Palmer, B. A. Hammel, D. L. Hansen, T. W. Hussey, W. W. Hsing, and R. J. Dukart
Sandia National Laboratories
Albuquerque, New Mexico, 87185

ABSTRACT

The Sandia x-ray laser program is based on the intense keV radiation produced by stagnating gas puff implosions. The laser scheme uses this radiation to photoionize Ne-like ions to F-like ions, and produce a 3p-3s population inversion via recombination. An annular stagnation shell is used to separate the pump from the lasant. Converter technology is being developed to try the Na-Ne line matching scheme. Experimental and computational results will be presented.

DISCUSSION

The Sandia x-ray laser program is based on the availability of large pulsed-power drivers developed for the light ion beam ICF program. When thin cylindrical foils are imploded in a pulsed-power diode, the foils are observed to be unstable. Calculations indicate that annuli of comparable mass/length, but greatly increased thickness (reduced density) would have improved stability properties[1]. But the same calculations indicated anomalously intense photon emission at 1 keV and above. The emission, nominally 10 kJ in 10 ns from a cylindrical stagnation region of radius 0.1 cm and length 2 cm, produces an intensity at the center plane of the cylinder of about 2 TW/cm^2 of 1 keV x-rays. To gauge the size of the pump output assume that the radiation pulse rises to its maximum in 5 ns, so that the rate of increase of pump power is 0.4 TW/cm^2-ns. In the first publication on x-ray lasers, Duguay and Rentzepis[2] estimated a requirement of 0.004 TW/cm^2-ns for 3s-2p lasing at 372 Å in Na^{1+}, and 0.25×10^{13} TW/cm^2-ns for a Cu K_α laser at 1.54 Å. Clearly, we have a respectable pump, but not one appropriate for a hard x-ray laser. The estimates in Ref.(2) suggest pump power scaling as inverse laser wavelength to the sixth power.

For low density thick-shelled "foils" we use annular gas puffs produced by high (4-8) Mach number nozzles. The gas is preionized and imploded. Figs.(1) and (2) show time integrated spectra from Ne and Xe gas puffs imploded on themselves. Because the nozzle does not produce a true annulus, "zippering" is observed in time-dependent output, i.e. the emission turns on earliest at the nozzle end of the diode and the onset of emission moves down the Z axis with time. A variety of schemes are being examined to reduce "zippering".

Because pulsed-power technology is inherently on a ns or longer time scale and because of "zippering", we want a laser medium that might show a population inversion on a ns or longer time scale. The $(2p)^5(3p)-(2p)^5(3s)$ transition in Ne-like ions is an appealing candidate. The population inversion is produced by photoionizing Ne-like ions with our pump, and developing the population inversion via recombination of F-like ions and the disparity of radiative lifetimes of the upper and lower laser levels. Fig.(3) shows the ionization energy of Ne-like ions as a function of Z, and the range of high energy photon emission observed for various gases. The intersection identifies suitable Z for laser media. But Fig.(3) does not involve pump power.

Fig.(4) shows an idealized target design, with a 1 mm radius gas puff stagnation layer surrounding a laser medium. The stagnation layer is introduced to delay shock wave disruption of the laser medium until after lasing. Experiments have shown that a stagnation layer made of CH foils or low density foam does not significantly change the high energy output from the stagnating gas puff. Given the geometry, and assuming that the laser medium is composed of ground state Ne-like ions, one asks what high energy output (J) would be required for an ionization time constant of 0.1 ns, 1.0 ns, and 10 ns. The required energies are called J_{100}, J_{10}, and J_1, respectively. If we assume that the high energy output is a delta function in energy located at the threshold for $(2p)^6$ photoionization, we calculate the dashed curves in Fig.(5). If we use realistic photoionization cross sections and spectral distributions for the various gases, we calculate the hooked shaped curves in Fig.(5) (i.e. there is an optimum Z for each gas puff spectrum). A J_1 criterion is unlikely to identify a useful system, while with J_{100} and a broad pump, one will rapidly photoionize F-like ions to O-like ions, etc. Thus J_{10} is a reasonable criterion. The results in Figs.(3) and (5) indicate a pump power scaling of inverse wavelength to the fourth power, neglecting the energy required to produce the Ne-like plasma. Also shown in Fig.(5) are outputs from various modifications of the Proto II accelerator, and a projected future driver. We are currently using a diode designed by PSI to optimize gas puff output, and have measured > 20 KJ at or above 1 keV with imploding Ne gas puffs. Thus we can examine laser media near Z = 30.

In modeling these systems with broadband pumps such as shown in Fig.(2), one finds that a J_{10} criterion produces a maximum F-like ion population in 2-3 ns, after which it decreases as the O-like and N-like populations increase because the intense broadband pump can photoionize F-like and O-like ions. To avoid this waste of both photons and F-like ion population we are developing low Z-converters to produce output spectra similar to Fig.(1), i.e. emission on the K_α lines from the H-like and He-like ions of the low Z-converter. This is done by coating the outer surface of the stagnation layer with the low-Z material. Since the laser medium does not exist as a gas, it initially is a coating on the inner surface of the stagnation layer. The laser medium is produced by thermally exploding this inner coating. Schematics of some current targets are shown in Fig.(6). With K_α radiation from H-like Al, one can photoionize the $(2p)^6$ shell of the Ne-like ion of both Cu and Ni. But while this pump wavelength can photoionize the $(2p)^5$ shell of F-like Ni, it cannot photoionize the $(2p)^5$ shell of F-like Cu. Thus, with an Al converter and a Cu target one can prevent the photoionization of F-like ions, and by varying the target from Cu to Ni, one can test the hypothesis experimentally. By varying the converter material one can examine other possible lasants, provided the conversion process is efficient. Preliminary measurements with Ne gas puffs and Al converters show about 25% of the output above 1 KeV in Al H- and He-like K_α lines. Fig.(7) shows a time integrated spectrum from a thin foil stagnation layer coated with Al on the outside and Cu on the inside. Emission from Ne-like Cu is observed. Use of a Na converter would allow experiments on the Na-Ne line matching laser scheme.[3]

The approach followed in the Sandia x-ray laser program is traditional in that we separate the pump (flashlamp) from the laser medium. Much of our effort so far has been on flashlamp development, but we plan on doing laser experiments in the next year.

REFERENCES

1) T. W. Hussey, Bull. Am. Phys. Soc. 27, 1065 (1982)
2) M. A. Duguay and P. M. Rentzepis, Appl. Phys. Lett. 10, 350 (1967)
3) J. P. Apruzese and J. Davis, Phys. Rev. A31, 2976 (1985)

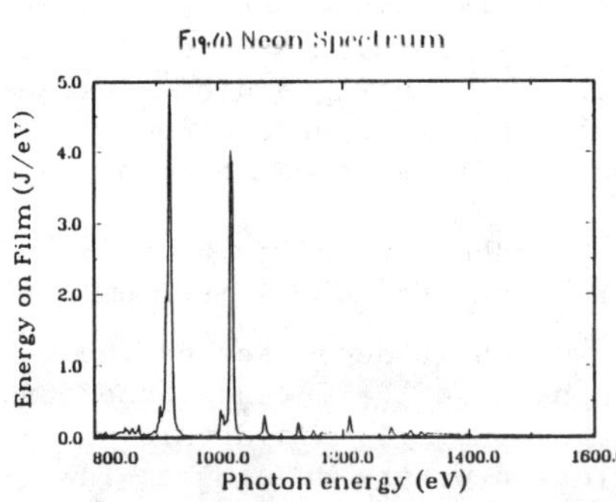

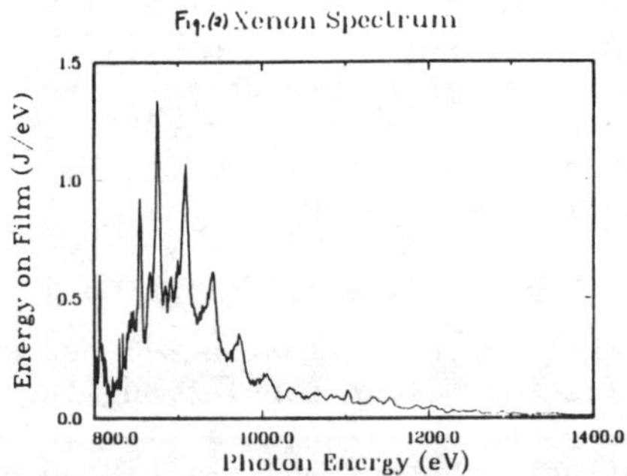

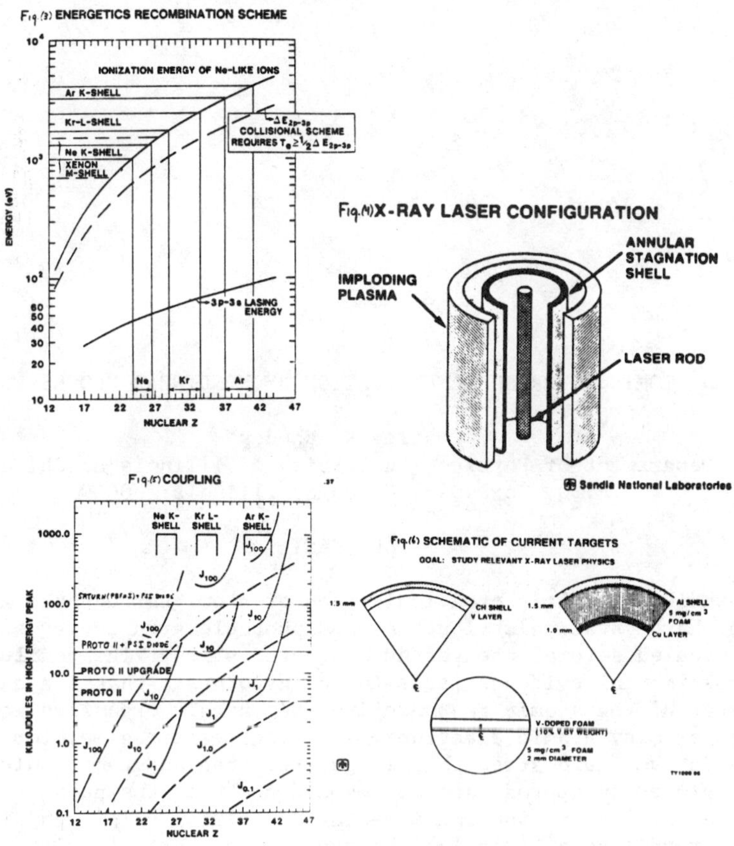
Fig.(3) ENERGETICS RECOMBINATION SCHEME
IONIZATION ENERGY OF Ne-LIKE IONS
Ar K-SHELL
Kr-L-SHELL
Ne K-SHELL
XENON M-SHELL
COLLISIONAL SCHEME REQUIRES $T_e \geq \frac{1}{2} \Delta E_{2p-3s}$
3p-3s LASING ENERGY
ENERGY (eV)
NUCLEAR Z
Fig.(4) X-RAY LASER CONFIGURATION
ANNULAR STAGNATION SHELL
IMPLODING PLASMA
LASER ROD
Sandia National Laboratories
Fig.(5) COUPLING
Ne K-SHELL
Kr L-SHELL
Ar K-SHELL
PROTO II + PSΣ DIODE
PROTO II UPGRADE
PROTO II
KILOJOULES IN HIGH ENERGY PEAK
NUCLEAR Z
Fig.(6) SCHEMATIC OF CURRENT TARGETS
GOAL: STUDY RELEVANT X-RAY LASER PHYSICS
CH SHELL
V LAYER
Al SHELL
FOAM
Cu LAYER
V-DOPED FOAM (10% V BY WEIGHT)
2 mm DIAMETER

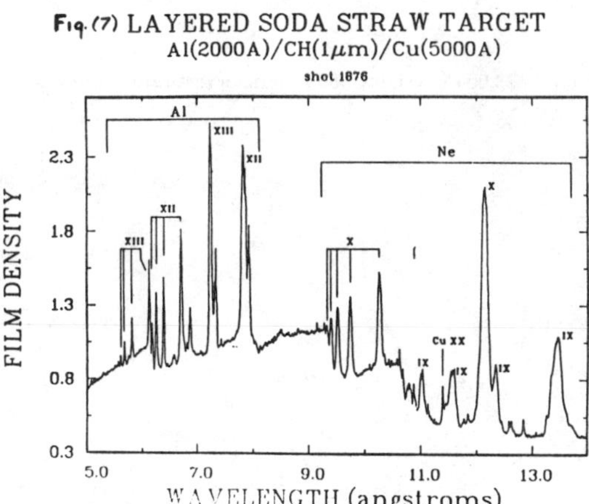
Fig.(7) LAYERED SODA STRAW TARGET
Al(2000A)/CH(1μm)/Cu(5000A)
shot 1876
Al
Ne
Cu XX
FILM DENSITY
WAVELENGTH (angstroms)

ATOMIC INNER-SHELL EXCITATION BY NONLINEAR PROCESSES

Charles K. Rhodes
Department of Physics, University of Illinois at Chicago
P. O. Box 4348, Chicago, Illinois 60680

ABSTRACT

Studies of multiphoton ionization of atoms through the analysis of ion charge state distributions and photoelectron energy spectra have revealed several unexpected characteristics. The confluence of the experimental evidence leads to the hypothesis that the basic character of the atomic response involves highly organized, coherent motions of many atomic electrons. The important regime, for which the radiative field strength E is greater than an atomic unit (e/a_0^2), can be viewed in approximate correspondence with the physics of fast ($\sim$ 10 MeV/amu) atom-atom and ion-atom scattering. This physical picture furnishes a basis for the expectation that stimulated emission in the x-ray range can be produced by direct highly nonlinear coupling of ultraviolet radiation to atoms.

DISCUSSION

A basic and long-standing problem in the field of coherent sources is that associated with the generation of coherent energy in the extreme ultraviolet and soft x-ray regions. During the last three years, picosecond rare gas halogen (RGH) excimer laser technology, on account of the very favorable scaling relationships governing the spectral brightness of these sources,[1,2] has emerged as a key factor in new techniques useful for generation of coherent radiation below 100 nm. Recently, the operation of RGH systems has been extended down into the femtosecond region,[3,4] a development that will enable sources with peak powers P in the 1 TW $\leq$ P $\leq$ 10 TW range to be used in basic physical studies. Light sources of this kind should permit the generation of focal intensities above $\sim 10^{20}$ W/cm^2.

Recent research findings[1,5,6] lead to the conclusion that the direct multiphoton excitation of appropriate amplifying media with high spectral brightness ultraviolet sources is a possible option for the generation of short wavelength radiation in the kilovolt range. In addition to satisfying the demanding energy density requirements generally called for to create amplification in the x-ray range, this method of excitation utilizes the coherence obtainable from RGH sources to enhance the coupling strength and provide selectivity in the energy flow.[7,8] The application of this technique to the x-ray region requires an extended study of the basic character of high order nonlinear processes in the ultraviolet in an intensity range corresponding to radiating field strengths E greater than an atomic unit (e/a_o^2).

This new experimental regime, in which the optical field strength E is considerably greater than an atomic unit (e/a_o^2), is now open for systematic study[1] and electric field strengths on the order of 100 (e/a_o^2) should be attainable. This range of field strengths, as noted above, has become accessible principally because of the availability of an ultraviolet laser technology capable of producing subpicosecond pulses[2] with pulse energies approaching the joule level in low divergence beams at high repetition rates. Clearly, a technology of this genre will make possible the creation of physical conditions unachievable with any other known experimental means.

Figure (1) illustrates the parameters of the physical regime, in terms of pulse intensity (I) and pulse time scale (τ) that characterize our experimental studies. It is seen that the ultraviolet laser technology will take us a considerable distance into the unexplored area, a zone associated with field strengths far greater than an atomic unit (e/a_o^2) and time scales that are advancing toward an atomic time τ_a.

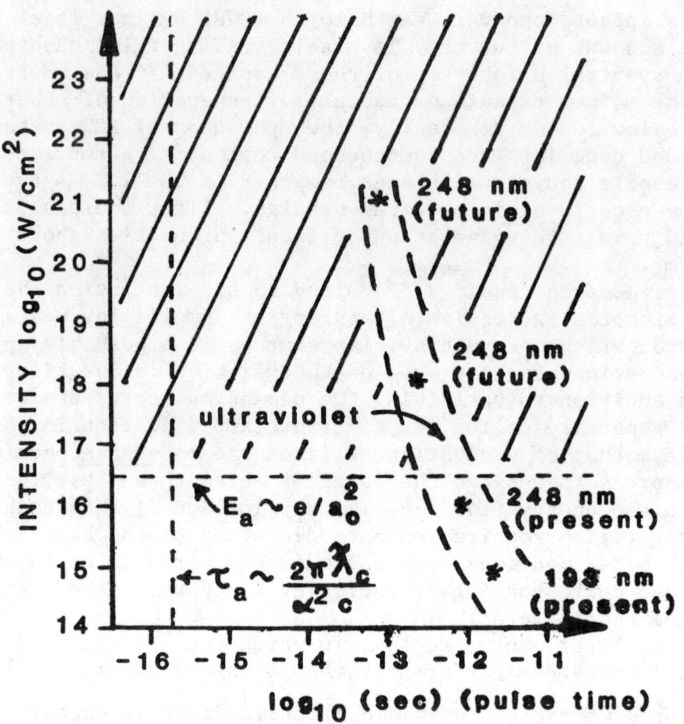

Fig. (1): Conditions of irradiation achievable with high brightness short pulse ultraviolet sources. The anticipated ultraviolet development path is shown as well as its relation to atomic field strengths (E_a) and atomic times (τ_a).

It has been conjectured[7,8] that fundamentally different atomic motions may be driven under such extreme conditions of irradiation. These involve organized coherent motions of many atomic electrons. Quantum mechanically, such states are multiply excited atomic levels that, on account of the large energy necessary to excite them, generally lie adjacent to a multiple continuum. It has been possible to describe the intra-atomic couplings involving these excited electrons in two ways. One utilizes an analogy[7] with atomic collisions,[9] while the other employs an analysis based on time-dependent Hartree Fock theory.[8] Both of these physical pictures[7,8] indicate that a substantial fraction of the energy absorbed by the atom can be channeled into inner-shell excitation, and thereby, generate population inversions in the x-ray region.[1] If the selectivity predicted by these models for inner-shell excitation is not greatly reduced,[10] the currently observed rates of energy transfer[11] are consistent with the production of stimulated emission in the 100 - 1000 eV range.

ACKNOWLEDGEMENTS

The author wishes to acknowledge fruitful discussions with T. S. Luk, U. Johann, A. P. Schwarzenbach, I. A. McIntyre, A. McPherson, H. Jara, and K. Boyer. This work was supported in part by the Office of Naval Research, the Air Force Office of Scientific Research under contract number F49620-83-K0014, the Air Force Office of Scientific Research, Department of Defense -- University Instrumentation Program under grant number USAF 840289, the Department of Energy under grant number DE-AC02-83ER13137, the Lawrence Livermore Laboratory under contract number 5765705, the National Science Foundation under grant number PHY-84-14201, the Defense Advanced Research Projects Agency, the Innovative Science and Technology Office of the Strategic Defense Initiative Organization, and the Los Alamos National Laboratory under contract number 9-X54-C6096-1.

REFERENCES

1. C. K. Rhodes, Science 229, 1345 (1985).
2. H. Pummer, H. Egger, and C. K. Rhodes in Topics in Applied Physics: Excimer Lasers, Vol. 30, Second enlarged edition, edited by C. K. Rhodes (Springer-Verlag, Berlin, 1984) p. 217.
3. A. P. Schwarzenbach, T. S. Luk, I. A. McIntyre, U. Johann, A. McPherson, K. Boyer, and C. K. Rhodes, "Subpicosecond KrF* Excimer Laser Source," submitted to Optics Letters.
4. J. H. Glownia, G. Arjavalingham, P. P. Sorokin, and J. E. Rothenburg, "Amplification of 350 fs Pulses in XeCl Excimer Gain Modules," (to be published).
5. T. S. Luk, H. Pummer, K. Boyer, M. Shahidi, H. Egger, and C. K. Rhodes, Phys. Rev. Lett. 51, 110 (1983).
6. T. S. Luk, U. Johann, H. Egger, H. Pummer, and C. K. Rhodes, Phys. Rev. A32, 214 (1985).
7. K. Boyer and C. K. Rhodes, Phys. Rev. Lett. 54, 1490 (1985).

8. A. Szöke and C. K. Rhodes, "A Theoretical Model of Inner-Shell Excitation by Outer-Shell Electrons," Phys. Rev. Lett., in press.
9. J. S. Briggs and K. Taulbjerg, in Structure and Collisions of Ions and Atoms, edited by I. A. Sellin (Springer-Verlag, Berlin, 1978) p. 105.
10. U. Johann, T. S. Luk, H. Egger, and C. K. Rhodes, "Rare Gas Electron Energy Spectra Produced by Collision-Free Multiquantum Processes," submitted to Phys. Rev. A.
11. U. Johann, T. S. Luk, I. McIntyre, A. P. Schwarzenbach, K. Boyer, and C. K. Rhodes, "Subpicosecond Studies of Collision-Free Multiple Ionization of Atoms at 248 nm," submitted to Phys. Rev. Lett.

STIMULATED SOFT X-RAY EMISSION IN A CONFINED PLASMA COLUMN

S. Suckewer, C.H. Skinner, C. Keane, D. Kim,
J.L. Schwob,* E. Valeo, D. Voorhees, and A. Wouters

Plasma Physics Laboratory, Princeton University, Princeton, NJ 08544

ABSTRACT

Measurements of high gain on the CVI 182 Å line are compared to theoretical predictions of a 1D code. Spectra of the emissions in axial and transverse directions, obtained with recently installed multichannel detectors, are presented. A new two-laser approach to generating gain at wavelengths below 100 Å is described.

INTRODUCTION

An enhancement of ~ 100 of stimulated emission over spontaneous emission of the CVI 182 Å line (one pass gain ~ 6.5) was measured in a recombining magnetically confined plasma column by two independent techniques using intensity calibrated monochromators.[1] In this experiment a commercially available TEA CO_2 laser of energy 300 J and pulse duration 80 nsec was incident on a carbon disc target in a strong B = 90 kG solenoidal magnetic field. After ionization by the laser pulse, the plasma cooled rapidly by radiation losses, and strong recombination created a population inversion between levels 3 and 2 in hydrogen-like carbon (CVI). It was possible to demonstrate amplification of the stimulated emission in a two-pass arrangement with a soft X-ray mirror. With a 12% measured reflectivity of the mirror a 120% increase in the stimulated emission at 182 Å was observed in the axial direction.

Experiments were also conducted with the CO_2 laser focussed onto the end of an axially oriented thick (35-300 μ) carbon fiber which generated gains of up to k ≈ 6 cm^{-1}.[2] More details of all the above work may be found in Refs. 1 and 2. In this article we will focus on the most recent results.

COMPARISON OF GAIN MEASUREMENTS WITH ONE DIMENSIONAL MODEL

A key element in the achievement of a high gain length G = kl ~ 6.5 was the realization, from measurements of the radial profiles of the CVI line radiation, that the most favorable conditions for maximum gain should exist in the off-axis regions of the plasma column. Gain was measured by recording the enhancement, E, of the axial CVI 182 Å emission; that is the ratio of the 182 Å stimulated plus spontaneous emission in the axial direction to the mostly spontaneous 182 Å emission in the transverse direction. The enhancement is related to the one pass gain, G, by:

$$E = (\exp G - 1)/G.$$

* Permanent address: Racah Institute of Physics, Hebrew University, Jerusalem

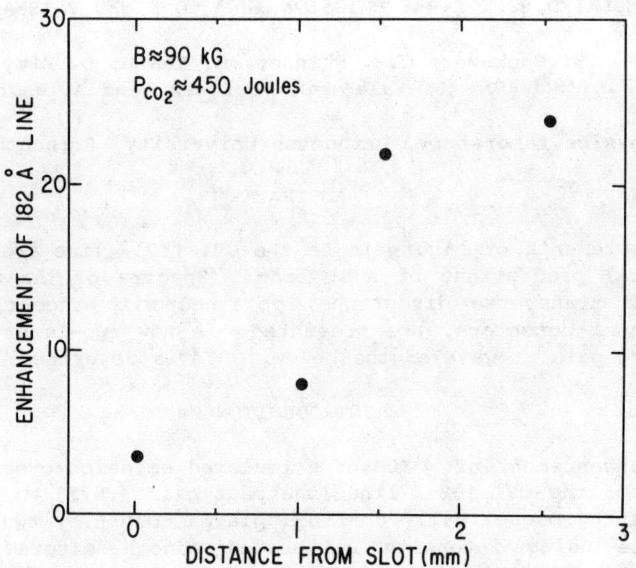

Fig. 1. Measurement of the enhancement of the CVI 182 Å emission as a function of radius in the plasma column. The slot limited the observation region of the spectrometers.

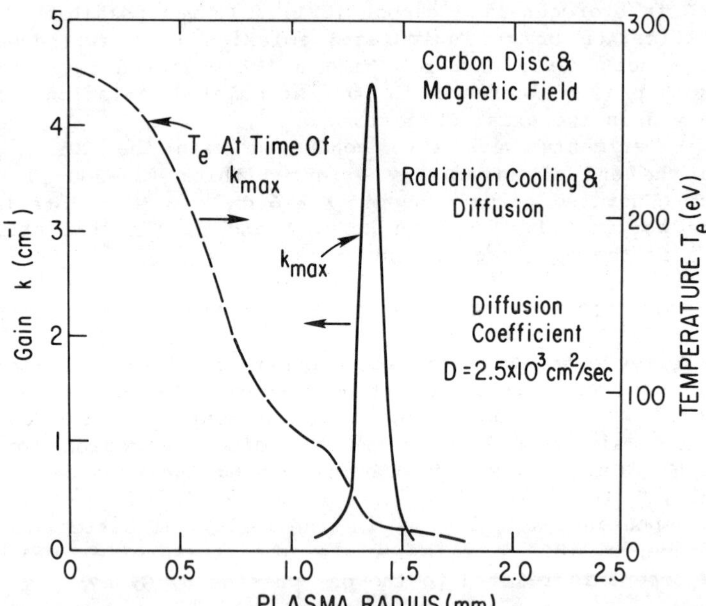

Fig. 2. Radial Profiles of CVI 182 Å gain, k_{max}, and electron temperature, T_e, versus radius in the plasma column as predicted by a 1D code.

By varying the position of the laser focus on the carbon disc target with respect to the observation volume of the axial and transverse monochromators it was possible to measure the gain as a function of radius in the plasma column. The results of one such scan are shown in Figure 1.

In this experiment the CO_2 laser energy was not optimal but it can be seen that there is a rapid rise of the enhancement in the region x = 1.5 to 2.5 mm off axis. Further experiments with optimal plasma conditions led to measurements of an enhancement of E ≈ 100 for x = 1.3 to 1.5 mm.[1]

A one-dimensional hydrodynamic plus atomic physics model has been developed to aid understanding these results.[3] In this model a single mean flow velocity was used to describe the ion mass motion. After solution of time-dependent equations for the ion density, momentum and electron energy, the gain was calculated by a post processor code from the electron density and temperature and the number density of the ground state populations of fully stripped and hydrogen like carbon. Because the laser pulse length is longer than the compressional Alfven transit time, radial pressure balance is quickly established in the plasma. Strong heating on the cylindrical axis of symmetry (at the laser focus) leads to a centrally peaked temperature profile with a corresponding electron density minimum. On the other hand, off axis, strong radiative cooling by CIV leads to low temperature, high density conditions conducive to a fast recombination rate and high 182 Å gain. With the introduction of an ion ion diffusion rate an order of magnitude greater than the classical value, totally stripped ions were transported from the center to the cold off axis region where fast recombination generated high gain.

Figure 2 shows the predicted gain versus radius and it can be seen that high gain occurs in a narrow 100 μ - 200 μ wide annulus at a radius of 1.4 mm. This is in excellent agreement with the experimental results.

AXIAL AND TRANSVERSE EMISSION SPECTRA

Recently, multichannel detectors, based on microchannel/reticon arrays, were installed in the axial and transverse spectrometers. These permit the recording of emission spectra in the axial and transverse directions in a single laser shot. One example is shown in Fig. 3.

In the transverse spectrum, the spontaneous CVI 182 Å emission is weak compared to the strongest line in the spectrum, OVI 173 Å. However in the axial direction, the stimulated 182 Å emission dominates the spectrum.

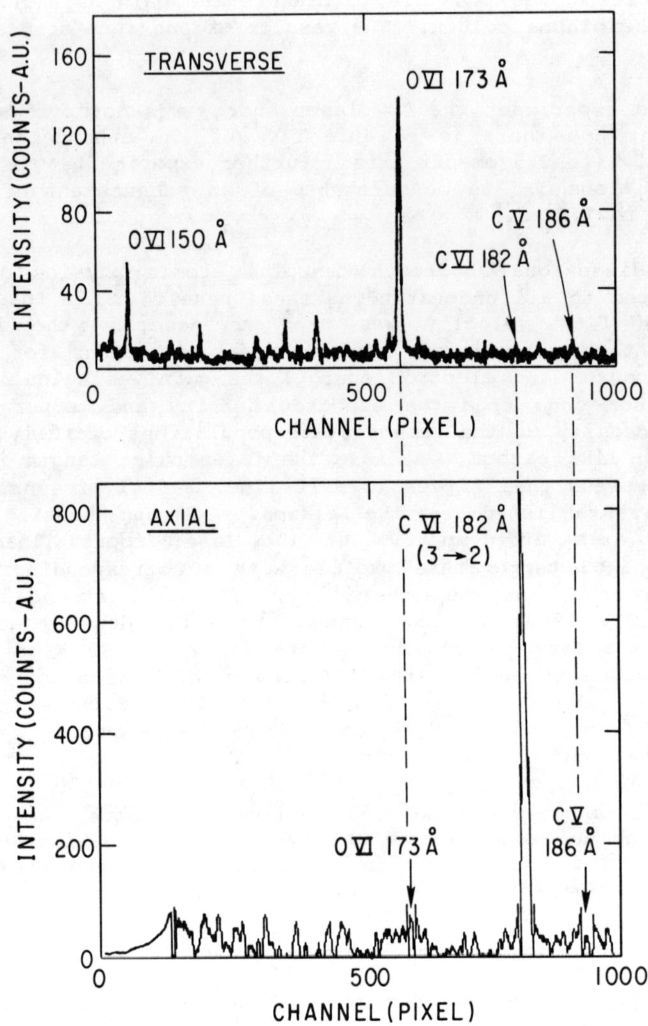

Fig. 3. Transverse and axial spectra in the region near 182 Å from a carbon disc target with four carbon blades. The laser energy was 500 J.

TWO LASER APPROACH TO X-RAY LASER DEVELOPMENT BELOW 100 Å

There is considerable interest in developing lasers in the wavelength region below 100 Å. We have begun constructing an experiment in which a 1.5 kJ CO_2 laser generates a highly ionized magnetically confined plasma column in which a powerful ($I \gtrsim 10^{15}$ W/cm^2) picosecond laser beam produces a population inversion and gain. The role of the CO_2 laser is to provide access to the high energy, short wavelength transitions of high Z ions which are then excited by the picosecond laser via multiphoton processes.

The feasibility of exciting either one subvalence electron or two valence electrons of an ion through multiphoton processes[4] was studied for the argon and krypton isoelectronic sequences excited by a powerful picosecond KrF* (2480 Å) laser.[5] The excitation of two valence electrons (e.g. $4s^2 4p^4 5s^2$ state) is especially attractive because of the faster progression to shorter wavelengths with increasing charge of the target ion. The potential lasing wavelength for e.g. Cd^{12+} is 89 Å. An additional advantage with the higher ionization stages is that competing processes such as photoionization or autoionization are reduced or eliminated. Multiphoton excitation is also expected to significantly increase the population inversion and gain at 182 Å in the current experiment.[1]

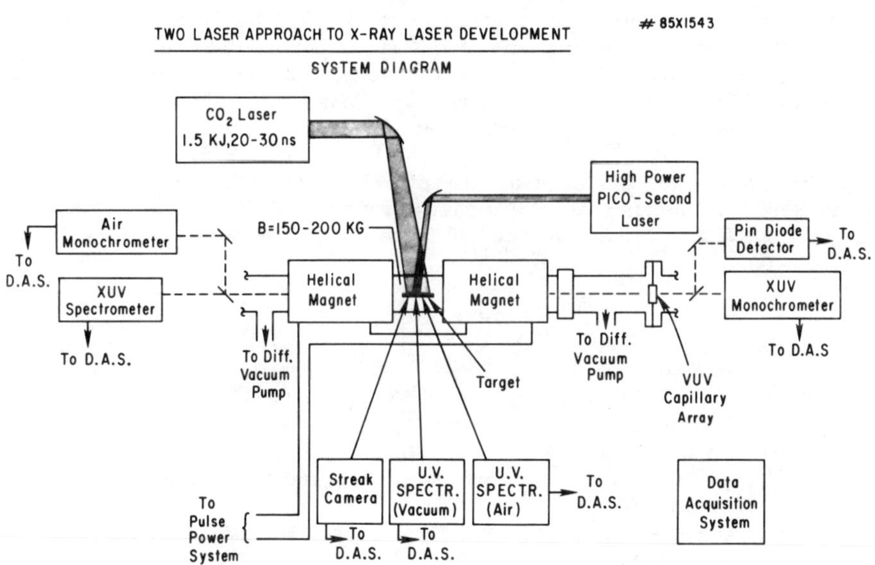

Fig. 4. Diagram of new experimental setup.

As shown in Fig. 4, the new experimental system incorporated two lasers; a 1.5 kJ CO_2 laser to produce the ionized medium and the powerful picosecond laser to generate the population inversion. The target is placed in a solenoidal magnet which will radially confine the plasma and produce the long, thin geometry suitable for laser action. Primary diagnostics will consist of multichannel soft X-ray spectrometers and an X-ray streak camera.

The powerful picosecond laser is expected to generate a 1 J, 1ps, 2480 Å laser pulse. The main oscillator is a YAG-laser pumped dye laser at 6470 Å which is amplified in a three stage dye amplifier producing an energy output of a few mJ. This is then frequency doubled and mixed with the 1.06 μ YAG-laser to produce a 1 ps, 2480 Å pulse with an energy of a few hundred μJ. In the final stage, two KrF^* amplifiers increase the pulse energy to the joule level with an (unfocussed) power in the terawatt range. Focussing by a suitable lens will produce intensities in excess of 10^{15} W/cm^2 and in addition to short wavelength laser applications, will produce new insights into the interaction of radiation with ions in a regime where the laser field is comparable to the couloumb field between the electrons and the nucleus.

ACKNOWLEDGMENTS

We would like to acknowledge support and encouragement from H. Furth and J.R. Thompson; assistance with the analysis of the multiphoton excitation scheme by C.W. Clark, M.G. Littman, T.J. McIlrath, and R. Miles; contributions to the design of the picosecond laser by L. Meixler, C.H. Nam, T. Srinivasan, and B. Tighe; and technical assistance from L. Guttadora and J. Robinson.

This work was made possible by financial support by the U.S. Department of Energy Basic Energy Sciences, Contract No. KC-05-01, and the U.S. Air Force Office of Scientific Research, Contract No. AFOSR-86-0025.

REFERENCES

1. S. Suckewer, C.H. Skinner, H. Milchberg, C. Keane, and D. Voorhees, Phys. Rev. Lett. 55, 1753 (1985).
2. H. Milchberg, C.H. Skinner, S. Suckewer, and D. Voorhees, Appl. Phys. Lett. 47, 1151 (1985).
3. E. J. Valeo, C. Keane, and R.M. Kulsrud, Bull. Am. Phys. Soc. 30, 1600 (1985).
4. T.S. Luk, H. Pummer, K. Boyer, M. Shahadi, H. Egger, and C.K. Rhodes, Phys. Rev. Lett., 51, 110 (1983).
5. C.W. Clark, M.G. Littman, R. Miles, T.J. McIlrath, C.H. Skinner, S. Suckewer. and E. Valeo, J. Opt. Soc. Am. March 1986.

TECHNIQUES FOR SOFT X-RAY SPECTROSCOPY

P. G. Burkhalter, D. J. Nagel, and M. Emery
Naval Research Laboratory, Washington, DC 20375-5000

P.D. Rockett and G. Charatis
KMS Fusion, Inc., Ann Arbor, MI 48106

ABSTRACT

Soft x-ray diagnostics were developed for acquiring spectra in the 6-60 Å wavelength region in rectangular-focused laser beam experiments. The line targets were prepared by a micro-lithographic method. Convex-curved crystal and grazing-incidence grating spectrographs collected spatially-resolved aluminum spectral data. A two-dimensional hydrodynamic model predicted plasma parameter profiles specific to the laser heating conditions. Emissivities were determined as a function of target dimensions.

INTRODUCTION

The methodology was developed for acquiring and measuring spectral emissivities from laser-produced line plasmas.

INSTRUMENTATION

Figure 1 shows the instrumentation used to acquire XUV spectral data from rectangular targets. The curved-diffraction-crystal x-ray spectrographs were positioned for on axis viewing of the line plasmas through 10 micron defining slits placed 1 cm from the aluminum target. The (013) diffraction plane in KAP[1] provides a means for collecting high resolution ($\lambda/\Delta\lambda$=3100) spectral data for the Al XIII and Al XII alpha lines and their satellites while PET crystals collected the entire Al spectra. The XUV spectrograph is a 1-m grazing incidence device originally designed at N.R.L. to collect solar spectra on rocket flights. The XUV spectrographs were secured to kinematic mounts and could be used to view the target either on or orthogonal to the plasma axis. The spectrographs were accurately aligned orthogonal to the direction of the Chroma laser beam, with the slits parallel to the Al target surface. This arrangement provided spatial definition of the plasma temperature and density profiles as has been reported using the tracer dot method developed at N.R.L.[2,3]

Fig. 1. Instrumentation used to acquire soft x-ray data from Al plasmas.

RESULTS

Figure 2 shows portions of the x-ray and XUV spectra collected from Al line plasma generated by 1.7×10^{13} w/cm^2 irradiance of 0.527 um laser light from the CHROMA laser that was focussed with a cylindrical lens to a 1.1 cm length. The plasma was viewed through the slits as illustrated to yield spectral lines that are an image of the collisionally-confined blow-off plasma emission in the direction of the laser beam.

The FAST2D fully two-dimensional hydrodynamic laser-matter-interaction code at N.R.L.[4] has been adapted to geometries and plasma conditions appropriate to opacity studies. Plasma temperature and density profiles are computed for the target and laser conditions. The code is valuable for experimental design and predicts both the temporal and spatial development of the rectangular plasma expansion.

The spectrograms are scanned by a digitizing microdensitometer in steps with 10 micron spatial resolution. The spectral densities are computer processed using known film calibrations for the direct exposure films.[5] Line intensities were determined as a function of distance from the original target surface and can be compared with collisonal-radiative equilibrium calculations[6] to interpret plasma conditions.

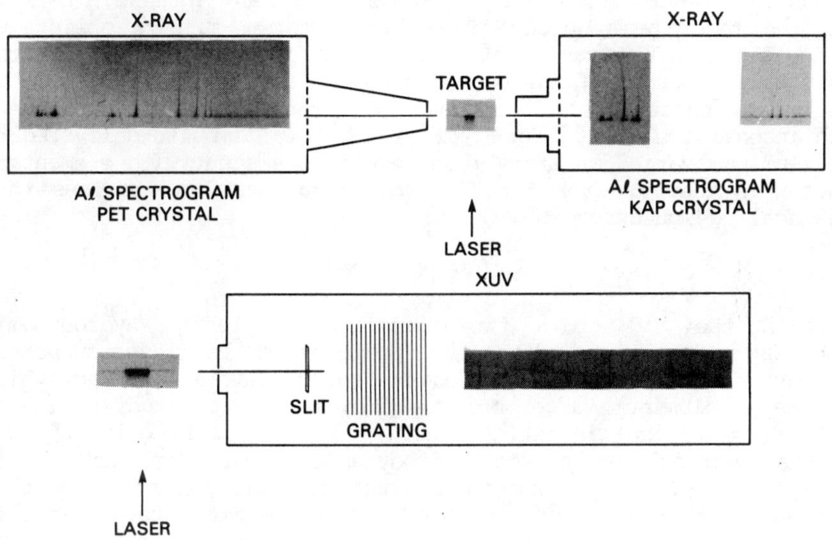

Fig. 2. X-ray and XUV spectrograms from rectilinear targets.

REFERENCES

1. P.G. Burkhalter, D.B. Brown, and M. Gersten, J. Appl. Phys. 52, 4379 (1981).
2. M.J. Herbst, P.G. Burkhalter, R.R. Whitlock, J. Grun and M. Fink, Rev. Sci. Instrum. 53, 1418 (1982).
3. P.G. Burkhalter, M.J. Herbst, D. Duston, J. Gardner, M. Emery, R.R. Whitlock, J. Grun, J. Apruzese, and J. Davis, Phys. Fluids 26, 3650 (1983).
4. M.H. Emery, J.H. Gardner, and J.P. Boris, Phys. Rev. Lett. 48, 677 (1982).
5. P.D. Rockett, C.R. Bird, C.J. Hailey, D. Sullivan, D.B. Brown, and P.G. Burkhalter, Appl. Optics 24, 2536 (1985).
6. J.P. Apruzese, D. Duston, and J. Davis, private communication.

LASER-SUPPORTED DETONATION WAVES IN AN OBLIQUELY INCIDENT BEAM

C. L. Bohn and M. L. Crawford
United States Air Force Academy, Colorado Springs, Co. 80840-5821

ABSTRACT

We conducted a computational study of laser-supported detonation waves (LSDWs) propagating up an intense, narrow, infrared laser beam which strikes an aluminum surface obliquely. The detailed hydrodynamics at the aluminum surface varied with the angle of incidence of the beam due to the asymmetric geometry, but the gross properties of the plasma and the LSDW propagation were both insensitive to θ. For $\theta \lesssim 60^{\circ}$, the total impulse delivered to the aluminum varied as $1/\cos\theta$, a result consistent with elementary blast-wave theory. For $\theta > 60^{\circ}$, the total impulse was less than this scaling would suggest.

INTRODUCTION

With the aid of a two-dimensional, Eulerian hydrodynamic computer code, we studied the propagation of laser-supported detonation waves (LSDWs) formed by an intense laser beam which strikes an aluminum target obliquely in air. This study builds on earlier theoretical investigations.[1,2] In particular, it uncovers to what extent the plasma hydrodynamics resembles that of an interaction at normal incidence, and to what extent elementary blast-wave theory might be used to model the total impulse delivered to the target. This paper contains a summary of the results; a complete account will appear elsewhere.[3]

ASSUMPTIONS

In every calculation, the laser beam was taken to be a spatially and temporally uniform slab of wavelength 10.6 μm, intensity 30 MW/cm^2, and half-thickness 2mm. A LSDW was made to form early, within 0.1 μsec, and it propagated until 5 μsec elapsed, after which the computation was terminated. Beam attenuation and energy deposition in the plasma were computed using three simplifying assumptions: Saha equilibrium ionization prevails, the electron and ion temperatures are equal, and the equation of state is that of an ideal gas. The hydrodynamics are insensitive to these assumptions.[2] In addition, re-radiation from the plasma is unimportant.[4]

RESULTS

Very early in the interaction, a planar LSDW forms oriented parallel to the target surface and propagates vertically away from the surface in response to the projected beam intensity. Once the center of this LSDW has propagated to the edge of the obliquely incident beam, a second LSDW oriented perpendicular to the beam has formed. This second shock absorbs the beam, thereby causing the first shock to dissipate. Subsequently, the second shock propagates up the beam in response to the beam intensity. The gross properties of the plasma and the shock propagation are found to be essentially independent of the angle of incidence of the beam. This is because the gas flow behind the shock is sonic.[5] Consequently, information about the target's orientation cannot propagate back to the shock. However, the detailed hydrodynamics at the target surface does depend on θ .

The total impulse I is the time-integrated pressure force. Once the transient effects associated with LSDW formation have passed, I(t) grows approximately linearly with time, regardless of θ. This computational result agrees with the prediction of elementary blast-wave theory that the pressure force is independent of time. In addition, I is found to vary approximately as $1/\cos\theta$, reflecting the geometric circumstance that the footprint of the plasma on the target grows as $1/\cos\theta$. For angles less than 60°, the data follow this scaling quite closely, but they fall below this prediction at larger angles. The physical interpretation is that small pressures acting over large areas dominate the total impulse. Accordingly, the gross structure of the plasma, which is insensitive to θ, dominates. For these reasons, we anticipate that elementary blast-wave theory will also provide reasonably accurate scaling rules when applied to a more realistic, fully three-dimensional interaction.

This work was sponsored by the Air Force Office of Scientific Research.

REFERENCES

1. A. N. Pirri, Phys. Fluids, 16, 1435 (1973).
2. P. E. Nielsen, J. Appl. Phys., 46, 4501 (1975).
3. C. L. Bohn and M. L. Crawford, to be published.
4. J. A. McKay, et. al., Paper 84-1586, presented at the AIAA 17th Fluid Dynamics, Plasma Dynamics, and Lasers Conference, Snowmass, Co, 1984 (unpublished).
5. Ya. B. Zeldovich and Yu. P. Raizer, Physics of Shock Waves and High Temperature Hydrodynamic Phenomena (Academic Press, New York, 1966), Vol. 1.

THE IODINE LASER

G. Brederlow
Max-Planck-Institut für Quantenoptik (MPQ), D-8046 Garching, FRG*

ABSTRACT

This paper will deal with the description of the design of the 2 kJ/7 TW single beam Asterix IV iodine laser and the steps necessary to obtain a good beam quality and a homogeneous intensity profile of the laser beam.

INTRODUCTION

The space available restricts the description of a few selected topics of those dealt with in the oral presentation of this paper. Therefore the survey of the basic features of an iodine laser[1], the measures for improving its pulse width flexibility[2 3], the generation of higher harmonics of iodine laser light[4], and the description of the setup and performance of the 300 J/1 TW Asterix III iodine laser[5] have to be referred to in the literature. In the following only the setup of the single beam 2 kJ/7 TW Asterix IV iodine laser, now under construction at MPQ Garching will be described. Also the measures required for maintaining a satisfying beam quality and intensity profile through a chain with saturated amplifiers will be reported.

THE ASTERIX IV LASER

The Asterix IV laser which will be used for laser plasma experiments is designed to deliver 2 kJ of energy at output pulse lengths of 1 ns and a maximum power of 7 TW at the shortest pulse length of 0.1 ns. The shot rate is 1/20 min. The basic for the design of this laser was a 1-dim. computer programme[6]. With this programme the energy extraction and the pulse shape variation along the amplifier chain was computed by solving the Maxwell-Bloch equations. According to the hyperfine splitting of the laser levels a six line code was employed.

The schematic of the Asterix IV laser is shown in Fig. 1. The pulse to be amplified is selected from an acousto-optically mode-locked oscillator pulse train by a pulse selection system or generated by the method of ultra-fast gain switching[2]. In this way oscillator pulses between 0.4 and 20 ns will be produced. Due to the non-linear gain or absorption in the saturated amplifiers resp. absorbers a pulse length reduction will take place, leading to an output pulse length between 0.1 ns and several ns.

The oscillator pulse is amplified by six amplifiers of successively increasing diameters, lengths and stored energies (A1-A6). The layout of the amplifier chain is such that the output energy density of an amplifier will not exceed $3 J/cm^2$ at $t_p \geq 1 ns$ resp. $1 J/cm^2$ at $t_p = 0.1$ ns. The output energies of the amplifiers A2,A3,A4 and A5 are chosen such that the energy density at the entrance of the

* Supported in part by Euratom

0094-243X/86/1460092-4$3.00

following amplifier is about 0.5 J/cm^2. Thus amplifier saturation is nearly achieved and the extraction efficiency for the stored inversion energy is about 50 %. The second amplifier (A2) is operated in the double-pass mode to obtain medium saturation already in the third amplifier. The entrance and exit beams are separated by a dielectric polarizer P after double pass through a $\lambda/4$ plate (QWP).

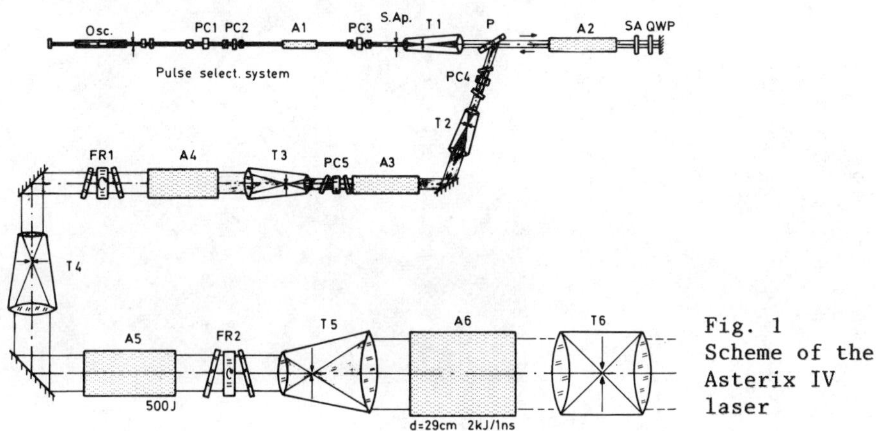

Fig. 1 Scheme of the Asterix IV laser

The amplifiers which are assembled from sections, have a new design. The quartz tube for containing the laser medium is now surrounded by two plastic half shells on which the flashlamps with their reflectors are mounted (fig. 2). Thus an easy exchange of the flashlamps will be possible. In fig. 3 a section of the end-amplifier is shown. The test of this section yielded the required stored inversion energy (400 J at 40 kV) and a nearly rectangular inversion profile. With shear-interferometric measurements it could be shown that after the completion of the pumping process (T=22 µs) there are no optical inhomogeneities in the volume of the laser medium.

The beam diameter will be expanded from 6 to 290 mm by an image relaying system consisting of vacuum spatial filter telescopes (T1-T5). This system relays an uniformly irradiated soft aperture (S.Ap.) through the chain. It provides a high fill factor for the amplifiers while maintaining a uniform intensity profile at critical positions of the system such as AR-coatings crystals, polarizers, etc.

The requirement of optimum energy extraction from the amplifier chain calls for a small signal amplification exceeding the actual amplification by several orders of magnitude. Therefore special measures have to be taken to restrict the prepulse power so that sensitive targets are not damaged before the main pulse strikes them. To reduce the leaking of oscillator pulses through the closed pulse selection system two selected Pockels cells/Glan prisms in line (PC1, PC2) with a high contrast ratio ($1 : 10^8$) are used. For amplifier decoupling and the reduction of amplified spontaneous fluorescence, three Pockels cells (PC1, PC2, PC3) and a saturable dye absorber (SA) (contrast ratio 1 : 20) are installed.

Only two Faraday rotators (FR1, FR2) are required to protect the laser system from light retroreflected from the target. This is due to the fact that the bandwidth of the spectrally broadened and Doppler shifted reflected light is much broader than the gain bandwidth of the amplifiers. Thus this light experiences only a small amplification by the residual inversion.

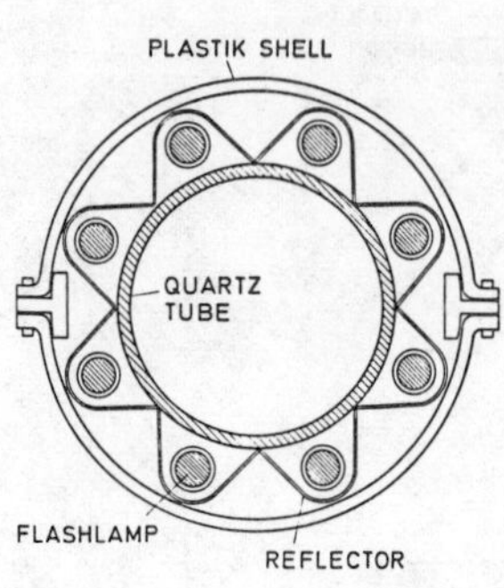

Fig. 2 Cross-section of an amplifier-section

Fig. 3 Section of the end-amplifier (d = 29 cm)

Laser plasma experiments require an excellent beam quality. Therefore measures have to be taken to keep the wave-front deformations along the beam line as small as possible. Measurements performed with test sections of the Asterix IV amplifiers revealed that the laser medium itself introduces only deformations less than $\lambda/10$. The main source for a beam degradation may be self-focussing effects. But since the iodine laser is a gas laser with a relatively small value of the non-linear refraction index n_2 of the laser medium, self-focussing effects are less pronounced than in solid state lasers. The main contribution to the B-integral, which is a measure of the beam quality, arises from optical glass components such as windows, polarizers etc. The critical value of $B = 4$ will be exceeded by the Asterix IV laser at maximum loading with pulses shorter than 1 ns. The value of the B-integral can, however, be reset by spatial filtering which will be achieved by stops placed in the focal plane of the telecopes of the image relaying system[7]. Since even at the highest output power of 7 TW/0.1 ns the increase of the B-integral per amplifier (including the telescope) is less than 1, the loading of the system is limited by damage thresholds of components.

Targets have often to be placed in the quasi near field of the lens ahead of the focus plane. The intensity profiles in this plane should then be as homogeneous as possible. To fulfill this requirement Fresnel diffraction has to be eliminated by image relaying of a uniformly illuminated iris placed ahead of the first amplifier through the laser system. Measurements performed with a combination of a single image relaying telescope and amplifier yielded, however, that a saturated amplifier changes the image properties such that a rectangular input profile shows an edge enhancement in the image plane[8]. Only an amplifier operated in the small-signal regime will exactly reproduce the input profile. This behaviour was confirmed by numerical calculations by computing the intensity profile in the image plane by the paraxial wave equation with the assumption that no coherent effects will occur. An example of results of the measured and calculated intensity profiles are shown in fig. 4 for a linear and a non-linear laser medium with homogeneously illuminated hard and soft apertures.

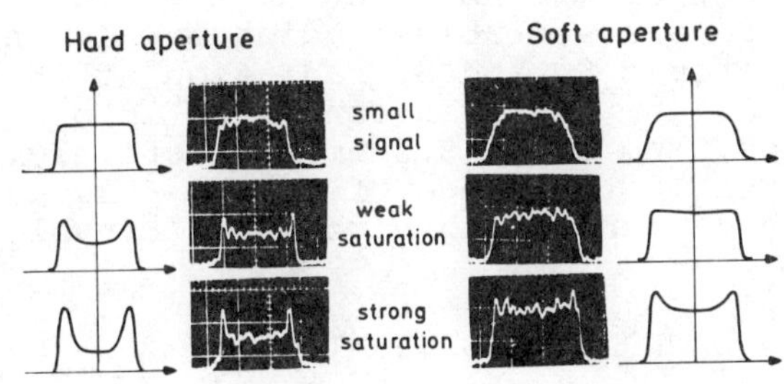

Fig. 4 Measured and calculated intensity profiles in the image plane. Fresnel number: F = 1.1

The edge enhancement of a rectangular intensity profile by a saturated amplifier can be explained by the fact that from the frequencies of the spatial spectrum forming the rectangular intensity profile the higher spatial frequencies of the edges have smaller amplitudes and therefore experience a higher amplification. The image thus shows an over-shoot at the edges. The edge enhancement, however, decreases with increasing Fresnel number. Since the Fresnel number F in an amplifier chain is ranging from $40 \leqq F \leqq 1400$, the overshoot is less dramatic as shown in Fig. 4 (F = 1.1) and can be eliminated by using a soft aperture in the object plane.

REFERENCES

1. G. Brederlow, E.E. Fill, K.J. Witte, The High Power Iodine Laser, Springer Verlag, Berlin, Heidelberg, New York, 1983.
2. E.E. Fill, W. Skrlac, K.J. Witte, Opt. Commun. 37, 123 (1981).
3. K.J. Witte et al., IEEE J. Quantum Electron. QE17, 1809 (1981).
4. E.E. Fill, Opt. Commun. 33, 321 (1980).
5. G. Brederlow et al., IEEE J. Quantum Electron. QE16, 122 (1980).
6. T.Uchiyama, K.J.Witte, IEEE J.Quantum Electron, QE18,885 (1982).
7. J.T. Hunt, J.A. Glaze, W.W. Simmons, P.A. Renard, Appl. Opt. 17, 2053 (1978).
8. E.E. Fill, Opt. Commun. 49, 362 (1984).

GENERATION OF VUV RADIATION BY THE ANTI-STOKES RAMAN LASER PROCESS

B. Wellegehausen and K. Ludewigt
Institut für Quantenoptik, Universität Hannover
3000 Hannover, Fed. Rep. of Germany

ABSTRACT

Frequency up-conversion by the anti-Stokes Raman laser process offers interesting perspectives for the generation of short wavelength coherent radiation. For the spectral range of 100 nm - 200 nm the group VI elements O, S and Se are suitable candidates. In initial experiments with atomic Se and S, radiation at 199.5 nm, 254.8 nm and 219.1 nm has been converted into radiation at 158.7 nm, 167.5 nm and 148.3 nm, respectively. Threshold pump energies in the range of 0.1 μJ - 20 μJ indicate the possibility of generating high power tunable vuv radiation.

GENERAL CONSIDERATIONS

The generation of coherent radiation in the vacuum ultraviolet or even in the x-ray spectral range requires special techniques and efforts. An interesting technique, which recently has been realized for the first time[1], is the frequency up-conversion by the anti-Stokes Raman laser process. This technique makes use of high lying metastable levels, which serve as energy storage levels. If such a metastable level (level 2 in Fig. 1) is selectively populated and inverted with respect to a lower lying level, then coherent pump radiation of frequency ω_p can be up-converted into coherent anti-Stokes radiation of frequency ω_{AS}. Depending on the energy difference ΔE between the levels 2 and 1, large frequency shifts in a single step are possible.

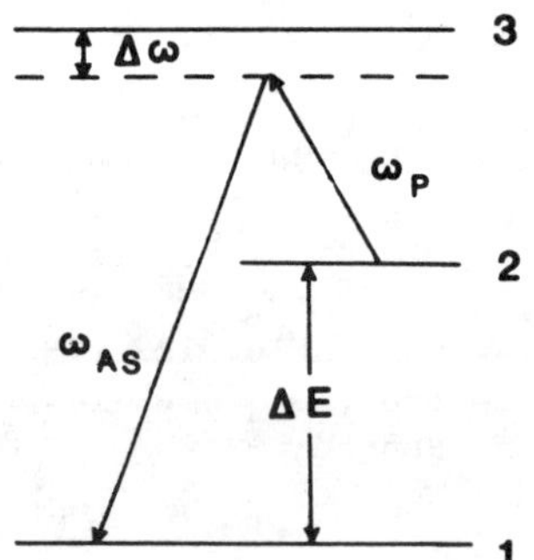

Fig. 1. Anti-Stokes frequency up-conversion scheme. ω_p : pump frequency, ω_{AS} : anti-Stokes frequency, $\Delta\omega$: frequency detuning. Population inversion between level 2 (metastable) and level 1.

This anti-Stokes technique allows tuning around resonance over a more or less extended range, depending on the pump intensity I_p and the inversion density ΔN, as the anti-Stokes gain coefficient α_{AS} is given by[2] $\alpha_{AS} = K \cdot \Delta N \cdot I_p / \Delta\omega^2$. ($\Delta\omega$ is the frequency detuning; the factor K contains the oscillator strengths of the involved transitions and the Raman linewidth. The formula is valid for detunings large compared to all relevant linewidths). This anti-Stokes process is non parametric and therefore requires no index matching, in contrast to many other nonlinear techniques used to generate short wavelength coherent radiation. Finally, this method also allows high output energies and large conversion efficiencies as has been demonstrated earlier[3].

Three main problems have to be solved in order to realize an anti-Stokes Raman laser.

First is the choice of a material with the appropriate metastable level and a suitable three level scheme as indicated in Fig. 1. Schemes using electronic levels in atoms, molecules[4], and ions[5,6] or even schemes involving nuclear levels[7,8] have been proposed and are presently investigated. However, so far only a few atomic systems (Tl[1,3], In[9], Br[10], I[11], Sn[2], Pb[3], Se[12] and S) have been realized.

The second problem is the generation of the necessary population inversion of the metastable level with respect to a lower lying level. For this, in principle, all excitation techniques used for normal lasers can also be applied. A very convenient technique, used with all atomic systems realized up to now, is photodissociation of suitable molecules. Thereby, the desired metastable atom can directly be produced.

Finally, the tunable pump radiation required for the up-conversion has to be generated. Most favorable, of course, would be anti-Stokes schemes where dye laser radiation can directly be converted. However, in many cases the pump transition needs a much shorter wavelength, and therefore already nonlinear techniques (frequency doubling, H_2-anti-Stokes Raman shifting of dye laser radiation) have to be used to produce the necessary pump radiation.

EXPERIMENTAL

For the vuv spectral range between about 100 nm and 200 nm, the group VI atoms O, S and Se with their high lying metastable $^1S_0$ level are suited[13]. Anti-Stokes Raman laser investigations on atomic Se and S have been started, and so far tunable vuv radiation at 167.5 nm and 158.7 nm in case of Se, and at 148.7 nm and 148.3 nm in case of S has been generated. Level schemes with the

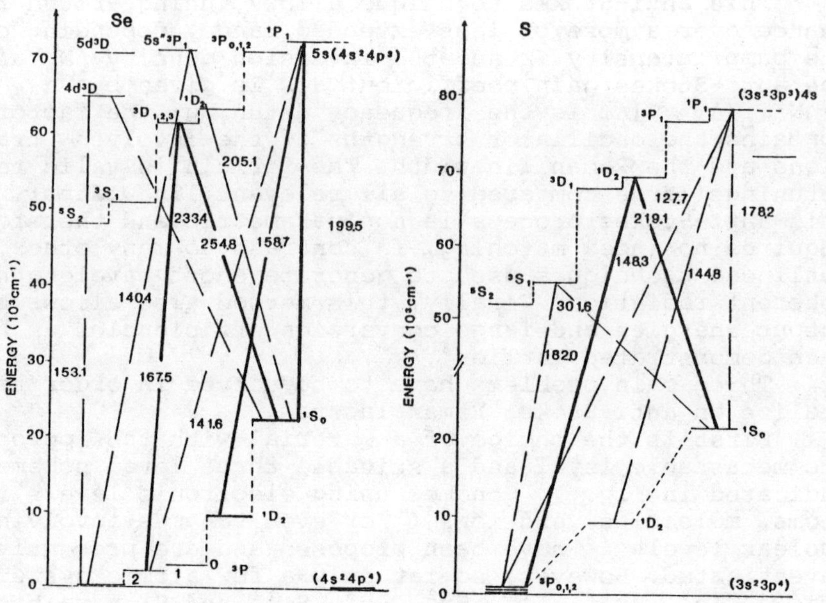

Fig. 2. Level schemes of atomic Se and S. Realized (——), possible (---) anti-Stokes laser cycles. Wavelengths in nm.

corresponding laser cycles are given in Fig. 2. The necessary population inversion between the initial and final level of the anti-Stokes laser cycle is, for Se, generated by photodissociation of the molecule COSe with ArF (193 nm) excimer laser radiation[14] and, for S, by photodissociation of COS with F_2 (157 nm) laser radiation[14]. Typical molecular vapor pressures are around 0.3 mbar and estimated inversion densities are about $2 \cdot 10^{15}$ cm^{-3}. The pump radiation was generated either by frequency doubling or H_2-Raman shifting of a pulsed dye laser, pumped by an excimer laser (EMG 150; Lambda Physics). This excimer laser simultaneously generates the photodissociation radiation. So far obtained experimental results are summarized in Table I. The data show that vuv anti-Stokes Raman lasers can be operated at low thresholds. It is expected that the output energies can be scaled by applying higher pump energies, by increasing the inversion density, the interaction length and the active volume. Further interesting cycles are possible with atomic Se and S. In some cases good coincidences to strong pump sources such as the ArF laser itself or its first Stokes/anti-Stokes components in H_2 or D_2 exist, allowing powerful anti-Stokes lasers at different wavelengths in the range of about 130 nm - 160 nm.

TABLE I DATA OF Se AND S ANTI-STOKES RAMAN LASER

	Pump transition	Laser transition	threshold	output energy	tuning range	max. pump energy
Se	^1S_o - $^1P_1$ 199.5 nm	$^1P_1$ - $^1D_2$ 158.7 nm	0.1 μJ	∿ 1 μJ	± 9 cm^{-1}	5 μJ
	^1S_o - $^3D_1$ 254.8 nm	$^3D_1$ - $^3P_1$ 167.5 nm	20 μJ	∿ 5 μJ	± 5 cm^{-1}	2 mJ
S	^1S_o - $^3D_1$ 219.1 nm	$^3D_1$ - $^3P_{1,O}$ 148.3 nm 148.7 nm	5 μJ	∿ 0.1 μJ	± 1 cm^{-1}	10 μJ

Operation conditions: pump pulse length 10 nsec, cell length 14 cm

Se: OCSe vapor pressure ∿ 0.4 mbar
Photodissociation with 193 nm (ArF-Laser),
energy 10 mJ, beam cross section ∿ 2 x 2 mm

S : OCS vapor pressure ∿0.2 mbar
Photodissociation with 157 nm (F_2-Laser),
energy 2 mJ, beam cross section ∿3 x 8 mm

REFERENCES

1. J. C. White, D. Henderson, Phys. Rev. A25, 1226 (1982).
2. K. Ludewigt, K. Birkmann, B. Wellegehausen, Appl. Phys. B33 (1984).
3. B. Wellegehausen, K. Ludewigt, H. Welling, Proc. Soc. Photo Opt. Instr. Eng. 492, 10 (1985).
4. J. C. White, D. Henderson, T. A. Miller, M. Heaven, Laser Spectroscopy VI, (Springer Series in Optical Sciences Vol. 40, Springer Verlag, N. Y. 1983) p. 407.
5. A. J. Mendelsohn, S. E. Harris, Opt. Lett. 10, 128 (1985).
6. R. G. Caro, J. C. Wang, R. W. Falcone, J. F. Young, S. E. Harris, Appl. Phys. Lett. 42, 9 (1983).
7. C. B. Collins, F. W. Lee, D. M. Shemwell, B. D. De Paola, S. Olariu, I. I. Popescu, J. Appl. Phys. 53, 4645 (1982).
8. B. D. De Paola, C. B. Collins, J. Opt. Soc. Am. B1, 812 (1984).
9. J. C. White, D. Henderson, IEEE Journ. Quant. Electr. QE-20, 462 (1984).
10. J. C. White, D. Henderson, Opt. Lett. 8, 520 (1983).
11. J. C. White, D. Henderson, Opt. Lett. 7, 204 (1982).
12. K. Ludewigt, H. Schmidt, R. Dierking, B. Wellegehausen, Opt. Lett. 10, 606 (1985).
13. J. C. White, Opt. Lett. 9, 38 (1983).
14. G. Black, R. Sharpless, T. Slanger, J. Chem. Phys. 64, 3985 (1976), J. Chem. Phys. 62, 4274 (1975).

INTRACAVITY GAIN DETECTION APPLIED TO THE OPTIMIZATION OF FLOW PARAMETERS IN A PULSED IF LASER

R. L. Williamson
Sandia National Laboratories, Albuquerque NM 87185

L. Hanko and S. J. Davis**
Air Force Weapons Lab, Kirtland AFB, New Mexico 87117

ABSTRACT

Intracavity gain detection was used to optimize flow conditions in an optically pumped, pulsed IF laser. This effort was initiated to evaluate the usefulness of the intracavity gain detection technique in this type of application.

The extreme sensitivity of the intracavity gain detection technique has recently been demonstrated by Truesdell and Keller[1] who detected optical gains estimated to be as small as 2×10^{-5} in a CW system. Though other highly sensitive gain diagnostics are available they invariably require that the probe laser wavelength exactly match the wavelength of the inverted transition under investigation. This requirement is eliminated when using intracavity gain detection due to the observation that the probe output pulls to and "locks" onto the gain transition wavelength. This wavelength "pulling" phenomenon[2] makes the intracavity technique an extremely versatile tool for gain detection and optimization studies in any medium which exhibits potential or known laser transitions within the wavelength range of the probe laser.

In this paper we describe the application of this technique to the study of a pulsed gain system. The technique was used to study gain conditions in optically pumped, transient inversions of the IF B→X system. The method provided detailed temporal data on gain magnitude as a function of pressures and flow rates of both reactant gases and carrier gases in the flow cell.

The intracavity gain detection technique involves placing the test medium within the cavity of an operating dye laser and using the dye gain to offset all steady-state losses in the cavity. In our experiments we used a Coherent Model CR-590 CW dye laser for the probe laser. The output coupler was extended about 0.8 m to allow room for a 65 cm longitudinal flow reactor to be inserted into the cavity. IF was produced in the flow cell by mixing I_2 and F_2. The carrier gas was He. The test medium was optically excited with a Phase-R Model DL-1400 pulsed dye laser at a wavelength of about 496.5 nm, which corresponds to pumping v'=3 of the IF $B^3\Pi(0^+)$ state. The pump laser pulse powers ranged from 20 to 50 mj/pulse. The pulse rate was typically 0.3 Hz. Gain pulses were obtained by

*This work performed at Sandia National Laboratories supported by U.S. Department of Energy under Contract Number DE-AC04-76DP00789.
**Current address: Physical Sciences Inc., Andover, MA 01810

monitoring the transient probe laser output "pulled" to the gain transition wavelength by the transient IF gain.

An example of the type of data collected in the pulsed gain detection experiment is shown in Figure 1. The gain was detected at 625.2 nm (1 nm bandpass) corresponding to the 0'-5" transition, and the probe laser was set at 620.5 nm. Thus, in this example the probe laser has been "pulled" nearly 5 nm onto the gain transition wavelength. Gain onset was delayed since gain from a collisionally pumped state of IF was being monitored. The delay decreased as total cell pressure was increased. Our studies showed that "strong" thermalized gain could be detected under a wide range of total pressure (10-100 Torr) provided the ratio of F_2 to I_2 was maintained at 40-50. Gain was also observed on either side of this optimum range. For example, at a total cell pressure of 25 Torr, gain was observed for F_2/I_2 ratios ranging from 11 to 160. Thermalized gain could only be detected on B-X transitions involving the v'=0 level which is evidence of very fast V-T transfer in this system, causing the residence time in any intermediate v' level to be short relative to the stimulated emission lifetime.

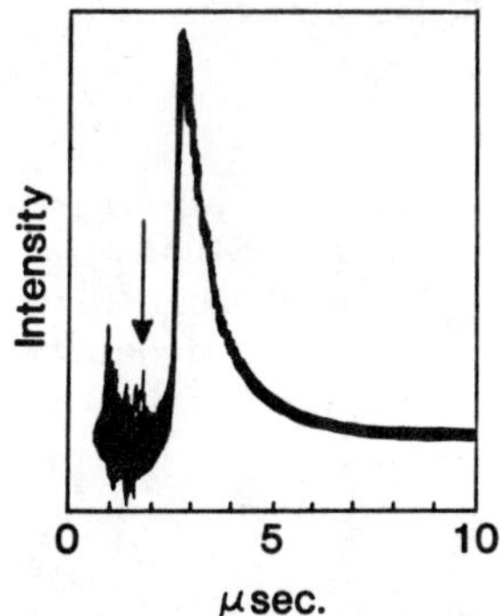

Fig. 1. IF gain pulse. The arrow marks the pump pulse.

Under normal conditions scanning the detection system revealed that the thermalized gain pulse was spread over a broad region (~2.5 nm) of the band. Although we were not able to resolve gain from individual rovibronic lines with the current experimental set-up it was apparent from these experiments that the gain was spread over many rovibronic transitions. This observation shows that monitoring gain yields more detailed information than simply monitoring IF laser output which originates from a single J' level.[3]

When the total cell pressure was lowered to less than about 7 Torr, the thermalized gain disappeared and gain from optically excited v',J' levels appeared. This "direct" gain was investigated primarily on the 3'-7" transition at 622.9 nm. The optically pumped gain was much weaker than the thermalized gain for the longitudinal cell configuration. Under normal conditions the optically pumped gain could only pull the probe output 0.5-1 nm. Efforts to lase IF on a direct transition failed, even at significantly higher pump powers, illustrating that the technique is sufficiently sensitive to detect subthreshold gain.

REFERENCES

1. K. A. Truesdell and R. A. Keller, Appl. Opt. 22, 339 (1983).
2. M. B. Klein, Opt. Commun. 5, 114 (1972).
3. S. J. Davis, L. Hanko, and R. F. Shea, J. Chem. Phys. 78, 172 (1983).

THE SERC BEAT - WAVE PROJECT

C. N. Danson, C. B. Edwards, R. G. Evans, R. W. W. Wyatt
SERC Rutherford Appleton Laboratory, Chilton, Didcot,
Oxfordshire, England OX11 OQX

ABSTRACT

In January 1986 the VULCAN[1] high power Nd Glass laser facility at the Rutherford Appleton Laboratory will be used in a series of experiments to investigate the generation of Beat-Waves in a Z-pinch plasma. The experimental campaign is being conducted in collaboration with the Plasma Physics Group at Imperial College, London.

INTRODUCTION

In order to make advances in High Energy Physics , it is necessary to have access in the laboratory to accelerated particles of extremely high energy. Over the past fifty years, there has been an exponential growth in the energy available to experimenters. However the increase has now started to saturate, and in order to achieve significant improvements in available energy a new generation of accelerators is required, as existing technology becomes inappropriate and prohibitively expensive. The Beat-Wave Accelerator[2] is a possible candidate for such a machine. The basic plasma physics upon which such an accelerator would depend is at present not well understood, and the experimental investigation of the important plasma wave growth and loss mechanisms provide an important incentive for the study to be described.

It has been recognised that electric fields of sufficient intensity for particle acceleration exist in the focus of a high power laser beam (GeV/cm). Such fields are transverse, and can therefore not be coupled directly to produce acceleration fields. The Beat-Wave concept relies on the coupling of electric fields within an electrically neutral plasma to give a longitudinal electric field by the generation of a Langmuir wave.

GENERATION OF BEAT-WAVES

If two electromagnetic waves of dissimilar frequency are propagated in a plasma such that the difference frequency between the waves is resonant with the plasma frequency, then a Langmuir or "Beat-Wave" is generated. Such a wave is longitudinally polarised and propagates with a **phase velocity** approximately equal to the **group velocity** of the incident waves.

The Langmuir wave arises by the following mechanism. The transverse electric field of one incident wave accelerates plasma electrons in the **transverse** direction. This motion couples with the magnetic field of the other incident wave to give a **longitudinal** (Ponderomotive) force which causes an electron density modulation in the direction of propagation of the incident waves.

The resulting electron distribution gives rise to a travelling corrugated electric potential in which charged particles may be **trapped** and **accelerated** if the appropriate injection conditions are fulfilled. It must be noted that the difference frequency between the input waves is at the plasma frequency, and can not propagate as an electromagnetic wave.

LASER DEVELOPMENT

The high power Nd Glass laser at RAL has been configured to produce coherent radiation at 1.064 μ m and 1.053 μ m in a 200 ps pulsewidth at a power of 1 TW. The wavelength of the beat for this case is 100 μm , and the plasma density required for resonance is 10^{17} /cc. Such densities are routinely available from Z-pinch plasmas. In order to produce laser output at the two wavelengths required, a novel front-end oscillator system has been constructed. The system is shown schematically in Fig.1. It consists of two actively modelocked oscillators operating with YAG at 1.064 μ m and YLF at 1.053 μ m.

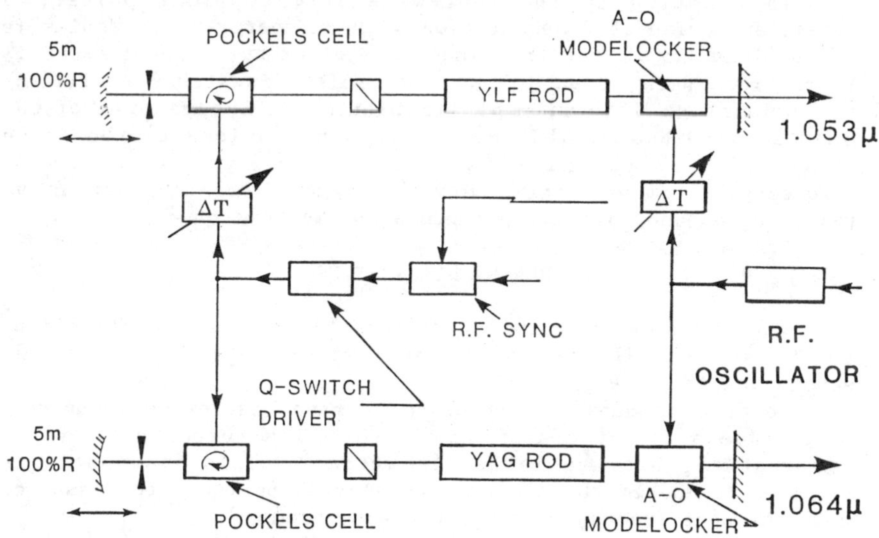

Fig.1 SCHEMATIC OF DUAL WAVELENGTH OSCILLATOR

Temporal synchronisation is achieved by powering the two resonantly matched modelocking crystals from a common RF power source, with the relative timing between the cavities adjusted by means of a phase shifting unit. Adjustment of the phase shifter has the effect of moving the modelocked pulse in one cavity relative to the position of the pulse in the other cavity. Both oscillators are Q-switched. Since the Q-switch build up time is different for the two cavities, it is necessary to provide relative delay between the Q-switch drive pulses. This is achieved by means of a cable delay as shown in Fig.1. The RF synchronisation circuit ensures that optical gates elsewhere in the laser chain are activated at the appropriate time. This is accomplished by gating the trigger circuits to the phase of the RF drive which effectively determines the position of the modelocked pulses in the oscillator cavities.

The laser output required is 100 Joules at each wavelength. The Vulcan laser facility uses a neodymium doped phosphate glass in a combination of rod and disc amplifiers[1], which exhibits higher gain at the 1.053 μ m wavelength. In order to achieve ballanced gain at both wavelengths silicate glass has been substituted in some rod amplifiers.

At an early stage in the rod amplifier chain, a small fraction of the short pulse output is coupled out from the system, is frequency doubled, and Raman shifted to provide a low power probe pulse at 632 nm for Raman-Nath scattering and Schlieren photography diagnostics.

In addition to the Beat-Wave drive optical pulses, a subsidiary pulse of 2 ns duration with an energy of 25 Joules is required at 0.5 μm for the Thomson scattering diagnostics. This is provided using a subsidiary long-pulse Q-switched oscillator in conjunction with an amplification chain independent of the main driving beams. This output is synchronised to the drive pulses using a pockels cell swithchout chain which is switched by an electrical pulse-forming network triggered from the main drive pulse in conjunction with a photoconductive switch[3].

PLASMA DIAGNOSTICS

A comprehensive suite of diagnostics has been developed by the plasma physics group at Imperial College, and is shown schematically in Fig.2.

Details of the time resolved Thomson scattering technique, which give a direct measurement of electron number density and temperature, are available elsewhere[4]. This diagnostic is required to enable the plasma frequency to be tuned to resonance with the beat frequency of the two pump beams.

The Beat-Wave interaction length is to be determined by Schlieren photography and by Raman-Nath Scattering. The interaction of the primary pump beams upon the Beat-Wave itself will generate sidebands at a spacing equal to the plasma frequency. These are to be detected by a spectrometer to give an absolute measure of the Beat-Wave amplitude when deconvolved with the interaction length measurements.

Raman and Brillouin backscattererd light will be detected with a streaked spectrometer arrangement as shown in Fig.2.

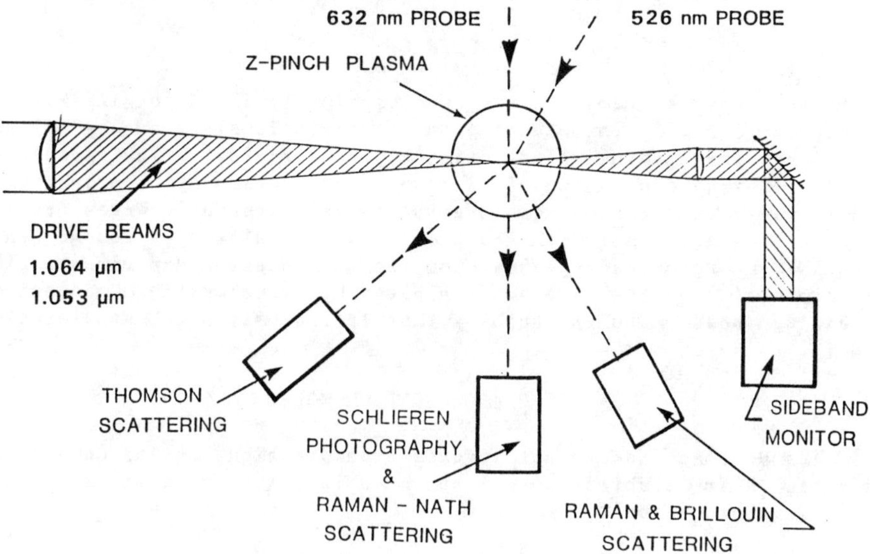

FIG.2 SCHEMATIC OF PLASMA DIAGNOSTICS

CONCLUSION

The VULCAN laser facility has been reconfigured in a novel way to provide the necessary output to drive an experiment designed to study the growth, saturation and loss processes of Beat-Waves in a laboratory Z-pinch plasma. A dual-wavelength synchronised oscillator system has been developed in order to provide the two wavelengths required. Synchronised, high power probe pulses have been provided to enable time-resolved measurements of important plasma parameters to be made.

REFERENCES

1. I. N. Ross, IEEE J. Quantum Electron., QE17, 1653, (1981)

2. T. Tajima and J. Dawson, Phys. Rev. Lett., 43, 267, (1979)

3. M. S. White, Optics Comm., 43, 53, (1982)

4. A. E. Dangor, Rutherford Laboratory, Laser Division, Annual Report 1985 p.A1.1

ELECTRON ACCELERATION IN A LASER-IRRADIATED PLASMA*

Sang-Hoon Kim and K. Wendell Chen
Center for Accelerator Science and Technology
The University of Texas at Arlington, Arlington, Texas 76019

ABSTRACT

Net inverse bremsstrahlung was previously found to give rise to a very large net dc force acting on relativistic electrons by a laser light in a longitudinal electrostatic plasma field. The predicted incident energy dependence of electron acceleration by co-propagating laser waves and the weaker longitudinal electrostatic waves is in qualitative agreement with the preliminary results of a recent Canadian INRS-Energie laser-plasma electron acceleration experiment. The experimental results are not explained by acceleration mechanisms requiring phase matching such as that in the beat wave acceleration models.

LASER ACCELERATION MODEL

The dc force due to net inverse bremsstrahlung acting on a relativistic beam electron in a laser wave interacting with a plasma field is shown to be[1]

$$\vec{F} = \left(\frac{\pi}{2}\right)^{1/2} \frac{\gamma^3 k_D (KT_p)^2 G^2}{mc^2 x^3} \exp(-1/2x^2)\hat{k}_o , \quad (1)$$

where $\vec{k}_o$ is the wave vector of the laser wave, KT_p and G are the electron thermal energy and turbulence parameter, respectively, $x = \Delta p/mc$ with Δp the mometum spread in the beam frame, and other notations have the standard meanings.[2] This force is not considered as the usual Lorentz force.

Equation (1) shows that the force is proportional to γ^3. Recently, Martin et al.[3] injected electrons into a plasma in which two laser waves and a plasma longitudinal wave (beat wave) co-propagate. No phase matching was attempted. The preliminary data show that when an electron beam of kinetic energy 1 MeV ($\gamma = 3$) was injected, electrons of 4.5 MeV ($\gamma = 10$) emerged from the plasma, while when an electron beam of kinetic energy 0.5 MeV ($\gamma = 2$) was injected, electrons of 1.5 MeV ($\gamma = 4$) emerged. The experimental evidence that the greater the electron beam energy, the greater is the acceleration, as well as the large observed acceleration without phase matching are those salient features consistent with the quantum-mechanical model.[1] We note that the apparent results are not consistent with the prevalent laser-electron acceleration models based on a classical treatment. Since the co-propagating laser wave in the z-direction is much stronger than the plasma wave, the electron trajectory must be affec-

*Supported in part by AFOSR 830368.

ted predominantly by the laser wave. In this case, the z-component of the Lorentz force acting on the electron is, from classical electrodynamics[4]

$$F_{1,z} = - \frac{e^2}{2mc^2\gamma} \frac{\partial A^2}{\partial z} , \quad (2)$$

where A is the vector potential of the laser wave. Since A is purely oscillatory, $F_{L,z}$ is also oscillatory in space and time with the same oscillation period of the electromagnetic wave. Thus in the presence of a far stronger laser wave such as in the INRS experiment, phase matching of the electron with respect to the plasma wave cannot be maintained over the wavelength of the laser wave. Moreover, even if one assumes that only the laser-induced plasma beat wave accelerates the electrons, it is difficult to explain the magnitude of the acceleration. This can be seen by noting that in beat wave acceleration the electron dynamics is described by

$$\frac{dv}{dt} = \frac{eE_o\lambda}{mc^2\gamma^3}\cos[2\pi(z - t)] \quad (3)$$

with the initial conditions: $z(0) = 0$ and $\gamma(0) = \gamma_i = \gamma_p = (1 - v^2/c^2)^{-1/2}$. Here z, t, and v are measured in units of the wavelength λ_p, the period T, and the phase velocity $v_p = \lambda/T$ of the plasma wave. Figure 1 shows the curves of the electron energy (γ) versus t for several different values of $W = eE_o\lambda/mc^2$. From the curves, one finds that the electron of $\gamma_i = 3$ does not emerge with $\gamma = 9$ for conditions as in INRS experiemt, $W = 10^{-2} \sim 10^{-3}$. $W = 10^{-2}$ is most likely far greater than that in the INRS experment.

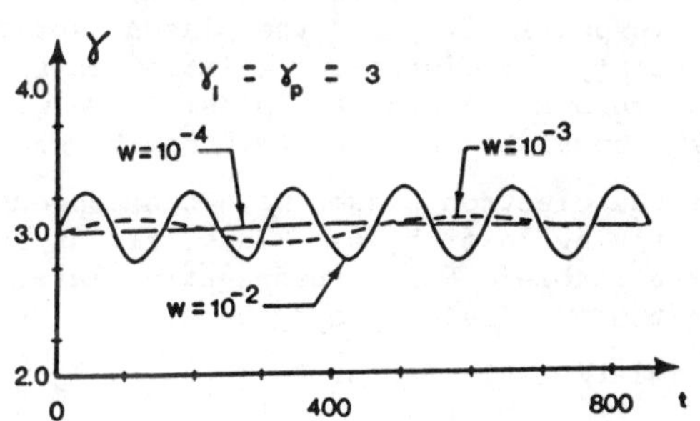

Fig. 1. Plots of γ vs t.

REFERENCES

1. S. H. Kim and K. W. Chen, Laser Acceleration of Particles, AIP Conf. Proc. No. 130, edited by C. Joshi and T. Katsouleas (AIP, New York, 1985), p. 190.
2. S. H. Kim, Phys. Fluids 27, 675 (1984).
3. F. Martin et al. APS Plasma Phys. Conf., 3-8 November, 1985, San Diego, California.
4. F. A. Hopf et al. in Novel Sources of Coherent Radiation, edited by S. F. Jocobs et al. (Addison-Wesley, Reading, 1978), Vol. 5, p. 41.

EXPERIMENTAL STUDIES OF THE PLASMA BEAT-WAVE ACCELERATOR

C. E. Clayton, C. Joshi, C. Darrow, D. Umstadter
and F. F. Chen
University of California, Los Angeles
Los Angeles, California 90024

ABSTRACT

This paper reviews a previous experiment on the excitation of relativistic electron plasma waves by collinear optical mixing and presents some plans for an upcoming upgrade of this experiment where an attempt will be made to observe the acceleration of injected test particles.

BASIC IDEA

The idea of using electron plasma waves excited by intense lasers to accelerate electrons was first proposed by Tajima and Dawson in 1979.[1] Recently, experiments have begun in a number of laboratories around the world.[2-4] In the plasma beat-wave scheme, two coherent beams of light of different frequencies (ω_0 and ω_1) are caused to interfere in a plasma. A force associated with the spatial gradients of the total laser electric field is exerted onto the plasma particles at the difference frequency and difference wavenumber of the two laser beams ($\Delta\omega$ and $\Delta\underline{k}$, respectively). If the plasma density is chosen such that $\Delta\omega$ and $\Delta\underline{k}$ satisfy the dispersion relation of a plasma wave, then this force can resonantly excite a plasma wave with frequency and wavenumber (ω_p, $\underline{k}_p$) equal to ($\Delta\omega$, $\Delta\underline{k}$). It can be shown that the longitudinal fields of the electron plasma wave can approch in magnitude the fields of the driving laser beams. Thus, fields in excess of 1 GeV/m are readily realizable. For co-propagating laser beams and $\omega_0 >> \omega_{pe}$, these longitudinal fields propagate at a relativistic phase velocity with a relativistic factor γ_ϕ given by $\gamma_\phi = \omega_0/\omega_{pe}$ where ω_{pe} is the plasma frequency.

CO_2 LASER REGIME

If one considers the "design equations" for plasma beatwave accelerator parameters, one finds that a CO_2 laser is perhaps the ideal laser to use for initial physics experiments. First, the growth rate and saturation amplitude of the electron plasma wave scale as $I_o\lambda^2$, where I_o is the intensity of the laser and λ its wavelength. Thus, CO_2 lasers, operating in the infrared (λ = 10 μm), need not be terribly intense to excite significantly large plasma waves. This allows interesting experiments to be done in relatively small laboratories. Another advantage is found in the required plasma parameters. Table I summarizes the parameters for the diffe-

Table 1. Parameters associated with collinear optical mixing using a CO_2 laser capable of running on various lines.

wave-lengths (μm)	density (10^{17} cm^{-3})	γ_ϕ	accel. length (mm)	energy gain (MeV)
9.6,10.6	1.1	10	3	3-9
9.6,10.3	0.6	15	15	16-72

rent CO_2 operating regimes. The main point is that plasmas with densities in the neighborhood of 10^{17} cm^{-3} and 1 cm in length are readily achieved in small laboratory devices (such as a "theta pinch" plasma source). We also note that the energy gains of injected electrons are so large (a factor of 10 or more) as to make an unambiguous observation of acceleration.

REVIEW OF RESULTS

In our previous experiment[2] a 16 J CO_2 laser (12 J at 10.6 μm and 4 J at 9.6 μm) was used to drive an electron plasma wave in a resonant plasma of about 10^{17} electrons/cm^3. The laser pulse duration was about 2 nsec and the intensity was 10^{13} W/cm^2. The plasma wave, driven in an arc-preionized hydrogen/nitrogen plasma, was found to grow to an amplitude of 9±6% of the cold wavebreaking amplitude in a time of about 500 psec, both measurements being in reasonable agreement with theory. The length of the electron plasma wave was found to be about 1.8 mm, limited by the inhomogeneity of the arc plasma. The above measurements were obtained by observing the Stokes and anti-Stokes radiation sidebands in the forward (CO_2) direction and from small-angle Thomson scattering performed at 90° to the CO_2 direction. The longitudinal electric field implied by these measurements is 1-5 GeV/m, more than an order of magnitude larger than the fields in state-of-the-art electron linacs.

FUTURE PLANS

Our planned upgrade of this experiment will differ from the original experiment in several ways. First, we will use a short laser pulse (50-200 psec) in order to suppress stimulated Brillouin scattering to a tolerable level and to improve the matching of the laser duration to the saturation time of the relativistic plasma wave. Table II shows a possible set of experimental parameters (alt-

Table II. A possible set of design parameters for future experiments. Parameters in first column are, in order: laser wavelengths, plasma density, plasma length, laser energy, pulse width, normalized electron quiver velocities, laser intensity, focal length, focal spot size. Second column: relativistic γ of the phase velocity, normalized wave amplitude, wave electric field, growth time to relativistic saturation, acceleration length, minimum injection energy, maximum energy gain, and pump depletion length. Numbers in parentheses are values at the relativistic saturation level.

Laser/Plasma		Wave/Particle	
λ_0, λ_1	9.56,10.27 μm	γ_ϕ	14.5
n_o	5.8 X 10^{16} cm^{-3}	ε	0.04(0.30)
L_p	20 cm	eE	1.0(6.9) GeV/m
E_ℓ	10 J	τ_{sat}	40 psec
τ_ℓ	50 psec	L_{acc}	1.7(1.0) cm
α_0,α_1	0.07	W_{inj}	2.4(0.3) MeV
I_o	7 X 10^{13} W/cm^2	ΔW	16(72) MeV
$2z_o$	6 cm	L_{dep}	18(0.6) cm
$2w_o$	620 μm		

hough the 50 psec pulse width is somewhat ambitious). A second major change is to use a preformed, fully ionized theta pinch plasma to reduce the plasma inhomogeneity. Finally, we will add an electron injector (linac) as a source of test particles to demonstrate, through detected acceleration and deflection, the qualities of the longitudinal and radial electric fields of the plasma wave.

REFERENCES

1. T. Tajima and J. M. Dawson, Phys. Rev. Lett. 43, 267 (1979).
2. C. E. Clayton et al., Phys. Rev. Lett. 54, 2343 (1985).
3. C. N. Danson et al., Bull. Am. Phys. Soc. 30, 1846 (1985).
4. F. Martin et al., Bull. Am. Phys. Soc. 30, 1613 (1985).

ACCELERATION OF ELECTRON BEAM BY A LARGE AMPLITUDE EM WAVE IN A UNIFORM MAGNETIC FIELD

S.P. Kuo
Polytechnic Institute of New York, Farmingdale, NY 11735
G. Schmidt
Stevens Institute of Technology, Hoboken, NJ 07030

ABSTRACT

Cyclotron resonance interaction between an electron beam and an electromagnetic wave for electron acceleration is analyzed. The Doppler shifted cyclotron resonance condition is shown able to be maintained over the entire interaction period of interest. We find that there is an upper bound, independent of the wave amplitude, on the transverse velocity of the electron. However, the result shows that the electron beam can be accelerated indefinitely in the axial direction if the wave amplitude is kept at constant.

INTRODUCTION

With the availability of high power lasers, the possibility of accelerating electrons to Gev energies in reasonable short distances (i.e. few Rayleigh lengths) has been studied extensively. A number of interesting concepts to achieve high-gradient accelerators have been reported.[1] Among them, some interesting features of a mechanism[2-3] which utilizes the cyclotron resonance interaction between an electron beam and a relativistically strong electromagnetic wave of right hand (R-H) circular polarization propagating colinearly along a dc magnetic field is discussed in the present work. We elaborate this cyclotron resonance interaction process for electron acceleration and show that the acceleration process is indeed effective and the produced energetic beam can have very small transverse energy spread.

ANALYSIS

We study the interaction between an electron beam and a large amplitude EM wave of R-H circular polarization propagating collinearly along a uniform magnetic field $B_o\hat{z}$. The equations governing the electron motion in the wave fields are given by

$$\frac{d}{dt}P^- + i\Omega P^- = \frac{e}{c}\frac{d}{dt}A_1^- \quad , \quad \frac{d}{dt}P_z = -\frac{e}{c}R_e[(\frac{\partial}{\partial z}A_1^-)v^{-*}]$$

$$mc^2\frac{d}{dt}\gamma = \frac{e}{c}R_e[(\frac{\partial}{\partial t}A_1^-)v^{-*}] \qquad (1,2,3)$$

where $P^- = P_x - iP_y, v^- = v_x - iv_y, A^- = A_x - iA_y = A_o^- + A_1^-$,

$$A_o^- = -i(B_o/2)(x-iy), A_1^- = -iA_1e^{i(kz-\omega t)}, \Omega = \Omega_o/\gamma, \Omega_o = eB_o/mc \quad ,$$

and R_e and $*$ stand for real part and complex conjugate respectively. A constant of motion is derived by combining (2) and (3) to be $\omega - \Omega_o/\gamma - kv_z \cong 0 = \omega - \Omega_o/\gamma_o - kv_{zo}$; a relation indicates that the Doppler shifted cyclotron resonance will be maintained by itself throughout the entire interation period of interest, i.e. a self-resonance effect, where $|d\ell n A_1/dz| << k$ is assumed. Using this resonance condition together with the definition of $\gamma = [1-(v_\perp^2+v_z^2)/c^2]^{-1/2}$, the relationship between v_z and $v_\perp$ are obtained as

$$v_z/c = \alpha \pm \sqrt{1-(1+\alpha)v_\perp^2/c^2}\,/(1+\alpha) \tag{4}$$

where $\alpha = \omega^2/\Omega_o^2 = 1/\gamma_o^2\,(1-v_{zo}/c)^2$. Since v_z has to be real, one constraint is deduced from (4), namely, $v_\perp^2/c^2 \leq 1/(1+\alpha)$, which sets an upper bound on the transverse velocity of the electron beam obtainable through the resonance interaction. It is noted that the upper bound is independent of the wave field amplitude. We now integrate (1) formally to $P^- = -\Omega_o(e/c)e^{i(kz-\omega t)}\int_o^t dt'[A_1/\gamma(t')]$, which is then substituted into (3) to yield

$$\gamma=\gamma_o+\frac{1}{2}\Omega_o\omega(e/mc^2)^2\,\{\int_o^t dt'[A_1/\gamma(t')]\}^2 \sim [\frac{3}{2}\sqrt{2\bar{v}_q^2\Omega_o\,\omega}\;\;t]^{2/3} \tag{5}$$

where $\bar{v}_q = eA_1(o)/mc^2$. Equation (5) shows that γ increases with time if the wave field amplitude A_1 is a constant. Since $v_\perp$ is bounded, it simply means that v_z can be increased indefinitely to approach c. Therefore, only the negative sign in (4) holds initially. However, it should be changed to the positive sign after $(v_\perp/c)_{max} = 1/(1+\alpha)^{1/2}$ is reached. Subsequently, $v_\perp$ decreases as v_z increases. This is because the increase of γ makes the electron become heavier.

To demonstrate the effectiveness of the acceleration mechanism, a numerical example is illustrated in the following. We use a CO_2 laser pulse having power flux $3x10^{13}$ w/cm^2 to accelerate the electron beam, it gives $\bar{v}_q = 0.032$. In order to match the resonance condition, a background magnetic field is needed. We choose $B_o = 50KG$. Since $\Omega_o = \gamma_o\omega(1-v_{zo}/c) = \omega/2\gamma_o$, it then requires $\gamma_o = 100$. This means that a 50 MeV beam is needed initially for satisfying the resonance condition. It is fortunately to be technologically feasible. From (5), we then find that for a 30 m interaction length, the electron beam is accelerated to $\gamma_f \sim 2x10^3$, i.e. 1GeV beam. We, therefore, have shown that the cyclotron resonance laser acceleration mechanism is very effective. This is mainly because the resonance condition can self-maintain throughout the entire interaction period. In addition, we have shown that the transverse velocity of the electron beam has an upper bound independent of the wave field strength. Therefore, the produced energetic beam will have very small transverse energy spread.

This work is supported by the Air Force Office of Scientific Research, under Grant No. AFOSR-85-0316.

REFERENCES

1. Laser Acceleration of Particles, AIP Conf. Proceedings No. 91, Ed. Paul J. Channell, AIP, NY (1982)

2. P. Sprangle and C.M. Tang, IEEE Trans. Nucl. Sci. NS-28, 3346 (1981); P. Sprangle, L. Vlahos, and C.M. Tang, IEEE Trans. Nucl. Sci., NS-30, 3177 (1983).

3. S.P. Kuo and G. Schmidt, J. Appl. Phys., 58, 3646 (1985).

LASER BEAM CONFIGURATIONS FOR CUMULATIVE INTERACTION WITH ELECTRONS IN A GAS*

J. R. Fontana
University of California, Santa Barbara, CA 93106

ABSTRACT

The index of refraction of a gas can be used to match the electron and wave phase velocities. Short laser pulses are required to avoid gas breakdown, and only high-energy electrons can overcome the molecular collision and matching condition problems. Appropriate field configurations are discussed as well as means of producing them. Optical waveguides may be used to enhance the net energy exchange or focusing effect for a given lase pulse energy and within the power density limitations at the guide boundary.

A laser field configuration already proposed[1] consists of a set of plane waves incident from all directions making the same angle θ with the electron beam axis. When θ, the gas index of refraction n, and the electron velocity β c fulfill the Cherenkov condition: $n\ \beta \cos\theta = 1$, each electron sees a constant field. To maintain the condition during acceleration, β must be close enough to unity; this high-energy requirement is also necessary to minimize the effect of collisions with gas molecules. This single angle, constant fields (SACF) pattern can produce energy gains of the order of 1 Gev/meter when short (picosecond) pulses are used in atmospheric pressure hydrogen[1,2].

Another configuration uses appropriate Gaussian modes presenting an axial electric field. One such mode is the $TEM^{c}_{1,0}$ which consists of an ordinary (1,0) mode configuration rotated around its axis, so the beam has an annular cross-section with an axial field at the center. Now the electrons see plane wave components distributed in angle and fields which decay away from the Gaussian beam waist. (Distributed angle, tapered fields: DATF). It is a property of these modes that the electron energy gain Δ W is proportional to $P^{\frac{1}{2}}$, where P is the total laser power, when the interaction occurs over a fixed fraction of the Rayleigh range, regardless of the beam convergence angle. However, the maximum field in the gas, E_m, does vary with this angle, so if E_m is limited by gas breakdown considerations, the angle θ of the most effective plane wave components is small.

A numerical example shows that with a laser of wavelength λ = 10.6 micrometers and pulse power P = 10^{14} watts, using $E_m = 1.5 \times 10^{10}$ volts/meter the SACF configurations yields Δ W = 26 Gev over a distance of 50 meters, while the DATF configuration with the $TEM^{c}_{1,0}$ mode only gives Δ W = 0.7 Gev over the same distance.

For a given P, the interaction length, and hence Δ W, in a SACF configuration can be increased by using a cylindrical optical wave-

* Work supported by DOE Contract DE-AS03-84ER40136.

guide to reflect the fields back towards the axis[2]. The new interaction length is larger for smaller guide radius a, but so are also the fields at the guide wall surface, and the wall composition determines a maximum acceptable value. The acceptable incident field E_1 depends on both the material and the angle θ.

The reflection loss on polished metal guides is large[3], and so lossless dielectric walls appear more promising. Each reflection allows a fraction of the incident power to be transmitted and hence lost, but a grazing angles this loss is small.

The following scaling laws apply. E_0 is the longitudinal field on axis and all other symbols are as defined above.

For given E_i, E_0 and λ: $a \propto \theta^{-3}$ and $P \propto \theta^{-6}$
For given E_i, λ and θ: $a \propto E_0^2$ and $P \propto E_0^4$
For given E_0, λ and θ: P and $a \propto E_i^{-2}$

A problem with waveguides is the velocity differential between the electrons and the laser pulse in the guide. The minimum pulse duration for an electron to remain within the pulse after N wall reflections is:

$$\tau = 2\,N\,a\,\sin\theta/c$$

Table I shows results for λ = 10.6 micrometers, θ = 20 milliradians, N = 5 and $E_i = 10^9$ volts/meter, probably safe for typical walls[4].

Table I Optical waveguide examples

E_0 (volts/meter)	10^9	0.5×10^9
a (cm)	3.2	0.8
Interaction length (m)	19.2	4.8
Length of waveguide (m)	16	4
Δ W (Gev)	16	2
τ (picoseconds)	21.3	5.3
P (watts)	1.65×10^{13}	10^{12}

REFERENCES

1. J. R. Fontana and R. H. Pantell, Jour. Appl. Phys. 54 (8), 4285-4288, August 1983.
2. J. R. Fontana, AIP Conf. Proc. No. 130, Laser Acceleration of Particles, 1985, (C. Joshi Ed.).
3. E. Garmire, T. McMahn and M. Bass, IEEE J. Quantum Electon. QE-15, 491 (1971).
4. S. Solimeno, AIP Conf. Proc. No. 91, 1982 (P. J. Channell Ed.), p. 176.

CONCEPT OF INFRARED LASER PARTICLE ACCELERATORS WITH OVERSIZED DBR AND HFB WAVEGUIDES

J. Arnesson, S. Gnepf, M. Nessi, W. Wölfli and F.K. Kneubühl
Physics Department, ETH, CH-8093 Zurich, Switzerland

ABSTRACT

We present an infrared-laser accelerator scheme which makes use of hollow oversized linear periodic and helical waveguide structures originally designed for distributed feedback (DFB) and helical feedback (HFB) lasers.

INTRODUCTION

The present high-power IR, VIS and UV lasers suggest that in particle accelerators [1] the microwave sources could be replaced [2-4] by powerful lasers. Their short wavelengths would increase field gradients and, therefore, either reduce the size or increase the particle energy of the accelerators. Here we propose a new scheme for particle acceleration with oversized periodic [5] or helical [6] waveguides driven by 10 μm CO_2 or submillimeter-wave free-electron lasers.

ACCELERATION OF ELECTRONS IN AN OVERSIZED PERIODIC WAVEGUIDE

In Fig. 1 we show an oversized cylindrical linear periodic hollow waveguide of mean radius a and a harmonic corrugation of amplitude $a_1 \ll a$ and period L, which is suited for particle acceleration as well as for DFB lasers [5]. This corrugation causes a coupling between forward and backward running modes [7]. For particle acceleration, the coupling between a TM_{01} forward and a TM_{01} backward mode has been found to be the most advantageous combination. This is due to the fact that the longitudinal electric field displays a single maximum on the waveguide axis where both, the transverse electric and the magnetic fields vanish. For helical corrugations of the waveguide as used in the HFB laser [6], all possible mode couplings exhibit off-axis field maxima. They display a corrugation dependent rotation of the field pattern along the axis which might imply synchrotron radiation by the accelerated particles.

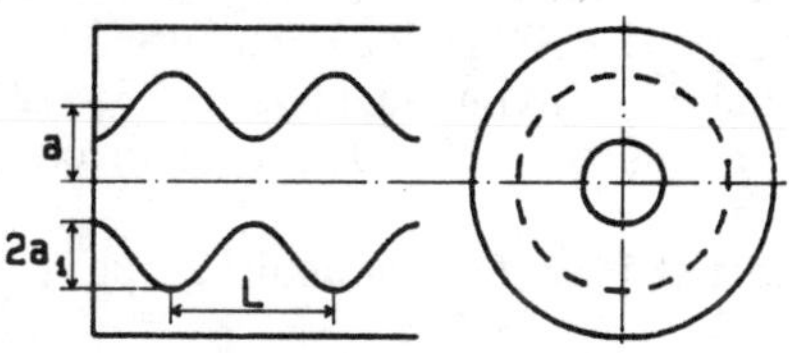

Fig. 1 Waveguide with linear corrugation

In order to investigate energy gain and transverse stability of a beam of accelerated electrons we have derived the electromagnetic field for an eigenmode of the TM_{01}/TM_{01} coupling of the linear periodic structure shown in Fig. 1 and calculated numerically the effect of this field on a beam of 6 MeV electrons entering the waveguide with velocities parallel to the waveguide axis. The laser power was assumed to be 5 MW at a wavelength of 10.6 μm. The resulting electron trajectories and energy gain as a function of distance from the waveguide axis are shown in Fig. 2. All trajectories remain stable with a slight focussing towards the waveguide axis and with a maximum energy gain ΔE of approx. 1.6 MeV for electrons on the axis. The present calculation does not include waveguide losses, parasitic modes, and material breakdown which we intend to cover in future investigations.

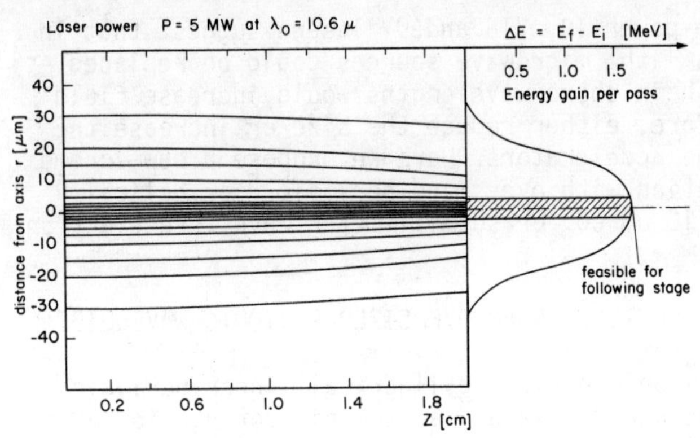

Fig. 2 Electron trajectories and energy gain

CONCLUSION

Oversized hollow metallic waveguide structures with linear periodic corrugation exhibit modes which allow a stable acceleration of charged particles with CO_2 or FIR free-electron laser radiation.

REFERENCES

1. P.M. Lapostolle, A.L. Septier, eds.: Linear Accelerators; North Holland Publ. Comp.; Amsterdam (1970)
2. K. Shimoda, Appl. Opt. 1, 33 (1962)
3. P.J. Channell, ed.: Laser Acceleration of Particles; Am. Inst. of Physics Conf. Proc. No91, NY (1982)
4. J. Arnesson, F.K. Kneubühl, Infrared Phys., 25, 121 (1985)
5. E. Affolter and F.K. Kneubühl, IEEE J. of Quantum Electron., QE-17, 1115 (1982)
6. H.P. Preiswerk, M. Lubanski, F.K. Kneubühl, Appl. Phys., 33, 115 (1984)
7. V.L. Bratman, G.G. Denisov, N.S. Ginzburg, M.I. Petelin, IEEE J. of Quantum Electron., QE-19, 282 (1983)

STRONG INTERNAL ELECTRIC FIELDS IN NONLINEAR FORCE PRODUCED CAVITONS: LASER ACCELERATION OF PARTICLES

F. Green, G. W. Kentwell, H. Hora and S. Tapalaga

Department of Theoretical Physics
University of New South Wales
Kensington, 2033
Australia

The presence of strong electric fields inside plasmas can lead to basically new and previously unexpected plasma phenomenon. These strong electric fields were predicted by hydrodynamic theory [1] [2] and confirmed experimentally in laser produced plasmas [3]. Though there are numerous observations of electric double layers in experimental laboratory plasmas and satellite observations of extraterrestrial space plasmas [4], the theoretical understanding was limited only to kinetic theory generally assuming collisionless plasmas and using the Vlasov equation. In macroscopic hydrodynamic plasmas, these phenomena could not be explained with the established theory of the space-charge quasi-neutral plasmas. While Alfven had reasons to assume these electric fields in plasmas [5] or to base the description of the plasma on electrical current densities as a dualistic substitute of the magnetic field, this description was widely criticized. Kulsrud's remark [6] that "Alfven's electric fields are intuitively not clear" is symptomatic.

The study of laser produced plasmas changed this situation. It was motivated by the knowledge of the nonlinear force of the electrodynamic interaction of lasar radiation with an irradiated target and with the generated plasma. The quasineutrality theory was able to derive the force density in a plasma as

$$\underline{f} = -\mathrm{grad}P + \underline{j}\mathrm{x}\underline{H}/c + \underline{E}\nabla.\underline{E}/4\pi + \nabla.\underline{E}\underline{E}(n^2-1)/4\pi \qquad (1)$$

with the thermokinetic pressure P in the plasma and with the laser produced electric and magnetic fields $\underline{E}$ and $\underline{H}$ and current density $\underline{j}$. n is the optical refractive index. This relation was shown to be perfect and complete if the laser irradiation is monochromatic (no temporal change) [7] [8] and was proved from momentum conservation and the absence of shear forces in collisionless plasmas at oblique incidence of the radiation on the plasma. For the more difficult transient non-monochromatic case, the following form has been suggested of the nonlinear force (after subtracting the thermokinetic term -gradP) [9]

$$\underline{f}_{NL} = \underline{j}\mathrm{x}\underline{H}/c + \underline{E}\nabla.\underline{E}/4\pi + [1 + (1/\omega)\partial/\partial t]\nabla.\underline{E}\underline{E}(n^2-1)/4\pi \qquad (2)$$

using the central radian frequency ω of the laser radiation. This result is somewhat controversial [9] and does not follow as rigorously as the time independent case.

An alternative derivation of the nonlinear force indicated that the laser radiation is acting essentially at the high density electron cloud in the plasma which is then pulled or pushed by the nonlinear, force while the ions follow only after some high electric field had been generated to attract them to the electrons. This was the motivation that, contrary to the established hydrodynamic quasi neutral plasma theory, a genuine two fluid model was required with a complete hydrodynamic description of the electron fluid and the ion fluid coupled by the generated internal electric field in one spatial dimension (in three dimensions the generated magentic fields will also result [10]).

The model is most realistic including collisions (viscosity, thermal transport, equipartition) and their nonlinear dependence on the laser intensity using the correct formula. The difficulty was that very small time steps were necessary resulting in long computations; numerical instabilities had to be suppressed and a new technique for a stable treatment of the boundary conditions for an Eulerian code was necessary [11]. It turned out that oscillating and damped dynamic electric fields in the plasma are created which in the most simplified static case are the ambipolar fields. The driving energy densities, however, are then of the high values of the laser and very high electric fields have been generated especially in the density minima of the plasma (cavitons) which are characteristic for the nonlinear force action [8].

This analysis led to the explanation of the measured inverted double layers in laser produced plasmas and to the discovery of a new second harmonic resonance [12]. The unexpected generation of homogeneous second harmonic emission from highly inhomogeneous plasmas with very strongly varying density during laser irradiation [13] had thus been explained [2]. Other conlusions are that the rotation of tokamak plasmas is a basic process, [1] and that the formerly unexplained inhibition of themal conduction in pellet fusion (which suppresses neutron generation) could now be predicted quantitatively from the result of the double layers [2].

Furthermore this theory predicts that the electric fields in the laser produced cavitons are very high. For the irradiation with a laser intensity of 10^{16} W/cm^2, dynamic electric fields above 10^8 V/cm are produced. These fields are nonconservative fields and cannot be described by potentials. Electrons moving through these fields can gain or lose energy depending on the phase of the motions.

Extensive numerical computations have beenperformed and it was confirmed that this mechanism for increasing the energy of relativistic electrons agrees with the simply expected values. The extension of this mechanism can be applied to accelerate electrons to energies of 10 GeV or much more using the nonlinear force produced caviton fields generated by available carbon dioxide lasers such as ANTARES (where 80 TW power can be produced with 150 psec steeply rising pulses). Combinations of these steps of acceleration can provide the desired TeV electrons by a technique which is less expensive than classical techniques by a factor of 10 or more [14].

REFERENCES

[1] P. Lalousis and H. Hora, Laser and Particle Beams 1, 283 (1983); H. Hora, P. Lalousis and D. A. Jones, Phys. Lett. 99A, 89 (1983).

[2] H. Hora, Laser and Particle Beams 3, 58 (1985).

[3] S. Eliezer and A. Ludmirsky, Laser and Particle Beams 1, 251 (1983); A. Ludmirsky et al. IEEE Transact. Plasma Sci. 13, 132 (1985).

[4] R. Schrittwieser and G. Eder, eds., 2nd International Symposium on Double Layers in Plasmas (Institute of Theoretical Physics, Innsbruck, 1984); N. Hershkowitz, Space Science Rev. 41, 351 (1985).

[5] H. Alfven, Cosmic Plasmas (Reidel, Dordrecht, 1981).

[6] R. Kulsrud, Physics Today 36, 56, (1983).

[7] H. Hora, Phys. Fluids 12, 181 (1969).

[8] H. Hora, Physics of Laser Driven Plasmas (Wiley, New York, 1981).

[9] G. W. Kentwell and D. A. Jones, Physics Reports (in print); H. Hora, Phys. Fluids 28, 3473 (1985).

[10] J. A. Stamper, K. Papadopoulos, R. N. Sudan, S. O. Dean, E. M. McLean and J. M. Sawson, Phys. Rev. Lett. 26, 1012 (1971).

[11] P. Lalousis, Ph.D. Thesis, University of New South Wales, 1983; P. Lalousis and H. Hora, Computer Techniques and Applications, J. Noye et al., eds. (Elsevier, North Holland, 1984).

[12] H. Hora, P. Lalousis and S. Eliezer, Phys. Rev. Lett. 53, 1650 (1984); H. Hora, A. K. Ghatak, Phys. Rev. 31A, 3473 (1985).

[13] I. V. Aleksandrova, W. Brunner, S. I. Fedotov, R. Guther, M. P. Kalashnikov, Yu. A. Mikhailov, S. Ploze, R. Riekher, and G. V. Sklizkov, Laser and Particle Beams 3, 197 (1985).

[14] P. J. Clark, S. Elieser, F. J. M. Farley, M. P. Goldsworthy, F. Green, H. Hora, J. C. Kelly, P. Lalousis, B. Luther-Davis, R. J. Stening and J.-C. Wang, Laser Acceleration of Particles, C. Joshi et al. eds., AIP Proceed. No. 130 (American Institute of Physics, New York, 1985), p. 380.

PLANAR METAL-GRATING FREE-ELECTRON LASERS

J. Walsh, T. Buller, B. Johnson, E. Garate,
P. Muhkopadhyay
Dartmouth College, Hanover, N.H. 03755

ABSTRACT

The predicted operating characteristics of a planar metal-grating free-electron laser (GFEL) are examined. It is shown that the kinematic parameters of the electron beam and the geometry of the grating determine the wavelength of the radiation and the relative coupling strength. The effect of additional constraints due to beam quality, such as emittance of energy spread, are also mentioned briefly.

INTRODUCTION

A free-electron laser which uses either a grating[1], or a thin dielectric film[2], to couple an electron beam to radiation fields can, in principle, operate at short (far-infrared - visible) wavelengths with a beam energy which is comparatively much lower than a device which uses a magnetic undulator for this purpose. Smith and Purcell[3] obtained spontaneous emission at visible wavelengths from a 300-350 KV, 5 μA beam, while the visible wavelength free-electron laser (FEL) operating on the ACO[4] storage ring used a 166 MeV, 16-100 mA beam. The comparison is striking and hence the practical difficulties associated with guided-wave FEL's are of interest.

In this note, the way in which grating and electron-beam-related constraints together determine the possible operating range of a grating-based free-electron laser (GFEL) will be reviewed. First, the general characteristics of the grating's dispersion relation will be examined and the way in which the operating wavelength is determined will be discussed. Following this, constraints due to coupling strengths will be considered and finally, a brief summary of some further constraints relating to beam quality will be presented.

DETERMINATION OF THE OPERATING WAVELENGTH

A typical grating geometry and its dispersion curve in the n = 0 Brillouin zone, for the lowest order TM mode, are shown in Fig. 1. It is assumed for simplicity that the tooth-and-slot widths are equal. In the region above the grating the axial electric field associated with this bound guided mode hs the form:

$$E_z = \sum_{n=-\infty}^{\infty} E_{no} e^{-q_n x} e^{i(k_n z - \omega t)} \qquad (1)$$

where the coefficients E_{no} are determined by the boundary conditions along the surface x = 0. Wavenumbers which appear in eq. (1) are defined by

$$k_n = k_o + \frac{2n\pi}{\ell} \tag{2a}$$

and

$$q_n = \sqrt{k_n^2 - \frac{\omega^2}{c^2}} \tag{2b}$$

The wavenumber, k_o, is that of the zeroth space harmonic. Other field components are determined by Maxwell's equations. The dispersion curves shown in Fig. 1 then follow from the application of the boundary conditions at the slot.

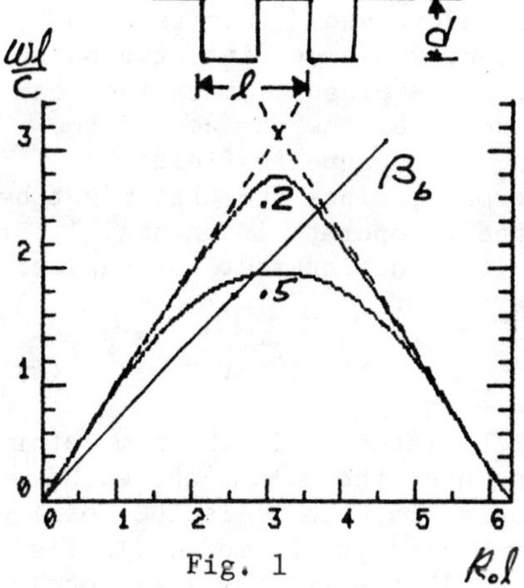

Fig. 1

Normalized dispersion relations for d/ℓ = 0.2 and 0.5

The dispersion curves displayed in Fig. 1 are plots of angular frequency ω measured in units of grating period, ℓ, divided by c, the speed of light vs. the wavenumber in the n = 0 zone. The latter is also displayed in dimensionless units. Variations in the shape of the curves are determined by the ratio of slot depth/period (d/ℓ). As this parameter increases, the relative value of the phase velocity ω/ck_o at the peak of the curve ($k_o\ell = \pi$) decreases. This parameter also determines the extent of the region over which the phase velocity of the wave is close to the speed of light.

In a GFEL energy is exchanged between an electron beam moving over the surface of the grating and a bound guided mode. The beam bunches in the retarding phase of the axial component of the field and the coupling strength peaks when the beam velocity is near, but slightly greater than, the mode phase velocity. Hence the operating frequency is determined by the condition

$$\beta_b = \frac{\omega}{ck} \tag{3}$$

where β_b is the relative beam velocity (V_b/c). Condition (3) relates the electron beam and grating parameters. This is illustrated by the line with slope β intercepting the dispersion curve at its peak, on Fig. 1. The dispersion curve and the condition of synchronism are self-similar functions of the grating geometry. Variation of d and ℓ by the same amount does not change the shape of the curve. Hence in principle it is always possible to choose d and ℓ in order to obtain operation at any wavelength with a given relative beam energy ($\gamma_b = 1/\sqrt{1-\beta_b^2}$). As γ_b^{-1} is varied. "tuning curves", $\omega\ell/c$ vs. γ_b-1, are obtained. Two examples are shown in Fig. 2. Note that as the relative slot depth decreases, the curve peaks at a higher value of the beam's relative kinetic energy.

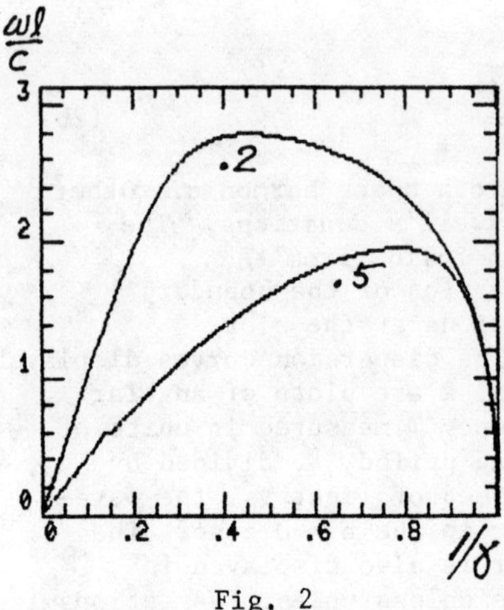

Fig. 2

Normalized tuning curve for the parameters of Fig. 1.

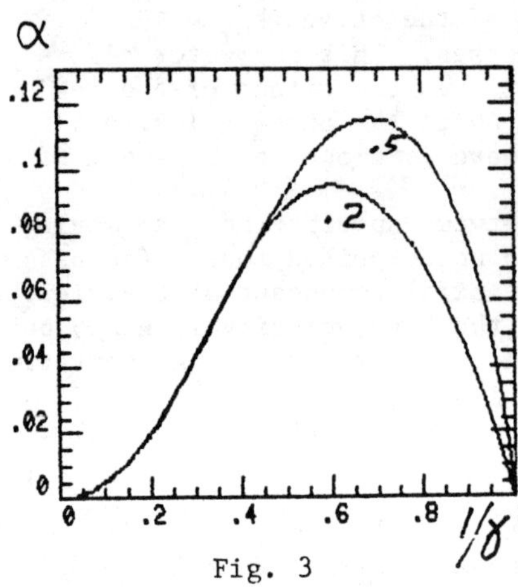

Fig. 3

The coupling factor α for the parameters shown in Fig. 1.

BEAM COUPLING IN A GFEL

The operating wavelength of a GFEL is determined by the beam velocity and the grating parameters. In order to proceed further with an assessment of practicality, the question of beam - mode coupling must be addressed. This is conveniently divided into two parts. First, it is presumed that the beam is guided over the surface of the grating by a magnetic field. Furthermore, since the GFEL might be expected to operate in an oscillator mode, it is of interest to consider the parameter

$$\alpha = \int |E_z|^2 \, dV/\mathcal{E} \qquad (4)$$

where the integral in the numerator is done over the volume above the grating surface and $\mathcal{E}$ is the total energy stored in the mode. In Fig. 3, α is plotted vs. the reciprocal of beam energy. Two values of the parameter d/ℓ are shown on the figure and it can be seen that as d/ℓ decreases, the peak is lowered and is shifted toward a higher beam energy.

As the operating wavelength of a GFEL is decreased, a second constraint also becomes important. It can be seen from eq. (1) that the space harmonics evanesce exponentially above the surface of the grating. At velocity synchronism

$$q_o = \sqrt{k_o^2 - \omega^2/c^2} \qquad (5a)$$

$$= k_o/\gamma_b \qquad (5b)$$

or

$$q_o = 2\pi/\lambda\beta_b\gamma_b \qquad (5c)$$

As the period ℓ decreases, $q_n \gg q_o$

for all n ≥ 1. In this limit the energy to a large extent is carried in the n = 0 space harmonic. The coupling is controlled by equ. (5) and the achievable beam focussing. A detailed development of a gain expression would include this constraint together with the energy dependence of the bunching process. In the limit of axially bunching this contributes an additional factor of γ^{-3} to the gain expression. This shifts the gain peak toward smaller values of $k_o\ell$.

CONSTRAINTS RELATED TO BEAM QUALITY

The complete gain expression for a GFEL is determined by Poynting's theorem. In the anticipated operating regime it is assumed that the field distributions are to first order unaffected by the presence of an electron beam. An initially unmodulated beam enters a resonator formed by a grating and suitable mirrors. The beam is then modulated by, and transfers energy to, the fields stored in the resonator. This process has a characteristic line shape given by

$$F'(\theta) = -\frac{d}{d\theta}\left(\frac{1-\cos\theta}{\theta^2}\right) \tag{6a}$$

where

$$\theta = (k_o v_b - \omega)L/V_b \tag{6b}$$

F' > 0 is the region of positive gain and L is the resonator length.

Gain is maximized when all the beam electrons act in concert and hence any beam inhomogeneities measured in units of θ must be less than π. In the case of beam energy inhomogeneity the relation

$$\frac{L}{\lambda}\,\frac{1}{\gamma_b^2}\,\frac{\delta\gamma_b}{\gamma_b} \lesssim 1 \tag{7}$$

where $\delta\gamma_b/\gamma_b$ is the relative energy spread is a quantitative measure of the constraint.

SUMMARY

In conclusion: application of kinematic constraints determine the operating wavelength of a GFEL. However, since the grating period may be small there is no need to use beams of very high energy to reach short wavelengths. Coupling (eqs. (3) and (5)) and beam quality constraints (eq. (7)) both require an increase in beam energy as the operating wavelength is decreased. Balancing these concerns implies that a beam in the 1-5 MeV range would be required for FIR-IR operation.

REFERENCES

1. J. Walsh *et al.*, Phys. Rev. Lett. 53(8), 779 (1984).
2. J. Walsh *et al.*, IEEE Journ. Quant. Elect. QE-21, 920 (1985).
3. S.J. Smith and E.M. Purcell, Phys. Rev. 92, 1069 (1953).
4. C. Bazin *et. al.*, Le Journ. de Phys. Lett. 41, 547 (1980).

Support of ARO Contract DAAG29-85-K-0176 and HDL Contract DAAL02-85-C-0110 are acknowledged.

THE MICROTRON FEL ČERENKOV:
theoretical and experimental aspects

F.Ciocci, G.Dattoli, A.De Angelis, A.Dipace, E.Fiorentino, G.P.Gallerano, I.Giabbai, G.Giordano, T.Letardi, A.Renieri, E.Sabia, and J.E.Walsh (*), G.Schettini (**)
ENEA, TIB, Divisione Fisica Applicata
P.O.Box 65 - 00044 FRASCATI (Rome), Italy

ABSTRACT

In this note we review the essential features of the Čerenkov Free Electron Laser (ČFEL) from both the theoretical and experimental point of view. In particular we discuss the ČFEL microtron joint ENEA Frascati-Dartmouth experiment.

So far a number of FEL experiment have been successfully operated[1]. They work with different e-beam sources ranging from Storage Ring, to Linacs and microtrons and to Van de Graaf. Their common feature is that the basic emission mechanism is provided by the interaction of a relativistic e-beam with an undulator. Coherent emission by free electrons can be however achieved using different processes as the Čerenkov and Smith-Purcell effects or the metal grating interaction[2].

Among this proposed FEL mechanisms it has been pointed out that a Čerenkov based FEL (ČFEL) may provide, in the FIR low gain region, a compact laser device with performances better than an Undulator based FEL (UFEL)[3].

An highly-simplified version of the ČFEL is the following. An e-beam moves near and parallel to the surface of a thin dielectric waveguide of length L. The beam couples to the axial component of a TM mode of the guide and thereby emits spontaneous Čerenkov radiation. This radiation is feedbacked in an optical cavity, reinteracts with the beam, is amplified and it is possible to realize an oscillator. Albeit the CFEL and UFEL appears rather different devices, it has been shown that the basic emission features can be reduced to the same mechanism, so that ČFEL gain formula reads

$$g = g_o \frac{\partial}{\partial\Theta} \left(\frac{1 - \cos\Theta}{2} \right) \qquad (1)$$

where $\Theta = (kv_o - \omega) \frac{L}{V_o}$ is the relative transit angle and g_o is

(*) Permanent address: Dept. Physics and Astronomy, Dartmouth College, Hanover (U.S.A.).
(**) ENEA Student.

the gain coefficient

$$g_o = \frac{2\pi}{\lambda} \cdot \frac{L^3}{\sigma_x \sigma_y} \cdot \frac{\hat{I}}{I_o} \cdot \frac{e^{-\alpha_o}}{\gamma^5} \qquad , \qquad (\gamma = E/_{mc^2})$$

$\sigma_{x,y}$ are the e-beam transverse dimensions, λ is the operating wavelength, $\hat{I}$ and I_o the peak and Alfvén current respectively and

$$\alpha_o = \frac{4\pi\delta}{\lambda\beta\gamma} \qquad , \qquad \beta = {}^V/_C$$

is fixed by the gap δ between the e-beam and the dielectric.

A simple comparison between the ČFEL and UFEL gain formulas has shown that the former offers larger gain in the FIR region.The gain formula [1] is relevant to the homogeneously broadened regime; the inhomogeneous broadening, due to the energy spread and emittances, have been evaluated too and turns out to be identical to those playing the same role in the UFEL. Finally the short pulse effects, due to the bunched structure of the e-beam, have been included and it has been proved that something similar to the well known UFEL lethargic behaviour arises [4].

A joint collaboration between the ENEA Frascati Center and the Darthmout College is in progress, and is aimed to realize a ČFEL with a microtron whose beam characteristics are summarized in tab.1 [5]. The emission wavelength is expected at $\lambda \sim 100$ μm with an average output laser power around 10 W at 50 Hz of repetition frequency.

Table I

Energy	5 MeV
Average current	0.2 A
Peak current	4 A
Macropulse duration	4 μs
Vertical emittance	8π mm-mrad
Horizontal emittance	24π mm-mrad
Energy spread	0.5 %

REFERENCES

1. See e.g. "Proceedings of the Lake Tahoe FEL Conference" (1985).
2. J.E.Walsh, F.T.Buller, B.Johnson, G.Dattoli and F.Ciocci, IEEE J.Q.E. Special issue on FEL, Ed.by C.W.Roberson and W.Granadstein (July 1985).
3. J.E.Walsh, B.Johnson, G.Dattoli and A.Renieri, Phys.Rev.Lett. 53, 779 (1984).
4. G.Dattoli and A.Renieri, FEL Handbook vol.4, Ed. by M.L.Stitch and M.Bass, North Holland Amsterdam (1985) pag.1.
5. G.Messina, L.Picardi and A.Vignati, Private communication.

PARTICLE SIMULATIONS VS SMALL SIGNAL FEL THEORY

C. J. Elliott and M. J. Schmitt
Los Alamos National Laboratory, Los Alamos, N. M. 87545

ABSTRACT

We compare particle simulations and small signal theory in the low gamma (< 4) regime with large magnetic fields. We extend theory for this regime and compare growth rates. These agree within 10%, and the differences are attributed to non-linearities and harmonics.

INTRODUCTION

We compare low γ high magnetic field particle simulations with analytical theory. Although the region we explore does not have large space charge effects, it is substantially more complicated than the high γ limit[1] often described. In particular, a new excitation mechanism leads to enhancement of harmonics. The transverse motion we consider has non-sinusoidal components and is given by

$$\sinh(k_w y) = \sqrt{\frac{m}{1-m}}\sin(k_w z)$$

where k_w is the static wiggler magnetic field wave number and m is an elliptical integral parameter for the time dependent parameterization of the electron motion that depends on the total energy of the electron $m_e c^2\gamma$, and the static wiggler field induction amplitude, B_w. It is the higher odd harmonics $(2n+1)k_w$ of the fundamental wavenumber in the transverse motion that mimic a superposition of static magnetic fields having those wavenumbers and generate harmonics we see in simulations. In addition, the longitudinal z motion is not steady and is a second source of harmonics. These effects are of special interest to the ELF experiment[2] and to proposed low-γ particle accelerators[3].

LOW DENSITY PARTICLE SIMULATIONS

We use WAVE[4] to compute the 1-D particle simulations having densities that are small, $\approx 10^{-5}$, fractions of the critical density. WAVE was modified by removing the $\partial\vec{A}/\partial t$ term that introduced a spurious slow drift of the saturated field. Small diamagnetic oscillations were removed by using a magnetic permeability slightly less than unity that properly balanced the initial fields and initial currents. First, third, and fifth harmonics were observed as strong components, and the total energy was fit by a straight line on an expotential plot of energy versus time. The particular run reported had a growth rate of 5.5 $10^{-3}\omega$ where ω is the angular frequency of the propagating electromagnetic component.

ANALYTICAL THEORY

The analytical theory begins with the Lorentz force equation for the electrons and the wave equation for the vector potential. The perturbations can be displayed on a triangular diagram that represents the dispersion relation in terms of the variations in $\vec{A}, \gamma$, and T, where $\vec{A}$ is the vector potential and T is the time required for an electron to reach a location. The resonant condition depends on the kinematics through the average velocity in the longitudinal direction, $c\bar{\beta}$. A substantial amount of algebra results in the cold-beam self-consistent dispersion equation

$$\omega^2 - c^2k^2 - \frac{\omega_p^2}{\gamma}\phi_a - \omega_p^2\beta\frac{\omega}{c}\sum_l \frac{d}{d\gamma}\left[\frac{e_l^2}{k + (2l+1)k_w - \omega/(c\bar{\beta})}\right] = 0,$$

$$e_l = \frac{1}{\lambda_w}\int_0^{\lambda_w} dz \frac{\beta_{x0}}{\beta_{z0}} \exp\left[-i\omega[T - z/(c\bar{\beta})] - i(2l+1)k_w z\right],$$

$T(z)$ is given by an elliptical integral with parameter m, $\omega_p^2\beta$ measures the current external to the wiggler, and the larger than unity quanity, ϕ_a, is given by

$$\phi_a = \int_0^{\lambda_w} \frac{\beta^3\, dz}{\beta_{z0}^3\, \lambda_w}.$$

This result reduces at high γ with a single frequency to the well known Bessel function form[1].

THE COMPARISON

The dispersion relation gain agreed with the linear fit to the total log field energy within 10%. There are a number of contributing factors that may explain this small discrepancy. First, the total field energy contains not only the fundamental energy but also the harmonic and electrostatic energies. These corrections were originally believed to be small, but estimates show that 10% of the energy is in the third harmonic. Thus, the agreement achieved is reasonable.

REFERENCES

1. R. C. Davidson and J. S. Wurtele, IEEE Trans. Plasma Sci. PS-13, 464 (1985).
2. T. J. Orzechowski, et al, Phys. Rev. Lett. 54, 889 (1985).
3. A. M. Sessler, AIP Conf. Proc. 91, 163 (1982).
4. D. Forslund, Space Science Reviews 42, 153 (1985).

THE HUGHES LOW-VOLTAGE FREE ELECTRON LASER PROGRAM*

R. J. Harvey and F. A. Dolezal
Hughes Research Laboratories, Malibu,CA 90265

ABSTRACT

The Hughes Free Electron Laser Program is based upon a previous low voltage oscillator experiment in which a 225-kV electron beam was used to generate 60 kW of 30-GHz radiation in a test model FEL[1]. The operation of these experiments and the impications for other low-voltage experiments are reviewed. The design of a second-generation low-voltage FEL and its incorporation into a two-beam, two-stage FEL are discussed.

INTRODUCTION

The need for compact, versatile mm-wave and infrared sources has driven the Hughes program toward a two-beam, two-stage FEL system. Previous FEL experiments[1] have demonstrated the feasibility of depressed collectors for use in E-beam collection to increase the net efficiency, and generation of low voltage FEL emission in the backward wave mode[2] Recently, first stage experiments[3] were carried out with a 225 kV, 12-A E-beam system. The results of these experiments have been folded into the design of a two-beam two-stage FEL to be carried out in conjuction with the University of California at Santa Barbara.

THE 30-GHz OSCILLATOR EXPERIMENT

The experimental apparatus consists of an electron beam, which is focussed and confined by a 400-G axial magnetic guide field and injected into a high-Q distributed-Bragg-reflector resonator and a "square" profile, 30-cm long, linear wiggler-magnet array. This device has been used to generate 60 kW of output power at approximately 30 GHz. The output is consistent with an efficiency of 3.5% which is the linewidth of the laser.

The wiggler is a "square" wiggler which provides two axis focussing to insure propagation of the electron beam through the wiggler. The two-dimensional focussing wiggler provides magnetic-pressure profiles near the axis which increase in intensity and become square in shape toward the magnets. The 400-G net magnetic field is polarized in a plane at 45 degrees to the magnets. This wiggler provides adequate beam focussing except at the magnet corners where the finite dimensions of the magnet allow leakage paths in the corners.

The measured FEL system gain is 4.6%, therefore a cavity Q of at least 10,000 is necessary for this FEL. Two distributed-

*Sponsored in part by Office of Naval Research under contract number N00014-82-0220 and by Hughes IR&D funds.

Bragg reflectors, located on either end of a cylindrical copper waveguide, provide a cavity with the necessary Q and provide a wide aperture for propagation of the E-beam. The cavity Q, defined by the ratio of center frequency to the half-width of the line in frequency space, has been measured by two different techniques. In one case a frequency swept klystron signal is injected into one end of the cavity and the output is collected with a crystal detector. In the other case a single-frequency, fast-risetime signal is injected into the cavity and the effect on the risetime of the throughput signal is monitored. A Q of approximately 12,000 is measured for each case. The Q of these cavities has also been measured as a function of the separation between the reflectors. The cavity behaves as a true Fabry-Perot resonator with two modes present. The high-Q modes correspond to the TE_{11} mode and lower Q modes correspond to the TE_{21} mode. While the lower Q modes can be excited on the bench, they do not present a high enough Q for the FEL to resonate.

THE TWO-BEAM, TWO-STAGE FEL EXPERIMENT

The experiments have provided a basis by which a new first-stage FEL has been designed. This FEL will have sufficient power to pump the second stage of a two-beam two-stage FEL. The second stage is driven by the 6-MeV beamline at UCSB. A schematic of this system is shown in Figure 1. The first stage output will be used to pump a second high-Q cavity to 100MW/cm^2 of standing wave mm-wave power. Large aperture distributed-Bragg reflectors will terminate the cavity, allowing passage of the second stage electron beam and infrared radiation. This system allows for independent optimization and alignment of both stages from the standpoint of efficiency, gain and stability.

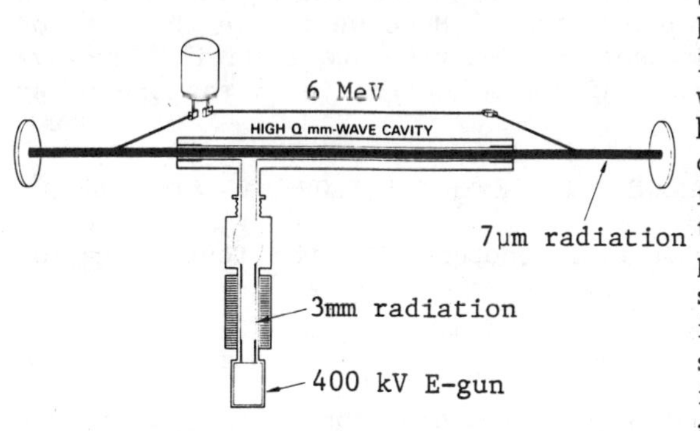

Figure 1. HRL/UCSB Two-Stage, Two-Beam FEL Concept.

REFERENCES

1. F. Dolezal, R. Harvey and C. Parazzoli, IEEE JQE. QE-19, 309 (1983).
2. F. A. Dolezal, R. J. Harvey, A. J. Palmer, and R. W. Schumacher, Proc. SPIE 453, 356 (1984).
3. R. J. Harvey and F. A. Dolezal, Proc. IEEE Conference on Lasers '84, San Francisco, CA , November 26-30, 1984.

A QUANTUM DESCRIPTION OF THE FREE-ELECTRON LASER WITH A UNIFORM MAGNETIC FIELD AND A CORRUGATED SLOW-WAVE STRUCTURE

Josip Šoln
U.S. Army LABCOM, Harry Diamond Laboratories, Adelphi, MD 20783

ABSTRACT

A method is proposed to describe quantum mechanically the free-electron laser with two explicit slow-wave structures: one due to the uniform magnetic field and the other due to the metal grating. The aim of this description is to characterize the free-electron laser capable of generating radiation in the far-infrared/microwave spectral region.

INTRODUCTION

When studying the physical characteristics of a single-particle free-electron laser (FEL), one usually assumes that the radiation is free-like in nature. While, by and large, classical and quantum treatments of such FEL's are expected to be physically equivalent (e.g., the wiggler-FEL[1-3]), the FEL in a uniform magnetic field does not follow this pattern; the classically evaluated gain[4] and the classical limit of the quantum gain[5] do not coincide. This justifies extending the quantum formulation to the FEL's with more than just one explicit slow-wave structure. Here we tackle the case of the uniform-magnetic-field[5]/orotron[6] FEL with two explicit slow-wave structures: one due to the uniform magnetic field[5] and the other due to the metal grating.[6]

EQUATION OF MOTION AND DYNAMICALLY GENERATED SLOW-WAVE STRUCTURES

For the uniform-magnetic-field/orotron FEL the force equation for the electron is (the velocity of light, c = 1)

$$M \frac{d}{dt}(\gamma\vec{v}) = -e\left[\vec{v} \times \vec{B} + \hat{z} E_o e^{-i2\pi z/\ell}\right] , \qquad (1)$$

where $(-e)$ is the electron charge, M the electron mass, $\vec{v}$ its velocity, and $\vec{B} = \hat{z} B$. The electric field in (1) is associated with the metal grating whose period is ℓ. Peak electric field E_o satisfies $E_o \ll |\vec{v} \times \vec{B}|$. This allows us to seek the solution for (1) exactly in terms of B and to the first-order in E_o. Since $z \simeq v_3 t + O(E_o)$, where v_3 is the velocity of the electron guiding center (EGC), we find from (1) that an electron will oscillate about the EGC with frequencies

$$\Omega = v_3 2\pi/\ell \quad , \quad \omega_c \quad , \; (v_3 2\pi/\ell) + \omega_c \quad , \; (v_3 2\pi/\ell) - \omega_c \quad , \qquad (2a,b,c,d)$$

where $\omega_c = eB/M\gamma$, $\gamma^2 = (1 - v_3^2)^{-1}$. The electron simultaneously oscillates along the z-axis (2a), and perpendicular to it (2b,c,d).

A Doppler-shifted radiation frequency, ν, is given as $\nu = \omega/2\pi$, $\omega = \Omega/(1 - v_3\cos\Theta)$, where $\cos\Theta = \hat{z}\cdot\hat{k}$, and $\vec{k}$ is the photon momentum. We see that for $(v_3 2\pi/\ell) - \omega_c > 0$, all ν's are different from each other. For example, for B = 1 T, $\gamma \approx 2.2$ ($v_3 \approx 1$), and $\ell = 1.7$ cm, we have $\nu(a) = 170$ GHz, $\nu(b) = 120$ GHz, $\nu(c) = 300$ GHz, and $\nu(d) = 50$ GHz.

For a given Ω from (2) we define the slow-wave structure with the wave vector $\vec{K} = \hat{z}\,K$, $K = \Omega/v_3$. K(a) and K(b) correspond to the metal grating[6] and the uniform magnetic field,[5] respectively. K(c) and K(d) are dynamically generated and are unique to this FEL. Since all K's are different from each other, each case (characterized with a given Ω) can be treated separately when a quantum formulation of this FEL is made where the quantum recoil involving the corresponding $\vec{K}$ is needed. Hence, the quantum mechanical description of this FEL is feasible and will result in characterization of a very attractive radiation source in the far-infrared/microwave spectral region.

REFERENCES

1. W. B. Colson, Phys. Lett. A 64, 190 (1977).
2. W. Becker, Opt. Commun. 33, 69 (1980).
3. J. Šoln, J. Appl. Phys. 52, 6882 (1981).
4. S. K. Ride and W. B. Colson, Appl. Phys. 20, 42 (1979).
5. J. Šoln, J. Appl. Phys. 58, 3314 (1985).
6. J. Šoln and R. P. Leavitt, J. Appl. Phys. 56, 29 (1984).

ULTRAVIOLET EXCITATION AND STIMULATED EMISSION IN CRYOGENIC RARE-GAS HALIDE SOLUTIONS

H. Jara, M. Shahidi, H. Pummer, H. Egger, and C. K. Rhodes
Department of Physics, University of Illinois at Chicago
P. O. Box 4348, Chicago, Illinois 60680

ABSTRACT

Cryogenic rare-gas halide solutions, excited optically with radiation at 351, 248 and 193 nm, exhibit fluorescence bands corresponding to rare-gas halide dimer and trimer species. For liquid Ar, Kr, and Xe hosts, these emissions display a systematic trend of wavelength shifts and state lifetimes. Due to their strong dipolar character, these excimers radiate in the liquid at wavelengths which are considerably red shifted[1] with respect to the gas phase values, an effect which can be explained on the basis of the solute-solvent interaction. In the case of XeF*, stimulated emission has been observed on the B1/2 → X1/2 transition at 404 nm. Energies of ∿ 110 μJ have been measured in 5 nsec pulses.

INTRODUCTION

In this paper we discuss the experimental results and theoretical implications of spectroscopic studies performed on a new type of optical material, namely, cryogenic rare gas-halogen (F_2 and Cl_2) liquid phase excimer systems, excited by high power ultraviolet laser radiation. In addition, the properties of stimulated emission on the XeF* (B-X) excimer band in liquid argon, are also analyzed.

These studies are well motivated by potential applications as well as by basic scientific considerations. Optical materials in the liquid phase, with its characteristic high density (on the order of 10^{22} atoms/cm^3), are not only attractive media candidates for the storage of optical energy, but could also be used themselves as the active media of new high photon flux laser systems. On the other hand, the study of the behavior of electronically excited impurities in simple cryogenic liquids is of basic physical interest, since the characteristic properties of certain excited states can furnish a fundamental understanding of the solute-solvent interactions.

EXPERIMENTAL RESULTS

A description of the experimental apparatus used in these experiments can be found elsewhere.[1] The cryogenic rare gas-halogen solutions investigated were Ar(F_2), Ar (Kr,F_2), Ar(Xe,F_2), Kr(F_2), Kr(Xe,F_2), Xe(F_2), Ar(Kr,Cl_2), Ar(Xe,Cl_2), Ar(Kr,Xe,Cl_2), Kr(Cl_2), Kr(Xe,Cl_2), and Xe(Cl_2), where the species given in parentheses designate the impurity materials. All the solutions investigated exhibited strong visible and/or near ultraviolet fluorescence under focussed laser irradiation for at least one of the excitation wavelengths at our disposal (193, 248, 351 nm). In addition, many solutions fluoresced strongly when irradiated at two or even three of the available excitation photon energies. Except for the Ar(F_2) liquid mixture, which exhibited only one ultraviolet emission when irradiated at 193 nm, all other cryogenic solutions exhibited visible fluorescence with a strength that permitted direct visual observation. Furthermore, most solutions displayed strong fluorescence under irradiation with the unfocussed laser beam, at intensities of less than 10^6 W/cm^2.

The recorded fluorescence spectra of these liquid solutions contained features which are very similar in shape and bandwidth to the ones displayed by the corresponding gas phase spectra. Generally, two different types of emission bands were observed, one broad with typical widths (FWHM) of $\sim$ 60 nm, and the other much narrower with corresponding widths of only $\sim$ 6 nm. Consequently, by using the characteristic widths of the emissions in the gas phase and assuming that the lower states of the rare gas-halide dimers (RgX*) and trimers (Rg_2X^*), with their characteristically weak interactions, are only slightly modified by the interaction with the host, it was then possible to assign the broad bands to the trimers and the narrow features to the B-X and D-X rare gas-halide dimer transitions. The C-A dimer emissions, due to the larger slope of the A state, were found to account for the additional broad bands (a fraction of our experimental observations are displayed in Table I). However, without exception, the wavelengths measured for emissions arising from liquid phase mixtures did not agree with the corresponding gas phase values. A systematic comparison between the liquid phase and gas phase data led to a consistent and clear picture for the net spectroscopic effects, which result from the interaction of these excited molecules with the surrounding atoms of the liquid host. Accordingly, we were able to conclude that the energies of the strongly dipolar rare gas-halide dimer and trimer states <u>decrease</u> as a result of the solute-solvent interaction.

Excimer Transition	Liquid Host	Center Wavelength (nm)	Shift (Exp) (eV)	Shift (Theor) (eV)
XeF* (B-X)	Argon	404	0.46	0.4
Kr_2F^*	Krypton	444	0.16	0.2
Xe_2Cl^*	Xenon	574	0.37	0.4

<u>Table 1</u>: Center wavelengths of some excimer emissions in the liquid. Experimental and theoretical red shifts are displayed in the third and fourth columns.

Using this experimental apparatus, but with a slightly modified[2] cold cell, it was possible to achieve laser action at 404 nm in the red-shifted XeF* (B-X) band in liquid argon, after transverse optical pumping at 351 nm. The energy of the emission was ∿ 110 μJ in a FWHM pulse of ∿ 5 nsec and a bandwidth of ∿ 60 cm^{-1}, in a solution containing ∿ 200 ppm of F_2 and ∿ 1% of Xe. This corresponds to a total inverted population of approximately 4.5×10^{14} molecules and a conversion efficiency of ∿ 1%. However, this liquid became optically poor due to the formation of clouds. The amount of clouds increased as the liquid was irradiated until the stimulated emission became weak, and eventually, extinguished.

SOLUTE-SOLVENT INTERACTION

Using our experimental findings as a guideline, we have also been able to develop a theoretical model,[3] based on Onsager's solute-solvent interaction[4] which reasonably reproduces the observed liquid-induced red shifts (see Table I). In Onsager's model,[4] the solute molecule is treated as a polarizable point dipole at the center of a spherical cavity, having a spatial scale on the order of molecular dimensions. The dipolar electric field of the solute system polarizes the cavity causing a non-uniform charge density to develop on its surface. This charge density, in turn, generates a uniform "reaction" electric field $\vec{E}_r$ which is parallel to the original dipole moment $\vec{\mu}$ (see Figure 1). It is, therefore, the electrostatic coupling of the dipole moment of the solute molecule with the reaction electric field which serves as the physical mechanism responsible for the lowering in energy of the solute excimer states. In this picture, Onsager's cavity is regarded as the volume of polarized matter over which the energy lost by the solute molecule is distributed.

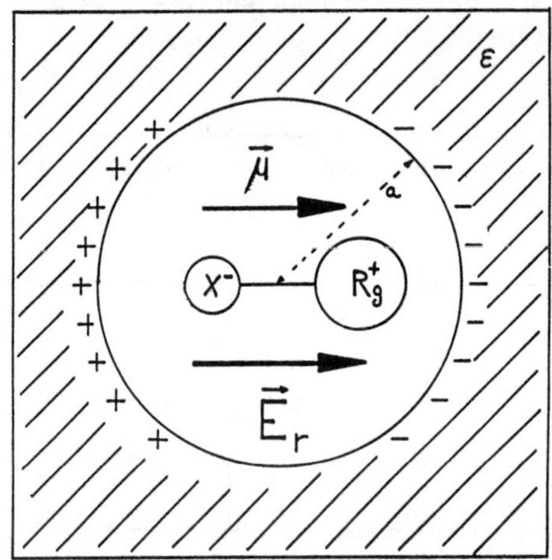

Fig. (1): Onsager's cavity (radius a), containing a permanent dipole moment $\vec{\mu}$. The reaction electric field $\vec{E}_r$ is also displayed.

CONCLUSION

In summary, we have performed spectroscopic studies on a new class of optical materials, namely rare gas-halogen excimers in the liquid phase. These excimers are generated in the liquid environment through direct optical excitation at moderate to low intensities.[1] This shows that new and efficient channels for the deposition of optical energy are available in the liquid phase. Also, these studies led us to the first observation of stimulated excimer emission in the liquid phase and to a deeper understanding of the basic solute-solvent interactions.

ACKNOWLEDGEMENTS

The authors wish to acknowledge R. Slagle, J. Wright, T. Pack, and R. Bernico for technical assistance. This work was supported in part by the Office of Naval Research, the Air Force Office of Scientific Research under contract number F49620-83-K0014, the Air Force Office of Scientific Research, Department of Defense -- University Instrumentation Program under grant number USAF 840289, the Department of Energy under grant number DE-AC02-83ER13137, the Lawrence Livermore National Laboratory under contract number 5765705, the National Science Foundation under grant number PHY-84-14201, the Defense Advanced Research Projects Agency, the Innovative Science and Technology Office of the Strategic Defense Initiative Organization, and the Los Alamos National Laboratory under contract number 9-X54-C6096-1.

REFERENCES

1. H. Jara, H. Pummer, H. Egger, and C. K. Rhodes, Phys. Rev. B30, 1 (1984).
2. M. Shahidi, H. Jara, H. Pummer, H. Egger, and C. K. Rhodes, Opt. Lett. 10, 448 (1985).
3. H. Jara, H. Pummer, H. Egger, M. Shahidi, and C. K. Rhodes, "Interaction of Rare Gas-Halide Excimers with Simple Cryogenic Liquids," to be submitted for publication.
4. L. Onsager, J. Am. Soc. 58, 1486 (1936).

POSSIBLE NEW LASERS BASED ON PLASMAS SIMILAR TO THERMIONIC CONVERTERS*

E. J. Britt
J. L. Lawless
Space Power Inc., 253 Humboldt Court, Sunnyvale, Ca. 94089
J. B. McVey
Rasor Associates, 253 Humboldt Court, Sunnyvale, Ca. 94089

ABSTRACT

This paper describes novel plasma recombination lasers that can be produced with conditions similar to the plasma in a thermionic convertor. Calculations have shown that a population inversion can be obtained by either time variation of the current in a thermionic convertor discharge or by gas dynamic expansion of plasma flow driven by heat pipe action. Sudden modulation of the current can cool the plasma in a thermionic convertor with electrons coming from the thermionically emitting electrode to produce an inversion of the 7p-7s line in cesium. Alternatively, if the inter-electrode plasma is made to flow through a supersonic expansion nozzle, a population inversion in the downstream plume may also be produced. Either of these approaches or a combination of them can be used to convert heat directly into laser output. Two laser lines in the cesium vapor at 2.93 and 3.10 microns are predicted. Two other novel laser concepts are also mentioned: a solar pumped atmospheric laser and a laser based on the space plasma around an orbiting vehicle.

INTRODUCTION

The direct conversion of ~1500 K heat to laser light will be considered theoretically. The laser will be produced using recombination kinetics in a plasma similar to that of a thermionic energy converter. Both unsteady and gas dynamic schemes for causing recombination will be considered. The predicted laser lines are infrared at 2.93 and 3.10 microns. Unlike most lasers which consume electrical power and produce heat as a byproduct, this laser consumes heat and may produce electricity as a byproduct. Two other more speculative novel laser concepts will be mentioned. One is pumped by naturally occurring atmospheric processes. The other converts the kinetic energy of orbital motion of a satellite in LEO to laser light.

*The thermionic laser research was supported by NSF Grant CPE-8204813. The atmospheric laser work was supported by NAS3-22239.

THERMIONIC PLASMA LASERS

The plasma in a thermionic energy converter is a two-temperature cesium plasma formed between two electrodes with an inter-electrode spacing typically less than 1 mm. One is hot (~1500 K) and emits electrons thermionically and the other is cool (<1100 K) and does not. The ionization fraction is typically 1% and the neutral pressure is of order 100 Pa (1 Torr). More information on the steady behavior of these plasmas can be found in Ref. 1, 2, and 3.

It was first predicted theoretically[4] and later demonstrated experimentally[5,6] that if a plasma with an initially high ionization fraction is rapidly cooled, the resulting rapid recombination of electrons and ions can produce a population inversion in the excited level of the resulting atom or ion. This is called a plasma recombination laser. Further informaiton on recombination laser kinetics may be found in Ref. 7.

In a thermionic converter, the important recombination reaction is:

$$e + e + Cs^{+} \rightarrow e + Cs \qquad (1)$$

Reaction (1) occurs as a series of elementary reactions involving the electronically excited levels of cesium. A simultaneous solution of the conservation equations for these levels has been performed and a population inversion between Cs(7P) and Cs(7S) is predicted to occur if recombination is sufficiently rapid.

To achieve sufficiently rapid recombination, the electrons must be cooled rapidly. Typically, the electron temperature in a steady thermionic plasma is ~3000 K. To achieve an inversion, calculations show that this must be cooled to ~1500 K within a few microseconds. Two methods for doing this have been calculated. The first is illustrated qualitatively in Fig. 1. A thermionic converter is initially operating in steady state at some high current. The current is suddenly reduced at time t=0. Calculations show that, because of the consequent reduction of ohmic heating, the electron temperature drops to near the cathode temperature in less than one microsecond. An inversion then occurs, peaking at 3.4×10^{10} cm^{-3} at t=2μs, and lasting for ~8μs.

A second method for rapid cooling of a thermionic plasma uses gas dynamic cooling. The gas flow is driven by heat-pipe action, preserving the heat driven nature of this laser. A conceptual diagram of this is shown in Fig. 2. To date, the calculated peak inversions for this method of cooling have been $O(10^{8}$ $cm^{-3})$, much less than for the unsteady method.

ATMOSPHERIC LASERS

Two schemes for creating atmospheric lasers will be considered. First, solar radiation and geophysical processes act to produce plasmas and excited species in the atmosphere of earth and other planets. In some cases these effects result in excited state population inversions which may be useful for extracting energy in the form of laser output. Population inversion of the CO_2 10.6μ transition has been observed experimentally on Mars. Similar inversions have been inferred from atmospheric models of Venus and other planets. The feasibility of an orbiting (or balloon-carried) optical system for an atmospheric laser in the earth's atmosphere has been studied. A conceptual diagram is shown in Fig. 3. The conclusions are:

- o At least one population inversion has been identified in the earth's atmosphere: $O_2(^1\Delta) \rightarrow O_2(V=1)$. A collision partner such as MgO would have to be dispersed to make a laser using this inversion.
- o Collisions with vibrationally excited nitrogen molecules may cause an inversion of the 10.6μ CO_2 line in the earth's atmosphere. The inversion density depends on temperature (≥200K), and probably is relatively small.
- o Beaming of energy into the atmosphere to "pump" a lasing region appears feasible. Intersection of two microwave beams in the ionosphere has been studied by the Russians and may be useful for ATLAS pumping.
- o A high altitude nuclear explosion may result in a temporary inversion. Compton electrons and gamma radiation can create excitation by processes similar to those responsible for EMP.

Secondly, recent observations of plasmas caused by collision with residual portions of the atmosphere, colliding with orbiting vehicles such as the space shuttle, have prompted a speculation that a tethered system may by used to form a laser from this plasma excitation. The collisional energy of an atom striking the orbiting vehicle is greater than the ionizational potential of most gases and, therefore, a "bow" shock and/or a plumed-type wake forms around many orbiting objects, particularly in LEO. Outgasing from the orbiting vehicle enhances this effect, and it may be deliberately used to increase the plasma density or change the type of constituents.

A long tether, perhaps shaped like a trough or scoop, could be deployed from an orbiting vehicle to produce a plasma column that may be lased. A possible candidate species for this type of laser might be the auroral green line of oxygen at 5577 Angstroms since atomic oxygen is one of the typical constituents of this type of plasma.

Better modeling of the conceptual space plasma laser and the atmospheric laser are needed. If further study indicates feasibility, a space shuttle experiment could be used to test either of the concepts.

REFERENCES

1. N. S. Rasor, Ch. 5 in Applied Atomic Collision Physics, edited by H. S. W. Massey, E. W. McDaniel and B. Bederson (Academic Press, New York, 1982).
2. E. J. Britt and J. B. McVey, U.S. Energy Rep. No. COO-2263-16, Rasor Associates, Sunnyvale Ca., 1979.
3. J. L. Lawless and S. H. Lam, J. Appl. Phys. 57 (1986)
4. L. I. Gudzenko and L. A. Shelepin, Sov. Phys.-JETP 18, 998 (1964).
5. W. T. Silvfast, L. H. Szeto, and O. R. Wood, Appl. Phys. Lett 36, 615 (1980).
6. E. M. Campbell, R. G. Jahn, W. F. von Jaskowsky, and K. E. Clark, Appl. Phys. Lett 30, 575 (1977).
7. J. L. Lawless, J. Appl. Phys. 55, 3226 (1984).

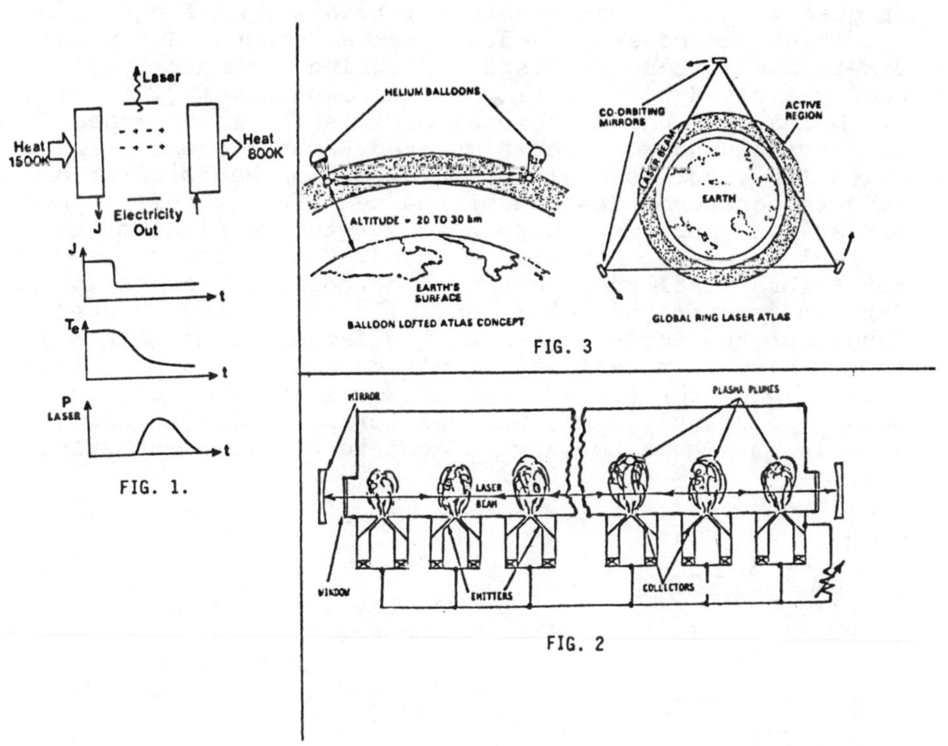

FIG. 1.

FIG. 3

FIG. 2

OPTICALLY PUMPED MOLECULAR BEAM LASERS

B. Wellegehausen
Institut für Quantenoptik, Universität Hannover
3000 Hannover, Fed. Rep. of Germany

U. Gaubatz and K. Bergmann
Fachbereich Physik, Universität Kaiserslautern

ABSTRACT

Investigations on Na_2 and I_2 dimer lasers using a supersonic molecular beam as laser active material have been performed. In the case of Na_2 a threshold pump power of only 17 μW has been obtained.

INTRODUCTION

Optically pumped lasers with diatomic molecules[1] operate between rotational-vibrational levels of different electronic states (Fig. 1). Thereby, pump radiation of frequency ω_p is down-converted into a set of frequencies ω_d, which may cover extended spectral ranges. A typical dimer laser set-up consists of a ring resonator collinearly excited by a continuous narrowband pump laser (Fig. 2). Normally, unidirectional oscillation is obtained, due to the influence of the stimulated Raman process[1]. The molecular vapor is enclosed in a cell or heatpipe at a certain temperature. Instead of the heatpipe or cell, a molecular beam streaming perpendicular to the plane of the optical resonator may also be used. Due to the reduced rot.-vibrational temperature of the beam, the operation conditions are strongly improved, allowing a much shorter length of the active material. So far, systems with Na_2 and I_2 have been realized[2,3]. In case of Na_2 a circular nozzle (diameter 0.5 mm) was used, yielding an active length of roughly 1 mm, whereas for I_2 a slit nozzle (50 μm x 14 mm) was necessary to achieve a sufficient gain.

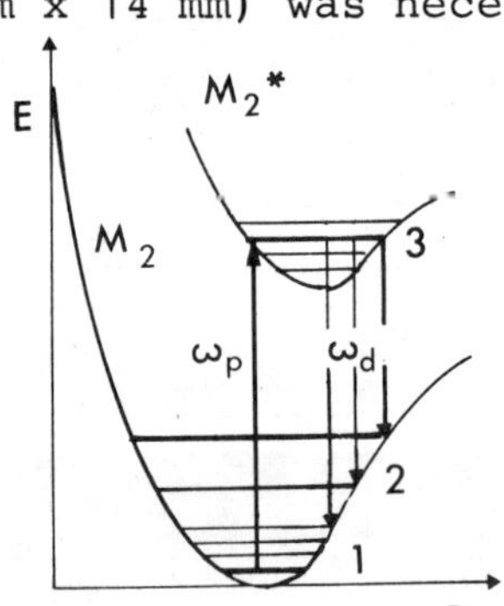

Fig. 1. Level scheme of a diatomic molecule M_2 with λ-type three level laser cycle. ω_p: pump frequency, ω_d: dimer laser frequencies.

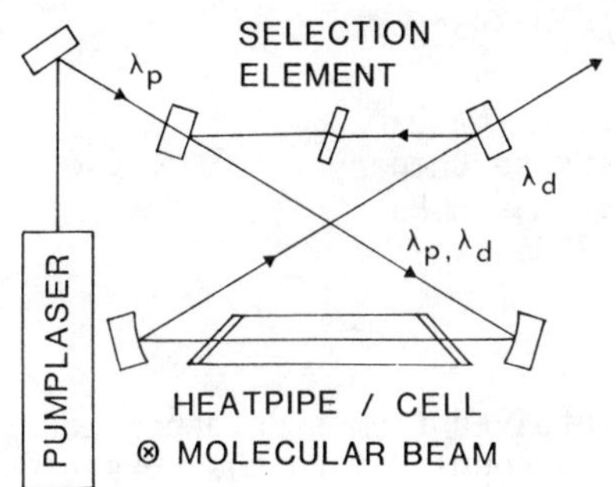

Fig. 2. Dimer ring laser set-up. λ_p: pump wavelength, λ_d: dimer laser wavelength. The selection element allows oscillation on a single laser line.

EXPERIMENTAL RESULTS

The current best laser data are summarized in Table I. For I_2 oscillation was only obtained for a single line, while for Na_2 multiline emission was possible. An important parameter is the distance Z of the active region to the nozzle, measured in units of a characteristic nozzle dimension D. Lowest thresholds are observed at Z/D values around 4. For Z/D > 10, the laser operates in an almost collision free regime. In the first experiments with Na , with slit nozzles or circular nozzles and seeded beams, lasing at Z/D values up to 100 have been achieved.

TABLE I DATA of Na_2 AND I_2 MOLECULAR BEAM LASER

	pump transition: $X^1\Sigma_g^+(0,28)\rightarrow B^1\Pi_u(6,27)$			oven temperature: 1000 K
	Ar^+ laser (476.5 nm)			beam velocity: 1500 m/s
	laser emission: B(6,27)→X(v",26/28)			$T_{vibr.} \sim 100$ K, $T_{rot} \sim 40$ K
Na_2	oscillation for $5 \leq v'' \leq 32$			nozzle: diameter D=0.5 mm
	threshold	output power	pump power	oscillation up to Z/D=24
	∿17 μW	15 mW^x	∿100 mW	^{x}multiline unidirectional oscillation
	pump transition: $X^1\Sigma(0_g^+)(0,14)\rightarrow B^3\Pi(0_u^+)(43,15)$			oven temperature: 120 K
	Ar^+ laser (514.5 nm)			beam velocity: 400 m/s
I_2	laser emission: B(43,15)→ X(v",14/16)			slit nozzle: 50 μm(D) x14 mm
	oscillation only for v"=83(1.34 μm)			oscillation up to Z/D=50
	∿20 mW	1 mW	500 mW	bidirectional oscillation forward:backward >10:1

REFERENCES

1. B. Wellegehausen, IEEE J. Quant. Electr., QE-15, 1108 (1979).
2. P. L. Jones, U. Gaubatz, U. Hefter, K. Bergmann, B. Wellegehausen, Appl. Phys. Lett. 42, 222 (1983).
3. U. Hefter, J. Eichert, K. Bergmann, Opt. Commun. 52, 330 (1985).

THE PHOTON AVALANCHE LASER

M. E. Koch and W. E. Case
LTV Aerospace and Defense Company
P. O. Box 650003, MS TH-85
Dallas, TX 75265

ABSTRACT

Laser output at 4.9 μm has been obtained from a Pr doped $LaCl_3$ crystal. Pump laser energy at 529 nm is coupled into the crystal via the photon avalanche process. This process involves metastable state ion absorption of a pump photon and subsequent ion-ion energy transfer. The latter step generates two metastable state ions (self-induced absorption). This highly nonlinear process is characterized by a sharp onset with pump intensity and, at higher intensities, depopulation of the ionic ground state. Experimental results will be compared with predictions from an existing rate equation model.

In this report we describe a new infrared cw laser based on a novel mechanism which we call photon avalanche. Laser action is achieved by optical pumping of Pr^{3+} ions in an $LaCl_3$ host crystal. This device differs from typical optically pumped lasers in two important respects. First, optically pumped lasers are usually energized by absorption of pump photons by species in their ground state or states within kT of the ground state. Thus, a large ground state population ensures a reasonable population of excited states. Second, for cw operation, the laser transition occurs usually between excited states. Neither of these conditions is true for the photon avalanche laser. In this device the photon avalanche is activated by an absorption involving only excited states of the Pr^{3+} ion; furthermore, the photon avalanche process drastically reduces the ground state population, producing a population inversion of excited state ions over ground state ions even for moderate cw pump intensities.

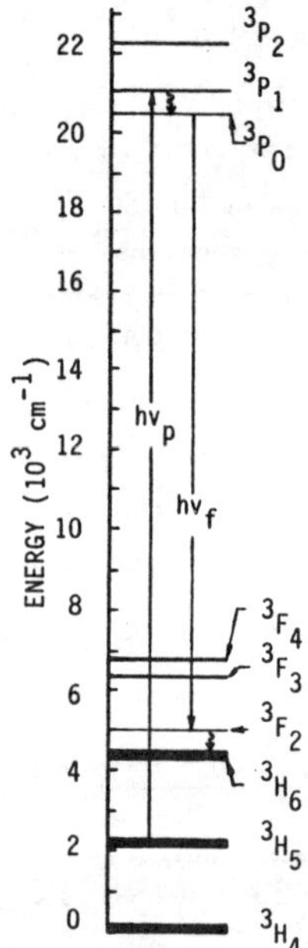

Fig. 1. Pr^{3+}:$LaCl_3$ Triplet States

The photon avalanche process has been observed in Pr doped $LaCl_3$ and $LaBr_3$[1] at liquid nitrogen temperatures and in Sm doped $LaBr_3$ at liquid helium temperatures.[2] A triplet energy level diagram for Pr^{3+} in $LaCl_3$ is shown in Figure 1. The pump laser frequency is resonant with a transition from the $^3H_5(2)$ excited state of the Pr^{3+} ion to the $^3P_1(1)$ optical state. Following absorption of a pump photon, both radiative and non-radiative processes result in de-excitation to a $^3H_6$ state. A few representative decay transitions are shown in the figure. The final step entails resonant energy transfer wherein a $^3H_6$ ion transfers half it's energy to a neighboring ground state ion leaving both in the $^3H_5$ state. There are now two ions ready to absorb pump photons. Each time a photon is absorbed, a high probability exists for creating an additional absorber. In this way, a chain reaction takes place with multiple production of $^3H_5$ ions within the lifetime of the initial $^3H_5$ state.

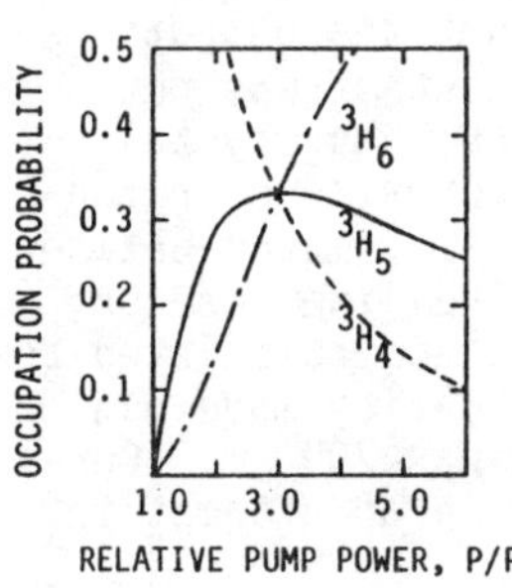

Fig. 2. Population Inversion (Theory)

The population of the $^3H_5$ level increases dramatically for pump intensities above a sharply defined critical value. Absorption becomes so strong that the avalanche is confined to within a few hundred microns of the entrance surface of the crystal. Beyond that distance only a few percent of the incident pump power is transmitted. A rate equation model has been developed[3] which predicts the various features of the avalanche including the sharp pump power onset, pump beam penetration depth, and the occurence of a population inversion of $^3H_6$ over $^3H_5$ and $^3H_5$ over $^3H_4$ at moderate pump intensities. A calculation of occupation probabilities for the first three electronic states is shown in Figure 2. The model will be the subject of a future report.

A schematic diagram of the laser is shown in Figure 3. Cylindrical lenses, L1 and L2, focus a beam from an Ar ion laser pumped ring dye laser to a line focus on the edge of the laser crystal, S. The dye laser is tuned to the $^3H_5$ (2) - $^3P_1$ (1) transition at 529 nm with 170 mW power. The crystal is oriented so that π-line radiation emitted parallel to the line focus exits at Brewsters angle to the crystal surfaces and normal to the cavity mirrors, Figure 3(c). The width of the line focus is typically 110 μm; the length of the gain region is 1.98 mm. The crystal was doped with 3.82% molar Pr. The laser cavity is hemispherical with a 190 mm radius high reflector and a 95% reflectivity flat output coupler located within a few mm of the crystal. The crystal is cooled to near liquid nitrogen temperatures (125K) by a dewar arrangement which provides an outer vacuum tight envelope, A, a radiation shield, B, and a sealed sample chamber, C, which is connected to the dewar cold finger, D. Light throughput is via fused silica pump windows, W3 and W4, and CaFl windows W1, W2, W5, and W6 Intracavity windows are mounted at Brewster's angle to the optical axis. The infrared beam was detected by a cooled InSb photovoltaic detector.

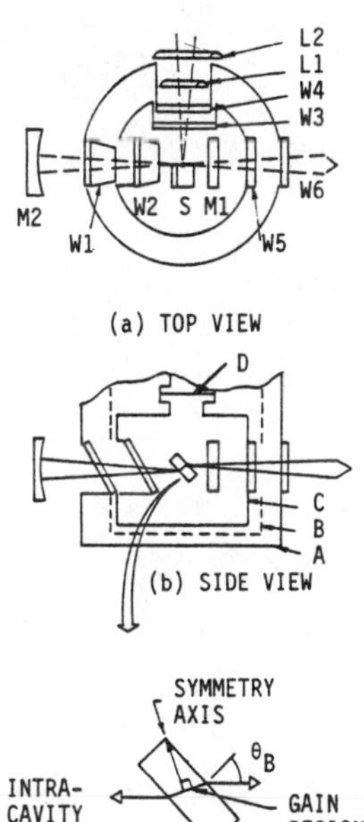

(c) MAGNIFIED SIDE VIEW

Fig. 3. Schematic Diagrams of Laser Apparatus

A well defined 5μW output beam was obtained from the laser. The beam had an oval cross section with the major axis normal to the polarization and propagation axes and the minor axis parallel to the polarization axis. Beam angles, between directions where intensity is reduced to $1/e^2$ of the intensity maximum, were 26 and 9 mR, respectively. The laser output spectrum was contained in a .4 μm wide (FWHM) band centered at 5 μm.

There are at least two reasons for believing that the output might be significantly increased. First, the laser output was not optimized with respect to variations in parameters such as crystal size (gain length), mirror reflectivities and curvatures, and Pr concentrations. Second, large losses were incurred by a mismatch between the cavity mode volume and the pump focal volume. The line focus of the pump laser at the crystal face was both measured and calculated to be about 110 μm in diameter. On the other hand the cavity mode diameter as inferred from the vertical beam divergence was 700 μm. Current experiments are aimed at exploring these areas in an attempt to improve laser output.

REFERENCES

1. J. S. Chivian, W. E. Case, D. D. Eden, Appl. Phys. Lett. 35, 124, 1979.
2. N. J. Krasutsky, J. Appl. Phys., 54, 1261, 1983.
3. W. E. Case, Vought Corporation Internal Report, 1977.

Trends in Chemical Oxygen-Iodine Lasers

V. I. Igoshin, V. A. Katulin, N. L. Kuprianov and M. V. Zagidullin
P. N. Lebedev Physical Institute
Kuibyshev Affiliate
Kuibyshev, USSR, 443020

Abstract

Potential resources of chemical oxygen-iodine lasers (COIL) are in the evaluation stage at present. The experimental and theoretical research is reviewed in this report. Principal attention is given to the authors' investigations directed at the improvement of COIL performance.

Introduction

The chemically pumped oxygen-iodine laser (COIL) is a new candidate for applications such as laser technology, laser chemistry, medicine, and laser fusion. The possibility of radical improvement of laser performance depends on the following questions:
- Can the pressure in the chemical reactor be raised?
- Can the singlet oxygen by transported at high pressure?
- Is there the possibility for increasing laser gain?
- Can the COIL operate without cold trap?
- Is it possible to obtain pulsed operation of the COIL?

Active Medium of CW COIL

The calculations show that we can obtain the laser gain in the range of 10^{-3} to 10^{-2} cm^{-1} at O_2 ($^1\Delta$) pressure higher than 4 torr. The principal means for increasing the gain is to raise the iodine concentration in the active medium. Unfortunately, the increase of I_2 concentration is accomplished by a decrease of the lifetime for stored energy. Within a wide range of parameters the following relation is valid:

$$K_{max} \cdot X_{0.5}/U_0 = \text{const} \simeq 10^{-5}\ cm^{-1} \cdot s,$$

where K_{max} is the maximum gain, $X_{0.5}$ is the length of the laser zone, and U_0 is the flow velocity. According to our calculations the allowed limit of water vapor concentration is 0.1 relative to the O_2 concentration and the allowed limit of Cl_2 is 0.2. One interesting idea is to cool the solution in a chemical reactor down to -40 - -50^oC. This cooling allows us to obtain lasing without a cold trap. The finite rate of hyperfine and translation relaxation decreases laser efficiency several times. The saturation intensity is approximately 5 kw/cm^2 at 1 torr of gas pressure.

Degradation of the Active Medium with Increasing Oxygen Pressure

Calculations indicate several possible reasons for medium degradation: 1) increase of relative content of water vapor at the entrance of the active zone, 2) increase of scattering from droplets emanating from the chemical reactor, 3) decrease of O_2 ($^1\Delta$) concentration due to its homogeneous relaxation before entering the laser cavity. For an adiabatic chemical reactor and a cylindrical cold trap the increase of oxygen pressure results in increasing water vapor concentration in accordance with Stephan's formula for condensation of vapor in the presence of other gases. According to our calculations, when gas pressure increases over two torr, the concentration of vapor for known experimental COILs is too high for effective laser operation. The operation of a chemical reactor at lower temperatures of solution allows us to overcome this difficulty and increase oxygen pressure up to 10 torr.

Pulsed Operation of COIL

For laser technology, the pulsed operation of a COIL is of interest. However in the presence of I atoms, the lifetime of stored energy is very small (less than 1 ms). During this time one can fill only a small volume with laser gas. A new scheme of operation of a COIL which is free from this disadvantage has been proposed by us[1] and later successfully realized[2]. The compound which contains I (e.g. CF_3I, HI, CH_3I) is injected into the flow of singlet oxygen. In the laser volume, photodissociation of this compound takes place and atomic iodine is produced. The time of laser pulse is determined by energy transfer from O_2 ($^1\Delta$) to I. One interesting idea is to use a solid aerosol of I_2 for production of I atoms. The calculations show that self-heating of the mixture $O_2^* + H_2 + I_2$-aerosol results in evaporation of iodine and the formation of an active laser medium. By combining within the laser volume the pulsed aerosol and singlet oxygen, it is possible to achieve operation of a pulsed COIL with photolytical production of iodine atoms at oxygen pressure as high as 100 torr.

References

1. M. V. Zagidullin et al., Kvantovaya Elektronika 11, 201 (1984).
2. N. G. Basov et al., Kvantovaya Elektronika 11, 1893 (1984).

REMARKS ON NONCOLLINEAR FREE-ELECTRON LASERS

W. Becker and J. K. McIver
Center for Advanced Studies, University
of New Mexico, Albuquerque, NM 87131

Off-axis schemes for free-electron lasers were first discussed by Fedorov and coworkers.[1] They were found to produce significantly higher gain than the standard on-axis setup if the FEL operated in the long undulator (warm beam) regime. This applies as long as overlap between the electron beam and the radiation field can be achieved throughout the undulator which will be taken for granted. In the long undulator regime, the width of the electron energy distribution is larger than the width of the spontaneous emission profile so that the gain is proportional to the derivative of the former rather than the latter. This applies whenever $\gamma^{-2}(L/\Lambda)(\Delta p/p) \gg 1$ where L and Λ denote the length and period of the undulator, $E = mc^2\gamma$ the electron energy and $\Delta p/p$ the relative width of the electron momentum distribution. In this regime the gains of the fundamental and the first two harmonics are[2]

$$G_i = (e^2 a)^2 \pi\rho L\Lambda\theta_c f'(p)/(m_* e) g_i((\theta/\theta_c)^2) \tag{1}$$

with $g_1(x) = (1-x)^{-3/2}(2x-1)^2$, $g_2(x) = 16(ea/m_*)^2 x(1-x)^{3/2}$, and $g_3(x) = 3(ea/m_*)^4(1-x)^{\frac{1}{2}}(6x - 9x + 1)^2$. Here $\sqrt{2}a$ is the amplitude of the vector potential of the linearly polarized magnetic undulator, ρ the electron density and $m_*^2 = m^2 + (ea)^2$. The quantity θ is the angle between the directions of the electron and the radiation field. Equation (1) specifies the gain at the frequency $i\omega$ for the ith harmonic where ω is kept constant, i.e. independent of the angle θ. In order to make up for this, the electron energy has to be increased off-axis so that $\gamma(\theta) = (1-x)^{-\frac{1}{2}}\gamma\ (\theta = 0)$ where $x = (\theta/\theta_c)^2$. This is only possible up to a critical angle $\theta_c = 2(\pi c/\Lambda)^{\frac{1}{2}}$. For $\theta > \theta_c$ the frequency ω can no longer be emitted. The expressions given above for the functions $g_i(x)$ have been evaluated to lowest order in the field strength a of the undulator. They are plotted in Fig. 1 and exhibit a

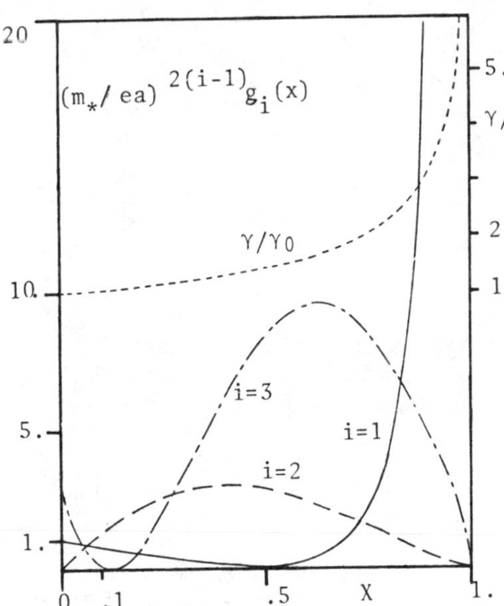

Fig. 1. The gain enhancement vs. x for the fundamental and first two harmonics

significant enhancement of the gain off-axis. In fact, the gain of the fundamental becomes infinite for $\theta = \theta_c$. This enhancement is particular to the long undulator regime: in the common short undulator (cold beam) regime the functions $g_i(x)$, have an additional factor of $(1-x)^3$ which deemphasizes the off-axis gain.[3]

A convenient way of calculating the small-signal gain on - or off-axis is by quantum mechanical perturbation theory of lowest order in the coupling between the electron and the radiation field. If one is only interested in the fundamental one may also expand with respect to the coupling to the undulator field. In this case, if one uses the Klein-Gordon equation, the three diagrams of Fig. 2 must be considered. Only the last one, the seagull term, contributes, on axis. Off-axis, the first two diagrams each are of order $mc^2/\hbar\omega$. However, these leading contributions cancel. Great care must be exercised in order to obtain properly the next-to-leading terms. Exceptionally high off-axis gains that have been reported are erroneous and can be traced back to this point.

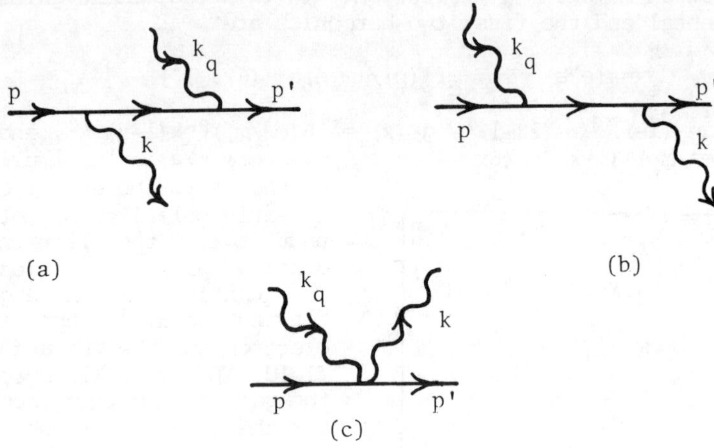

Fig. 2. The lowest order Feynman diagrams in scolar QED that contribute to spontaneous emission in a free-electron device. k_q (k) is the wavevector of the undulator (emitted radiation).

REFERENCES

1. D. F. Zaretskii, E. A. Nersesov, and M. V. Fedorov, Zh. Eksp. Teor. Fiz. 80, 999 (1981) [Sov. Phys. JETP 53, 508 (1981)]; M. V. Fedorov, Progr. Quant. Electr. 7, 73 (1981); M. V. Fedorov, Sov. Phys. SP 24, 801 (1981).
2. W. Becker and J. K. McIver, to be published.
3. W. B. Colson, G. Dattoli, anfd F. Ciocci, Phys. Rev. A31, 828 (1985).

FIRST DETECTION OF HIGHER MODES OF A HELICAL FEEDBACK (HFB) GAS LASER

J. Arnesson, S. Gnepf and F.K. Kneubühl
Physics Department, ETH, CH-8093 Zurich, Switzerland

ABSTRACT

Measurements of resonant modes of a HFB 496 µm CH_3F laser with an oversized metal-waveguide structure are presented. Higher HFB modes have been observed for the first time.

INTRODUCTION

Helical feedback (HFB) was first introduced in lasers by Preiswerk et al. [1-3] in 1982 as an extension of the concept of distributed feedback (DFB) which is widely used in various types of lasers [4-6]. This new type of feedback exhibits a longitudinal-mode selectivity higher than the standard linear DFB because of strong polarization effects. Preiswerk et al. [1-3] observed only one longitudinal HFB mode. In this paper we report on first observations of additional HFB mode.

DESIGN AND EXPERIMENTAL ARRANGEMENT

The waveguide used in our laser is a 30 cm long initailly smooth brass pipe in which a thread has been cut with a nominal pitch close to half the guide wavelength of the dominant emission at 496 µm of methyl fluoride (CH_3F). This permits operation close to the first quasi Bragg resonance [1-3], where the threshold for HFB oscillation is lower. The ratio between the mean diameter of the waveguide cross section and the emitted wavelength is close to eight. The experimental setup (Fig. 1) shows the waveguide surrounded by a heating jacket filled with silicone oil to allow homogeneous heating. For a temperature range of about 200°C and a thermal expansion coefficient $\eta = 1.85\ 10^{-5}\ {}^{o}C^{-1}$ for brass we obtain a tuning range exceeding two guide wavelengths λ_g. The CH_3F gas is optically pumped with linearly polarized light from a hybrid CO_2 laser tuned to the 9P(20) line and operating in the TEM_{00} mode. Before entering the FIR laser cavity the CO_2 beam is reduced to a spotsize comparable to the waveguide aperture in order to obtain optimal pumping. A quartz window assures that only the FIR emission is detected. The pyroelectric detector is covered with a thin Teflon plate to eliminate background noise.

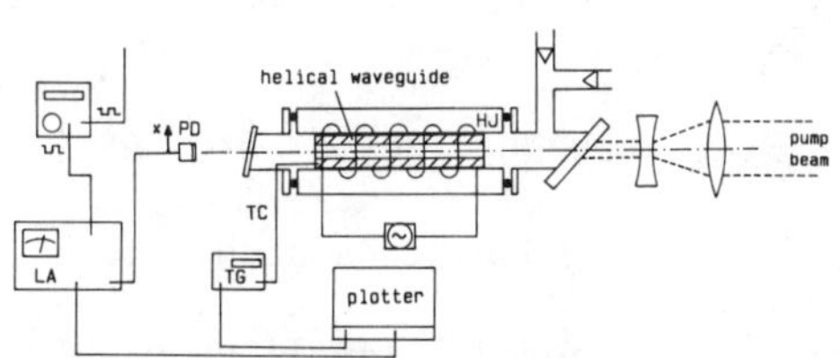

Fig. 1. Experimental setup.

EXPERIMENT

An optimal adjustment of the helical waveguide colinear to the pump beam results in a symmetric far field pattern of the HFB laser emission. With this adjustment, scans under different conditions have been performed between room temperature and 220°C. In a first experiment we have examined the FIR laser emission with a conical waveguide in front of the detector. Under this condition and with the detector centered on the symmetry axis, two resonances with a separation of approximately 80°C were detected. This represents the first observation of more than one resonance of an HFB laser. In an attempt to identify the modes we have removed the conical waveguide and again scanned over the whole temperature range for different radial positions of the detector. The sensitive area of the detector is 2x2 mm^2. This allows us to pick out and study any selected part of the spatial beam profile. Two scans corresponding to 0 mm (position A) and 3 mm (B) displacement from the axis respectively are shown in Fig. 2. We now find that the resonance at 115°C is no longer a single resonance but resolved into a fundamental mode (f.m.) at T_1 = 109°C and a higher mode at T_2 = 124°C. The resonance at T_3 = 196°C, on the other hand, remains single. It is interpreted as a second fundamental HFB mode. The separation between the two fundamental modes is almost λ_g, which is twice the separation found from the standard resonance condition for weak-gain DFB and Fabry-Perot laser resonators. This is a confirmation of the higher longitudinal mode selectivity of HFB lasers.

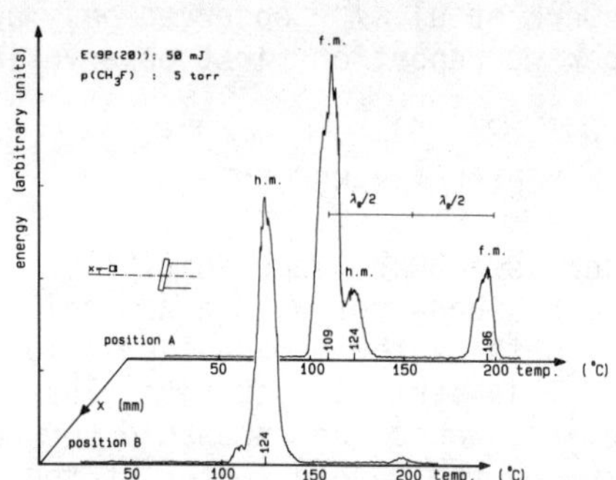

Fig. 2. Observed HFB modes

REFERENCES

1. H.P. Preiswerk, G. Küttel and F.K. Kneubühl, Phys. Lett. 93A, 15 (1982)
2. H.P. Preiswerk, M. Lubanski, S. Gnepf and F.K. Kneubühl, IEEE J. QE-19, 1452 (1983)
3. H.P. Preiswerk, M. Lubanski and F.K. Kneubühl, Appl. Phys. B33, 115 (1984)
4. H. Kogelnik and C.V. Shank, Appl. Phys. Lett. 18, 152 (1971)
5. M. Nakamura, A. Yariv, H.W. Yen, S. Sanekh and H.L. Garvin, Appl. Phys. Lett. 22, 515 (1973)
6. E. Affolter and F.K. Kneubühl, Phys. Lett. 74A, 407 (1979)

RYDBERG-BREMSSTRAHLUNG-MASER

A. Jay Palmer
Hughes Research Laboratories, Malibu, Ca. 90265

ABSTRACT

Stimulated emission of bremsstrahlung radiation by electrons in high Rydberg states is analysed theoretically. This approach to bremsstrahlung masers is shown to overcome some of the previous difficulties in realizing a practical value for the bremsstrahlung gain in the millimeter-wave regime by allowing volumetric production of mono-energetic electrons within a time period which is short compared to the thermalization time of the electron energy distribution, and by relaxing the criteria on the energy dependance of the collision cross-section for achieving a positive bremsstrahlung gain coefficient.

INTRODUCTION

The major roadblock to achieving bremsstrahlung maser action in the millimeter-wave and infrared regimes has been the requirement for achieving a high electron collision frequency to produce an adequate gain on the one hand with the competing requirement for a low electron collision frequency to maintain a monocromatic and beamed electron distribution on the other hand. In addition the electron momentum transfer cross section must fall faster than $1/E^2$ for an isotropic electron distribution or faster than $1/E$ for a beamed electron distribution[1]. Coulomb collisions and collisions with atoms exhibiting a Ramsaucr miminum in their cross-sections are the only examples of collision processes where bremsstrahlung instabilities have been observed[2]. In these cases signal frequencies remained well below the millimeter-wave regime due to the inability to meet all of the above requirements.

A new approach to achieving bremsstrahlung maser action is presented here. It is based on producing electrons in a mono-energetic, high-quantum-number Rydberg state using pulsed dye lasers, rather than using beamed or plasma electrons as has been the case in all previous approaches to bremsstrahlung masers. The required high-density, mono-energetic electron distribution can, in this way, be established in a time interval which is short compared to the collisional relaxation time of the distribution. Also, because of the different form of the gain coefficient, the criteria on the energy dependance of the collision cross section required for positive bremsstrahlung gain is less restrictive for Rydberg electrons than the criteria for free electrons listed above.

STIMULATED EMISSION CROSS-SECTION

Rydberg-state bremsstrahlung emission is an example of collisionaly aided radiation emission (CARE) in the impact collisional regime[3]. In this regime, the net stimulated emission cross-section for transitions from a given Rydberg state with principle quantum number, n to nearby states with principle quantum number, m, can be written:

$$\sigma_{st}(n,\omega) = 4/(\hbar c)\ \{\sum_{m<n}\gamma_c(n)\ \omega(n,m)\ \mu^2(n,m)\ m^2 \times 1/[\Delta^2(n,m) + \gamma_c^{\,2}(n) + \chi^2(n,m)] - \sum_{m>n}\gamma_c(n)\ \omega(n,m)\ \mu^2(n,m)\ m^2 \times 1/[\Delta^2(n,m) + \gamma_c^{\,2}(n) + \chi^2(n,m)]\ \} \quad (1)$$

where, $\gamma_c(n)$ is the collision frequency for the excited Rydberg state, $\omega(n,m)$ is the transition frequency, $\mu(n,m)$ is the corresponding transition dipole moment, $\hbar$ is Planck's constant divided by 2π, c is the speed of light and χ is the Rabi frequency for the transition.

DEMONSTRATION REQUIREMENTS

When hydrogenic approximations are used for $\omega(n,m)$ and $\mu(n,m)$, Eq. (1) predicts that broad, bremsstrahlung gain bands in the 100 - 1000 GHz portion of the spectrum can be produced by exciting high Rydberg states to densities of on the order of a few times 10^{11} cm^{-3} in the presence of a few tens of Torr of rare gas perturbers. To assure that mirrorless maser action at the Rydberg transition line centers does not compete with the bremsstralung maser emission it will be necessary to tune the resonant cavity to the desired frequency in the gain band, and to expose the gain medium to a few milliwatts/cm^2 of millimeter-wave radiation near the desired frequency. The spectral signatures and threshold behavior of the Rydberg bremsstrahlung maser emission will provide a new measurement of Rydberg level collision cross-sections through Eq. (1) which can be compared with theory and with the other measurements of the cross-section based on radiation quenching. In common with the collisionless Rydberg masers, the efficiency of the Rydberg Bremsstrahlung maser is low due to the poor quantum efficiency of the process. However, the greatly improved power and tunability of the Bremsstrahlung maser will likely offer benifits to a number of applications in a portion of the spectrum where there is still a dirth of tunable coherent sources.

REFERENCES

1. A. Rosenberg, Y. Ben-Aryeh, J. Politch, and J. Felsteiner, Phys. Rev. A., **25**, 1160 (1980); A. J. Palmer, Appl. Phys. Lett. 42, 1011 (15 June 1983).
2. G. Bekefi, "Radiation Processes in Plasma", Wiley & Sons, New York, 1966, p. 306.
3. S. Yeh and P. R. Berman, Phys. Rev. A, 19, 1106 (1979)

HIGH EFFICIENCY KINETIC ENERGY INJECTION AMPLIFIER

G. W. Kentwell[1)] and H. Hora[1)2)]

[1)] Department of Theoretical Physics
University of New South Wales
Kensington 2033, Australia

[2)] Iowa Laser Facility and
Department of Physics and Astronomy
University of Iowa
Iowa City, Iowa 52242

Earlier studies of a free electron laser amplifier without a wiggler field [1] have shown the technique to be limited to very long wave lengths. We suggest, as an alternate approach, injection of solid particles (clusters or pellets) into the laser pulse, where the kinetic energy of the solid particle is transferred into optical laser energy [2].

A free electron laser without wiggler field was derived from the inversion of the experiments [3] where a neodymium glass laser beam was focused to intensities of 10^{15} W/cm^2 in a low density gas of about 10^{-4} Torr pressure. The laser ionizes the gas according to the tunnel type Keldysh-process [4] and the electrons are accelerated radially by the nonlinear force [5] which is a generalization of the ponderomotive force [6]

$$f_{NL} = (d/dr)\ (E^2 + H^2)/8\pi \tag{1}$$

where E and H are the electric and magnetic field of the laser radiation and r is the radial coordinate for the focus. The theory is complicated because the laser field of the beam of finite diameter has a longitudinal component [6]. For low laser intensity, the multiphoton ionization of the gas was analyzed with a van der Wiel analyser [7].

The radial acceleration converts half of the oscillation energy of the electrons into kinetic energy of the electrons. This was about 100 eV in the first measurement [3] and was 1000 eV [4] when using ten times higher laser intensity at maximum. The ejection process for electrons was due to nonlinear collisionless absorption because the translative kinetic energy of the electron had to be taken from the optical energy of the laser. The idea was to invert this process by injecting electrons into the laser beam, to slow them down while changing their energy into oscillation energy and to take this out as optical energy of the laser beam. The electron is finally at rest and has under these optimized conditions converted its translative kinetic energy by 100% into optical energy.

0094-243X/86/1460153-3$3.00

The similar mechanism, as described by Kibble [8] for time independent cw laser beams, had to be generalized. It was necessary to combine this with the time dependent switching-on and switching-off process of the laser beams [9]. This results in translative motion of the electrons with the laser beam [9] which has been evaluated. It was found that for the lateral injection, a certain axial component of the motion of the injected electrons is necessary. This component corresponds to the tranfer of optical momentum to the amplified pulse [10] and also confirms then the model of the switching process [9] and of the laser amplifier [1].

The amplification of this lateral injection free electron laser (LIFEL) is limited by the highest electron beam current densities available today of Megaamperes/cm^2. This results in amplifications of only 10^{-3} at 10.6 μm, and is of interest, e.g. for wave lengths around 0.4 mm if then laser beams in the GW range are available. A further problem is the high electric field of the electrons after slowing them down within the area of the laser pulse.

All these problems of limited density and space charge can be overcome if instead of working as a free electron laser, the scheme of injection is changed to using neutral solid particles like clusters or small pellets for the injection. In this case however, the velocity of the injected particles has to be in the range of about 10^7 cm/sec to fit with the maximum laser intensity corresponding to an oscillation energy of the electrons of at least few electron volts. This energy is required in order to highly ionize the injected pellet. The plasmatized pellet is then slowed down to come to rest at the center of the laser pulse. If all paramters are fitted correctly, as they were in the former case of the LIFEL, the pellet will yield a much higher density without space charge problems, and will have the much higher inertia of the ions, as long as the coupling accoring to the Debye lengths is fulfilled. This is known from the recent investigations of electric fields in laser produced plasmas [11].

Under these optimized conditions, the efficiency of the laser amplifier is very high, e.g., 80% or more, since the only losses are the ionization of the solid material and losses by bremsstrahlung of the generated plasma. The amplification per interaction, A, has a similar dependence on the square of the wave length as known from the wiggler FEL and from LIFEL [1][10], but it is applicable to much shorter wave lengths. The amplification is independent of the ion mass for pellet densities of the solid hydrogen and is given by

$$A = (3.64/E_{ph})^2 \qquad (2)$$

where E_{ph} is the photon energy of the laser pulse in electron volts. This high amplification in the visible and UV range makes this scheme most interesting for building a highly efficient laser for laser fusion and for other high power and high energy applications.

The cluster injection free electron laser can be understood as an inversion of the well known process of plasma ejection from the axis of a laser beam due to nonlinear (ponderomotive) force at self-focusing [12]. The acceleration of the plasma in the radial direction is a nonlinear absorption process converting optical energy into kinetic energy of plasma motion. Reversing this process, operation in a pulsed mode, and optimizing of injection velocity, pellet (cluster) size, timing, and pulselength will result in amplification. This mechanism of inversion by self focusing gives evidence about the correct momentum transfer and it shows how the wave front of the laser pulse can be manipulated by appropriate synchronization of the pellets from different radial directions. This scheme works at wave lengths as short as 100 Angstroms or less and might have applications for x-ray lasers. The manipulation of the phase fronts is also important for bending or focussing of beams and also for correcting and improving the quality of a laser pulse during the amplification process.

References

[1] H. Hora, Conference on Energy Storage, Compression and Switching, Venice, December 1978, V. Nardi *et al*., eds. (Plenum, New York, 1982), p. 131.
[2] H. Hora, J.-S. Wang, P. C. Clark and R. L. Stening, Laser and Particle Beams **4**, 83 (1986).
[3] B. W. Boreham and H. Hora, Phys. Rev. Lett. **42**, 776 (1979).
[4] K. G. H. Baldwin and B. W. Boreham, J. Appl. Phys. **52**, 2627 (1981); B. W. Boreham and B. Luther-Davies, J. Appl. Phys. **50**, 2533 (1979).
[5] H. Hora, Phys. Fluids **12**, 181 (1969).
[6] H. Hora, *Physics of Laser Driven Plasmas* (John Wiley, New York, 1981).
[7] P. Kruit, *et al*., Phys. Rev. A **28**, 248 (1983); M. J. van der Wiel (private communication).
[8] T. W. B. Kibble, Phys. Rev. Lett. **16**, 1054 (1966).
[9] R. Klima and V. P. Petrzilka, Cz. J. Phys. **22B**, 869 (1972); R. Dragila and H. Hora, Phys. Fluids **25**, 1057 (1982), G. W. Kentwell and D. A. Jones, Physics Report (submitted); H. Hora, Phys. Fluids **28**, 3473 (1985).
[10] H. Hora and G. Viera, *Laser Interaction and Related Plasma Phenomena*, H. Hora and G. H. Miley, eds. (Plenum, New York, 1984), Vol. 6, p. 373.
[11] H. Hora, P. Lalousis and S. Eliezer, Phys. Rev. Lett. **53**, 1650 (1984).
[12] H. Hora, Z. Phys. **226**, 159 (1969); J. S. Bakos, *et al*., J. Appl. Phys. **52**, 627 (1981); S. P. Kuo, *et al*., Phys. Fluids **26**, 2529 (1983); J. Limpouch, *et al*., Kvantovaja Elektronika **11**, 1416 (1984); M. Luccesi, *et al*., Nuovo Cim. **B84**, 111 (1984); I. Ursu, *et al*., Physica B&C **123**, 379 (1984).

HOT-CELL-FREE 4.3-μm CO_2 LASER

A.S.Solodukhin, B.I.Stepanov, S.A.Trushin
Institute of Physics, Minsk 220602, USSR

ABSTRACT

A simple sealed off hot-cell-free CO_2 laser with longitudinal dc discharge oscillating in the region of 4.3-μm (10^01-00^01 band) is described.

INTRODUCTION

The development of laser sources in the 4-5 μm region is of interest for applications such as atmospheric monitoring, laser photochemistry, and gas media diagnostics.

The possibility of 4.3-μm lasing in CO_2 (10^01-10^00 band) under the conditions of combined electric and optical excitation has been shown by B.I.Stepanov et al.[1] Such oscillation was obtained later by T.A.Znotins et al.[2]. An intracavity cell with heated CO_2 was used in[2] to separate the sequence band line. Here we describe a simple hot-cell-free technique for 4.3-μm oscillation in a longitudinal dc discharge low-pressure CO_2 laser.

EXPERIMENTAL SETUP AND RESULTS

In our experiments 10.6-um (00^02-10^01 band) and 4.3-um (10^01-10^00 band) radiation pulses were produced in the same active medium. The sealed off three-electrode discharge tube we used had an active length of 1.5 m and a 1.5 cm bore and was sealed with ZnSe Brewster windows. The optical scheme of 4.3-um CO_2 laser is shown in Fig.1. The 10.6-um laser cavity was formed by a copper mirror, a grating, and a rotating Q-switch mirror. The possibility of reliable separation of the 00^02-10^01 sequence band transitions which are 0.3 to 1 cm^{-1} from the 00^01-10^00 regular band transitions and have smaller gain values was provided in our case by the use of mixtures with reduced content of carbon dioxide where sufficiently high values of vibrational temperature of the CO_2 molecule asymmetric mode are realized.

The grating was set in the non-autocollimation regime to increase the wavelength discrimination and served for both tuning over individual lines and separating 10.6- and 4.3-μm radiation. The 4.3-μm cavity was formed by the copper mirror, the grating, and the ZnSe mirror.

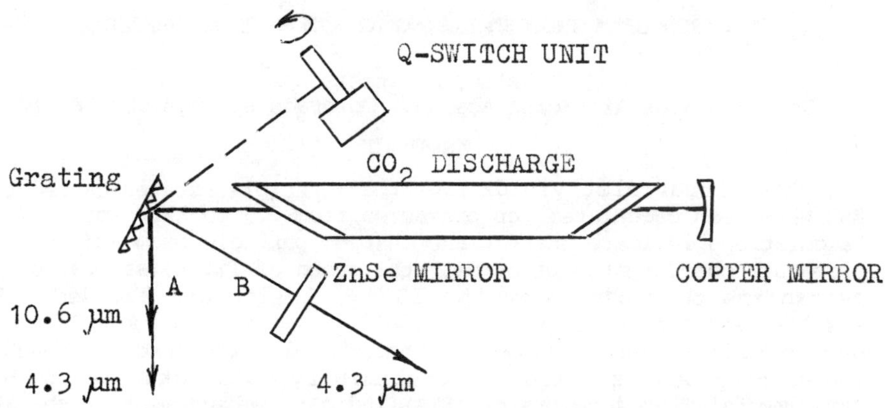

Fig.1. Schematic of the experimental setup

As shown in Fig.1. 4.3-μm radiation is output in two directions: through the grating zeroth order (A) and the partially transmitting mirror (B).

The results of investigating the output parameters have shown that for effective operation of 4.3 μm CO_2 laser active mixtures containing 4-5% of carbon dioxide with 20% of nitrogen must be used[3]. In the conditions close to optimum, namely, with $Xe:CO_2:N_2:He$ = 3:4:20:73 mixture at a pressure of 16 Torr, current of 30 mA, repetition rate of 30 Hz in the above system, the pulse peak power on the line P(26) of the 10^01-10^00 band was 60 W with a 300 ns FWHM.

The action of a strong radiation pulse in the 00^02-10^01 band on the electric-discharge excited active medium creates, under specific conditions, total inversion in the 10^01-10^00 transition. Consequently, when oscillating on an individual line of the 00^02-10^01 band, oscillation on many lines of the P- and R-branches in the 10^01-10^00 band can be obtained. For example, in our scheme we obtained lasing on 17 lines of the P-branch in the 10^01-10^00 band with J from 8 to 40 when oscillating in the 00^02-10^01 band was only on the P(25) line. In this case tuning over the lines of the 4.3 μm band was performed by rotating the ZnSe mirror.

When using a saturable absorber cell inside the cavity instead of a mechanical Q-switch unit a pulse repetition rate up to 20 kHz with an average power 60 mW was obtained.

1. B.I.Stepanov et al., Sov.Phys.Dokl.23, 910 (1978).
2. T.A.Znotins et al., Appl.Phys.Lett. 39, 199 (1981).
3. A.S.Solodukhin et al., Opt.Quant.Electr. 18, to be published (1986).

ELECTRON BEAM INDUCED LASING AT ATMOSPHERIC PRESSURES

Walter L. Atchison
United States Air Force Academy, Colorado Springs, Co. 80840

ABSTRACT

The ultraviolet yields from the N_2 triplet states caused by a REB have been calculated for pressures from 0.5 to 1000 torr. These calculations indicate that as the charged particle beam and UV pulse propagate concurrently at or near the speed of light, several of the mechanisms that shut down the 3371 Å lasing are avoided. The results also predict significant yields at pressures above 100 torr due to collisional activation from the electron cascade caused by the primary beam electrons. Preliminary data taken on the LLNL Experimental Test Accelerator (ETA) indicate enhancement of the 3371 Å line by greater than 10^3 above that expected from spontaneous emission alone at a pressure of 1 atmosphere.

INTRODUCTION

While studying the emissions caused by high energy electron beams in N_2/O_2 mixtures, intense pulses of UV light are usually observed[1]. The wavelengths of these emissions indicate that they are from the 3371 Å band of N_2 and the 3914 Å band of N_2^+. The intensities of these emissions are greater than can be explained by simple spontaneous emission. It is the intent of this study to analyze and predict these emissions as a collateral effect of the propagation of charged particle beams. The basic thrust of this effort is therefore to analyze the axial non-equilibrium UV emission caused by the interaction of a relativistic electron beam (REB) with air.

This effort includes both a computational and experimental study of the phenomenon. There were two primary tasks: 1) Develop an axially symmetric one dimensional model of both the charged particle beam and the co-propagating UV pulse, 2) Attempt to measure both the transverse and forward directed axial intensity of UV emissions caused by the Lawrence Livermore National Laboratory's (LLNL) Experimental Test Accelerator(ETA).

COMPUTATIONAL MODEL

In order to calculate the forward directed axial UV intensity resulting from the beam-gas interaction, it is necessary to model three basic processes. The first process to model is the environment produced by the incident electron beam. The quantities of interest are the number density of the cascade electrons, the number density of the plasma electrons, and the induced electric field caused by the dynamics of the electron beam. The second process to model is the collisional activation of N_2 excited state densities. Finally, it is necessary to calculate the initialization and growth of the UV intensity as the entire system propagates forward in space. The majority of the calculations done used a fourth order Runge-Kutta approach. To insure stability, several of the results were compared to solutions found using Gear's technique[2]. Several analytic solutions were also used as a test to check convergence.

MEASUREMENTS

The calculations were compared with measurements made using the LLNL Experimental Test Accelerator. The ETA produces a 10 kiloamp electron beam with an average energy of 5 Mev and an emittance of 120 mrad-cm. While this beam is not stable above 100 torr due to the hose instability, it is possible to get 1 to 2 meters of straight propagation by expanding the radius and reducing the current to decrease the hose growth rate. It is also possible to monitor both the beam position and the net current along the line of propagation.

The ultraviolet intensities were measured at three locations using PM tubes and interference filters. Two PM tubes were used to view the emissions in the transverse direction. They were located 0.20 and 1.20 meters downstream from where the beam entered the test chamber. The third PM tube was located 8.55 meters down stream looking axially upstream.

Table I. Power Into Axial Viewing PMT on ETA

P(torr)	Peak Spont. Emission(W/cm)	Measured Axial Power(W)	Calculated Axial Power(W)	Ratio
2	810	0.810	1.0	1.2
5	2250	0.880	3.2	3.6
10	3040	0.860	43.5	50.6
50	1010	0.130	140.0	1077
100	517	0.065	65.0	1000
200	225	0.029	31.0	1069

CONCLUSIONS

In general the results of the calculations indicate a strong amplification of the UV intensity in the forward direction of the REB. For a 10 kiloamp beam at 5 Mev in a 1 atm. pressure the linear saturated gain is predicted to be approximately 1.8 kw/cm^2 per cm of REB/UV beam propagation. This number scales linearly as the energy deposited by the electron beam, and linearly with pressure above 100 torr. The experimental measurements were insufficient to verify the pressure dependence of pulse shape and linear gain, but do show significant enhancement of the forward directed ultraviolet intensity over that of the spontaneous emission observed tranverse to the direction of beam propagation.

[1] T.J. Fessenden, "Optical Emissions from an Ionized Channel Produced by an Electron Beam", UCID-17647, Lawerence Livermore National Laboratory, Livermore, Ca., 1977.

[2] A.C. Hindmarsch, "GEAR.. Ordinary Differential Equation System Solver", UCID-30001 Rev. 3, Lawrence Livermore National Laboratory, Livermore, Ca., Dec 1974.

Electronic Assignments of the Violet Bands of Sodium

G. Pichler
Iowa Laser Facility and Department of Physics, University of Iowa, Iowa City, Iowa 52242 and Institute of Physics, University of Zagreb, P. O. Box 304, 41001 Zagreb, Yugoslavia

J. T. Bahns, K. M. Sando and W. C. Stwalley
Iowa Laser Facility and Department of Chemistry, University of Iowa, Iowa City, Iowa 52242

W. Müller
Fachbereich Chemie, Universitaet Kaiserslautern, 6750 Kaiserslautern, West Germany

D. D. Konowalow
Department of Chemistry, State University of New York, Binghamton, New York 13901

L. Li and R. W. Field
Department of Chemistry, Massachusetts Institute of Technology, Cambridge, Massachusetts 02139

Abstract

The puzzling violet bands of sodium (425-460 nm), known since 1932[1], have been shown conclusively to arise from the superposition of two distinct continuum emission bands - one singlet ($2^1\Sigma_u^+$-$X^1\Sigma_g^+$) and one triplet (primarily $2^3\Pi_g$-$1^3\Sigma_u^+$). Each continuum emission system resulting from bound-free transitions shows complex interference structure (arising from multiple branches of the corresponding Mulliken difference potential (MDP))[2].

Introduction

Since the first observation there have been over thirty publications dealing with the interpretation of these bands (to be reviewed by two of us (GP and WCS)). However, no completely satisfactory electronic assignment (or mechanism) has been presented.

Results

Figure 1 shows the singlet oscillatory continuum fluorescence spectrum which results from populating a <u>single</u> upper level (v' = 41, J' = 60) of the $2^1\Sigma_u^+$ state[3] (from the v" = 3, J" = 59 level of the ground $X^1\Sigma_g^+$ state[4]) using a single longitudinal mode of the 351.1 nm line from an argon ion laser.

The triplet oscillatory continuum fluorescence spectrum (Figure 2) results from populating a <u>single</u> level (in this case v' = 44, J' = 12) of the $2^3\Pi_g$ state[5] via optical-optical double resonance, with resulting fluorescence to the $1^3\Sigma_u^+$ state[6].

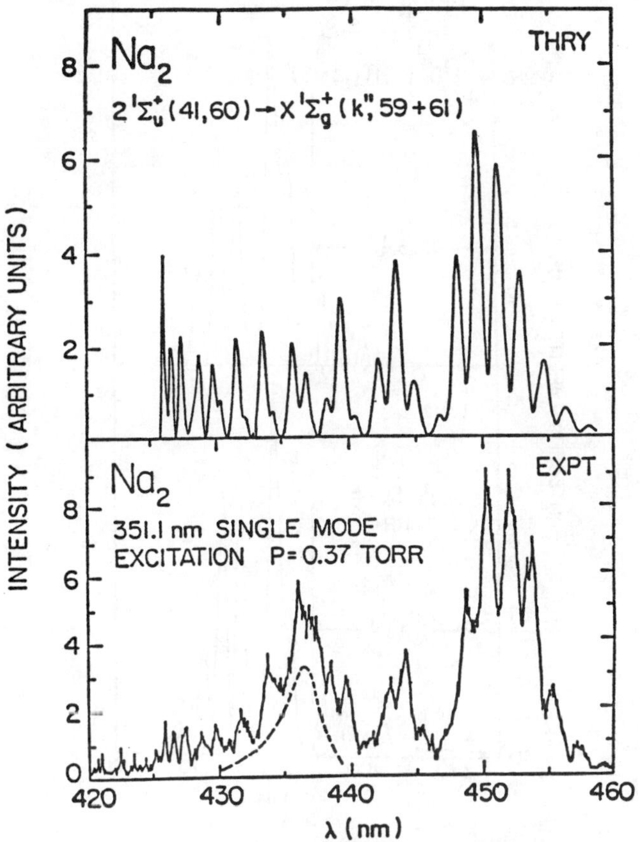

Figure 1. The bound-free emission spectrum for the Na_2 singlet transitions $2^1\Sigma_u^+$ (v' = 41, J' = 60) - $X^1\Sigma_g^+$ (v" or K", J" = 59 and 61). The lower figure is experimental (total P = 0.37 torr) while the upper figure is the quantum mechanical calculation (see text). All regions of strong emission are shown.

Conclusions

The violet bands of sodium have been unambiguously assigned to transitions between electronic states of Na_2 through direct comparison of theoretical and experimental evidence[7].

Acknowledgements

Helpful communications with P. D. Kleiber, S. P. Heneghan, A. M. Lyyra, W. T. Luh, S. Rice, J. Vedder and J. Tellinghuisen are gratefully acknowledged, as is support from AFOSR and NSF.

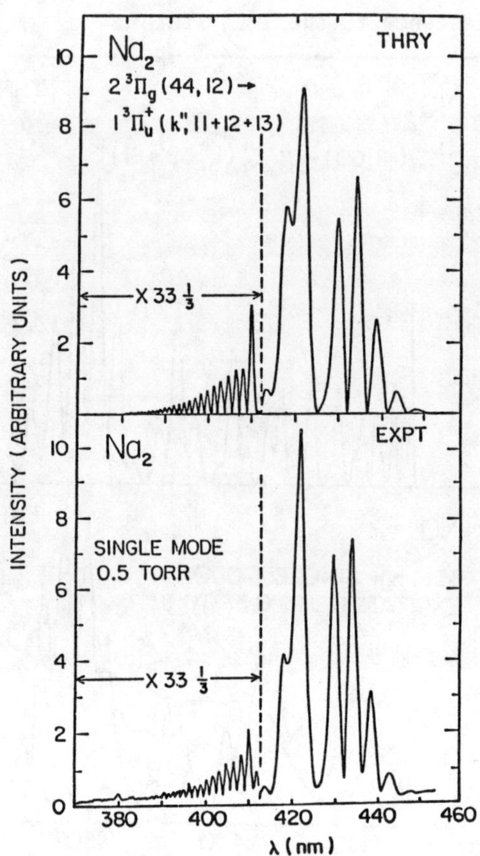

Figure 2. The bound-free emission spectrum for the Na_2 triplet transitions $2^3\Pi_g$ (v' = 44, J' = 12) - $1^3\Sigma_u^+$ (K", J" = 11, 12 and 13). The lower figure is the experimental (total P = 0.55 torr) while the upper figure is the quantum mechanical calculation (see text). All regions of strong emission are shown.

References

1. H. Bartels, Z. Physik 73, 203 (1932).
2. R. S. Mulliken, J. Chem. Phys. 55, 309 (1971).
3. J. Verges, C. Effantin, J. d'Incan, D. L. Cooper and R. F. Barrow, Phys. Rev. Lett. 53, 46 (1984).
4. K. K. Verma, J. T. Bahns, A. R. Rajaei-Rizi, W. C. Stwalley and W. T. Zemke, J. Chem. Phys. 78, 3599 (1983).
5. L. Li and R. W. Field, private communication.
6. D. D. Konowalow, M. E. Rosenkrantz and M. L. Olson, J. Chem. Phys. 72,2612 (1980).
7. G. Pichler, J. T. Bahns, K. M. Sando, W. C. Stwalley, W. Muller, D. D. Konowalow, L. Li and R. W. Field, submitted to Phys. Rev.A.

LINEWIDTH OF ULTRAVIOLET RADIATION GENERATED IN NONLINEAR PROCESSES

C.Y. Robert Wu, Sui Xu and Tai-Sone Yih
University of Southern California, Los Angeles, Ca. 90089-1341

ABSTRACT

Extremely broad linewidths of ultraviolet coherent radiation generated in a four-wave mixing process in sodium vapor have been observed. When the wavelength of the input visible laser increases it is found that (1) the linewidth of the UV output generated in the 333nm region correspondingly increases and (2) the integrated intensity of the UV output initially increases and then slowly decreases.

INTRODUCTION

The linewidth of the tunable infrared generation, e.g., SERS and SHRS, has been known to be significantly broader than that of the input laser if an intermediate state is resonantly involved in the nonlinear processes.[1] We have carried out systematic measurements of linewidths of the ultraviolet (UV) coherent radiation generated in four-wave mixing (4-WM) processes in sodium vapor. In the present work we report an interesting result of linewidths of the UV radiation generated at 333nm.

RESULTS AND DISCUSSION

The experimental setup used in the present work has been previously reported.[2] The linewidth of the dye laser is $0.3cm^{-1}$. The spectral bandwidth of the monochromator used in the analysis of the UV output was 0.04nm ($3.6cm^{-1}$ at 333nm).

By tuning the dye laser wavelength λ_ℓ near the 3s-4d two-photon resonance transition, a highly collimated output beam from the sodium vapor was observed in the forward direction along the dye laser beam. Typical UV output spectra in the 333nm region obtained using various dye laser wavelengths are shown in Fig. 1. It is quite clear that the UV output is tunable as its position shifts toward long wavelength side when the dye laser wavelength increases. The measured positions of the tunable UV output agree very well with those calculated according to a 4-WM scheme of $\omega_{UV}=2\omega_\ell-\omega_{4p-4s}$ (λ=2.20837 for $4P_{3/2}-4S_{1/2}$ and 2.20564μm for $4P_{1/2}-4S_{1/2}$). Our instrumental bandwidth is quite sufficient to resolve these two components which can be clearly seen in the spectrum taken by using λ_ℓ=578.7nm. However, when the laser wavelength was tuned away from the 3s-4d two-photon resonance transition the UV output occurred in a single well defined spectral line. We found that it is contributed from the fine transition of $4P_{3/2}-4S_{1/2}$. Similar effects in the generation of SERS have also been observed.[1]

As shown in Fig. 1 the broadening of the linewidth of the generated UV output obviously increases as λ_ℓ increases. In Fig. 2a, we have plotted the linewidth (FWHM) as a function of detuning parameter Δ, which is defined as $\Delta(cm^{-1}) = \omega_{4d-3s} - 2\omega_{\ell}$, where ω_{4d-3s} is the energy separation between the atomic sodium 4d and 3s levels in unit of cm^{-1}. As one can see the linewidth appears to broaden linearly with Δ. The most striking effect is that the FWHM reaches $26cm^{-1}$ at $\Delta = 52.6cm^{-1}$. To qualitatively explain this extraordinary broadening a theoretical calculation is currently in progress.

In Fig. 2b, we have plotted the integrated intensity of the tunable UV output as a function of Δ. The output intensity increases to a maximum at $\Delta \sim 15cm^{-1}$ and then slowly decreases to the limit of the present work. The large shift in the maximum can be explained as due to Stark shift in which the DC electric fields are generated by the charge produced through multiphoton ionization of the sodium vapor. As the 4d level is approached, the degree of ionization increases, so as the DC electric fields.[3] Thus, the 4d level is removed from the two-photon resonance.

ACKNOWLEDGMENT - This work was partially support by the USC Faculty Research and Innovation Fund.

1. D.C. Hanna, M.A. Yuratich and D. Cotter, Nonlinear Optics of Free Atoms and Molecules (Springer-Verlag, Berlin, New York, 1979), p.196 and 231.
2. J.K. Chen, C.Y.R. Wu, C.C. Kim and D.L. Judge, Appl. Phys. B 33, 155(1984).
3. B. Dai and P. Lambropoulous, to be published.

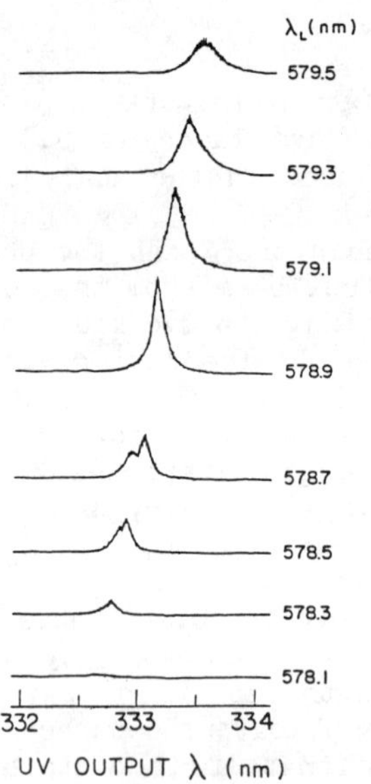

Fig. 1. The coherent UV output spectrum in the 333nm region as a function of input dye laser wavelength λ_{ℓ}.

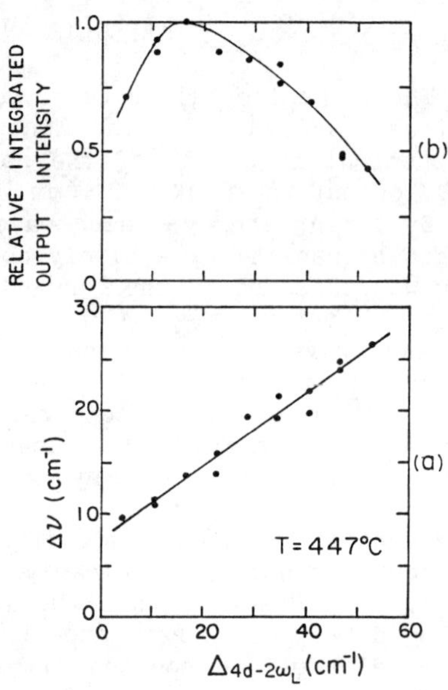

Fig. 2. The linewidth (FWHM) and the integrated intensity of the UV coherent radiation as a function of the detuning parameter Δ.

NEW HEAT PIPE OVEN DEVICES FOR BROAD BAND EXCITATION LASER STUDIES

Mark A. DeFaccio, Steven J. Davis, David I Rosen, William C. Stwalley* and David O. Ham
Physical Sciences Inc., P.O. Box 3100, Andover, MA 01810
*Iowa Laser Facility, University of Iowa

We have investigated heat pipe oven type devices which allow excitation or observation over the full length of the vapor zone. Such a device is necessary for a flashlamp or solar pumped metal vapor laser. These broadband excitation sources can excite a vapor effectively only if a clean window can be maintained close to the vapor. Metal vapors, especially alkali vapors, coat and discolor windows making them opaque. We have successfully operated two devices with sodium at temperatures up to 450°C for tens of hours with no significant deposit on the window that was within 2 to 3 cm of the sodium vapor. Additionally, the sodium vapor was sufficiently uniform and loss free in our most recently studied device to obtain lasing as readily as in a conventional heat pipe device.

So far we have tested two heat trough type designs with large side windows. We have operated a more conventional crossed heat pipe oven (borrowed from the Iowa Laser Facility) for temperature calibration and comparison as a standard. Diagrams of the three devices are shown in Figure 1; the trough shown in Figure 1(b) was hollowed from single piece of stainless steel and the device shown in Figure 1(c) was fabricated by welding a ss hopper onto a ss pipe. Each of these devices is very different from presently used heat pipe devices.[1-4] We have operated each device with several different wick configurations varying the mesh size (usually 70) of the ss screen , the number of layers, the height of the wick on the side, and whether the top was open or covered. We have also used either He, Ar or N_2 as a buffer gas at pressures from the vapor pressure of sodium to tens of Torr. In all cases the large window stayed clean up to fairly high temperatures and with an open topped, multi-layered wick

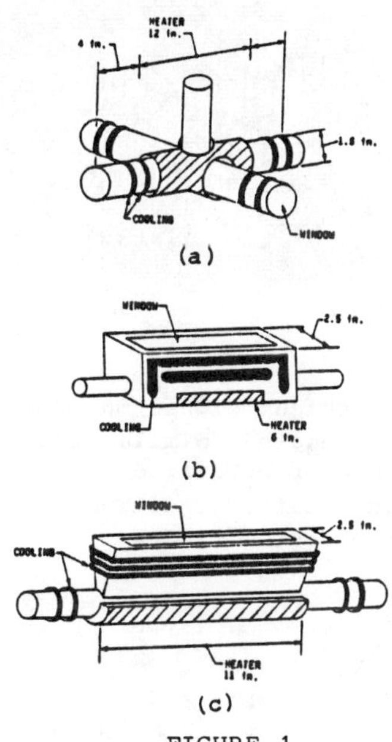

FIGURE 1

The three heat pipe oven devices studied; (a) a crossed heat pipe oven (b) a rectangular heat trough with open top, and (c) a pipe with hopper to support the window. Wide black areas indicate cooling coils, cross-hatched areas indicate heaters.

0094-243X/86/1460165-2$3.00

we have operated for hours at high buffer gas pressure with temperatures above 500°C maintaining a clean window.

We have measured temperature distributions in each of these devices by measuring intensity ratios of flourescence lines from Na_2 excited by a HeNe laser. The HeNe laser is absorbed by V" = 2, 4, 6 and 8 of $Na_2(X)$[5]. Using this technique calibrated in the crossed heat pipe oven, we measured the temperature in the device in Figure 1(c) to be constant within about 10°C over a length greater than 10 cm. This temperature distribution was insensitive to pressure.

The uniformity of the sodium vapor in the Figure 1(c) device was good enough to achieve Na_2 A → X laser operation pumped by a dye laser over the full tuning range of Rhodamine B dye. Laser performance was essentially the same as in the crossed heat pipe oven. A portion of the result of this laser result in the Figure 1(c) device is shown in Figure 2.

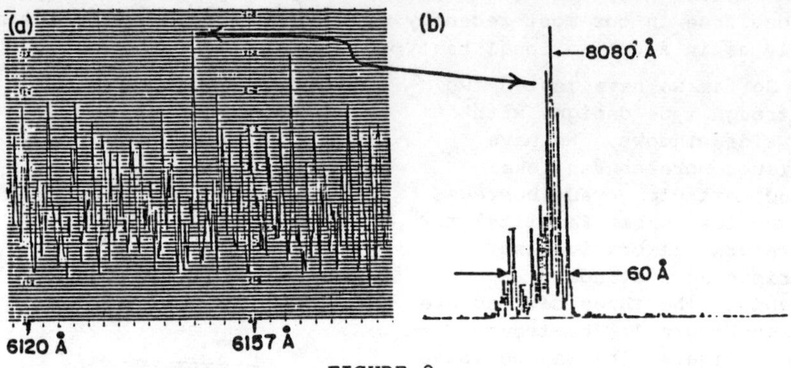

FIGURE 2

(a) OPL output signal on photodiode as a function of pump laser wavelength. Similar signals were observed over entire tuning range of Rhodamine B dye from 6000 A to 6340 A.

(b) Laser output spectrum pumped by the dye laser at the indicated wavelength.

REFERENCES

1) C.R. Vidal and J. Cooper, J. Appl. Phys. 40, 3370 (1969)

2) H. Scheingraber and C.R. Vidal, Rev. Sci. Instrum. 52, 1010 (1981)

3) a) R.W. Boyd, J.G. Dodd, J. Krasinski, and C.R. Stroud, Jr., Opt. Lett. 5, 117 (1980); b) R. W. Boyd and D. J. Harter, Appl. Opt. 19, 2660 (1980)

4) M.A. Cappelli, P.G. Cardinal, H. Herchen, and R.M. Measures, Rev. Sci. Instrum. 56, 2030 (1985)

5) a) K.K. Verma, T.H. Vu, and W.C. Stwalley, J. Mol. Spectrosc. 85, 131 (1981)

b) W.T. Zemke, K.K. Verma, T. Vu, and W.C. Stwalley, J. Mol. Spectrosc. 85, 150 (1981)

Acknowledgement: This work has been supported by AFOSR.

HIGH ENERGY LASER INTERFACES WITH NUCLEAR FUEL ENERGY SOURCES

M. A. Prelas and J. F. Kunze
Dept. Nuclear Engr., University of Missouri, Columbia, MO 65211

F. P. Boody
Princeton Plasma Physics Laboratory, Princeton, NJ 08544

ABSTRACT

Methods of interfacing nuclear fuels with high energy lasers can be categorized by either an indirect interface or a direct interface. An example of an indirect interface is using the nuclear fuel as a low grade heat source to drive a cycle which produces electricity and the electricity in turn drives a laser. There of course are many many proposed cycles which can produce electricity--MHD, thermoelectric, thermionic, etc. A direct interface, conversely attempts to directly use the products of nuclear reactions to drive a laser medium. Lasers of this type have been studied and there are some examples of such systems commonly referred to as nuclear-pumped lasers[1]. Work discussed in this paper involves a new method of direct interfacing using a nuclear driven flashlamp and an aerosol core reactor.

INTRODUCTION

The concept of a nuclear-pumped laser dates back nearly to the discovery of the first laser[1]. A nuclear-pumped laser utilizes the energetic products of nuclear reactions to "pump" the laser medium. Fifteen lasers which capitalize on this technique of "pumping" have been discovered (see Table 1). Three of the nuclear-pumped lasers outlined in Table 1 have good efficiencies (>1%). One of the reasons for the low number of nuclear-pumped lasers discovered thus far is due to the low power density inherent in using controlled nuclear reactions (<10 kW/cm^3). (It should be noted that this is a limitation on pulsing methods for controlled nuclear reactions: the energy density for controlled nuclear reactions is reasonably high [< 100 J/cm^3].) Consequently, many efficient, high power lasers which characteristically require very high power densities (>100 kW/cm^3) can not be driven with this pumping method. A way of interfacing the nuclear fuel and the laser which can achieve higher power densities than conventional nuclear-pumping techniques[1] and can take advantage of natural atomic resonances was introduced by the authors[1,3]. This method incorporates a nuclear driven flashlamp, a non-imaging optical concentrator and an aerosol core reactor.

TECHNICAL DISCUSSION

The use of an aerosol medium and fluorescer coupled to a laser has been discussed in Reference 3. Current research indicates that such a design can easily be assembled to form a critical core and will have a relatively small mass (1 Tonne/MJ laser energy)[4]. The major improvement being examined is the use of reflective materials on the aerosol fuel[4].

An essential aspect in the design of a reactor/laser system using an aerosol fuel is to form a crtical assembly of fuel elements, each containing a laser (See References 3 & 4). It is therefore important to combine the small beams from the laser array to form a single diffraction limited beam for power transmission over long distances. In order to generate a single diffraction limited beam far field, it is necessary that a method of phase locking the array of lasers (as many as 1000) be available. This task has been done on a small scale with CO_2 lasers and appears to be technically feasible for a large array[5].

Table 1. Nuclear-Pumped Lasers (Ref. 1,2)

Laser	λ(nm)	So	P_{th}	η%	Laser	λ(nm)	So	P_{th}	η(%)
Xe_2*	170	γ	NA	NA	HF	2500	γ	NA	NA
CO	5100	ff	1860	1	Xe	3500	ff	2	.002
N	860 & 939	α	4.5	10^{-5}	Hg^+	615	α	45	.0001
Ar	1790 & 1270	p	980	.001	Xe	2026, 3508 & 3652	p	160	.0016
Kr	2520 & 2190	p	520	10^{-5}	Ar	2397 1190 & 1150	ff	370	.02
Xe	2630 & 2480	ff	37	1.33	Cd	534 & 538	p	3.11	.008
Cl	1586	p	110	.0002	C	1450	α	.04	.006
CO_2	10600	α	27	1					

So-Source (γ-gamma thermonuclear explosion, ff-fission fragments from controlled nuclear reactions, α-alpha from $B^{10}(n,\alpha)Li$, p-proton from $He^3(n,p)T$), P_{th}-threshold power density (W/cm^3), η-laser energy out/energy deposited (%).

CONCLUSION

New methods of interfacing nuclear fuels to lasers hold promise for light, compact multi-megajoule lasers.

REFERENCES

1. Prelas M. A. and Loyalka S. K. (1981), Progress in Nuclear Energy, **8**, 35-52.
2. Jalufka N. W., "Direct Nuclear-Pumped Lasers", NASA Technical Paper 2091, Langley Research Center, Hampton, VA.
3 Prelas M. A., Boody F. P. and Zediker M. S. (1985), Space Nuclear Power Systems 1985, M. El-Genk and M . Hoover eds., Orbit Book Co., Malabar, FL.
4. Prelas M. A., Kunze J. F. and Boody F. P. (1986), Lasers and Related Plasma Phenomena Vol. 7, Plenum Press, NY.
5. Hayes C. L. and Davis W. C. (1979), Appl. Opt., **18**(24).

THE APPLICATION OF THE KINETICS OF Si AND SiF IN SPECIFIC ELECTRONIC STATES TO AN INVESTIGATION OF THE PUMPING OF A POPULATION INVERSION IN THE MOLECULE, SiF.

D.R.HARDING AND D.HUSAIN
CAMBRIDGE UNIVERSITY, ENGLAND CB2 1EP

ABSTRACT

The feasibility of constructing a chemical laser operating in the pulsed mode on the transition SiF($a^4\Sigma^- \rightarrow A^2\Sigma^+$)+h$\nu$ pumped by the reaction of Si+F_2 is investigated on the basis of absolute kinetic data for Si and SiF and determined by (a) both time-resolved atomic and molecular absorption spectroscopy on the three low lying states of Si and the ground state of SiF and (b) chemiluminescence of SiF(A-X), following pulsed irradiation of $SiCl_4$+RF+He mixtures. The kinetic behaviour of SiF($X^2\Pi_r$) and SiF($A^2\Sigma^+$) is quantitatively consistent with the rate data for atomic silicon and intra-cavity stimulated emission signals detected in these investigations are accounted for in terms of the generation of a population inversion between SiF($a^4\Sigma^-$) and SiF($A^2\Sigma^+$) from the reaction of Si(^1D, ^1S)+F_2. Calculations of wavelengths of SiF(a-X) and SiF(a-A) emission and lifetimes are presented.

INTRODUCTION

A pulsed chemically pumped i.r. laser operating on an electronic transition, SiF($a^4\Sigma^- \rightarrow A^2\Sigma^+$) has earlier been proposed and a series of related experiments were undertaken to estimate its feasibility and conditions for operation. In this study the pumping mechanism has been investigated further and extended. This work has shown a photolytic pulsed discharge through a $SiCl_4/F_2$/He mixture to be a viable technique for generating 1) an appreciable population of the product species, SiF, and 2) within the restrictions of the radiative and non-radiative lifetimes, a population inversion.

REACTION KINETICS OF Si AND SiF IN THE PRESENCE OF DIFFERENT FLUORINATING AGENTS

Atomic silicon was monitored as a function of time by the method of time-resolved resonance absorption spectroscopy. The attenuation of the signal is least squares fitted to the modified Beer-Lambert law where [Si] follows a pseudo first order decay. This principle and technique was extended to examine the experimental collisional properties of SiF+F_2/CF_4/SF_6 and SiF_4 and permitted the detection of the formation and subsequent decay of SiF($X^2\Pi_r$). These rates of reaction were used to determine the optimised concentration ratios for the lasing investigation. Chemiluminescence experiments on the formation and decay of SiF($A^2\Sigma^+$) gave an estimate for the non-radiative lifetime of SiF($A^2\Sigma^+$).

THEORETICAL ANALYSIS OF THE RADIATIVE TRANSITIONS INVOLVING SiF($X^2\Pi_r$,$A^2\Sigma^+$,$a^4\Sigma^-$) AND INVESTIGATION FOR STIMULATED EMISSION

Calculations of the transition wavelengths, relative transition intensities and lifetimes of vib-electronic states were performed. Although the $a^4\Sigma^- \rightarrow A^2\Sigma^+$ transition is disallowed for Hund's case (b) transitions, breakdown from Hund's case (b) to Hund's case (c) by increased vib/rot distribution when forming the SiF($a^4\Sigma^-$) states as well as the weight of the molecule, would permit a transition. Hence it is impossible to calculate the radiative lifetime of the $a^4\Sigma^-$ state though it is reasonable to infer that it would be considerably greater than that for the $A^2\Sigma^+$ state. Within this context a population inversion is a viable possibility providing a sufficient density of the $a^4\Sigma^-$ state can be attained by the pumping mechanism.

CF_3I/He mixtures were used to detect I($5^2P_{1/2} \rightarrow 5^2P_{3/2}$) (1.315$\mu$) stimulated emission to confirm the functionability of the optical resonator and detection equipment. This was then extended to look for SiF emission from $SiCl_4/F_2$/He mixtures at 1.4μ .

Whilst it is believed that stimulated emission could be observed for $SiCl_4/F_2$/He as the profiles took a similar shape to those for the iodine atom laser, the absolute profile of the emission could not be absolutely attained because of the distortion of the detection system due to the electrical discharge arising for the duration of the pulse discharge. Nonetheless a signal similar in shape and a fraction of the intensity of the I lasing signal could be observed in a tuned laser cavity. When the cavity was decoupled or attempts to output couple the signal were made no emission could be observed.

Fig. 1. Intra-cavity emission profiles

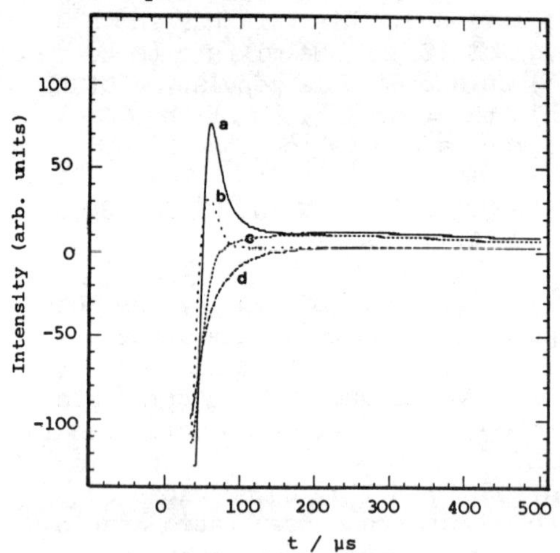

(a) CF_3I/He mixture-I atom lasing pulse
(b) $SiCl_4/F_2$/He mixture
(c) empty reactor (d) decoupled cavity

Table 1. Lifetimes and Transition wavelengths (μ) for SiF(a-A)

		$A^2\Sigma^+$ v'' 0	1	2
$\tau \times 10^{-6}$s		0.18	0.185	0.19
$a^4\Sigma^-$	0	1.424	1.582	1.773
v'	1	1.270	1.394	1.540
	2	1.147	1.248	1.363
	3	1.048	1.131	1.225
	4	0.965	1.035	1.113
	5	0.895	0.955	1.021
	6	0.836	0.887	0.944

INCOHERENT OPTICAL PUMP SOURCES FOR LASERS BASED ON SURFACE DISCHARGES

R. E. Beverly III
R. E. Beverly III and Associates, Columbus, Ohio 43221

ABSTRACT

The inherent low inductance of surface discharges permits large-area optical sources to be constructed which produce short-pulsewidth, intense radiation suitable for pumping or preionizing laser media. The technical efficiencies for pumpbands in the soft-uv through visible spectral regions are generally inferior to those for optimized, short-pulse flashlamps (typically by 20-50%). The specific radiated energy and radiant excitance produced by high-power planar surface discharges, however, can easily exceed the corresponding capabilities for flashlamps.

INTRODUCTION

An investigation was undertaken of the electrical, gasdynamic, and radiative properties of surface discharges intended as intense optical pump sources for lasers. Large-area (~20-300 cm^2), short-pulse (~1 μsec) planar surface discharges were produced across various ceramic and polymer substrates and operated in all of the rare gases, N_2, and air at absolute pressures up to 4 atm.

ELECTRICAL BEHAVIOR

Studies of the discharge electrical parameters and source dimensions show that the surface-discharge resistance R_{sd} and discharge efficiency η_d decrease with increasing charging voltage V_c and energy storage capacitance C_s, while these same parameters increase with increasing discharge gap d and atomic weight for atmospheres consisting of the rare gases. This behavior illustrates the close relation between plasma expansion velocity, which determines the cross-sectional area, and R_{sd}. The plasma expands faster for the light rare gases, the cross-sectional area is larger, and R_{sd} is smaller. The plasma expands slowly in the heavy rare gases with a concomitant increase in R_{sd}. The characteristic operating time (time interval between application of high voltage and onset of appreciable current flow) in Ar is <100 nsec for $V_c \geq 20$ kV and d = 10 cm, consistent with breakdown by the single-stroke leader mechanism. The time required for recovery of gap electrical strength (≈100 μsec) is much longer than the characteristic times for plasma expansion and recombination, and it is believed that continued surface ablation, occurring well after the discharge pulse, produces a flux of low-ionization-potential atoms which is responsible for this delay.

GASDYNAMIC BEHAVIOR

The presence of a conductive backplane insures that surface-discharge plasmas are tightly pressed against the surface of the dielectric substrate. Surface discharges do not have an explosion limit analogous to flashlamps, although a shock wave is launched normal to the surface, detaches from the plasma, and travels through the cold, surrounding gas atmosphere. Measured wave velocities can be correlated with the theory of a Chapman-Jouguet detonation despite the fact that the physical processes occurring in a surface discharge depart significantly from those in an optical detonation such as a laser-supported absorption wave. The shock velocity D was determined to obey the theoretical proportionality $D \propto (G/\rho)^{1/3}$ for high-pressure operation, where G is the discharge power density and ρ is the unperturbed gas density. If the electrical circuit parameters are fixed, then $D \propto \rho^{-0.22}$ and the proportionality is weaker than predicted by the optical detonation model. This behavior can be explained by noting that R_{sd}, η_d, and, hence, G decrease with decreasing ρ. For operation in the rare gases, measured shock velocities decrease with increasing gas atomic weight for conditions of constant ρ.

RADIATIVE PROPERTIES

Spectroscopic studies show that planar surface discharges radiate as Lambertian sources which cannot be characterized as ideal blackbodies [effective brightness temperatures $T_b \sim$ 10,000-20,000 K (visible wavelengths) for specific input energies of ~1-4 J/cm^2]. The plasma is not opaque for $\lambda \lesssim 500$ nm and shows considerable structure due to line emission from neutral and singly-ionized species originating from the gas atmosphere and vaporized substrate material. The presence of line transitions from vaporized and excited substrate species can greatly enhance the radiative output, especially in the uv and vuv spectral regions. Discharges across perovskite ceramic substrates, such as $SrTiO_3$, yielded a 70% improvement in soft-uv output compared with discharges across plastics. The specific radiated energy for the rare gases is almost directly proportional to the atomic weight of the gas, consistent with the hypothesis that the current and optical densities of the plasma are proportional to its cross-sectional area which, in turn, is related to the gasdynamic expansion characteristics of the plasma sheet.

The conditions which lead to formation of a high-conductivity plasma sheet on the surface of a dielectric are complex and interdependent, and no comprehensive model has been devised. Experimentally, the degree of spatial homogeneity depends upon the peak discharge voltage, the initial rate of voltage rise, the specific capacitance of the substrate, the gas atmosphere and pressure, and the ablation mode of the substrate. Uniform, coalesced discharges are quite easy to form in the heavy rare gases even at superatmospheric pressure and low voltage. Discharges in helium and attaching gases such as air are usually multichannel, and fast circuitry operating at a higher voltage is required to maximize the extent of surface coverage by plasma.

FAST PLASMA MIXING - A NEW EXCITATION METHOD FOR CW GAS LASERS

G. Schaefer
Texas Tech University, Lubbock, TX 79409-4439

ABSTRACT

Fast Plasma Mixing is proposed as a method to generate population inversions for continuous lasers. Selective excitation is accomplished by preparing collision partners for excitation in two or more separate beams. Excitation in a well defined reaction volume and a fast flow will allow the fast removal of decay products from the active volume. This pumping method will allow the operation of new CW laser systems not possible with conventional discharges and CW operation of laser systems, which at this time can only be operated as pulsed lasers. A test device has been used to study the proposed excitation method. CW operation as a Fluorine laser has demonstrated the general feasibility of this method. Of special interest are molecular lasers pumped by charge or energy transfer from atomic species or by chemical reactions and excimer lasers.

INTRODUCTION

The method of "Fast Plasma Mixing" is proposed to overcome two general limitations present in gas discharges for the operation of efficient CW-lasers:
(1) In a gas discharge, there is always a large number of competing processes, and only one or a few are utilized for the excitation of the intended upper laser level. This is true also for laser systems in gas mixtures using energy transfer or charge transfer processes. In continuous gas discharges energy transfer from atomic metastables or ions to molecules can not be used efficiently, since a significant fraction of the energy transfered through electron collisions will always go into excitation and ionization of the molecular gas. This has two disadvantages. First, the energy transferred to the molecules through electron collisions is lost, and second, the molecules are now distributed over a large number of excited states.
(2) A large number of laser systems are self terminating. The reasons can be that the lifetime of the lower laser level is too long, that the starting species for excitation are depleted, or that absorbing species are produced as a result of the excitation mechanism. Typical examples are excimer lasers.

For such systems fast plasma mixing is, therefore, proposed as an excitation method.

PRINCIPLE OF FAST PLASMA MIXING

A general scheme of Fast Plasma Mixing is shown in Figure 1. Two different particle beams (beam 1 in x-direction and beam 2 in y-direction) are crossed within the laser volume. The direction of the laser resonantor is in z-direction. To generate an excitation volume with gain over a sufficient length, the cross sections of the two beams can be rectangular, as produced by slit nozzles extended in z-direction (see Figure 2), or by an array of nozzles. At least one of the two beams has to carry excited species at high densities (preferably long living species such as metastables, radicals, and/or ions). These species can be produced in different ways, such as with electron beams, with a DC or microwave plasma torch, with a fast flowing hollow cathode discharge, with a hollow cathode arc, or with a laser plasmatron. The species excited into the intended level are produced through collisions of particles from different beams. Such a device will combine two advantages: (1) selective excitation of specific states with high pump rates and (2) a fast removal of reaction products from the active volume.

EXPERIMENTAL SETUP AND TEST RESULTS

A test device using a slit type copper hollow cathode was constructed: HCD dimensions (see Fig. 2): $d_y = 0.08$ mm, $d_z = 20$ mm, $l_x = 0.8$ mm. This device was tested with helium in the operation parameter range: inlet pressure of HCD: $p_{in} = 100 - 1200$ mbar; outlet pressure of HCD: $p_{out} = 1 - 200$ mbar; current: $I = 0.02 - 2.0$ A; voltage: $V = 220 - 230$ V. We achieved CW laser action on the lines 703.75 nm, 712.78 nm, and 731.06 nm. The laser threshold current was 50 mA. The line 731.06 nm was operated CW for the first time.

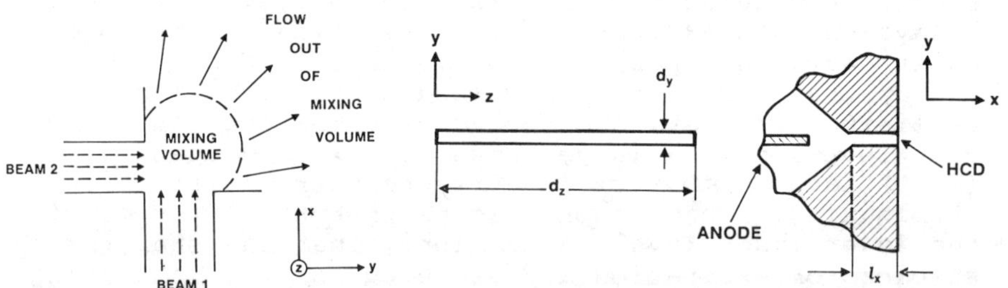

Fig. 1. Schematic of fast plasma mixing

Fig. 2. Slit hollow cathode

MULTIWAVELENGTH EXCIMER LASER STUDIES*

Y. Zhu, R.A. Sauerbrey, F.K. Tittel, and W.L. Wilson
Department of Electrical and Computer Engineering
Rice University, Houston, TX 77251-1892

W.L. Nighan
United Technologies Research Center, East Hartford, CT 06108

ABSTRACT

The simultaneous generation of both UV and blue-green coherent radiation in electron beam pumped high pressure argon buffered Kr, Xe, NF_3, F_2 mixtures is reported. In the case of the XeF(B→X) at 351 nm and XeF(C→A) centered at 480 nm the relative output of the two lasers can be adjusted by variation of the krypton partial pressure. In these studies, a total output of more than 0.5 J/l was realized, corresponding to an efficiency of ~0.5%.

DISCUSSION

For various applications such as materials processing, multicolor laser spectroscopy, and in medical laser systems, it is desirable to have a multiwavelength laser output. The simultaneous efficient operation of XeF(B→X) and XeF(C→A) laser action in the same rare gas halide mixture is reported.

Fig. 1 shows the XeF(C→A) peak gain coefficient measured at 488 nm as a function of XeF(B→X) intracavity flux. The peak gain for the aligned XeF(B→X) cavity, g_a, is normalized to the peak gain for the misaligned cavity, g_m. The solid line is the result of a modelling calculation for the XeF(C→A) gain dependence on the XeF(B→X) flux using the kinetic data given in Ref. 1. Both experiments and theoretical analysis showed that the XeF(C→A) peak gain does not decrease considerably until the XeF(B→X) intracavity flux exceeds several MW cm^{-2}. Consequently a XeF(B→X) and a XeF(C→A) laser could operate simultaneously in the same device, both exhibiting high photon fluxes.

The experimental set-up used in this work has been described in detail in Ref. 2. Various multicomponent gas mixtures comprised of F_2, NF_3, Xe, Kr, and Ar were excited by an intense electron beam (1 Mev, 200 A cm^{-2}, 10 ns FWHM). A special dual wavelength resonator was employed using mirrors that were highly reflective at both 350 nm and ~480 nm.

The temporal and spectral behavior of the XeF(B→X) and XeF(C→A) laser output was investigated.[3] The output energies at 350 nm and around 475 nm were studied as a function of the Kr pressure as shown in Fig. 2. A combined energy of more than 0.5 J/l, corresponding to an efficiency of ~0.5%, was obtained. By variation of the Kr pressure, the relative output energy at the two wavelengths could be

---

*This work was supported in part by the Office of Naval Research, the National Science Foundation, and the Robert A. Welch Foundation.

0094-243X/86/1460175-2$3.00 Copyright 1986 American Institute of Physics

varied conveniently.

The addition of Kr plays an important kinetic role in the Kr_2F production possibly leading to improved XeF(C→A) performance especially on the short wavelength side of XeF(C→A) spectrum. Furthermore, it was shown that the presence of Kr reduces considerably the transient absorption for XeF(C→A) emission. On the other hand, Kr_2F is a strong absorber at 350 nm, which explains the reduction of XeF(B→X) output with increasing Kr pressure.

Further improvement in efficiency by optimum resonator design and mixture refinement appears possible. Simultaneous dual wavelength operation has also recently been achieved in commercial discharge excited rare gas halide excimer lasers.[4]

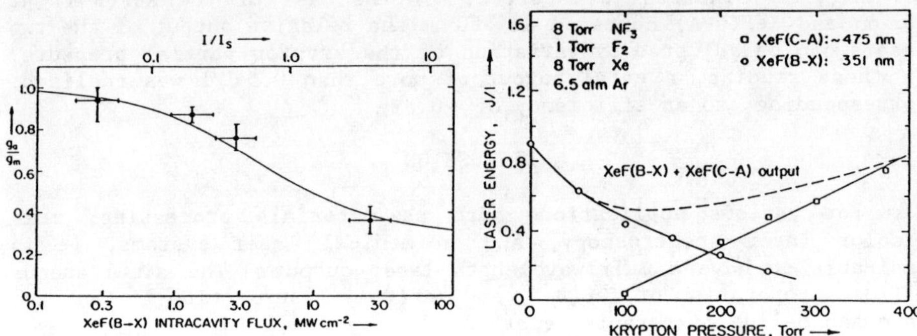

Fig. 1. Normalized XeF(C→A) peak gain coefficient at 488 nm vs. XeF(B→X) intracavity flux.

Fig. 2. XeF(B→X) and XeF(C→A) dual laser output energy vs. Kr pressure.

REFERENCES

1. R. Sauerbrey, W. Walter, F.K. Tittel, and W.L. Wilson, Jr., J. Chem. Phys. 78, 735 (1983).
2. Y. Nachshon, F.K. Tittel, W.L. Wilson, Jr., and W.L. Nighan, J. Appl. Phys. 56, 36 (1984).
3. R. Sauerbrey, Y. Zhu, F.K. Tittel, W.L. Wilson, Jr., N. Nishida, F. Emmert, and W.L. Nighan, IEEE J. Quantum Electron. QE-21, 418 (1985).
4. R. Sauerbrey, W.L. Nighan, F.K. Tittel, W.L. Wilson, Jr., and J. Kinross-Wright, accepted by IEEE J. Quantum Electron., 1986.

AURORA: MULTI-KILOJOULE KrF LASER SYSTEM FOR ICF STUDIES

L. A. Rosocha, P. S. Bowling, M. D. Burrows, J. A. Hanlon, M. Kang, J. McLeod, J. R. Ratliff, D. O. Whitcomb, and G. W. York, Jr.
University of California, Los Alamos National Laboratory
P. O. Box 1663, MS E548, Los Alamos, New Mexico 87545

Aurora is a high power krypton fluoride laser system that serves as an end-to-end technology demonstration prototype for scaling lasers of interest to short wavelength inertial confinement fusion (ICF). The system employs optical angular multiplexing and serial amplification by electron beam driven KrF laser amplifiers to deliver multi-kilojoule, 5 ns, 248-nm laser pulses to fusion targets.

The main amplifiers range in aperture size from 10 cm x 12 cm to 100 cm x 100 cm and range in stage gain from 10 to 50. The system functions by replicating a 5 ns portion of the front end TEA laser oscillator pulse by means of aperture and intensity division. A 480 ns long pulse train that consists of 96 separate beams, each of 5 ns duration results. These spatially separated distinct pulses are then angularly encoded (given different path angles of a few milliradians) and sent through the entrance pupil of an optical relay system. This relay system, which provides for efficient filling of the amplifiers, sends the beams through two single-pass amplifiers and one double-pass amplifier, which amplify the pulse train energy to more than 10 kJ.

The system is being built in two phases: the first phase includes the multiplexing and serial amplification, while the second phase includes the additional demonstration of demultiplexing and delivery of pulses to target. In the first phase, the 480 ns laser pulse train will be delivered to a diagnostic station. In the second phase, the pulse train will be compressed into a single 5 ns pulse that will be focused onto a fusion target.

Aurora will serve to address the critical technologies involved in scaling KrF lasers to high power required for fusion. In particular, it will examine: uniform e-beam pumping of large laser volumes, efficient energy extraction from the KrF medium, optical angular multiplexing at the multi-kilojoule level, control of amplified spontaneous emission, large amplifier staging, and alignment of multibeam systems.

Figure 1 is an illustration of the second phase Aurora KrF fusion laser prototype.

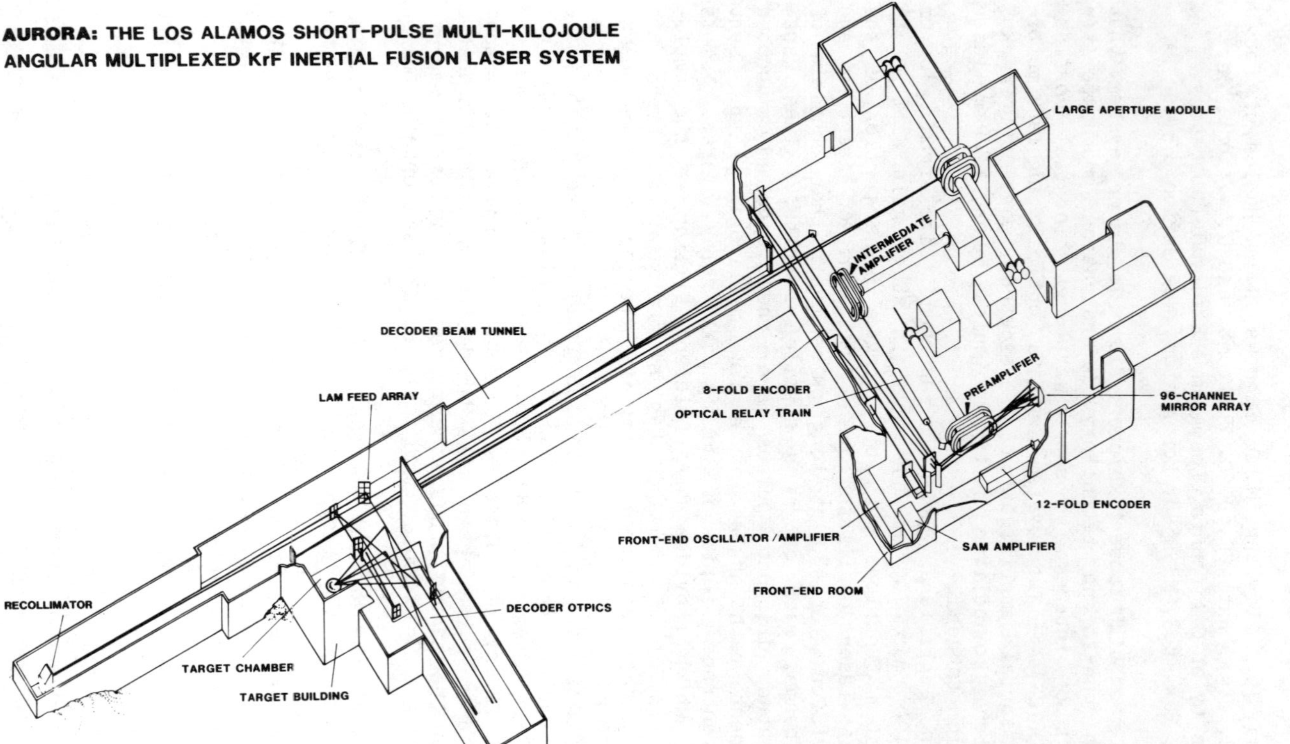

Fig. 1. An artist's conception of the full Aurora laser system. As configured, the system is intended to operate as a 96 beam angular multiplexed prototype. The laser system, which consists of the four serial electron beam driven amplifiers and optical encoder is ont he right. The decoder and target system is shown on the left; 48 of the 96 beams will be stacked and delivered to fusion targets at the 5 to 8 kJ level.

SOLAR-PUMPED PHOTODISSOCIATION IODINE LASER*

J. H. Lee, W. R. Weaver, D. H. Humes,
M. D. Williams and M. H. Lee+

NASA Langley Research Center, Hampton, Virginia 23665-5225

ABSTRACT

The scientific feasibility of a solar-pumped iodine photodissociation laser for space applications is under investigation. Recently, a 2-watt CW output for more than one hour was achieved using n-C_3F_7I vapor as the laser material and a vortex-stabilized argon arc as the light source.

INTRODUCTION

Direct solar-pumped lasers for use in space communications and power transmissions have been discussed since 1966.[1,2] Development of a solar-pumped iodine photodissociation laser has been reported by Lee et al. in 1981[3] and obtained a maximum laser output of 4 W in excess of 10 ms for a single static filling of n-C_3F_7I. Among the laser materials available at the present time, the iodides such as C_3F_7I or C_4F_9I have the advantages of broad-band absorption of solar radiation and low threshold. This presentation reports high-power CW operation of an iodine laser and is part of the NASA program to develop a solar-pumped laser for space applications.[4]

EXPERIMENT AND RESULTS

The solar simulator used is an argon arc contained in a single quartz tube which is cooled by a thin layer of spiraling water on its inside surface. The total continuous radiative power obtainable is 45 kW. Figure 1 compares the spectral irradiance of the solar simulator (solid line) on the laser tube with that of the air-mass-zero solar spectrum (dotted line). When the arc current is 200 A the pumping power integrated over the absorption band (broken line) of the iodide n-C_3F_7I is approximately that obtainable by 760x solar concentration.

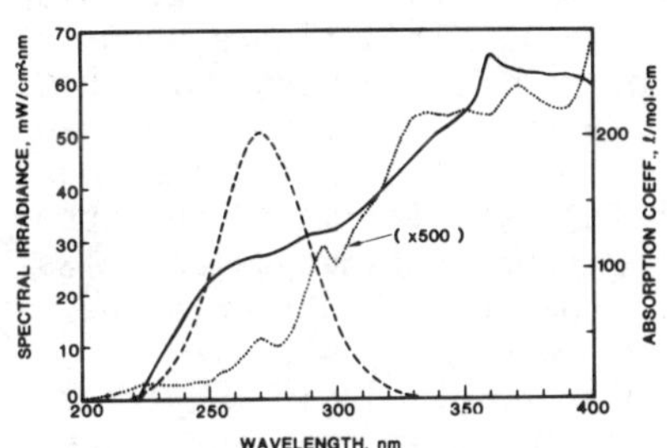

Fig.1 Spectral irradiance of simulator(——), solar spectrum(•••) and absorption curve of C_3F_7I(---).

* Supported in part by NASA Grant NAG-1-441.
+ Hampton University, on leave from Inha Univ., Inchon, Korea.

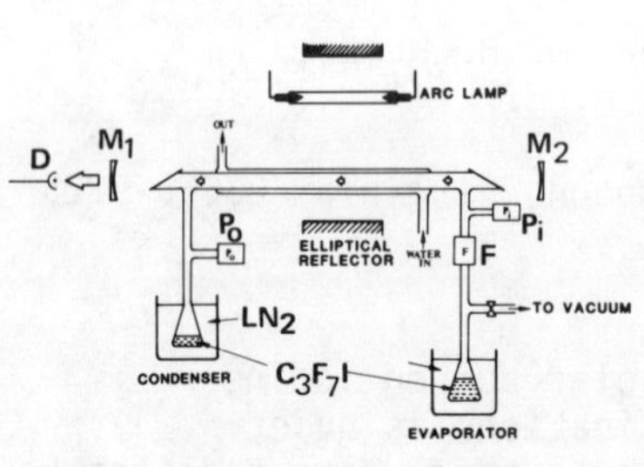

Fig.2 Experimental setup.

The laser resonator is formed by 10 m radius-of-curvature mirrors placed 0.8 m apart. (Fig. 2) Reflectances of these mirrors are 85% and 100%. The solar simulator lamp is placed 17.5 cm from the laser tube in one focal line of an elliptical cylindrical reflector. The effective gain length is 15 cm. The laser tube which is cooled by deionized water is a 12 mm ID Suprasil quartz tube with Brewster windows. Iodide flow was obtained with an evaporator-condenser unit and the flow rate was controlled by varying the temperature of the evaporator and adjusting a valve in the line, while the condensing bottle was immersed in LN_2. The inlet (P_i) and the outlet (P_o) pressures were measured to determine the velocity of iodides in the laser tube. The laser output power was measured with a Ge photodiode and a pyroelectric power meter.

Fig. 3 shows the laser output power for the various flow rates of n-C_3F_7I. The flow rate is expected to control the total photodissociation rate and consequently the total inversion in the laser tube. Two factors operate against each other for the flow rate: (1) supply of new iodide and removal of photodissociation wastes which demand a fast flow, and (2) irradiation time required for absorption of the pump light which is best served by a slow flow or static fill. This figure shows a flow rate of 1,740 sccm (0.4 g/s) is optimum for the system. This rate corresponds to a C_3F_7I flow velocity of 18 m/s for a pressure of 1.3 kPa. The effective irradiation time or the dwell time for the lasant is 8 ms which is comparable to the lifetime of the upper laser level I*. With the 15% T output mirror continuous lasing was obtained for over one hour with the CW power of 2 W at the higher arc current.

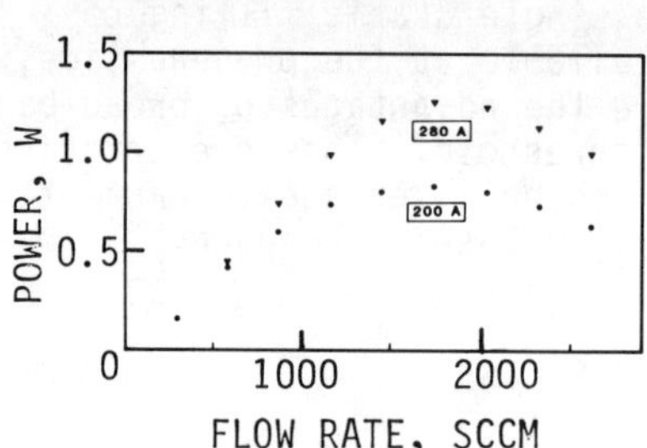

Fig.3 Laser power vs flow rate.

These results indicate that the solar-pumped iodine CW laser can be scaled to a high power level that is significant for space applications.

REFERENCES

1. C. G. Young, Appl. Opt. 5, 993 (1966).
2. H. Arashi et al., Japanese J. Appl. Phys. 23, 1051 (1984).
3. J. H. Lee and W. R. Weaver, Appl. Phys. Lett. 39, 137 (1981).
4. E. J. Conway, "Potential New Solar Laser Concepts," NASA Langley Research Center (unpublished), 1985.

ENHANCEMENT OF BLUE-GREEN LASER EFFICENCY BY A SPECTRUM CONVERTER*

K. S. HAN, C. H. OH** and J. H. LEE
Hampton University, Hampton, Va. 23668

ABSTRACT

In order to enhance the efficiency of a blue-green laser through spectrum conversion of the pumping light, a converter dye, BBQ, was mixed in the laser dye solutions. The laser was pumped with a plasma radiation source. The maximum increase of laser output at the dye mixture of LD490+BBQ or coumarin 503+BBQ was about 80%. The enhancement is mainly due to the abundance of near uv in the pumping source, the fairly good match of the fluorescence band of converter dye with the absorption band of the laser dye, and a small overlap of fluorescence band of laser dyes with triplet-triplet absorption band of converter dye.

INTRODUCTION

In order to achieve the more efficient utilization of the near uv-abundant HCP light source for pumping dye lasers, the laser dye mixture method was used. Various concentrations of laser dye mixture LD490+BBQ in ethanol and coumarin 503+BBQ in p-dioxane were employed. Description of the HCP plasma source is reported elsewhere.[1] As shown in figure 1, the converter dye BBQ absorbs light in the region 286-333nm of the HCP pumping source where the laser dye LD490 or coumarin 503 absorbs little. On the other hand the converter dye BBQ fluoresces near 380nm which lies in the absorption band 360-416nm of the laser dye LD490 or coumarin 503. The irradiance of the pumping light which is calibrated with a standard W-Halogen source shows a maximum at 300nm in the converter dye absorption band. The intensity at this peak is six times as high as that at 390nm. Thus, as the concentration of converter dye BBQ increases in the dye mixture, the fluorescence light near 380nm increases and is absorbed by laser dye LD490 or coumarin 503. An enhancement of laser output is expected as the concentration increases. However, there is a small overlap of the fluorescence band

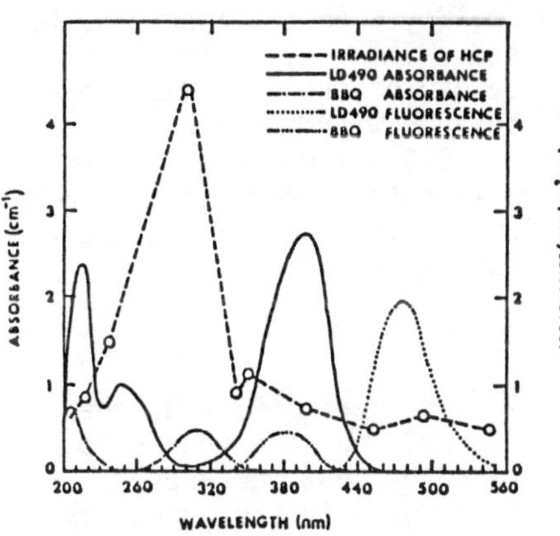

Fig. 1 Absorption curves of dyes and spectrum of HCP pumping source.

* Research supported in part by ARO grant DAAG-85-G-0107 and in part by ONR contract N0001-80-C-0957.

** On leave of absence from Kyungpook National Univ. Taegu, Korea of the laser dye and the triplet-triplet absorption band (λ_c = 530 nm) of the converter dye.[2] Consequently the enhancement of laser output is affected at high concentration.

EXPERIMENTAL RESULTS AND DISCUSSION

A small HCP, which was designed for use as a laser pumping source, was used under the optimum condition of argon fill gas pressure of 1 Torr and an applied energy of 0.9kJ stored at 30kV. Figure 2 shows laser output of the LD490 dye as a function of the BBQ concentration. The laser output increases as the BBQ concentration increases until it reaches 5x10^{-6}mol/ℓ where the energy enhancement is 80%. This increase is expected by the converter effect mentioned earlier. Figure 2 also shows there are smaller enhancements for the concentrations of laser dye less than optimum (lower curves). For the smaller concentration of LD490, the optimum concentration of the converter dye is also small. However, if the BBQ concentration with 4x10^{-4}mol/ℓ of LD490 is more than 10^{-5} mol/ℓ, then the laser energy decreases as BBQ concentration increases. This may be due to the increased triplet-triplet absorption by BBQ of the fluorescence of the laser dye. Similar results are obtained with the dye mixture of 8x10^{-5}mol/ℓ coumarin 503 and 4x10^{-6}mol/ℓ BBQ which are not shown here. The results are generally agree with the expectation from a spectrum converter theory but the effect of the radiationless energy transfer[3] in the dye mixture will be investigated in future.

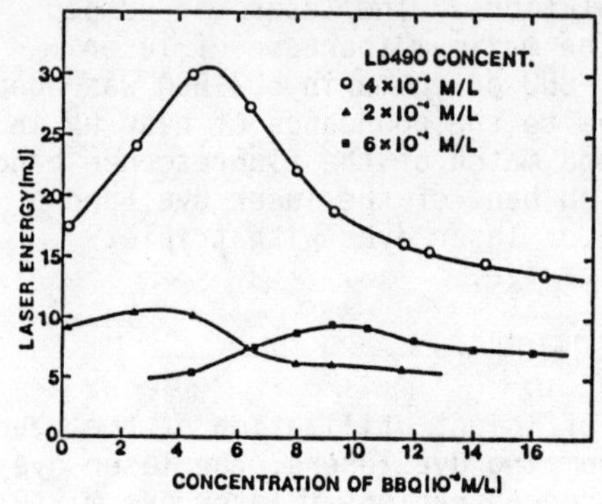

Fig. 2 Laser output of LD490 vs converter dye BBQ concentration.

CONCLUSION

Using the dye mixture of LD490+BBQ or coumarin 503+BBQ, the efficiency of a blue-green laser with the HCP pumping source is increased by about 80%. The result indicates that the energy transfer method can be applied to pumping not only blue-green dye lasers but also pumping uv dye lasers as well.

REFERENCE

1. K. S. Han, S. H. Nam, and J. H. Lee, J. Appl. Phys. 55, 4113 (1984)
2. T. G. Pavlopoulos, Opt. Comm. 24, 170 (1978).
3. Th. Förster, "Transfer Mechanisms of Electronic Excitation," Discuss. Faraday. Soc., 27, 7 (1959)

LINEWIDTH REDUCTION OF SEMICONDUCTOR LASERS*

J. Harrison and A. Mooradian

Lincoln Laboratory, Massachusetts Institute of Technology
Lexington, Massachusetts 02173-0073

The semiconductor diode laser (SL) has a number of properties that make it a desirable light source for spectroscopy. Aside from the advantages in size, cost and complexity, SL's cover a broad spectral range and may easily be modulated at rates up to several gigahertz. However, coherence remains an issue in many applications.

Optical feedback can be used to achieve dramatic linewidth reduction. By placing a SL in an external cavity, the linewidth may be reduced by as much as the square of the increase in the photon lifetime (i.e., several orders of magnitude). The linewidth of the external cavity SL, $\Delta\nu_{xcav}$, is related to that of a solitary device at the same pump level, $\Delta\nu_{sol}$, by the following relation

$$\Delta\nu_{xcav} = \Delta\nu_{sol}\cdot[n\ell/n\ell+L]^2\cdot[G_{xcav}/G_{sol}]^2 \quad , \qquad (1)$$

where the first term in brackets is the ratio of the active optical path length to the sum of the active and passive optical path lengths of the external cavity. The second bracketed term in Eq. (1) is the ratio of the net gains at threshold of the external cavity and solitary lasers. In our experiments, the square of the length ratio is typically about 10^{-4} while that of the gain term is about three. The external cavity SL is broadly tunable when a dispersive element is included in the cavity. By employing compact, stable cavity structures and carefully isolating the laser from acoustic sources, one can construct a very narrow line, tunable source that can deliver tens of milliwatts. High power, pulsed SL's can also be used in external cavities to produce much higher peak powers in a single longitudinal mode for efficient frequency mixing.

We have constructed two compact external cavity lasers using Hitachi HLP-1400 (GaAl)As lasers. AR coatings were applied to the internal facets with residual modal reflectivities of less than 1%. The lasers were imaged through a thin etalon onto 5% output couplers using commercial microscope objectives that transmitted 70% of the incident light at 833 nm. Line coincidence was achieved by a combination of current and etalon tuning, and no effort was made to thermally stabilize the lasers. The heterodyne signal was derived from a Ge avalanche PIN photodiode and observed on a spectrum analyzer. The data obtained on a logarithmic scale (Fig. 1a) is slightly broader than the heterodyne lineshape that is calculated to be about 5 kHz FWHM in this experiment. In order to obtain a measure of the linewidth jitter due to acoustic sources, Figs. 1b and 1c show the heterodyne signal on a linear scale at sweep rates of 17 MHz/sec and 100 kHz/sec, respectively. It is evident that in several milliseconds, the acoustically driven peak-to-peak jitter is within a factor of two of the intrinsic heterodyne

*This work was sponsored by the Department of the Air Force.

0094-243X/86/1460183-4$3.00 Copyright 1986 American Institute of Physics

linewidth. After a moderate warmup period, the maximum frequency variation of the heterodyne signal was observed to be less than 2 MHz over 10 minutes.

With such stable, narrow-line devices, it is straightforward to phase-lock two external cavity lasers. The experimental schematic is included in Fig. 2a. The heterodyne signal is mixed with a local oscillator set to the heterodyne center frequency. The mixer output is then amplified in order to supply a feedback current that is proportional to the phase shift of the heterodyne signal. Note that because the tuning rate of the external cavity laser was 25 MHz/mA, the peak-to-peak variation in the feedback current over several seconds was less than 5 μA. Figure 2b shows the heterodyne signal under phase-locked conditions. The observed signal indicates that the phase-locked heterodyne linewidth is less than the 15 Hz linear resolution halfwidth of the spectrum analyzer.

In conclusion, optical feedback may be employed to obtain ultranarrow, tunable spectroscopic sources. We have demostrated phase-locking between two very stable external cavity semiconductor diode lasers.

The views expressed are those of the authors and do not reflect the official policy or position of the U.S. government.

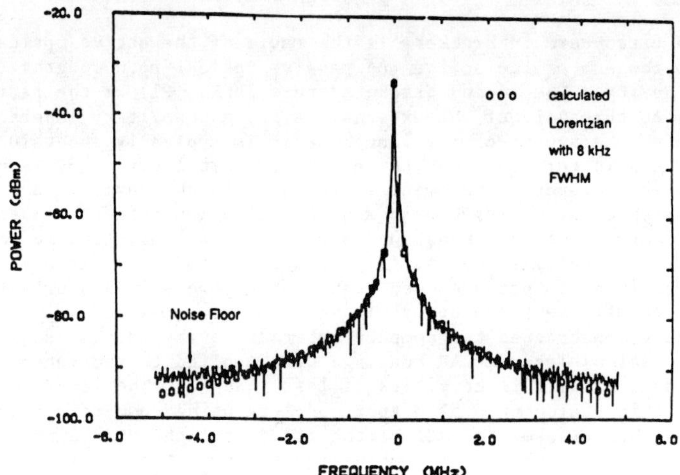

Fig. 1. Heterodyne signal of two free-running external cavity lasers.
a) log (10 dB/div) vertical scale: 1 s sweep, 30 kHz resolution bandwith; circles indicate 8 kHz FWHM Lorentzian curve.

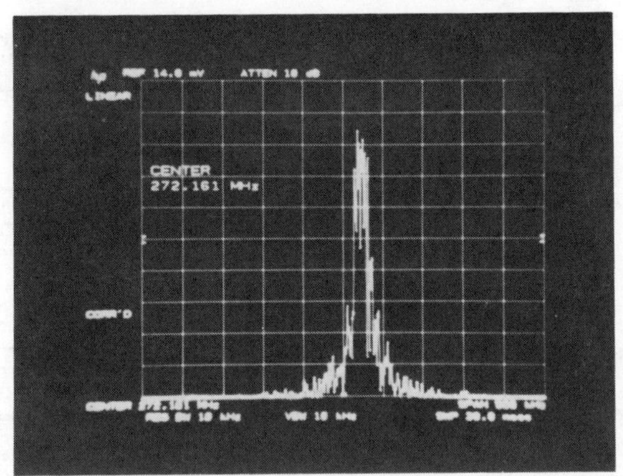

Fig. 1. b) Linear vertical scale: 50 kHz/div, 30 ms sweep, 10 kHz resolution bandwidth.

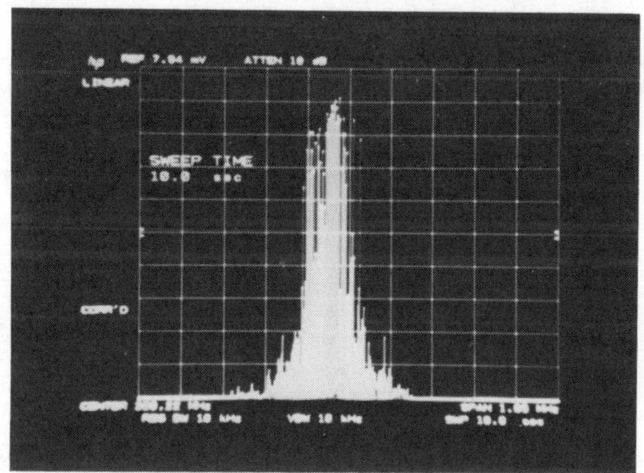

Fig. 1. c) Linear vertical scale: 100 kHz/div, 10 s sweep, 10 kHz resolution bandwidth.

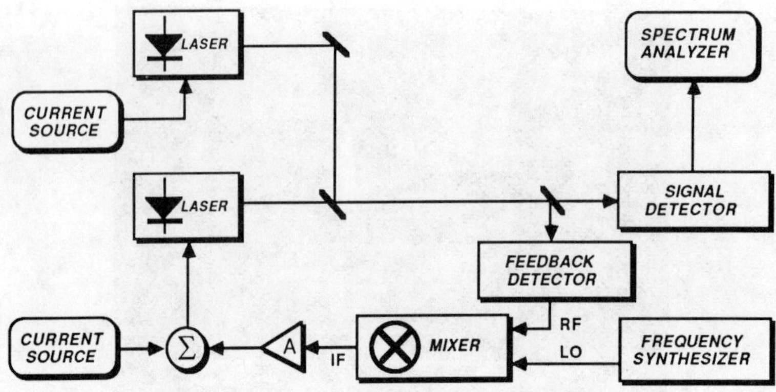

Fig. 2. a) Schematic of phase-locking experiment.

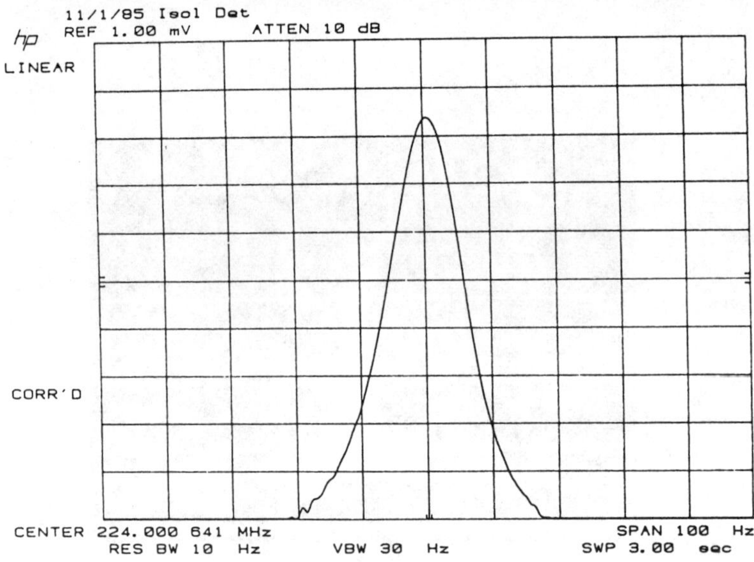

Fig. 2. b) Heterodyne signal of two phase-locked external cavity lasers. Linear vertical scale: 10 Hz/div, 10 Hz resolution bandwidth (i.e., 15 Hz resolution halfwidth).

HIGH POWER SEMICONDUCTOR LASER DIODE ARRAYS

Peter S. Cross, Spectra Diode Laboratories, San Jose, CA 95134

ABSTRACT

The cw optical power obtainable from semiconductor laser diodes has been extended to unprecedented levels in recent years through the use of multistripe arrays. By spreading out the optical power with more than 100 stripes, single-facet, cw output in excess of 5 Watts has been demonstrated, and 500 mW cw is now commercially available. Recent improvements to array performance include: arrays up to 1 cm wide that generate quasi-cw (150 usec pulse) output in excess of 11 Watts, and a novel device structure which produces up to 215 mW cw in a single diffraction limited lobe.

INTRODUCTION

Multiple stripe lasers are rapidly maturing as efficient sources of high power, continuous wave (cw) optical radiation. Single stripe diodes are typically limited by heating and catastrophic facet damage phenomena to about 50-100 mW single facet output power. By spreading out the emitting region of the laser, diodes with ten stripes have been reported to produce in excess of 800 mW cw output [1], and up to 2.6 Watts cw has been demonstrated with 40-stripe devices [2]. Current areas of active development include: increasing power output, improved beam quality and higher reliability.

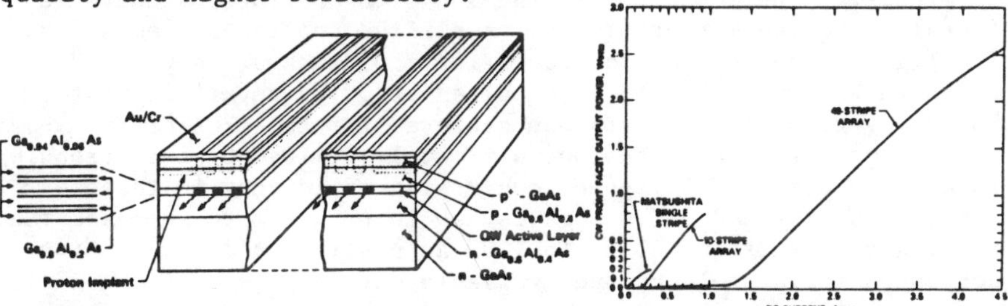

Fig. 1 - Laser Diode Array

Fig. 2 - cw Light versus Current Characteristics

HIGH CW POWER OUTPUT

A schematic diagram of such a laser is shown in Figure 1. The device consists of a series of layers of AlGaAs grown on a GaAs substrate to provide confinement of both photons and injected carriers in the active region. In order to obtain both high efficiency and high output power, the arrays are fabricated using metalorganic chemical vapor deposition (MOCVD) for the growth of very thin (quantum well) active layers. The MOCVD process provides a high degree of uniformity of critical layer properties such as thickness, composition and carrier concentration [3]. The light versus current characteristics of 10 and 40 stripe arrays are compared in Figure 2 with the highest power single stripe laser reported in the literature [4]. It should be noted that the peak powers indicated in Figure 2 represent the catastrophic limit at which the device fails almost instantaneously. Reliable operation can only be achieved at power levels about a factor of four below the catastrophic limit. For example, commercially available 40-stripe lasers are rated at 500 mW cw power output [5].

Very recently [6], even larger arrays with greater than 100 stripes have been investigated and have achieved cw output power in excess of 5 Watts. Preliminary lifetests of these devices being carried out at 1 Watt cw show no signs of rapid degradation after several hundred hours of operation.

VERY HIGH POWER, QUASI-CW ARRAYS

The size of the multistripe arrays (number of stripes) can in principle be increased indefinitely, but the increased heat load of very large arrays precludes true cw operation. However, long pulse (several hundred microseconds), quasi cw operation can still be achieved if the duty factor is kept sufficiently small. As an example of work currently being conducted, a 1 cm wide bar with 200 stripes arranged as shown in Figure 3 has been fabricated [7]. The active stripes are placed in twenty clusters of ten in order to minimize problems with heating and with spurious, transverse (parallel to the mirror facets), amplified spontaneous emission. The light versus current characteristic of such a laser "bar" is shown in Figure 4 for 150 usec pulses at 50 Hz repetition rate. Power in excess of 11 Watts was achieved for this 20% stripe packing density configuration, ultimately limited by catastrophic facet breakdown. Total power conversion efficiency in these arrays has exceeded 30%. Work is now in progress to increase the packing density to as close to 100% as possible and correspondingly scale up the available output power.

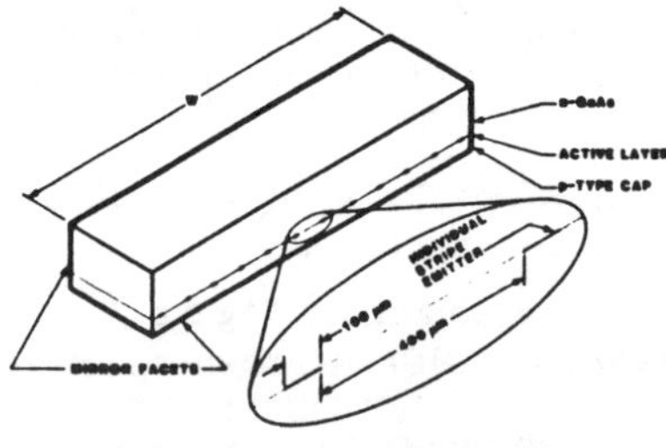

Fig. 3 - Schematic Diagram of a 1 cm Laser Bar

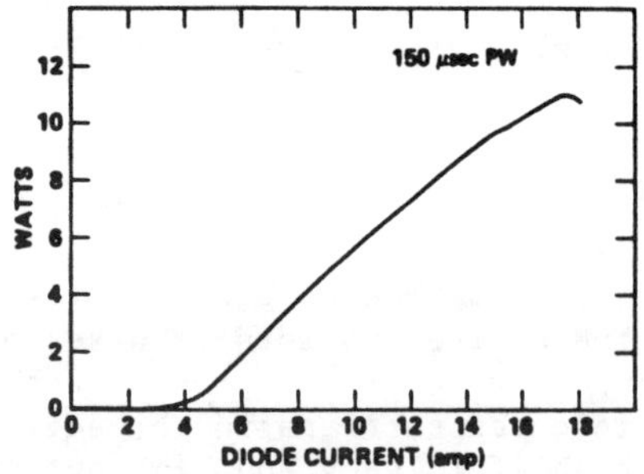

Fig. 4 - Quasi-cw Light vs. Current Curve

IN PURSUIT OF DIFFRACTION LIMITED BEAMS

The stripes in high power arrays are typically placed in such close proximity that phase locking occurs among the stripes which causes the array to oscillate in certain "supermodes" [8]. Unfortunately, in nominally uniform arrays there is usually higher gain in the active stripes than there is between the stripes, a situation that tends to favor oscillation in a high order supermode that emits a double lobed radiation pattern.

In order to force the desired single lobed mode to oscillate, several structural variations have been investigated over the past few years. The structure that has demonstrated the highest cw power in a diffraction limited beam is shown in Figure 5 [9]. Instead of using identical, straight lines, the active region consists of stripes that are offset at two locations in order to effectively equalize the gain in all paths between the mirror facets. This is equivalent to putting gain in between the lasing stripes which has been shown theoretically to enhance the gain of the single lobe mode [10]. In addition, the width of the gain stripes in monotonically chirped to provide a further discrimination in favor of the single lobed output [11]. The cw far-field radiation pattern for such a device is shown in Figure 6. The fundamental mode was maintained up to 350 mW cw, with 215 mW being in the 0.7 degree wide (diffraction limited) central lobe. Research into other device structures to achieve still higher power diffraction limited beams is being actively pursued in several laboratories.

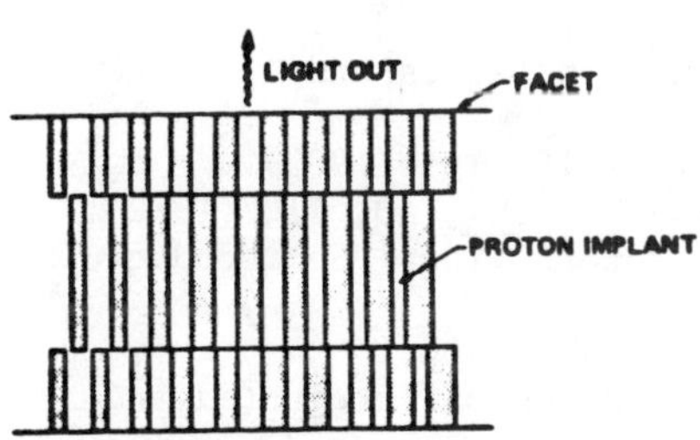

Fig. 5 - Contact Pattern of an Asymmetric, Offset Stripe Laser

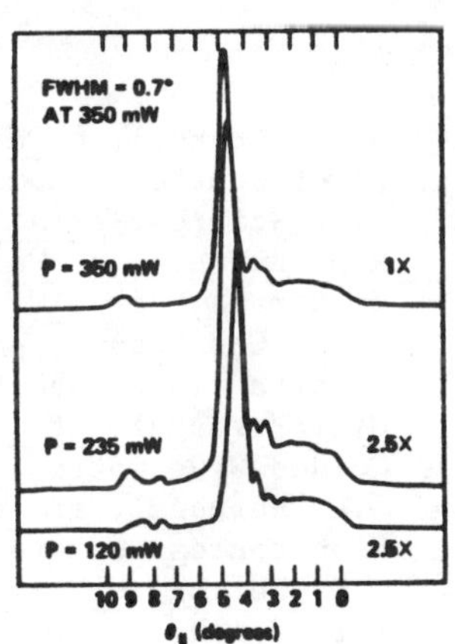

Fig. 6 - Radiation Pattern of Device in Fig. 5

SUMMARY AND CONCLUSIONS

In conclusion, the power available from small, efficient diode lasers has increased dramatically in the past several years through the development of monolithic, multistripe arrays. CW power output in excess of 5 Watts has been achieved, and 500 mW cw is available commercially. Quasi-cw power levels using very wide arrays (1 cm) have now exceeded 11 Watts, and substantial increases in power are anticipated in the next few years. Furthermore, 10 stripe arrays have achieved reliability levels sufficient for many commercial applications. Specifically, arrays rated at 100 mW and 200 mW cw have median lifetimes at room temperature of 26,000 and 5000 hours, respectively.

A large number of applications of high power arrays are now under development in commercial, medical and military areas. Some examples include: pumping of Nd:YAG and other solid state lasers, illumination for machine vision and night vision, fiber power transmission for remote sensing and pyrotechnic ignition, and medical applications such as laser scalpels. When devices become available with diffraction limited radiation patterns, the number of applications will expand significantly to include optical data storage, laser printing and point to point communication links. The inherent small size, high efficiency and potential for very low cost should make semiconductor diode laser arrays one of the fastest growing segments of the optoelectronics market in the next few years.

REFERENCES

1. G.L. Harnagel, et.al., Conference on Lasers and Electrooptics, paper ThZZ1 (1985).
2. D.R. Scifres, et.al., Elec. Lett., 19, 169 (1983).
3. R.D. Burnham, et. al., Appl. Phys. Lett., 40, 118 (1982).
4. K. Hamada, et. al., Ninth IEEE International Semiconductor Laser Conference, paper C1 (1984).
5. Spectra Diode Laboratories, Model SDL-4450.
6. G.L. Harnagel, et.al., Topical Meeting on Integrated and Guided Wave Optics, (1986).
7. G.L. Harnagel, et. al., submitted for publication to Electronics Letters.
8. J.K. Butler, et. al., Appl. Phys. Lett., 44, 293 (1984).
9. D.F. Welch, et.al., Appl. Phys. Lett., 47, 1134 (1985).
10. W. Steifer, et.al., Elec. Lett., 21, 118 (1985).
11. E. Kapon, et.al., Appl. Phys. Lett., 45, 200 (1984).

SEMICONDUCTOR LASER RESEARCH IN JAPAN

T. Yabuzaki
Radio Atmospheric Science Center, Kyoto University
Uji, Kyoto 611, Japan

ABSTRACT

Recent developments of semiconductor lasers in Japan are overviewed from the user's point of view, together with the improvements of spectral characteristics by using external optical systems.

INTRODUCTION

Recently various types of semiconductor lasers have been extensively developing in many Japanese companies, national laboratories, and universities. These developments have been mainly motivated by the practical applications, but the newly developed lasers are quite attractive in their use in the fundamental researches such as laser spectroscopy. In this paper we describe briefly on the most advanced and the most interesting researches and developments done in recent years, from the user's point of view, the user wanting to get good semiconductor lasers in fundamental researches.

LASERS DEVELOPED FOR OPTICAL COMMUNICATIONS (λ=1-2μm)

The lases operating at λ=1.3μm and 1.5μm, at which the single-mode fiber shows respectively the minimum dispersion and minimum loss (now 0.2 dB/km), have been simultaneously developed for the optical communications. The materials used is mostly GaInAsP/InP.

High Output Power Lasers : High output power can be realized by the FP(Fabry-Perot) type BH(Buried Heterostructure) lasers, in which a thin active layer is buried in the semiconductor with larger bandgap, in order to confine well the injection current and light wave into the active region. The maximum output power of a single laser (excluding a phase-locked laser array) at 130μm is now 140 mW (quantum efficiency:61%).[1]

Single-Mode Lasers : As well known the FP type laser behaves with mode-hopping when the injection current or laser temperature is varied, or the current is modulated at high frequency. To remove this disadvantage, the DFB(Distributed FeedBack) laser was developed,[2] in which the optical feedback is made by a grating located just below the active layer. So, even when the temperature or current is changed in a wade range, the DFB laser operates on a stable single-longitudinal mode, whose frequency shifts continuously without mode-hopping. The DFB laser may become a powerful tool in the field of laser spectroscopy, because the frequency tunable range exceeds 1000 GHz. In addition, the output power is high, the maximum is now 95mW at 1.3 μm, with low threshold current 10-20 mA.[3] Several types of integrated DFB lasers were also reported recently. Among them, the interesting one for laser spectropists may be the five-wavelength DFB laser.[4] This laser consists of five DFB lasers fabricated on the common substrate, each laser having a grating with slightly different pitch. Then these lasers emit the light with different frequencies, the shifts of which are continuous and parallel when the temperature is changed. So, it is possible to cover cover the wavelength range over about 30 nm. Another important single-mode laser is the <u>DBR</u>(Distributed Bragg Reflector) laser,[5] where a grating is monolythically fabricated out of the active layer as a frequency selective reflector. The temperature region of the DBR laser for single-mode operation is not large compared with that of the DFB laser, but it has an advantage to be easily fabricated in an opto-electric integrated circuit. An interesting integrated DBR laser was proposed,[6] where a current can be injected to the grating region to control the pitch of the grating. So it is possible to tune or modulate electrically the laser frequency in a wide frequency range, without significant change of the output power.[6]

SHORTER WAVELENGTH LASERS (λ<1μm)

In this wavelength region, the researches and developments have been directed mainly to (1) increase of output power, (2) realization of stable single-longitudinal-mode lasers, (3) shortening of wavelengths, (4) reduction of noise, and (5) and developments of new type lasers with various functions, such as surface emitting lasers, bistable lasers, optical transistor, electrically beam-scanable lasers, etc. Here we will focus our attention to the directions (1)-(3).

<u>High Power Lasers</u> : To increase the output power, improvements of FP type lasers have been done on optimization of facet reflectivities, efficient current and light wave confinements, and avoidance of COD

(Catastrophic Optical Damage) by using nonabsorptive windows and/or by expanding the spot-size (i.e. by decreasing the radiation density) at the facet. The maximum output power attained is about 200 mW,[7] which is from a GaAlAs laser with the very thin active layer enough to expand the spot-size, and with optimization facet reflectivities done by multi-layered coating.

Stable Single-Mode Lasers : There was a technical difficulty to fabricate a DFB GaAlAs laser, which comes mainly from the oxidation of the surface of the grating. But, recently this difficulty was partially removed, and a DFB laser with low threshold current (40 mA) could be realized at the wavelength range of 0.8 μm.[8]

Shortening of Wavelengths : The shortening of the oscillation wavelength has been done mostly by usingh new materials having large energy band-gap, such as AlGaInP/GaAs and InGaAsP/GaAsP. The shortest wavelength on cw operation at room temperature is now 671 nm, which is from a InGaAsP/GaAsP DH laser.[9] With respect to the pulsed operation, the shortest wavelength obtained so far is 621.4 nm at room temperature,[10] and 580 nm at 77K.[9]

IMPROVEMENTS OF SPECTRAL CHARACTERISTICS

The spectral of above mentioned lasers are generally very wide (tens of MHz), and the frequency is very sensitive to the variations of temperature and injection currents. The improvements of the spectral characteristics, which is particularly important in coherent optical communications, have been extensively doing by using external optical systems.

Spectral Narrowing by Using an External Mirror : The reduction of the spectral width is possible by placing an external mirror, which increases the Q-value of the laser cavity. Several groups reported the spectral narrowing by following compound cavity systems.[11] The spectral width of about 1MHz was obtained by for the DFB laser with monolythically integrated waveguide and mirror, where the distance L between the mirror and the active region being about 1.5 mm. When an external mirror was placed 36 cm away from a DFB laser, the spectral width was reduced to 40 kHz. Further narrowing down to 20kHz was observed for a FP type laser with an external etalon and a grating.

Frequency-Stabilization : For the frequency-stabilization, atomic and molecular absorption lines, most of which are Doppler-broadened, have been used as external frequency references. The narrow atomic spectral lines were observed by using the techniques of saturated absorption spectroscopy, and obtained Doppler-free spectra of Cs-D_2 line have been used in very high frequency-stabilization.[12]

CONCLUSIONS

Recent researches and developments of semiconductor lasers in Japan have been overviewed, from the user's point of view. The progress is so rapid in these years, various values for the most advanced lasers may be soon replaced by new ones. The present progress is mainly motivated by its use in the practical applications, but semiconductor lasers become undoubtedly one of th most important lasers in fundamental sciences in near future.

ACKNOWLEDGEMENTS

I would like to express my deep thanks to Profs. Y. Suematsu, K. Iga, M. Ohtsu, Drs. H. Yajima, Y. Yamamoto, R. Lang, I. Teramoto, K. Ito, S. Yano, W. Susaki, T. Niina, and H. Katsuta for their kind cooperation to give me many informations of newly developed lasers.

1. M. Yamaguchi et al., presented at the meeting of IECE of Japan (1985).
2. K. Utaka et al., Electron. Lett. 17, 961 (1981).
3. I. Mito et al., presented at OFS'85, PD-09 (1985).
4. H. Okuda et al., Jpn. J. Appl. Phys. 23, L904 (1984).
5. T. Tanbun-ek et al., Electron. Lett. 17, 967 (1981).
6. M Yamaguchi et al., Electron. Lett. 21, 63 (1985).
7. K. Hamada et al., IEEE JQE,21, 623 (1985).
8. S. Noda et al., Tech. Rep. of IECE of Japan, OQE85-67, 47 (1985), (in Japanese).
9. A. Usui et al., Jpn. J. Appl. Phys. 24, L163 (1985).
10. A. Fujimoto et al., Jpn. J. Appl. Phys. 21, L729 (1982).
11. S. Murata et al.: J. Omoto et al: S.Tai et al.: and T. Uno et al., all presented at meeting of Jpn. Soc. of Appl. Phys. (1985).
12. T. Yabuzaki et al., Proc. Int. Conf. on LASER'S 83, p93 (1983).

LAUNCHING LIGHT FROM SINGLE-MODE SEMICONDUCTOR LASERS INTO SINGLE-MODE OPTICAL FIBERS

Harish R. D. Sunak

University of Rhode Island
Department of Electrical Engineering
Kingston, RI. 02881-0805

ABSTRACT

Single-mode optical fibers are extremely important for long-haul broadband optical communications systems. The coupling of light into these fibers is hence of utmost importance, to obtain a high coupling efficiency and negligible optical feedback from the fiber to laser. This invited paper reviewed and compared techniques to achieve these requirements.

INTRODUCTION

Single-mode fibers are now being widely used for long-haul optical communication systems and are also being considered for local area networks. The optical attenuation of these fibers is extremely low, of the order of 0.2-0.3 dB/km at the wavelength of 1.55 μm. They can also be designed to have zero chromatic dispersion in the 1.3-1.55 μm range. Laser-to-fiber coupling is a significant performance factor as a 3 dB increase in the coupling efficiency implies a 10-15 km increase in the repeater separation. Since repeaters are a major fraction of the cost of any communication system, important cost saving can be achieved. A 5 dB increase, which has been achieved in practice, implies an increase of 17-25 km in the repeater span.

The objectives of coupling from the single-mode semiconductor laser into the single-mode optical fiber are to (i) maximize the coupling efficiency (η) and (ii) minimize optical feedback from fiber to laser. If the latter objective is not fulfilled, the laser's dynamic, static and spectral characteristics are severely degraded and the system performance also deteriorates.

COUPLING PROBLEMS

There are many problems to be considered when solving the above coupling problems. These are (i) electrical drive current to laser and its control, (ii) optical output from laser and its characteristics compared to fiber characteristics, (iii) fabrication and handling of miniature optical components or (iv) suitable modification of the fiber end; (v) positioning of the components very accurately, (± 0.1 μm) (vi) fixing the components without alteration of the coupling efficiency, (vii) long-term stability with respect to temperature shock, time effects and mechanical vibrations. I shall restrict myself to the optical coupling problems in this paper.

The optical coupling problem is entirely a mode-matching problem. In general, the laser emits a non-symmetric laser beam with ellipticity $\varepsilon = \omega_{1}/\omega_{11}$ where ω_{1} and ω_{11} are the spot-sizes perpendicular and parallel to the lasing junction. The fiber fundamental mode (LP_{01} or HE_{11}) has a symmetric intensity distribution which is approximately gaussian and having a spot-size W_F. Most good coupling experiments reported to date in the literature have used buried heterostructure (BH) lasers which have the following characteristics: (i) linear light output versus current curve, (ii) single-transverse or spatial mode to achieve high η, (iii) single-longitudinal mode to achieve low chromatic dispersion, and (iv) stable and symmetric field pattern. This last consideration is important for high η; but even then, there is a large mismatch in the spot-sizes, e.g., W_L for the laser is 1.7 μm and W_F for the fiber is 3.9 μm (typical value at 1.55 μm). Hence poor η is obtained when butt-coupling the laser to a plane-ended optical fiber, typically < 10%.

RESULTS

Methods have to be developed to match W_L to W_F. This is done in the following ways: (i) modify the fiber end, (ii) introduce a micro-optic component between the laser and fiber, and (iii) a combination of (i) and (ii) above. In category (i) we have the following options: hemispherical taper end, hemispherical tipped lens, quadrangular end, high-index microbead lens, high-index taper end, and conical and etched lenses. In category (ii) the following have been reported: cylindrical lenses; ball and graded-index (GRIN) lens; plane convex and GRIN lens; cylindrical and GRIN lens, etc.

The best bench experiment has been reported by the method [1] of using high-index microbead lens where a coupling loss of 1.6 dB was achieved, corresponding to a 70% launching efficiency. In package design, this value was 2.1 dB and hence a < 3 dB loss is realistic in future package design. For more details, the reader is referred to Ref. [1] and the references cited therein, and also to Ref. [2].

More recently, a novel lens design has been reported in Ref. [2] in which typical η = 40%, and best η = 70% (1.6 dB loss as well), have been achieved with extremely low optical power feedback into the laser (of the order of 10^{-7}). Not many researchers have addressed themselves to this important problem. The main feature of this coupling optics is an aspherically shaped lens modeled from the end of a silica thread of an appropriate length. The focal length of the lens exceeds 100 μm and hence the large distance between laser and fiber greatly reduces the optical feedback. Loss due to spherical aberration was minimized by the aspherical shape of the lens. For further details, the reader is referred to Ref. [2]

CONCLUSION

The coupling of single-mode semiconductor lasers into single-mode optical fibers presents many difficult problems. Many options are available to attain the mode matching condition and hence increase the launching efficiency. It is reasonable to expect a < 3 dB coupling loss in well-designed packages. To decrease the optical feedback, the distance between the laser and the fiber should be as large as possible, and approximately 100 μm.

REFERENCES

[1] G.D. Khoe et.al., "Progress in monomode optical-fiber interconnection devices", IEEE/OSA J. Lightwave Technology, Vol. LT-2, 217 (1984).

[2] W. Bludau, R. H. Rossberg, "Low-loss laser-to-fiber coupling with negligible optical feedback", IEEE/OSA J. Lightwave Technology, Vol. LT-3, 294 (1985).

LOW-TEMPERATURE PRESSURE-DEPENDENT MAGNETO-OPTIC MEASUREMENTS IN STRAINED-LAYER SUPERLATTICES

E. D. Jones
Sandia National Laboratories, Albuquerque, NM 87185

ABSTRACT

Low-temperature magneto-optic data for $In_{0.2}Ga_{0.8}As$/GaAs and GaAs/$GaP_{0.2}As_{0.8}$ strained-layer superlattice systems are discussed. Also presented are low-temperature, pressure-dependent magneto-optic data for the InGaAs/GaAs SLS system.

INTRODUCTION

Magneto-optical measurements in semiconductors provide important information about band structures and band parameters. As an example of the utility of these kinds of measurements, we will describe results obtained in the strained-layer superlattice (SLS) systems $In_{0.2}Ga_{0.8}As$/GaAs and GaAs/$GaP_{0.2}As_{0.8}$.

Strained-layer superlattices are multilayer structures grown from lattice mismatched materials where the layers are thin enough so that the mismatch is accommodated by coherent strain. This lattice mismatch and the attendant biaxial strain has been used to create light-hole InGaAs/GaAs conductors.[2] The built-in strain provides a fundamental modification of the semiconductor band structure which can be exploited to tailor the optical properties of these kinds of materials. The theoretical advantages and design flexibilities of these SLS structures have been discussed by Osbourn.[3,4] Preliminary magneto-optic data for the InGaAs/GaAs and GaAs/GaPAs SLS have been previously presented.[5-7]

EXPERIMENTAL

The $In_{0.2}Ga_{0.8}As$/GaAs SLS structures used in these studies were grown by computer-controlled molecular beam epitaxy[8], while the GaAs/GaPAs SLS samples were prepared by metal-organic vapor deposition.[9] For all the structures discussed here, there are nominally 50 cycles of alternating layers with well thicknesses of about 100 Å. Typical 77 K mobilities are 4000 cm^2/Vs and two-dimensional carrier concentrations are 10^{12} cm^{-2}. Complete information regarding these structural and electrical parameters for the samples can be found in Refs. 5-7.

The experimental apparatus and techniques for performing hydrostatic-pressure-dependent magneto-optic measurements at low temperatures and high magnetic fields have been described in Ref. 6. For all the data discussed here, the magnetic field direction was normal to the SLS layers.

RESULTS

The magnetic-field-induced energy shift ΔE, relative to the band-gap energy E_g is given by (cgs units)

$$\Delta E = (n + \frac{1}{2}) (e\hbar B/\mu c) \qquad (1)$$

where n is the index of the Landau level, e is the electronic charge, $\hbar$ is Planck's constant over 2π, B is the magnetic field, and μ is the reduced effective mass given by $1/\mu = 1/m_e + 1/m_h$, where m_e and m_h are respectively the conduction-band mass and the valence-band mass. All masses are expressed in terms of the free electron mass m_o.

For each SLS structure listed in Table I, the reduced effective mass μ was determined by measuring the magnetic field dependence of the interband Landau-level transition energies. For the GaAs/GaPAs n-type SLS sample (R833) at atmospheric pressure and T = 1.6 K, a fit of Eq. (1) to the data yielded $1/\mu = 18.8$. Table I lists the values for $1/\mu$ for the various samples. For both GaAs/GaPAs SLS structures (R830 and R833), an analysis of the magnetic field dependence of a conduction-band to impurity (acceptor level) transition gave a value for the conduction-band effective mass $m_e = 0.073$, which is in excellent agreement with the value $m_e = 0.074$ obtained by Worlock, et. al.[10] for the GaAs/AlGaAs superlattices. Thus, with a knowledge of μ and m_e, a value of $m_h = 0.20$ is calculated.

Table I Magneto-optic parameters at atmospheric pressure for various strained-layer superlattices

Sample	E_g(meV)	$1/\mu$	m_e	m_h
InGaAs/GaAs, p-type (T0030)	1308	19.9	0.069	0.19
InGaAs/GaAs, n-type (M316)	1318	20.7	0.069	0.16
GaAs/GaPAs, n-type (R830)	1530	18.3	0.073	0.22
GaAs/GaPAs, n-type (R833)	1533	18.8	0.073	0.20

The results of similar measurements for the InGaAs/GaAs SLS structures (M316 and T0030) at atmospheric pressure and T ≦ 4 K are summarized in Table I. In these samples, we also observed a magnetic-field-dependent conduction-band to impurity (acceptor level) transition which gave an conduction-band effective mass $m_e = 0.069$. This value for m_h is consistent with our measured conduction-band effective mass value $m_h = 0.073$ obtained in the GaAs/GaPAs SLS structures when it is scaled to $In_{0.2}Ga_{0.8}As$. Thus, values of m = 0.16 and $m_h = 0.19$ are calculated for these two samples. These magneto-optic determined values for the in-plane valence-band masses are in good agreement with the magneto-transport determined values of $m_h \approx 0.15$ obtained in similar p-type InGaAs/GaAs SLS structures.

The band-gap energy pressure-coefficients for the InGaAs/GaAs SLS samples (M316 and T0030) are summarized in Table II. A more detailed description can be found in Ref 6. As can been seen from Table II, dE_g/dP = 12.2 and 10.2 meV/kbar in the n- and p-type samples, respectively. These results are in qualitative agreement with the experimental values obtained for the GaAs/AlGaAs superlattices.[6]

Table II Pressure Coefficients for InGaAs/GaAs SLS ($T \leq 4K$)

Sample	dE_g/dP (meV/kbar)	dlog(μ)/dP (percent/kbar)
InGaAs/GaAs, p-type (T0030)	10.2	3.2 ± 0.9
InGaAs/GaAs, n-type (M316)	12.2	1.8 ± 1.0

The pressure coefficients of the reduced-effective mass μ for the two InGaAs/GaAs SLS structures were determined by measuring the magnetic-field dependence of the interband Landau-level transition energies at each pressure. The results for these measurements have been described in Ref. 6 and are summarized in Table II. The experimental uncertainties of the reduced effective mass are about 3%. As can be seen in Table II, the measured pressure coefficients dlog(μ)/dP are (1.8 ± 1.0) percent/kbar for the n-type sample (M316) and (3.2 ± 0.9) percent/kbar for the p-type sample (T0030).

Using magneto-transport measurements, Schirber, et. al.,[11] have measured the pressure dependence $d\log(m_e)/dP$ of the conduction-band effective mass in a similar n-type InGaAs/GaAs SLS sample to be (1.3 ± 0.3) percent/kbar. Thus, the average experimental pressure coefficient $d\log(m_h)/dP$ for the in-plane valence-band mass can be calculated to be (5.0 ± 3.5) percent/kbar. While the error estimates are large, the indications are that the pressure dependence for the in-plane valence-band mass m_h may be larger than for the conduction-band mass m_e. However, an estimate for the pressure coefficient for this mass, based on simple $\vec{k} \cdot \vec{p}$ theory gives a value for $d\log(m_h)/dP$ of about 1 percent/kbar. At the present time, these differences are not understood. Meanwhile, we are proceeding with pressure-dependent magneto-optic measurements on the GaAs/GaPAs SLS structures and also with pressure-dependent magneto-transport measurements on p-type InGaAs/GaAs SLS in order to resolve these disparities.

ACKNOWLEDGMENTS

This work was done in collaboration with R. Biefeld, L. Dawson, T. Drummond, I. Fritz, P. Gourley, G. Osbourn, and J. Schirber.

This work was supported by the U.S. Department of Energy under contract number DE-AC04-76DP00789.

REFERENCES

1. J. E. Schirber, I. J. Fritz, and L. R. Dawson, Appl. Phys. Lett. 46, 187 (1985).
2. P. L. Gourley and R.M. Biefeld, Appl. Phys. Lett. 45, 751 (1984).
3. G. C. Osbourn, J. Appl. Phys. 53, 1585 (1982).
4. G. C. Osbourn, Phys. Rev. B27, 5126 (1983).
5. E. D. Jones, H. Ackermann, J. E. Schirber, T. J. Drummond, L. R. Dawson, and I. J. Fritz, Solid State Commun. 55, 525 (1985)
6. E. D. Jones, H. Ackermann, J. E. Schirber, T. J. Drummond, L. R. Dawson and I. J. Fritz, Appl. Phys. Lett. 47, 492 (1985).
7. E. D. Jones, J. E. Schirber, I. J. Fritz, P. L. Gourley, R. M. Biefeld, L. R. Dawson, and T. J. Drummond, to be published - Mat. Res. Bull.
8. I. J. Fritz, L. R. Dawson, and T. E. Zipperian, Appl. Phys. Lett. 43, 846 (1983).
9. R. M. Biefeld, G. C. Osbourn, P. L. Gourley, and I. J. Fritz, J. Electron. Mater. 12, 903 (1983).
10. J. M. Worlock, A. C. Maciel, A. Petrou, C. H. Perry, R. L. Aggarwal, M. Smith, A. C. Gossard, and W. Wiegmann, Surf. Sci. 142, 486 (1984).
11. J. E. Schirber, T. E. Zipperian, G. C. Osbourn, L. R. Dawson, and J. B. Snelling, Mat. Res. Bull. 20, 871 (1985).

PICOSECOND LASER INDUCED NONLINEAR EXCITATION OF SEMICONDUCTORS

W. E. Bron
Indiana University, Bloomington, IN 47405

ABSTRACT

A review is presented on the results of recent investigations of laser excitation, and subsequent decay, of a coherent optical phonon state in III-V semiconductors.

RESULTS AND DISCUSSION

We have reported[1,2] the generation, and subsequent decay, of near-zone-center, coherent longitudinal optical (LO) phonons in GaP and ZnSe. Generation of coherent optical phonons is achieved through coherent Raman excitation (CRE) and detection is (for the most part) obtained through time resolved coherent anti-Stokes Raman scattering (TRCARS). The general experimental techniques are discussed in references 2 and 3. The coherent phonon state most closely resembles a classical harmonic oscillator. Thus the generation process may be described by

$$\mu(\ddot{Q} + \Gamma\dot{Q} + \omega^2 Q) = q\cdot(\overleftrightarrow{R}\vec{E}_\ell\cdot\vec{E}_s - \frac{4\pi e^*}{\varepsilon_\infty}\vec{P}_{NL}). \qquad (1)$$

In eq. 1, μ is the reduced lattice mass, Γ a phenomenological decay rate, ω the frequency associated with the harmonic oscillator (that of LO phonons in the present case), q is a unit vector oriented along the LO phonon propagation direction, $\bar{R}$ the appropriate Raman tensor, $\vec{E}_\ell$ and $\vec{E}_s$ are fields associated with the two laser beams involved in CRE,[1] e* the effective lattice charge, and ε_∞ the high frequency dielectric constant. Thus, eq. (1) implies that the coherent amplitude, ⟨Q⟩, of the coherent phonon state is driven by forces involving a Raman interaction and by forces associated with the longitudinal component of the nonlinear polarization, $\vec{P}_{NL}$.

It follows that the intensity of the TRCARS signal can be written as[4]

$$I_c(\Delta t) = AS\int_{-\infty}^{\infty} dt\,|\vec{E}_p(t+\Delta t)\times[N(\bar{R}\cdot q)Q(t) + \bar{\chi}^{(3)}\vec{E}_\ell(t)\vec{E}_s(t)]|^2. \qquad (2)$$

The parameters A and S are defined in reference 4, $\vec{E}_p$ refers to the intensity of a probe laser pulse which can precede or be delayed in time by arbitrary amounts, Δt, from the time of creation of the coherent phonon state at $\Delta t = 0$, and, $\bar{\chi}^{(3)}$ is an effective third-order nonlinear electronic susceptibility.

It is clear from eq. (2) and Fig. 1 that there are also two contributions to $I_c(\Delta t)$; one due to the response of the lattice, $I_c^l(\Delta t)$ and one due to the response of the electronic system, $I_c^e(\Delta t)$, of the solid. As has been demonstrated in ref. (1), the contribution due to I_c^e is observed only as long as the temporal overlap exists among the three laser beams (~ 50 ps) (circles in Fig. 1), whereas I_c^l can be observed for times of the order of hundreds of picoseconds (crosses in Fig. 1). Hence, the lattice component represents free fall dephasing of the coherent phonon state. In the case of GaP and ZnSe, it is observed that $I_c^l(\Delta t) \propto \langle Q\rangle^2 \propto \exp(2t/T_2)$. Moreover, the temperature dependence of the dephasing time, $T_2/2$, of LO phonons is observed[1,2] to correspond to three-phonon anharmonic processes within a temperature range from 0 to 150K. In addition, it is observed[1,2] that the temperature dependence of $T_2/2$ as measured via $I_c^l(\Delta t)$ is the same as that observed in traditional measurements of the linewidth of the spontaneous (incoherent) Raman scattering intensity. This result was foreshadowed by earlier theoretical predictions of the dephasing through anharmonic decay of a coherent phonon state[5] and by incoherent phonons.[6]

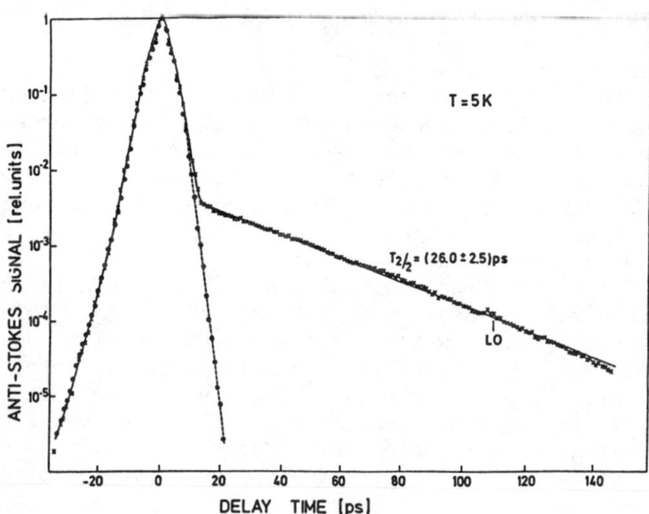

Fig. 1 CARS signal as a function of delay time. Data obtained for a high-quality sample of GaP held at a temperature of 5K.

It is also clear from eq. (2) that once $Q(t)$ and $\bar{R}$ are known from experiment, the elements of $\chi^{(3)}$ can be extracted from $I_c(\Delta t)$ provided the temporal profiles of the laser pulses are determined. The profiles can be obtained through independent measurements of the autocorrelation and the crosscorrelation signal, plus $I_c^e(\Delta t)$[2,4]. The analysis, though tedious, can indeed be carried out,[2,4] and yields all the real χ' and imaginary χ'' components of $\bar{\chi}^{(3)}$. For example, the components for GaP under our experimental conditions are found (in units of 10^{-10} esu) to be: $\chi'_{1221} = 2.3 \pm 0.3$, $\chi''_{1122} = -0.8 \pm 0.8$, $\chi'_{1221} = 2.9 \pm 0.2$, $\chi''_{1221} = -0.7 \pm 0.3$, $\chi'_{1111} = 1.7 \pm 0.5$ and $\chi''_{1111} = -0.3 \pm 1.1$. The corresponding values for ZnSe (in units of 10^{-11} esu) are: $\chi'_{1122} = 9.1 \pm 1.6$, $\chi''_{1122} = -4.0 \pm 1.4$, $\chi'_{1221} = 6.9 \pm 0.5$, $\chi''_{1221} = -1.7 \pm 2.0$, $\chi'_{1111} = 9.2 \pm 1.9$ and $\chi''_{1111} = -4.6 \pm 3.0$.

Nonzero values of χ'' (the imaginary components) imply that real absorption of laser light must have occurred. Since the laser energies $\hbar\omega_\ell$, $\hbar\omega_s$, $\hbar\omega_p$ are all less than the electronic gap energy, E_g, of the semiconductor, excitation across the gap must proceed by multiphoton absorption. It is now known, however, that "hot" electron-hole pairs created in this way relax to an electron-hole plasma at some minimum of the conduction band in a time short[7] compared to our temporal resolution of approximately one picosecond.[3] The LO phonons under observation possess an energy $\hbar\omega \sim 50$ meV and wavevector $|\vec{q}| \sim 3000\ \text{cm}^{-1}$. Thus electrons which scatter from these phonons can gain or lose energy and momentum of these amounts. But this turns out not to be possible for the following reasons. The important relevant parameter is the gradient of the conduction band surface. For example, if one assumes a very simple parabolic conduction band for, say, GaP and an effective mass of $m^* = 0.29\ m_\ell$[9] one finds that LO phonons which can interact with an electron must possess $|\vec{q}| \sim 10^5\ \text{cm}^{-1}$ or higher. This holds for electronic states which are of the order of 1eV above the band minimum (~ 20 LO phonon energies). This is, of course, particularly so for the electron-hole plasma with energies near the band minimum. Thus, $T_2/2$ is not influenced by the presence of the plasma, as has indeed been observed by us. However, the plasma does interfere with the $\langle Q\rangle^2$ through its influence on the second term on the right hand side of eq. (1). We have shown experimentally, as well as theoretically,[8] that if either $(E_\ell)^2$ or $(E_s)^2$ become sufficiently strong, then higher order terms in P_{NL}; particularly, free electron contributions to $\chi^{(3)}$, may change the sign of this term. The resulting decrease in the total generating force on the coherent phonon state (right side of eq. (1)), then results in decrease in the LO phonon concentration.

The examples, presented above, of nonlinear excitation in semiconductors due to interaction with laser light, are but a few which illustrate the potential of these techniques. More investigations on the possible existence of stimulated phonon decay, phonon breakdown, phonon renormalization effects, and free-electron-phonon interactions are either underway or planned for the future.

The author acknowledges the collaboration of Drs. B. K. Rhee and J. Kuhl, and financial support through ARO DAAG29-83-K-0091.

REFERENCES

1. J. Kuhl and W.E. Bron, Physica 117B/118B, 532 (1983), Solid State Commun. 49, 935 (1984).
2. W.E. Bron, J. Kuhl, B.K. Rhee, submitted for publication.
3. J. Kuhl and D. von der Linde, "Picosecond Phenomena III", Ed. K.B. Eisenthal, R.M. Hochstrasser, W. Kaiser and A. Laubereau. (Springer, Berlin, 1982) 201-4.
4. B.K. Rhee, W.E. Bron, and J. Kuhl, Phys. Rev. B30, 7358 (1984).
5. P. Carruthers and K.S. Dy, Phys. Rev. 147, 214 (1966).
6. A.A. Maradudin and A.E. Fein, Phys. Rev. 128, 2589 (1962).
7. See J.A. Kash, J.C. Tsang and H.V. Hvam, Phys. Rev. Lett. 54, 2151 (1985) and references cited therein.
8. B.K. Rhee and W.E. Bron, submitted for publication elsewhere.
9. S.A. Abagyan, G.A. Ivanov, A.P. Izergin and Yu. E. Shanurin, Sov. Phys. Semiconductors 6, 985 (1972).

PICOSECOND TIME RESOLVED STUDIES OF NONRADIATIVE RELAXATION IN RUBY AND ALEXANDRITE

S. K. Gayen, W. B. Wang, V. Petričević and R. R. Alfano
Institute for Ultrafast Spectroscopy & Lasers
Physics Department
City College of New York
New York, NY 10031

ABSTRACT

Dynamics of the nonradiative transitions between the $^4T_2$ pump band and the ^2E storage level of the trivalent chromium ion in ruby and alexandrite crystals is studied using the picosecond excite-and-probe absorption technique. A 527-nm picosecond pulse excites the $^4T_2$ state of the Cr^{3+} ion, and an infrared picosecond probe pulse monitors the subsequent growth and decay of population in the excited states as a function of pump-probe delay. An upper limit of 7 ps is determined for the nonradiative lifetime of the $^4T_2$ state in ruby. A vibrational relaxation time of 25 ps for the $^4T_2$ band in alexandrite is estimated. The time to attain thermal equilibrium population between the ^2E and $^4T_2$ levels of alexandrite following excitation of the $^4T_2$ band is estimated to be ~ 100 ps.

INTRODUCTION

Two technologically important solid state laser crystals are ruby ($Cr^{3+}:Al_2O_3$) and alexandrite ($Cr^{3+}:BeAl_2O_4$). Although the characteristics of the absorption, excitation and emission spectra of these crystals are well documented,[1-4] there is a paucity of any such detailed investigation of the nonradiative relaxation processes among the excited states of the laser-active Cr^{3+} ion in these hosts. However, knowledge about nonradiative processes is crucial for selecting and formulating the design criteria for solid state laser materials. In this report, we present first, direct picosecond time-resolved measurements of the excited state decay dynamics in these systems using the picosecond excite-and-probe absorption technique.

The laser-active ion in both ruby and alexandrite is trivalent chromium. The absorption spectrum of both the systems is characterized by two broad bands attributed to the $^4A_2 \rightarrow ^4T_2$ and $^4A_2 \rightarrow ^4T_1$ transitions in Cr^{3+}. These broad bands serve as the pump states which relax nonradiatively to the metastable ^2E level. Ruby is a three-level laser system where the laser action arises from the $^2E \rightarrow ^4A_2$ transition, and the emission is over two narrow lines centered at 694.3 and 692.9 nm, the well-known R lines. The $^2E \rightarrow ^4T_2$ energy gap in alexandrite is only 800 cm^{-1} as compared to 2300 cm^{-1} in ruby (Fig. 1). This smaller energy gap allows thermal repopulation of the $^4T_2$ level from the ^2E. The metastable ^2E level in alexandrite thus acts as a storage level for vibronic

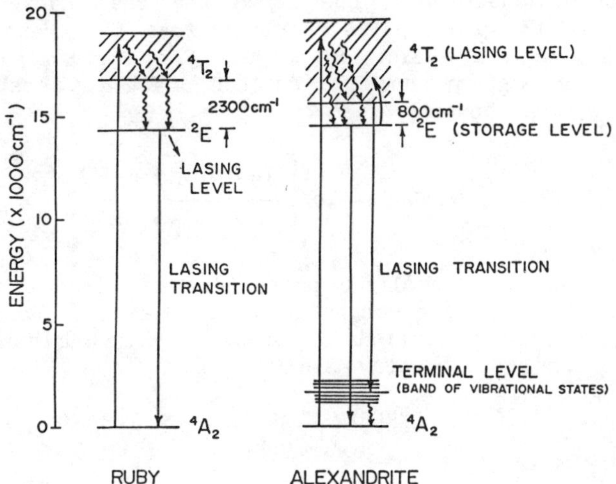

Fig. 1 A simplified energy-level diagram of Cr^{3+} in ruby and alexandrite. Only the levels directly associated with laser action are shown.

transitions from the $^4T_2$ state to the vibrational levels of the $^4A_2$ ground state. In this mode of operation, alexandrite is a four-level laser system and the emission is wavelength tunable over the continuous range from 701 to 818 nm. In addition to this phonon-terminated vibronic mode alexandrite, like ruby, has been demonstrated to lase on the R-line at 680.4 nm as well. The objective of the present work is to understand the transition dynamics between the $^4T_2$ and the ^2E levels as well as the intra-$^4T_2$ vibrational relaxation in these two crystals.

EXPERIMENTAL APPARATUS AND TECHNIQUES

The picosecond excite-and-probe absorption arrangement was used to study transitions from the $^4T_2$ and ^2E states in ruby and alexandrite. A 7-ps, 527-nm pump pulse excites higher lying vibrational states of the broad $^4T_2$ band. These excited vibrational states relax predominantly via nonradiative transitions resulting in a growth of population in the zero- and lower lying vibrational states of the $^4T_2$ and the metastable ^2E state. This growth of population is monitored by an infrared probe pulse which may be wavelength tuned from 2-5 μm. Both the pump and the probe pulses are derived from a single, amplified 1054-nm pulse generated by a passively mode-locked Nd:Glass laser. Time delay between the pump and the probe pulses may be varied continuously from 0-400 ps. Details of the experimental arrangement, detection scheme, signal averaging and processing technique are presented elsewhere.[4]

In this measurement, one looks for the change in optical absorption of the probe pulse resulting from the changes initiated in the sample by the exciting pulse. For a given pump-to-probe delay, the absorption from the excited state is measured by the change in optical density given by

$$\Delta OD(t) = \log\left[\frac{(I_s / I_r)_u}{(I_s / I_r)_p}\right], \tag{1}$$

where I_s/I_r is the normalized probe intensity, and u and p stand for unpumped and pumped conditions of the sample. The kinetics of the excited-state transitions is studied by measuring $\Delta OD(t)$ as a function of pump-probe delay time.

EXPERIMENTAL RESULTS

A. Ruby

The literature dealing with the relaxatior times of the nonradiative transitions in ruby and transitions from the $^4T_2$ state

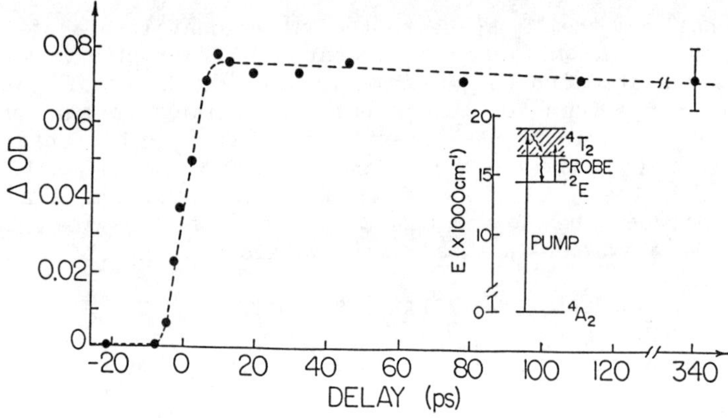

Fig. 2 Rise and decay of population in the excited ^2E state of ruby probed by a 3.4-μm pulse at room temperature. The sample is in the form of a cylinder 4.5 mm long and 9 mm in diameter with c-axis parallel to the axis of the cylinder, and contains 0.04% Cr_2O_3 by weight. Measurements are taken along 4.5-mm path length of the sample. Both the pump and the probe pulses are linearly polarized at right angles to the c-axis of the ruby sample. Inset shows the relevant energy level diagram of $Cr^{3+}:Al_2O_3$, and the pump, probe and relaxation transitions. The zero time is accurate within 5 ps.

in particular, covers a period of more than two decades. The theoretical and experimental estimates of the lifetime of the $^4T_2$ state in ruby also have a very wide range, from 10^{-6} to 10^{-13} s. While the theoretically predicted value is as low as 10^{-13} s, experiments have only provided upper limits to the lifetime, the lowest so far being 0.3 ns.[4]

The result of the present picosecond excite-and-probe measurement in ruby showing the time evolution of the optical density at 3.4 μm in the ^2E state at room temperature is displayed in Fig. 2. The salient features of the curve are a rapid rise followed by a long decay. The long-lifetime decay reflects the depopulation of ^2E state, which shows no appreciable change over the time scale of this measurement. This is expected since the lifetime of the ^2E state at room temperature is 3 ms. The risetime (time for growth of population from 10% to 90%) is ~ 10 ps, limited by the duration of the pump pulse. This leads to an experimentally determined shortest-so-far upper limit of ~ 7 ps for the $^4T_2$ lifetime in ruby.

B. Alexandrite

The situation in alexandrite is much different and more complex. Figure 3(a) displays the time evolution of the optical density in the excited states at 3.4 μm at room temperature. The curve is characterized by a ~ 25-ps risetime, longer compared to that in ruby; followed by a multicomponent decay. The faster component has a decaytime of ~ 80 ps, whereas the longer component does not exhibit any appreciable change within the timescale of this measurement.

To further understand this complicated behavior, we extended the present measurement to a probe wavelength of 2.4 μm, and the result is presented in Fig. 3(b). The salient features of the curve are a laser-pulse-width-limited sharp rise followed by a ~ 25-ps lifetime decay. The key to the different behavior at the two probe wavelengths lies in the difference in energy of the two photons and the different transitions which may be initiated by those photons. With 3.4 μm as the probe wavelength, transitions from both ^2E to $^4T_2$, and zero vibrational level to higher vibrational levels of $^4T_2$ manifold are energetically possible. However, a 3.4-μm photon is not energetic enough to cause a transition even from some higher-lying vibrational state of $^4T_2$ to the $^4T_1$ band. On the other hand, a 2.4-μm photon has sufficient energy to initiate transitions from ^2E to some higher vibrational states of $^4T_2$, and from higher vibrational states of $^4T_2$ to $^4T_1$ band. However, this energy is greater than the width of $^4T_2$ band, so any transition from zero vibrational level of $^4T_2$ to higher vibrational levels is ruled out.

We attribute the change in optical density at 2.4 μm to absorptive transition from higher-lying vibrational states of $^4T_2$ to the $^4T_1$ state. The resolution-limited risetime indicates the increase of population as long as the pump pulse is on. The ~

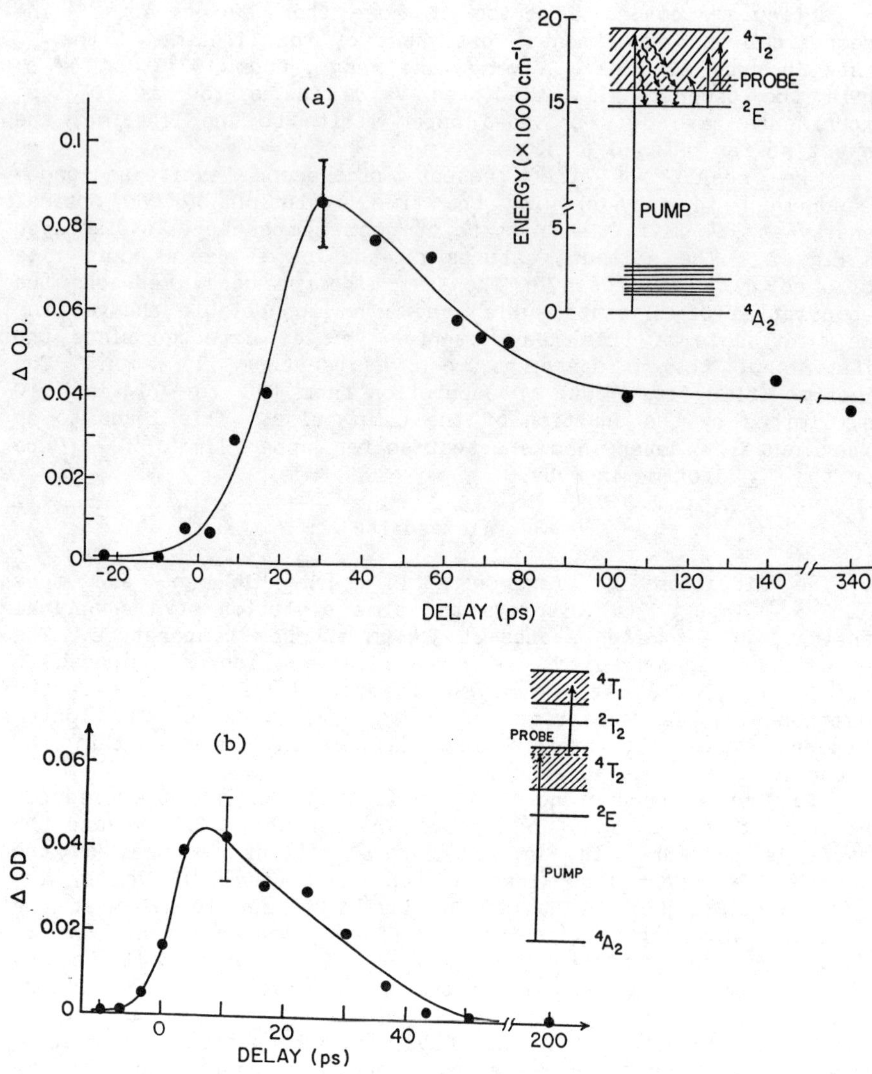

Fig. 3 Time evolution of the optical density (a) at 3.4 μm and (b) at 2.4 μm in the ^2E and $^4T_2$-zero-vibrational states of alexandrite at room temperature. The sample is in the form of a rectangular parallelepiped 12mm X 10mm X 2.4mm in dimension and contains 0.4 at.% Cr^{3+}. Both the pump and the probe pulses are linearly polarized parallel to the b-axis of the alexandrite crystal. Measurements were taken through the 2.4-mm path length of the sample. Insets show relevant energy level diagram and transitions of interest.

25-ps decay time reflects the vibrational relaxation time of higher-lying vibrational states of $^4T_2$ band which are excited by the pump pulse. This relaxation leads to growth of population in the zero vibrational level of the $^4T_2$ band and in the metastable ^2E level. The change in optical density at 3.4 μm is determined by the population of the two levels and the absorption cross sections for 3.4-μm photon from the two levels. The risetime of 25 ps for the optical density at 3.4 μm (Fig. 3(a)) is consistent with the 25-ps vibrational relaxation time that populates the two levels. The flat region at long time in 3.4-μm curve indicates that a thermal equilibrium has been reached between the ^2E and $^4T_2$ level. Its depopulation is due to radiative transitions. Since the room temperature fluorescence lifetime of alexandrite is 262 μs, no appreciable change is observed in the time scale of this measurement. The 80-ps lifetime component reflects the nonequilibrium population kinetics among the ^2E and $^4T_2$ levels and is related to the "effective thermalization time". A more detailed and quantitative treatment of these features is outside the scope of the present report.

SUMMARY

Picosecond excite-and-probe absorption measurements in ruby yield a new, shortest-so-far upper limit of 7 ps for the nonradiative lifetime of the $^4T_2$ state. In alexandrite, a vibrational relaxation time of ~ 25 ps is estimated for the $^4T_2$ band. Thermal equilibrium in population between the ^2E storage level and $^4T_2$ band in alexandrite is attained within about a 100 ps.

This research is supported by Army Research Office.

REFERENCES

1. The literature dealing with optical and spectroscopic properties of ruby is very extensive. For a brief review of optical and laser properties see W. Koechner, Solid-State Laser Engineering, (Springer-Verlag, 1976), pp. 44-52 and references therein.
2. Richard C. Powell, Lin Xi, Xu Gang, Gregory J. Quarles and John C. Walling, Phys. Rev. B32, 2788 (1985), and references therein.
3. John C. Walling, Otis G. Peterson, Hans P. Jenssen, Robert C. Morris and Wayne O'Dell, IEEE J. Quantum Electron. QE-16, 1302 (1980), and references therein.
4. S. K. Gayen, W. B. Wang, V. Petričević, R. Dorsinville and R. R. Alfano, Appl. Phys. Lett. 47, 455 (1985). References 2-16 in this paper constitute a list of research performed on nonradiative relaxation in ruby over the last two decades.

A New CW, Tunable Laser in the 1.05-1.08 µ Region

L.D. Schearer
University of Missouri-Rolla, Rolla, Mo. 65401

M. Leduc
Laboratoire de Spectroscopie Hertzienne, Paris, FRANCE

ABSTRACT

The CW laser properties at room temperature of lanthamide hexa-aluminate doped with Nd (LNA) were investigated and the tuning characteristics in the 1.05-1.08 µ region obtained. When a 1 cm long ´c´ axis crystal is pumped by an Ar laser(514nm) or a Kr laser(752nm), CW emission is obtained with slope efficiencies of 10% and 25% respectively. A Lyot filter within the laser cavity forces the LNA crystal to oscillate in either of two major bands centered at 1082nm and 1054nm. A thin etalon permits tuning over 12nm.

INTRODUCTION

There have been continuing efforts to find other Nd doped crystalline hosts which retain the advantages of YAG as to performance without showing some of its disadvantages such as Nd sepregation during crystal growth, low solubility of Nd in the YAG lattice, and limited tuning in an important region of the spectrum. Earlier work[1,2] with LNA has confirmed interest in this new material as a new, high-power, solid state Nd laser[3].

Among the advantages of LNA is the ability to obtain Nd concentrations more than 6 times greater than Nd in YAG without severe concentration quenching or severe ion segregation. This earlier work and that reported here demonstrate that LNA has an efficiency and threshold comparable to Nd:YAG. The broad fluorescence spectrum also suggested the possibility of obtaining efficient, CW tunable emission in a region of the spectrum which is generally devoid of tunable sources.

We report here the characteristics of the LNA laser and its tuning characteristics when pumped by either the green output of an Ar ion laser or the IR output of a Kr ion laser.

EXPERIMENTAL RESULTS

The laser cavity used in these experiments consists of a meniscus which acts both a lens for the pump radiation and as a curved mirror forming one end of the cavity. The pump radiation is focussed on the crystal. The meniscus is transparent to the pump radiation but totally reflecting at the laser wavelengths. An internal lens of focal length 2.75 cm is located beyond the crystal and focusses the fluorescence to points between a plane, 90% transmitting output mirror and infinity. The total cavity length is 25 cm. This configuration was used for fluorescence, gain, and threshold measurements.

The fluorescence spectrum from the LNA when pumped by the Kr ion laser is shown in fig. 1 . The spectrum was obtained by replacing the output mirror by an optic fiber. The fluorescence was then examined with a 1/2 m grating spectromenter and detected with a germanium detector. The fluorescence shows 2 broad, principal peaks, centered at 1054nm and 1082nm. The full-widths at half height of the two peaks are 4.4 and 7.4nm, respectively. Fluorescence widths in LNA are considerably broader than the corresponding transitions in YAG.

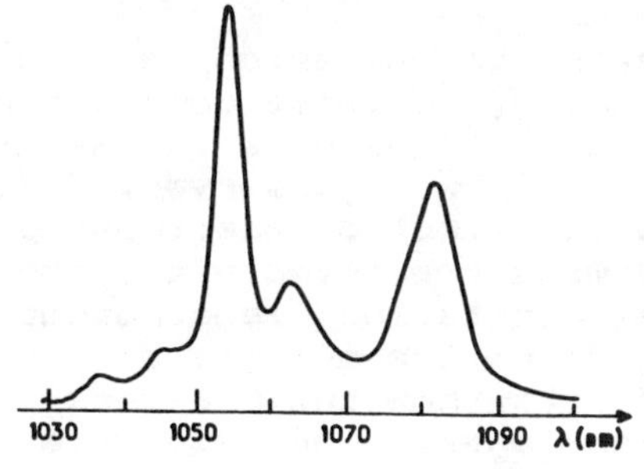

Figure 1. LNA fluorescence.

With Kr laser pumping(752nm) the threshold for the 1 cm crystal was less than 90 mW. The slope efficiency was 25% with an output mirror transmission of 10%.

With a 4 plate Lyot filter and a 0.2 mm uncoated etalon in the cavity the output could be tuned over the range shown in fig. 2. A 1mm thick etalon with a 50% reflective coating yielded single mode power of 150 mW at 1083nm when pumped with 1.9W of 752nm power from the Kr laser.

We also obtained CW laser emission from a 3.2cm LNA rod when pumped by Kr arc lamps in a conventional YAG laser cavity. Substantial improvement is expected with longer LNA crystals.

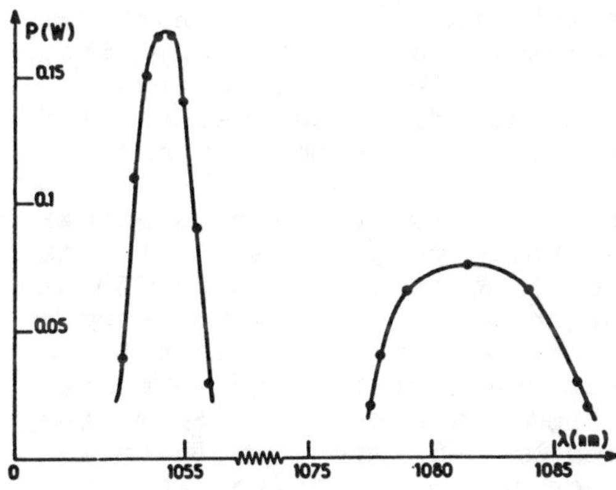

Figure 2. Tuning Curves, LNA.

REFERENCES

1. D. Vivien, A.M. Lejus, J. Thery, R. Collongues, J.J. Aubert, R. Montcorge, and F. Auzel, C.R. Acad. Sci. Paris, 298, 195(1984).
2. Kh. S. Bagdasarov, L.M. Dorozhkin, A.M. Kevorkov, Yu. I. Krasilov, A.V. Potemkin, A.V. Shestakov, and I.I. Kuratev, Sov.J. Quantum Electr. 13, 639(1983).
3. L.D. Schearer, M. Leduc, D. Vivien, A.M. Lejus, and J. Thery, IEEE J. Quantum Electr. to be published, May 1986.

SPECTROSCOPIC PROPERTIES OF $La_3Lu_2Ga_3O_{12}$:Nd^{3+} CRYSTALS

D.K. Sardar*, G.J. Quarles, and R.C. Powell
Oklahoma State University, Stillwater, OK 74078

M.R. Kokta
Union Carbide Corp., 750 S. 32 St., Washougal, WA 98671

ABSTRACT

Laser-pumped, single-pass transmission measurements were performed on a $La_3Lu_2Ga_3O_{12}$ crystal containing 3.3 at.% Nd^{3+} to determine the usefulness of this material as a laser. No optical gain was observed. To ascertain why, time-resolved, site-selection spectroscopy measurements were made to determine the effects of ion-ion interaction and two-photon excitation spectroscopy measurements were made to determine the effects of excited state absorption. The results show the presence of very weak energy transfer between ions in non-equivalent crystal field sites and the presence of very strong two-photon absorption transitions.

We report here the results of spectroscopic measurements obtained on Nd-doped $La_3Lu_2Ga_3O_{12}$ (LLGG) crystals relevant to the use of this material in laser applications. Neodymium may be substituted completely into gallium based garnet systems with minimum distortion of the oxygen polyhedra and maximum distance between Nd lattice positions which minimizes ion-ion interaction problems.

For low power excitation in the visible spectral region, the fluorescence emission originates from the $^4F_{3/2}$ metastable state at wavelengths longer than 850 nm as seen in other Nd-doped materials. The fluorescence lifetime of this emission decreases from 290 µs at 10 K to 205 µs at 300 K. Attempts were made to observe single pass gain at 1059 nm using high power pumping from the doubled output of a modelocked Nd-YAG laser. No gain was observed but the fluorescence emission shifted to the visible spectral region between 400 and 780 nm. This demonstrates the presence of multiphoton excitation processes and subsequent emission from higher energy metastable states similar to that observed in other Nd-doped systems.[1] The emission is found to be due to transitions from the $^2P_{3/2}$ state with a lifetime of 0.32 µs and a rise time of 317 ns, and the $^2(F2)_{5/2}$ state with a lifetime of 2.52 µs and a rise time of 200 ns.

The temperature dependence of the fluorescence life-

*Permanent Address: University of Texas at San Antonio, San Antonio, TX 78285.

time after lower power excitation can be fit by expression of the form $\tau_f^{-1}=\tau_r^{-1}+C\{\exp[\Delta E/(k_BT)]-1\}^{-1}$, where τ_f and τ_r are the fluorescence and radiative lifetimes, respectively. The last term describes the quenching of the lifetime due to radiationless processes involving the absorption of phonons of energy ΔE. C is a constant containing the matrix element for these transitions. The values obtained from fitting the data are τ_r=295 μs, C=1538 μs^{-1}, and ΔE=30 cm^{-1}.

Dye laser, time-resolved spectroscopy techniques were used to investigate the characteristics of energy transfer between Nd^{3+} ions in nonequivalent crystal field sites. The time evolution of the fluorescence emission in the 880 nm spectral region was monitored at temperatures of 10 K and 100 K. At higher temperatures thermal broadening of the lines prevented accurate measurements. Each transition has two distinct peaks associated with ions in two nonequivalent crystal field sites. No energy transfer was detected between ions in these two types of sites at the two temperatures investigated. However the energy difference of their transitions is the same as the activation energy for lifetime quenching which may mean that transfer does occur at higher temperatures and is associated with the lifetime quenching.

The presence of energy transfer across the inhomogeneously broadened spectral band for each transition could be observed at 100 K and was analyzed using the technique developed for studying energy transfer in doped glasses.[2] The results give an average energy transfer rate of α=392 μs.

These results show that compared to the same concentration of Nd-ions in other host materials, Nd-doped $La_3Lu_2Ga_3O_{12}$ has a higher quantum efficiency, weaker nonradiative decay, and weaker energy transfer processes. However, multiphoton excitation processes prohibited the observation of single pass gain for the type of experimental conditions used here.

ACKNOWLEDGMENTS: The OSU part of this research was sponsored by the U.S. Army Research Office. One of the authors (DKS) gratefully acknowledges financial support from from the University of Texas at San Antonio for this work.

REFERENCES

1. G.J. Quarles, G.E. Venikouas, and R.C. Powell, Phys. Rev. B <u>31</u>, 6935 (1985).
2. S.A. Brawer and M.J. Weber, Appl. Phys. Lett. <u>35</u>, 31 (1979).

OPTICAL PROPERTIES OF $LaMgAl_{11}O_{19}$:Cr,Nd*

M. D. Shinn, W. F. Krupke, J. A. Caird, and H. W. Newkirk
Lawrence Livermore National Laboratory,
P. O. Box 5508, Livermore, CA 94550

ABSTRACT

The optical properties of Cr-doped $LaMgAl_{11}O_{19}$ (LMAO:Cr) and LMAO:Cr,Nd have been measured. Evidence of Cr to Nd radiationless energy transfer was observed. Since LMAO:Nd may become a useful laser material, the prospect of further improving laser performance through Cr-sensitization is discussed.

INTRODUCTION

Researchers in the Soviet Union have demonstrated laser action in LMAO:Nd with an efficiency about 20% higher than YAG:Nd.[1] Increased efficiency was anticipated as the optical quality improved. It is known that Cr sensitization doubles the laser efficiency of GSGG:Cr,Nd.[2] The LMAO lattice has three six-fold coordinated Al sites,[3] and readily accepts Cr ions. The optical properties of LMAO:Cr, and the observed energy transfer in LMAO:Cr,Nd is the topic of this study.

EXPERIMENTAL RESULTS

Two doped LMAO crystals were grown using the Czochralski technique. The first crystal (LMAO:Cr,Nd) contained about 9×10^{20} Cr ions/cc and 3×10^{19} Nd ions/cc as determined by the ICP technique. The second crystal (LMAO:Cr) contained about 2.3×10^{20} Cr ions/cc. The segregation coefficients for Cr and Nd are 1.2 and 0.5 ± 0.2, respectively.

The polarized absorption and emission spectra of the LMAO:Cr crystal is shown in Fig. 1. The positions of the absorption bands and the presence of $^2E\rightarrow^4A_2$ line emission near 700 nm are indicative of an intermediate strength crystal field at the Cr site, like alexandrite or YAG:Cr. The fluorescence decay after laser excitation (λ_{ex}=640 nm) is the sum of three single exponentials, with lifetime values of 0.28 ms, 1.0 ms, and 2.7 ms. From this data it appears that Cr is incorporated in all three six-fold coordinated Al sites. There are also two four-fold coordinated Al sites in LMAO, however there is no evidence, based on absorption spectra analysis, which indicates that Cr enters these sites.

When the LMAO:Cr,Nd sample was excited at 632.8 nm, Nd fluorescence was observed. This is indicative of Cr to Nd energy transfer.

*Work performed under the auspices of the U.S. Department of Energy by Lawrence Livermore National Laboratory under Contract No. W-7405-ENG-48.

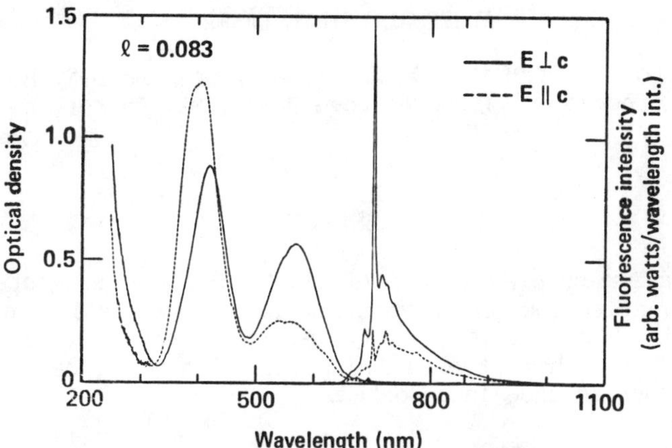

Figure 1. Absorption and emission (λ_{ex}=632.8 nm) of $LaMgAl_{11}O_{19}$:Cr at room temperature.

The Nd fluorescence peaked about 40 μs after excitation of the Cr ions and exhibited a decay that was the sum of two exponentials with lifetimes of 225 μs and 700 μs. There are two La sites in LMAO and the lifetime data indicates that Nd ions occupies both sites.[4] The Cr decay was nonexponential and decayed more rapidly than in the LMAO:Cr sample. At this time it is not known how much of the increased nonradiative rate is due to energy transfer to Nd ions and how much is due to Cr ion concentration quenching. Assuming the increased decay rate is entirely due to energy transfer to Nd ions, the Cr to Nd energy transfer efficiency may be calculated:[5]

$$\eta = 1 - \int_0^\infty I_{s'}(t)dt / \int_0^\infty I_s(t)dt \quad (1)$$

where $I_s(t)$ is the intensity of the Cr ions alone and $I_{s'}(t)$ is the intensity of the Cr ions in the presence of Nd. Application of (1) to the decay data yields η=0.87, a value comparable to GSGG:Cr,Nd. The 40 μs value for the Nd ion maximum intensity after Cr ion pumping is also comparable to GSGG:Cr,Nd. However, it should be stressed that the degree of concentration quenching of the Cr ion lifetime will greatly affect these results. Based on this data, it would appear that the Cr-sensitization of Nd in LMAO will be about as effective as found for GSGG. A similar increase in laser efficiency should also be found. Further experiments are planned to check the energy transfer efficiency.

REFERENCES

1. Kh. S. Bagdasorov, et.al., Sov. J. Quantum Electron. 13, 1082 (1983).
2. E. V. Zharikov, et.al., Sov. J. Quantum Electron. 13, 82 (1983).
3. A. Kahn, et.al., J. Appl. Phys. 52, 6864 (1981).
4. M. Gasperin, et.al., J. Solid State Chem. 54, 61 (1984).
5. M. J. Taylor, Proc. Phys. Soc. 90, 487 (1967).

ATHERMAL Nd:BEL LASERS

R.C. Morris, T. Chin, O. Kafri, M. Long, and D.F. Heller
Allied Corporation, 7 Powder Horn Drive, Mt. Bethel, NJ 07060

ABSTRACT

Recent measurements of dn/dT in Nd:BEL have suggested that a judicious selection of crystallographic orientation can yield laser rods that exibit greatly reduced thermal lensing. Such "athermal" rods have been fabricated. Their optical properties and laser performance is reported.

INTRODUCTION

Neodymium doped $La_2Be_2O_5$ (Nd:BEL) is a monoclinic, congruently melting compound which has emerged as an attractive, intermediate gain, single crystal laser material. Considerable development of this material, in the areas of crystal growth, spectroscopy, physical property characterization, and laser physics has already taken place at Allied and elsewhere[5,6]. Recent results demonstrate effective thermal lens compensation and improved laser performance from certain crystallographic cuts of Nd:BEL. These results together with the previously accumulated body of information suggest that Nd:BEL may substantially outperform more conventional Nd^{+3} laser materials, including Nd:YAG and Nd:Glass, in a number of important applications.

THERMO-OPTIC AND STRESS-OPTIC EFFECTS

As a consequence of its monoclinic structure, BEL is optically biaxial. The relation between the mutually orthogonal optical principal vibration directions, $\underline{X}$, $\underline{Y}$ and $\underline{Z}$, and the crystallographic directions $\underline{a}$, $\underline{b}$ and $\underline{c}$, is shown in Figure 2.1. The optical vibration direction $\underline{Y}$ coincides with the crystallographic 2-fold rotation axis $\underline{b}$. The optical $\underline{X}$ and $\underline{Z}$ directions (along with the optic axes) lie in the crystallographic $\underline{a}$-$\underline{c}$ plane (a mirror plane). In this plane, $\underline{a}$ and $\underline{c}$ are related by the angle β, which is 91°33', while c and $\underline{Z}$ are related by $\rho = 31.7°$ at $\lambda = 1.0$ µm. The refractive indices n_1, n_2 and n_3 are 1.964, 1.997 and 2.035 at $\lambda = 1.0$ µm. More complete refractive index data from $\lambda = 600$ nm to 2000 nm are given in reference (2).

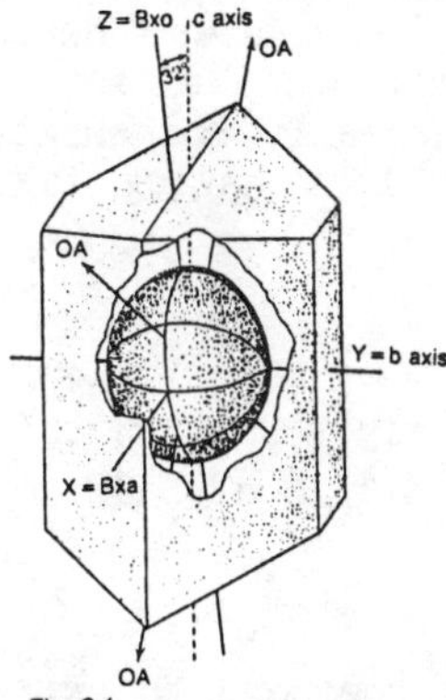

Fig. 2.1
Monoclinic Nd:BeL Showing the Orientation of the Indicatrix for Sodium Light. Optic Plane is Parallel to (010). Bxa and Bxo denote the acute and obtuse optical axis bisectrices, respectively.

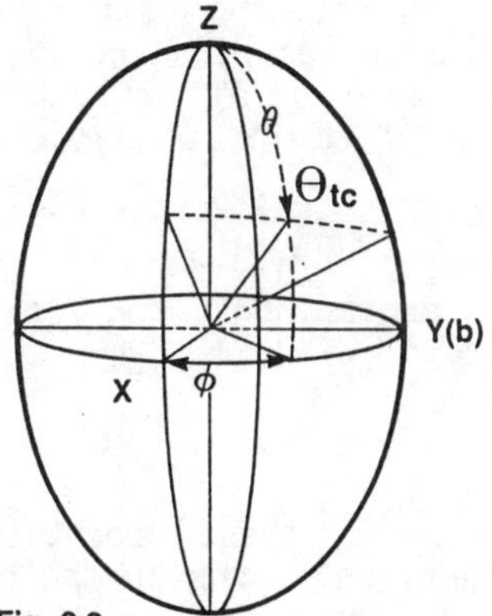

Fig. 2.2.
Locus of zero dn/dT vibration directions on $La_2Be_2O_5$ indicatrix.

The first order temperature coefficients of the three principal refractive indices have recently been determined at λ = 1.15 µm in the temperature range of 25-50°C. (The temperature coefficient of the angle ρ has not been measured). The measured dn/dT values are shown in Table 1.

Table 1. - Thermo-optic Effects in $La_2Be_2O_5$
@ λ = 1.15 µm, 25-50°C

Polarization	(n)	α_n*(ppm/°C)	dn/dT(ppm/°C)
E‖X	1.96	+1.46	+2.86 ±0.5
E‖Y(b)	1.99	+0.77	+1.53 ±0.5
E‖Z	2.03	-3.06	-6.23 ±0.5
≈E‖a		+1.10	+2.19
≈E‖c		-1.85	-3.68
E‖[Z-63.0°-Y]		0	0
E‖[Z-54.4°-$\overline{X}$]		0	0

*$\alpha_n \equiv 1/n\ \delta n/\delta T$

The existence of positive and negative values for dn/dT implies that dn/dT = 0 for a locus of non-principal vibration directions as shown in Figure 2.2. Knowing the values for the refractive indices n_1, n_2 and n_3, and their temperature dependences dn/dT, the locations of the zero dn/dT vibration directions on the indicatrix can be calculated. The thermally compensated (i.e. zero dn/dT) orientations correspond to loci having fixed values of the polar angle $\theta = \theta_{tc}$ given by:

$$\theta_{tc} = \arctan \left[n_3^{-3} \frac{dn_3}{dT} / [n_1^{-3} \cos^2 \phi \frac{dn_1}{dT} + n_2^{-3} \sin^2 \phi \frac{dn_2}{dT}] \right]^{-1/2} \quad (3)$$

In general, these vibration directions will correspond to extraordinary ray paths in the crystal.

THERMAL LENSING

Thermal lens formation in laser rods arises from the combined effects of the thermooptic, stress-optic and elastic responses of the material to the temperature and stress fields resulting from thermal loading. An expression for the focal length f in a cylindrical laser rod of isotropic material is:(3)

$$f^{-1} = \frac{P}{KA} \left[\frac{1}{2} \frac{dn}{dT} + \alpha C n^3 + \frac{2r(n-1)}{L} \right] \quad (4)$$

where K is the thermal conductivity, A is the rod cross sectional area, P is the heat dissipated in the rod, α is the thermal expansion coefficient, n is the refractive index, r is the rod cylinder radius, L is the rod length, and C is a measure of the stress-optic effect.

The last term, arising from rod end elastic distortions, can be made negligibly small by mounting the rod such that the ends are not pumped. Generally, of the remaining two terms, the dn/dT contribution is larger by about an order of magnitude than the stress-optic contribution.

A full theoretical description of thermally induced optical distortion in an asymmetric crystal such as BEL should account for the anisotropy of thermal expansion and thermal conductivity as well as the more complex elastic, thermo-optic and stress-optic effects. However, since, dn/dT is generally the dominant contribution we undertook a preliminary experimental investigation of rod geometry thermal lensing in BEL, hoping to locate first order temperature compensated crystal cuts.

In this study, moire deflectometry,[4] using HeNe (λ = 632.8 nm and 1.15 µm) probe beams, was used to measure the strength of thermally induced lensing as a function of input power in a series of 6.35 mm diameter x 90 mm long BEL rods of varying crystallographic orientation in the Y-Z plane. Average pump powers in the range of 0-1 kW were used. The results are shown in Figure 2-3. Thermal lensing compensation is found very near the predicted Z-63°-Y vibration direction.

Figure 2-4 shows the thermal lens strength as a function of input power for the nearly compensated Z-62°-Y (vibration direction) cut for two wavelengths, λ = 632.8 nm and λ = 1.15 µm. The variation of thermal lensing rate with wavelength while significant is not enormous over this wavelength region. The moire deflectometry lens power measurement is directional and measures one cylindrical component of the lens at a time. The symbols ∥ and ⊥ in Figure 2-4 refer to the cylindrical lens components with axes ∥ and ⊥ to the extraordinary ray vibration direction, respectively. It can be seen that there is a residual astigmatic lensing rate of about 0.1 diopters/kW for this cut at λ = 1.15 µm, probably the result of the (anisotropic) stress-optic effect.

Nd:Bel Thermal Lensing Rate vs. Polarization Direction @632.8 nm

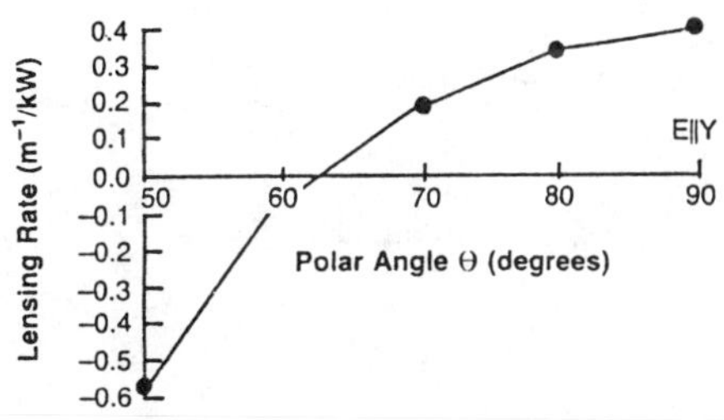

Figure 2-3. Rate of Thermal Lensing as a Function of Vibration Direction in Bel YZ Plane.

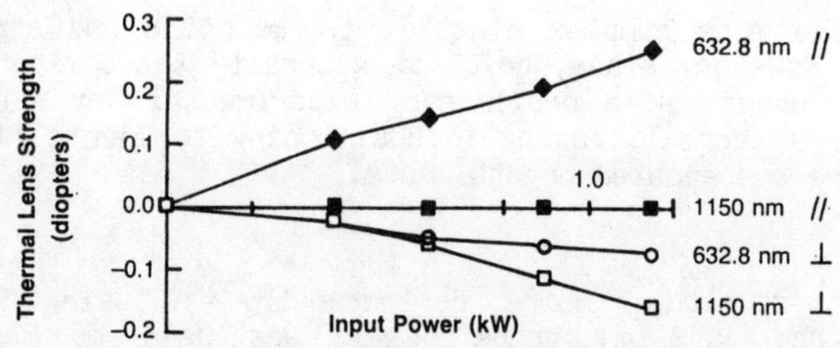

Figure 2-4. Thermal Lensing vs. Input Power for Z-62°-Y Cut Showing Wavelength Dispersion of Astigmatism.

One method for reducing or eliminating the residual thermal astigmatism for Nd:BEL relys on the use of a slab shaped gain element with a rectangular cross section, as shown schematically in Figure 2.5. Here laser beam propogation is straight through, parallel to the long axis x. Pumping and heat extraction are through the major faces parallel to the xz plane. The crystal cut is selected for zero lensing rate for the cylinder axis z parallel to the long edge of the rectangular aperture cross section. This is the axis of maximum (cylindrical) thermal lensing for this thermal geometry. Since there will be little or no thermal gradient in the z direction, the cylindrical thermal lens with axis y will be the product of two small numbers and therefore exceedingly weak.

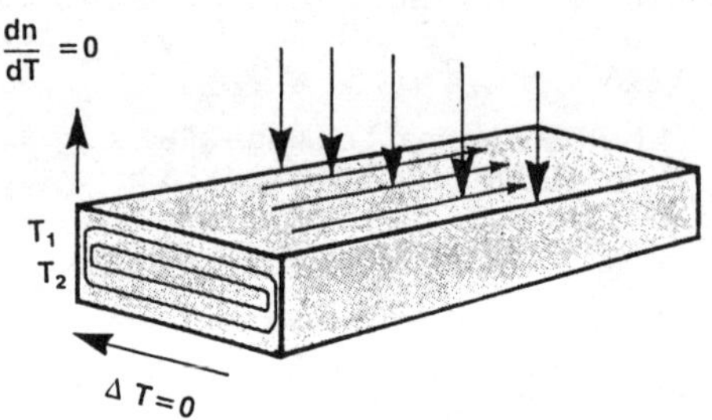

Fig. 2.5. Straight-Thru Nd:Bel Slab

LASER STUDIES

A Z-62°-Y Nd:BEL rod with dimensions (6.35 mmϕ x 80 mm long) was fabricated and a series of experiments were performed with this rod. Using a double ellipse pump chamber, the thermal-lensing was studied using moire deflectometry and results are summarized in Fig. 2-4.

An oscillator-amplifier set-up with the Z-62°-Y rod in the amplifier was used to measure the relative small signal gain at various wavelengths. The relative gain was then compared with gain predictions from emission cross-section measurements. The results are described by Table 2. It is worth noting that one can conclude from the results that excited state absorption is not a major concern for any transition studied in the 1.07, 1.08, or 1.36 μm band in Nd:BEL.

Table 2
Comparison of Small Signal Gain and Emission Cross Section

λ μm	$\sigma(\Theta,\lambda)^{+}$ 10^{-20} cm^2	$G(\lambda)/E^{*}$ J^{-1}	$G(\lambda)/G(\lambda=1.070)$	$\sigma(\Theta,\lambda)/\sigma(\Theta,\lambda=1.070)$
1.070	10.2	0.014	1	1
1.079	7.34	0.010	0.714	0.714
1.365	1.89	0.0025	0.179	0.185

*From Small Signal Gain Measurement, all values are ± 10%.
+From Emission Spectra (H.P. Jenssen et al., unpublished):

$$\sigma(\Theta,\lambda) = \sigma_y(\lambda)\sin^2\Theta + \sigma_z(\lambda)\cos^2\Theta$$

where $\sigma_{y,z}(\lambda)$ are the emission cross sections for y and z polarized light and Θ = 62° was the electric field polarization direction in the athermal rod.

Subsequently, the Z-62°Y rod was placed in a simple resonator and we obtained lasing at 1.07 μm, 1.079 μm, and 1.36 μm wavelengths. It was found that precautions were required to prevent the higher cross-section lines (1.07, 1.079 μm) from lasing when operating at 1.36 μm. Recent results, are summarized in Fig. 2.6. Furthermore, a 9.5 mm x 100 mm Z-62°Y rod produced 1 J per 20 nsec pulse at λ = 1.079 μm and 10 Hz repetition rate, i.e., 10 watts average power, 50 MW peak power.

SUMMARY

Athermal Nd:BEL is a material that has a good mix of laser parameters: high efficiency (especially Q-switched), high average power, thermal lens correction, and wavelength flexibility; these properties make Nd:BEL an excellent candidate for high average power operation.

REFERENCES

1. R.C. Morris, C.F. Cline, R.F. Begley, M. Dutoit, P.J. Harget, H.P. Jenssen, T.S. La France and R. Webb, Appl. Phys. Lett. 27, 444 (1975).
2. H.P. Jenssen, R.F. Begley, R. Webb and R.C. Morris, J. App. Phys. 47.
3. W. Koechner, Solid State Laser Engineering, Springer-Verlag, New York, p353, (1976).
4. O. Kafri, Opt. Lett., 5, 5455 (1980).
5. Birnbaum et al, "Engineering Design of Repetitvely Q-switched Solid State Lasers for Precision Ranging Applications," first report, NASA Contract Number NAS5-23698 (1979).
6. M. Birnbaum et al, "Repetitively Q-switched Nd:BEL Lasers", final report, NASA Contract Number NAS 5-25098 (1979).
7. T.S. Lomheim and L.G. DeShazer, Phys. Rev. B 20, 4343 (1979).

GSGG Crystal Growth and Quality: A Status Report

S. Stokowski, J. Caird, M. Shinn, L. Smith, and R. Wilder*
Lawrence Livermore National Laboratory, Livermore, CA 94550

ABSTRACT

Gadolinium scandium gallium garnet (GSGG) co-doped with Cr and Nd is the crystalline-laser material with the highest measured efficiency (5% absolute) under flashlamp pumping. It is presently being developed for a variety of laser applications. We report measurements of the present crystal quality of 1.5 to 2-inch diameter boules, finding 0.24 m^{-1} loss coefficient; $\lambda/50$ homogeneity in a 1/4-inch by 3-inch rod; and 0.4 nm/cm, birefringence. We have evidence that a one-micron absorption band, which appears frequently in Cr doped GSGG, is caused by Cr^{4+}, which is generated to compensate for Ca^{2+} impurities in the crystal. Scale-up of boule growth to 5-inch diameter is now underway.

SUMMARY

Gadolinium scandium gallium garnet (GSGG) co-doped with Cr and Nd is the crystalline material of choice for new, efficient lasers. The basis for this choice of GSGG:Nd,Cr is its high laser efficiency (due to Cr sensitization), good optical quality, resistance to solarization, acceptable thermal shock figure-of-merit, and the high degree of industrial experience in large GGG growth, which is a similar material. We have measured the laser performance of GSGG: Nd, Cr rods, the spectroscopic, optical, and thermo-mechanical properties of GSGG [1] and report here our measurements of the present crystal quality of 1.5 to 2-inch diameter boules.

GSGG:Nd,Cr laser rods have almost twice the efficiency of YAG:Nd because of Cr sensitization. We measured a 5.0% absolute and 7.0% slope efficiency from 1/4" dia. by 3" long rods compared with 2.7% absolute and 3.7% slope from YAG:Nd rods in the same cavity. These rods were cored from 1.5"-diameter boules grown by Airtron, Allied, or Material Progress Corp.

The efficiency of a laser and its beam quality are dependent on the optical quality of the laser material. Optical quality for this application is defined by the measured values for transmission loss, wavefront distortion, and birefringence. We measured these properties on about 30 GSGG:Nd,Cr 1/4"-diameter rods and compared them with values obtained from YAG:Nd rods (Fig. 1). To measure both absorption and scattering loss at 1.06 µm, our apparatus rejects light scattered beyond 250 microradians (the diffraction limit) by the test sample.

*Work performed under the auspices of the U.S. Department of Energy by Lawrence Livermore National Laboratory under Contract No. W-7405-ENG-48.

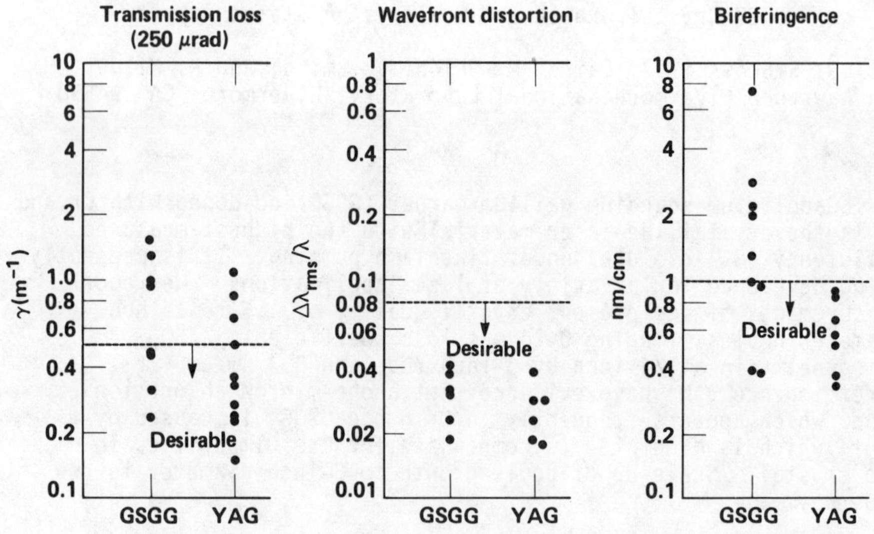

Fig. 1. Measured quality of 1/4" diameter by 3" GSGG and YAG rods.

Some GSGG:Nd,Cr boules have extremely high (200 m^{-1}) values of transmission loss at 1.06 microns caused by an 1100 nm band (Fig. 2) which is found only in GSGG and GGG crystals containing Cr. Through chemical analysis, we find the strength of this absorption band is proportional to the Ca impurity content in the crystal.

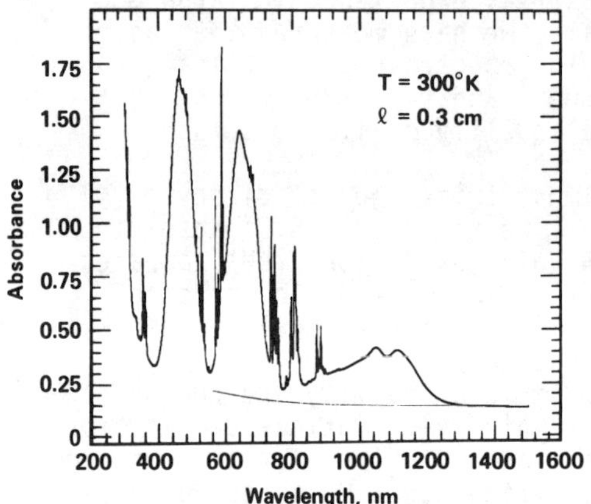

Fig. 2 A 1100-nm absorption band in GSGG:Nd,Cr; possibly due to Cr^{4+}.

We suggest then that Ca^{2+} impurities are charge-compensated by the formation of Cr^{4+}.

To demonstrate a high-efficiency, medium-power (1 KW) laser using GSGG:Nd,Cr as the laser material, we have contracted with Allied Corp. to grow 5-inch diameter boules from which we can cut up to five 10 x 20 x 1 cm plates. These crystal boules will be grown by the Czochralski technique. The iridium crucible containing the melt is 9 inches in diameter by 9 inches high. Approximately 30 kg. of high purity Sc_2O_3 will be used in this program.

REFERENCE

1. W. F. Krupke, M. D. Shinn, J. E. Marion, J. A. Caird, and S. E. Stokowski, "Spectroscopic, Optical, and Thermo-Mechanical Properties of GSGG:Nd,Cr", to be published in the J. of the Opt. Soc. of Am.

Cr^{3+} IN K_2NaScF_6: CRYSTAL GROWTH AND SPECTROSCOPY

L. J. Andrews and B. C. McCollum*
GTE Laboratories, Waltham, Ma. 02254

R. H. Bartram
University of Connecticut, Storrs, Ct. 06268

ABSTRACT

Crystal growth of the candidate tunable laser material $K_2NaScF_6:Cr^{3+}$ by a vertical Bridgman technique is described along with studies of its radiationless relaxation and excited state absorption.

INTRODUCTION

The ordered perovskite K_2NaScF_6 substitutionally incorporates Cr^{3+} into octahedral Sc^{3+} sites where it has a $^4T_{2g}$ lowest excited state, unit fluorescence quantum efficiency, and a peak stimulated emission cross section of 7×10^{-21} cm^2 at 768 nm. Because of these desirable spectroscopic characteristics, attempts were made to grow this material as high quality single crystals and to more fully characterize it for tunable near infrared lasers.

CRYSTAL GROWTH

The growth of K_2NaScF_6 single crystals at 1000 C was done in a vertical Bridgman apparatus using graphite and platinum crucibles sealed to prevent KF evaporation. It was found that batches containing stiochiometric mixtures of the component fluorides invariably resulted in marginal optical quality crystals in which a second, ordered perovskite phase could be detected by weak satellite lines in x-ray diffraction powder patterns and by inhomogeneity in the Cr^{3+} fluorescence lifetime. An examination of the ternary KF-ScF_3-NaF phase field demonstrated that off-stiochiometric batch mixtures lean in KF minimized the problem but could not eliminate it. Nevertheless, it was possible to produce sufficiently homogeneous material to be useful for spectroscopic studies.

EXCITED STATE ABSORPTION

Excited state absorption (ESA) measurements of $K_2NaScF_6:Cr^{3+}$ have provided the first example of absorption from the $^4T_{2g}$ state

*Present address: Polaroid Corp., Cambridge, Ma.

0094-243X/86/1460227-2$3.00

of Cr^{3+}. Only one ESA band was found in the visible, the $^4T_{1g}(b) \leftarrow {}^4T_{2g}$ transition at 19,500 cm^{-1} with a peak $\sigma = 1.4 \times 10^{-20}$ cm^2. No other features were detected between 15,000 and 40,000 cm^{-1} which shows that spin forbidden doublet $\leftarrow$ quartet transitions are not significant. Comparison of these results with prior work[1] on high field Cr^{3+} shows a substantial difference between absorption from the $^4T_{2g}$ and ^2E_g states of Cr^{3+}, the latter state leads to a far more congested ESA spectrum due to the higher density of terminal (doublet) states.

RADIATIONLESS RELAXATION

Thermal induced radiationless relaxation of the $^4T_{2g}$ fluorescence of $K_2NaScF_6:Cr^{3+}$ is well described as Arrhenian with a rate (s^{-1}) given by $1.2 \times 10^{13} \exp[-7270\ cm^{-1}/kT]$. Attempts to calculate this behavior with single-mode[2] or multimode[3] linear coupling models failed by 6-8 orders of magnitude. A new model was developed within the harmonic approximation incorporating the essential features of linear coupling to a_{1g} and quadratic coupling to t_{2g}. This new approach was found to provide order of magnitude agreement between calculated and measured rates when the degree of t_{2g} quadratic coupling (i.e., frequency change) is fixed by crystal field theory.

REFERENCES

1. W. M. Fairbank, G. K. Klauminzer, and A. L. Schawlow, Phys. Rev. B 11, 60 (1975).
2. K. Huang and A. Rhys, Proc. Roy. Soc. (London) A204, 406 (1950); K. Huang, Sci. Sinica 24, 27 (1981).
3. M. H. L. Pryce, in Phonons, ed. R. W. H. Stevenson (Plenum, N. Y., 1966), p. 403.

ACKNOWLEDGMENT

The authors wish to thank Dr. D. Gabbe of M.I.T. for providing a sample of $K_2NaScF_6:Cr^{3+}$ and NVEOL (DAAK20-82-C-0134) and ARO (DAAG29-82-K-0158) for providing support.

MECHANISM OF SCATTERING CENTERS FORMATION IN TITANIUM SAPPHIRE

Jaroslav L. Caslavsky
U.S. Army Materials Technology Laboratory, Watertown, MA 02172

INTRODUCTION

Since the Ti^{3+}:sapphire exhibits a broad vibronic emission band, it is considered a promising material for solid-state tunable lasers.[1] Although the Ti^{3+}:sapphire single crystals have been grown for several years by numerous methods,[2] they all have a common denominator: a high level of scattering centers. Recent research on the origins of the scattering centers in Ti^{3+}:sapphire single crystals indicates that the origin of the scattering centers is of an intrinsic nature. The problem is further amplified by the lack of suitable materials, namely those for crucibles and thermal shields needed for the growth. Nonreactivity of a protective gaseous atmosphere also becomes a pressing problem at these temperatures.

Ti_2O_3, also known as titanium sesquioxide, belongs to the same symmetry group as sapphire. It can be grown in the form of single crystals with the nonstoichiometric ratio of Ti:O richer in oxygen.[3] Ti_2O_3 reacts easily with nitrogen and or oxygen to TiN or TiO_2, respectively. Ti_2O_3 melts at approximately 1750°C. It forms with the Al_2O_3, a limited solid solution with eutectic at 1700°C at which point the solubility of Ti_2O_3 in Al_2O_3 is approximately 2.8 mol%.* This entails that the distribution coefficient has to be well below one. Consequently, a significant concentration gradient of the Ti ions shall be present in the Ti^{3+}:sapphire single crystals along the growth axis. TiO_2 is unstable in the Al_2O_3 melt. It loses oxygen and forms Ti_3O_5, which is a mixed valence oxide. It is stoichiometric and appears to be the most stable titanium oxide over a wide range of temperatures as well as partial pressures of oxygen. In the available literature, nothing is known about the Ti_3O_5 solid solubility in Al_2O_3. When a small amount of Ti_3O_5 is added to Al_2O_3 and melted under an atmosphere of argon, the melt solidifies in large transparent crystals of a color typical for the TI^{3+}:sapphire. However, such a material has an abundance of black inclusions. When an attempt was made to determine the composition of this second phase, no X-ray pattern from the second phase could be obtained because the amount of the second phase in the Al_2O_3 is too low. Raman spectroscopy also does not give any meaningful answer. Microscopic examination of thin sections of Ti_3O_5-doped sapphire revealed an apparent similarity with the scattering centers generally found in thin sections of Ti^{3+}:sapphires. In order to investigate the behavior of Ti_3O_5 in the Al_2O_3 melt, the following experiment was performed.

*The authors for the solubility paper are A. J. Strauss, M. M. Stuppi, and R. E. Fahey.

Ti^{3+} ions in the Ti^{3+}:sapphire single crystals grown by the Vertical Solidification of the Melt (VSOM) technique[2] are introduced by doping Al_2O_3 with Ti_2O_3. In this particular experiment, Ti_3O_5 was used as an alternative dopant material, while the other growth and charge parameters remained the same. The crystal obtained was free of cracks and of the usual pinkish color, but the amount of black inclusions was unusually high. From this crystal, a disk 10 mm thick, 75 mm in diameter was cut perpendicular to the growth axis and polished on both sides. This polished disk revealed that the scattering centers are situated in $(10\bar{1}2)$ planes. In other words, a pink "star sapphire" was obtained (see Fig. 1). After the crystal was annealed for 50 hours in 760-mm Hg pressure of oxygen at 1550°C, all colors disappeared. The pink transparent regions became water clear and the scattering centers changed from black to white.

This experiment suggests that the Ti_3O_5 oxide as such is not soluble in the Al_2O_3 melt. Ti_3O_5 is a mixed valence oxide which dissociates at the melting point of Al_2O_3 to TiO_2 and Ti_2O_3. Ti_2O_3 dissolves in the Al_2O_3 melt while TiO_2 loses oxygen and forms any of the Magnoly phases characterized by the formula Ti_xO_{2x-1} where x = 4 to 10. Since Ti^{3+}/Ti^{4+} ratio is controlled by the partial pressure of oxygen above the Al_2O_3 melt, which of the Magnoly phases formed can be inferred from Figure 2. There shall always be some oxygen partial pressure above the Al_2O_3 melt since the molten Al_2O_3 decomposes to a suboxide according to the reaction (1).

$$Al_2O_3 \Rightarrow Al_2O_{(g)} + O_{2(g)}. \tag{1}$$

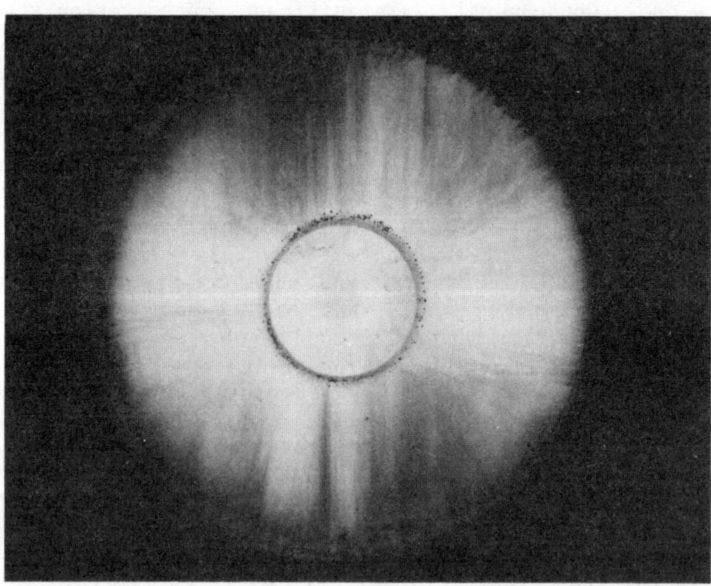

Figure 1. Ti^{3+} "star" sapphire $[10\bar{1}0]$ axis is perpendicular to the plane of the picture.

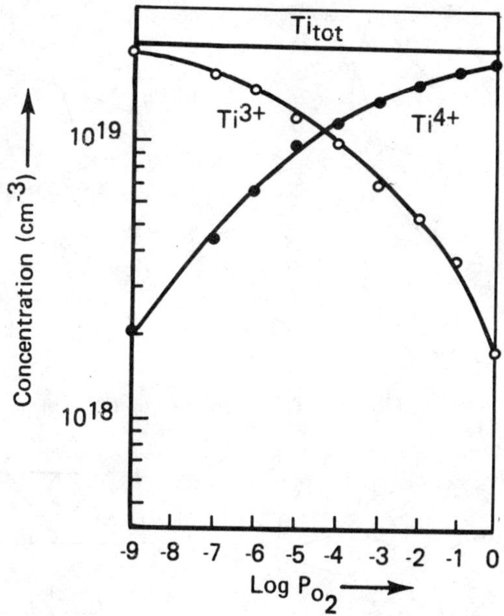

Figure 2. Relative concentration of Ti^{3+} and Ti^{4+} in sapphire melt as a function of P_{O_2}.

Kinetics of reaction (1) is controlled by total pressure which consequently will affect oxygen partial pressure and thereby various Magnoly phases are formed. All the Magnoly phases invariably segregate on $(10\bar{1}2)$ planes and if annealed at elevated temperatures, in an oxygen atmosphere, they oxidize to TiO_2. A study of the sapphire structure revealed that in most directions the stacking sequences follow the scheme where layers of oxygen atoms alternate with layers of aluminum atoms filling 2/3 of the octahedral sites. On the other hand, the stacking along the the $\langle 50\bar{5}\ 10\rangle$ directions is characterized by octahedral sites which are all filled with aluminum atoms (see layer A in the composite, Fig. 3). This sequence is followed by a layer B where the oxygen octahedra axes are rotated; nevertheless, all the octahedral sites are again occupied with aluminum atoms. In the next sequence, the axis of oxygen octahedra is further rotated and all of its octahedral sites marked 0 are empty. It is apparent that the negativity in the 0 layer in contrast to the A and B layers is enhanced, therefore the Ti^{4+} ions are attracted to these planes, consequently giving the "star" appearance to such sapphire single crystals. This conclusion can be even further substantiated by the fact that if large diameter undoped sapphire single crystals are cooled fast, they show perfect cleavages along the $(10\bar{1}2)$ planes.* On the other hand, "star"

*J. L. Caslavsky, to be published.

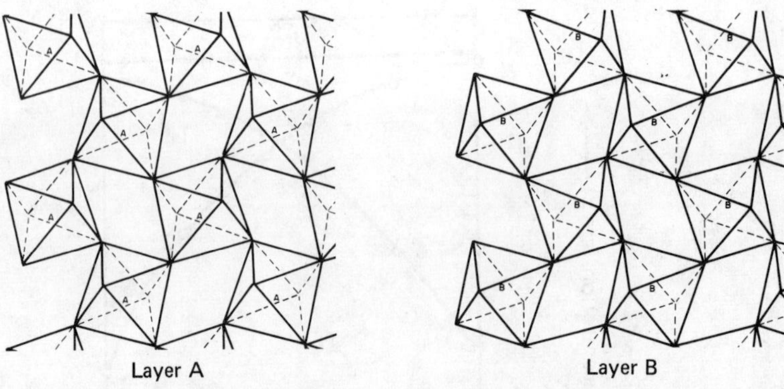

Layers A, B, O, are projected to (0001) plane

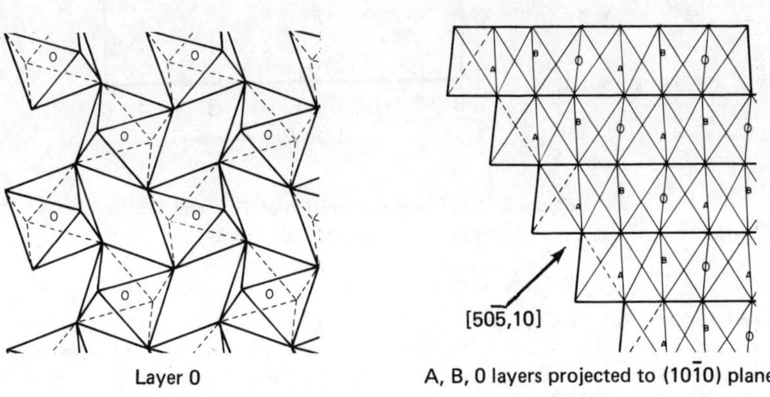

Figure 3. Computer image of Al_2O_3 structure.

sapphire is much less prone to crack formation during rapid cooling. This may explain the enhanced negativity of the O layer manifests itself in repulsive forces in (10$\bar{1}$2) planes which makes them the weakest planes in the sapphire structure. That would also explain why the presence of Ti^{4+} ions in sapphire crystals highly diminishes the formation of cracks.

CONCLUSION

Reaction (1) implies that certain partial pressure of oxygen will be present in any sapphire melt. This oxygen causes some of the Ti^{3+} ions to oxidize, consequently forming scattering centers. It is apparent that the sapphire melt shall always contain some amount of oxygen, thus imposing restrictions on the growth of the scattering centers free Ti^{3+}:sapphire single crystals. However, if crystals of large diameter are grown on a very flat interface, the scattering centers can be eliminated in about 2/3 of the inner core of the crystals. This suggests that the VSOM crystal growth technique is suitable for this purpose.

The absorption spectra of both the Ti^{3+}:"star" sapphire and the "star" sapphire obtained through the annealing do not show the anomalous absorption at approximately 800 nm.[4] This indicates that the Ti ion in any of the possible oxidation stages is not responsible for such an absorption.

REFERENCES

1. P. F. Moulton, Opt. News 8, 9 (1982).
2. J. L. Caslavsky, "The Growth Large Diameter Single Crystals by Vertical Solidification of the Melt," U.S. Army Materials Technology Laboratory, AMMRC TR 84-34 (1984); Proc. ISSCG 5, ed. E. Kaldis (North Holland, Amsterdam, 1984) Ch. 10.
3. T. B. Reed, R. E. Fahey, and J. M. Honig, Mat. Res. Bull. 2, 561 (1967).
4. R. C. Powell, J. L. Caslavsky, Z. AlShaieb, and J. M. Bowen, J. Appl. Phys. 58, 6 (1985).

DEVELOPMENT OF HIGH STRENGTH SOLID STATE LASER MATERIALS*

J. E. Marion,
Lawrence Livermore National Laboratory, Livermore, CA 94550

ABSTRACT

The threat of laser material fracture limits the average power of many lasers to modest levels. Rupture occurs when the tensile surface stress, which results from the temperature gradient within the component, exceeds its strength. To increase the power output potential we have focussed on methods to strengthen the amplifier slabs. Two basic approaches are used; the subsurface damage from machining, and inducing a compressive stress at the slab surface. We report results on several strengthened systems including GSGG, GGG and YAG crystalline hosts, and LHG-5 phosphate glass (Hoya Glass Corp.).

SUMMARY

Of the thermo-mechanical properties which determine the power handling capability of a high average power laser component, strength is the parameter most conducive to significant improvement by materials engineering.[1] Fracture mechanics relate the tensile strength, σ_f, to the flaw depth, a, by

$$\sigma_f = Y\,K_c(a)^{-1/2} \qquad (1)$$

where Y is near unity and K_c is the fracture toughness.[2] In most optical components, particularly laser rods and slabs, the highest tension and the largest flaws are both located at the surfaces, the flaws resulting from the fabrication process.

Optical surface preparation typically commences with a grinding operation which produces relatively shallow machining grooves, as well as a characteristic set of cracks: the lateral and radial crack systems.[3] The radial cracks are the most serious because they extend perpendicular to the surface to surprisingly large depths. The flaws are typically not optically detectable because displacement of the crack faces is usually less than the wavelength of light. Subsequent lapping and polishing may not remove sufficient material to eliminate these cracks.

There are two basic strategies for improving strength: reduce the depth of the flaws or induce a compressive stress that incorporates the damaged layer. Several methods for reducing the depth of the flaws exist. The simplest method is by modifying the existing grinding and polishing schedule so that large amounts of material are taken off at each step, removing the damage from the previous step and replacing it with lighter damage.[4] Alternately,

*Work performed under the auspices of the U.S. Department of Energy by Lawrence Livermore National Laboratory under Contract No. W-7405-ENG-48.

the entire damaged layer can be removed by acid etching. Various thermal treatments can lead to crack healing. All of these methods improve the strength by reducing the flaw size, a, in Eq. 1.

In contrast, inducing a surface compressive stress leads to strengthening because the applied stress must overcome the residual compression before the tension can build up to the level required to propagate the flaws. Chemical ion exchange, in which large ions replace smaller ones, giving a compressive surface stress is the most common adaptation of this strategy. Ion implantation also appears feasible. We use a surface coating which is in residual compression to give the required stress distribution.

The two basic strategies have fundamental differences. Removing all subsurface defects will result in a component that is as strong as the inherent theoretical strength (5-20 GPa). These components, while strong, are also fragile because any flaw introduced by abrasion for instance, will immediately reduce the strength in accordance with Eq. 1. The surface compression approach has less potential for very high strengths but gives more durable components; unless the flaws introduced during handling extend beyond the compressive layer, the component strength is largely unaffected. A final note on these methods, is that for low and moderate strength components, these two approaches can be used in combination to give superior strength components (i.e., see Fig. 2).

Our development of strengthened laser materials has focussed on candidate high average power laser materials: LHG-5 phosphate glass and the garnet crystalline hosts, yttrium aluminum garnet (YAG), gadolinium gallium garnet (GGG), and gadolinium scandium gallium garnet (GSGG). A factor of three strengthening has been demonstrated (Fig. 1) for large material-removal special polishes on the garnets, as well as a fifteen times improvement by an acid etching treatment. Strength is increased by about 4X while retaining a high quality optical surface, by repolishing the surface after etching. These results are the focus of a separate publication.[4]

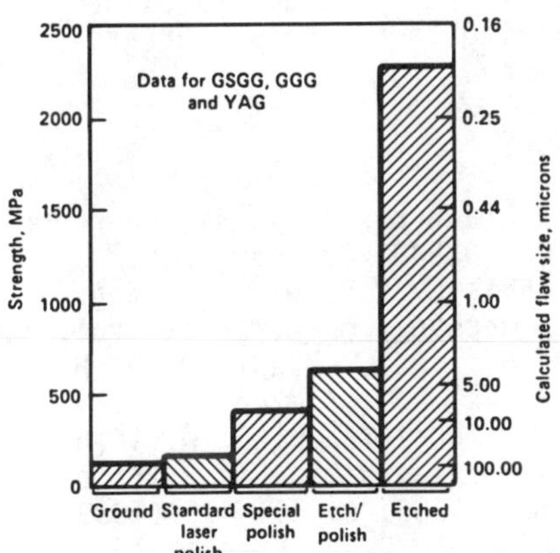

Fig. 1. Mean strength of laser garnets with various surface treatments. Special polish refers to the samples in which large amounts of material were removed during each grinding operation; 250 µm of material was removed during the 30-µm loose abrasive grind, 100 µm removed by 12-µm grit, and 75 µm removed by 3-µm grit, followed by polishing. Etching is for 20 min. in 200°C orthophosphoric acid.

The beneficial effect of a surface compressive layer on LHG-5 phosphate glass is illustrated in Fig. 2. Also, the effect of careful polishing is evident. Here, ion beam sputtered coatings of silicon nitride and silica (about 0.5 μm thick) were applied giving a significant surface compression. These coatings have the added advantage of being adaptable for evanescent wave suppression and perhaps for preventing the strength reduction which results from the interaction of water at crack tips under tensile stress.[1]

To reduce the threat of slab fracture, these strengthening methods must be combined with appropriate fracture testing and statistical analysis. Prudent design with brittle materials incorporates a Wiebull extreme value type fit of the fracture data with a scaling factor which is dependent on the flaw characteristics in order to estimate a safe operating stress level for a particular laser design.[1] In combination, these strategies can significantly increase the power output potential of high average power slab lasers.

Garnet specimens were finished by Nacin Optics, Monrovia, PA and by E. Prochnow, LLNL. The LHG-5 specimens were finished by Zygo Corp. Middlefield, CT. The ion beam coatings were applied by Optical Coating Laboratory, Santa Rosa, CA. Specimens were fractured in conjunction with S. M. Winfree.

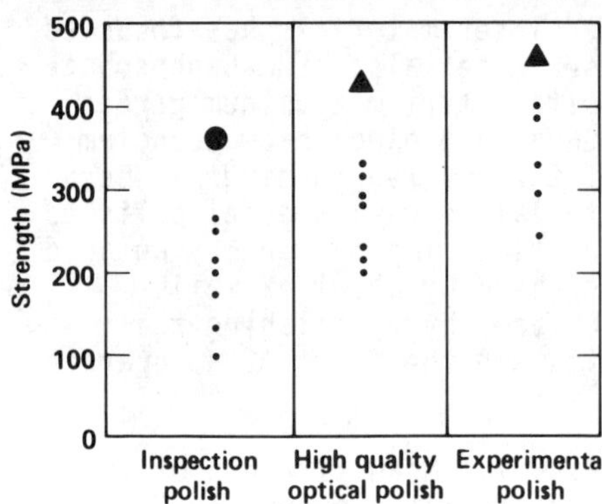

Fig. 2. Strength of LHG-5 phosphate glass as a function of surface finish, and with a surface compression coating. Small dots are data for uncoated samples, triangles are for silicon nitride (1500 MPa compression) and large dots are for silica coatings (200 MPa compression). The experimental polish consists of large material removal and sodium hydroxide etching to remove subsurface damage.

REFERENCES

1. J. E. Marion, "Fracture of Solid State Laser Slabs", Lawrence Livermore National Laboratory UCRL 93543, October 1985.
2. Handbook der Physik, Vol. 6, p. 551, G. R. Irwin, Springer, Berlin (1958); also see Fracture Mechanis of Ceramics, Vol. 1: Concepts Flaws and Fractography, eds. R. C. Bradt, D. P. H. Hasselman, F. F. Lange, Plenum Press, N.Y. (1974).
3. D. B. Marshall, A. G. Evans, B. T. Khuri-Yakub, J. W. Tien and G. S. Kino, "The Nature of Machining Damage in Brittle Materials", Proc. Royal Soc. London A385, p. 461-475 (1983).
4. J. E. Marion, "Strengthened Solid State Laser Materials", Appl. Phys. Lett. 47, no. 7, p. 694-696 (1985).

Sm^{2+} SENSITIZATION OF Nd^{3+} SOLID STATE LASERS

Stephen A. Payne and L. L. Chase
Lawrence Livermore National Laboratory,
P. O. Box 5508, Livermore, CA 94550

ABSTRACT

We have measured the $Sm^{2+} \rightarrow Nd^{3+}$ energy transfer efficiency in CaF_2 and have found that the transfer occurs through the dipole-dipole interaction. Although efficiencies of 0.9 can be achieved with reasonable levels of Nd^{3+} doping, the Sm^{2+} luminescence begins to quench at T>220K.

INTRODUCTION

We show that the transfer efficiency can be obtained by three independent methods, each of which will be discussed in turn below.[1]

RESULTS AND DISCUSSION

Method I: For an electric dipole-dipole interaction between the sensitizer (Sm^{2+}) and the activator (Nd^{3+}), the transfer rate can be described as[2]

$$P = \frac{R_o^6}{R^6 \tau_s} = \left(\frac{3}{4\pi}\right)\left(\frac{\hbar c}{n}\right)^4 \frac{1}{R^6 \tau_s} \int \frac{F_s(E) \cdot \sigma_a(E)}{E^4} dE \qquad (1)$$

where F_s is the area normalized emission spectrum of the sensitizer, σ_a is the lineshape of the activator absorption expressed in terms of the absolute value of the cross-section, τ_s is the sensitizer lifetime, and E is the photon energy. We find for the $CaF_2:Sm^{2+},Nd^{3+}$ system at T=200K that the critical radius R_o=11.4Å. The critical concentration, defined by $C_o=(4/3\ \pi\ R_o^3)^{-1}$, is then equal to 0.66 mole% Nd^{3+}.

Method II: The energy transfer process from Sm^{2+} to Nd^{3+} shortens the observed emission lifetime of Sm^{2+}. By fitting the transients to an equation of the form

$$I_s(t) = \exp(-t/\tau_s) \cdot \exp[-\sqrt{\pi} \cdot (C/C_o) \cdot (t/\tau_s)^{1/2}] \qquad (2)$$

an average value of C_o=0.57 at T=200K is determined, in good agreement with the result of Method I. The transfer efficiencies

*Work performed under the auspices of the U.S. Department of Energy by Lawrence Livermore National Laboratory under Contract No. W-7405-Eng-48.

calculated from the measured values of C/C_0 are shown in Fig. 1. The energy transfer is efficient at T<220K, above which it is known that the Sm^{2+} luminescence non-radiatively quenches. This non-radiative quenching at T>220K is an important problem which must be addressed.

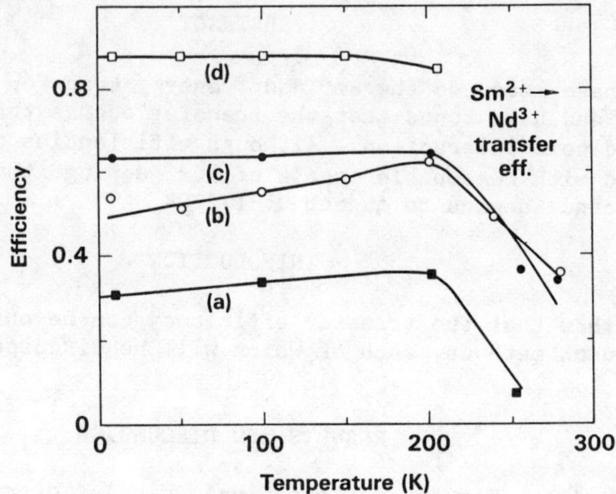

Figure 1. Energy transfer efficiency as a function of temperature. The Nd^{3+} mole% for samples a-d is 0.16, 0.17, 0.39 and 1.1; for Sm^{2+} it is 0.0020, 0.022, 0.0038 and 0.0037.

Method III: The energy transfer efficiency can be determined most directly, but perhaps not most accurately, with the relation

$$\eta = [A(Sm,exc)/A(Sm,abs)]/[A(Nd,exc)/A(Nd,abs)] \qquad (3)$$

where A is the area of a Sm or Nd band derived from the absorption or excitation spectra, as indicated paranthetically. For example, this method yields efficiencies for sample (a) of 0.27, 0.59 and 0.71 at T=10K, 100K and 200K, respectively, while Method II yields the values 0.31, 0.34 and 0.35.

REFERENCES

1. S. A. Payne and L. L. Chase, J. Opt. Soc. Am. B, submitted.
2. T. Foerster, Z. Naturforsch. 4a, 321 (1949); D. L. Dexter, J. Chem. Phys. 21, 836 (1953).

SPECTROSCOPIC, OPTICAL, AND THERMO-MECHANICAL PROPERTIES OF GSGG AND ITS LASER PERFORMANCE*

William F. Krupke
Lawrence Livermore National Laboratory, Livermore, CA 94550

ABSTRACT

Spectroscopic, optical, and thermo-mechanical properties central to the design of high average power Nd:Cr:GSGG lasers are presented. The status of crystal growth and achieved laser performance levels are briefly reviewed.

INTRODUCTION

The new high-performance solid state laser crystal, Gadolinium Scandium Gallium Garnet (GSGG), co-doped with trivalent neodymium (Nd) and chromium (Cr) ions, has become of interest worldwide due to the pioneering work of two groups: that of Prof. I. Shcherbakov and E. Zharikov (Moscow) and that of Prof. G. Huber (Hamburg). These researchers recognized[1,2] that Cr-doped gallium garnets, with large lattice constants, exhibit intense broadband fluorescence at 750 nm. This emission is resonant with several absorption bands of co-doped Nd ions, leading to efficient nonradiative sensitization of Nd luminescence and to an increase in laser efficiency by over a factor of two, compared to Nd:YAG. Gadolinium scandium gallium garnet was chosen for development over others because its near unity distribution coefficients for Cr(1.0) and Nd(0.7) ions permit the growth of uniformly-doped, large diameter boules by the Czochralski method. Additionally, gallium garnet melts have viscosities which allow for a flat growth interface, yielding optically-uniform boules free of an axial refractive index distortion, or "core."

To design Nd:Cr:GSGG lasers, particularly high average power lasers, values for an array of certain spectroscopic, optical, and thermo-mechanical properties are needed. These property values have been measured[3] and are presented in this paper. This is followed by brief reviews of the quality of Nd:Cr:GSGG crystals grown and the laser performance levels observed to date.

PARAMETER VALUE SUMMARY

The values of salient spectroscopic parameters of GSGG (and of YAG[4] for comparison) are given in Table 1. This data shows that the overall radiative transition rates of Nd:GSGG are very similar to those of Nd:YAG, as represented by Judd-Ofelt parameter values,

*Work performed under the joint auspices of U.S. Department of Energy under Contract No. W-7405-Eng-48 and DARPA Order #5358.

Table 1. Spectroscopic and laser parameters of Nd:GSGG and Nd:YAG

	Parameter	Unit	Nd:GSGG [3]	Nd:YAG [3,4]
	Judd-Ofelt parameters $\Omega_2, \Omega_4, \Omega_6$	10^{-20} cm^2	0.35; 2.35; 3.23	0.2; 2.7; 5.0
	Branching ratios $\beta_{J'}$, J' = 9/2, 11/2, 13/2	—	0.39; 0.51; 0.10	0.37; 0.50; 0.13
	J-O calculated radiative lifetime, $\tau_r(^4F_{3/2})$	μsec	284	259
	Zero concentration fluorescence lifetime, τ_o	μsec	281	~ 260
880 nm band	$\sigma_p(^4F_{3/2} \rightarrow {}^4I_{9/2})$	10^{-20} cm^2	3.0	3.1 (946 nm)
	$\sigma_{936}(^4F_{3/2} \rightarrow {}^4I_{9/2})$	10^{-20} cm^2	2.1	3.1 (946 nm)
	$\sigma(R_1 \rightarrow Z_5)$; $\lambda(R_1 \rightarrow Z_5)$	10^{-20} cm^2	3.7 (936 nm)	5.1 (946 nm)
	$\lambda_p(^4F_{3/2} \rightarrow {}^4I_{9/2})$	nm	883	946
1060 nm band	$\sigma_p(^4F_{3/2} \rightarrow {}^4I_{11/2})$	10^{-19} cm^2	1.3	2.8
	$\sigma(R_2 \rightarrow Y_3)$	10^{-19} cm^2	3.1	6.5
	$\lambda_p(^4F_{3/2} \rightarrow {}^4I_{11/2})$	nm	1061.1	1064.1
	$^*\Gamma_{sat}(^4F_{3/2} \rightarrow {}^4I_{11/2})$	J cm^{-2}	1.0	0.5
1320 nm band	$\sigma_p(^4F_{3/2} \rightarrow {}^4I_{13/2})$	10^{-20} cm^2	3.1	5.3
	$\sigma(R_2 \rightarrow X_4)$	10^{-20} cm^2	7.3	13.4 ($R_2 \rightarrow X_2$)
	$\lambda_p(^4F_{3/2} \rightarrow {}^4I_{13/2})$	nm	1335.5	1318.8
	$^*\Gamma_{sat}(^4F_{3/2} \rightarrow {}^4I_{13/2})$	J cm^{-2}	3.3	2.0

*Assumed homogeneously saturating population for pulse duration, τ_L, in range: 10s psec $\lesssim \tau_L \lesssim$ 10 nsec

transition decay rates, fluorescence branching ratios, and the radiative lifetime. However, the stimulated emission cross sections of the strong 1061 and 1335 nm transitions of Nd:GSGG are about half those of the analogous transitions of Nd:YAG. This implies that Nd:GSGG can store about two times as much energy as Nd:YAG when pumped to the onset of superfluorescence. The gain saturation fluences of the transitions in Nd:GSGG, calculated in the homogeneous saturation approximation, are about twice those of YAG. This implies that the extraction beam fluence must be twice as large in Nd:GSGG as in Nd:YAG for efficient extraction. Since the saturation fluences of Nd:GSGG (1.0 J cm^{-2} at 1061 nm and 3.3 J cm^{-2} at 1335 nm) are small compared to the damage fluences found for iridium-free Nd:Cr:GSGG crystals (19 J cm^{-2} surface and 28 J cm^{-2} bulk), efficient extraction of stored energy is assured.

Zharikov, et al.[1] have determined the microscopic energy transfer parameters for Nd:Cr:GSGG and find that a very efficient sensitization of Nd luminescence and laser output can occur for reasonable values of doping. An excitation quantum transfer efficiency in excess of 80% is realized for GSGG doped with 1 and 2 x 10^{20} Cr and Nd ions/cc, respectively.

Table 2 gives the values of optical, thermal, and mechanical properties of GSGG[3,5] and YAG.[3,6] The thermo-mechanical properties of these two crystals are sufficiently similar that their resistance to fracture under thermal loading is also very similar. The larger index of refraction and spectral dispersion of GSGG, compared to YAG, indicates a 2.5-fold larger nonlinear index of refraction for GSGG than YAG, thus limiting the use of GSGG for extremely high intensity laser applications.

Table 2. Optical, thermal, and mechanical properties of GSGG and YAG

Parameter		Unit	GSGG		YAG
			ref. 3	ref. 5	refs. 3,6
Index of refraction	n (1060 nm)	–	1.942	1.943	1.815
Index temperature derivative	$\partial n/\partial T$ (1150 nm)	10^{-6} °C^{-1}	10.5	–	8.9
Nonlinear refractive index (est.)	n_2 (1060 nm)	10^{-13} esu	~8	–	~3
Elasto-optic coefficients	p_{11} (633 nm)	–	-0.012	-0.097	-0.029
	p_{12} (633 nm)	–	+0.019	-0.040	+0.009
	p_{44} (633 nm)	–	-0.067	-0.066	-0.062
Elastic constants	C_{11}	GPa	277.4	269	333
	C_{12}	GPa	104.9	102	111
	C_{44}	GPa	80.4	77.4	115
Heat capacity	C_p	J g^{-1} K^{-1}	0.40	0.45	0.59
Thermal diffusivity	λ	cm^2 sec^{-1}	0.022	0.024	0.037
Thermal conductivity	κ	W m^{-1} K^{-1}	5-7	7	10-13
Thermal expansion coefficient	α	10^{-6} K^{-1}	7.5	–	8.2
Young's modulus	E	GPa	212	205	282
Poisson's ratio	ν	–	0.28	0.28	0.25
Fracture toughness	K_C	MPa m$^{-1/2}$	1.2	–	1.4
Breaking strength	σ_f	MPa	240-2400	–	280-2800
Thermal resistance parameter	R_T	W m^{-1}	660-6600	–	770-7700

QUALITY OF Nd:Cr:GSGG CRYSTALS

Optical quality measurements on 6.3 x 75 mm laser rods cored from the first two dozen 30 mm diameter boules of Nd:Cr:GSGG grown in a recent study indicate the following: the wavefront uniformity of the GSGG rods is about equal to that exhibited in production Nd:YAG laser rods ($\lambda/10$); the transmission loss at 1060 nm (0.0025 cm^{-1}) and the stress birefringence of GSGG rods (1 nm/cm) are somewhat larger than those of production YAG.

LASER EXPERIMENTS

Nd:Cr:GSGG lasers have now been operated long-pulse, Q-switched, and mode locked at a wavelength of 1061 nm. An efficiency of 5% (two times larger than for Nd:YAG) was achieved in the long pulse mode, consistent with expectations based on microscopic energy transfer studies. Laser action was also achieved at 1335 and 936 nm, the latter at room temperature for the first time.

ACKNOWLEDGMENTS

The data presented in this paper derives largely through the efforts of several colleagues at the Lawrence Livermore National Laboratory. I would particularly like to acknowledge the efforts of Drs. M. D. Shinn (spectroscopy), J. E. Marion and S. E. Stokowski (materials properties), and J. A. Caird (laser experiments).

REFERENCES

1. E. V. Zharikov, V. V. Laptev, V. G. Ostroumov, Yu. S. Privis, V. A. Smirnov, and I. A. Shcherbakov, Sov. J. Quantum Electron., 14, 1056 (1984).
2. D. Pruss, G. Huber, A. Beimowski, V. V. Laptev, I. A. Shcherbakov, and E. V. Zharikov, Appl. Phys., B28, 355 (1982).
3. W. F. Krupke, M. D. Shinn, J. E. Marion, J. A. Caird, and S. E. Stokowski, J. Opt. Soc. Am., B3, 102 (1986).
4. W. F. Krupke, IEEE J. Quantum Electron., 7, 153 (1971).
5. E. V. Zharikov, V. F. Kitaeva, V. V. Osiko, I. R. Rustamov, and N. N. Sobolev, Sov. Phys. Solid State, 26, 922 (1984).
6. A. A. Kaminskii, Laser Crystals (Springer-Verlag, Berlin, 1976), p 320.

ROOM TEMPERATURE $SrAlF_5:Cr^{3+}$ LASER EMISSION TUNABLE FROM 825 nm TO 1010 nm*

J. A. Caird, W. F. Krupke, M. D. Shinn and P. R. Staver
Lawrence Livermore National Laboratory, Box 808, Livermore, CA 94550

H. J. Guggenheim
AT&T Bell Laboratories, 600 Mountain Avenue, Murray Hill, N. J. 07974

ABSTRACT

The tunable $SrAlF_5$:Cr laser operates in a wavelength range from which frequency doubling can produce the blue wavelengths of interest for submarine laser communications (SLC). The principal fluorescence lifetime of 95 μs is long enough to allow for efficient flashlamp pumping. The estimated emission cross section of 1.9×10^{-20} cm^2 at 910 nm is high enough to allow efficient extraction of stored energy at a modest saturation fluence of 11.5 J/cm^2. To date, however, the output lasing slope efficiency has been limited to 15%, indicating high losses in the medium. The experiments are seen as a demonstration of the fact that Cr^{3+} ions doped in appropriate host media can provide tunable laser output in this important wavelength range.

SUMMARY

Crystals of $SrAlF_5$ nominally doped with 1 or 2 at.% Cr^{3+} were grown by horizontal zone melting. The tetragonal crystal has four inequivalent octahedral aluminum ion sites on which Cr^{3+} can substitute[1]. The fluorescence decay exhibited a double exponential time dependence. About 80% of the time integrated fluorescence was characterized by the longer lifetime of 95 μs at room temperature in the 1 at.% doped sample. Another 20% of the luminescence exhibited a shorter lifetime of 40 μs. Both lifetimes increased on the order of 10% on cooling to near liquid helium temperature, indicating that the decays are primarily radiative at room temperature.

The absorption spectrum exhibited a long wavelength "tail" which extended beyond 900 nm, and was much stronger in the σ-polarization. A lack of reciprocity between the absorption and emission spectra may indicate that the centers absorbing in this wavelength range have a low (perhaps zero) fluorescence quantum yield. The presence of a non-radiative site which absorbs in this wavelength range has been reported by Jenssen and Lai[2]. Assuming that the measured fluorescence lifetime is purely radiative, and using the fluorescence line shape function (assuming that it is approximately the same for all radiative centers), a peak emission cross section of 2.2×10^{-20} cm^2 was found for the primary emitting center (τ=95 μs) at 870 nm. The cross section drops to 1.9×10^{-20} cm^2 at 910 nm.

A crystal with 2 at.% Cr^{3+}, 1.93 mm thick, was pumped longitudi-

*Work performed under the joint auspices of the U.S. Department of Energy under Contract No. W-7405-ENG-48 and the U.S. Department of Defense under Defense Advanced Research Projects Agency order #5358.

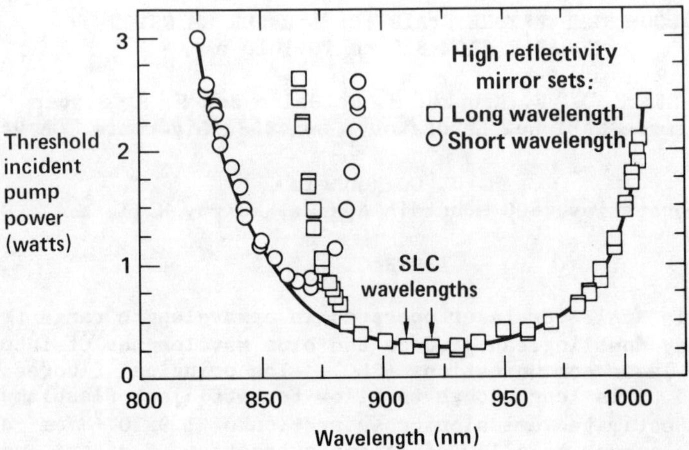

Figure 1. Threshold Kr ion pump power vs. $SrAlF_5$:Cr laser wavelength.

nally in a nearly concentric laser resonator by a krypton ion laser. The pump laser was chopped to yield a 2% duty factor in order to limit thermal loading of the sample. A single plate birefringent tuner was used along with two different sets of high reflectivity mirrors to vary the output wavelength. Laser emission was obtained with output polarization parallel to the c-axis only. The threshold pump power is plotted as a function of output wavelength in Fig. 1. In these experiments the pump beam waist diameter was about 100 μm, and increased somewhat with increasing pump power. The rise in threshold at the longer wavelengths is due to declining emission cross section, while at shorter wavelengths it is primarily due to increasing absorption loss. The tuning range obtained here, 825-1010 nm, is considerably broader than reported by Jenssen and Lai[2].

Output power measurements at 910 nm were made using a silicon photodiode which was calibrated by reference to a thermopile. With 1% output coupling a slope efficiency of 14.7% was obtained. In the standard model the slope efficiency is given by[3] $\eta_s=\eta_o T(L+T)^{-1}$, where T is the output coupling fraction, L is the round trip resonator loss, and η_o is the quantum defect limited efficiency (71% in this case). If this equation is used to estimate the loss coefficient in the laser material a value of 10% cm^{-1} is obtained. Measurements of the absorption spectrum indicate that absorption loss is about an order of magnitude smaller. The nature of the increased losses required to explain the low output efficiency is not understood at present. Substantial excited stated absorption, or strong absorption by nonradiative "dark" sites are possible explanations.

REFERENCES

1. S. C. Abrahams, et. al., J. Appl. Phys. 52, 4740 (1981).
2. H. P. Jenssen, and S. T. Lai, J. Opt. Soc. Am., B3, 115 (1986).
3. H. G. Danielmeyer in Lasers Vol. 4, A. K. Levine and A. J. DeMaria eds., Marcel Dekker, New York, 1976, p. 28.

TRANSIENT OPTICAL GRATING IN HgCdTe

J.R. Lindle, F.J. Bartoli, J.R. Meyer, and C. A. Hoffman
Naval Research Laboratory, Washington, DC 20375

ABSTRACT

The transient grating technique has been used to study ambipolar diffusion and free carrier recombination in HgCdTe. The grating is created by the interference of two coherent 1.06 μm, 1 nsec pulses from a Nd:YAG laser. Decay of the grating is monitored by observing the first-order diffraction of a 1 μs, 10.6 μm laser pulse which is synchronized with the pump pulses. Ambipolar diffusion dominates the decay at small grating spacings and low excitation levels, while recombination is more important when the grating spacing is larger.

DISCUSSION

Previous studies[1] have demonstrated that the transient grating technique can be used to accurately measure ambipolar diffusion coefficients and recombination lifetimes in wide-gap semiconductors. Here, we have adapted this technique for use with HgCdTe, a narrow-gap semiconductor of great current interest because of its infrared detector applications. In the present experiment the transient grating is generated by irradiating the sample with two coherent 1 nsec, 1.06 um laser pulses derived from the same mode-locked Nd:YAG laser. These pulses are spatially and temporally overlapped on the sample so that they interfere and spatially modulate carrier density, which leads to a modulated Drude contribution to the refractive index. The decay of the free carrier grating is monitored by irradiating the grating with a 10.6 μm probe and observing the first-order diffracted light. The probe pulse, which is produced by a CO_2 TE laser, is 1 μsec in duration and is electronically synchronized with the 1.06 μm excite pulse. For sufficiently small grating spacings, decay is dominated by homogenization of the carrier distribution due to ambipolar diffusion. Conversely, for large grating spacings carrier recombination will dominate the decay. The grating spacing is determined by the angular separation between the two 1.06 μm excitation laser pulses.

The present experiment was performed at 295K on a $Hg_{1-x}Cd_xTe$ sample of composition X=0.30. Fig. 1 shows the time dependence of the grating efficiency for a grating spacing Λ of 28 μm and a pump intensity of 147 W/cm^2. The diffusion coefficient can be obtained from the expression

$$D = \frac{\Lambda^2}{8\pi^2\tau_g} \qquad (1)$$

where τ_g is the 1/e grating lifetime. From a grating lifetime in Fig. 1 of 20 nsec, the diffusion coefficient is determined to be 5 ($\pm$ 1) cm^2/sec.

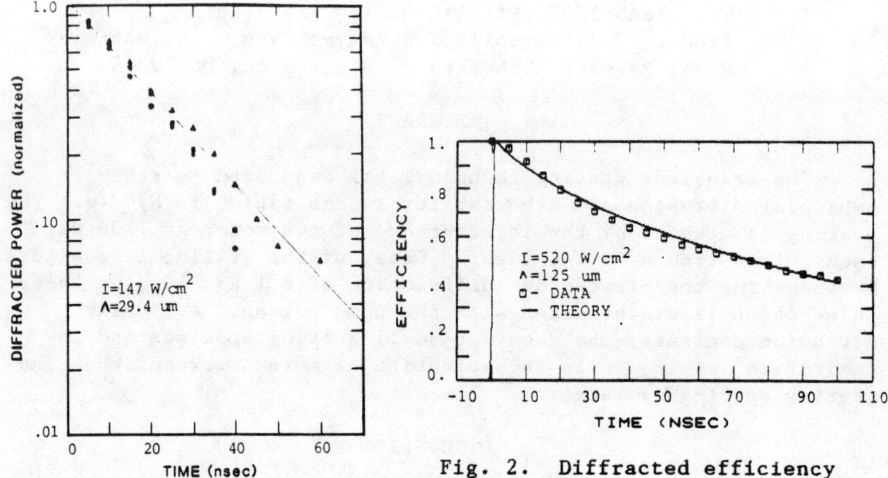

Fig. 1. Semilog plot of diffracted power versus time.

Fig. 2. Diffracted efficiency versus time.

The results obtained on the same sample for a grating spacing of 125 μm and 520 W/cm^2 are shown in Fig. 2. In this case the effects of recombination and diffusion cannot strictly be decoupled and we employ a numerical solution to the carrier rate equations. The curve of Fig. 1 shows the best fit to the experimental data, which was obtained using a linear recombination coefficient of 2×10^6/sec, an Auger coefficient of 4 $\times10^{-27}$ cm^6/sec and a diffusion coefficient of 5.0 cm^2/sec. These values are consistent with previous results from the literature[2,3].

The present work demonstrates that the transient optical grating technique is a valuable tool for characterizing the transport and recombination properties of narrow-gap semiconductors such as HgCdTe. In addition to allowing the measurements to be performed non-destructively (without applying contacts), the transient grating technique also allows the material properties to be determined as a function of position on the sample surface. This advantage over conventional measurement techniques should prove to be useful in screening materials used in IR detector arrays.

REFERENCES

1. For example, see A. L. Smirl, S. C. Moss, and J. R. Lindle, Phys. Rev. B 25, 2645 (1982).
2. W. Scott, E. L. Stelzer, and R. J. Hager, J. Appl. Phys. 47, 1408 (1976).
3. F. Bartoli, R. Allen, L. Esterowitz, and M. Kruer, J. Appl. Phys. 45, 2150 (1974).

LIGHT SCATTERING FROM OXYGEN IN ZnTe*

Harold P. Hjalmarson, E. D. Jones, and C. B. Norris
Sandia National Laboratories
Albuquerque, New Mexico 87185

ABSTRACT

Resonant light scattering experiments and calculations have been performed on substitutional oxygen impurities in ZnTe.

INTRODUCTION

The nature of optical emission following absorption of light depends on the vibronic coupling. For a case of strong vibronic coupling, such as an impurity embedded in a solid, the rate of vibronic relaxation following excitation is very rapid, and the resultant optical emission consists mainly of ordinary luminescence (OL). In general, however, a small component of this secondary emission consists of resonant Raman scattering (RRS) and hot luminescence (HL)[1-4]. In this paper, we measure and calculate this secondary emission for Te-site O in ZnTe.

EXPERIMENTS AND THEORY

The experiments utilized a tunable pulsed dye laser to pump the absorption bands of ZnTe:O samples at 10K. The total secondary emission was analyzed by a double monochromator[5].

The theory uses a Kramers-Heisenberg approach[3,4,6] to generalized light scattering which includes OL, HL, and RRS. The theory is formulated so that the complete secondary emission spectrum is calculated in terms of the electron-phonon spectral density of states (EPSDOS) $D(\omega)$. This EPSDOS is derived from the OL spectrum using an inversion technique[7].

DISCUSSION

The theory[7] and data[5,8] for OL at 10K are shown in Fig. 1. The theoretical phonon spectrum $D(\omega)$ is a sum of contributions from longitudinal optical (LO) and transferse acoustic (TA) modes. These components are centered at ω_{LO} = 26.1 meV and ω_{TA} = 7.1 meV and they are characterized by dimensionless couplings $S_{LO} = 3.5$ and $S_{TA} = 2.0$[6].

Having considered the OL, we turn our attention to RRS and HL for incident light frequencies ω_i lying within the ZnTe:O absorption band. Data (dotted line) and theory (solid line) for the secondary emission spectrum above the zero phonon line at 1.986 eV and the case $\omega_i \sim 4\omega_{LO} \sim 104$ meV are shown in Fig. 2. Near the laser line, the spectrum becomes dominated by the laser lineshape which is approximated by the dashed line.

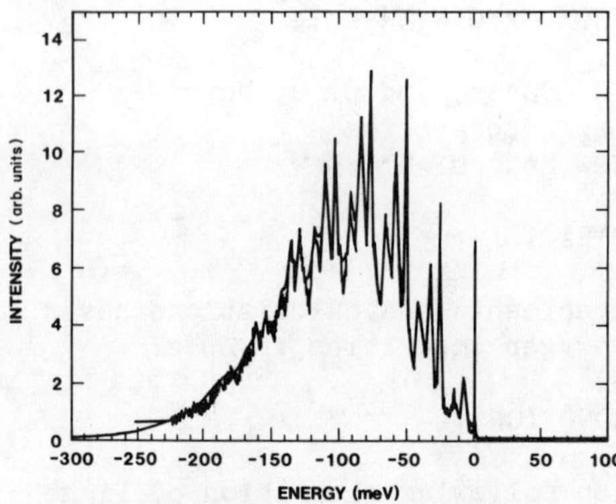

Fig. 1 Superimposed theoretical and experimental ordinary luminescence spectra.

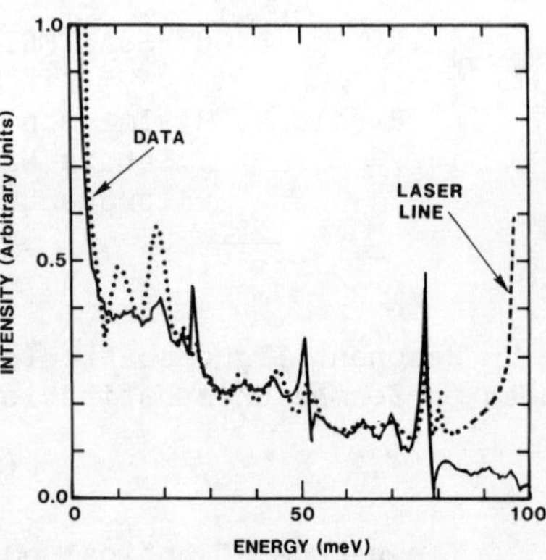

Fig. 2 Secondary emission for the incident laser at 4 LO energy units above the zero phonon line.

In all cases, one observes generally good agreement between the theory and experiment[6]. In particular, we point out that the slowly varying background as well as the peak positions and heights are predicted by the theory. Furthermore, we observe that at $\omega = \omega_{LO}$ and $\omega = 2\omega_{LO}$ the trend to domination by TA mode contributions is predicted by the theory.

For this choice of ω_i, the HL and RRS cannot be distinguished. Other theory and data, however, show RRS peaks which "follow" the laser line and HL peaks which "follow" the zero phonon line.

In conclusion, we have successfully performed resonant light scattering from ZnTe:O. We have obtained the EPSDOS D(ω) from the OL data and then used D(ω) to predict the complete secondary emission spectrum.

*This work was supported by the U. S. Department of Energy under contract number DE-AC04-756DP00789.

REFERENCES

1. Y. R. Shen, Phys. Rev. B 9, 622 (1974).
2. K. Rebane and P. Saari, J. Lum. 16, 223 (1978).
3. V. Hizhnyakov and I. Tehver, Phys. Stat. Sol. 21, 755 (1967).
4. Y. Toyozawa, J. Phys. Soc. Japan 41, 400 (1976).
5. E. D. Jones, H. P. Hjalmarson, and C. B. Norris, J. Lum. 31/32, 436 (1984).
6 H. P. Hjalmarson, E. D. Jones, and C. B. Norris, to be published.
7. H. P. Hjalmarson, L. A. Romero, D. C. Ghiglia, E. D. Jones, and C. B. Norris, Phys. Rev. B32, 4300 (1985).
8. R. E. Dietz, D. G. Thomas, and J. J. Hopfield, Phys. Rev. Lett. 8, 391 (1962).

Czochralski Growth and Laser Performance of Alexandrite Crystals

Xing-an Guo, Bang-xing Zhang,
Lu-sheng Wu and Mei-Ling Chen

Anhui Institute of Optics and Fine Mechanics
Academia Sinica, P. O. Box 25, Hefei, Anhui
People's Republic of China

Abstract

Alexandrite ($BeAl_2O_4:Cr^{3+}$) crystals have been grown by the Czochralski technique and continually tunable laser output with energy of 304 mJ and slope efficiency of 0.46% in the wavelength range from 735 to 786 nm has been obtained using c-axis rods. Tunable Q-switch pulse output and $LiIO_3$ double-frequency have been also obtained.

Optical homogeneous alexandrite crystals ($BeAl_2O_4:Cr^{3+}$) have been grown by the Czochralski technique using high-frequency induction heating of an indium crucible in a N_2 atmosphere. Before growing single crystals, the alexandrite phase was synthesized by solid-phase reaction. Oxides of beryllium, aluminum and chromium were used as the initial reagents. Cr^{3+} concentrations in charges were 0.05-0.3 at.%. Crystallization was carried out on seeds oriented along [100], [010] and [001] directions (a = 5.470 A, b = 9.404 A, c = 4.427 A). Growth habits have been studied and it has been found that the anisotropy of growth rates is in the order [100] > [010] > [001]. This result is different from that obtained by G. V. Bukin _et al_.[1]. The effective distribution coefficient of Cr^{3+} ions in alexandrite was determined to be 2.1 by spectrophotometer.

The optical quality and spectroscopic characteristics of alexandrite crystals were measured. The fluorescence lifetime was 260 μs. The high-angle scattering loss coefficient was about 0.2% cm^{-1}. Optical homogeneity was measured by a Twayman-Green interferometer and the interference pattern was 1 fringe/inch. With a sample of 20 mm thickness we have measured the absorption and luminescence spectra of alexandrite at room temperature. The spectra are similar to those in reference 2.

We have obtained tunable output of 304 mJ with a tuning range from 735 to 786 nm, a slope efficiency 0.46%, a linewidth of 0.8 nm, and a pulse width of 170 ns at room temperature, using a 6.8 mm diameter, 65 mm long, c-axis alexandrite rod and Xe flashlamp pumping. Tunable Q-switched output (pulsewidth 30 ns and peak power 10 MW) and $LiIO_3$ frequency (tunable range from 368 to 383 nm and frequency doubling conversion efficiency of 9%) have been also obtained.

References

1. G. V. Bukin _et al_., Journal of Crystal Growth _52_, 537 (1981).

2. J. C. Walling, IEEE Journal of Quantum Electron. _QE-16_, 1302 (1980).

NOVEL DESIGN FOR RUBY LASER OSCILLATOR

K. A. Arunkumar, J. D. Trolinger, D. Lee
Spectron Development Laboratories, Inc.
3303 Harbor Blvd., Ste. G-3
Costa Mesa, CA 92626

ABSTRACT

A novel approach to the design of a Q-switched ruby laser oscillator has resulted in increasing its output energy for the same input energy. This was achieved by doubling the length of the gain medium. To make that possible, the back mirror was replaced by a 90 degree roof prism and the output coupler by a special mirror which has 100% R coating on one-half and X% R coating on the other half. Laser oscillations occurred between the 100% and X% R mirror coatings and the laser beam traveled along a direction parallel to the axis of the rod, but shifted from it by 2 mm. The output pulse has TEM_{oo} mode profile.

OBJECTIVE

The aim of this work is to develop a Q-switched, TEM_{oo} oscillator more efficient than the commercially available ones. It is also intended that the resulting design will lead to the development of a portable/compact laser.

APPROACH

The approach taken was to utilize the extra active volume normally unused in a commercially available TEM_{oo} mode, Q-switched laser oscillator. This was achieved by folding the cavity as shown in Figure 1

85-1981

FOLDED CAVITY RUBY LASER RESONATOR

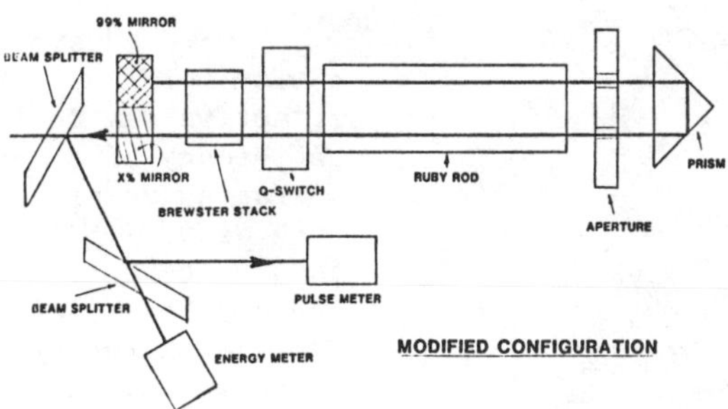

This arrangement basically doubles the length of the gain medium and since the gain ∝ exp (gl) the oscillator output goes up substantially. Figure 2 compares the output of a modified ruby oscillator against that from the standard cavity. For the same input energy, the output energy was increased by 100% - 500%. This result suggests that this configuration can help cut down the physical length of the laser resonator without dropping the output energy. Hence, laser cavity simplification/portability is one of the major consequences of this result.

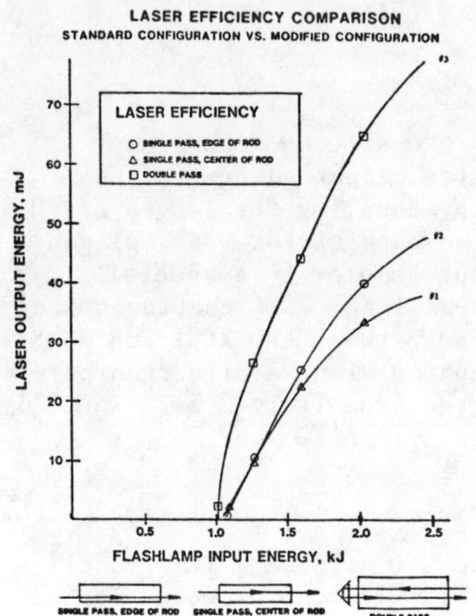

The 40 nanosecond pulse duration originally obtained from the commercial unit was reproduced in the modified version. The pulse duration was reproducible from shot to shot with no measurable variation. The spatial profile of the beam was examined to check whether the beam path shifting from the rod axis affected the TEM_{oo} mode structure. For that, the oscillator output after proper beam expansion was recorded on a photographic plate. The processed plate was then scanned using a He-Ne laser and a photo detector. The resulting beam profile shown in Figure 3 confirms the transverse field variation to be indeed gaussain.

Also the measured beam divergence of 230 μ radiaus compared well with the 200 μ radiaus measured for the standard oscillator.

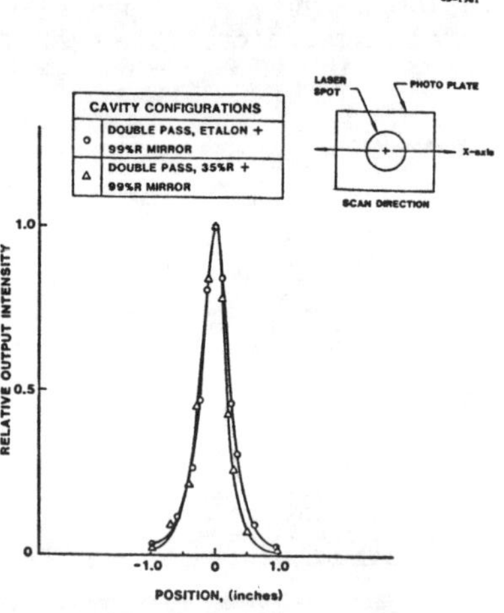

Laser beam intensity profile on the horizontal X-axis. Flashlamp input energy is 1.07 kJ.

In conclusion, we find that cavity folding to be a simple and effective way to improve the efficiency of a Q-switched laser operating under TEM_{oo} mode by as much as 500%. A corollary to this concept is the halving of the physical size of the laser oscillator/rod without any energy increase. Thus cavity folding can lead to the production of a compact/portable laser resonators.

Tuning Characteristics and New Laser Lines in an Nd:YAP CW Laser

L.D. Schearer
University of Missouri-Rolla, Rolla, Mo. 65401

M. Leduc
Laboratoire de Spectroscopie Hertzienne, Paris, FRANCE

ABSTRACT

We have investigated the CW laser properties of Nd doped Yttrium-Aluminum-Perovskite (YAP) in a Kr arc-lamp, pumped cavity. Seven laser transitions within the $^4F_{3/2}-^4I_{11/2}$ multiplet have been observed with thresholds and power outputs comparable to the corresponding transitions in Nd:YAG. A simple Lyot filter in the cavity is used to select the transition. The addition of a thin, uncoated etalon enabled us to obtain the tuning characteristics of the transitions at 1079.5nm and 1084.5nm. Tuning widths of 3nm are obtained.

INTRODUCTION

Interest in Nd:YAP is based on early reports of high average power in pulsed systems and high CW power output with thresholds and efficiencies comparable to YAG[1,2]. Interest in this system was also motivated by the fact that the orthorhombic symmetry yields polarized laser output. The YAP laser is then free from the depolarizing thermal birefrigence which affects YAG performance when polarizing elements are inserted in the YAG cavity

EXPERIMENTAL RESULTS

We report here CW laser emission from 7 lines in the 1064-1099nm range. We also report the tuning characteristics for the transitions centered at 1079.5nm and 10845nm. We were primarily motivated by the possibilty of obtaining tunable, CW laser emission at 1083.2nm corresponding to the resonance transition in He-3.

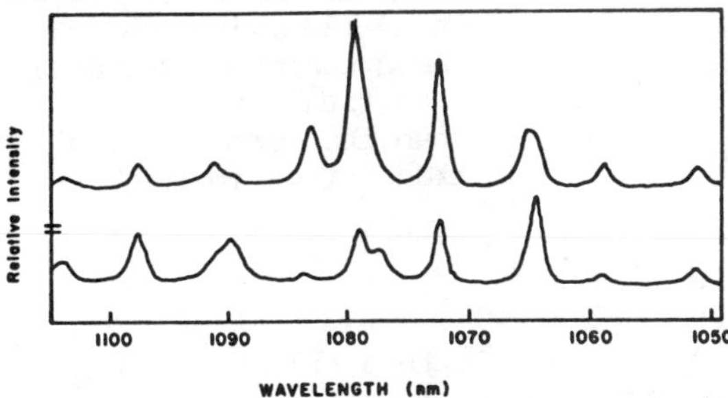

Fig. 1. YAP Fluorescence Polarization Along a and c axes

The $YAlO_3$:Nd crystal was 90mm in length and 5mm diameter with a Nd concentration of 0.7% with the crystalline ´a´ axis along the rod length. The crystal was mounted in a conventional ventional arc-lamp pumped YAG laser cavity, replacing the Nd:YAG crystal. The $YAlO_3$ is optically biaxial with the laser

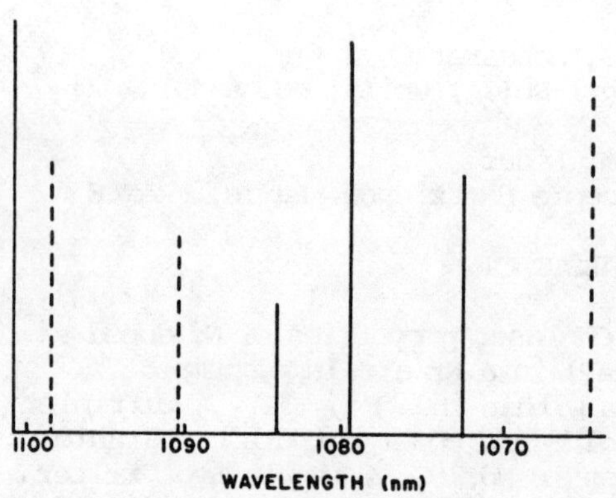

Fig. 2. Laser Intensity vs. Wavelength

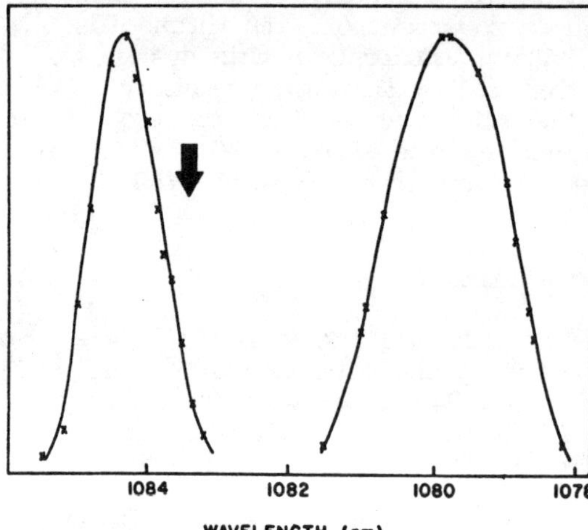

Fig. 3. Tuning Curves for 1084.5nm and 1079.8nm.

emission polarized either along the b or c crytalline axes (depending on wavelength). Figure 1 illustrates the biaxial properties of the material. The relative intensities of the two curves suggests how any tuning elements such as Lyot filters or Brewster prisms should be oriented within the cavity for a particular wavelength. With no polarizing elements within the cavity, the output of the laser will be centered at 10798.5nm and polarized along the c axis. In general, output powers are comparable to that obtained with YAG crystals of the same size although thresholds are slightly higher.

The addition of a Lyot filter within the cavity, properly oriented, enabled us to tune the transitions shown in fig. 2. A thin, uncoated etalon added to the cavity permitted us to tune within each transition. With cavity mirrors that were both nominally totally refelecting, we obtained the tuning curves shown in fig. 3. These tuning curves are significantly broader than the corresponding transitions in YAG[3].

REFERENCES

1. M. Bass and M.J. Weber, Appl. Phys. Letters, 17, 395(1970)
2. G.A. Massey and J.M. Yarborough, Appl. Phys. Letters, 18, 576 (1971).
3. J. Marling, IEEE J. Quantum Electr. 14, 56, (1978).

FREE JET SPECTROSCOPY BY COHERENT RAMAN METHODS

G. A. Pubanz, M. Maroncelli and J. W. Nibler
Oregon State University, Corvallis, OR 97331

ABSTRACT

Coherent Raman techniques can serve as useful probes of the structural and dynamic properties of monomers and small aggregates formed under the cold, nonequilibrium conditions of free jet expansions. The results of recent measurements on hydrogen bonded systems such as HCN dimer and trimer and on weak van der Waals complexes such as $(CO_2)_n$ are presented. The utility of high resolution Raman spectroscopy in the time domain is also discussed briefly.

INTRODUCTION

In recent years the study of gas phase spectroscopy has been greatly enhanced through the use of free jet expansions. The cooled samples yield simpler spectra with reduced Doppler linewidths. For those techniques which can probe close to the nozzle, studies of the expansion dynamics can provide valuable information about the conversion of vibrational and rotational energy into translational energy. The low temperature conditions of these expansions also favors the formation of weakly bound complexes, systems enjoying much current research activity.

Recently, coherent anti-Stokes Raman spectroscopy (CARS), inverse Raman spectroscopy (IRS) and Raman gain spectroscopy (RGS) have proven useful as probes of jet expansions. These methods are attractive because of their inherent sensitivity and the small volume from which signal is actually produced. The latter feature enables probing very close to the nozzle and permits an accurate characterization of the local temperature and composition of gases in the free expansion zone.

COMPLEXES

There is considerable interest in the study of small molecular clusters. As the first step in the nucleation process, dimers and higher polymers provide insight into the fundamental processes of aggregation and, with mixed species, solvation. Structural information such as bond lengths, angles and force constants in these gas phase species is extremely useful in refining models of the attractive part of the potential energy surface. In addition, linewidths can provide information on lifetimes and reveal details of vibrational predissociation pathways.

We have used CARS and photoacoustic Raman spectroscopy (PARS) to probe clusters of HCN molecules. In the case of static HCN samples, PARS and FTIR spectra of dimer and trimer species were obtained. Improved resolution and species discrimination was achieved with jet expansions probed by the CARS technique.[1] The experimental details are outlined elsewhere.

The series of spectra in Fig. 1 gives evidence for the growth of HCN clusters in cold free jets. Under relatively mild expansion conditions the dimer is the most evident species, however, as higher pressures of argon are added, the trimer and higher polymers begin to dominate the spectrum. Note that with a stagnation pressure of 6 atm. and 10% HCN in Ar the monomer peak is almost lost as a shoulder on the polymer peak. Further work[2] indicates that both the dimer and the trimer are linear in the gas phase and normal coordinate analysis shows that one hydrogen bond in the trimer is significantly stronger than the other.

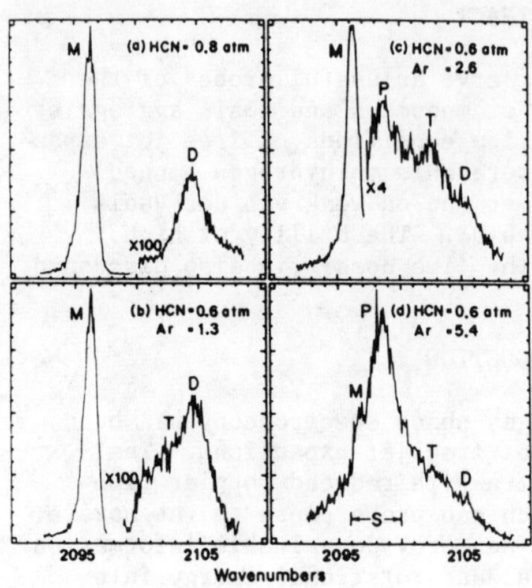

Fig. 1 CARS spectra of the ν_1 region in supersonic expansions of HCN showing bands due to complexes. Peaks are labeled as M = monomer, D = dimer, T = trimer and P = higher polymers. In spectrum (d) the region labeled S denotes the position and width of ν_1 in crystalline HCN at 78 K.

The structural problem of the CO_2 dimer has been investigated using a variety of techniques since the IR spectrum was first observed by Mannick et al.[3] in 1971. Based upon PQR structure it was concluded that the CO_2 dimer had a polar, T-shaped conformation. In the intervening fourteen years additional gas phase and matrix infrared studies have been done as well as electric deflection and infrared vibrational predissociation experiments. In 1974 Koide and Kihara[4] proposed an alternative, non-polar structure based upon empirical calculations. This structure, having an offset parallel geometry, contains an inversion center and belongs to the C_{2h} point group. More recent _ab initio_ calculations by Brigot et al.[5] suggest that this non-polar form would be the more stable form but that the energy difference is small so that some T-shaped dimer could be present.

We have used CARS to study the CO_2 clusters formed in dilute He/CO_2 expansions from a cooled nozzle.[6] Figure 2 compares the CARS spectrum of the ν_1 symmetric stretching band with an infrared spectrum obtained by Kopec.[7] The key observation is that there is no evidence in the Raman spectrum of the dimer Q branch seen in the infrared at 1284.7 cm^{-1} even though, for a T-shaped structure, this mode should have high Raman intensity. There is also no peak in the infrared correlating with the dimer Q branch observed at 1281.3 cm^{-1} in the Raman. Hence the rule of mutual exclusion appears to apply to the CO_2 dimer, a result favoring the offset parallel structure with C_{2h} symmetry. The data do not, however, exclude the possibility of up to 25% of a second, higher energy, polar form which could account for electric deflection results.[8]

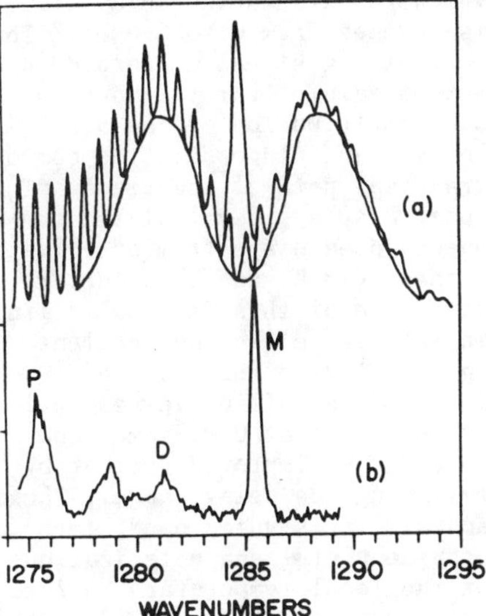

Fig. 2 Comparison of CARS spectrum of CO_2 (b) with a reproduction of the infrared spectrum (a) reported by Kopec.[7] Kopec has drawn a smooth line under the rotational R branch peaks of $^{12}C^{16}O^{18}O$ to emphasize the dimer P-R contours and has interpolated the monomer structure in the Q branch region of the dimer. As before M = monomer, D = dimer and P = polymer.

FOURIER TRANSFORM RAMAN SPECTROSCOPY

Great effort has gone into developing high resolution methods for CARS and other nonlinear Raman techniques. The motivation for this is to resolve the closely spaced lines which characterize the cooled spectra of many monomers as well as dimers and higher polymers. The most advanced pulsed experiments feature transform limited linewidths of about 0.002 cm^{-1}. Picosecond CARS carried out in the time domain can be considered as an alternative to the high resolution methods in the frequency domain.

Time domain CARS utilizes picosecond pulses and a variable optical delay between the initial pump and Stokes beams and the following probe beam which produces the anti-Stokes signal. Polarization beats in the latter signal result from interferences among the various vibration-rotation modes excited by the spectrally broad pump beams. Analysis by Fourier methods allows these time domain spectra to be presented in the conventional frequency domain.

Graener et al.[9] demonstrated this technique on methane in free jet expansions using a 24 psec initial pump and a broad Stokes pulse tuned to excite the dominant Q branch peaks of the symmetric stretch. Figure 3 shows the time domain spectrum of neat methane which results as the delay time is scanned from 0 to 2 nsec. The very strong signal at zero delay arises mainly from the nonresonant contribution of the air, lenses, and windows in the common traversal path of the beams. The inset displays a calculated frequency domain spectrum of cold methane (78 K) and the Fourier transform of this (the solid line in the figure) is in excellent agreement with the measured signal. The effective resolution is estimated at 0.003 cm^{-1} and this can be improved further by extending the delay range. These and similar studies on N_2 jets[10] provide useful characterizations of the local temperature in free jet expansions.

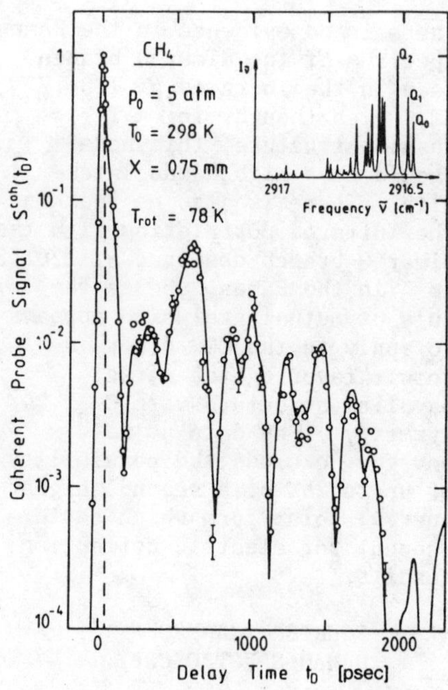

Fig. 3 Coherent probe scattering signal vs. delay time between excitation and probing pulses for the ν_1 Q band of pure CH_4.

REFERENCES

1. G. A. Hopkins, M. Maroncelli, J. W. Nibler and T. R. Dyke, Chem. Phys. Lett. 114, 97 (1985).
2. M. Maroncelli, G. A. Hopkins, J. W. Nibler and T. R. Dyke, J. Chem. Phys. 83, 2129 (1985).
3. L. Mannick, J. C. Stryland and H. L. Welsh, Can. J. Phys. 49, 3056 (1971).
4. A. Koide and T. Kihara, Chem. Phys. 5, 34 (1974).
5. N. Brigot, S. Odiot, S. H. Walmsley and J. L. Whitten, Chem. Phys. Lett. 49, 157 (1977).
6. G. A. Pubanz, M. Maroncelli and J. W. Nibler, Chem. Phys. Lett. 120, 313 (1985).
7. R. L. Kopec, Ph.D. Thesis, Indiana University, Bloomington (1981).
8. J. M. Lobue, J. K. Rice and S. E. Novick, Chem. Phys., Lett. 112, 376 (1984).
9. H. Graener, A. Laubereau and J. W. Nibler, Opt. Lett. 9, 165 (1984).
10. S. A. Akhmanov, N. I. Koroteev, S. A. Magnitskii, V. B. Morozov, A. P. Tarasevich and V. G. Tunkin, J. Opt. Soc. Am. B 2, 640 (1985).

TIME-RESOLVED STIMULATED RAMAN EXPERIMENTS

M.P. van Exter and A. Lagendijk
Nat.Lab., Univ. of Amsterdam, Amsterdam, The Netherlands

ABSTRACT

We have performed time-resolved stimulated Raman measurements on CS_2, benzene and diamond. These time-resolved SRS measurements produce more information than ordinary Raman or even non-time-resolved (two beams) SRS experiments because the signal is sensitive to the optical phases of the laser beams and registers therefore the so-called electronic contribution.

Finally we discuss the (im)possibility to resolve an inhomogeneously broadened line in its homogeneous components using the time-resolved Raman technique. Computer simulations support our theoretical analysis.

EXPERIMENT

Using two lasers, at least one of them tunable in frequency, it is possible to obtain a frequency scan of the Raman active excitations of the sample.[1] The quantity one measures in this two-beam experiment is the gain of the "Stokes beam" and is proportional to the imaginary part of $\chi^{(3)}(-\omega_s,\omega_s,\omega_\ell,-\omega_\ell)$. An alternative technique is a time-resolved spectrocopic method first proposed by Heritage.[2] Two pairs of laser pulses are formed, each one containing (5 ps) pulses of both frequencies (ω_s,ω_ℓ). The first pair is used to excite an internal mode of the system while the second pair, after a variable delay, is used to observe the relaxation of this mode. We have demonstrated through calculations and experiments[3] that a time-resolved experiment produces information about the absolute value, and in principle also about the phase, of $\chi^{(3)}(-\omega_s,\omega_s,\omega_\ell,-\omega_\ell)$ convoluted with the pulse profiles. An important consequence is the appearance of the electronic contribution (being almost purely real) superimposed onto the nuclear contributions.[4]

An experimental run is depicted in Fig.1. The envelope of the plotted curve represents the overall amplitude relaxation of the coherently excited vibrational mode (T_2-relaxation) which is related to the absolute value of $\chi^{(3)}(-\omega_s,\omega_s,\omega_\ell,-\omega_\ell)$. The appearing fringes demonstrate the importance of the phases of the optical beams and provide us in principle with the phase information of $\chi^{(3)}$.

An interesting situation arises when we tune the frequency difference of our lasers away from the eigenfrequency of the internal mode. The mode gets of course less excited, but due to interference effects it is possible that the strength of the time-resolved SRS signal first decreases sharply, as the time delay between the probe-pair and the pump-pair is increased, and at lager delays recovers to an "ordinary" value. The appearance of this

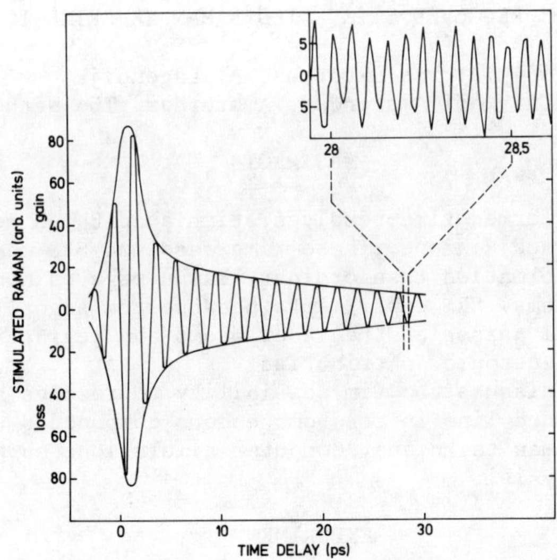

Fig.1. A time-resolved SR measurement of the 656 cm^{-1} ν_1-mode of liquid CS_2. The fringes shown are only a schematic representation of the more rapid oscillations experimentally measured. In the insert we have pictured part of the experimental data. Notice that the rapid oscillations exactly correspond to the inverse Raman frequency of this mode.

intensity dip as a function of the time delay depends critically on the pulse shapes of the excitation pulses and more importantly on the mistuning. After having found this effect in our theory we have demonstrated its presence experimentally.

Analytical calculations with an inhomogeneously broadened line, consisting of uncoupled homogeneously broadened components, demonstrate the impossibility to observe these separate components. Although some modes get more excited than others (because they are closer to resonance) one always measures the total behavior of the ensemble of modes. Only when coupling between the components (e.g. spectral diffusion) is taken into account it may become possible to subtract more information.

REFERENCES

1. B.F. Levine and C.G. Bethea, IEEE QE-16, 85-8 (1980).
2. J.P. Heritage, Appl.Phys.Lett. 34, 470-2 (1979).
3. M. van Exter and A. Lagendijk, Opt. Comm. 56, 191 (1985).
4. R.W. Hellwarth, Prog.Quant.Electr. 5, 1-68 (Pergamon Press, 1977).

MEASUREMENTS OF THE NONRESONANT THIRD-ORDER SUSCEPTIBILITY

G. J. Rosasco and W. S. Hurst
National Bureau of Standards
Gaithersburg, MD 20899

ABSTRACT

We advance evidence for the validity of a long-known dispersion formula for the electronic contribution to the third-order nonlinear susceptibility.

INTRODUCTION

Recent advances in measurement techniques have resulted in increased accuracy in determinations of the nonresonant third-order susceptibilities of gases. As a result, there now exists a reliable body of data from which one can estimate the dispersion in the electronic contribution to the susceptibility $\chi^{NRE}(-\omega_1,\omega_2,\omega_3,\omega_4)$. Dispersion refers to the dependence of χ^{NRE} on the frequencies of the fields involved in the nonlinear interaction, and from this dependence a prediction of the value of χ^{NRE} for one nonlinear process can be made from a measurement by another nonlinear process. Accurate values of the electronic contribution are generally important for harmonic generation experiments and for accurate deconvolutions of CARS diagnostic spectra. In the following, we present evidence to support the validity of a simple, long-known dispersion formula for the electronic portion of the susceptibility. We begin with the definition of the third-order nonlinear susceptibility used in the present work, then introduce a representation due to Hellwarth[1] for this susceptibility. This representation separates the electronic and nuclear framework contributions under the Born-Oppenheimer approximation and allows us to discuss the electronic portion in terms of the electronic hyperpolarizability. The dispersion formula for this quantity then is introduced and briefly discussed. Values of the electronic hyperpolarizability determined by means of dc-Kerr effect (dc-KE), field-induced second harmonic generation (FISHG), third harmonic generation (THG), coherent anti-stokes Raman spectroscopy (CARS), and ac-Kerr effect (ac-KE) are intercompared by means of the proposed dispersion formula. Two gases are considered, Ar and H_2, since data are available from a variety of nonlinear techniques that cover a large frequency range.

THE THIRD-ORDER SUSCEPTIBILITY

The third-order susceptibility used in the present work is implicitly defined by

$$P_{1,i} = D\chi_{ijkl}(-\omega_1,\omega_2,\omega_3,\omega_4)\ E_{2,j}E_{3,k}E_{4,l} \tag{1}$$

where D=6, 3, or 1 depending on the number (3,2,1) of distinguishable field components. Summation of the repeated indices (j,k,l) is implied. The polarization, P, and the fields, E, are expressed in terms of their Fourier components. The representation of the susceptibility[1] chosen for this discussion is

$$\chi_{1111}(-\omega_1,\omega_2,\omega_3,\omega_4) = \frac{3\sigma}{24} + \frac{1}{24} \sum_{\substack{j,k=2 \\ j\neq k}}^{4} [A(\omega_j+\omega_k)+B(\omega_j+\omega_k)] \ . \qquad (2)$$

In Eq. (2), the contributions of the vibrational and orientational nuclear motions are represented by the terms A and B, generally complex, and the electronic contribution is given by the real electronic hyperpolarizability, σ. As presented in Eq. (2), the hyperpolarizability is explicitly dispersionless. Hellwarth[1] and Owyoung[2] have advanced a simple dispersion formula based on an anharmonic electronic oscillator. From this model they find that the electronic third-order susceptibility can be expressed in terms of a product of first-order susceptibilities, one for each frequency, i.e.

$$\chi^{NRE}(-\omega_1,\omega_2,\omega_3,\omega_4) \propto \chi(\omega_1)\chi(\omega_2)\chi(\omega_3)\chi(\omega_4) \ . \qquad (3)$$

The first-order susceptibility function χ is assumed to have a pole at $\omega_0 >> \omega_i$, and thus the dispersion relation takes the simple form

$$\chi^{NRE}(-\omega_1,\omega_2,\omega_3,\omega_4) \propto \chi_o \ [1 + K\ (\omega^2)_{eff}] \qquad (4)$$

with K a constant and

$$(\omega^2)_{eff} = \omega_1^2 + \omega_2^2 + \omega_3^2 + \omega_4^2 \ . \qquad (5)$$

Dispersion relations have been derived from formal perturbation series solutions for the third-order susceptibility by Dawes[3] and Finn[4] for the cases of dc-KE, FISHG, and THG. Their results predict relative rates of dispersion (versus the frequency of the fundamental) for these three effects in the ratio respectively, 1:3:6, as does Eq. (4). It also has been observed[5,6] that <u>ab initio</u> calculations for He are approximately in this ratio. Recently, Shelton[6], using an approach similar to that of Dawes, has derived a dispersion relation for χ_{1111}^{NRE} in the form of Eq. (4). In order to verify the dispersion relation we plot the various values of the nonresonant electronic susceptibility versus the effective squared frequency defined in Eq. (5). We present the data in terms of $\chi_{1111}^{NRE}=3\sigma/24$ and use electrostatic units.

In a recent paper[7] we have given the formulae which relate the values of the susceptibility determined by various nonlinear techniques and for a variety of definitions of the susceptibility. Also in that paper, we collected data from a wide number of sources. In terms of direct measurements of dispersion, only for FISHG have there been systematic studies[5,6,8,9]. All these data are presented in Fig. 1. The solid lines for Ar and H_2 were determined from the data provided by Mizrahi and Shelton[10]. These data were presented in the form of ratios to the susceptibility of He. We first apply the dispersion formula to plot the calculations of Klingbeil[11] for THG in He; we then calculate the magnitudes for Ar and H_2 from the data of Ref. 10. A small correction to the H_2 nonresonant susceptibility values to account for the far off-resonance vibrational contributions was also made (see also Ref. 9). Although our method of data reduction based upon the dispersion formula (Eq. 4) is different from that of Mizrahi and Shelton (Ref. 8), the agreement in general is within 1%. However, the most reliable values for Ar and H_2 should be obtained from Ref. 8. Our measurements, labeled ac-KE, are derived from a phase modulated version of stimulated Raman spectroscopy which is similar to OHD-RIKES[12]. A small (+3%) correction has been applied to the dc-KE measurement of H_2 to account for far off-resonance Raman contributions. All other data are taken from the original references and corrected to our definitions and units (cf. Ref. 7).

CONCLUSIONS

We see from the figure that the large (>15%) variations in the values of the susceptibility obtained from the various techniques are now revealed as systematic variations associated with dispersion. The variation from the dispersion curve is on the average less than 5%, the uncertainty associated with each point being assigned, somewhat arbitrarily, as ±6%. We emphasize that the various techniques represented in the figure derive their quantitative basis from very different references. For FISGH and THG, *ab initio* calculations for He are the reference. For CARS and for our measurements, various Raman cross-sections serve as reference. The dc-KE measurements rely on an absolute determination of birefringence. Thus, we conclude that the level of agreement with the dispersion formula displayed in the figure is good evidence for its utility in predicting the electronic hyperpolarizability.

REFERENCES

1. R.W. Hellwarth, Prog. Quant. Electr. 5, 1(1977).

2. A. Owyoung, Ph. D. Thesis, California Inst. of Technology (1971).

3. E.L. Dawes, Phys. Rev. 169, 47(1968).

4. R.S. Finn, quoted in D.S. Elliott and J.F. Ward, Mol. Phys. 51, 45(1984).

5. V. Mizrahi, Ph. D. Thesis, Univ. of Toronto (1985).

6. D.P. Shelton, J. Chem. Phys. 84, 404 (1986).

7. G.J. Rosasco and W.S. Hurst, Phys. Rev. A32, 281(1985).

8. V. Mizrahi and D.P. Shelton, Phys. Rev. Lett. 55, 696(1985).

9. V. Mizrahi and D.P. Shelton, Phys. Rev. A32, 3454(1985).

10. Data kindly provided by D.P. Shelton and V. Mizrahi prior to publication.

11. R. Klingbeil, Phys. Rev. A7, 48(1973).

12. G.J. Rosasco and W.S. Hurst, J. Opt. Soc. Am. B2, 1485(1985).

13. R.L. Farrow and L.A. Rahn, J. Opt. Soc. Am. B2, 903(1985).

14. T. Lundeen, S-Y. Hou, and J.W. Nibler, J. Chem. Phys. 79, 6301(1983).

15. G.J. Rosasco and W.S. Hurst, submitted for publication.

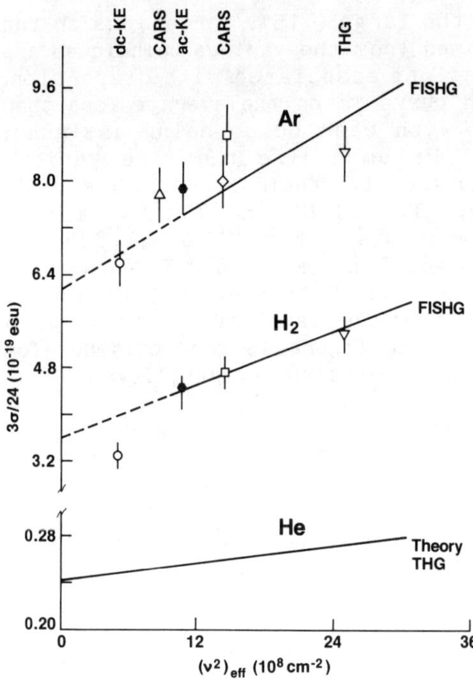

Figure 1. Values of $3\sigma/24$ from various nonlinear techniques plotted versus the effective squared frequency defined in Eq. 5. The solid lines are explained in the text. The CARS measurement for Ar with the diamond symbol is from Ref. 13. The CARS measurement for H_2 with the square symbol is taken from the value for D_2 given in Ref. 14 adjusted to H_2 by the ratio given in Ref. 9. All other data are taken from the summary given in Table IV of Ref. 7.

TWO-PHOTON EXCITATION STUDIES IN XENON

L. C. Bobb, M. B. Rankin, and J. P. Davis
Naval Air Development Center, Warminster, PA 18974-5000

ABSTRACT

A frequency-doubled dye laser was used to excite ground state xenon into the 6p states and the subsequent fluorescence to the 6s states was measured. Xenon pressures from 0.01 to 10.0 torr were used with buffer gases of oxygen and nitrogen at pressures from 0 to 760 torr. Both spontaneous and stimulated fluorescence were observed in the forward and backward directions.

INTRODUCTION

Two-photon excitation to the 6p states of xenon has been accomplished by using both excimer lasers and dye lasers. One of the early studies[1] used a KrF laser to populate the 6p(1/2,0) state of xenon; subsequent formation of XeO* in the presence of N_2O was observed. Resonance ionization spectroscopy was studied using a dye laser by three-photon ionization of xenon atoms through the 6p(1/2,0) and the 6p(3/2,2) states[2]. Three-photon ionization of xenon through the 6p(1/2,0) state was accomplished also by the simultaneous use of an ArF (193 nm) and a XeF (351 nm) excimer laser[3]. Lasing on the 6p(1/2,0) → 6s(3/2,1) at 828 nm was reported as well[3]. Two-photon laser spectroscopy studies of xenon collision pairs over a pressure range of a few torr to 10,000 torr were reported recently[4]. In our studies we have populated three of the 6p states individually and observed the subsequent spontaneous fluorescence and superradiant emission to the 6s-states. Figure 1 shows the relevant energy level diagram.

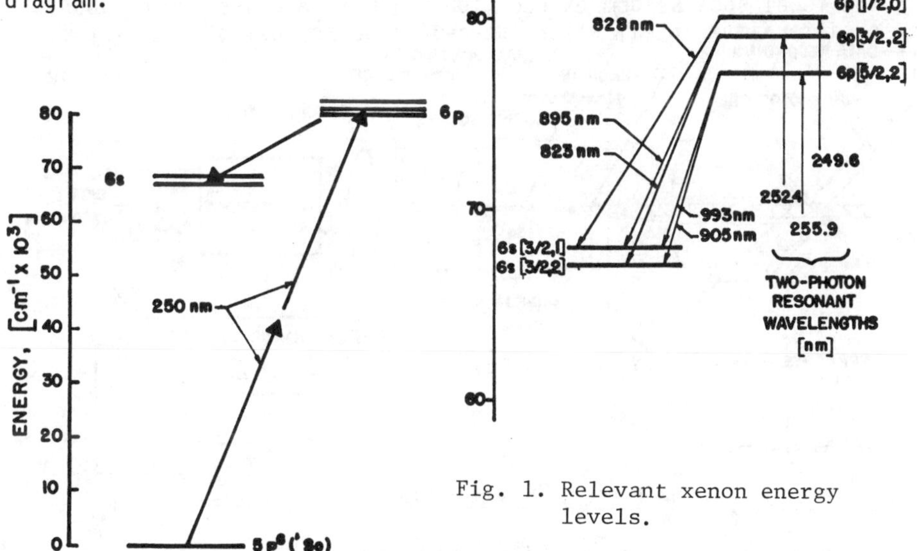

Fig. 1. Relevant xenon energy levels.

EXPERIMENTAL APPARATUS

A Q-switched Nd:YAG oscillator/amplifier drives nonlinear crystals which generate the second and third harmonics of the 1.06 μm fundamental wavelength. The 355 nm third harmonic is isolated and used to pump a three-stage dye laser using coumarin 500 dye. The system runs at 10 pulses/second. The dye oscillator is a Hansch-type design[5] using both a grating and an intracavity etalon as tuning elements. The grating alone limits the linewidth of the oscillator to approximately 8 GHz. The air-spaced intracavity etalon (FSR of 37.5 GHz, finesse of about 20) further narrows the linewidth to a few longitudinal cavity modes, approximately 1.5-2 GHz. Longitudinal mode spacing for the oscillator is about 500 MHz. The tuning elements are enclosed in an airtight chamber to allow pressure scanning. To generate the required UV wavelengths, the dye laser is tuned between 499.2 and 511.8 nm, and its output is frequency-doubled in lithium formate monohydrate (LFM) to produce wavelengths ranging from 249.6 to 255.9 nm, spanning the range of two-photon resonance wavelengths for the xenon 6p states. A maximum conversion efficiency of ∿1% from 500 to 250 nm was obtained with the unfocussed dye laser beam. After doubling, the mixed beam transits a water solution of $CoSO_4$ (1 mole/liter concentration, 3 cm path length) which strongly absorbs the visible light while transmitting the 250 nm radiation. The UV beam has a diameter of about 2 mm, a divergence of about 0.8 mr and a maximum pulse energy of 1 μJ. Each pulse is composed of 1-2 spikes with a 1.6 ns spacing; the duration (FWHM) of the individual spikes is 600 ps, the detection system limit.

Details of the UV focussing and fluorescence collection optics are shown in Figure 2. Lens L1 (suprasil, 2.5 cm diameter, 5 cm focal length) focusses the UV beam into the cell to a minimum diameter of approximately 40 μm. The maximum average power density in the focal spot seldom exceeds 100 MW/cm^2. A glass 800 nm long-pass filter immediately after the cell absorbs most of the UV pump

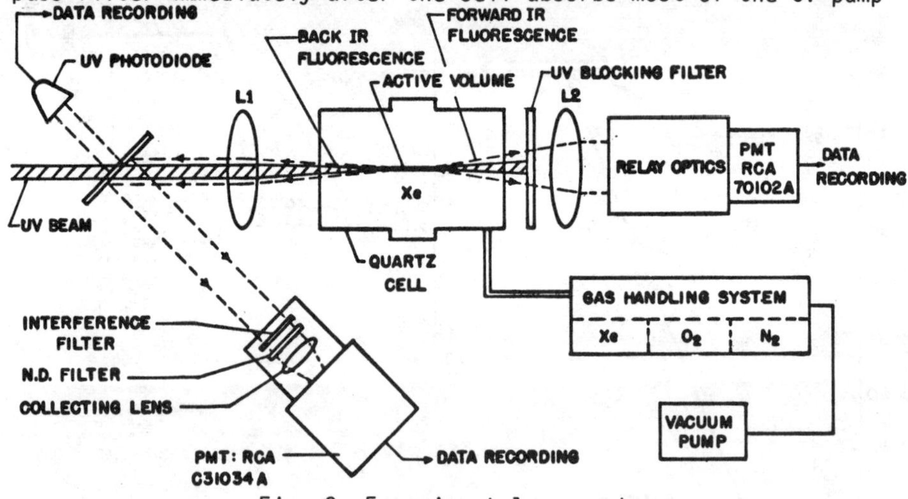

Fig. 2. Experimental apparatus.

light and visible wavelengths shorter than 800 nm while transmitting the near-IR fluorescence light. Glass lens L2 (50mm diameter, 50mm focal length) collects and collimates fluorescence light from the vicinity of the UV beam waist. Apertures of variable diameter inserted after L2 are used to study the angular distribution of fluorescence energy. Interference filters for fluorescence line isolation can also be inserted into the collimated section after L2.

EXPERIMENTAL RESULTS

For UV pulse energies below a threshold value, the fluorescent energy increases quadratically with input energy. The fluorescence is also isotropic and characterized by the normal fluorescence decay time (28 ns for the 6p(1/2,0) state). Above the threshold energy, the fluorescent radiation becomes highly bi-directional, and is characterized by a much shorter radiation time ($\sim$0.6 ns). Using the calibrated system shown in Figure 2, it was determined that the stimulated fluorescent energy is equal within experimental uncertainty in the forward and backward directions. The angular distribution of the fluorescence was measured; fifty percent of the energy in the forward direction was within a cone of 4°. And, the stimulated fluorescence is unpolarized on a time-averaged basis. An image intensifier was used to view the fluorescent radiation in the forward direction. It was found that when the pump laser is detuned from line center to the high frequency side, the radiation occasionally appears as a ring. This ring varies in diameter with xenon pressure. These observations so far are consistent with four-wave mixing in the xenon. However, with the laser tuned to line center the observations are characterized by amplified spontaneous emission (ASE).

In Figure 3, the ASE energy in the forward direction is shown plotted against the UV pulse energy. The same general behavior is observed for xenon pressures from 0.01 to 10 torr and nitrogen pressures from 0 to 760 torr. The data are characterized by large fluctuations in the ASE signal, a linear growth in fluorescent energy with input energy and a threshold value below which only spontaneous emission is observed. We have not been successful in explaining the linear growth of the ASE energy. Three-photon ionization and the optical Stark effect, both of which would cause departures from the quadratic growth in the fluorescent energy, were calculated to be too small at the measured input powers. Additionally, our super-radiance model predicts a threshold energy value 50 times larger than experimental values.

The threshold energy versus xenon pressure is shown in Figure 4. The data points in this figure were assembled from various measurements taken over a one-year interval. The best fit to the data is a curve for which the threshold energy grows as $1/p^{0.65}$. A simple model for the growth of the threshold energy with decreasing xenon pressure predicts that E_t grows as $1/p^{0.5}$. This data will be extended to lower pressures to provide a better understanding of the threshold behavior.

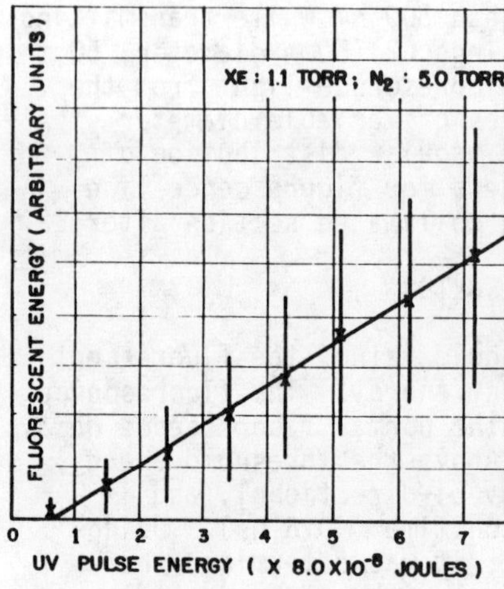

Fig. 3. Superradiant pulse energy vs UV pulse energy.

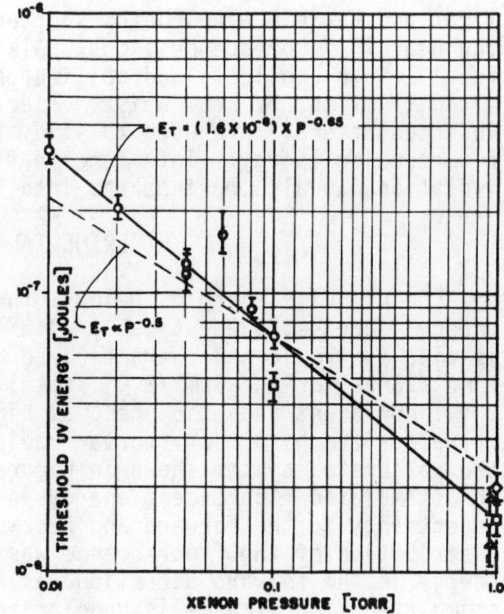

Fig. 4. Threshold energy vs xenon pressure.

ACKNOWLEDGEMENTS

The contributions of Tim Berkoff, Bob Starry, Chris Giranda, and Tom Curran of NADC and Dr. Lorenzo Narducci of Drexel University are gratefully acknowledged.

REFERENCES

1. D. Kligler, D. Pritchard, W. K. Bischel, and C. K. Rhodes, J. Appl. Phys. 49(4), 2219 (1978).
2. C. H. Chen, G. S. Hurst, and M. G. Payne, Chem. Phys. Lett. 75(3), 473 (1980).
3. A. W. McCown, M. N. Ediger, and J. G. Eden, Phys. Rev. A 26(4), 2281 (1982).
4. T. D. Raymond, N. Bowering, Chien-Yu Kuo, and J. W. Keto, Phys. Rev. A 29(2), 721 (1984).
5. T. W. Hansch, Appl. Optics 11(4), 895 (1972).

APPLICATION OF NONLINEAR RAMAN SPECTROSCOPY IN COMBUSTION AND PLASMA DIAGNOSTICS

Alan C. Eckbreth
United Technologies Research Center, East Hartford, Ct. 06108

INTRODUCTION

Nonlinear Raman spectroscopy has emerged as a very powerful analytical technique due to its extremely high spectral resolution, superior signal/noise ratios and sensitive detection levels. Of the myriad of nonlinear Raman approaches,[1,2] CARS (coherent anti-Stokes Raman spectroscopy) and SRGS (stimulated Raman gain) are seeing the most utilization in measurement and spectroscopic investigations, respectively. In most diagnostic applications, CARS is preferable to SRGS for a number of reasons to be detailed. In spectroscopic investigations, SRGS is generally preferred. In this overview, we will highlight these distinctions, review some important diagnostic developments in CARS and tabulate practical applications of CARS in combustion and plasmas.

COMPARISON OF CARS AND SRGS

Most practical combustion and plasma environments are time-varying in nature being either transient or unsteady. Due to the nonlinear dependence of CARS/SRGS on temperature and density, time averaging over parameter fluctuations can lead to serious measurement errors. Thus, single pulse, "instantaneous" measurements are required. Furthermore, fluctuation magnitudes are often as equally important as mean values. CARS, being a signal generative process, is easily multiplexed with broadband Stokes sources and spectrally recorded with optical multichannel detectors. Although SRGS was first demonstrated with a broadband source,[3] i.e., in its loss variant, namely, inverse Raman, it is difficult to measure the small gas phase gains amidst the background and noise with multichannel detectors. Turbulence effects in practical environments generally cause some beam steering and defocussing due to refractive index effects leading to signal loss. Thus, the signal has to be normalized and this generally cannot be done accurately externally. With CARS, the presence of the background nonresonant susceptibility permits in situ normalization either explicitly or implicitly through spectral signature analysis. A corresponding normalization scheme is not inherent in SRGS. Both techniques are critically dependent on beam overlap at the measurement location. Although the beams can be arbitrarily oriented in SRGS, the phasematching requirements of CARS are not particularly disadvantageous. The spectral complexities of CARS, i.e. constructive and destructive interference effects, are tolerated diagnostically but best avoided for spectroscopic studies. For spectroscopy, SRGS is the method of choice due to the purity of the individual lines and is providing much of the fundamental data base necessary for the accurate modelling of CARS spectra.

DIAGNOSTIC HISTORY OF CARS

After the discovery of anti-Stokes rings in the early sixties,[4] CARS remained in the province of nonlinear optics until Taran's pioneering applications to flames in the early seventies in which quantitative species measurements[5] and thermometry[6] were demonstrated. In 1974, Nd:YAG lasers were applied to CARS[7] and much of the progress in this field was accelerated by the commercial emergence of reliable, scientific Nd:YAG lasers in the mid seventies. Temporal and spatial precision followed with the introduction of broadband Stokes sources for single pulse recording[8] and crossed-beam phase matching[9] for pointwise measurements respectively. In the late seventies, electronic resonance enhancement in gases and flame radicals was observed,[10] the incoherent convolution for multimode pump lasers was introduced,[11] and nonresonant background suppression using polarization approaches was demonstrated in flames.[12] In addition, the first few practical device demonstrations appeared which will be reviewed shortly. At the beginning of this decade, collisional narrowing of CARS spectra at elevated pressures was reported[13] and three-dimensional phase matching was introduced to observe pure rotational Raman transitions.[14,15] Shortly thereafter, three-dimensional phase matching with just two beams was introduced[16] and explicit in-situ referencing was employed for quantitative flame measurements.[17] In the mid-eighties, it was shown that the incoherent spectral convolution for multimode pumps was not accurate over all concentration ranges and that a partially-coherent formulation was necessary.[18,19] The effect of laser field statistics on CARS spectra was seen with multimode pump lasers[20] and single pulse spectral noise was observed to decrease with single-mode pump lasers.[21] These three effects, i.e. convolution, statistics and single pulse quality, argue strongly for use of a single-mode pump laser for quantitative CARS work. Recently, dual broadband Stokes approaches have been advanced to permit simultaneous measurement of a multiplicity of species through a combination of two and three-color wave mixing.[22]

PRACTICAL UTILIZATION OF CARS

CARS has been demonstrated in many practical environments in the propulsion, automotive, energy conversion and plasma fields. CARS has been employed for measurements in gas turbine combustors or simulations thereof,[23-27] in jet engine exhausts,[28] in supersonic combustion[29] and over burning solid propellants.[30] It has been applied to numerous internal combustion engine studies[16,31-33] as well as to an actual diesel engine.[34] In the energy arena, it's been successfully utilized in furnaces,[35,36] chemical reactors,[37] a coal gasifier[38] and a simulated MHD exhaust.[39] In the plasma field, CARS has been used to investigate nonequilibrium glow discharges,[40,41] a low pressure H_2 plasma[42] and silane CVD plasmas[43] for semiconductor device fabrication.

SUMMARY

After more than a decade of development, CARS is maturing as an important and powerful nonintrusive diagnostic tool and should contribute to increased fundamental and applied understanding in combustion and plasma science.

REFERENCES

1. A. B. Harvey, Ed., Chemical Applications of Nonlinear Raman Spectroscopy (Academic Press, New York, 1981).
2. G. L. Eesley, Coherent Raman Spectroscopy (Pergamon, Oxford, 1981).
3. W. J. Jones and B. P Stoicheff, Phys. Rev. Lett. 13, 657 (1964).
4. P. D. Maker and R. W. Terhune, Phy. Rev. 137A, 801 (1965).
5. P. R. Regnier and J. P. E. Taran, Appl. Phys. Lett. 23, 240 (1973).
6. F. S. Moya, S. A. J. Druet and J. P. E. Taran, Opt. Comm. 13, 169 (1975).
7. R. F. Begley, A. B. Harvey and R. L. Byer, Appl. Phys. Lett. 25, 387 (1974).
8. W. B. Roh, P. W. Schreiber and J. P. E. Taran, Appl. Phys. Lett. 29, 174 (1976).
9. A. C. Eckbreth, Appl. Phys. Lett. 32, 421 (1978).
10. S. A. J. Druet, B. Attal, T. K. Gustafson and J. P. Taran, Phys. Rev. A 18, 1529 (1978).
11. M. A. Yuratich, Mol. Phys. 38, 625 (1979).
12. L. A. Rahn, L. J. Zych and P. L. Mattern, Opt. Comm. 39, 249 (1979).
13. R. J. Hall, J. F. Verdieck and A. C. Eckbreth, Opt. Comm. 35, 69 (1980).
14. Y. Prior, Appl. Opt. 19, 1741 (1980).
15. J. A. Shirley, R. J. Hall and A. C. Eckbreth, Opt. Lett. 5, 380 (1980).
16. D. Klick, K. A. Marko and L. Rimai, Appl. Opt. 20, 1178 (1981).

17. R. L. Farrow, P. L. Mattern and L. A. Rahn, Appl. Opt. 21, 3119 (1982).
18. H. Kataoka, S. Maeda and C. Hirose, Appl. Spect. 36, 565 (1982).
19. R. E. Teets, Opt. Lett. 9, 226 (1984).
20. L. A. Rahn, R. L. Farrow and R. P. Lucht, Opt. Lett. 9, 223 (1984).
21. D. R. Snelling, R. A. Sawchuk and R. E. Mueller, Appl. Opt. 24, 2771 (1985).
22. A. C. Eckbreth and T. J. Anderson, Appl. Opt. 24, 2731 (1985).
23. A. C. Eckbreth, Comb. Flame 39, 133 (1980).
24. G. L. Switzer, et al., J. Energy 4, 209 (1980).
25. R. Bedue, P. Gastebois, R. Bailly, M. Pealat and J. P. Taran, Comb. Flame 57, 141 (1984).
26. D. A. Greenhalgh, F. M. Porter and W. A. England, Comb. Flame 49, 171 (1983).
27. S. Fujii, M. Gomi and K. Eguchi, J. Fluids Eng. 105, 128 (1983).
28. A. C. Eckbreth, G. M. Dobbs, J. H. Stufflebeam and P. A. Tellex, Appl. Opt. 23, 1328 (1984).
29. T. J. Anderson, I. W. Kay and W. T. Peschke, Proceedings 22nd JANNAF Combustion Meeting (CPIA, Silver Spring, MD. 1986) in press.
30. K. Aron and L. E. Harris, Chem. Phys. Letts. 103, 413 (1984).
31. I. A. Stenhouse, D. R. Williams, J. B. Cole and M. D. Swords, Appl. Opt. 18, 3819 (1979).
32. L. A. Rahn, S. C. Johnston, R. L. Farrow and P. L. Mattern, Temperature (American Institute of Physics, New York, 1982) Vol. 5, Part I, p. 609.
33. G. C. Alessandretti and P. Violino, J. Phys. D. Appl. Phys. 16, 1583 (1983).
34. K. Kajiyama, K. Sajiki, H. Kataoka, S. Maeda and C. Hirose, SAE Technical Paper 821036 (1982).
35. A Ferrario and C. Malvicini, in Analytical Laser Spectroscopy, S. Martellucci and A. N. Chester, Eds., NATO Adv. Study Instit. Ser. B119, 89 (1985).
36. M. Alden and S. Wallin, Appl. Opt. 24, 3434 (1985).
37. W. A. England, J. M. Milne, S. N. Jenny and D. A. Greenhalgh, Appl. Spect. 38, 867 (1984).
38. D. J. Taylor, Los Alamos Scientific Laboratory Report LA-UR-83-1840 (1983).
39. D. L. Murphree, et al., AIAA Paper 82-0377 (1982).
40. A. B. Harvey, Anal. Chem. 50, 905A (1978).
41. T. Drier, U. Wellhausen, J. Wolfrum and G. Marowsky, Appl. Phys. B 29, 31 (1982).
42. M. Pealat, et al., J. Appl. Phys. 52, 2687 (1981).
43. N. Hata, et al., Jap. J. Appl. Phys. 22, L1 (1983).

CARS APPLICATIONS IN CHEMICAL REACTORS, COMBUSTION AND HEAT TRANSFER

D.A. Greenhalgh and F.M. Porter
Harwell Combustion Centre, B10.4, Harwell Labs., Oxfordshire U.K.

ABSTRACT

This paper illustrates the use of the CARS technique in the fields of Chemical Reactor engineering, combustion and Heat Transfer. Examples of recent results from a catalytic chemical reactor, an operating production petrol engine and an oil spray furnace are given. The experimentally determined accuracy of CARS nitrogen thermometry for both mean and single pulse measurements is presented.

APPLICATIONS

The CARS spectroscopy presented in this paper utilises equipment either identical or equivalent to earlier work[1]. Either folded BOXCARS or in the case of the i.c. engine, colinear phase matching is employed. Spectral simulation uses the latest analytical methods of Greenhalgh and Hall[2]. The application of the CARS technique to Chemical Reactors has been pioneered by England[3] and we present an example of some recent work on the methanation reaction ($CO+3H_2 \rightarrow CH_4 + H_2O$) using a tubular catalytic reactor[4]. In simple isothermal, isobaric, situations England[3,4] has shown that accurate concentration measurements can easily be made from a direct estimate of the non-resonant background and species specific resonant signals of the required species. Figure 1 illustrates a plot of CARS measurements made, on-axis, inside a 100mm by 6.3mm i.d. catalytic reactor tube. The dots are CARS measurements and the continuous curve is taken from a theoretical model of the reactor[4]. The excellent quality of the "in-situ" CARS data has enabled a careful evaluation of kinetic models, for an alumina supported nickel catalyst, to be performed[4].

A more demanding application for CARS is the internal combustion engine. Here we illustrate recent data obtained by Williams and others, from a production, 2 litre petrol engine, operating under typical driving conditions of operation. Figure 2 shows an example of a spectrum taken before combustion of the fuel/air charge at the point of measurement. Approximately 50 single shot spectra have been averaged, from several engine cycles, but conditionally sampled for the same in cylinder pressure at the selected crank angle. The key to the excellent analysis lies in the choice of the correct relaxation model. In this instance rotational relaxation rates (γ_{jk}) and linewidths (Γ_j) are given by a new model, the Differential Energy Gap model which is given below:

$$\Gamma_j = \gamma_{jk} = P \cdot T^{-A} \cdot K \cdot \rho_j^{-B} \cdot \rho_k^{+C}$$

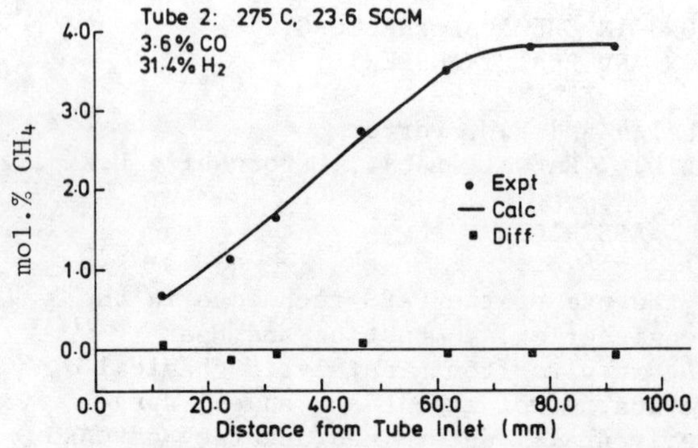

FIG. 1 Fitted Methane Conc. Profile

where P is pressure, T is temperature, P_j is the nomalised population of the jth state, the constants are K = 33.3226, A = 1.2807, B = 1.63, C = 1.4 the above is for up rates (assuming no dephasing Γ_i is thus $\sum_k \gamma_{jk}$) and down rates follow from detailed balance.

This model has been derived from the combination of some preliminary data of Rahn[6] and spectra similar to that shown in Figure 2.

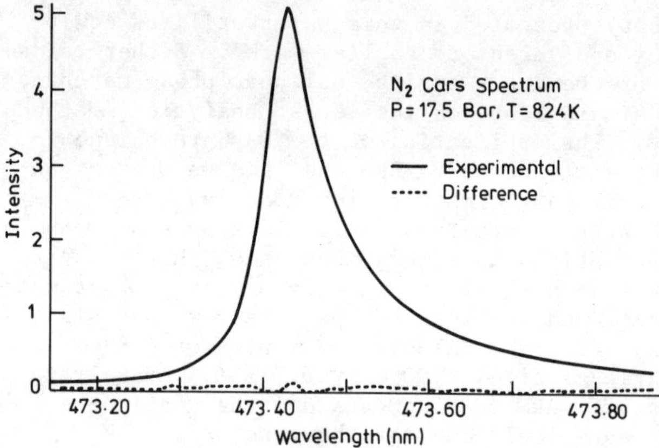

FIG. 2 I.C. Engine Spectrum

The use of CARS in turbulent oil-spray combustion depends primarily on the capability of CARS to make single pulse measurements. We have undertaken an extensive study of the use of CARS for 1 atmosphere thermometry in oil-spray combustion. The results of this work are extensively detailed elsewhere[7]. However two typical p.d.f's or temperature histograms are shown in figure 3. This data was taken around the burning fuel spray. A key question concerns the accuracy of the temperature. In the temperature region 0-1,100K we have compared averaged CARS and thermocouple measurements taken inside an isothermal tube furnace of known temperature (±1K). Figure 4 therefore illustrates the accuracy for our multimode, broad CARS system for data taken on 3 separate days. This data was processed by least squares fitting a theoretical spectra calculated using an early linewidth model[8]. Recently, we have found that the DEG model reduces the small systematic error, near 800K is reduced. Clearly an accuracy of ±20K, at worst, is possible.

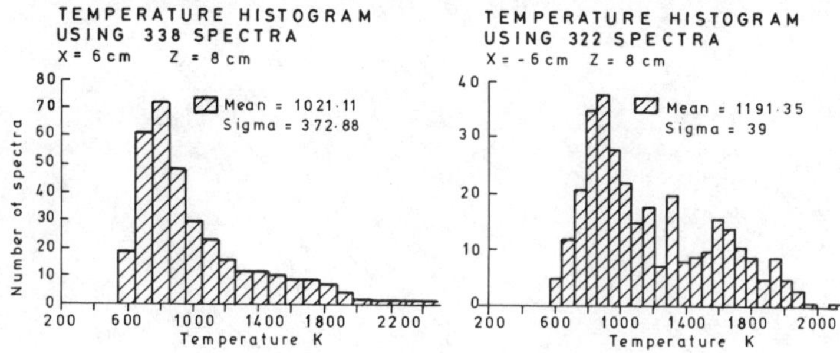

FIG. 3 Temperature Histograms for Oil Spray Flame

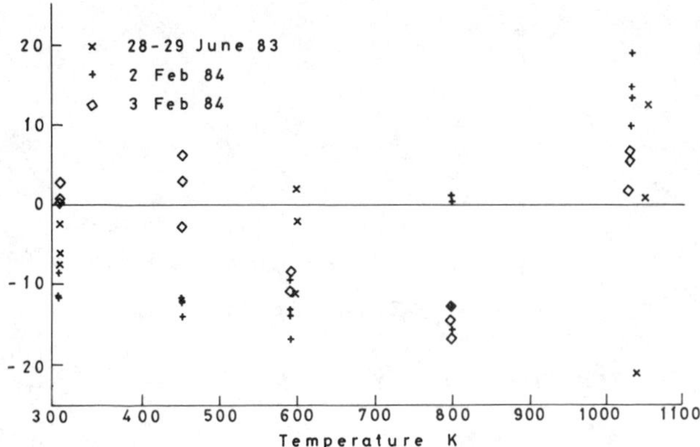

FIG. 4 Temperature Errors for Averaged CARS Spectra

Lastly we have also assess the accuracy of our single-shot CARS thermometry, also by comparison with air in an isothermal tube furnace. This data is presented in figure 5 and is plotted for 3 gas temperatures as a function of the inverse of the square root of the counts peak channel on the multi-channel, intensified diode array detector. Two points are clear. Firstly detector noise, presumably arising from the intensifier/photocathode, contributes a major part of the measurement uncertainty for signal level below 2,500 counts (maximum detector range 16,383 counts). Dye laser noise, which has been described elsewhere is dominant only for strong CARS signals. The marked contribution of detector counting statistics to the measured temperature is surprising. However it should be noted that (i) the counting statistics are correlated by cross-talk between channels generated in the intensifier (typically 3-4 channels) and (ii) this noise affects different intensity portions of the spectrum unequally. The latter will be particularly devastating for an analysis method which places an undue weight on the lower intensity portions of a spectrum.

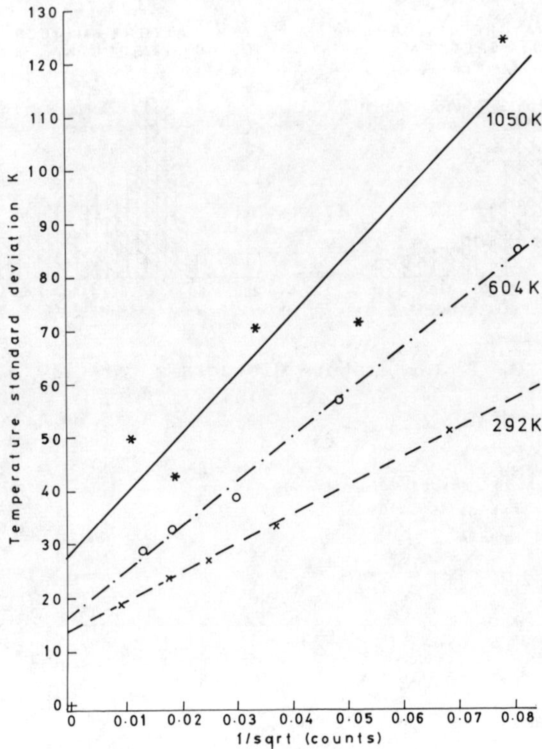

FIG. 5 CARS Temperature Uncertainty (1σ)

REFERENCES

1. D.A. Greenhalgh and S.T. Whittley, Appl. Opt. 24, 907 (1985).
2. D.A. Greenhalgh and R.J. Hall, Opt. Comm. to be published (1986).
3. W.A. England, J.M. Milne, S.N. Jenny and D.A. Greenhalgh, Appl. Spectrosc. 38, 867 (1984).
4. W.A. England, D.H.W. Glass, G. Brennan and D.A. Greenhalgh, J. Catalysis, to be published (1986).
5. D.A. Greenhalgh, D.R. Williams and C.A. Baker, "CARS Thermometry in a Firing, Production Petrol Engine", I.Mech.E. AUTOTECH '85, Birmingham, UK.
6. L.A. Rahn - private communication.
7. F.M. Porter and D.A. Greenhalgh, "Application of the Laser Optical Techniques CARS to Heat Transfer and Combustion", Harwell Report AERE-R 11824 (1985).
8. D.A. Greenhalgh, F.M. Porter and S.A. Barton, J. Quant. Spect. & Rad. Trans. 34, 95 (1985).

PLASMA DIAGNOSTICS USING COHERENT ANTI-STOKES RAMAN SPECTROSCOPY

W. F. Lynn and P. P. Yaney
University of Dayton, Dayton, Oh. 45469

L. P. Goss
Systems Research Laboratories, Inc., Dayton, Oh. 45440

S. W. Kizirnis
USAF Aero Propulsion Laboratory, WPAFB, Oh. 45433

ABSTRACT

Non-Boltzmann population distributions between the rotational and vibrational levels of N_2 in a transverse hollow-cathode discharge were observed using CARS and a scanned probe laser. A planar, crossed beam geometry was used which allowed high spatial resolution. The apparent rotational and vibrational temperatures were determined to be near 900 K and 1500 K, respectively, with a clear dependence upon position. The estimated uncertainty is ±75 K. Deviations from previously reported temperatures in positive-column discharges is attributed to differences in electron energy distribution.

EXPERIMENT AND ANALYSIS

The rotational-vibrational populations in low-current, positive-column N_2 discharges were measured using colinear CARS and found to be non-Boltzmann.[1,2] Vibrational temperatures were reported to be > 2500 K, rotational temperatures near 400 K and the non-equilibrium condition was attributed to electron impact.

In this study a planar crossed beam geometry (BOXCARS) was used. The essential characteristics of the apparatus were the pump beam energy, pulsewidth and linewidth which were ~40 mJ, 10 ns and 1 cm^{-1}, respectively, and the probe beam parameters, ~ 0.5 mJ, 7 ns and 0.3 cm^{-1}, respectively. The anti-Stokes beam was detected by a PMT and the output was time averaged at 10 Hz using a boxcar integrator.

A cross section of the hollow-cathode discharge and the CARS data are presented in Figure 1. The cathode length was 9.95 cm. The pressure and current were 3.7 torr and 150 mA, respectively. The N_2 flow rate was ~80 $cm^3 min^{-1}$. The CARS data were recorded on an X-Y recorder and are presented in 0.26 cm^{-1} intervals. Position I represents measurements made ~1.5 mm from the cathode. Closer measurements were not made due to an interaction between the pump beams and the cathode.

Rotational and vibrational temperatures were determined by fitting convolved theoretical CARS spectra to the data. The Yuratich convolution[3] was used as opposed to the more accurate method of Teets.[4] The error was negligible based on estimates that $\geq$ 30% of the plasma was ground state N_2. Stark effects[5] and saturation[6] were also neglected based on estimated intensity and spectral intensity.

The results of the analysis revealed a clear temperature

difference between the cathode and cathode slot where the rotational and vibrational temperatures differ by 125 K and 200 K, respectively. The non-equilibrium condition is attributed to low-energy electron scattering where the maximum scattering cross section occurs with 2-3 eV electrons.[7] Electron energy distributions were reported for positive-column and transverse hollow-cathode He discharges.[8] The two energy distributions appear to intersect at ~2 eV. The hollow-cathode has a larger number of electrons with energies < 2 eV. However, the positive-column has a more even distribution and for electron energies > 2 eV the electron population exceeds that in the hollow-cathode by as much as two orders of magnitude. This difference may explain the higher vibrational temperature previously reported.[1,2] The higher rotational temperature observed is a result of the higher operating current (150 vs. 7 mA). Hence electron scattering may be one mechanism for non-equilibrium and the previously reported non-Boltzmann vibrational population for v > 2.

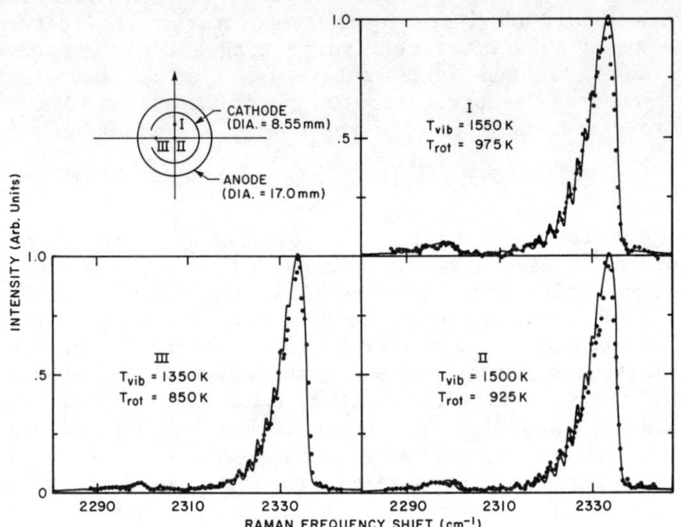

Fig. 1. Hollow-cathode cross section, data locations, CARS data (pts.), and fitted theoretical spectra with temperatures.

REFERENCES

1. W. M. Shaub, et al., J. Chem. Phys. 67, 1883 (1977).
2. V. V. Smirnov, V. I. Fabelinskii. JETP Lett. 28, 427 (1978).
3. M. A. Yuratich, Mol. Phys. 38, 625, (1979).
4. R. E. Teets, Opt. Lett. 9, 226 (1984).
5. L. A. Rahn, et al., Phys. Rev. Lett. 45, 620, (1980).
6. R. L. Farrow, R. P. Lucht, First ILS Conf.,pap. TM53, (1985)
7. B. I. Schneider, et.al., Phys. Rev. Lett. 43, 1926, (1979).
8. G. Medicus, R. A. Olson, XIth Int. Conf. on Phenomena in Ionized Gases,pap. 5.1.2.2., 1973.

MULTIPHOTON EXCITATION TECHNIQUES FOR COMBUSTION DIAGNOSTICS

J. E. M. Goldsmith
Combustion Research Facility
Sandia National Laboratories, Livermore, California 94550

INTRODUCTION

Several species that are of considerable interest to combustion research (e.g., atomic hydrogen and oxygen) are not amenable to conventional optical detection in flames because the vacuum-ultraviolet radiation required to excite their first electronic resonances is not transmitted by combustion gases. Resonant multiphoton excitation of atomic or molecular transitions using wavelengths longer than 200 nm provides a convenient method for gaining spectroscopic access to these species. Following the multiphoton excitation step, the atom or molecule can then be observed by sensitive optical detection techniques that include monitoring the ion/electron pairs produced by subsequent photoionization (optogalvanic detection), or monitoring the photons emitted from radiative decay to lower-lying electronic states (fluorescence detection). This paper will summarize and, where feasible, compare the various combustion diagnostic techniques based on multiphoton excitation for detecting atomic hydrogen and oxygen in flames, with references cited for applications of these methods in combustion environments.

ATOMIC HYDROGEN

The excitation schemes that have been successfully used to detect atomic hydrogen in flames are shown in Fig. 1; we have successfully worked with all seven schemes. Scheme 1 was the first method reported in the literature.[1] This technique used two distinct wavelengths for resonant two-photon excitation of the n=1 - n=2 transition, with a crossed-beam configuration providing excellent spatial resolution. Photons from either beam provided sufficient energy to photoionize the n=2 state. The 266-nm beam was supplied by the fourth harmonic of the Nd:YAG laser; as in the other schemes, a tunable UV beam was produced by shifting a tunable, visible dye laser beam to the UV by the appropriate combination of frequency doubling, frequency mixing, and Raman shifting. This scheme worked well in flames that did not contain nitrogenous species (e.g., hydrogen-oxygen flames), but difficulties were encountered in flames that used air as the oxidant due to interferences with NO produced in the flame.

Scheme number 3 in Fig. 1 is an alternative scheme for two-photon resonant, three-photon ionization of hydrogen that uses the single laser wavelength 243 nm. This method is simpler, since the fourth-harmonic output of the Nd:YAG laser can be dispensed with, but at the expense of some loss in spatial resolution, since the point defined by the intersection volume of the two laser beams in the two-wavelength method is replaced by the longer volume limited in extent only by the nonlinear intensity behavior of the multiphoton ionization process. This loss in spatial resolution is more than offset by far superior immunity to NO interference. Scheme number 3 has also been used to measure a relative atomic hydrogen concentration profile in a hydrogen/oxygen flame as a first step toward making quantitative concentration measurements with this method. Excellent agreement was obtained between the relative profile and an absolute concentration profile calculated from a measured OH

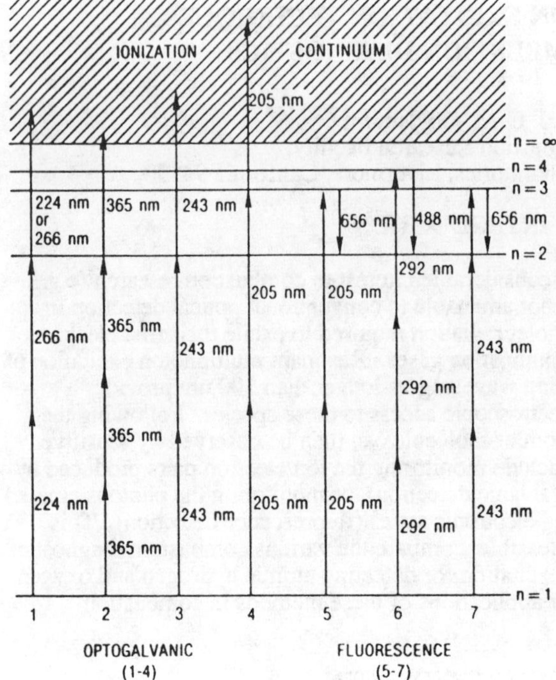

Fig. 1. Multiphoton excitation schemes that have been used to detect atomic hydrogen in flames.

concentration profile,[2] indicating a sensitivity on the order of 100 ppm even with the very modest laser energy of 50 μJ/pulse; substantially higher sensitivity should be achievable with higher pulse energies, although care must be taken that the signal does not become large enough to affect the signal collection efficiency.[3]

Scheme 2, three-photon resonant (n=1 - n=2) four-photon ionization with 365-nm radiation, was first demonstrated in a low-pressure (75 Torr) hydrogen-oxygen-argon flame.[4] We were able to observe this hydrogen resonance in an atmospheric-pressure hydrogen-oxygen diffusion flame only with great difficulty; interferences from NO completely masked the atomic hydrogen signal in hydrogen-air flames. We obtained good signals using scheme 4 (two-photon-resonant, three-photon ionization using 205-nm radiation) in rich hydrogen-oxygen flames, but encountered interferences in other flames, and did not pursue this scheme further since it had no advantages relative to scheme 3.

Fluorescence detection of atomic hydrogen in flames requires excitation to n=3 or higher states, since the only spontaneous emission pathway from n=2 is in the vacuum ultraviolet at 122 nm, which is absorbed by the flame gases. This technique was first demonstrated by Lucht et al. using scheme 5 (two-photon n=1- n=3 excitation with 205-nm radiation, with detection of 656-nm radiation emitted from n=3 - n=2 decay).[5] We have repeated this experiment, comparing a relative atomic hydrogen concentration profile with one measured in the same flame using scheme 3 and calculated from a measured OH concentration profile.[2] In this flame, we found that more atomic hydrogen was created by photolysis of water by the 205-nm beam than was naturally present in the flame, resulting in an invalid measurement; this photochemical difficulty is treated fully in a recent publication.[6]

Two other fluorescence methods have appeared in the literature. Three-photon-resonant excitation with fluorescence detection using scheme 6 has been reported.[7] As with scheme 2, this method suffers somewhat from weak signal levels due to the use of a three-photon excitation without resonant enhancement. A three-photon scheme that takes advantage of a two-photon resonance to form a two-step excitation process, scheme 7, has also been described[8] and used for the first demonstration of imaging of atomic hydrogen.[9] This latter scheme appears to provide the strongest signal level of the three fluorescence methods, although with some cost in complexity.

ATOMIC OXYGEN

The demonstrated techniques based on multiphoton excitation processes for detecting atomic oxygen in flames are shown in Fig. 2. All share in common resonant two-photon excitation of the 2^3P-3^3P transition using 226-nm radiation. Optogalvanic detection takes advantage of subsequent photoionization from the 3^3P state.[10] Fluorescence detection was first demonstrated by observing the 845-nm radiation emitted from spontaneous decay of the 3^3P state down to the 3^3S state.[11] Emission at 778 nm from 3^5P-3^5S decay (following the 3^3P excitation) has also been observed, presumably due to collisional energy transfer from the 3^3P state to the 3^5P state.[12] Linear imaging of atomic oxygen in flames by using a linear diode array, rather than a photomultiplier, to detect the spontaneous emission from oxygen atoms following two-photon excitation has recently been reported[13] and repeated in our laboratory.[9] Although not yet observed, near-coincidences between atomic and molecular oxygen excitation wavelengths suggests that measurements of atomic oxygen concentrations using these transitions could produce artificially high values because of predissociation of the excited molecules.[14]

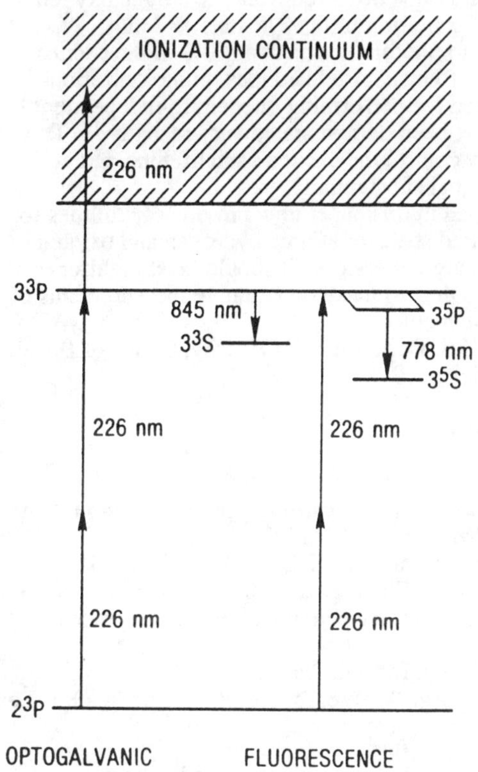

Fig. 2. Multiphoton excitation schemes that have been used to detect atomic oxygen in flames.

As described earlier in this paper for atomic hydrogen, optogalvanic detection has also been used to measure a relative atomic oxygen concentration profile in a hydrogen-oxygen flame as a first step toward making quantitative concentration measurements with this method. Excellent agreement was obtained between the measured profile and a profile calculated from a measured OH concentration profile.[2] The calculated oxygen concentration indicates a sensitivity on the order of 100 ppm for this flame even with the very modest laser energy of 100 μJ/pulse; substantially higher sensitivity should be achievable with higher pulse energies, although care must be taken that the signal does not become large enough to affect the signal collection efficiency.[3]

DISCUSSION

The development of detection schemes for combustion studies based on multiphoton excitation are still in their infancy. In addition to the references cited above for atomic hydrogen and oxygen, two-photon excitation has also been used to detect OH and CO in flames,[15-17] with linear imaging of CO demonstrated in the latter. All these methods have had substantial success in detecting the desired species in flames. In a real combustion application, however, the ability to make quantitative measurements, of relative if not absolute concentration profiles, becomes an overriding issue. The difficulty we have had in using scheme 5 for measuring the atomic hydrogen concentration profile in a lean atmospheric-pressure demonstrates one of several kinds of problems that can occur,[6] and points out the need to apply all of these techniques with considerable care to be certain that they really are "non-intrusive."

The effects of collisional quenching remain one of the major difficulties in measurements of species concentrations in flames using resonant optical (i.e., non-Raman) methods. This is especially a problem with fluorescence detection, as demonstrated by the fact that 778-nm emission is observed from the atomic oxygen 3^5P-3^5S transition following two-photon excitation of the 3^3P state.[12] With optogalvanic detection, it may be feasible to photoionize nearly 100% of the excited atoms, completely eliminating quenching effects.[1] A related idea, photoionization-controlled loss spectroscopy, was proposed for performing quenching-independent H-atom fluorescence measurements,[5] and has been demonstrated very recently.[18] This method uses photoionization to effectively control the excited-state lifetime, but at the expense of decreasing the fluorescence signal substantially.

Alternatively, the methods described in this paper now provide capabilities for studying collisional quenching of the excited states of atomic hydrogen and oxygen in flames. Once the quenching mechanisms are understood, it should be straightforward to correct for the effects of quenching, and thus to make quantitative measurements of atomic concentrations in flames with these methods.

This work was supported by the U.S. Department of Energy, Office of Basic Energy Sciences, Division of Chemical Sciences.

REFERENCES

1. J. E. M. Goldsmith, Opt. Lett. **7**, 437 (1982).
2. J. E. M. Goldsmith, *Twentieth Symposium (International) on Combustion* (The Combustion Institute, Pittsburgh, 1984), pp. 1331-1337.
3. J. E. M. Goldsmith, Digest of Technical Papers, CLEO '84, p. 92 (1984).
4. P. J. H. Tjossem and T. A. Cool, Chem. Phys. Lett. **100**, 479 (1983).
5. R.P. Lucht, J.T. Salmon, G.B. King, D.W. Sweeney and N.M. Laurendeau, Opt. Lett. **8**, 365 (1983).
6. J. E. M. Goldsmith, Opt. Lett. (submitted for publication).
7. M. Aldén, A. L. Schawlow, S. Svanberg, W. Wendt, and P.-L. Zhang, Opt. Lett. **9**, 211 (1984).
8. J. E. M. Goldsmith, Opt. Lett. **10**, 116 (1985).
9. J. E. M. Goldsmith and R. J. M. Anderson, Appl. Opt. **24**, 607 (1985).
10. J. E. M. Goldsmith, J. Chem. Phys. **78**, 1610 (1983).
11. M. Aldén, H. Edner, P. Grafström, and S. Svanberg, Opt. Commun. **42**, 244 (1982).
12. A. W. Miziolek and M. A. DeWilde, Opt. Lett. **7**, 390 (1984).
13. M. Aldén, H. M. Hertz, S. Svanberg, and S. Wallin, Appl. Opt. **23**, 3255 (1984).
14. J. E. M. Goldsmith and R. J. M. Anderson, Opt. Lett. **11**, (1986), in press.
15. D.R. Crosley and G.P. Smith, J. Chem. Phys. **79**, 4764 (1983).
16. J. E. M. Goldsmith and N. M. Laurendeau, Appl. Opt. **25**, 276 (1986).
17. M. Aldén, S. Wallin, and W. Wendt, Appl. Phys. B **33**, 205 (1984).
18. J. T. Salmon (private communication).

THEORY OF PHASE CONJUGATION IN FOCUSED BEAM WITH PARTIALLY-CORRELATED COMPONENTS

J.T. Lin
Litton Systems, Inc., Laser Systems Division, Orlando, FL 32854

ABSTRACT

The degree of noise discrimination and the quality of the phase-conjugated Stokes fields are numerically analyzed via an ensemble-averaged Maxwell equation for a focused pump beam consisting of two uncorrelated components: the plane wave and the speckle (noise). Effects of parameters such as focus-waist, cell length and the focal length on the discrimination of these two uncorrelated components are shown.

INTRODUCTION AND EQUATION OF MOTION

Consider a focused pump field consisting of a plane wave (PW) and a speckle (SP) component

$$E_L(r,z) = f_0^{\frac{1}{2}} E_0 + f_1^{\frac{1}{2}} E_1, \tag{1}$$

where the focusing functions are related to the confocal length of the PW (z_0) and of the SP(z_1) by

$$f_{0,1} = 1/[1 + (z/z_{0,1})^2] \tag{2}$$

We shall find the Stokes fields correlated with the pump in the conjugated form of $E_s(r,z) = a_0(z)E_0^* + a_1(z)E_1^*$. Taking the ensemble average of the random transverse coordinate, r, with a joint Gaussian correlatin and assuming that $E_iE_j = 0$, for $i \neq j = 0,1$, we obtain a set of coupled equations for the amplification of the PW and SP:

$$\frac{da_0}{dz} = \left(\frac{gI_{L0}}{2}\right)\left[F(z)a_0 + (f_0f_1)^{\frac{1}{2}}\rho a_1\right]$$
$$\frac{da_1}{dz} = \left(\frac{gI_{L0}}{2}\right)\left[(f_0f_1)^{\frac{1}{2}}(1 - \rho)a_0 + [F(z) + f_1\rho]a_1\right] \tag{3}$$

where $F(z) = f_0 + (f_1 - f_0)\rho$, and ρ is the fraction of the SP component defined by:

$$\rho = \langle |E_1|^2 \rangle / I_{L0}, \qquad I_{L0} = \langle |E_0|^2 \rangle + \langle |E_1|^2 \rangle , \tag{4}$$

Equations (3) have been studied for the interacting, unfocused case with $f_0=f_1=1$, where the pure SP gain is reported to be twice that of the pure PW gain.[1] This may be readily seen for the limiting cases of

ρ=0 (no SP) and ρ=1 (no PW). For focused but non-interacting case, results have also been reported, where the scattered fields are governed by an exponential growth with an integrated gain given by $\arctan(2L/z_R)$, where L is the cell length and Z_R is the Rayleigh range or twice of the confocal parameter.

RESULTS

Given an initial speckle fraction (SF), the output SF may be solved analytically only for the unfocused case whose results are shown in Figure 1. We see that due to the interaction between PW and SP and the double-gain feature of SP, perfect conjugated scatterings are available only in the pure-component limits, i.e., when the pump field is pure SP(ρ=1) or pure PW(ρ=0).

For the focused case, the interaction of the fields can only be solved numerically. Figure 2 shows the results for the focused case with confocal length ratio $z_1=10z_0$, i.e., the SP component has a divergence angle ten times that of the PW. In Figures 1 and 2, the input SF is defined by Equation (4) whereas the output SF is defined by $1/(1+1/R)$, where the speckle reflectivity is given by:

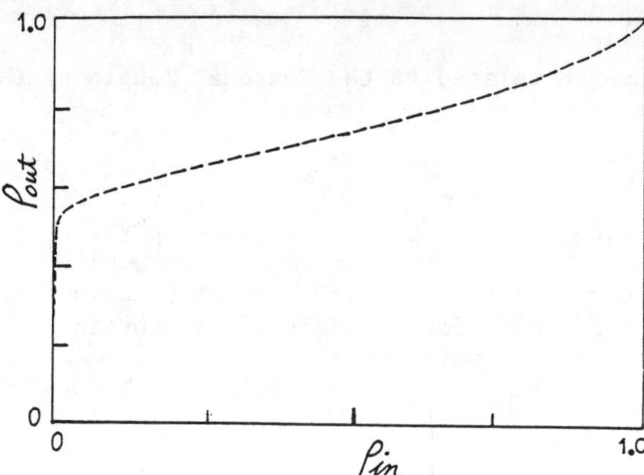

Fig. 1. The output versus input speckle component for the interacting unfocused case.

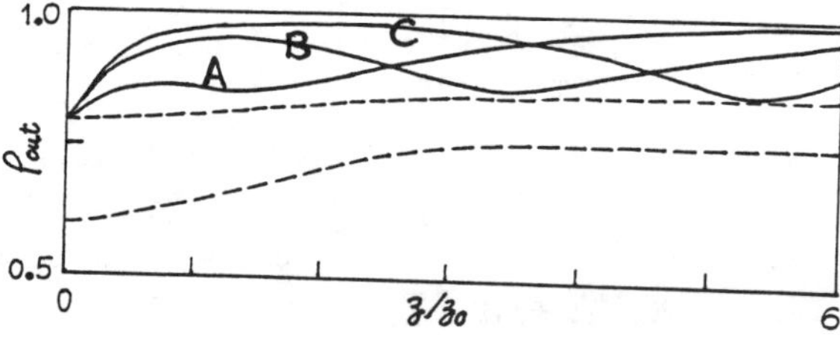

Fig. 2. Propagation z-dependence of the speckle components for the unfocused cases (dashed curves) and the focused cases for focal length f = (A)5L/6, (B) L/2 and (C) L/6.

In conclusion, we note that for systems with injected seed-Stokes, the pump beam will be severely depleted and a complete model including the z-dependence of E_0 and E_1, should be considered.[2]

REFERENCES

1. B. Ya. Zedovich, et.al., Principles of Phase Conjugation (Springer-Verlag, N.Y. 1985), Ch. 3 and 5.

2. J.T. Lin, Raman Laser Sources via Stimulated Raman and Brillouin Processes in Gas Mixtures, J. Opt. Soc. Am. (1986), in press.

PHASE CONJUGATION REVISITED

D. C. Brown and K. K. Lee
General Electric Company, PO Box 5000 Binghamton, NY 13902

ABSTRACT

We have proved, by using nonlinear wave equation among three independent methods, that: complete correction of phase distortions due to an optical device (or a system of several optical devices) by phase conjugation is possible if and only if the device (or the system) which produces such phase distortion is gainless, lossless, and time-order invariant. Effects of saturated gain, finite coherence, and system symmetry are also discussed briefly.

INTRODUCTION

The capability of optical phase conjugation mirrors (PCM) to correct phase distortion due to linear refractive index inhomogeneities has been demonstrated both theoretically and experimentally.[1] It has been suggested[1,2] that PC could be a valuable means of eliminating phase inhomogeneities associated with high power laser systems where phase distortions arise from a number of sources including wavefront errors associated with optical components, turbulence, and active-medium index of refraction variations across amplifier apertures.

With the success of many experiments and in theoretical discussions, there is an entrenched perception in both the optical and optical users communities that optical phase conjugation (OPC) can in principle undo <u>all</u> the phase distortions in an optical system. Thus, OPC is perceived to be a method which can correct all the phase errors in an optical train. We would like to analyze and explicitly state the domain of validity for such OPC approaches by stating and proving a fundamental theorem.

As is well-known, in order that the retro-reflected wave from the PCM compensates exactly the wavefront distortion of the incoming wave, the system has to be "time-reversal" invariant. The degree of incomplete compensation depends on the degree of symmetry breaking of the "time-reversal" invariant.

Next, we consider the simplest system. Let a plane wave propogate from the left through an optical device, which has neither gain nor loss, and it causes a phase change of $\Delta\Phi$. If the reflectivity of the PCM is unity, then clearly the phase change can be compensated completely. If the device has a gain (or loss), but the phase change due to the device is independent of the intensity (e.g., in an unsaturated region of the gain and the intensity is below the threshold of any nonlinear effect) then one can still completely compensate the phase distortion due to optical devices by the PCM[3].

Although variuos methods of achieving PC such as degenerate four-wave mixing and stimulated Brillouin scattering are themselves nonlinear processes, until recently most theoretical and experi-

0094-243X/86/1460286-3$3.00

mental work on phase correction by PC have focussed on phase distortions which are independent of intensity. For a medium with a nonlinear index of refraction, such as a cubic nonlinearity, the phase distortion depends on the intensity along the propagation path. Thus, for a gain medium with a cubic nonlinearity, the phase conjugation process is not able to completely correct the phase distortion due to the device.[4,5] Indeed, we will prove the following theorem:

Complete correction of phase distortion due to an optical device (or a system of several optical devices) by phase conjugation is possible in principle if and only if the device (or system) which produces such phase distortion is gainless, lossless, and time-order invariant.

PROOF

Assuming a slow variation of E, the non-linear wave equation for a cubic medium with gain in the cgs system is

$$\nabla^2 E - \frac{\mu}{c^2}[\varepsilon_o + \varepsilon_2 |E|^2]\frac{\partial^2 E}{\partial t^2} + i\beta E = 0 \tag{1}$$

where $\beta = \beta(E^2)$ is the intensity dependent gain coefficient. Consider a monochromatic incident electromagnetic field

$$E_{inc} = \psi(\vec{r})\exp[i(\omega t - kz)] + c.c. \tag{2}$$

After the field passes through a nonlinear (e.g., cubic) medium and is incident upon a PCM which procudes a "reflected" field

$$E_{ref} = \alpha(E^2)\, f(\vec{r})\exp[i(\omega t + kz)] + c.c. \tag{3}$$

where α is the gain term. It should be emphasized that <u>in the following derivation no detailed knowledge of the relationship between the total gain α and the gain coefficient β is required</u>. Thus, the total field is

$$E = [\psi(\vec{r})\exp(-ikz) + \alpha f(r)\exp(ikz)]\exp(i\omega t) + c.c. \tag{4}$$

physically, this means that the pulse is long so that the nonlinear medium is simultaneously subjected to both the incident and the phase conjugated light. For much shorter pulse such that the incident and the phase conjugated light do not overlap in the nonlinear medium where the phase distortion took place, then the coupled terms $2\alpha^2|f|^2\psi$ in Eq. (5) and $2|\psi|^2 f$ in Eq. (6) drop out. Substituting (4) into (1), we obtain

$$\nabla^2\psi + [\frac{\omega^2\mu\varepsilon_o}{c^2} - k^2]\psi + \frac{\omega^2\mu\varepsilon_2}{c^2}(|\psi|^2 + 2\alpha^2|f|^2) - 2ik\frac{\partial\psi}{\partial z} + i\beta\psi = 0, \tag{5}$$

$$\alpha\nabla^2 f + [\frac{\omega^2\mu\varepsilon_o\alpha}{c^2} + \nabla^2\alpha - k^2\alpha]f + \frac{\omega^2\mu}{c^2}\varepsilon_2(\alpha^2|f|^2 + 2|\psi|^2)f + 2\frac{\partial\alpha}{\partial z}\frac{\partial f}{\partial z} + 2ik\frac{\partial\alpha}{\partial z}f + 2ik\frac{\partial f}{\partial z} + i\alpha\beta f = 0. \tag{6}$$

The nonsynchronous terms, exp ($\pm 3ikz$), have been dropped. Comparing Eq. (6) with the complex conjugate of Eq. (5), we note that the conjugated amplitude of the forward wave $\psi^*(\vec{r})$ and the amplitude of the backward wave obey the same propagation equation iff $\alpha = 1$, $\beta = 0$, and $|\psi| = |f|$. Here $\alpha = 1$ and $\beta = 0$ definitely require that the medium do not have any gain. Thus, the theorem is proven. The same conclusion follows for the PCM. One can also prove the theorem by using PC operator, or by constructing examples with gains in nonlinear media[6].

DISCUSSION

The effect of an intensity dependent phase accumulation is important for high-power solid state laser systems but is considerably less important in gas lasers where the nonlinear coefficients are two to four orders of magnitude lower than solid-state laser media. Nonetheless, for high-power operation, where PC processes are utilized to remove device (or system) phase distortion, either gas or solid-state amplifiers often operate in the saturated gain regime. It is well-known that the laser equation for saturated gain is $a^+ = (\alpha - \beta)\ a^+ - \Gamma\ (a^+a)\ a^+$. Since a^+ is directly related to the field amplitude, it is clear that the <u>nonvanishing of the saturated gain coefficient acts like the nonlinear index of refraction of a nonlinear medium</u>. Thus, even if the device has a negligible nonlinear coefficient, the saturated gain coefficient Γ "replaces" the nonlinear coefficient γ for a saturated device. Similar conclusions can be reached by appealing to the Frantz-Nevdik saturated gain equations. Numerically, we have shown such non-correctable phase distortion due to saturated gain as well as nonlinear mdeia with gain[8].

To summarize, we have shown that only phase distortion due to passive devices can be completely conjugated. PC devices utilizing nonlinear inelastic scattering processes, (such as SRS, SBS, SRWS,) and elastic scattering processes (such as FWM) are themselves gain devices. Consequently, our theorem and our analysis apply[8]. In other words, such PC devices do introduce uncorrectable phase distortions (however small) by themselves, in addition to the uncorrectable phase distortion introduced by other gain media.

From an information theory or thermodynamics viewpoint, phase distortions can be viewed as increase in entropy. Should PC eliminate all phase distortions due to a system, then we have an optical example of a constant of motion of entroyp macroscopically. What we have shown is that globally in order to have gain, the "optical entropy" must increase.

REFERENCES

1. B.Ya Zel'dovich, et al, JETTP Lett 15, 109 (1972); VD Sidorovich, Sov Phys Tech Phys 21, 1270 (1976); DM Pepper, Opt Eng, 21, 156 (1982).
2. E Bergman, I Bigio, B. Feldman, R Fisher, Opt Lett 3, 82 (1978).
3. VI Kryzhanovskii, et al, Sov Tech Phys Lett 7, 170 (1982).
4. LA Bol'shov, et al, Sov Phys JETP Lett 31, 287 (1980).
5. M Nazarathy, J Shamir, J Opt Soc Am 73, 910 (1983).
6. DC Brown, KK Lee, to be published
7. See e.g. H Haken in Encyclopedia of Physics Vol 25/2c, (Springer, 1970).
8. DC Brown, W Gehm, KK Lee, in preparation.

APPLICATIONS OF STIMULATED RAMAN SCATTERING IN LASER SYSTEMS

H.Komine, W.H. Long, Jr., S.J. Brosnan and E.A. Stappaerts
Northrop Research and Technology Center, Palos Verdes, CA 90274

ABSTRACT

The property of Raman amplifiers for producing good beam quality Stokes output is investigated experimentally. The theoretical basis of this effect is discussed.

INTRODUCTION

Stimulated Raman scattering in molecular and atomic media has been actively investigated for potential application in advanced laser systems in the near UV and visible regions.[1] In addition to its frequency shifting property, stimulated Raman scattering provides a means for improving the spatial coherence of laser beams[2] and for coherently summing the power from several laser beams.[3] Wave-Optics code simulations predict these effects in Raman conversion systems based on oscillator-amplifier configurations which satisfy certain conditions on parameters such as laser beam aberration, spectral bandwidth, and amplifier Fresnel number. This paper presents some of the results of initial experimental beam clean-up investigations that demonstrate efficient generation of nearly diffraction limited Stokes beam in a high gain Raman amplifier pumped by aberrated beams. The results are discussed in terms of a theoretical model.

EXPERIMENT

The experimental apparatus consisted of a XeCl discharge laser as a pump source and a hydrogen Raman converter which was configured as an oscillator-amplifier system.[4] Figure 1 shows a schematic layout of the experiment, including diagnostics for pulse waveform and beam wave front. Two self-referencing interferometers are used to measure the input pump and output Stokes wave fronts. The input pump beam phase front was distorted by inserting an aberrator to simulate poor spatial coherence. Experimental data were taken, using various aberration strength and spatial scale sizes.

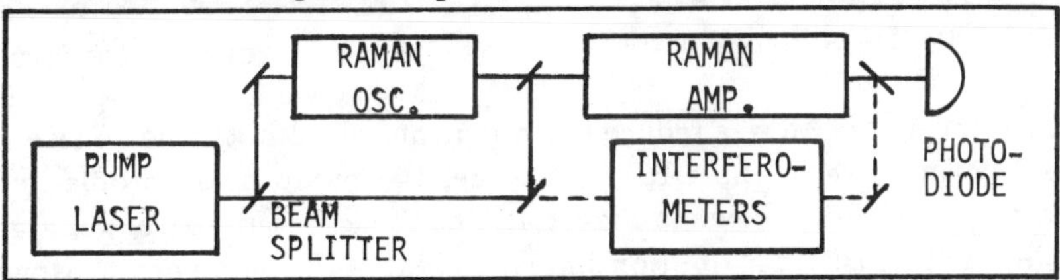

Fig. 1. Experimental Schematic

The pump laser was operated as a master oscillator/power amplifier system with energy output of about 0.2 joule in 60 nanosecond pulses. The power amplifier output was linearly polarized and had a spectral bandwidth of 3 GHz at 308.2 nm. A small fraction of this output was used to drive a Raman oscillator which generated a nearly diffraction limited injection beam.

The aberrator was a thin column of flowing helium gas which introduced refractive index variations across the pump beam. By changing the flow rate and column width the aberrator produced phase distortions of up to about one wavelength (peak-to-valley) with a scale size of 0.3 to 0.5 cm. Smaller scale sizes and larger amplitudes resulted in significant pump intensity nonuniformity as the beam propagated into the Raman amplifier.

The Raman amplifier was pumped by the aberrated beam with an unsaturated peak gain of approximately G=14 in a 2-meter long collimated geometry. The injected Stokes intensity level was about 0.1% of the input pump intensity. Thus, the saturation gain, $G_{sat}=\ell n\ (P_{in}/S_{inj})$, is reached in about 1 meter.

Under these conditions peak power depletion exceeding 90% was observed with converted power going into the amplified first Stokes beam. Generation of other Stokes and anti-Stokes components was insignificant. Also amplified spontaneous emission was not detected when the Stokes injection beam was blocked. Therefore, controlled amplification factors of several hundred were realized in a strongly saturated amplifier.

A representative wave front interferogram of the aberrated pump beam is shown in Figure 2. The aberration scale size is comparable to the beam width in this case. The corresponding interferogram for the amplified Stokes beam is shown in Figure 3.

The rms value of the analyzed optical path difference (OPD) is about 0.1 wavelength for the Stokes as compared to 0.2 wave for the input pump beam.

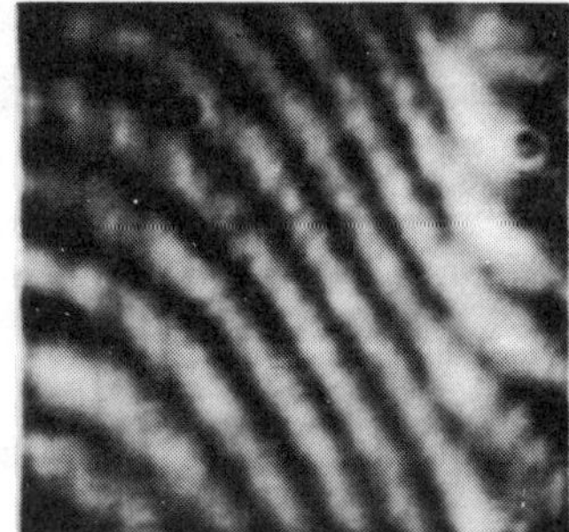

Fig. 2. Pump Interferogram

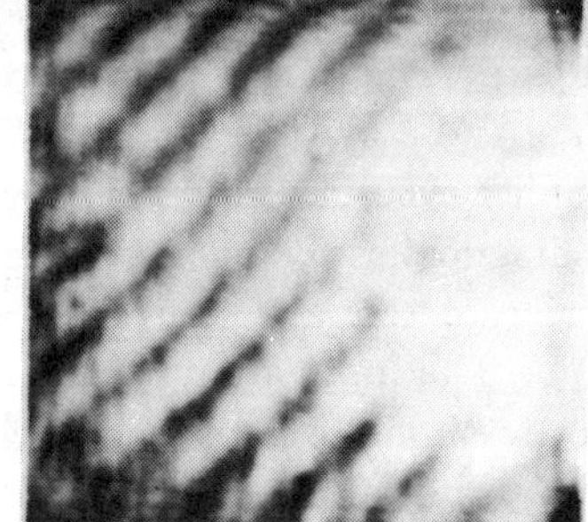

Fig. 3. Stokes Interferogram

When the scale size was reduced to about one third of the beam size along with a 70% increase in the OPD rms value, the pump beam profile became somewhat distorted, as shown in Figure 4. The Stokes interferogram in Figure 5 also shows an intensity distribution similar to that of the pump beam. However, the interference fringes remain relatively unaffected by the increased pump aberration.

Evidently the pump wave front is not transferred directly onto the amplified Stokes beam.

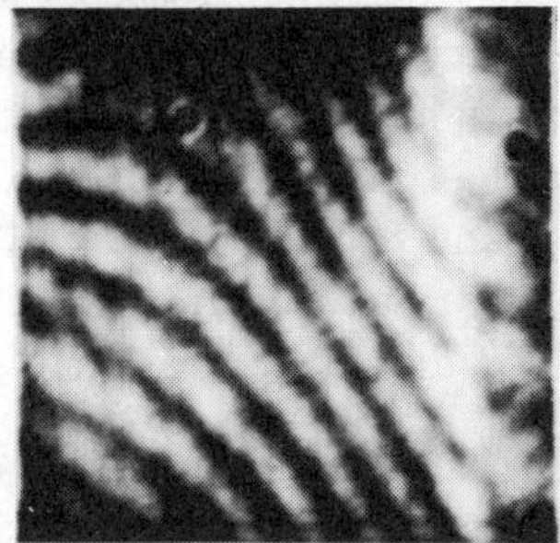

Fig. 4. Pump Interferogram (Reduced scale size)

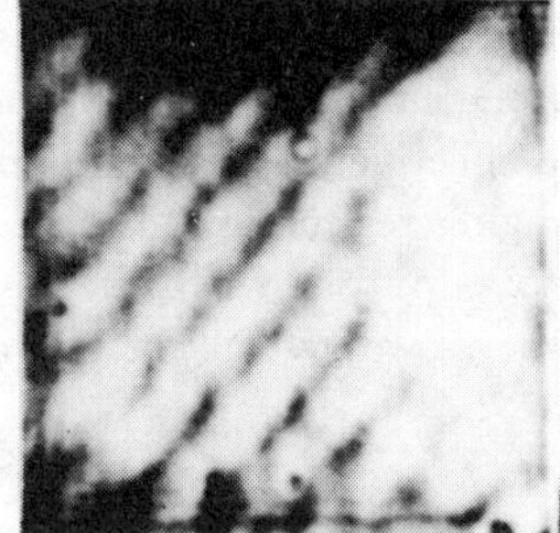

Fig. 5. Stokes Interferogram (Reduced scale size)

A more vivid example of this phenomenon is shown in Figures 6 and 7, which show respectively, the pump and amplified Stokes beam interferograms obtained when a weak negative spherical lens was inserted in place of the aberrator.

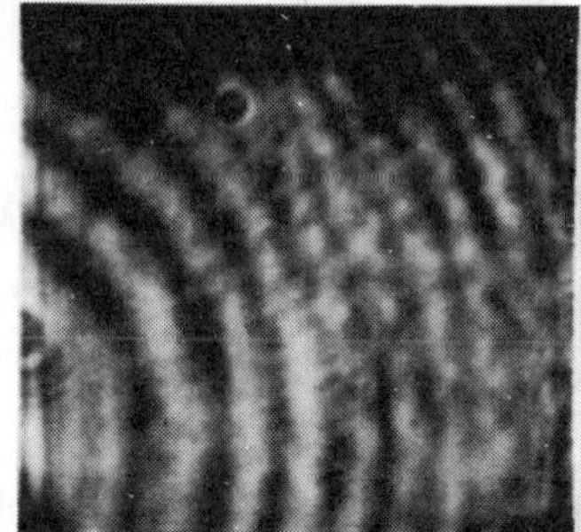

Fig. 6. Pump Interferogram (Spherical lens)

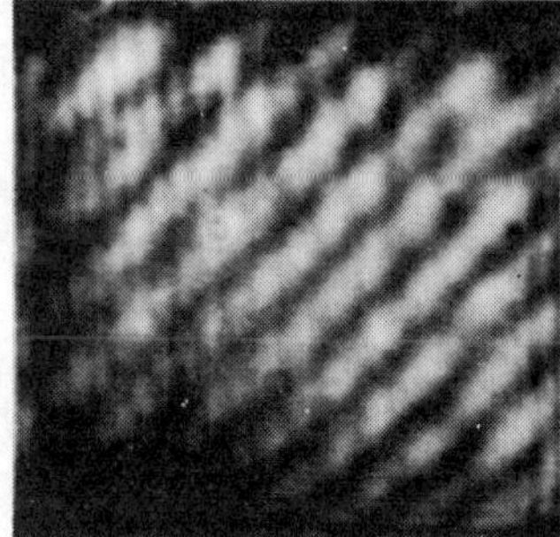

Fig. 7. Stokes Interferogram (Spherical lens)

The experimental data and observations can be understood in terms of a model, developed by Long and Stappaerts,[5] for Raman beam clean-up with moderately aberrated pump beams. In this model the phase aberrations are considered to cause intensity modulations of the beam profile as the beam propagates along the Raman cell. Since the Raman gain is higher over more intense regions of the pump beam, significant amplification of the Stokes beam occurs in these regions. Thus, the amplified Stokes beam profile tends to become a replica of the pump beam after it has propagated to the saturation gain length. In the weak aberration limit, the input pump profile essentially determines the Stokes output. As the aberrations become stronger, the Stokes output may develop intensity nonuniformities; this can lead to reduced energy in the central spot of the far field pattern. If the pump beam breaks up into several locally intense beams, the far field pattern can be expected to show substantial energy outside the central spot. While the interferogram measurements are unsuited to the latter case, the far field intensity patterns have indicated the effect of beam breakup. Up to this point, the Stokes beam coherence tends to decrease as the pump aberrations become stronger. However, for sufficiently large apertures, strong small scale aberrations may cause rapid intensity fluctuations which can be effectively averaged along the amplifier. This is the so-called intensity averaging regime in which severely aberrated pump beams are converted into a Stokes output with low beam divergence.[6] Hence, the Raman beam clean-up process works well for two distinct ranges of pump wave front aberrations.

CONCLUSIONS

The experimental results of Raman beam clean up demonstrate the feasibility of producing a spatially coherent beam by converting aberrated pump beams in a high gain Raman amplifier. High power conversion efficiencies realized in these Raman converters make them practical for many applications with stringent beam quality requirements. Our theoretical model of the beam clean-up process can be utilized to guide the design of such subsystems as an important part of advanced laser systems.

REFERENCES

1. R. Burnham and N. Djeu, Opt. Lett. 3, 215 (1978)

2. R.S.F. Chang, R.H. Lehmberg, M.T. Duignan, and N. Djeu, IEEE J. Quant. Electron. QE-21, 477 (1985)

3. N.G. Basov, A.Z. Grasyuk, Yu. L. Karev, L.L. Losev, and V.G. Smirnov, Sov. J. Quant. Electron. 9, 780 (1979)

4. H. Komine and E.A. Stappaerts, Opt. Lett. 7, 157 (1982)

5. Theoretical model is discussed in a paper to be published.

6. J. Goldhar and J.R. Murray, IEEE J. Quant. Electron. QE-18, 399 (1982)

A CONTINUOUSLY TUNABLE SEQUENTIAL RAMAN LASER: WITH APPLICATION TO SURFACE SPECTROSCOPY

P. Rabinowitz, B. N. Perry, F. M. Hoffmann and N. Levinos
Exxon Research and Engineering Company, Annandale, NJ 08801

ABSTRACT

Continuously tunable coherent radiation in the 1-12μm range has been obtained via sequential Raman scattering of pulsed-dye-laser radiation in hydrogen. A multiple-pass-cell configuration was used to enhance the Raman gain and produce an overall quantum efficiency of 45% to a wavelength of 5μm. A nonlinear wave model of the process has been developed that includes the effects of diffraction, four-wave mixing and temporal pulse shape and gives numerical outputs in agreement with experiment. Additionally, the output radiation has been used in 5μm range for surface spectroscopy studies of sub-monolayers of CO adsorbed on an Ru(001) single crystal.

In these proceedings we report a frequency conversion technique using stimulated Raman Scattering (SRS) which maintains broad continuous tunability while relaxing the usual requirements for high pump power. This is done by increasing the net Raman gain through the use of repeated focusing in the active medium with a multiple pass cell (MPC).[1,2]

We have employed the MPC to obtain a diffraction limited continuously tunable infrared source through sequential Raman scattering in H_2. This configuration has provided a factor of eighteen reduction in the required pump laser power compared with single-pass scattering, allowing the use of a commerical dye-laser as the pump. With this configuration, an output energy of 1mJ per pulse in a single transverse mode (TEM_{oo}) has been obtained out to 5μm in a bandwidth of 0.2 cm^{-1} and with a quantum efficiency of 45%. The dye pyridine 1 was used in these measurements. Continuous tuning over the range from 5.9μm to 12μm, the cutoff of the MPC window, was obtained using Styrl 7 dye. At the long wavelength end of this range both dye-laser power and conversion efficiency fell sharply so that a fraction of a microjoule per pulse was produced at 12μm. The shorter wavelength range was obtained with DCM and Rhodamine dyes.

The experimental apparatus is diagrammed in Figure 1. A frequency-doubled Quanta Ray DCR 201 Nd:Yag laser pumps a Lambda Physik FL 2002 dye laser. This arrangement produces pulses of 7 nsec width and 0.2 cm^{-1} bandwidth, with 40-60 mJ of energy using pyridine 1. The tunable radiation passes through a first aperture which reduces the pulse energy by about 15 mJ and produces a transverse Airy-like intensity pattern in the far-field. The radiation is then mode-matched to a confocal parameter of 28 cm with a Galilean telescope and passes through a second aperture, adjusted in diameter to coincide approximately with the first zero of the Airy-like intensity pattern. This arrangement provides more than 98% of the transmitted energy in the TEM_{oo} mode. The MPC, containing 14 atm of H_2 at ambient temperature, encloses two spherical copper mirrors of 7cm clear aperture and 1M radius of curvature, held at 1.73M separation by an invar structure to form an off-axis spherical interferometer[3,4] having 22 passes to ray path closure. The measurements reported here were done with 21 passes.

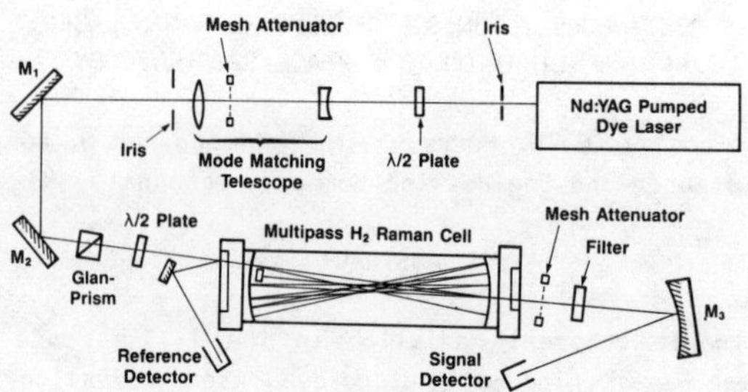

Figure 1. Experimental Apparatus

The experiments consisted of measuring the conversion of the dye laser pump radiation to first, second, and third Stokes wavelengths as a function of input energy. The combined output radiation was passed through various filters to eliminate all but one Stokes component. An interference filter passing 3.7μm to 6μm was used, as well as filters of Si, fused quartz and glass (Corning 7-69 and Schott KG3). Energies were measured using pyroelectric detectors with digital read-outs (Laser Precision).

Figure 2 shows the measured quantum efficiency (photons out/photons in) for production of first Stokes (0.97μm), second Stokes (1.62μm) and third Stokes (5.0μm) radiation, as a function of injected pump energy. The injected pump energy is evaluated just after the input window, and the output radiation is evaluated just before the exit window, by taking account of Fresnel reflection losses. Each point in the figure is the average value of several groups of 100 pulse averages; the size of the dots represents about one standard deviation of a 100 pulse average.

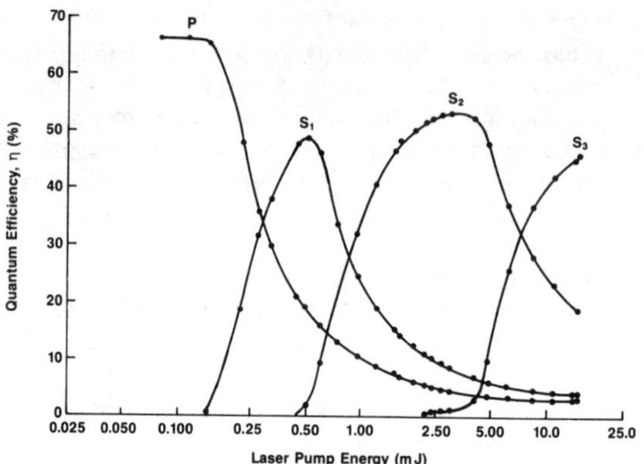

Figure 2. Measured quantum conversion efficiencies η (photons out of MPC/pump photons into MPC) including mirror losses for pump (0.69μm), 1st Stokes (0.97μm), 2nd Stokes (1.62μm) and 3rd Stokes (5.0μm) radiation vs. input pump energy.

In addition to the efficient production of vibrational Stokes orders, spaced by 4,155.2 cm^{-1}, a number of parasitic Raman processes were observed. However, these accounted for only a small fraction of the converted energy. Pure rotationally shifted Stokes and anti-Stokes components surrounding pure vibrational Stokes were evident when the radiation was dispersed with a prism. These components were spaced by the 587 cm^{-1} rotational energy of ortho-H_2. The rotational Stokes energy was estimated as, at most, about 10% of the corresponding pure vibrational Stokes energy. This maximum occurred when the pure vibrational Stokes radiation was overdriven and strongly depleted. The rotational energy was typically less than 1% of the optimized vibrational Stokes energy.

In order to understand the optimization and wavelength scaling of MPC Raman conversion, we developed a non-linear model of the scattering process. Previous non-linear analyses of multiple Stokes generation[5-7] have described scattering of local plane wave pump and multiple Stokes fields. We adopted a model incorporating both a focussed Gaussian pump field and parametric processes to retain the effects of diffraction, non-uniform pump intensity, and four-wave mixing. To produce a numerically tractable non-linear model, we were guided by our previous results for coupled stokes and anti-Stokes growth with a prescribed pump field.[8,9] We found that growth takes place predominantly in the lowest order mode if the product of background dispersion and pump confocal parameter exceeds a value of 12. The experimental conditions were arranged such that this condition was satisfied for generation of all Stokes orders. As a result, we were able to use a one-dimensional model in which only the lowest order mode in each field is retained, with diffraction losses accounted for through mode projection coefficients. Six fields were retained: anti-Stokes, pump, and four Stokes fields.

Figure 3 shows the expected degree of quantum conversion to each Stokes order as a function of (average) input energy, and is to be directly compared to the experimental results of Figure 2. This comparison shows good agreement between the model and the experimental results. The discrepancies are attributed to the physical phenomena left out of the model, mainly transient effects and parasitic rotational Raman scattering. However, the main features of the experiment, namely the threshold energies and peak conversion efficiencies, are well reproduced by the model.

We have used this tunable radiation source for surface spectroscopy of CO adsorbed on Ru(001). Radiation from the source illuminated a Ru sample at grazing incidence in a UHV chamber. The light was plane polarized at 45° with respect to the plane of incidence providing a means for ratio detection of the absorption spectrum since only the p-polarization is absorbed by the CO molecule. A ZnSe prism with surfaces aligned near Brewsters' angle was used as the polarization analyzer. Detection was accomplished with a pair of pyroelectric detectors. The experimental results are shown in Figure 4 for the C-O stretch on Ru(001) with a surface coverage of θ=0.65 at 300K and at 85K. The spectra were taken after first raising the substrate temperature to 300K (a) and then recooling to 85K (b). This produces two distinct reversible changes: a 2.5 cm^{-1} blue shift and a narrowing of the linewidth from 7.5 to 5.5 cm^{-1}. The 5.5 cm^{-1} linewidth is smaller than previously observed[10] and gives a lower limit of 1.9 psec for the vibrational dephasing time. A number of mechanisms can contribute to the observed line shift and broadening,[10-12] however, our signal-to-noise-ratio was insufficient to distinguish among them. With improved signal to noise it should be possible to obtain a more detailed understanding of mechanisms which exchange energy with the surface. The use of lasers for surface spectroscopy, particularly at wavelengths longer than 10μm, offers the possibility of higher resolution, higher signal to noise, and time resolved measurements, all with the suppression of thermal background.

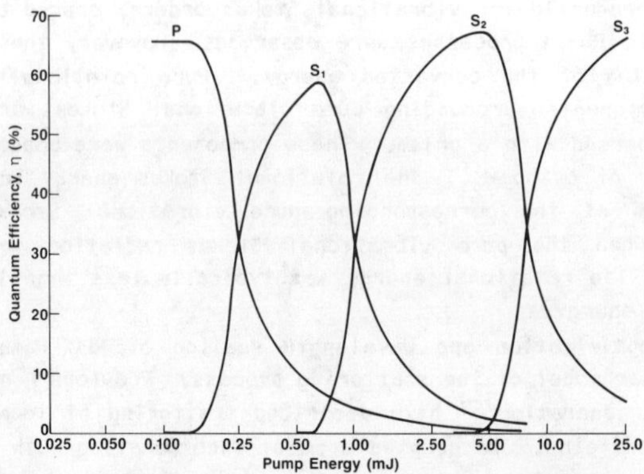

Figure 3. Calculated quantum conversion efficiencies η for pump and three Stokes orders vs. pump energy for measured average intensity distribution (i.e., measured effective pulse shape).

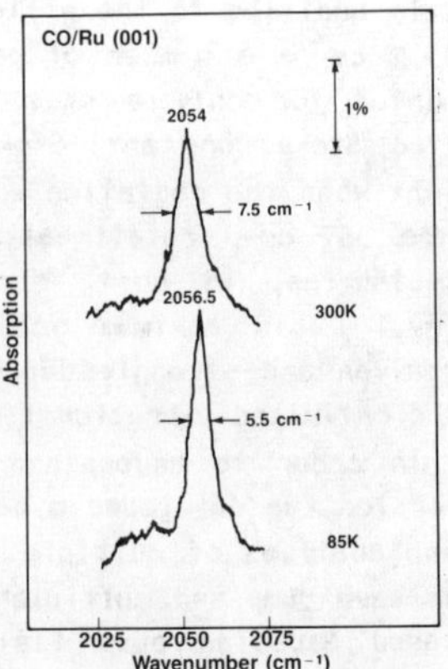

Figure 4. Scanned absorption spectra of CO/Ru(001) at 85 and 300K for θ=0.65.

In conclusion, we have demonstrated that sequential stimulated Raman scattering in an MPC is a useful technique for generating high brightness (TEM_{oo}) broadly tunable infrared radiation with moderate power pump sources. We have demonstrated its use for surface spectroscopy. We have also developed a theoretical model which adequately describes the process and allows its optimization. To extend the tuning range further into the infrared, a significant increase in the power of the visible pump laser is necessary because the exponential Raman gain falls approximately quadratically with increasing wavelength. However, tunable visible sources of higher power are available and could provide a source of high power tunable infrared to 30μm.

REFERENCES

1. P. Rabinowitz, A. Stein, R. Brickman and A. Kaldor, Opt. Lett., Vol 3, pp. 147-8 (1978).
2. R. L. Byer and W. R. Trutna, Opt. Lett., Vol. 3, pp. 144-6 (1978).
3. D. Herriott, H. Kogelnik and R. Kompfner, Appl. Opt., Vol. 3, pp. 523-6, (1964).
4. B. N. Perry, R. O. Brickman, A. Stein, E. B. Treacy and P. Rabinowitz, Opt. Lett., Vol. 5., pp. 288-90 (1980).
5. Y. R. Shen and N. Bloembergen, Phys. Rev., Vol. 137A, pp. 1787-1805 (1965).
6. J. H. Newton and G. M. Shindler, Opt. Lett., Vol. 6, pp. 125-7 (1981).
7. D. Von derLinde, M. Maier and W. Kaiser, Phys. Rev., Vol. 178, pp. 11-17 (1969).
8. B. N. Perry, P. Rabinowitz and D. S. Bomse, Opt. Lett., Vol. 10, pp. 146-8 (1985).
9. B. N. Perry, P. Rabinowitz and M. Newstein, Phys. Rev., Vol. 27a, pp. 1989-2002 (1983).
10. H. Pfnur, D. Menzel, F. M. Hoffmann, A. Ortega and A. M. Bradshaw, Surface Sci., 93 (1980) 431.
11. J. W. Gadzuk and A. C. Luntz, Surface Sci., 144 (1984) 429.
12. B. N. J. Persson and R. Ryberg, Phys. Rev. Lett., 54, 2119 (1985).

RAMAN SUSCEPTIBILITY OF LIQUID NITROGEN IN THE MID-INFRARED SPECTRAL REGION

C.L. Marquardt, M.E. Storm*, and N.P. Barnes†
Naval Research Laboratory, Washington, DC 20375-5000

ABSTRACT

We have measured first Stokes conversion efficiency as a function of pump power for stimulated Raman scattering in liquid nitrogen at pump wavelengths of 0.532, 1.06 and 2.05 μm. An anomolously large Raman susceptibility was observed in the case of 2.05 → 3.92 μm conversion. The mechanism of this effect is tentatively identified as a preresonance enhancement of the Raman cross-section in the vicinity of a collision-induced absorption band associated with the first vibrational overtone of nitrogen.

EXPERIMENT

In a single-pass, focused-beam Raman shifter the first Stokes energy conversion efficiency, E_s/E_p, is related to the peak pump power, P, by

$$E_s/E_p \approx (\lambda_p/\lambda_s) \quad F(P) \quad (e^{GP} - 1)/(e^{GP} + AP), \tag{1}$$

$$F(P) = \text{erf } (\ell n(P/P_t))^{1/2}, \; A = nb\lambda_p/hc^2, \tag{2}$$

where b is the confocal parameter and P_t is the threshold power. By fitting conversion efficiency curves to these equations we have determined experimental values of the Raman gain coefficient, G, for liquid nitrogen at three wavelengths and a variety of pumping configurations. Calculated values of G for the same wavelengths and pumping configurations were obtained from the relations:

$$G = 4\mu_o\epsilon_o c\lambda_s\lambda_p NLQ \; (d\sigma/d\Omega)/\pi nh\Delta\nu_r, \tag{3}$$

$$Q = [2\alpha/(1+\alpha^2)] \; [1+\alpha^2 - \alpha^2/U^2]^{1/2} \text{ Arctan } (UZ/b), \tag{4}$$

$$U = [\alpha^2 + (1 + \nu_r/\nu_s)^2/\alpha^2(1+\alpha^2)]^{1/2} \tag{5}$$

Here Z is the length of the medium, α the ratio of Stokes beam waist to pump beam waist, ν_r and $\Delta\nu_r$ the frequency and width of the Raman mode, N the number of molecules per unit volume, $(d\sigma/d\Omega)$ the differential cross section, and L the local field correction.

Comparison between experimental and calculated values of G is summarized in Fig. 1, where each point represents a series of experiments employing various pumping configurations

*Sachs/Freeman Associates, Landover, Md 20785
†Permanent Address: Los Alamos National Laboratory

at a fixed pump wavelength. Each series is seen to be internally consistent within the experimental accuracy of ± 10%. Values of G measured for λ_p = 0.532 μm and 1.06μm agree well with corresponding calculated values; i.e. G(exp)/G(calc) = 1. But for λp = 2.05 μm, G(exp)/G(calc) = 2.7 ± 0.3, indicating a threefold enhancement which is clearly outside experimental errors.

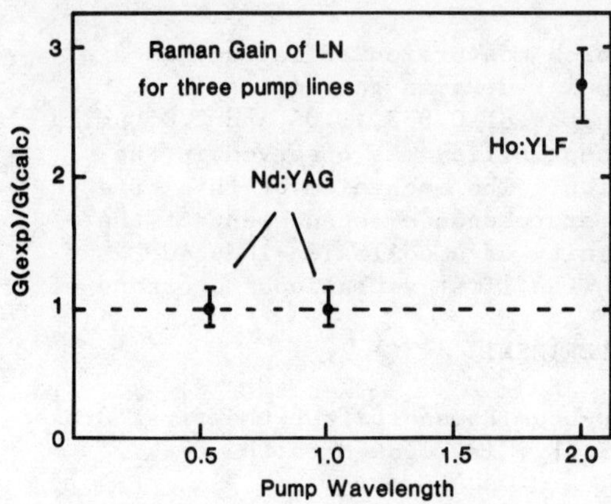

Fig.1 Comparison of experimental and calculated Raman gain coeffficients of liquid nitrogen at three pump wavelengths: 0.532 μm, 1.06μm and 2.05 μm.

DISCUSSION

We have considered the possibility that the observed gain enhancement could be a spurious effect due to systematic measurement errors, failure of the calculational model, self-focusing or optical feedback. We have performed experiments to test each of these hypothesized mechanisms, and have demonstrated conclusively that the observed enhancement is not due to any of these spurious effects. It is evident from Eq. 3 that potential sources for the wavelength dependence of G are contained implicitly in the dispersion of the refractive index (primarily through its effect on the local field correction) and in the wavelength dependence of the Raman cross-section. Using the IR absorption spectra of our liquid nitrogen samples we have demonstrated that the observed gain enhancement cannot be explained by anomolous dispersion of the refractive index. The peak of the collision-induced absorption by the first overtone of the N-N stretch mode occurs at 4656 cm^{-1} (2.14μm) in our ultrapure liquid nitrogen samples, with FWHM ≈ 96 cm^{-1} and peak absorption coefficient ≈ 3 x 10^{-2} cm^{-1}. Enhancement of the Raman cross section due to proximity of this collision-induced dipole absorption thus appears to be the most likely source of the effect illustrated in Fig. 1. If this understanding is correct, similar effects may occur in other diatomics used in tunable IR Raman lasers (e.g. in hydrogen for λp ≈ 1.2 μm).

ACKNOWLEDGMENTS

The authors are indebted to Prof. J.S. Shirk and Dr. J.F. Reintjes for many useful discussions.

SYNCHRONOUSLY PUMPED RAMAN OSCILLATOR OF HYDROGEN GAS FOR GENERATING PICOSECOND STOKES PULSES

Lin Lihuang
Shanghai Institute of Optics and Fine Mechanics
Academia Sinica, Shanghai, People's Republic of China

N. Morita and T. Yajima
Institute for Solid State Physics
University of Tokyo, Tokyo, Japan

Abstract

A Raman oscillator of compressed hydrogen gas synchronously pumped by a frequency-doubled mode-locked Nd:YAG laser has been used to generate 6 psec Stokes pulses as short as 1/6 of the pumping pulsewidth.

The picosecond stimulated Raman scattering in hydrogen gas was studied by using the second harmonic of a mode-locked Nd:YAG laser as an incident pulse[1]. In order to generate shorter Stokes pulses, we constructed a Raman oscillator in hydrogen gas to compress the Stokes pulse temporally. The Raman oscillator shown in Figure 1 is longitudinally pumped by the second harmonic of an actively-passively mode-locked Nd:YAG laser in the form of a train, and the optical length of the oscillator cavity is adjusted such that a single Stokes pulse is amplified by the successive pumping pulses. In this synchronously-pumped way the group velocity mismatch can be corrected with each passage through the active medium, and gain narrowing of the Stokes pulse can be achieved[2].

The Raman oscillator was a Raman cell filled with hydrogen gas, two 20 cm focal-length lenses (separation ~40 cm) and two plane mirrors of fused quartz, M_1 and M_2. M_1 was highly reflecting and M_2 was partially transmitting (70%) at the Stokes wavelength, and both were transparent at the pumping wavelength. The Raman cell was a 30 cm long stainless steel pipe of 2 cm inner diameter with 2 cm thick fused quartz windows at the ends, filled to pressures of up to 100 atm. The cavity length was about 150 cm.

As shown in Figure 1 the pulsewidths of the input pumping and output Stokes beams were monitored by the same two-photon fluorescence cell using a solution of D-POPOP in toluene. The TPF traces were recorded with an ISIT camera system and an optical multi-channel analyzer, and then read out on a pen recorder. The pumping and Stokes energies were measured simultaneously using two energy meters.

We have experimentally studied some opearting characteristics of such a Raman oscillator by means of changing the parameters of the hydrogen pressure and/or the pumping pulse train. It is found that the temporal character of the Stokes pulses (683 nm) is critically dependent upon setting the Raman oscillator cavity circulation time 2L/c equal to the pumping interpulse interval. When the condition above is achieved, the Stokes pulsewidth is the shortest at a given hydrogen pressure. Figure 2 shows that the output Stokes pulsewidths gradually shorten with increasing hydrogen pressure and/or the number of pumping pulses in the pulse train. A Stokes pulsewidth (assuming a Gaussian pulse shape) of about 6 ps, as short as 1/6 of the pumping pulsewidth, has been recorded for a hydrogen pressure of 100 atm and about 16 pumping pulses in the half width of the pulse train envelope. The ratio of the Stokes pulsewidth to pump pulsewidth is expected to decrease further with increasing hydrogen pressure and/or the number of the pumping pulses.

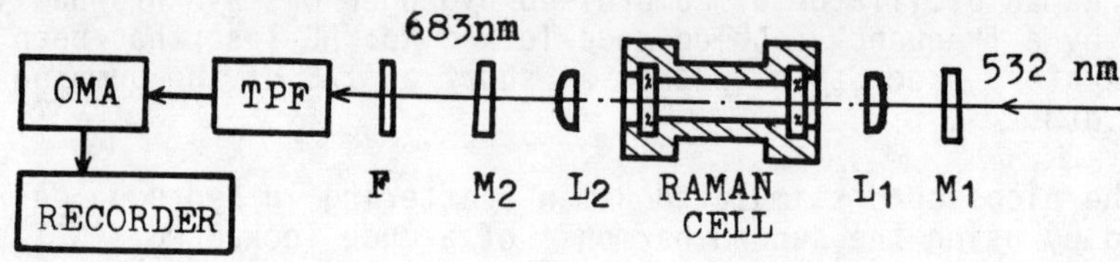

Fig.1 Schematic diagram of experimental setup.

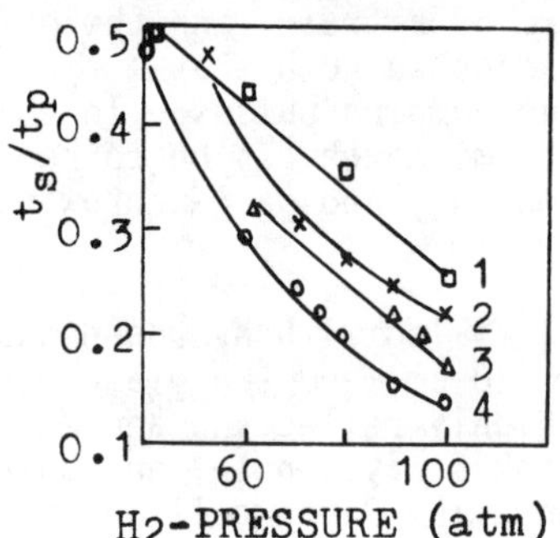

Fig.2 Ratio of t_s to t_p as functions of H_2 pressure.
1) Pumped by a single pulse (from Ref.1).
2) Train width: 80--100 ns.
3) Train width: 140--160 ns.
4) Train width: 200--250 ns.
t_s: Stokes pulsewidth;
t_p: Pumping pulsewidth.

References

1. N. Morita, L. H. Lin and T. Yajima, Appl. Phys. B 31, 63 (1983).
2. M. J. Colles, Appl. Phys. Lett. 19, 23 (1971).

NONEQUILIBRIUM TRANSITIONS IN THREE-WAVE NONLINEAR-OPTICAL PROCESSES

E.P. Gordov and A. L. Zhiliba
The Institute of Atmospheric Optics SB USSR Acad. Sci, Tomsk, 634055, U.S.S.R.

ABSTRACT

Statistical properties of interacting quantized field modes involved in three-wave nonlinear-optical processes have been studied. Considered are nonequilibrium transitions that may occur in second harmonic and subharmonic generation and parametric amplification under resonatorless conditions.

INTRODUCTION

Despite the well-known problems involved in the quantum electrodynamical calculations there appears to be an enhanced interest in accurate description of the nonlinear optics processes from the quantum standpoint[1-3]. In this work the semiclassical representation method (SRM) is used for a quantum-electrodynamical analysis of the nonlinear optics problems. SRM introduced elsewhere [4] enables one to calculate quantum averages through the use of the semiclassical theory solutions.

THREE-WAVE PROCESSES. GENERAL APPROACH

The SRM results in a convenient representation of the interaction between a finite number of waves in a nonlinear medium. The description desired can be provided by eliminating the variables for the modes not involved in the interactions considered. Physically it implies the case where the laser source field as well as that emitted in the coarse of the interaction appear to be superposition of a finite number of waves. For three-wave processes it yields for the field [5]

$$E = A_1(r,t)e^{i\omega_1 t - ik_1 r} + A_2(r,t)e^{i\omega_2 t - ik_2 r} + A_3(r,t)e^{i\omega_3 t - ik_3 r} + \text{c.c.} \qquad (1)$$

Here $E_i(r,t)$ $(i = 1,2,3)$ are slowly varying field amplitudes as a function of time, t, and coordinate, r. When the field amplitudes are independent of t and r, the rest of the modes can be eliminated in a closed form. In the case of the time and coordinate dependent amplitudes there appears an interaction between the modes considered and those eliminated. However, in the discussion below this interaction will be neglected subject to the restriction that the properties of the source and of the nonlinear medium result in the absence of the vacuum (eliminated) mode excitation.

Taking into account the above remarks one can write down the master equation for the three-wave processes as

$$\frac{\partial R}{\partial t} = \hbar^{-1}\sum_{n=1}^{3}\left(\frac{\partial}{\partial \bar{a}_n(t)}\langle d^c(t)\rangle_x - c.c.\right)R(t) + \\ + \hbar^{-2}\sum_{n,k=1}^{3}\int_{t_0}^{t} dt_1\Big[\langle d^c(t)d^c(t_1)\rangle_x \frac{\partial^2}{\partial a_n(t)\partial a_k(t)} + \\ + \langle d^c(t_1)d^c(t)\rangle_x \frac{\partial^2}{\partial \bar{a}_n(t)\partial a_k(t_1)} + c.c.\Big]R(t_1), \tag{2}$$

where $d_j(r)$ is the dipole moment density at point r; $d_c(r\ ;\ t) = c^+ d\ c$; l_j is the polarization vector of the field E_j. The evolution operator C obeys the Schrödinger equation of the form

$$i\hbar\frac{\partial C}{\partial t} = \sum_j d_j E(r,t)C, \tag{3}$$

where E is given by Eq.(1) and obeys the relevant set of the Maxwell equations.

DYNAMICS OF THREE-WAVE INTERACTIONS

Considered below is the dynamics of the parametric amplification (PA), the second harmonic generation (SHG) as well as the subharmonic generation (SG).

The Maxwell equations for PA are of the form

$$\frac{\partial A_{1,2}}{\partial z} + \frac{1}{u_{1,2}}\frac{\partial A_{1,2}}{\partial t} = \beta A_p \bar{A}_{2,1} e^{-i\Delta K\cdot z} - \gamma_{1,2}A_{1,2}, \\ \frac{\partial A_p}{\partial z} + \frac{1}{u_p}\frac{\partial A_p}{\partial t} = -\beta A_1 A_2 e^{i\Delta K\cdot z} - \gamma_p A_p. \tag{4}$$

Here ΔK is the phase mismatch, $\gamma_{1,2,p}$ is for the losses and $u_{1,2,p}$ is the group velocity.

The density matrix for the quantum problem does not depend on the coordinates. Thus, only the time dependence of the coupled waves appears to be important for further analysis. To obtain the relevant equations in the semiclassical approximations Eqs.(4) are to be averaged over the spatial variables. The following equations describe the time behavior of the waves.

$$\frac{1}{u_{1,2}}\frac{\partial a_{1,2}}{\partial t} = \beta a_p \bar{a}_{2,1} + E_{1,2} - \gamma_{1,2}\, a_{1,2} \ ,$$

$$\frac{1}{u_p}\frac{\partial a_p}{\partial t} = \beta a_1 a_2 - \gamma_p a_p \ , \quad E_{1,2} \equiv \frac{A_{1,2}(0) - A_{1,2}(z_0)}{z_0} \qquad (5)$$

where $a\ (t) = z_0^{-1}\int_0^{z_0} A(z)dz$. In deriving Eqs.(5) the spatial dependence of A_p is neglected. In the steady state the dependence of a_1 on E_1 takes the form

$$E_1 = a_1\left[\gamma_1 + \frac{\beta^2}{\gamma_2^2\gamma_p}\ \frac{|E_2|^2}{\left(1+\frac{\beta^2}{\gamma_2^2\gamma_p}|a_1|^2\right)^2}\right]. \qquad (6)$$

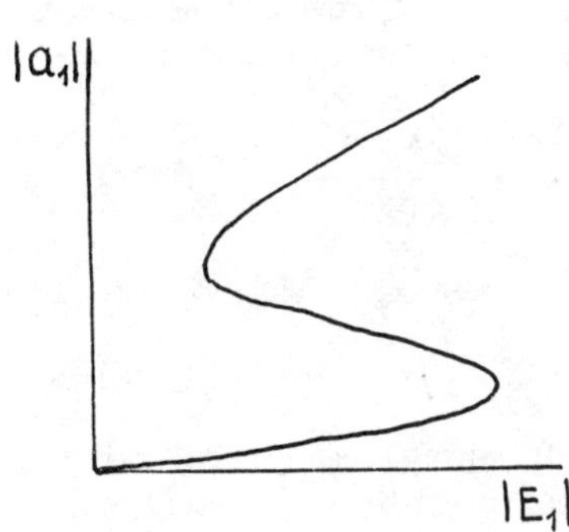

Fig.1. $|a_1|$ versus $|E_1|$.

Figure 1 shows this dependence for real amplitudes of a_1 and E_1 for a fixed values of E_2. For SHG and SG the following set of equations is obtained after averaging over z.

$$u_1^{-1}\dot{a}_1 = -\gamma_1 a_1 + i\beta \bar{a}_1 a_2 \ , \qquad (7)$$
$$u_2^{-1}\dot{a}_2 = -\gamma_2 a_2 + i\beta a_1^2 - E_2 \ .$$

It should be noted that for SG involved in a resonator the same set of equations is valid [7]. For $\gamma_2 \gg \gamma_1$ (the adiabatic case) a_1 obeys the equation

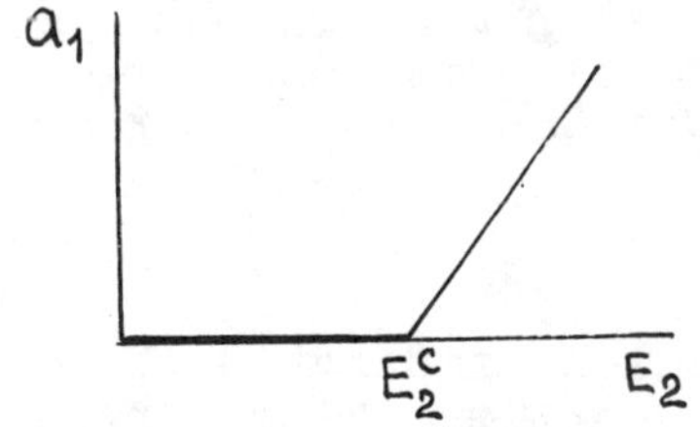

$$u_1^{-1}\dot{a}_1 = -\gamma_1 a_1 - i\beta\frac{E_2}{\gamma_2}\bar{a}_1 + \frac{\beta^2}{\gamma_2}|a_1|^2 a_1 \ , \qquad (8)$$

which shows that there is a critical point at $E_2 = E_2^c = \gamma_1\gamma_2/\beta$.

a_1 E_2^c E_2

Fig.2. a_1 versus E_2 .

From the physical standpoint the process under consideration can be described as follows: in SHG E_2 amounts to E_2^c at point z_0, then it becomes unstable, after the transition the SG process begins. Fig.2 shows a_1 versus E_2 .

THE QUANTUM-STATISTICAL PROPERTIES OF THE FIELD

The master equation (2) for SHG and SG in the adiabatic case ($\gamma_2 \gg \gamma_1$) is

$$\frac{\partial R}{\partial t} = \Big[\frac{\partial}{\partial a_1}\Big(\gamma_1 a_1 - i\beta\frac{E_2}{\gamma_2} + \frac{\beta^2}{\gamma_2}|a_1|^2 a_1 + \frac{1}{2}\gamma_1(\bar{n}_1+1)\frac{\partial^2}{\partial a_1 \partial \bar{a}_1} +$$

$$+\frac{1}{2}\frac{\partial^2}{\partial a_1^2}\left(\frac{\beta^2}{\gamma_2}a_1^2+\frac{i\beta E_2}{\gamma_2}\right)+c.c.\Big]R\ , \qquad (9)$$

$$R=\langle a|\rho^F|a\rangle\ , \qquad \bar{n}_1=\left(\exp\left(\frac{\hbar\omega_1}{kT}\right)-1\right)^{-1} .$$

Using the real amplitude and phase variables and linearizing the phase variables near the steady state $2\varphi_1-\varphi_2=\pm\pi/2$ one can write the stationary solution of resulting equation

$$R=N\left[\left(x^2-\frac{|E_2|}{\beta}\right)+1\right]^{\alpha-1}\exp(-x^2)\ , \quad \alpha\equiv\frac{\gamma_1\gamma_2\bar{n}_1}{\beta^2} . \qquad (10)$$

In this case $\delta=\langle a^{+2}a^2\rangle/\langle a^+a\rangle^2$ equals

$$\delta=1-\beta^2 z_0\left(\beta|A_2(z_0)|-\gamma_1\gamma_2\bar{n}_1 z_0-\beta^2 z_0\right)^{-1} . \qquad (11)$$

The degree of antibunching grows with the length of nonlinear medium, while the thermal noise presence ($\bar{n}_1\neq 0$) decreases it. For $\bar{n}_1=0$ $\delta=1+(-|A_2(z_0)|+1)^{-1}$, if $|A_2(z_0)|>1$, then $\delta\approx 1-1/x^2$.

It should be noted that Eq.(11) leads to a conventional result for the well-known limiting case. It follows from Eq.(11) that for $\gamma_1=\gamma_2=0$, $E_2=0$, $\delta=2$. On the other hand, under above conditions there exist only one stationary point at $x=0$ which is the Gaussian statistics of the field with $\delta=2$.

Concluding Remarks

The performed quantum-statistical consideration based on SRM takes into account space and temporal dependences of SHG and SG. To make it we imply that the nonlinear medium is divided into a set of volumes V_k, $k=1,N$, whose dimensions l_k satisfy the conditions $\lambda < l_k \ll z_0$, where is the wavelength. This allows us to treat the problem as the interaction V_j with the external field generated in previous considered volumes.

References

1. Yu.M.Golubev, V.N.Gorbachev, P.N.Zanadvorov. Opt. Spektrosk., 1982, v.53, p.876.
2. F.Casagrande, A.Lugiato. Acta Phys.Pol.,1984,A66,No.4.
3. J.Peřina, Quantum Statistics of Linear and Nonlinear Optical Phenomena, D.Reidel Publ.Co., 1984.
4. E.P.Gordov, S.D.Tvorogov, Phys.Rev.,1980, D22,p.908.
5. N.Bloembergen, Nonlinear Optics.A.A.Benjamin,Inc., New-York-Amsterdam, 1965.
6. P.D.Drummond, K.J.McNeil, D.F.Walls, Optica Acta,1980, v.27,p.321.

THE ROLE OF PUMP DEPLETION AND PHONON TRANSIENCY IN STIMULATED RAMAN AMPLIFICATION

J.T. LIN
Litton Systems, Inc., Laser Systems Division, Orlando, FL 32854-7300

ABSTRACT

Transient features of stimulated Raman amplification are studied theoretically for systems of single and multiple-pulse trains. Pulsation of the pump and the Stokes intensity profiles is studied analytically and numerically via the dynamic polarization. Energy feedback from the Stokes to the pump occurs when the pump field is depleted and the phonon amplitude has a 180° phase shift. Steady-state and transient results are compared through the symmetry of the temporal profile of the pump. For systems of pulse trains, the roles of pulse separation on the accumulated phonon excitation and the amplified Stokes intensity are discussed.

SINGLE PULSE

Transient stimulated Raman scattering (TSRS) has been reported for the case of finite dephasing time (T_2) but without the pump depletion,[1] and for systems with pump depletion but in the zero T_2 limit.[2] In the present paper, we shall study the transient stimulated Raman amplification (TSRA) where the pump is severely depleted by the injected seed-Stokes. One advantage of using TSRA is that the significant competing processes (backward SRS, SBS, and four-wave-mixing) in the TSRS are now suppressed, since the intensity of the injected seed Stokes is much higher than that of the spontaneous noise.

In the retarded time frame, the reduced amplitudes (assumed to be real) for the phonon (Q), the pump (E_p), and the Stokes (E_s) are governed by:

$$Q = E_p E_s - (\partial Q/\partial t)/\Gamma \tag{1}$$

$$\partial E_p/\partial z = -gE_p Q/2 \tag{2}$$

$$\partial E_s/\partial z = gE_p Q/2 \tag{3}$$

For a long pulse with duration much longer than that of the dephasing time, i.e., $t_p >> 1/\Gamma$, the steady-state Stokes intensity, including the pump depletion,[3] may be obtained exactly as follows:

$$I_s^0 = E_s^2 = I_{p0} f(t) Z/(1 + Z) \tag{4}$$

$$Z = (I_{s0}/I_{p0}) \exp[Gf(t)], \qquad G = gz, \tag{5}$$

where I_{s0} and I_{p0} are the initial intensity of the seed-Stokes and the pump, with $I_{s0} \ll I_{p0}$, f(t) is their initial profiles. We readily see that the steady-state phonon, the first term in Eq. (1), has an instant response to the product $E_s(t)E_p(t)$, which leads to the same time symmetry as that of the pump, i.e., $E_s(-t) = E_s(t)$ and $Q(-t) = Q(t)$ when $f(-t) = f(t)$.

In the slightly transient regime, we may treat the second term in Eq. (1) as a perturbation. For analytical results, we consider f(t) to be a hyper-secant square function which leads to the expression of the phonon amplitude

$$Q^{(1)} = Q^{(0)}\ [1 + \delta(t)], \tag{6}$$

where $Q^{(0)}$ is the zeroth-order (or steady-state) value and $\delta(t)$ is the first-order correction given by:[3]

$$Q^{(0)} = [I_s^0\ (1 - I_s^0)]^{\frac{1}{2}} \tag{7}$$

$$\delta(t) = \left(\frac{1}{\Gamma t_p}\right)\tanh\left(\frac{t}{t_p}\right)\left[2 + Gf(t)(1 - Z)/(1 + Z)\right]. \tag{8}$$

The important features of TSRA extracted from the above results are:

a) the transient correction term (TCT) breaks the symmetry of f(t), since $\tanh(-t) = -\tanh(t)$; b) the TCT contribution is governed by $G/\Gamma t_p$ rather than Γt_p only; and c) when $\delta(t) < -1$, the phonon amplitude becomes negative and causes the feedback mechanism [see Eqs. (2) and (3) with $Q<0$].

The above analysis are in agreement with our computer exact solutions shown in Figures 1 and 2, for a steady-state total gain G = 9 at various pulse duration. We note in Figure 1 that even when $\Gamma t_p = 10$, the transient feature is still found, since $G/\Gamma t_p = 0.9$ in this case. Figure 2 shows the energy feedback or the pulsation due to the dynamic phonon polarization (Q), which has a delayed response and changes phase when the pump is depleted to zero intensity. These effects are more significant for higher transient gain (G_T). The transient gain is defined by the very short pulse case, $t_p \ll 1/\Gamma$, and is given by $G_T \propto (G\Gamma I_{p0} t_p)^{\frac{1}{2}}$, which is pump energy dependent rather than intensity dependent.

MULTIPLE PULSES

Numerical results for the multiple-pulse systems are shown in Figures 3 and 4, where a train of 6 pulses is considered for the seed-Stokes and the pump (perfect overlap and equally separated pulses are assumed). Figure 3 shows the effects of the pulse separation (t_s) on the accumulated phonon excitation (Q^2). We see that higher Q^2 is achieved when the pulses are less separated. Pulsation (for the t_s = 0.4 n.s. case) and saturation effects (for the t_s = 0.8 n.s. case) are found in these multiple pulse systems as in that of the single pulse case. Finally, we remark that these features of TSRA have been observed experimentally, but only for the single pulse case.[4]

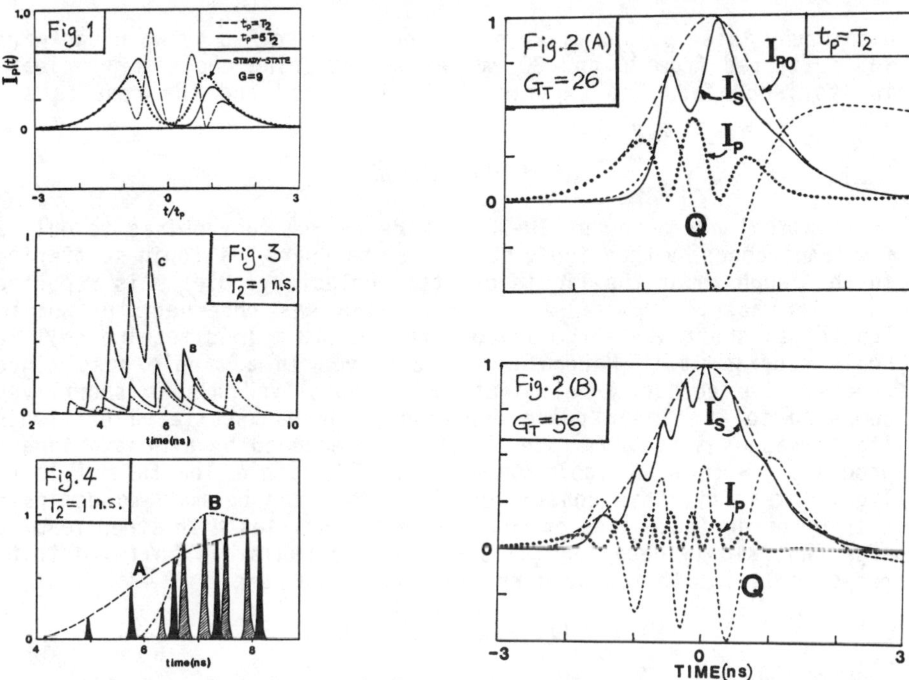

Fig 1 shows the intensity profiles of the depleted pump for steady and transient states. Fig. 2 shows profiles of the initial pump (I_{po}), output pump (I_p), Stokes (I_s) and the phonon complitude (Q) at various transient gains (G_T). The phonon excitations are shown in Fig. 3 at various t_s = (A) 0.8 (B) 0.4 (n.s.).

REFERENCES

1. R.L. Carman, F. Shimizu, C.S. Wang and N. Bloembergen, Phys. Rev. A2, 60 (1970).

2. J.N. Elgin and T.B. O'Hare, J. Phys. B12, 159 (1979).

3. J.T. Lin, J. Opt. Soc. Am. (B), in press (1986).

4. G.I. Kachen and W.H. Lowdermil, Phys. Rev. A14, 1472 (1976).

OPTICAL PULSE NARROWING BY STIMULATED SBS

D.V.G.L.N. Rao and D.N. Ghosh Roy
Physics Department, University of Massachusetts
Boston, Massachusetts 02125

ABSTRACT

Giant ruby laser pulses of 20 ns are narrowed to 2-5 ns when backward SBS scattered from 10 and 30 cm cells containing cholesteryl oleate in its isophase. The experiment is described and the results are discussed.

INTRODUCTION

Optical narrowing of 20-30 ns wide 6943 Å ruby pulses to only a few nanoseconds by transient, stimulated backward Brillouin scattering in the isophase of the liquid crystal Cholesteryl Oleate is reported in this paper. Optical pulse narrowing was observed in sample lengths as short as 3-5 cm as compared to 1.3 m long methane cell in Hon's experiment.[1] Narrowing was achieved in a single pass since regenerative feedback was eliminated. For shortlived phonons (shortlived compared to the laser pulse duration), the backscattered Brillouin light reaches a steady state condition preceeded by a nonstationary process. With a suitable combination of sample length and laser light intensity, this nonstationary process can be made to dominate thereby producing pulses of reasonable intensities with steep leading edges and slowly decaying tails.[2] This mechanism is believed to be responsible for the pulse narrowings reported here.

EXPERIMENT AND RESULTS

The laser used was a passively Q-switched, water cooled, ruby laser operating in a single mode with peak power in the range of 1-20 MW and a pulse-width of 20 ns. The scattering medium was the liquid crystal Cholesteryl Oleate (Aldrich Chemical, 97% pure, cholesteric-isotropic transition at 49°C) in its isophase. A fast pulse photodetector (S-1 photocathode, 0.5 ns response time) and a Tektronix 519 Oscilloscope (0.3 ns response) were used for signal monitoring. Pulse energy was measured by a ballistic thermopile connected to a Keithly microvoltmeter. The cell was kept sufficiently distant from the laser to avoid any regenerative feedback. The damping rate t_p of the phonon amplitudes was 0.2 ns and the corresponding steady state gain was estimated to be 7.5×10^{-3}cm/MW. Two cell lengths were used -- 10 cm and 20 cm. For a Gaussian incident pulse of 20 ns width, the backscattered pulses were narrowed to 2-3 ns for the 10 cm cell and 5-6 ns for the 30 cm cell. SBS conversion was estimated to be roughly 25 to 30%. The incident laser and the experimental backscattered pulse at the entrance window of the 30 cm cell are shown in Fig. 1 below.

By Laplace transforming the fundamental equations for the backward SBS process and inverting the integral for the backscattered field at the Stokes pole, an approximate expression was obtained for the pulse shape:

$$F(\tau ;z = L) = 2 \exp [(G-\tau)/2] \sqrt{G/\tau} \cdot I_1(\sqrt{G\tau})$$

$$-[\exp(-\tau)](G/\tau)[I_1(\sqrt{G\tau})]^2 \qquad (1)$$

where L is the cell length, G the exponential gain over L, $\tau = t/tp$, and I_1 = the modified Bessel function of order one. Using this expression, the following approximate estimates were obtained for τ max = time to reach peak intensity $\approx$ **G**, $\Delta\tau$ = FWHM $\approx 4G/\sqrt{G-3}$ for large G. Equation (1) and the above approximations agreed with the experimental results reasonably well.

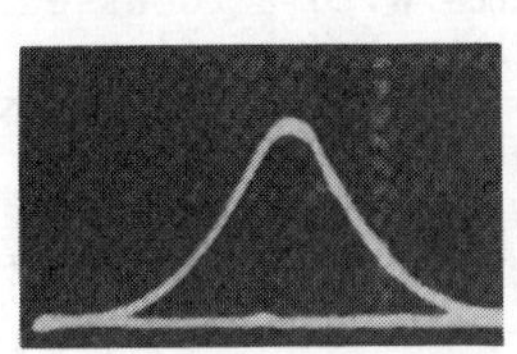

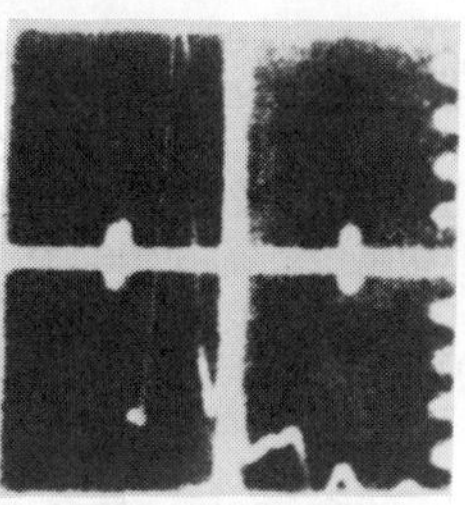

Fig.1 a) Incident laser pulse. Scale: 20 ns/cm
b) Backscattered pulse from 30 cm cell. Scale: 10 ns/cm

CONCLUSION

Narrow, intense optical pulses can be generated using SBS by enhancing only the transient part of the scattering for shortlived phonons by judicious control of cell length and laser power. High conversion can be achieved if Brillouin active media of small linear absorption are used.

REFERENCES

[1] D. Hon, Opt. Let. 5, 516 (1980)
[2] M. Maier, Phys. Rev. 166, 113 (1968)

AMPLIFIED SPONTANEOUS EMISSION STUDIES IN XENON

M. B. Rankin, L. C. Bobb, J. P. Davis, C. Giranda
U.S. Naval Air Development Center, Warminster, PA 18974-5000

ABSTRACT

Measurements of the spectral width of 828 nm amplified spontaneous emission (ASE) in xenon give line widths of 1.0-2.5 GHz, larger than the Doppler value 397 MHz expected for spontaneous emission of the fluorescence. Maxwell-Bloch model simulations of ASE events are compared with experimental results. Agreement to within a geometry-related scale factor is found.

INTRODUCTION

ASE is created in xenon by two-photon-resonant excitation of the 6p(1/2,0) state from the ground state with a focussed sub-nanosecond pulse of UV radiation at 250 nm. Optical gain on the 6p(1/2,0) 6s(3/2,1) transition (828 nm) results. The ensuing bidirectional emission is fast, intense, and contained in 70 milliradian cones coaxial with the UV beam. Paper MG8 of these proceedings describes this research more broadly. Here we highlight recent results.

LINEWIDTH MEASUREMENTS AND NUMERICAL SIMULATIONS

We measured ASE linewidth with a scanning Fabry-Perot (FP) etalon whose spectral resolution was about 530 MHz. Figure 1 shows

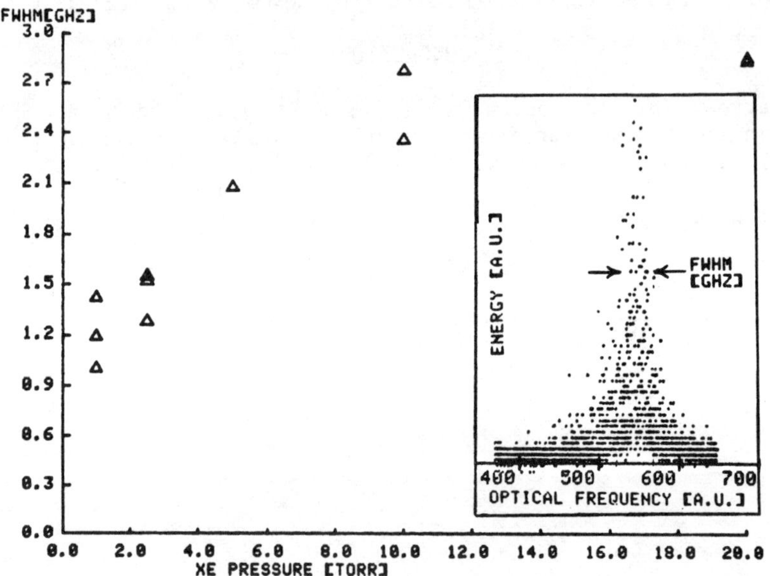

FIG 1. 828 nm xenon ASE linewidths are plotted versus xenon pressure

0094-243X/86/1460310-2$3.00

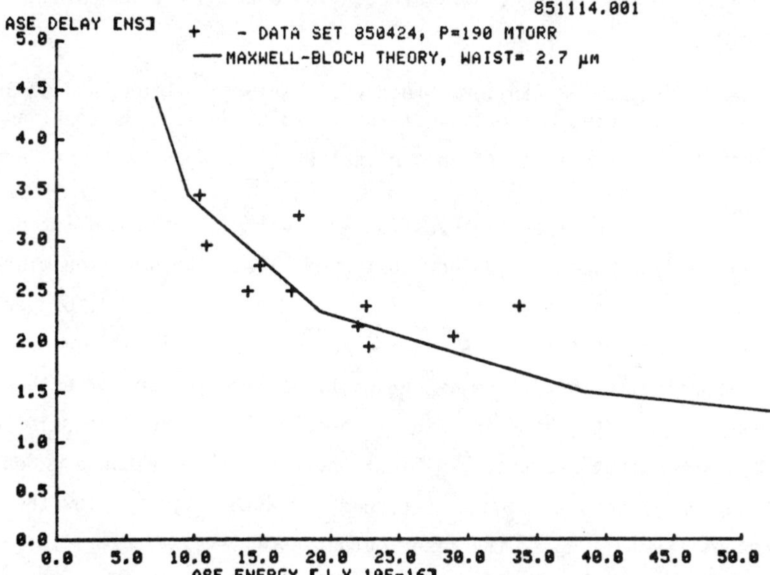

FIG 2. Experimental and simulated delay times plotted versus energy

raw data (inset), and the growth of linewidth with xenon pressure. The linewidth exceeds the Doppler- or collision-broadened widths expected of spontaneous emission[1], as well as the minimum width required by the observed 600 ps duration of the ASE pulse.

We model ASE as superradiant emission[2] which can begin during excitation. Figure 2 compares experimentally measured ASE pulse energies and delay times with simulations in which the number of excited xenon atoms matches the number of photons emitted bidirectionally. The diameter of the active region in the simulation is 5.4 μm, smaller than the 40 μm divergence-limited spot size of the UV pump beam. We are studying this discrepancy.

ACKNOWLEDGEMENTS

We thank Tim Berkoff, Robert Starry and Tom Curran of NADC and Dr. Lorenzo Narducci and the Drexel University Quantum Optics group for their contributions.

REFERENCES

1. McCown, A. W., M. N. Ediger, and J. G. Eden, Phys. Rev. A26, 2281(1982).
2. Feld, M. S. and V. S. Letokhov (eds.), Coherent Nonlinear Optics: Recent Advances (Springer Verlag, New York, 1980), pp 7-57; Bonifacio, R. (ed.), Dissipative Systems in Quantum Optics, (Springer-Verlag, New York, 1982), pp 4-6, 111-145.

A MODULATED NMR LASER NEAR TO BIFURCATION USED AS A SMALL SIGNAL DETECTOR

B. Derighetti, M. Ravani, R. Stoop, P.F. Meier, E. Brun, and G. Broggi
Physics Institute, University of Zürich, 8006 Zürich, Switzerland

and R. Badii, Institute of Theoretical Physics, University of Zürich

We performed experimental investigations with a doubly modulated NMR laser working as a small-signal detector [1,2]. The NMR line width was modulated with a pump signal $A_p \sin\omega_p t$ leading to a cascade of period-doubling instabilities as A_p is increased. A small input signal $a_i \sin\omega_i t$ was supplied to the system as an additional modulation of either the Q-factor or of the NMR line width. The effects of the 2 signals on the NMR laser output are shown in Fig.1 where the time response is depicted. In (a) only the pump signal is turned on. With A_p just below the first instability ($\omega_p \rightarrow \omega_p/2$) the laser output oscillates with ω_p. In (b) both signals are present. The nonlinear coupling of the input signal with the pump signal leads to a response where the frequencies ω_i, $\omega_p/2$ and their combination frequencies are showing up.

Instead of analysing the continuous time behavior of the NMR laser, it is possible to give a good account of the phenomenon by sampling the output in phase with the pump signal. The strobed data display two marked features: i) the superposition of 2 beating signals of frequency ω_i and $\omega_p/2n$ (n=1,2,3 according to the instability), ii) the peaked frequency response $\Delta S(\omega_i)$ where ΔS is the amplitude of the beating i.e., the maximum difference between two consecutively strobed values. Fig.2 shows the measured response $\Delta S(f_i=\omega_i/2\pi)$ for n=1. Only one peak appears at $\omega_i = \omega_p/2$. The absence of peaks at ω_i (mod ω_p) indicates that n=1 is not a bifurcation. Figure 3 shows the measured response $\Delta S(f_i)$ for n=2 with ω_p as in Fig.2. The expected peaks at ω_i and ω_i (mod $\omega_p/2$) are present as expected for a period-doubling bifurcation.

The records of Fig.2 and Fig.3 suggest the use of this double-modulation technique to investigate the nature of instabilities in nonlinear systems.

1) K. Wiesenfeld and B. McNamara, Phys.Rev.Lett. 55 (13) 1985

2) B. Derighetti, M. Ravani, R. Stoop, P.F. Meier, and E. Brun Phys.Rev.Lett. 55 (1746) 1985

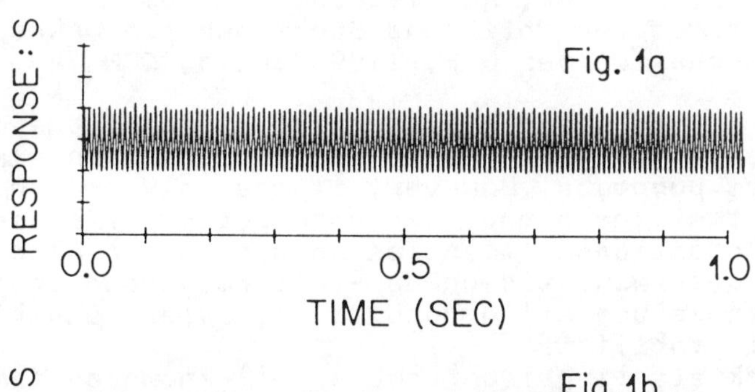

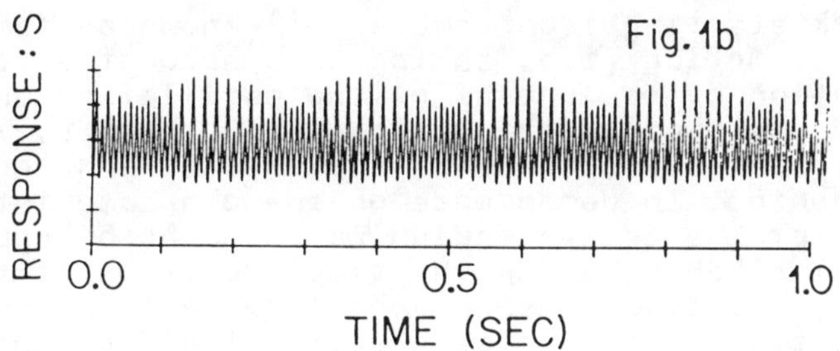

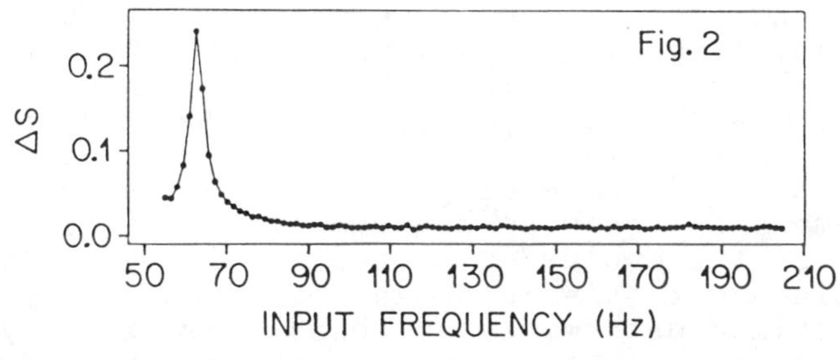

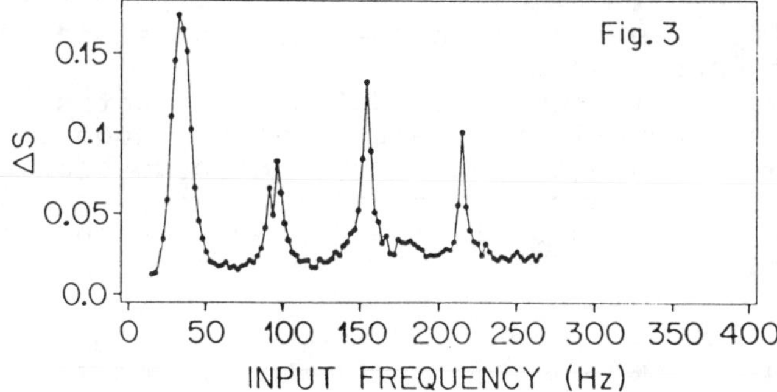

TIME EVOLUTION OF THE TOTAL ELECTRIC FIELD STRENGTH IN MULTIMODE LASERS

W. Brunner, R. Fischer, H. Paul
Zentralinstitut für Optik und Spektroskopie, Akademie der Wissenschaften der DDR, 1199 Berlin, GDR

The spectral and dynamic properties of lasers have always attracted the interest of reseachers using lasers for various purposes. However, it was only in the last few years that the output characteristics of lasers have attracted additional attention in connection with the so-called deterministic chaos [1]. It was shown that, under suitable operating conditions, lasers exhibit specific instabilities.

In our work starting from Lamb's well-known system of equations of motion (i.a. taking into account also the time variation of the phases of the oscillating modes [2]). we studied numerically the build-up of standing-wave multimode gas laser spectra in the case of inhomogeneous line broadening. In dependence on the characteristic parameters of the active medium we found both regular and irregular behaviour in the time evolution of the amplitudes, relative phases, secondary beat frequencies, the intensity, and the total electric field strength. The irregular behaviour is characterized by strong fluctuations of all these quantities.

The results for a special set of parameters are shown in Fig. 1 for $g/\varkappa$ = 1,3. Here, $\Delta\omega$ is the mode spacing, γ the homogeneous line width, Γ the inhomogeneous line width, $\varkappa$ - the cavity loss, and $g/\varkappa$ the relative excitation. Varying the parameter $\Delta\omega/\gamma$ at a given ratio Γ/γ = 1,39 we found a single-pulse regime, a three-pulse regime and irregular behaviour. The sensitive dependence of the evolution on the initial conditions is a strong hint that this irregular behaviour corresponds to the so-called deterministic chaos. Lowering Γ/γ at given $\Delta\omega/\gamma$ = 0,3125, the system changes over from the chaotic state to a stable three-mode regime and, finally a stable one-mode regime.

From our results we conclude that a chaotic behaviour will be found in the case of strong mode competition in the presence of some degree of mode coexistence With decreasing mode coexistence, the system turns over in a regular regime even in the case of strong mode competition.

REFERENCES

1. J.Opt.Soc.Am. B2, January 1985, Special Issue on Instabilities in Active Optical Systems, ed. N.B. Abraham, L.A. Lugiato, L.M. Narducci
2. W. Brunner, R. Fischer, H. Paul, Sov.J.Quantum Electronics 13, 58 (1983)

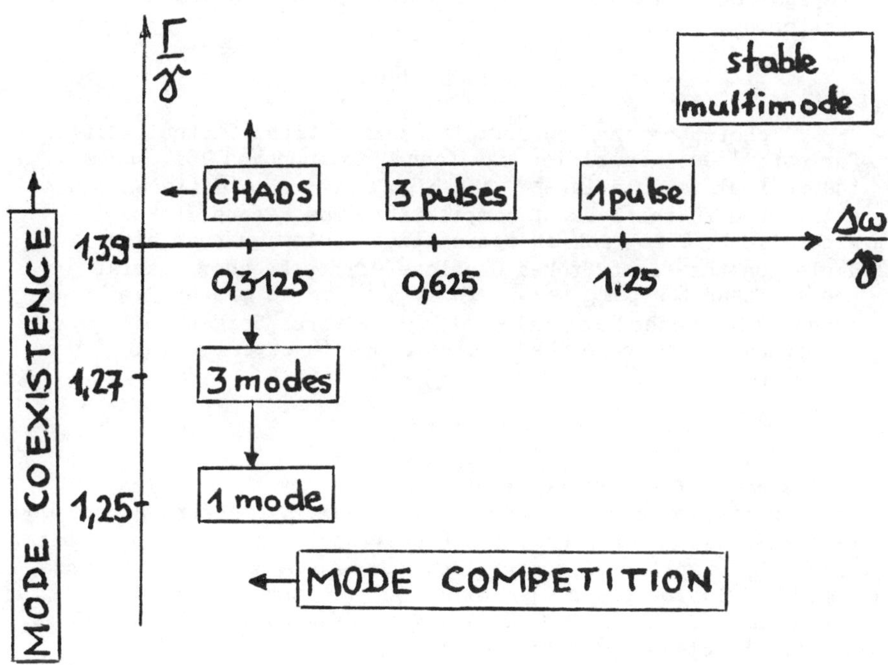

Fig. 1

SPATIAL MODE STRUCTURE OF STIMULATED STOKES EMISSION FROM A RAMAN GENERATOR

J.L. Carlsten, J. Rifkin and D.C. MacPherson
Physics Dept., Montana State University, Bozeman, MT 59717

ABSTRACT

The spatial mode structure of the stimulated Stokes emission from an H_2 Raman generator has been accurately measured. These measurements were compared with predictions of Yariv's theory of propagation in a quadratic gain media applied to Raman scattering.

INTRODUCTION

A photon conversion efficiency into first Stokes radiation for soft focus geometries was found[1] to approach 90%. From theoretical calculations[2,3] the high conversion efficiency was attributed to the lack of competition from second Stokes generation. This lack of competition indicates that after depletion the first Stokes develops with a broader spatial profile than the pump beam. In this paper we present an approximate method for calculating the first Stokes mode structure and compare its predictions with experimental measurements.

THEORY

Growth of the Stokes field is driven by the nonlinear polarization produced by a pump laser field. The transverse mode of the Stokes results from the transverse mode of the pump beam. Since the transfer of energy from pump to Stokes field depends exponentially on the pump intensity, only the transverse dependence of the most intense part of the pump beam should affect the Stokes mode structure.

If the pump intensity's transverse dependence is expanded in a Taylor series and terms of order greater than r^2 are neglected, the important central portion of the beam can be fit very well for distributions like Gaussians and Airy patterns. The treatment of the Stokes field in a region where the gain is fit by a quadratic is identical to Yariv's[4] treatment of a quadratic index media.

From Yariv's treatment we find that a Gaussian Stokes input will eventually reach a steady state mode structure. Steady state is reached when gain narrowing and diffraction have balanced each other.

Since depletion occurs over a short distance, the pump intensity can be approximated by a step function. Yariv's treatment gives us the complex beam radius at the end of the gain region. From the complex beam radius the effective location and size of the Stokes waist can be determined. The effective waist location may be quite different from the location of the pump focus.

EXPERIMENTAL RESULTS

In our experiment a 4.5 mm top hat distribution from an XeCl laser was focused with a 191.6 cm focal length lens into the center of a 50 cm cell filled to 1500 psi with Hydrogen. The spatial patterns were measured using a Reticon linear photo diode array.

From spatial profiles of the pump it was found that the width of the central peak was fairly constant through the cell. At low intensities the Stokes beam was close to Gaussian. The measured size and location of the effective waist agreed fairly well with the quadratic theory. However, at high pump intensities the Stokes output developed non-Gaussian structure. Nonetheless, the theory appears to predict the correct profile in the spatial wings.

ACKNOWLEDGEMENTS

We would like to thank K. Manes in particular for suggesting the application of Yariv's theory to stimulated Raman scattering. In addition, we would like to thank J. Cooper, K. Druhl, M. Raymer, A. Szoke, J. Telle and R. Wenzel for illuminating discussions about this experiment and M. Babb and M. Sangarakumaran for technical help with the data acquisition and digitization. This research has been supported in part by Los Alamos National Laboratory, Lawrence Livermore National Laboratory and the Innovative Science and Technology Office of the Strategic Defense Initiative Organization administered by the Office of Naval Research.

REFERENCES

1. J.L. Carlsten, J.M. Telle and R.G. Wenzel, Opt. Lett. 9, 353 (1984)
2. B. Bobbs and C. Warner, "Absence of second Stokes in a Raman generator with no four-wave mixing", submitted to Optics Letters.
3. K. Druhl, "Stokes beam parameters at large gain for focused pump beams", submitted to Optics Letters.
4. A. Yariv, Quantum Electronics, New York: Wiley, 1975.

FIRST-PASSAGE-TIME STATISTICS FOR THE GROWTH OF LASER RADIATION

Marvin Young and Surendra Singh
Physics Department, University of Arkansas, Fayetteville, AR 72701

ABSTRACT

We report on the measurements of the first-passage-time distributions for the growth of laser radiation in a Q-switched laser near threshold.

The dynamics of a single-mode laser has been studied by a number of workers both theoretically[1] and experimentally.[2-4] Here we report on the measurements of yet another aspect of this problem, namely, the passage-time distributions for the laser intensity to grow from an initial value zero to a predetermined value for the first time. The time evolution of the complex field amplitude of the laser is governed by the nonlinear Langevin equation

$$\dot{E} = -\frac{\partial U}{\partial E^*} + \eta(t) \ , \tag{1}$$

where the potential $U(|E|^2)$ is given by

$$U(|E|^2) = -\tfrac{1}{2}a|E|^2 + \tfrac{1}{4}|E|^4 \ . \tag{2}$$

The Langevin noise term $\eta(t)$ represents spontaneous emission fluctuations and is taken to be a delta-correlated Gaussian random process with zero mean. The pump parameter a is proportional to the net gain. If the pump parameter is suddenly changed from a large negative value to some positive value (Q-switching), the light intensity grows from the initial value zero toward the nonzero steady-state intensity $I = |E|^2 = a$. The time taken by the light intensity to reach a predetermined threshold is called the first-passage-time (FPT). The FPT will fluctuate each time the laser is Q-switched reflecting the fact that the growth is initiated by the random spontaneous emission fluctuations.

Measurements of the FPT distributions were performed on a single-mode He:Ne laser operating at $\lambda = 633$ nm near threshold. The laser is turned on by applying a voltage pulse to an intracavity acousto-optic modulator. At the same time a gate allows pulses from a clock to reach a scaler. The growth of the light intensity is monitored by a photomultiplier tube (PMT) whose output is fed to a discriminator. When the PMT output voltage crosses the discriminator threshold (corresponding to some intensity) an output pulse is triggered which stops the gate. The number stored in the scaler is then a measure of the FPT. This number is transferred to a computer memory. By repeating this process a histogram of the FPT is built up. Figure 1 shows measured FPT distributions for several different pump parameters and for a fixed threshold $I_{th} = 2$. It is seen that with increasing pump parameter the peak of the distribution shifts to smaller values and its width decreases. These

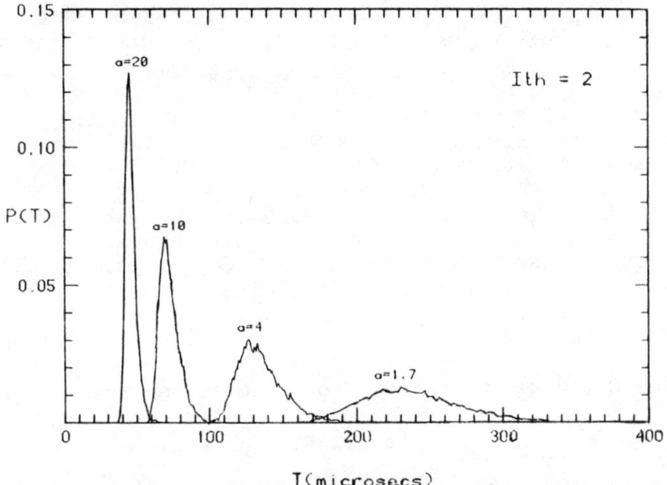

Fig. 1. Examples of the measured FPT distributions.

observations are in qualitative agreement with the predictions of Eq.(1). Detailed comparison between the predictions of Eq.(1) and our measurements is in progress and the results will be published elsewhere.

This work was supported by NSF grant PHY-8501359 and by Joseph H. DeFrees grant of Research Corporation.

REFERENCES

1. J. P. Gordon and E. W. Aslaksen, IEEE J. Quant. Electron. QE-6, 428 (1970); F. Haake, Phys. Rev. Lett. 41, 1685 (1978); M. Young and S. Singh, Phys. Rev. A31, 888 (1985).
2. F. T. Arecchi and V. DiGiorgio, Phys. Rev. A3, 1108 (1971).
3. D. Meltzer and L. Mandel, Phys. Rev. A3, 1763 (1971).
4. Measurements of the FPT distribution with multiplicative noise in a dye-laser operating far above threshold (a ≳ 1000) have been performed by R. Roy, A. W. Wu and S. Zhu, Phys. Rev. Lett. 55, 2794 (1985).

PHYSICAL INSIGHTS ON THE CW ON-RESONANCE WHOLE BEAM ABSORPTIVE AND OFF-RESONANT DISPERSIVE SMALL-SCALE SELF-FOCUSING IN PROPAGATION OF LASER BEAMS IN SATURABLE ABSORBERS+

J.P. Babuel-Peyrissac*, J.P.Marinier*, and C. Bardin**
Department de Physico-Chimie* and Informatique Internationale**
Centre d'Etudes Nucleaires de Saclay, 91191 Gif/Yvette, France
F.P. Mattar
Department of Physics, New York University, New York, NY 10003
and George R. Harrison Spectroscopy Laboratory,
Massachusetts Institute of Technology, Cambridge, MA 02139
J. Teichmann* and Y. Claude**
Departement de Physique* and Departement d/Informatique**
Universite de Montreal, Box 6128, Montreal, P.Q., Canada
and B.R. Suydam
Los Alamos National Laboratory, Los Alamos, New Mexico 87544

ABSTRACT

Analytical and numerical calculations to elucidate the propagation of very intense CW laser beams in saturable absorbers are presented. The beam profile can be either (i) smooth thus experiencing whole beam collapse for the case of on-resonance, or (ii) corrugated with small-scale ripples which leads to off-resonant small-scale self-focusing.

SUMMARY

Boshier et al.'s surprising prediction[1] of CW on-resonance self-focusing for beams more intense $|e_o|^2$ than the saturation intensity I_s in two-level systems is understood through the interplay of the radially dependent nonlinear absorption (defined in plane wave analysis by the Beer reciprocal length α), and the forward free space diffraction (characterized by the Rayleigh/diffraction length z_d), towards the beam center.

A Fresnel number defined in terms of α, namely $F=\alpha z_d$; perfectly characterizes the competitive effects of the nonlinear medium and beam geometry. This new phenomenon had not occurred previously, as reviewed by Marburger[2] where less intense beams were studied using a cubic nonlinearity which is obtainable from the saturable one by a power series development. The saturable absorption, which varies from one concentric shell to another, attenuates the low-intensity 'wings' of the beam more than the high-intensity 'center', abruptly stripping the beam edges which form a shrinking aperture as outlined by Teichmann et al.[3] The inward intensity flow does not experience a nonlinear boosting amplification as discussed by Mattar et al.[4a] for the coherent self-focusing in Self-Induced-Transparency (SIT)[4b] since there is not any population inversion in shells closer to the axis. This fact was recognized by LeBerre et al.[5] when they developed their model in which an abrupt aperture stop approximates the stripping of the beam by the nonuniform absorption and then allowed for free space propagation. Concomitantly, Teichmann et al. developed a soft shrinking aperture which elucidates (see Fig. 1) stripping over the propagation distance by expanding the field profile at the edges into a power series in terms of exponential decay and which is followed by free space propagation. The latter results in Fresnel diffraction causing on-axis minima and maxima. The stripping and the early stages of an inward flow of intensity can be described using a perturbative treatment. Such treatment, however, only describe the onset of the enhancement. The calculation must be followed by free space propagation so that the same on-axis self-focusing obtained by rigorous calculations occurs. If the stripped beam profile is approximated by a parabola the propagation in free space may be solved analytically (see Fig.2) as an equivalence to the aperture model. However, the on-axis enhancement of the free space beam is larger than what it would have been had the absorbing medium effect been maintained. A small corrective nonlinear field is obtained as a perturbation driven solely by the free space parabolic beam. This simple physical picture has been confirmed by both numerical and laboratory experiments.[6]

If the intense beam profile is not perfectly smooth then it suffers some small-scale self-focusing (SSSF) as in high-power Nd Lasers and it collapses before the whole beam self-focusing length. (See Fig.3) Suydam's[7] small-scale linearized perturbation calculation for a cubic non-linearity has been generalized[8] to describe a saturable medium. Both amplitude and phase perturbations are considered in our calculation and lead to excellent agreement for short propagation length with a plane wave analytical treatment. Conditions leading to the maximum enhancement SSSF have been determined as well as those leading to non-growing ripples. Additional computations on perturbed Gaussian and super-Gaussian beams have also been realised. Parametric dependence of the growth rate of the transverse perturbational on the detuning, saturation intensity and Fresnel numbers are reported.

REFERENCES

1. M.G. Boshier and W.J. Sandle, Optics Comm. 42, 371 (1982).
2. J.H. Marburger, Progress in Quantum Electronics, Vol. 4, eds. J.H. Sanders and S. Stenholm (Pergamon Press 1975), p. 35.
3. (a) J. Teichmann and F.P. Mattar, Abstracts Digest of the Annual Meeting of the Canadian Association of Physicists and Canadian Astronomical Society, Victoria, B.C., Canada (June 1983); (b) J. Teichmann, Y. Claude and F.P. Mattar, Opt. Commun. 54, 33 (1985).
4. (a) F.P. Mattar and M.C. Newstein, IEEE J. Quantum Electron QE-13, 507 (1977); (b) H.M. Gibbs, B. Bolger, F.P. Mattar, M.C. Newstein, G. Forster and P.E. Toschek, Phys. Lett. 37 1743 (1976) and (c) J.J. Bannister, H.J. Baker, T.A. King and W.G. McNaught, Phys. Rev. Lett. 44, 1062 (1980).
5. (a) M. Leberre, E. Ressayre and A. Tallet, Phys. Rev. A-25, 1604 (1982); (b) M. Leberre, F.P. Mattar, E. Ressayre and A. Tallet, Proc. Max Born Centenary Conf., publ. by SPIE,

Vol. 369, 270 (1983); (c) M. LeBerre, E. Ressayre, A. Tallet, H.M. Gibbs, M.C. Rushford and F.P. Mattar, Coherent and Quantum Optics V, ed. L. Mandel and E. Wolf (Plenum Press), p. 347; and (d) M. LeBerre, E. Ressayre and A. Tallet, Coherent and Quantum Optics V, eds. L. Mandel and E. Wolf (Plenum Press, 1984), p. 331 and Phys. Rev. A-29, 2669 (1984).

6. K. Tai, H.M. Gibbs, M.C. Rushford, M. Peyghambarian, M. LeBerre, E. Ressayre, A. Tallet, W.J. Sandle, J. Teichmann, F.P. Mattar and P.D. Drummond, Optics Lett. 9, 243 (1984).
7. B.R. Suydam, IEEE J. Quantum Electron. 10, 837 (1970) and 11, 225 (1975).
8. J.P. Babuel-Peyrissac, J.P. Marinier, C. Bardin, J. Teichmann, Y. Claude, F.P. Mattar and B.R. Suydam, Refereed Proc. Southwest Conf. on Optics, ed. R.S. McDowell and S.C. Stotlar, Publ. by SPIE, Vol. 540, 560 (1985).

FIGURE CAPTIONS

Fig. 1: Soft shrinking aperture model with the beam edges being stripped out due to saturable absorption. The free space propagation which follows the soft aperaturization is illustrated with the z-dependence of the on axis and the total energy.

Fig. 2: Rigorous calculation shows (a) on-axis energy vs. propagational distance (b) 3-dimensional energy distribution vs. radius and propagational distance shows on-axis enhancement after stripping of edges.

Fig. 3: The small-scale self-focusing distance is shorter than that of the whole beam collapse length.

Figure 1.

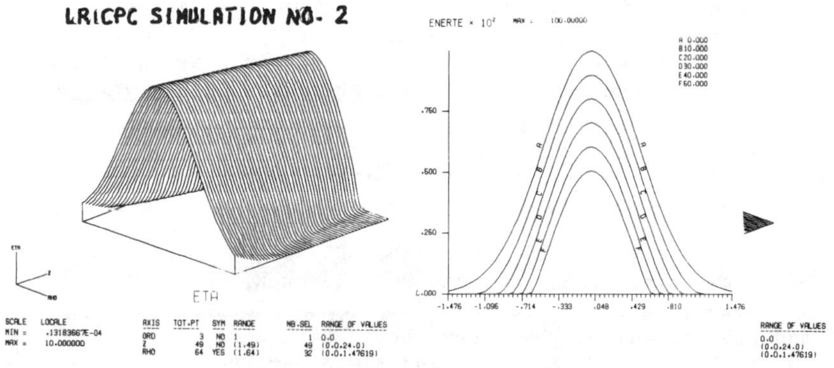

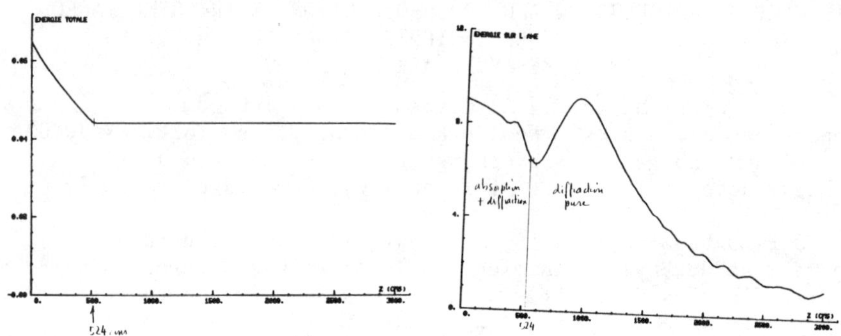

Figure 2.

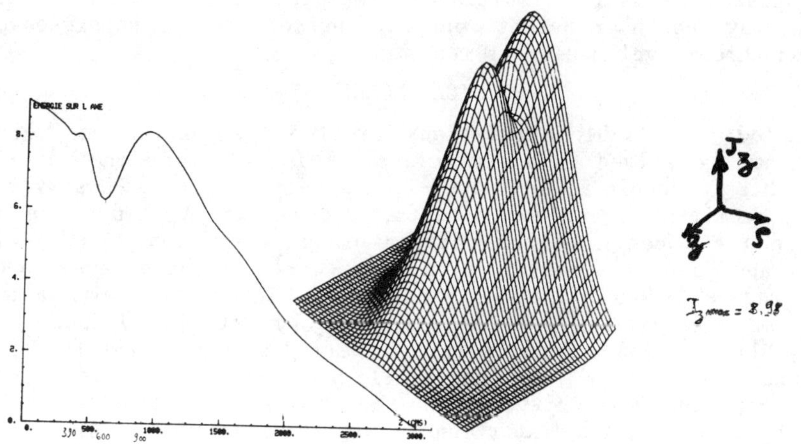

Figure 3.

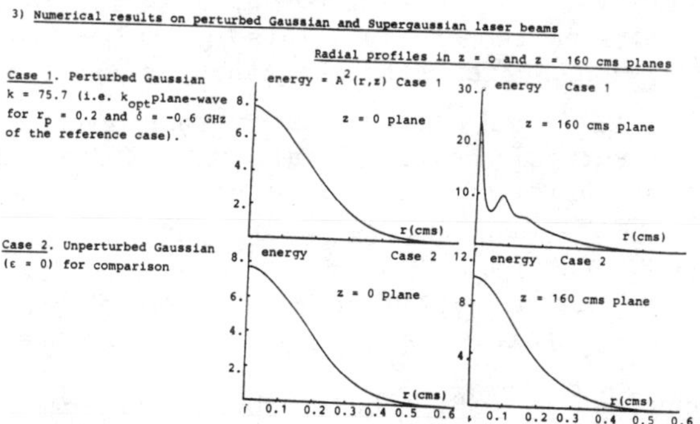

DIFFRACTION, DISPERSION AND CROSS-DEPLETION IN THE SIMULTANEOUS PROPAGATION OF COHERENT DIFFERENT-WAVELENGTH PULSES IN THREE-LEVEL ABSORBERS†,††

F.P. Mattar□,# A. Matos□ and A. Richard□
Department of Physics,□ New York University, New York, NY 10003
and George R. Harrison Spectroscopy Laboratory,#
Massachusetts Institute of Technology, Cambridge, MA 02139
and
J.P. Babuel-Peyrissac*, J.P. Marinier* and C. Bardin**
Departement de Physico-Chimie* and Informatique Internationale**
Centre d'Etudes Nucleaires de Saclay, 91191 Gif/Yvette, France

ABSTRACT

The effect of diffraction and dispersion on longitudinal reshaping associated with the simultaneous resonant co-propagation of two different-wavelength coherent optical pulses in a homogeneously broadened three-level medium is reported.

PHYSICAL MODEL

Coupled Maxwell-Bloch equations for two frequencies and three-level atoms are solved in a semi-classical formalism in the paraxial limit. This is double Self-Induced-Transparency (SIT)[1]. Phase variations, concomitant frequency offsets, and relaxation times are included in the model. In the uniform plane wave regime (UPWR) with Brewer's and Hahn's[2] symmetrization, the two-laser three-level interactions reduce to an effective single field interacting with a two-level atom. The latter evolves, as shown by McCall and Hahn, as a solitary SIT solution. Konopnicki and Eberly[3] used the UPWP to obtain on-resonance analytical and numerical predictions of simultaneous solitons of equal velocity, which they called simultons. The two beams propagate without distortion and at the same speed. The effect occurs only for restricted sets of conditions for the partial initial population, oscillator strengths and Rabi frequencies. Stroud and Cardimone[4] reported an analytical generalization which includes detuning. Newstein et al[5] predicted on-resonant SIT self-focusing (that was ascertained by independent observations[6] in Na, Ne and I). Therefore, it is expected that in research the interplay of diffraction and the medium inertial response inevitably will redistribute each beam energy both spatially and temporally. This interplay is characterized by F_g, the Beer's length Fresnel number that is associated with the Beer length α^{-1}. One would thus expect that even for identical transitions and identical beams distortion occurs. However, the two-beam energy synchronicity remains unaltered.

† The algorithm development was funded by DOE at the University of Rochester and by ARO and ONR at the Polytechnic Institute of New York. The code vectorization for the Cray-XMP was realized at Saclay under the auspices of the French Atomic Energy Commision CEA.

†† Partially supported by ONR, AFOSR, ARO and NSF; part of this work was performed at the MIT Regional Laser Center which is an NSF Regional Instrumentation facility.

NUMERICAL RESULTS

We studied the case of equal Beer's length nonlinearities $\alpha_a=\alpha_b$ $\mu_a\neq\mu_b$ with $\lambda_a\neq\lambda_b$ but different Beer-length Fresnel number so that $\mu_a^2\lambda_a^{-1}=\mu_b^2\lambda_b^{-1}$. Subsequently, we studied different Beer's lengths $\alpha_a\neq\alpha_b$ but with equal Beer-length Fresnel numbers $F_{ga}=F_{gb}$. As a result of the combined effect of free space diffraction and the nonlinear absorption the two beams remain synchronized in spite of the different characteristics.

In Fig. 1, we display the field energy for each beam versus τ at different propagation distances to illustrate the simultaneous distortion-free co-propagation with equal velocity of two unequal laser beams in a three-level cascade medium partially populated initially. In Fig. 2, the diffraction reshaping takes place synchronously in non-identical transitions with equal Beer's lengths. The temporal coincidence of the peaks prevails even though the beam transverse structure becomes distorted. Both beams experience self-focusing or self-defocusing simultaneously. This result is a generalization of the KE simulton condition since the peak of the two pulse keeps occurring at the same instant for the different z planes. However, for both equal Beer-length-Fresnel numbers for strong diffraction (i.e., both Fresnel numbers F_a and F_b are smaller than one-half and unequal) and for the most general transverse variations considered, both the shape preservation and the synchronization characteristics are eliminated. Figure 3 exhibits the general transverse problem.

In the first graph, the on-axis time-integrated energies and associated beam effective radii are plotted as a function of z. They acquire a radial energy redistribution and experience unequal on-axis energy magnification. The unequal beams of different color with unequal F_g goes through a different self-focusing while propagating in transitions of unequal α. In graph b, the spatial distribution of the Energy of each Laser and the population of the two lowest levels are shown as intervals for different τ in isometric plots versus z and δ. The temporal non-coincidence was also analyzed in the UPWR as a function of the ratio of the initial delay to the pulse length.

Calculations with non-Gaussian radial profiles have been carried out to investigate how diffraction can be enhanced or diminished and how desynchronization can be inhibited. The simulton evolution was also studied in both the Lambda and Vee atomic configurations. Moreover, the effect of the input temporal shape on the simulton was also investigated for both the UPWR and diffraction calculations. Universal curves displaying the generalization of the simulton were obtained[7].

REFERENCES

1. S.L. McCall and E.L. Hahn, Phys. Rev. 183, 457 (1969).
2. R.G. Brewer and E.L. Hahn, Phys. Rev. A11, 1641 (1975).

3. M.J. Konopnicki and J.H. Eberly, Phys. Rev. A24, 2567 (1981).
4. C.R. Stroud, Jr. and D.A. Cardimone, Opt. Commun. 37, 221 (1981).
5. N. Wright and M.C. Newstein, Opt. Commun. 1, 8 (1973); F.P. Mattar and M.C. Newstein, IEEE J. Quantum Electronics QE-13, 507 (1977).
6. H.M. Gibbs, B. Bolger, F.P. Mattar, M.C. Newstein, G. Forster and P.E. Toschek, Phys. Rev. Lett. 37, 1743 (1976); and J.J. Bannister, H.J. Baker, T.A. King and W.G. McNaught, Phys. Rev. Lett. 44, 1062 (1980).
7. F.P. Mattar, J.P. Babuel-Peyrissac, J.P. Marinier and C. Bardin, to be published in J. Opt. Soc. Am. B (1986).

FIGURE CAPTIONS

Fig. 1 The simultaneous distortion-free co-propagation with equal velocity of two unequal laser beams in a three-level cascade medium partially populated initially.

Fig. 2 Diffraction reshaping effects for temporarily-coincident two laser co-propagation in non-identical transitions with equal Beer lengths.

Fig. 3 General transverse effects associated with unequal beams, unequal Beer lengths, unequal geometric Fresnel numbers and unequal Beer-length-Fresnel numbers lead to unsynchronized on-axis energy enhancement and beam narrowing.

Figure 1

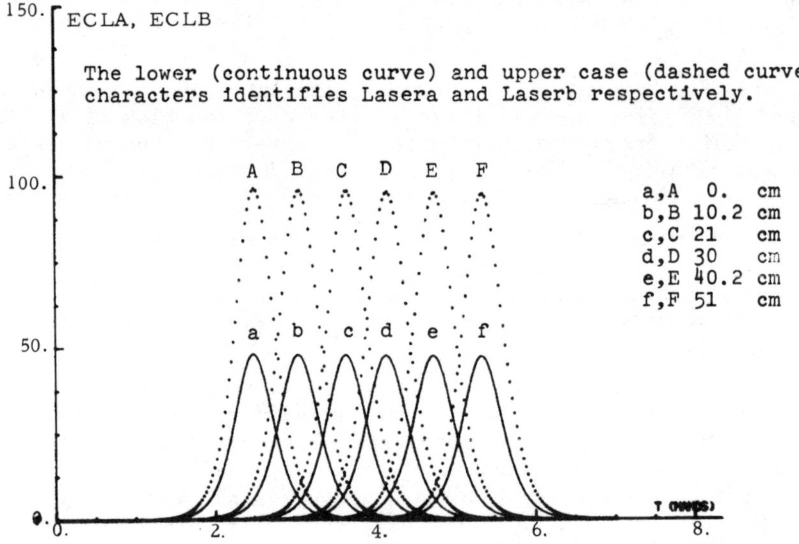

Figure 2

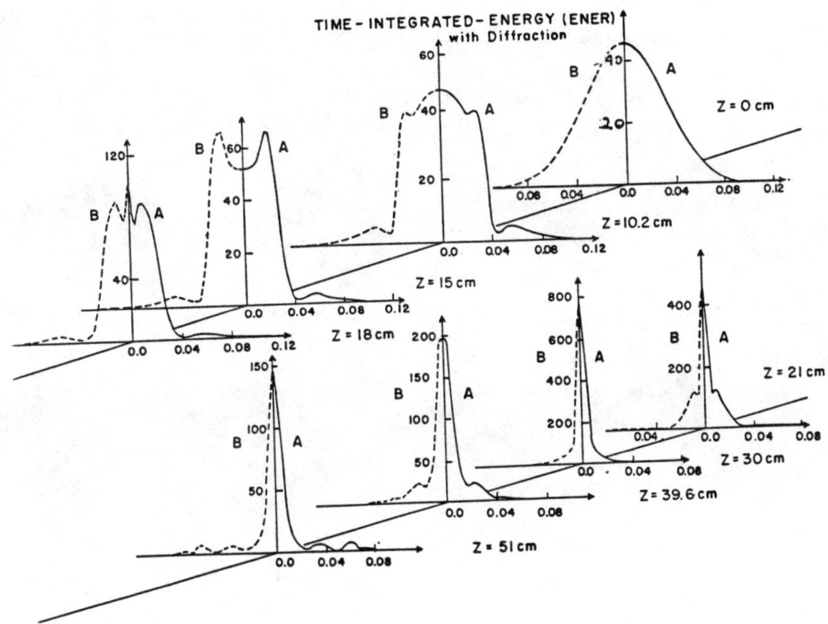

Figure 3, graph a

Diffraction simultaneous

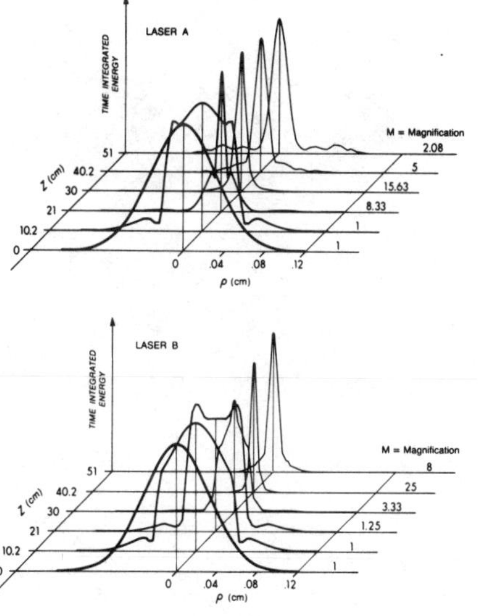

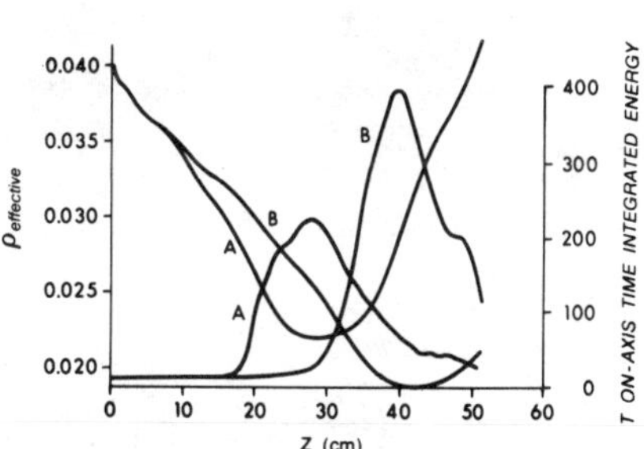

Diffraction simultaneous

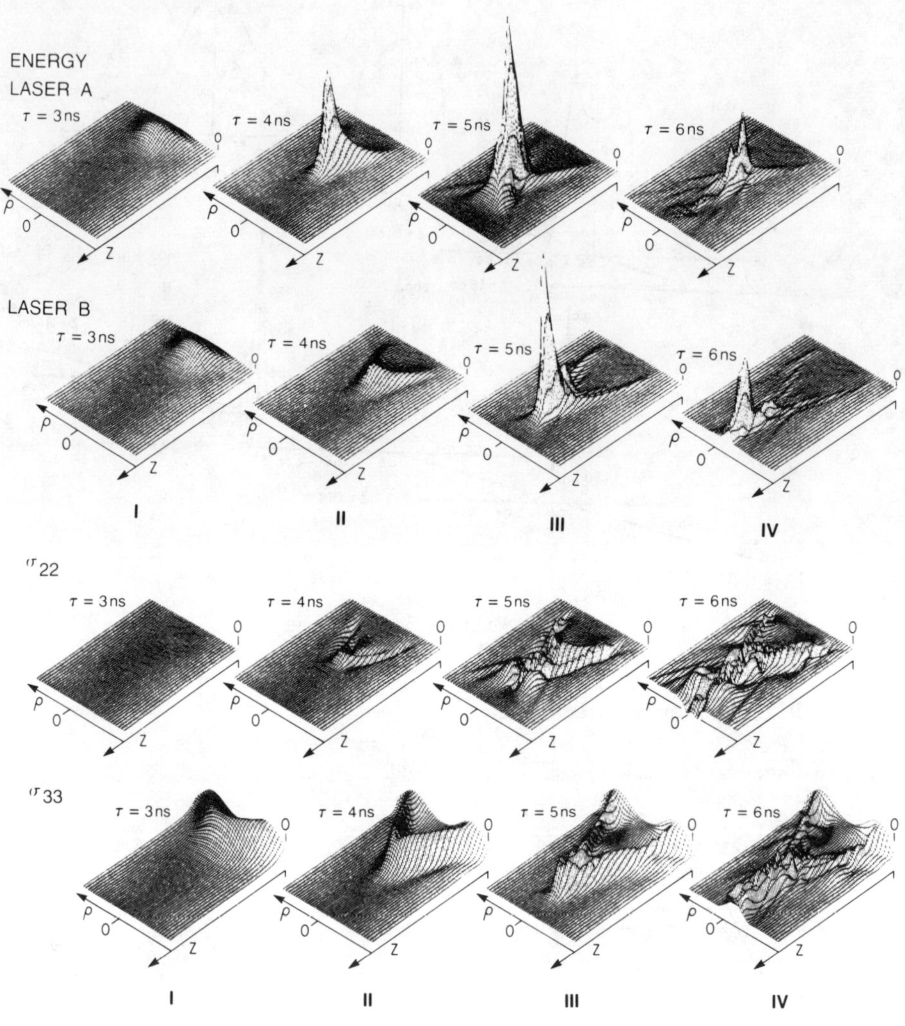

Fig. 3 - Graph b Simultaneous Different Wavelength Propagation

HIGH REFLECTIVITY SBS OF 2.9μ HF LASER RADIATION IN XENON

Michael T. Duignan,* W.T. Whitney and B.J. Feldman
Laser Physics Branch, Code 6540
Naval Research Laboratory, Washington, DC 20375-5000

ABSTRACT

We have successfully demonstrated, for the first time, stimulated Brillouin scattering of single-line HF laser radiation in high pressure xenon. The temporal behavior of the input, transmitted and SBS reflected laser pulses was monitored. Energy and power reflectivities were also measured as a function of input laser power, xenon pressure, and f-number of focusing optics. Brillouin reflection could be unambiguously distinguished from much weaker reflections due to occasional optical breakdown. Power reflectivities of greater than 50% have been observed.

INTRODUCTION

Stimulated Brillouin scattering (SBS) has frequently been employed to compensate for phase distortions in a refracting medium.[1] It thereby holds great promise for applications requiring high power laser beams of near diffraction-limited spatial quality. However, efficient SBS in the mid-IR is impeded by a host of technical limitations.[2] Most high-gain SBS media, suitable for use in the UV through near IR, absorb in the mid-IR. Furthermore, SBS gain tends to be lower, but attempts to compensate with increased laser intensity are limited by self-focusing and optical breakdown.

RESULTS

We have demonstrated, for the first time, high reflectivity SBS in gaseous media beyond 1.3μ and have observed power reflectivities of ~ 50% at 2.91μ. High pressure (~ 40 atm) Xe gas at room temperature was used as the SBS medium.

The apparatus employed is diagramed in Fig. 1. Briefly, a grating tuned pulsed chemical HF laser operating on the 2P8 line at 2.91 μ supplied 1.5-2 μs pulses of up to 2 J total energy. Beam diameter was controlled by means of an intracavity iris and was typically 30-40 mm. The beam was focused into the xenon cell by means of a 50 or 100 cm f.l. CaF_2 lens. Input, throughput and reflected power and energy were monitored on each shot.

*Potomac Photonics, Inc.
Bldg. 335, University of Maryland, College Park, MD 20742

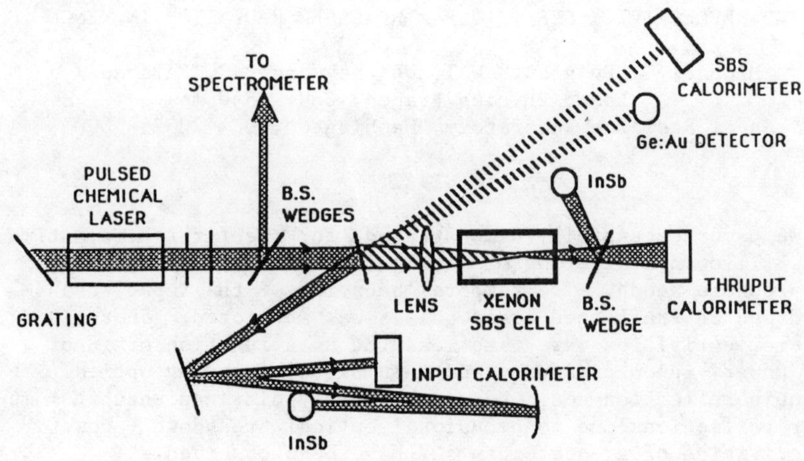

Fig. 1 - Experimental set-up. B.S. - wedged CaF_2 flat used as beamsplitter; InSb - Indium Antimonide photovoltaic detectors; cross-hatched-input beam; hatched - SBS reflection.

SBS reflection was detected at an input threshold of 0.7 J and increased linearly with input power to 0.33 J at 1.6 J input energy. SBS typically did not "turn on" until ~ 1.5 μs after the start of the input pulse. Considering only the time the reflection was active, > 50% power reflectivities were measured. SBS was not observed when Xe pressure was < 30 atm. Over the observed range f/12 - f/30, no aperture effects on SBS reflectivity were noted. Indications of optical feedback into the laser cavity were noted and complicated measurements on some shots, especially throughput power depletion, but calorimetric energy measurements were self-consistent.

Experiments are currently underway to measure the phase conjugate properties of the stimulated Brillouin reflection.

REFERENCES

1. B. Ya. Zel'dovich N.F. Pilipetskii and V.V. Shkunov, in "Optical Phase Conjugation," R.A. Fisher, Ed., Academic Press, N.Y., 1983, p. 135.

2. V. Yu. Zalesskii and A.M. Kokushkin, Sov. J. Quantum Electron. 15, 541 (1985).

LASER-INDUCED DISSOCIATION DYNAMICS IN TRIATOMIC MOLECULES

W.-K. Liu
Physics Department, University of Waterloo, Ont., Canada N2L 3G1

W. H. Fletcher
Chemistry Department, University of Tennessee, Knoxville, TN 37916

D. W. Noid
Chemistry Division, Oak Ridge National Laboratory, TN 37830

The study of intramolecular vibrational energy transfer in isolated molecules has received much attention recently[1], and the classical trajectory method has been found to be a useful tool. There has been a number of classical studies of molecular excitation and dissociation by infrared lasers[2]. One of the early work on multiphoton dissociation of a polyatomic system was by Noid et al.[3] on CD_3Cl. They have demonstrated that (i) the energy absorbed is a sharply peaked function of the excitation frequency, (ii) the dissociation threshold depends on the fluence rather than the intensity of the laser, and (iii) the translational energy distribution follows a statistical distribution upon dissociation. Good agreement with statistical theory was also obtained by Martin and Wyatt[4] recently in a similar study of a model two-dimensional system. It should be noted that in these studies, the molecule is initially in its ground vibrational state.

In a recent study of the dynamics of excited vibrational states of triatomic molecules, Hedges and Reinhardt (HR)[5] found that exceptionally long lived states exist with energy well above the dissociation limit. Statistical theories assume that for a given energy E, all states of the same energy are equally accessible and hence the lifetime is a decreasing function of E. The existence of long lived states thus suggests non-statistical behavior. It is the purpose of this paper to investigate the effect of a laser on the dissociation dynamics of molecules initially prepared in such excited states.

We shall use the model Hamiltonian of HR for a non-rotating, non-bending triatomic molecule ABA:

$$H_m = p_1^2/2\mu + D(1 - e^{-r_1})^2 + p_2^2/2\mu + D(1 - e^{-r_2})^2 - p_1 p_2/m_B \qquad (1)$$

where r_1, r_2 are the AB bond lengths and p_1, p_2 their conjugate momenta. $\mu = m_A m_B/(m_A+m_B)$ is the reduced mass with m_A, m_B being the effective masses of A and B. D is the dissociation energy of a single bond. The interaction between the bonds is due to the momentum coupling through the central atom B. The parameters chosen for the present study are $m_A=1$, $m_B=64$ and D=51, which correspond to the weak coupling case of a H_2O molecule[5]. In the presence of a laser, we must add to the Hamiltonian (1) the coupling term

$$H_i = - d(r_1,r_2)\, \mathcal{E}_o \cos \Omega t \tag{2}$$

where $d(r_1,r_2)$ is the dipole moment of the molecule, and $\mathcal{E}_o$, Ω are the laser electric field strength and frequency, respectively. We use for our model H_2O molecule[6]

$$d(r_1,r_2) = d_o\, (r_1\, e^{-r_1/r^*} - r_2\, e^{-r_2/r^*}) \tag{3}$$

where $d_o = 6.23$ debye/Å and $r^* = 0.6$ Å. The classical trajectory for given initial conditions is then obtained by integrating numerically the Hamilton's equations

$$\begin{aligned} \dot{r}_i &= \partial H/\partial p_i = \partial H_m/\partial p_i \\ \dot{p}_i &= -\partial H/\partial r_i = -\partial H_m/\partial r_i + (\partial d/\partial r_i)\, \mathcal{E}_o \cos \Omega t \end{aligned} \tag{4}$$

where $H = H_m + H_i$.

The initial state of the molecule is specified by the classical actions (N_1, N_2) of each bond in the zeroth order local mode description (i.e. in the absence of the $-p_1p_2/m_B$ term in eq.(1)). The initial Cartesian coordinates and momenta are then obtained from a uniform grid of angle variables conjugate to N_1 and N_2. The procedure of Woodruff and Thompson[7] is then employed to calculate the distribution of trajectory lifetimes, from which the dissociation probability f_d and the dissociation rate R can be extracted. Typically, 1600 trajectories are used for each calculation.

We shall focus our attention on the state $(N_1,N_2) = (6,6)$ for which its total energy E is above the dissociation threshold (E/D=1.68). In the absence of a laser, it is found that f_d=9%, and hence this state corresponds approximately to a long-lived excited state discussed in HR. Coupling a laser to the system would induce dissociation. In Fig. 1 we present the frequency dependence of f_d for different laser intensities. It is clear that f_d peaks at a laser frequency Ω_{max} for a given intensity I, and that Ω_{max} increases slightly with I. Similar behavior is also observed for the dissociation rate R. Fig. 2 shows the dependence of f_d and R (at Ω_{max}) on the laser intensity I, and as expected, both are increasing functions of I. It should be noted that in these studies, the average energy absorbed ΔE is very small: for I = 1.6 TW/cm^2, $\Delta E < 1$ photon energy while for I = 8 TW/cm^2, $\Delta E < 2$ photon energy.

Finally, we compare in Table I the dissociation behavior of the state (3,9) with the state (6,6). It is clear that, contrary to the

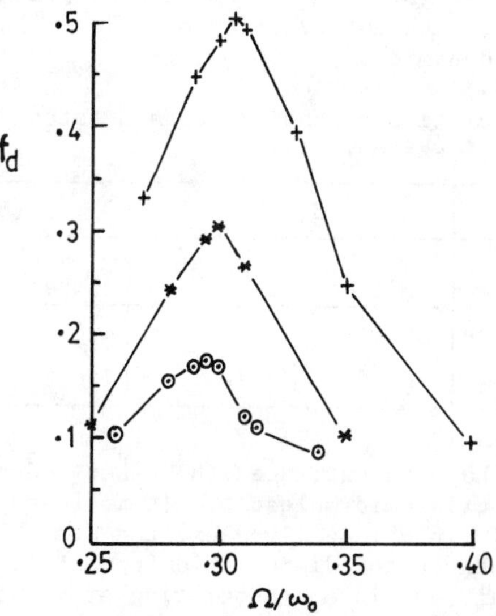

Fig. 1. Fraction of molecules dissociated f_d as a function of laser frequence Ω for different intensities I. $\omega_0=\sqrt{2D/\mu}$ is the harmonic frequency of the Morse oscillator. + -- I=8 TW/cm^2, * -- I=4 TW/cm^2, ⊙ -- I=1.6 TW/cm^2.

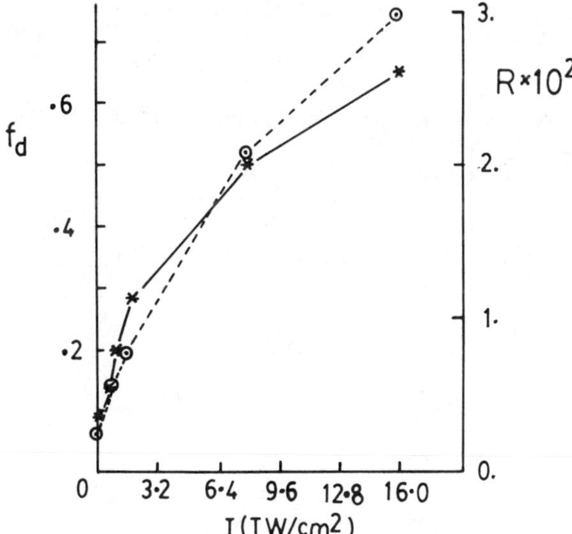

Fig. 2. Fraction of molecules dissociated f_d (*) and dissociation rate R (⊙) as a function of laser intensity I at the optimal laser frequencies.

long-lived state (6,6), (3,9) is a dissociating state even in the absence of the laser, and introduction of a laser has little effect on its dissociation dynamics.

Table I. Dissociation probabilities and rates for different initial states.

State	(3,9)		(6,6)	
I (TW/cm^2)	f_d	$R\times10^2$	f_d	$R\times10^2$
0	0.71	2.76	0.09	0.23
0.8	0.72	2.93	0.20	0.58

In summary, we have investigated the laser-induced dissociation dynamics of a model triatomic molecule. It is found that when the molecule is initially in a long-lived excited state, interaction with a laser will enhance the dissociation probabilities. However, if the molecule is already in a dissociating state, the laser will have little effect on its dynamics.

This work was supported by NSERC of Canada and U.S. DOE under contract DE-AC05-84OR21400 with Martin Marietta Energy Systems, Inc. The use of the facilities of the University of Tennessee Computing Center and the University of Waterloo Science VAX under the WATDEC project #S34 is acknowledged.

1. D.W.Noid, M.L.Koszykowski and R.A.Marcus, Ann. Rev. Phys. Chem. 32,267(1981); S.A.Rice, Adv. Chem. Phys. 47,117(1981).
2. S.K.Gray, J.R.Stine and D.W.Noid, Laser Chem. 5,209(1985), and references therein.
3. D.W.Noid, M.L.Koszykowski, R.A.Marcus and J.D.McDonald, Chem. Phys. Lett. 51,540(1977).
4. D.L.Martin and R.E.Wyatt, Chem. Phys. 64,203(1982).
5. R.M.Hedges and W.P.Reinhardt, J. Chem. Phys. 78,3964(1983).
6. R.T.Lawton and M.S.Child, Mol. Phys. 40,773(1980).
7. S.B.Woodruff and D.L.Thompson, J. Chem. Phys. 71,376(1979).

POSSIBLE STOCHASTIC EFFECTS IN MICROWAVE IONIZATION OF HIGHLY EXCITED STATES OF ATOMIC HYDROGEN

James E. Bayfield
University of Pittsburgh, Pittsburgh, PA 15260

ABSTRACT

The classical theory of the nonlinear dynamics of a bound electron in strong external time-oscillating fields predicts stochastic motion of the electron when the field strength and frequency are appropriate. Experiments have been performed to test various predictions of the classical theory. Quantum mechanical calculations have been carried out that exhibit features expected classically and that predict photon absorption probabilities in accord with the experiments. The present quantum picture of classically stochastic motion involves a simultaneous and rapid occurence of a multitude of many-photon absorption processes involving many states of the field-free atom. A scaling of the observed microwave phenomena to optical frequencies is discussed.

CLASSICAL NONLINEAR DYNAMICS OF THE DRIVEN BOUND ELECTRON

Classical nonlinear systems are classified by the topology of the system's trajectories in phase space. A charged particle in combined Coulomb and time-oscillating electric fields is considered to be a conservative, deterministic nonlinear system. As the problem is not integrable, Poincare sections of phase-space (1) and classical probabilities for energy absorption by the atom from the oscillating field (2) are obtained numerically. The Poincare sections exhibit several important features, namely an uppermost KAM surface which is the boundary between regions of phase space that are predominantly quasiperiodic and stochastic, and islands of quasiperiodic motion within the stochastic region. A transition from quasiperiodic motion to stochastic motion occurs as the strength of the oscillating field is increased through threshold values. The stochastic trajectories exhibit a correspondence between sequences of random numbers and sets of time-samplings of the system time-development in some subspace of phase space. As time develops, there is an exponential divergence of "adjacent" trajectories in phase space, associated with an exponential dependence on initial conditions. If the system is started out in a stochastic region, the time-averaged absorption of energy from the oscillating field is "diffusive" in the sense of satisfying a Fokker-Planck equation.

MICROWAVE STUDIES WITH ELECTRICALLY-POLARIZED HYDROGEN ATOMS

Classical calculations using the one-dimensional surface-state electron (SSE) model (1) can be compared with experiments using quasi one-dimensional atoms electrically polarized by an additional external electric field that is static

(3). State-selective laser excitation prepares the atom in a quantum state with parabolic quantum numbers such as n, n_1, n_2, m = 63, 0, 62, 0. This state is maximally stretched out along the static field direction, with the mean electron position relative to the nucleus being about 4000 Angstroms. If the oscillating field is a microwave field polarized along the static field direction, then to a good approximation the many-photon microwave absorption leaves the electron quantum-mechanically in a linear combination of stretched atom states with n_1=m=0 and different n (4), consistent with experimental observations using state-selective static field ionization for quantum-number analysis of the atoms after exposure to the microwave field (3,5). Classically the electron motion is almost completely along a line, and almost stays on that line. Thus experimental measurements of probabilities for n-changing and for ionization can be compared with classical values obtained within the SSE approximation and with quantum values obtained within the stretched-atom (SHA) approximation (5,6). So far there is qualitative agreement for microwave frequencies a fraction of the initial electron orbit frequency. Some theoretical predictions indicate a quantum suppression of classical diffusion when these frequencies are about equal (2), a region yet to be experimentally investigated.

QUANTUM SIGNATURES OF THE BOUNDARIES OF CLASSICALLY STOCHASTIC REGIONS IN PHASE SPACE

In the process of carrying out the numerical many-state quantum calculations of n-changing transition probabilities (6) a quantum threshold for basis-set instability in the final-state probability distribution was noticed (5). The probabilities below some value of quantum number n were basis-set independent for any sufficiently large set, while for higher values of n this was not so. A correlation with the values of action for the location of the uppermost KAM surface in phase space was discovered, leading to an identification of quantum thresholds for basis-set instability with classical thresholds for stochastic motion. Above the quantum threshold, wave functions for the atom in the field are delocalized (6). Table I shows the correlation in quantum number for a number of values of microwave field strength F_μ, static field strength F_s, microwave frequency ω, and initial state quantum number (classical action) n_o. One case showing the variation of thresholds with the initial phase ϕ of the microwave field oscillation relative to the initial conditions for the electron motion is included. The observed significant dependence on ϕ is physically expected classically as the trajectory should depend upon which side of the nucleus the electron starts out on and whether the electron velocity is initially directed towards or against the initial total electric field direction. Such a phase dependence exists for sudden-switching of the microwave field and also when the microwave field is gradually switched on over the time of one hundred field oscillations. This dependence is in keeping with the sensitivity to initial conditions characteristic

TABLE I. Comparison between (a) classical action thresholds for onset of stochastic motion and (b) quantal quantum number thresholds for basis-unstable delocalized behavior.

F_μ(V/Cm)	F_s(V/cm)	ω(GHz)	n_o	(a)	(b)
9.1	8.2	7.1047	63		
		$\phi=0$:		68	69
		$\phi=\pi/2$:		62	62
9.1	0.0	7.1047	63	70	70
17.3	0.0	7.1047	63	67	66
16.0	5.45	7.0061	60	64	65
11.0	0.0	57.0	66	58	58
9.8	0.0	31.58	63	55	54

of classically stochastic motion.

Not only is the classical threshold region for stochastic motion a feature of the quantum results explaining the n-changing probability data, but also the island features representing classical resonances embedded in the stochastic region of the phase space trajectory plots have been observed in experiments (7) and in model quantum calculations of microwave ionization (8). Thus the qualitative features of transition probabilities obtained by classical theory, by quantum theory and by experiment so far have been the same, although a number of classical features such as the n-scaling laws (1) remain to be investigated. The question of whether this experimentally-tested agreement between quantum and classical theory implies the existence of chaos in quantum systems remains open. A deeper investigation may be necessary, requiring the measurement, quantum definition, and quantum calculation of direct measures of "stochasticity" that correspond to classical measures such as the information dimension and the characteristic exponent (9). Questions to be resolved by quantum measurement theory may well arise during such an effort.

SCALING TO OPTICAL FREQUENCIES

A question of interest to the laser science community concerns the possibility that the phenomena discussed above also occur at high intensities for short-pulse laser radiation interacting with more strongly bound electrons in atoms, molecules, and wells in semiconductors. For instance, is there relevance to the atom ionization experiments at 10 microns (10) and at 193 nm (11)? The likely answer is yes, provided that enough photons are absorbed via a sufficiently dense manifold of quantum energy levels, and that quantum bottlenecks inhibiting the absorption do not interfere. As is believed to occur in the infrared dissociation of molecules, laser ionization of ground state atoms will usually involve initial bottlenecks; many-photon ionization of excited atoms has a better chance of being predominately diffusive.

In the limit of no quantum bottlenecks, an approximate

formula for a lower bound for the field intensity threshold for ionization can be obtained from simple average energy-absorption requirements for exposure to a given number of oscillations of the external field. After normalization to the microwave data for highly excited hydrogen, the semiempirical curve (a) in Figure 1 is the result. This is to be compared with curve (b), which is an estimated intensity threshold curve for diffusive n-changing for the case of nearly equal frequencies of field oscillation and electron orbiting. Curve (b) sets the Rabi-flopping frequency equal to the energy level spacing, a condition equivalent classically to the electron's orbit being altered on average every orbital period. The figure indicates a real possibility for interesting work at frequencies much above the microwave region, with past optical ionization experiments being at or above predicted intensity values. The high value shown for the infrared ionization of xenon (10) is explained by a bottleneck arising from the large minimum excitation energy of this atom.

The author acknowledges stimulating discussions with R. V. Jensen, J. N. Bardsley, B. Sundaram and G. Casati, who have provided unpublished information. This work was supported by the U. S. National Science Foundation.

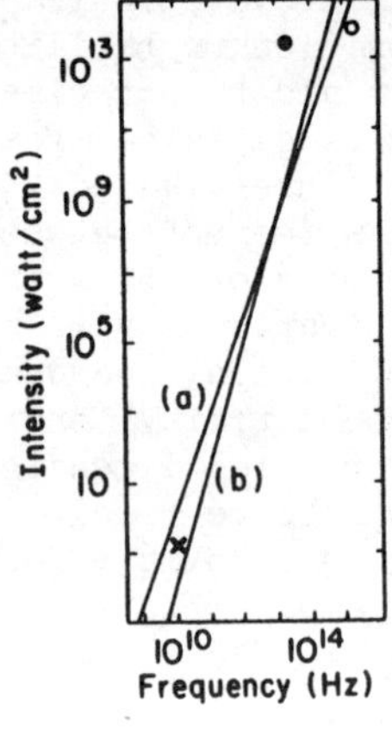

Figure 1.

● Ref. 10
O Ref. 11
X Refs. 3,7

REFERENCES

1. R. V. Jensen, Phys. Rev. Lett. 54, 2057 (1985).
2. G. Casati et al, Phys. Rev. Lett. 53, 2525 (1984).
3. J. E. Bayfield et al, Phys. Rev. Lett. 54, 313 (1985).
4. J. N. Bardsley and B. Sundaram, Phys. Rev. A 32, 689 (1985).
5. J. E. Bayfield, in "Fundamental Aspects of Quantum Theory", V. Gorini and A. Frigerio, eds., Plenum Press, to be published.
6. J. N. Bardsley, B. Sundaram, L. A. Pinnaduwage and J. E. Bayfield, sumbitted for publication.
7. K. A. H. van Leeuwen et al, Phys. Rev. Lett. 55, 2231 (1985).
8. B. Sundaram, Thesis, University of Pittsburgh, 1986.
9. J.-P. Eckmann and D. Ruelle,, Rev. Mod. Phys. 57, 617 (1985).
10. S. L. Chin et al, J. Phys. B <u>18</u>, L213 (1985).
11. T. S. Luk et al, Phys. Rev. Lett. <u>51</u>, 110 (1983).

BISTABILITY AND CHAOS OF LASER-DRIVEN MOLECULES

J. M. Yuan
Drexel University, Philadelphia, PA 19104

M. Tung
Swarthmore College, Swarthmore, PA 19081

G. C. Lie
IBM, Kingston, NY 12401

ABSTRACT

Multiphoton vibrational excitation of molecules in contact with some dissipative mechanisms has been studied by analyzing a classical, semiclassical, and quantum model. Classical solutions of a driven damped Morse oscillator show bistable behavior for a wide range of damping constants, as laser intensity or frequency is adiabatically varied. In the case of a large damping constant and an intense field, period-doubling bifurcations leading to chaotic solutions appear on both the upper and lower branches. Biustable behavior has also been found in analyzing a semiclassical model of Narducci, et al. But in this model instabilities have not been seen unless a field amplitude modulation was introduced. Both classical and semiclassical solutions show that the steady-state values of the average vibrational coordinate (or energy) can be represented by a cusp catastrophe when plotted against laser intensity and frequency. A quantum mechanical study of a Morse oscillator coupled to a reservoir and driven by a classical field has also been carried out. Molecular response shows sudden increases which can be attributed to molecular bistability.

INTRODUCTION

All morning talks so far have been concerned with conservative systems. My talk serves as a turning point to dissipative dynamical systems in this Chaos Symposium. Afternoon talks will be mainly concentrated on dissipative optical systems. Mine still deals with molecular problems.

Most realistic physical, chemical, biological and engineering systems belong to dissipative, nonlinear ones. But only until recent years, a large number of researchers have taken the challenge of solving the nonlinear problems directly.[1,2] New mathematical methods and techniques have to be developed for their solutions, because most linear methods are not directly useful.[3] In general, even small nonlinear dissipative terms in an equation can yield steady-state solutions which are qualitatively different from those of the linear ones. On the other hand, the phenomenology of such nonlinear systems is extremely rich and fascinating, as exemplified by the solutions of the logistic equations.[4] And the same kind of mathematics is applicable to problems of many different fields, such as chemical reactions, lasers, fluids, power systems, biological and medical systems.[2]

Molecular vibrational motion with its well-known nonlinear (or anharmonic) contribution should exhibit some interesting dynamical behavior, characteristic of nonlinear problems. The purpose of this article is to summarize the results that we have obtained in the studies of driven, damped molecular systems. We shall describe in the following sections results obtained by analyzing three models: one classical, one semiclassical, and one quantum mechanical.

CLASSICAL MODEL

We have studied numerical solutions of the following driven damped equation of a Morse oscillator,[4,5]

$$\ddot{x} + \gamma \dot{x} + (1 - e^{-x})\, e^{-x} = A \cos \omega t. \tag{1}$$

To simulate multiphoton excitation of an HF gas under high pressure we have integrated Eq.(1) for γ=0.001, A=0.01 and changing ω. We have found a wide range of ω (from 0.36 to 0.94) for which two different types of steady oscillations exist; one with larger amplitude than the other. The exact steady oscillation that a trajectory is asymptotic to depends on the initial condition chosen. For processes taking place in a condensed phase, we have solved Eq.(1) with γ=0.4. The results are shown in Fig. 1, where the maximal value of the amplitudes of the steady-state oscillations(X_{max}) are plotted as a function of ω for five values of A. Starting from the curve with the smallest A value, A=0.15, we observe a broad peak with a maximum, located close to its intersection with the natural frequency curve (the dashed one). The A=0.25 curve shows similar behavior, except that the peak is more pronounced and a smaller peak appears at about half of the frequency of the first one. At A=0.3 a small bistable region, which implies hysteretic behavior as indicated by arrows, appears near the maximum. This region grows as A increases to 0.45 and at the same time the second peak becomes more pronounced. For A>0.25 X_{max} increases drastically as ω approaches zero, which originates from the fact that a molecule can be dissociated by a DC field with A>0.25. Thick segments of the A=0.45 curve correspond to chaotic behavior and they exist on both upper and lower branches. The scenario to chaos is via period-doubling bifurcations.

SEMICLASSICAL MODEL

Narducci et al.[6] studied the multiphoton vibrational excitation of an anharmonic oscillator whose Hamiltonian is given by

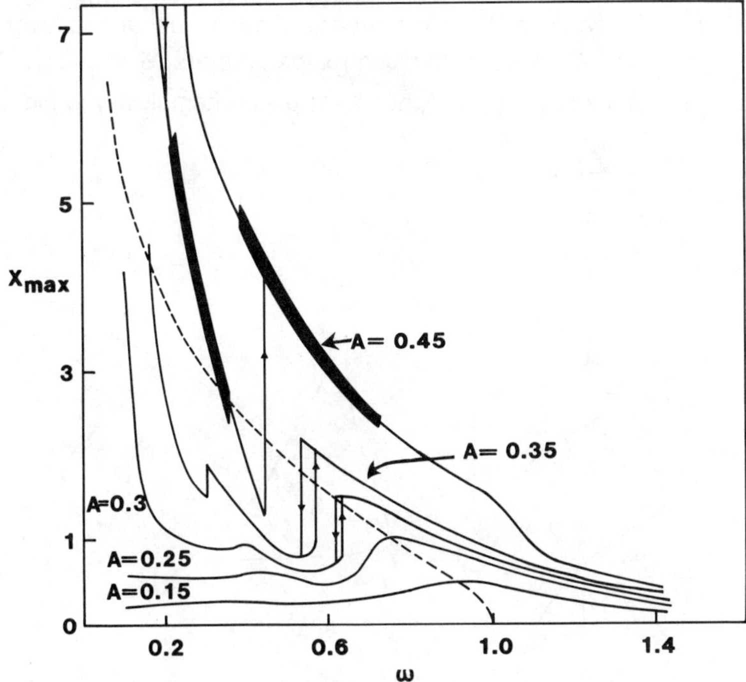

Fig. 1 Classical vibrational amplitude as a function of the driven frequency for several A values.

$$H = \hbar\omega_0 a^\dagger a - \hbar\varepsilon\, a^\dagger a(a^\dagger a + 1) \tag{2}$$

where $a^\dagger$ and a are the ordinary creation and annihilation operators. Based on a generalized master equation with energy and phase relaxation terms they derived a closed set of equations of motion for $\langle a^\dagger\rangle$, $\langle a\rangle$, $\langle a^\dagger a + 1\rangle$ by introducing a factorization ansatz. The model thus derived can be rewritten in the following form

$$\begin{aligned} \dot{X} &= -X + \delta Y - \alpha YZ \\ \dot{Y} &= -\delta X - Y + \alpha XZ - \Omega_R \\ \dot{Z} &= -2\,\Omega_R Y - \lambda Z \end{aligned} \tag{3}$$

where X,Y are the real and imaginary parts of $\langle a\rangle$, namely, quantities proportional to coordinate and momentum. $Z=\langle a^\dagger a\rangle$ is the average vibrational excitation and δ, α, Ω_R and λ are the scaled detuning, anharmonicity, Rabi rate and the longitudinal relaxation time. Steady states of Eq.(3) can be represented graphically by plotting the steady state value of the average vibrational excitation, Z_0, versus the control parameters, δ and Ω_R. The resulting surface forms a cusp catastrophe[7] as shown schematically in Fig. 2.

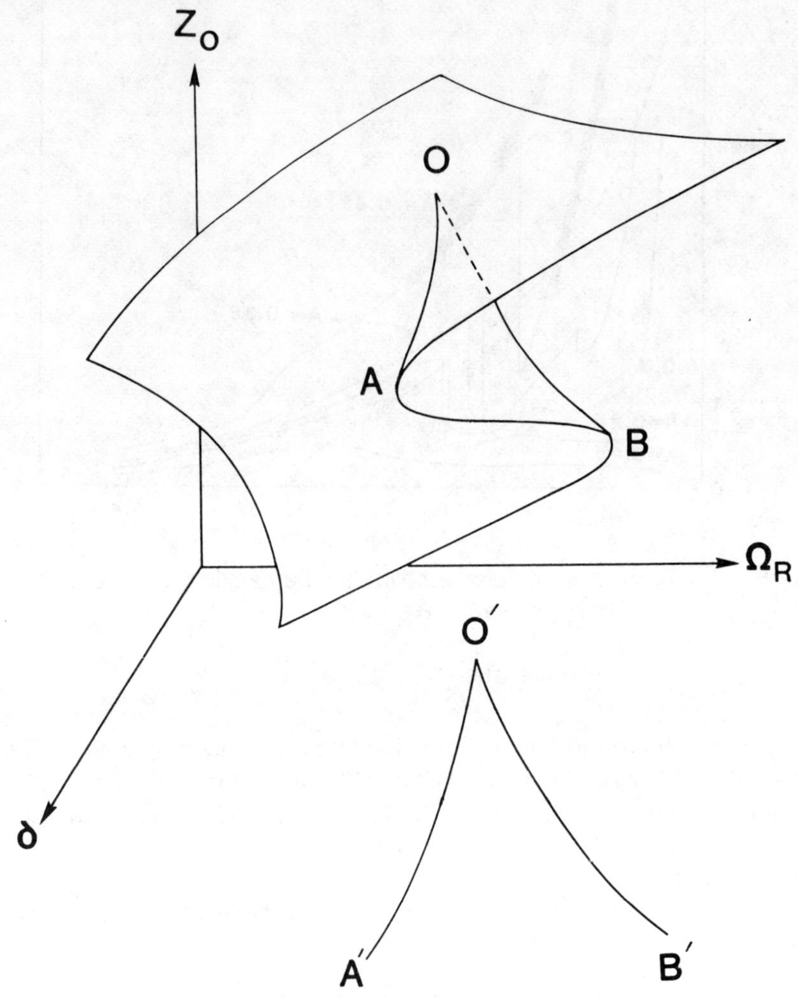

Fig. 2 Average vibrational excitation plotted as a function of the detuning and Rabi rate.

We observe that the λ-shaped curves around the bistable domains in Fig. 1 are similar to those obtained by cutting the cusp catastrophe surface with a constant Ω_R-plane. The distinct difference is that both upper and lower branches remain stable and bifurcations or chaos has not been found in the semiclassical model. However, the upper branch does exhibit instabilities,[8] similar to the kind found in the classical solutions, when a sinusoidal modulation is introduced to Ω_R or the field amplitude.

QUANTUM MECHANICAL STUDY

We have also investigated the dynamical behavior of a quantum model of a Morse oscillator, whose Hamiltonian can be expresssed as[9]

$$H = \hbar\omega_0(A^+A^- + I_0/2) \ , \tag{4}$$

where A^+, A^- and I_0 are the generators of an SU(2) Lie algebra. Dynamical equations of motion have been derived for this oscillator in coupling to a bath of harmonic oscillators[10] and driven by a classical field.[11] Sudden change in the oscillator energy has been found in the solutions obtained by numerically integrating the time evolution equations of the reduced density matrix elements directly.[11] This sudden change can be attributed to the molecular bistabiltiy seen in the classical and semiclassical models. However, quantum solutions do not exhibit hysterisis found in the classical and semicalsssical solutions. Details of this study will be published later.

REFERENCES

1. E. Ott, Rev. Mod. Phys. 53, 655 (1981).
2. N. B. Abraham, J. P. Gollub, and H. L. Swinney, Physica 11D, 252 (1984).
3. J. P. eckman and D. Ruelle, Rev. Mod. Phys. 57, 617 (1985).
4. M. J. Feigenbaum, J. Stat. Phys. 19, 25 (1978); 21, 669 (1979).
5. G. C. Lie and J. M. Yuan, J. Chem. Phys., Submitted.
6. L. M. Narducci, S. S. Mitra, R. A. Shatas, and C. A. Coulten, Phys. Rev. A16, 287 (1977).
7. J. M. Yuan, E. Liu, and M. Tung, J. Chem. Phys. 79, 5034 (1983).
8. E. Liu and J. M. Yuan, Phys. Rev. A29, 2257 (1984).
9. R. O. Levine, Chem. Phys. Lett. 95, 87 (1983).
10. M. Tung, E. Eschenazi, and J. M. Yuan, Chem. Phys. Lett. 115, 405 (1985).
11. M. Tung, Ph.D. Dissertation, Drexel University, 1985.

SINGLE AND MULTI-MODE INSTABILITIES IN LASER SYSTEMS

L.A. Lugiato
Dipartimento di Fisica, Universita' di Milano
Milano, 20133 Italy

D.K. Bandy, J.R. Tredicce and L.M. Narducci
Physics Department, Drexel University
Philadelphia, PA 19104

The appearance of random spiking in the output of a ruby laser[1] is one of the best known examples of the propensity of this device for developing noisy and uncontrollable behavior, even under the action of a nearly continuous pump. This aspect of laser operation has recently become the focus of systematic investigations by a large segment of the laser community which is now actively engaged in the study of effects that were originally viewed as mere nuisances[2,3].

One of the main reasons for the resurgence of interest in the anomalous behavior of continuously pumped lasers is the growing belief that the output pulsations are rooted in and governed by the same nonlinear dynamics that is responsible for laser action. Added impetus is provided by the recently acquired tools and perspectives on how to deal with nonlinear dynamical systems, and by the realization that the emergence of both spatial and temporal patterns is not only typical of lasers, but also of innumerable other systems in physics, chemistry, biology and the engineering sciences[4].

Thus, it is not surprising that after Casperson's pioneering studies of the 1970's[5] many other groups have invested their best efforts in the study of laser instabilities[6].

From a theoretical viewpoint, the situation is still far from being resolved. In fact, it is not even clear to what extent this is the fault of the theoretical models. Over the years, we have come to accept the notion that an adequate understanding of these problems requires the development of a theory of laser action that is based on the coupled dynamics of the atomic polarization and population variables, in addition to a suitable field equation. This is not a universal requirement because theories of the laser with a saturable absorber, for example, have been rather successful in describing bistable and unstable behavior, even in the rate equation approximation[7].

In dealing with the free running laser, on the other hand, the coherent equations are presently viewed as the most promising, or at least the most natural, starting point. In this brief review, we focus on a class of semiclassical models in which the active atoms behave as identical two-level systems (homogeneously broadened medium) and the laser field is confined by a unidirectional resonators, in the plane-wave approximation[8].

1. THE SEMICLASSICAL LASER EQUATIONS - MULTIMODE INSTABILITIES

We consider a ring laser resonator of length $\mathcal{L}$, confined by two mirrors of reflectivity R and by additional ideal reflectors to close the loop; the active atoms are contained in a cylindrical region of length L. For convenience, the longitudinal cavity mode (ω_C) that lies nearest the center of the atomic line (ω_A) will be called the resonant mode; all the others will be identified as nonresonant modes.

The coherent semiclassical laser equations are the well known Maxwell-Bloch equations for the space-time dependent amplitude $\mathcal{F}(z,t)$ of the electric field and for the atomic polarization $\mathcal{P}(z,t)$ and population difference $\mathcal{D}(z,t)$

$$\frac{\partial \mathcal{F}}{\partial z} + \frac{1}{c}\frac{\partial \mathcal{F}}{\partial t} = - \alpha \mathcal{P} \tag{1.1a}$$

$$\frac{\partial \mathcal{P}}{\partial t} = \gamma_{\perp} [\mathcal{F}\mathcal{D} - (1+i\delta_{AC})\mathcal{P}] \tag{1.1b}$$

$$\frac{\partial \mathcal{D}}{\partial t} = \gamma_{\parallel}\left\{\frac{1}{2} (\mathcal{F}^*\mathcal{P} + \mathcal{F}\mathcal{P}^*) + \mathcal{D} + 1\right\} \tag{1.1c}$$

In Eqs. (1.1) $\gamma_{\parallel}$ and $\gamma_{\perp}$ are the relaxation rates for the atomic population difference and polarization, respectively, α denotes the unsaturated gain per unit length, δ_{AC} is the detuning between the atomic line and the resonant mode, in units of $\gamma_{\perp}$. The evolution equations are supplemented by the boundary conditions

$$\mathcal{F}(0,L) = R\mathcal{F}(L,t-(\mathcal{L}-L)/c) \tag{1.2}$$

The search for the stationary solutions of the Maxwell-Bloch equations shows the existence of a multiplicity of possible solutions, each characterized by its own output intensity

$$|\mathcal{F}_j(L)|^2 = 2/(1-R^2)\ \{\alpha L + (1+\Delta_j^2)\ \ell n(R)\} \tag{1.3}$$

and by the operating frequency

$$\omega_j = \{(\omega_C + j\alpha_1)\ \gamma_{\perp} + \omega_A \kappa\}/(\gamma_{\perp}+\kappa) \tag{1.4}$$

where $\Delta_j = (\omega_A-\omega_j)/\gamma_{\perp}$, κ is the linewidth of the cavity resonances and $j=0,\pm1,\pm2\cdots$.

In the past, it has been common to assume that the steady state with the highest gain would be the one to be actually excited[9]. This view is not always correct for homogeneously broadened lasers, as we show in this paper.

If the intermode spacing is larger than the power broadened gain profile, only one steady state solution at a time satisfies the threshold condition. Hence, a detuning scan leads to periodic and continuous variations of the output intensity and operating frequency, as adjacent cavity modes fall under the gain curve and become active. This fact is intuitively obvious, and is confirmed by Eq. (1.3).

If one performs a linear stability analysis of the steady state configurations and records the real parts of the eigenvalues of the linearized problem, one finds that the system is stable for low enough values of the gain. At higher values of the pump parameter, one of the sidebands of the resonant mode may become unstable and, in this case, it will grow and interfere with the freely running steady state field. After a transient evolution, steady oscillations develop with a pulsing frequency approximately equal to the spacing between the unstable sideband and the resonant mode. This instability was discovered, independently, by Risken and Nummedal[10] and by Graham and Haken[11] in a ring laser whose unsaturated gain per pass and mirror transmittivity T satisfy the conditions

$$\alpha L \rightarrow 0, \quad T \rightarrow 0, \quad \alpha L/T \equiv 2C = \text{const} \tag{1.5}$$

(For a more general formulation, see Ref. 12.) Because the eigenvalue that develops a positive real part in this case can be associated with the field amplitude in the linearized equations, it will be convenient to label this phenomenon amplitude instability. Its threshold is quite high in the sense that the gain parameter is usually many times higher than the ordinary laser threshold value. Upon increasing the detuning parameter δ_{AC} the pulsing instability threshold grows also[13].

If the intermode spacing is made comparable to or even smaller than the power broadened atomic line, coexisting steady states develop as shown by the detuning scans of the steady state output intensity and laser operating frequency, Eq. (1.3,4). In this case, (Fig. 1a,b), we have two possible scenarios[14]: *(1) the active laser mode retains control of the operation well into the coexistence region; (2) the two coexisting steady states compete with one another*. In case (1), one observes intensity and frequency switching between different steady states, and hysteretic behavior on reversing the direction of the detuning scan. Case (2), instead, gives rise to self-pulsing. Both mode-switching and self-pulsing are caused by the development of a positive real part in the phase eigenvalue; for this reason, we label this effect phase instability[14]. This type of behavior is much more readily accessible to experimental verification because the gain requirements are much more modest than in the case of an amplitude instability, and in fact, it has been seen in a CO_2 laser operating only slightly above the ordinary laser threshold[15].

Note that both amplitude and phase instabilities are multimode phenomena. The amplitude instability is a high gain effect that displays a continuous behavior. The phase instability can be a low threshold effect and produce both discontinuous and hysteretic behavior (bistability) or, continuous variations. Lasers such as CO_2 and Nd:YAG, for example, whose ratio $\gamma_{\parallel}/\gamma_{\perp}$ is much smaller than unity, are predicted to be sources of discontinuous mode switching, while systems for which $\gamma_{\parallel} \approx \gamma_{\perp}$ are typically sources of mode-mode competition leading to self-pulsing.

2. SINGLE MODE INSTABILITY

Probably the simplest model of a laser system is that of a single mode unidirectional ring laser containing a two-level resonant medium. For many years, this model has been used as the starting point for the description of homogeneously and inhomogeneously

FIGURE 1a. Modulus of the steady state field amplitude as a function of the detuning parameter δ_{AC}. Two different steady states coexist over a finite range of operating conditions. A priori, one cannot predict what kind of dynamical evolution will develop in the coexistence region. For the selected ratio $\gamma_{\parallel}/\gamma_{\perp}$ this calculation predicts discontinuous mode switching and hysteresis. The intermode spacing is $\alpha_1 = 2$.

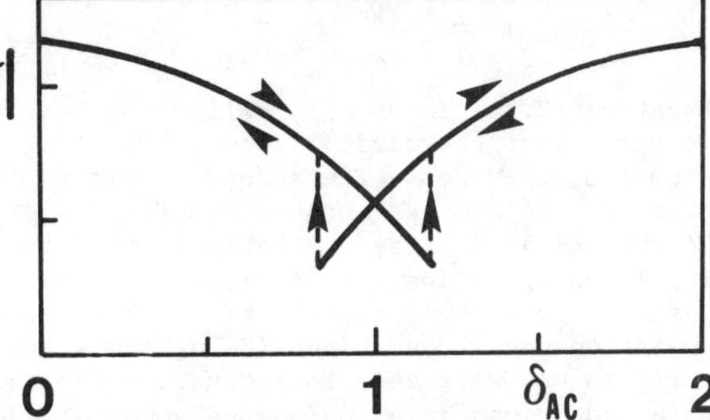

FIGURE 1b. Frequency offset $\delta\omega$ of the operating laser field from the reference cavity mode as a function of the detuning parameter δ_{AC}. For the selected parameters, this calculation predicts discontinuous frequency switching and hysteresis.

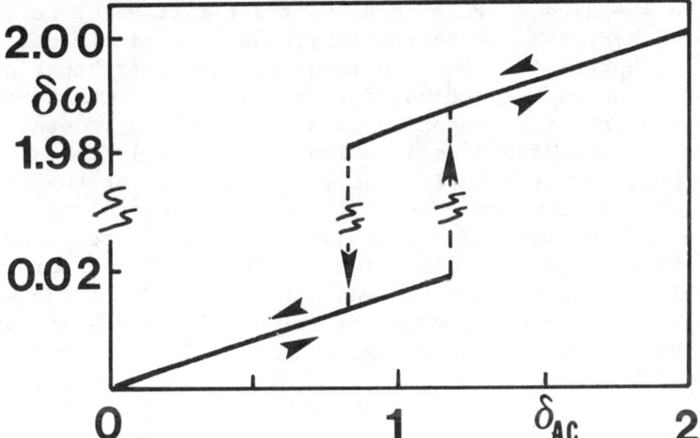

broadened active media, resonant or nonresonant laser cavities, with plane or Gaussian internal field profiles. The earliest studies of homogeneously broadened unidirectional resonators were carried out by Uspenskii[16], Grazyuk and Oraevskii[17] and Haken[18] who discovered the existence of operating conditions leading to instabilities and pulsations. The relation between the single mode model and the Maxwell-Bloch equations (1.1) and, in particular, the conditions for the validity of the single mode equations are[19]: *(i) the mean field limit must be a valid approximation, (ii) all the modal amplitudes must be initially zero, except for the resonant one, and (iii) the nonresonant modes must be stable.* In this case, the single mode equations take the form (in resonance)

$$\dot{f} = -\kappa\ (f + 2Cp) \tag{2.1a}$$

$$\dot{p} = \gamma_{\perp}(fd - p) \tag{2.1b}$$

$$\dot{d} = -\gamma_{\parallel}(fp + d + 1) \tag{2.1c}$$

where f,p,d are the single mode amplitudes of the field, polarization and population difference, respectively. An instability develops if the following two conditions are met[16-18]

i. $\kappa > \gamma_{\perp} + \gamma_{\parallel}$ (2.2a)

ii. $2C > 1 + (\kappa + \gamma_{\parallel} + \gamma_{\perp})(\kappa + \gamma_{\perp})/(\kappa - \gamma_{\parallel} - \gamma_{\perp})$ (2.2b)

Equation (2.2a) is usually called the bad cavity condition: it requires that the cavity linewidth be larger than the sum of the atomic decay rates. The second condition (2.2b) requires, as in the case of a multimode instability of the amplitude type, that the gain be sufficiently large. Because 2C = 1 is the threshold condition for laser operation, it is clear that one is faced with a high gain instability. The first experimental evidence of single mode laser operation under conditions (2.2a,b) was provided by Weiss and collaborators who were able to match both the bad cavity and the high gain conditions in a CO_2-pumped ammonia laser, while, at the same time, keeping the inhomogeneous Stark broadening effects to a minimum[20].

From a theoretical viewpoint, the dynamical behavior of Eqs.(2.1) has been investigated in great detail. These equations are isomorphic[21] to the so-called Lorenz model[22], first introduced to describe the onset of a hydrodynamic instability, and extensively analyzed in a monograph by Sparrow[23]. For the range of parameters that are typical of a laser and as a function of increasing gain, the time-dependent field, at first, approaches a steady state monotonically, then it develops relaxation oscillations, and finally falls prey to a strange attractor. For sufficiently small values of the ratio $\gamma_{\parallel}/\gamma_{\perp}$, the chaotic attractor is replaced by ordinary limit cycles and by a complicated sequence of period doubling bifurcations and hysteresis effects between coexisting steady states and pulsing solutions[24]. This regime of operation appears to be consistent with the setting employed by Weiss[20], whose experimental results are qualitatively similar to some of these predictions.

3. CONCLUSIONS

The relevance of this survey to real laser systems is still an open question. Apart from Weiss' single mode experiments[20] and Tredicce's CO_2 laser work[15], there is still limited experimental confirmation of the theoretical predictions of the Maxwell-Bloch equations for a homogeneously broadened laser. A spectacular multimode instability was discovered by Hillman and collaborators[25], not far above the ordinary laser threshold in a cw-pumped dye laser operating in the good cavity limit. Under certain conditions, this instability is characterized by discontinuous changes of the average dye laser output power, by the simultaneous emergence of symmetric sidebands, and by the disappearance of the laser resonant spectral component. It is difficult to explain these results in terms of our current understanding of the Risken-Nummedal[10] and Graham-Haken[11] models because the dye laser gain required to trigger this effect is only slightly larger than what is required to produce normal laser action, and because the resonant laser mode is apparently suppressed discontinuously by the emerging sidebands for increasing values of the pump parameter.

One is led to believe that essential physical features have been omitted from the traditional Maxwell-Bloch equations. A common approximation in nearly all studies of this problem is the plane-wave assumption. Attempts to include transverse effects in the Maxwell-Bloch model for an active medium[26] have met thus far with very limited success, apart from the recognition that the pattern of unstable behavior is altered in the presence of a transverse field profile.

This research has been partially supported by a contract with the US Army Research Office, and a grant from the Italian National Research Council (CNR).

REFERENCES

1. R.J. Collins, D.F. Nelson, A.L. Shawlow, W. Bond, C.G.B. Garrett and W. Kaiser, Phys. Rev. Lett. 5, 303 (1960).
2. N.B. Abraham, L.A. Lugiato and L.M. Narducci, J. Opt. Soc. Am. B2, 7 (1985).
3. Special issue on Instabilities in Active Optical Media, J. Opt. Soc. Am. B2, Jan. 1985, edited by N.B. Abraham, L.A. Lugiato and L.M. Narducci.
4. H. Haken, Synergetics: An Introduction, 3rd Ed. (Springer-Verlag, Berlin:1983).
5. L.W. Casperson and A. Yariv, IEEE J. Quant. Elect. QE8, 69 (1972); L.W. Casperson, IEEE J. Quant. Elect. QE14, 756 (1978). For a recent update and new results, see L.W. Casperson, J. Opt. Soc. Am. B2, 62 and 73 (1985).
6. R.B. Boyd, M. Raymer and L.M. Narducci, Eds., Proceedings of the International Conference on Instabilities and Dynamics of Lasers and Nonlinear Optical Systems (Cambridge University Press)[in press].
7. E. Arimondo, F. Casagrande, L.A. Lugiato and P. Glorieux, Appl. Phys. B30, 57 (1983).
8. L.A. Lugiato and L.M. Narducci, Phys. Rev. A32, 1576 (1985).
9. See, for example, O. Svelto, Principles of Lasers, 2nd Ed. (Plenum Press, NY:1982).
10. H. Risken and K. Nummedal, J. Appl. Phys. 39, 4662 (1968).
11. R. Graham and H. Haken, Z. Phys. 213, 240 (1968).
12. L.A. Lugiato, L.M. Narducci, E.V. Eschenazi, D.K. Bandy and N.B. Abraham, Phys. Rev. A32, 1563 (1985).
13. J. Zorell, Opt. Comm. 38, 127 (1981).
14. L.M. Narducci, J.R. Tredicce, L.A. Lugiato, N.B. Abraham and D.K. Bandy, Phys. Rev. A [to be published].
15. J.R. Tredicce, L.M. Narducci, N.B. Abraham, D.K. Bandy and L.A. Lugiato, Opt. Comm. [to be published].
16. A.V. Uspenskii, Radio Eng. Electron. Phys. (USSR) 8, 1145 (1963); ibid. 9, 605 (1964).
17. A.Z. Grazyuk and A.N. Oraevskii in Quantum Electronics and Coherent Light, P.A. Miles, Ed. (Acad. Press, NY:1964) p.162.
18. H. Haken, Z. Phys. 190, 327 (1966).

19. L.A. Lugiato, L.M. Narducci, D.K. Bandy and J.R. Tredicce, Phys. Rev. A [to be published].
20. See, for example, C.O. Weiss, J. Opt. Soc. Am. B2, 137 (1985) and references quoted therein.
21. H. Haken, Phys. Lett. 53A, 77 (1975).
22. E.N. Lorenz, J. Atm. Sci. 20, 130 (1963).
23. C.T. Sparrow, The Lorenz Equations: Bifurcation, Chaos and Strange Attractors (Springer-Verlag, Berlin:1982).
24. L.M. Narducci, H. Sadiky, L.A. Lugiato and N.B. Abraham, Opt. Comm. 55, 370 (1985).
25. L.W. Hillman, J. Krasinsky, R.B. Boyd and C.R. Stroud, Jr., Phys. Rev. Lett. 52, 1605 (1984). See also C.R. Stroud, Jr., K. Koch and S. Chakmakjian, in Proceedings of the International Conference on Instabilities and Dynamics of Lasers and Nonlinear Optical Systems, R.B. Boyd, M. Raymer and L.M. Narducci, Eds., (Cambridge University Press) [in press].
26. L.A. Lugiato and M. Milani, Z. Phys. B50, 171 (1983); L.A. Lugiato, R.J. Horowicz, G. Strini, L.M. Narducci, Phys. Rev. A30, 1366 (1984).

INSTABILITIES AND CHAOTIC EMISSION OF FAR-INFRARED NH_3-LASERS

C. O. Weiss, W. Klische
Phys.-Techn. Bundesanstalt, 3300 Braunschweig, F.R.G.

ABSTRACT

We have used optically pumped NH_3-far-infrared lasers to investigate self pulsing instabilities and transition to chaotic emission, theoretically predicted to occur for homogeneously broadened lasers high above threshold when the 3 characteristic times of the laser: decay time of population, polarization, and field, are of comparable magnitude. We have observed pulsing instabilities and chaotic emission from these systems for the first time and find qualitative agreement with the laser dynamics predicted from the simplest laser model (Lorenz equations).

For homogeneously broadened lasers with comparable decay times of polarization, population and field the three basic laser equations show an instability of the cw solution at high pump strengths and transition to periodic or chaotic pulsing [1, 2]. It has been argued in the past that this instability will not occur in real lasers since the comparability of polarization and field decay times is equivalent for common lasers (like CO_2, YAG Ruby) to a very high resonator loss [3].

We have reconsidered the possiblity of occurrence of these instabilities in a real laser. Very large polarization decay times (small homogeneous linewidths) can occur in low pressure gas lasers. Homogeneous broadening conditions can in these cases be created by narrow band optical pumping (i.e. laser pumping). Lasers in the far infrared with a few thousand laser transitions between 30 μm and 1000 μm wavelength are of this type.

Self-pulsing of FIR lasers had already been reported before (termed "relaxation oscillation") [4]. We have verified first that self-pulsing occurs in a popular FIR laser (CH_2F_2) [5, 6]. Additionally, period doubling and continuous spectra of the laser power output were observed indicating chaotic emission. Since the parameters describing the laser in this first experiment were badly defined (strongly spatially inhomogeneous conditions in the laser resonator), we have set up FIR laser systems more closely matching the simplest theoretical model of a laser [1] (Lorenz model). NH_3 (spectroscopically well studied) was used as the laser medium, which shows high gain due to its small partition function.

Fig. 1 shows a ring resonator used for measurements on the (aR 7,7) laser transition of $^{14}NH_3$ (81 μm wavelength). Optically pumped gas lasers can work unidirectionally in a ring if pumped off resonance[7]. The resonator has none of the usual coupling holes of FIR lasers and therefore achieves a maximum of homogeneity of pump field and generated FIR laser field. If homogeneous broadening in conditions are to be realized in a laser pumped gas laser, coherence effects by the pump laser must be avoided. This means that the pumpfield must not exceed a value at which roughly the pump Rabi frequency equals the homogeneous linewidth[8].

With this resonator, we have observed a classical period doubling transition to chaos[9] (Fig. 2) when the pump laser frequency was fixed (pumping off $^{14}NH_3$ line center) and the ring laser resonator was tuned towards the line center. This behaviour is also predicted for the Lorenz equation, however we have to be careful in our case since these instabilities occured for pump fields where the pump Rabi frequency was roughly equal to the homogeneous linewidth, thus strictly homogeneous broadening was not achieved.

The well known "period 3 window" in the chaotic range was also observable (Fig. 3).

In order to reduce pump thresholds to achieve more homogeneously broadened conditions we have then used the aR(2,0) $^{15}NH_3$ laser transition at 374 μm wavelength, which has higher gain at comparable pump Rabi frequency.

For unidirectional operation we have in this case found only periodic self pulsing, which is consistent with calculations[2] since the relaxation constants of this laser transition are different from the values which had been used before in the Lorenz equation.

In the case of bidirectional emission, interesting phase-locking effects were observed with subsequent period-doubling to chaos (including observation of clear period 3 and 5 windows)[11] in complete analogy with results on Benard convection cells[12].

Further measurements have shown[13] that under homogeneous broadening conditions indeed the instability behaviour as predicted from the Lorenz-equation is observed: We find the corresponding high thresholds (Fig. 4), and the twofold splitting of the laser line, predicted as an unusual feature of the solutions of the Lorenz laser equations, was found by heterodyne measurements using a 70 GHz Klystron harmonic.

Although the lasers used do not correspond exactly to the basic laser-Lorenz equations, since there is unavoidably a spatial inhomogeneity of all parameters present (Gaussian structure of pump and laser field and the fact that the pump beam is absorbed in the laser gas). We find nevertheless that the instabilities of the Lorenz model predicted 10 years ago do exist and the experiments show the qualitative features predicted theoretically.

REFERENCES

1. H. Haken, Phys. Lett. 53A, 77 (1975).
2. L. M. Narducci, H. Sadiky, L. A. Lugiato, D. Bandy, N. Abraham, Opt. Comm 55, 370 (1985).
3. L. W. Casperson in Lecture Notes in Physics, Vol. 182, Springer Verlag 1983, p. 88.
4. see e.g. N. M. Lawandy, G. Koepf, IEEE Journ. Quant. El. QE-16, 701 (1982).
5. C. O. Weiss, Journ. Opt. Soc. Amer. B2, 137 (1985).
6. E. J. Danielewicz, C.O. Weiss, IEEE Journ. Quant. El. QE-14, 705 (1978).
7. J. Heppner, C. O. Weiss, Appl. Phys. Lett. 33, 590 (1978).
8. J. Heppner, C. O. Weiss, U. Hübner, G. Schinn, IEEE Journ. Quant. El. QE-10, 392 (1980).
9. C. O. Weiss, W. Klische, P.S. Ering, M. Cooper, Opt. Comm. 52, 405 (1985).
10. H. Zeghlache, P. Mandel, J. Opt. Soc. Amer. B2, 18 (1985).
11. W. Klische, C. O. Weiss, Phys. Rev. A-31, 4049 (1985).
12. J. Maurer, A. Libchaber, J. Phys. (Paris) Lett. 40, 419 (1979).
    A. Libchaber, J. Maurer, J. Phys. (Paris) Colloque 41, C3-51 (1980).
13. E. M. H. Hogenboom, W. Klische, C. O. Weiss, A. Godone, Phys. Rev. Lett. 55, 2571 (1985).

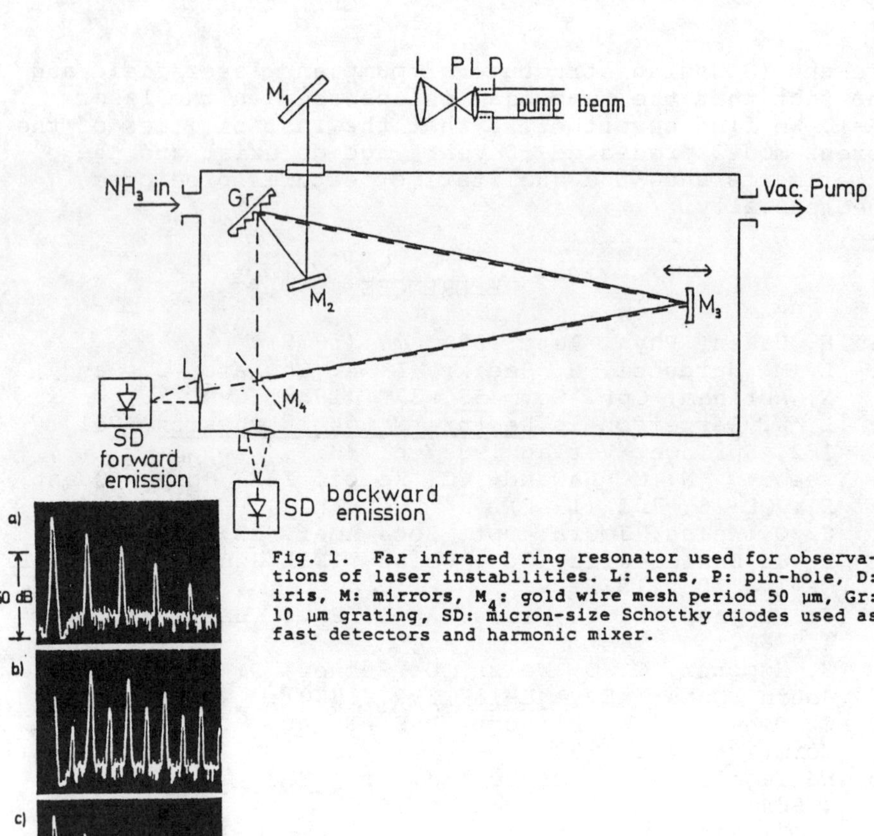

Fig. 1. Far infrared ring resonator used for observations of laser instabilities. L: lens, P: pin-hole, D: iris, M: mirrors, M_4: gold wire mesh period 50 µm, Gr: 10 µm grating, SD: micron-size Schottky diodes used as fast detectors and harmonic mixer.

a) b) c) d) e) f)

30 dB

0 1 2 3 4 MHz

Fig. 2. Period-doubling sequence as aR(7,7) 81 µm NH_3 laser resonator, emitting unidirectionally, is tuned towards gainline center (Intensity spectra) a: self-pulsing, b, c, d period-doublings. A further period doubling seems to be resolved in e. f: spectrum of chaotic laser emissions: the residual lines in the noise spectrum reflect the fact that only the laserpulse intensity is varying chaotically, the pulse frequency is fairly stable (see Fig. 3).

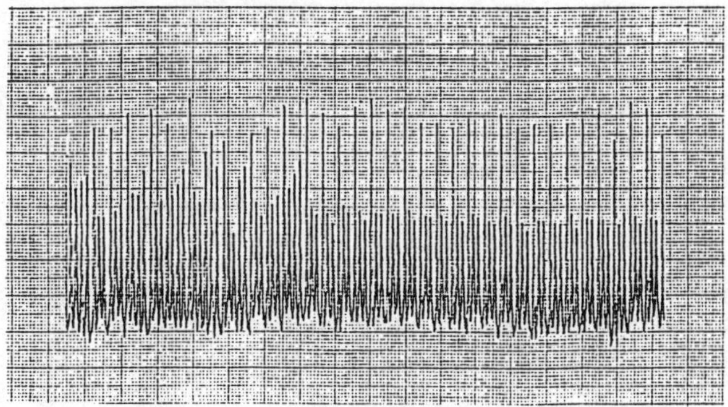

Fig. 3. Transition of the laser from chaotic emission to regular period-3 pulsing ("period-3 window" in the chaotic range) Laser intensity vs. time representation. Fundamental pulsing frequency is ~ 1.2 MHz.

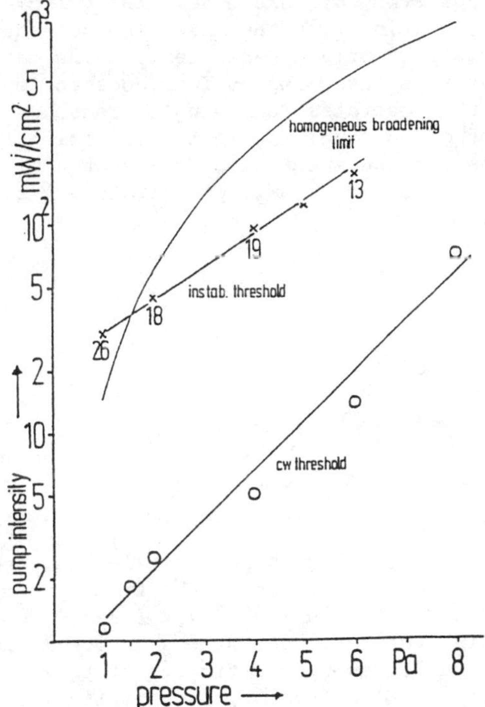

Fig. 4. Instability threshold and cw-laser threshold as a function of pressure. The cw threshold increases exponentially with laser gas pressure due to back-ground absorption in the gas.

The ratios of instability pump threshold to cw pump threshold (noted with the instability threshold curve) are in the right range for the Lorenz equations. Limit of homogeneous broadening is indicated.

THE SINGLE-MODE LASER MODEL: AN IMPROVED APPROXIMATION

L.A. Lugiato
Istituto di Fisica, Universita' di Milano, Milano 20133 Italy

D.K. Bandy, E.P. Dougherty, L.M. Narducci and J.R. Tredicce
Physics Department, Drexel University, Philadelphia, PA 19104

The single-mode model of a homogeneously broadened ring laser is the oldest and one of the most widely analyzed approximations for the description of laser action[1]. This model predicts stable operation above the so-called first laser threshold and allows the possibility of spontaneous self-pulsing if the cavity linewidth (κ) exceeds the sum of the polarization ($\gamma_\perp$) and population ($\gamma_\parallel$) relaxation rates, and if the gain of the active medium is larger than a critical value (second laser threshold).

In spite of the popularity of this model, the validity conditions and, especially, its connection with the Maxwell-Bloch equations have not been clarified in sufficiently general terms. The main purpose of this note is to address this issue and to introduce an improved description for a laser that operates as a single frequency device.

The starting point of our analysis is the traditional set of Maxwell-Bloch equations for the space-time dependent amplitude $\mathcal{F}(z,t)$ of the electric field and for the atomic polarization $\mathcal{P}(z,t)$ and population difference $\mathcal{D}(z,t)$

$$\frac{\partial \mathcal{F}}{\partial t} + \frac{1}{c}\frac{\partial \mathcal{F}}{\partial t} = -\alpha \mathcal{P} \tag{1a}$$

$$\frac{\partial \mathcal{P}}{\partial t} = \gamma_\perp [\mathcal{F}\mathcal{D} - (1 + i\delta_{AC})\mathcal{P}] \tag{1b}$$

$$\frac{\partial \mathcal{D}}{\partial t} = -\gamma_\parallel \left\{ \frac{1}{2}(\mathcal{F}^*\mathcal{P} + \mathcal{F}\mathcal{P}^*) + \mathcal{D} + 1 \right\} \tag{1c}$$

supplemented by the boundary conditions

$$\mathcal{F}(0,t) = R\mathcal{F}[L, t-(\mathcal{L}-1)/c] \tag{2}$$

In Eqs. (1), α is the unsaturated gain per unit length, and δ_{AC} is the frequency separation between the center of the atomic resonance and a selected reference cavity mode in units of $\gamma_\perp$; $\mathcal{L}$ and L represent the length of the entire ring resonator and of the active medium, respectively, while R is the reflectivity coefficient of two of the mirrors, the remaining ones being ideal reflector, for simplicity. As shown in Ref. 2, a rigorous modal expansion of Eqs. (1) can be carried out in the new reference system defined by the coordinates $z'=z$ and $t'=t + (\mathcal{L}-L)z/cL$ and in terms of the new field and atomic variables:

$$F(z',t') = \mathcal{F}(z',t') \exp(\ln(R)z'/L) \tag{3a}$$
$$P(z',t') = \mathcal{P}(z',t') \exp(\ln(R)z'/L) \tag{3b}$$
$$D(z',t') = \mathcal{D}(z',t') \tag{3c}$$

The resulting coupled modal equations become equivalent to the traditional single-mode equations (in resonance)

$$\dot{f} = -\kappa(f + 2Cp) \quad (4a)$$
$$\dot{p} = \gamma_{\perp}(fd - p) \quad (4b)$$
$$\dot{d} = -\gamma_{\parallel}(fp + d + 1) \quad (4c)$$

where f,p,d are the single-mode amplitudes of the field, polarization and population difference, respectively, (i) in the double limit

$$\alpha L \rightarrow 0,\ R \rightarrow 1 \text{ with } 2C = \alpha L/|\ln(R)| = \text{const}; \quad (5)$$

(ii) under the assumption that all the modal amplitudes other than those of the resonant mode are initially zero; and (iii) if multimode instabilities are ruled out for the selected operating conditions.

The steady state longitudinal field profile $F_{st}(z')$ tends to remain quite uniform, even for parameter values that are rather removed from the ideal limit (5), but this is not the case for the modulus of the atomic polarization $P_{st}(z')$ and for the population $D_{st}(z')$. For this reason, we maintain the single-mode approximation for the field, but do not carry out the modal expansion for the atomic variables. In this case, after setting

$$P(z',t') = e^{i\delta\Omega t'}\ p(z',t') \quad (6a)$$
$$D(z',t') = d(z',t') \quad (6b)$$

the new coupled equations of motion take the form[3]

$$\frac{df}{dt'} = i\delta\Omega f - \kappa\left(f + 2C\,\frac{1}{L}\int_0^L dz'\ p(z',t')\right) \quad (7a)$$

$$\frac{dp}{dt'} = \gamma_{\perp}\left\{df - [1 + i(\delta_{AC} - \delta\Omega)]p\right\} \quad (7b)$$

$$\frac{dd}{dt'} = -\gamma_{\parallel}\left[\frac{1}{2}\,(f^*p + fp^*)\ e^{2(z'/L)|\ln(R)|} + d + 1\right] \quad (7c)$$

The main differences between the original Maxwell-Bloch equations and Eqs. (7) are evident by inspection: we have neglected the longitudinal dependence of the field, so that the field modal amplitude satisfies an ordinary, rather than a partial differential equation; the spatial dependence of the atomic variables is maintained, but the variable z' appears only parametrically in the equations of motion.

The strategy employed in this approach is responsible for a significant improvement relative to the predictions of the conventional single-mode equations, even for values of αL of the order of a few units. This is quite unlike the behavior of the "mean field approximation" (Eqs. (4)) which loses accuracy quickly as αL becomes larger than the first laser threshold.

The situation remains quite satisfactory, even for parameter values where the Maxwell-Bloch equations predict the emergence of unstable behavior. In general, instabilities of the single-mode type tend to produce chaotic pulsations; in this case, of course, it is impossible to attempt a quantitative comparison between the different formulations. On the other hand, as shown in Ref. 4, periodic oscillations do develop, even for a single-mode laser, if the ratio $\gamma_{\parallel}/\gamma_{\perp}$

Figure 1. *A comparison between the gain dependence of the output field modulus* $|\mathcal{I}_{st}|$ *for (1) the mean field model; (2) the improved single-mode approximation; and (3) the exact solution of the Maxwell-Block equation. The parameters used in these calculations are* $R=0.7$, $\tilde{\kappa}=\kappa/\gamma_{\perp}=4.257$, $\tilde{\gamma}=\gamma_{\parallel}/\gamma_{\perp}=0.1$, $\alpha_I=75$ *and* $\delta_{AC}=0$.

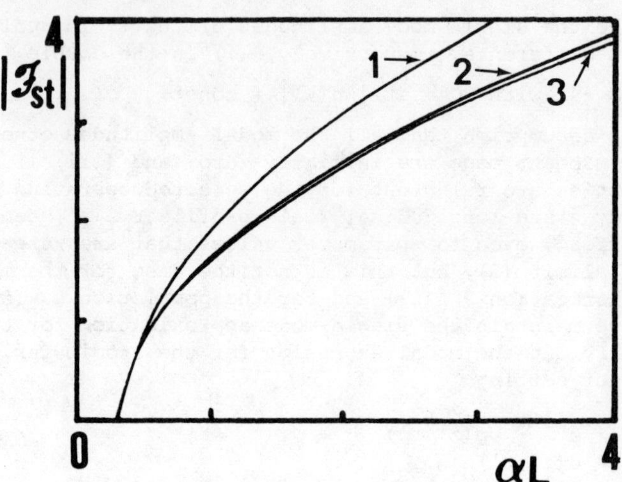

Figure 2. *Comparison between the steady state oscillations corresponding to unstable solutions of equations for the three models and conditions used in Fig. 1. The gain,* $\alpha L=4$.

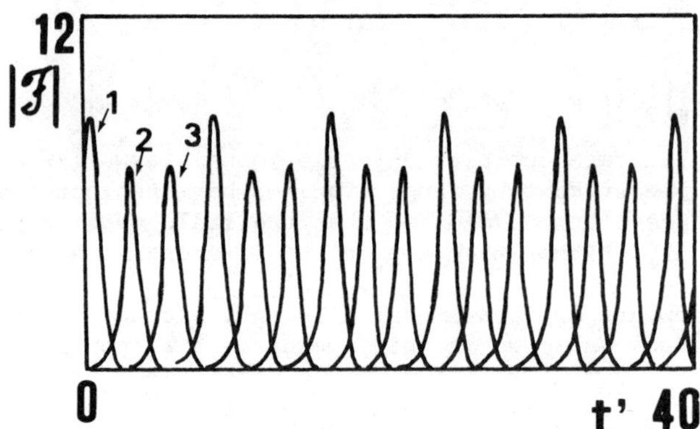

is sufficiently small. In Fig. 2 (above) we have assembled the steady state oscillations corresponding to unstable solutions of Eqs. (1), (4) and (7). Again, apart from an irrelevant shift along the time axis, the exact Maxwell-Bloch equations and the improved single mode approximation yield essentially identical results.

ACKNOWLEDGEMENTS. *This research has been partially supported by a contract with the U.S. Army Research Office and a grant from the Italian National Research Council (CNR).*

REFERENCES

1. For an overview of single mode lasers and references therein, see Section II of N.B. Abraham, L.A. Lugiato, L.M. Narducci, Journ. of Opt. Soc. of Am. B2, 7, 1985,
2. L.A. Lugiato, L.M. Narducci, E.V. Eschenazi, D.K. Bandy, N.B. Abraham, Phys. Rev. A32, 1563 (1985).
3. L.A. Lugiato, L.M. Narducci, D.K. Bandy, J.R. Tredicce, Phys. Rev. A (to be published).
4. L.M. Narducci, H. Sadiky, L.A. Lugiato, N.B. Abraham, Opt. Comm. 55, 370, 1985.

PHOTON-DRESSED DISCRETE STATES IN THE CONTINUUM

N. K. Rahman
Sezione di Chimica Fisica, Dipartimento di Chimica e Chimica Industriale, Università di Pisa, Via Risorgimento 35, Pisa (Italy)

A. Lami
Istituto di Chimica Quantistica ed Energetica Molecolare del C.N.R., Via Risorgimento 35, 56100 Pisa (Italy)

ABSTRACT

Generation of two or more photon-dressed discrete states in the atomic continuum by coupling a resonance to atomic bound states by external e.m. field and its experimental consequences are discussed.

The problem of coupling the atomic continuum in the vicinity of a resonance with a bound state by external e.m. field is an active issue due to some tantalizing possibilities offered with tunable and moderately powered lasers. Consider the coupling in Fig.1. If the "detuning" δ is fixed to be

$$\delta = q(\gamma_2 - \gamma_1) \quad (1)$$

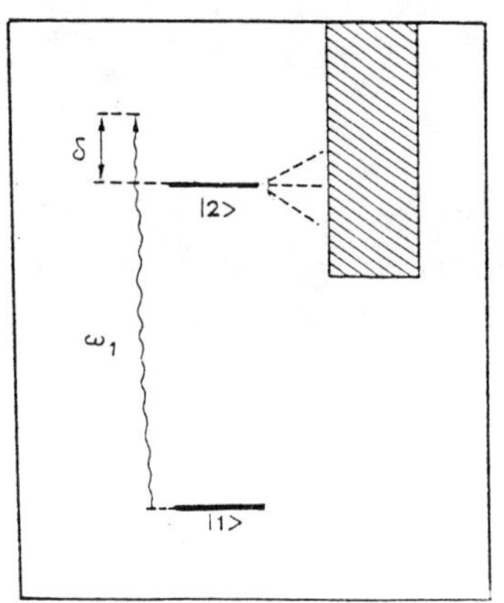

Fig. 1

where $\gamma_1 = \pi|\langle 1|\mu E_o|1\rangle|^2$, q is the Fano parameter, γ_2 is the natural width of the resonance, one obtains a partial freezing of the ionization. The phenomenon is equivalent to the generation of a photondressed discrete state in the continuum (to the extent that one can model the system with two discrete states and a continuum) as well as confluence in the parlance of those interested in the photoelectron spectrum[1].

If one couples the same resonance with another bound state by means of a second e.m. field (Fig.2), one may view this in a variety of ways. Generally, it is a setup of a double resonance experiment with the continuum. If one of the fields is held rather weak, that field may be considered as a probe field which among others may be utilized to probe the discrete state (confluence) that may be formed by the other field. One may, however, extend the concept further where the two fields have to be treated on an equal

0094-243X/86/1460360-2$3.00

footing. A particular situation arises when one fixes

$$\delta = q_1(\gamma_2 - \gamma_1) - \gamma_3(q_3 - q_{13}) \qquad (2)$$

$$\delta' = q_3(\gamma_2 - \gamma_3) - \gamma_1(q_1 - q_{13}) \qquad (3)$$

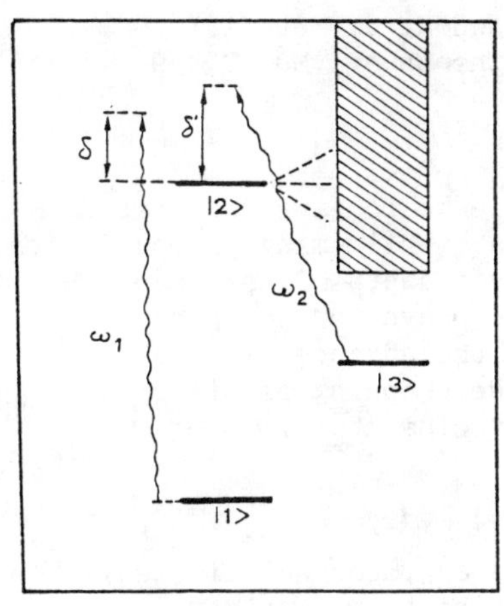

Fig. 2

where now all the terms have obvious meanings with the additional quantity q_{13} which is the ratio of the real to imaginary part of the matrix element between the two bound states of the atom [2]. When Eq.(2) and Eq.(3) hold, the ionization not only freezes partially but oscillates about some mean value after a transient regime has passed. Viewed again as a three level system interacting with continuum, this happens due to the generation of two photon-dressed discrete states in the continuum. The oscillation is due to the quantum beats. Concomitant with the oscillations of the ionization, the two bound states of the atom are populated even at long times which concurrently undergo oscillations. These oscillations are phenomena which persist with time.

Lately, we have examined the case in which N bound states of an atom are all coupled to a single resonance by N different e.m. fields. The question that arises is if it is at all possible to produce N photon-dressed discrete states in the continuum. We have found the answer to be positive and have obtained the corresponding conditions of the detunings [3]. This and other questions obviously remain as matter under active consideration of research.

REFERENCES

1. P. Knight, Comments At. Mol. Phys. 15, 193 (1984). This reference quotes most of the literature in this area.
2. A. Lami and N. K. Rahman, Phys. Rev. A. (Rapid Communications), 33, 782 (1986) and in Collisions and Half-Collisions with Lasers ed. N.K.Rahman and C.Guidotti (Harwood Academic, Chur-New York 1984).
3. A. Lami and N. K. Rahman, Phys. Rev. A. (submitted).

LASER-COOLING AND ELECTROMAGNETIC TRAPPING† OF NEUTRAL ATOMS

W. D. Phillips, A. L. Migdall, and H. J. Metcalf*
National Bureau of Standards, Gaithersburg, MD 20899

ABSTRACT

Until recently it has been impossible to confine and trap neutral atoms using electromagnetic fields. While many proposals for such traps exist, the small potential energy depth of the traps and the high kinetic energy of available atoms prevented trapping. We review various schemes for atom trapping, the advances in laser cooling of atomic beams which have now made trapping possible, and the successful magnetic trapping of cold sodium atoms.

INTRODUCTION

The thermal motion of atoms presents an important limitation to a number of physical measurements. The highest resolution spectroscopy, from rf to optical frequencies, is limited in precision and accuracy by motional effects. These include first and second order Doppler effects, transit time effects and motion induced field effects. Collisional studies are often complicated by thermal spreads in the relative velocity of the collision partners. Many atomic physics experiments, such as those involving deflection of atoms, long-time interactions between atoms and fields, or measurement of long atomic lifetimes would benefit from reduction of the thermal velocity and its spread.

For many years it has been possible to confine charged particles, and in particular atomic ions, in various kinds of ion traps.[1] More recently, laser cooling[2,3] has been used to reduce the temperature of ions in Penning[4] and rf[5] traps to below 20 mK.[6] Single ions have even been confined to dimensions less than a wavelength of light,[7,8] the Lamb-Dicke condition for suppression of first order Doppler broadening.

This ability to first trap and then cool atomic ions has provided a solution to many of the experimental problems associated with atomic motion. We, along with other groups, have been working toward achieving the same benefits of cooling and trapping for neutral atoms.

NEUTRAL ATOM TRAPS

A major obstacle to cooling and trapping neutrals is the very small depth of any of the proposed traps for neutral atoms. While ions with energies equivalent to room temperature or above can be easily contained in ion traps, practical neutral atom traps have typical depths of only a few kelvin or less. As a result, neutral

0094-243X/86/1460362-4$3.00

atoms must be cooled first, before being trapped. We shall see below that this represents a major difficulty.

Why are traps for neutral atoms so shallow compared to those for ions? Ions are trapped by electric or magnetic fields which apply strong Coulomb or Lorentz forces. Forces on neutral atoms require an interaction between an atomic dipole or higher order moment and a gradient in an applied electric or magnetic field. Such forces are necessarily rather small.

Consider, for example, an atom with a permanent magnetic dipole moment trapped in an inhomogeneous magnetic field. This is possible only for a quantum state where the projection of the magnetic moment on the field is negative, so that the magnetic energy, $W = -\vec{\mu}\cdot\vec{B}$, increases as the magnitude of the field increases. Then a field having a local minimum in the magnitude of B acts as a trap. (A field with a local maximum in B is not possible.[9]) If superconducting coils are used, one might make such a configuration where the available change in field magnitude is several tesla. A typical magnetic moment for an atom is a Bohr magneton, so the energy associated with a 2 T field change is only 1.3 K.

Laser dipole traps derive their potential energy from the interaction between the oscillating dipole moment induced in an atom by the oscillating electric field of a laser and the gradient in that same electric field. The simplest configuration for such a trap is a single, focussed, Gaussian laser beam, tuned below the atomic resonance, so that the atom is attracted to the maximum in the oscillating electric field at the center of the focus.[10] The depth of such a trap is roughly the energy by which the laser can be detuned from resonance and still saturate the transition. For a laser power of 1 W focussed to some tens of micrometers, tuned to a strong transition such as the 3S → 3P in sodium, the depth is about one kelvin. Various compromises which are needed to stabilize such a trap will reduce the achievable depth.[11]

Other proposed traps include radiation pressure traps,[12] hybrid laser-magnetic traps,[13] and electrostatic traps.[9] The first two of these have maximum depths on the order of one kelvin. The electrostatic trap, which relies on the interaction of an inhomogeneous static electric field with the electric dipole moment of an excited atom, can be considerably deeper if highly excited Rydberg atoms with large dipole moments are used. Unfortunately, the excited atoms decay and the trap is fundamentally unstable.

LASER COOLING ATOMIC BEAMS

Because there are no stable, deep traps for neutral atoms, most of the experimental effort in this area has been directed toward cooling free atoms, particularly atomic beams. To cool an atomic beam one simply directs a laser beam opposite to the direction of the atomic beam. As the atoms absorb photons, reradiating them in random directions, they slow down. The absorption-reradiation process must be repeated as rapidly as possible, so as to cool the atoms in the short time available before they leave the apparatus. By contrast, the cooling of trapped ions need not be so fast, since the time available is much longer.

Two processes interfere with such rapid cooling. First, as the atoms absorb photons and slow down, their velocity changes enough to Doppler shift them out of resonance with the laser while remaining much too fast to allow trapping. Furthermore, only a small fraction of the atoms in a thermal atomic beam have the right velocity to be in resonance with the laser. Second, the alkali atoms used in all experiments to date can be optically pumped by the cooling laser into a hyperfine state which cannot be re-excited by the laser.

Various groups have used different methods of overcoming these problems. The laser frequency may be rapidly increased to compensate for the changing Doppler shift of the decelerating atoms while coming into resonance with most atoms in the atomic beam.[14-17] Alternately, the resonance frequency of the atoms can be changed, using the changing Zeeman shift from a spatially varying magnetic field to achieve the same result.[18,19] The problem of optical pumping is solved by careful attention to selection rules for excitation and emission,[18] or by providing another laser frequency to re-excite atoms which decay to the wrong hyperfine state.[20,16,17] Using combinations of these techniques, several groups have recently been able to produce samples of atoms with average velocities at or near zero, often with velocity spreads of a few meters per second (kinetic energies of tens of millikelvin) or less.[16,17,21,22]

In addition to the one dimensional cooling described above, additional laser beams can be used to produce two[23] or three[17] dimensional cooling. Three dimensional cooling is so effective that atoms have been brought to energies as low as 240 μK, with their kinetic motion so strongly damped that they behave as if they were diffusing in a viscous fluid.[17,24]

A MAGNETIC QUADRUPOLE TRAP

Laser cooling of atomic beams has now made it possible to trap neutral atoms. We have achieved the first such trapping using a magnetic quadrupole trap.[25] The trap consists of a coaxial pair of coils, separated by 1.25 radii, carrying opposed currents. This produces a magnetic field whose magnitude is zero at the center of the trap and increases linearly to 0.025 T at the edge of an ellipsoidal region 20 cm^3 in volume. Laser cooled sodium atoms are stopped in the center of the coils and the current is then turned on. Atoms with energies as high as 17 mK (velocity of 3.5 m/s) are confined by the quadrupole field. The mean trapping time is greater than 0.8 s and is limited mainly by collisions with fast background gas atoms.

We expect that longer trapping times, cooling in traps, and various kinds of laser traps will be achieved in the near future by a number of groups. Details of experiments on laser cooling and trapping of neutral atoms and of proposals for neutral atom traps can be found in a recent review.[26]

REFERENCES

† This work was supported in part by the Office of Naval Research.
* Permanent address: Department of Physics, S.U.N.Y., Stony Brook, NY 11790.

1. D. J. Wineland, Science 226, 395 (1984).
2. T. Hansch and A. Schawlow, Opt. Commun. 13, 68 (1975).
3. D. Wineland and H. Dehmelt, Bull. Am. Phys. Soc. 20, 637 (1975).
4. D. Wineland *et al.*, Phys. Rev. Lett. 40, 1639 (1978).
5. W. Neuhauser *et al.*, Phys. Rev. Lett. 41, 233 (1978).
6. W. Nagourney *et al.*, Proc. Nat. Acad. Sci. U.S.A. 80, 643 (1983).
7. W. Neuhauser *et al.*, Phys. Rev. A22, 1137 (1980).
8. W. Nagourney *et al.*, Bull. Am. Phys. Soc. 30, 1831 (1985).
9. W. Wing, Progr. Quantum Electron. 8, 181 (1984).
10. A. Ashkin, Phys. Rev. Lett. 40, 729 (1978).
11. J. Dalibard *et al.*, Opt. Commun. 47, 395 (1983).
12. A. Ashkin, Opt. Lett. 9, 454 (1984).
13. W. Stwalley, Progr. Quantum Electron. 8, 203 (1984).
14. V. S. Letokhov *et al.*, Opt. Commun. 19, 72 (1976).
15. W. Phillips and J. Prodan, in Coherence and Quantum Optics V, L. Mandel and E. Wolf, eds. (Plenum, New York, 1984), p. 15.
16. W. Ertmer *et al.*, Phys. Rev. Lett., 54, 996 (1985).
17. S. Chu *et al.*, Phys. Rev. Lett., 55, 48 (1985).
18. W. Phillips and H. Metcalf, Phys. Rev. Lett., 48, 596 (1982).
19. J. Prodan *et al.*, Phys. Rev. Lett., 49, 1149 (1982).
20. S. Andreev *et al.*, Pis'ma Zh. Eksp. Teor. Fiz. 34, 463 (1981).
21. J. Prodan *et al.*, Phys. Rev. Lett. 54, 992 (1985).
22. V. Balykin *et al.*, Zh. Eksp. Teor. Fiz. 86, 2019 (1984).
23. V. Balykin *et al.*, Pis'ma Zh. Eksp. Teor. Fiz. 40, 251 (1984).
24. A. Migdall *et al.*, J. Opt. Soc. Am. A 2 (13), P51 (1985).
25. A. Migdall *et al.*, Phys. Rev. Lett. 54, 2596 (1985).
26. W. Phillips *et al.*, J. Opt. Soc. Am. B2, 1751 (1985).

HIGH RESOLUTION LASER SPECTROSCOPY OF ATOMIC HYDROGEN*

R. G. Beausoleil, B. Couillaud, C. J. Foot,[1] T. W. Hänsch, E. A. Hildum, and D. H. McIntyre[2]

Department of Physics,
Stanford University, Stanford, CA 94305, USA

Although spectroscopy of hydrogen has played a central role in the development of atomic theory and quantum mechanics, the resolution of optical spectral lines remained limited by Doppler broadening to about 1 part in 10^5 until 1971. Since then, methods of Doppler-free laser spectroscopy have achieved major improvements.[1-7] Thanks to recent experimental advances,[8-10] the rate of progress has quickened dramatically in 1985, as illustrated in Fig. 1, promising unprecedented opportunities for precision measurements of fundamental constants and for stringent tests of basic physics laws.

The 1S-2S two-photon transition with its natural linewidth of only 1.3 Hz has long been recognized as one of the most intriguing transitions to be studied by high resolution laser spectroscopy.[3] Until recently, however, intense monochromatic ultraviolet light at the required wavelength of 243 nm was only available from pulsed laser systems with considerable instrumental linewidth. Nonetheless, even the earliest crude experiments (C) achieved a resolution comparable to that of Balmer-α.[3] Wieman (E) was able to measure the Lamb shift of the 1S ground state to within better than 30 MHz by

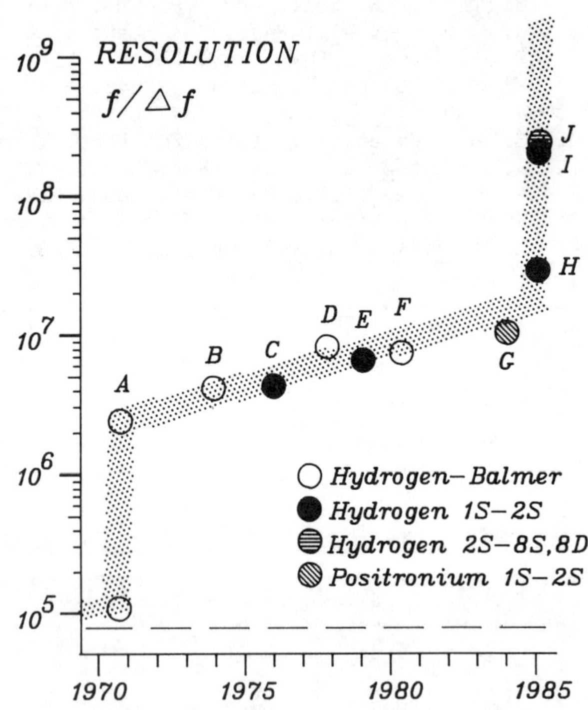

Fig. 1. Resolution achieved by Doppler-free laser spectroscopy of atomic hydrogen and positronium. The letter symbols refer to the following References: A=1, B=2, C=3, D=4, E=5, F=6, G=7, H=8, I=9, J=10

comparing 1S-2S with the n=2→4 Balmer-β line at 486 nm, even though the accuracy was limited by frequency chirping in the pulsed dye laser amplifiers.[5]

Hildum et al.[8] have achieved a much improved resolution of 3 parts in 10^9 (H) in a new experiment designed to reduce earlier systematic errors. The 243 nm light is produced with a sophisticated pulsed laser system, consisting of a 486 nm cw dye laser oscillator, an excimer-laser-pumped pulsed dye amplifier, a confocal filter interferometer servo-locked to the cw oscillator, and a urea frequency doubler. The atoms are excited in an atomic beam, and the signal is observed via photoionization of the metastable 2S state. With an interferometrically calibrated absorption line of $^{130}Te_2$ near 486 nm as the reference,[11] the absolute frequency of the 1S-2S interval has been measured as f(1S-2S) = 2 466 061 395.6(4.8) MHz. A new value of the Rydberg constant, R_∞ = 109 737.314 92(21) cm^{-1}, derived from this result, is not in good agreement with the most recent previous value derived from the frequency of Balmer-α.[6]

A more important breakthrough (I) was achieved when C.J.Foot et al.[9] observed the 1S-2S transition by continuous-wave two-photon spectroscopy. The 243 nm radiation is generated by summing the frequency of a 351 nm argon laser and a 789 nm cw dye laser in a KDP crystal heated to the critical phase matching temperature of 62 °C. A servo-locked build-up cavity enhances the dye laser radiation at the crystal, and a standing wave cavity further enhances the ultraviolet intensity at the final interaction region to 10-100 mW. A resolution of 5 parts in 10^9 has been obtained by simply observing the hydrogen atoms in a gas cell at room temperature. The linewidth is limited by transit broadening, but collisions and laser bandwidth also contribute. A new absolute frequency measurement of 1S-2S using this setup is presently underway at Stanford. The reference is provided by a 486 nm cw dye laser which is servo-locked to a calibrated line of $^{130}Te_2$. An external urea crystal produces a few nW of second harmonic radiation in near coincidence with the 243 nm radiation from the sum frequency generator, so that an rf beat signal can be used for a precise frequency comparison.

By using straightforward means, we expect to improve the resolution of the hydrogen 1S-2S transition by another 3 orders of magnitude in the near future. The laser linewidth can be much reduced by internal or external frequency stabilizers with fast servo response,[12] and pressure broadening can be eliminated with an atomic beam. In order to reduce transit broadening, the interaction time can be extended by sending the atoms along the light waves, or by Ramsey spectroscopy with two spatially separated interaction regions. Considerable improvement is gained if the hydrogen beam is cooled to liquid helium temperature by mounting the escape nozzle on a helium cryostat. In this case, transit broadening will be reduced to 7 kHz, and second order Doppler shifts remain smaller than 1 kHz.

For experiments in the more distant future, we are exploring means to cool the hydrogen atoms far below liquid helium temperature. Once the kinetic energy has been reduced to a few mK, the earth's gravitational field can be used to further slow the atoms in an atomic "fountain." Two-photon optical Ramsey spectroscopy is possible by letting freely falling atoms traverse a single standing wave laser field twice on their parabolic trajectories.[13,14] When averaging over a broad velocity distribution, we predict a simple near-Lorentzian lineshape with just the natural linewidth. A resolution of better than 1 part in 10^{15} may thus ultimately be achievable.

Although quantum electrodynamics is believed to provide a very good theory for the hydrogen atom, the accuracy of calculations[15] is limited by the uncertainties of fundamental constants, by unknown nuclear size and structure effects, and by computational approximations.

At present, the dominant contribution to the uncertainty of the predicted 1S-2S interval (2.5 MHz) is the uncertainty of the Rydberg constant (10^{-9}), and a measurement of the 1S-2S frequency can improve the Rydberg constant about tenfold, until we approach the uncertainties due to the electron mass (5×10^{-8}: 70 kHz), the charge radius of the proton (10^{-1}: 130 kHz), and approximations in the computation of QED corrections (60 kHz). The uncertainty of the 1S-2S hydrogen-deuterium isotope shift is dominated by nuclear size effects (180 kHz), with the proton/electron mass ratio contributing only 35 kHz.

More interesting measurements become possible if the 1S-2S resonance is compared with other sharp hydrogenic transitions, such as two-photon transitions from the metastable 2S state to long living highly excited nS Rydberg states.[10] The excitation of levels with $n \approx 100$ would require a laser wavelength near 729 nm, slightly more than 3 times larger than the 243 nm of the 1S-2S resonance, and it should be feasible to compare the two frequencies with extreme precision by detecting a microwave beat signal between the ultraviolet light and the third harmonic of the red laser. The small difference frequency df = f(1S-2S) - 3 f(2S-nS) depends critically on the Lamb shifts of the participating levels, and a precision measurement can provide accurate new values for the charge radii of the proton and deuteron, if the calculations of electron structure corrections and higher order QED corrections[15] are improved.

It is possible to construct composite frequencies which no longer depend on energy shifts which scale with the inverse cube of the principal quantum number.[9] Such frequencies are thus independent of nuclear size, and they no longer depend on uncalculated higher order QED effects. One interesting example is a composite of the two isotope shifts, $\Delta f(2S - nS) - 1/7\ (1 - 8/n^3)\ \Delta f(1S\text{-}2S)$. A measurement of this frequency can serve to determine a much improved value for the electron/proton mass ratio. Once this ratio and the proton size are better known, a much improved value of the Rydberg constant can be determined from the frequency f(1S-2S). By taking advantage of the composite frequency $f(1S\text{-}2S) - 7/(4+24/n^3)$ df, which is again independent of nuclear size effects, we can actually determine the Rydberg constant without having to find the proton size explicitely.

Very stringent tests of basic physics laws will become possible, if accurate values for these same constants become available from independent experiments.

*The work reported here has been supported by the National Science Foundation under Grant No. NSF PHY83-08721 and by the U.S. Office of Naval Research under Contract ONR N00014-C-0403.

[1]Permanent address: Clarendon Labs, Oxford, England
[2]National Science Foundation Predoctoral Fellow

REFERENCES

1. T. W. Hänsch, I. S. Shahin, and A. L. Schawlow, Nature 235, 63 (1972)
2. T. W. Hänsch, M. H. Nayfeh, S. A. Lee, S. M. Curry, and I. S. Shahin, Phys. Rev. Letters 32, 1336 (1974)
3. S. A. Lee, R. Wallenstein, and T. W. Hänsch, Phys. Rev. Letters 35, 1262 (1975)
4. J. E. M. Goldsmith, E. W. Weber, and T. W. Hänsch, Phys. Rev. Letters 41, 1525 (1978)
5. C. Wieman and T. W. Hänsch, Phys. Rev. A22, 192 (1980)
6. S. R. Amin, C. D. Caldwell, and W. Lichten, Phys. Rev. Lett. 47, 1234 (1981)
7. S. Chu, A. P. Mills, and J. L. Hall, Phys. Rev. Lett. 52, 1689 (1984)
8. E. A. Hildum, U. Boesl, D. H. McIntyre, R. G. Beausoleil, and T. W. Hänsch, Phys. Rev. Letters, submitted for publication
9. C. J. Foot, B. Couillaud, R. G. Beausoleil, and T.W. Hänsch, Phys. Rev. Letters 54, 1913 (1985)
10. F. Biraben and L. Julien, Opt. Commun. 53, 319 (1985)
11. J. R. M. Barr, J. M. Girkin, A. I. Ferguson, G. P. Barwood, W.R.C. Rowley, and R. C. Thompson, Opt. Commun. 54, 217 (1985)
12. J. L. Hall and T. W. Hänsch, Optics Letters 9, 502 (1984)
13. R. G. Beausoleil and T. W. Hänsch, Optics Letters 10, 547 (1985)
14. R. G. Beausoleil and T. W. Hänsch, Phys. Rev. A, submitted for publication
15. G. W. Erickson, J. Phys. Chem. Ref. Data 6, 831 (1977)

ONE-DIMENSIONAL GIANT DIPOLE ATOMS

M. H. Nayfeh, D. Yao*, Y. Ying†, D. Humm, K. Ng, and T. Sherlock
Department of Physics
University of Illinois at Urbana-Champaign
1110 W. Green Street
Urbana, Illinois 61801

ABSTRACT

An external dc electric field F imposed on highly excited atomic hydrogen is used to construct nearly one-dimensional atoms whose electronic distribution are highly extended along the field, and which may have enormous dipole moments ("giant dipole" atoms). The nuclear charge Z_1 that defines the energy and other properties of the "new" atom is a fraction of the proton charge. The fractions 0, 1/4, 1/2, 3/4 and 1 define four quarters that classify some properties of the atoms. The dipole moment is found to be opposite to the field in the first and third, and in its direction in the second and fourth quarter, and zero at the boundaries. Those atoms are unstable against ionization, however at $E < 0$ but $E > -2\sqrt{F}$ one can populate "giant dipole" atoms whose potential barriers are large enough (tunneling small enough) to render their lifetimes quite long. Hydrogen is a special case because these states are not mixed with less stable ones and so their "giant dipoles" survive long enough to be studied. We have used what we call "charge shape tuning" utilizing a three-photon process to excite them without the excitation of the overlapping continuum.

Recently it has been shown that in the presence of a strong external dc electric field, atoms have quantized energy levels for all energies including positive energies.[1] The motion of the electron in these states is nearly one-dimensional with the electronic charge distribution highly extended and resembling a cigar whose axis has specific directions.[2,3] The atoms have enormous size and may have enormous electric dipole moments ("giant dipole" atoms), but they spontaneously ionize in about 10^{-12}s in the positive energy region. Moreover, in general, one cannot exclusively prepare these types of states in the $E \geq 0$ region without preparing the highly excited normal state of the atom (continuum state) since, first of all, the excitation has to start from the ground state of the normal atom which is only weakly affected by electric fields and secondly both fields Coulomb and Stark will have to compete. For example, the efficiency of excitation or the "visibility" of the giant dipoles, which is a measure of how much they rise above the accumulated smooth continuum, tends to be very small (4% at 5 kV/cm). Moreover, for $E < 0$ but $E > -2\sqrt{F}$ other effects arise due to the scattering of the active electron from the rest of the electrons. This causes mixing and smearing of the Coulomb-Stark interaction to the degree of preventing selective excitation of the giant dipoles, causing mixing among states of short and long lifetimes, resulting in shortening of

0094-243X/86/1460370-5$3.00

the lifetimes of the long lived ones.[1]

Considering the shortness of their lifetimes, and the low efficiency of excitation it is clear that experimentation with these "new atoms" will not be easy unless these two properties are enhanced. Our recent theoretical and experimental work on atomic hydrogen has shown that it is possible by using multi-step excitation via resonant intermediate Stark states to selectively excite and enhance the giant dipole without the excitation of the continuum states.[3] Also we were able to prepare a variety of these giant dipoles with a variety of ionization lifetimes ranging from 10^{-7} - 10^{-12}s and to find systematic classification of some of their properties thus making them easier to work with.

The scheme we devised for this purpose relies on a process we call multi-stage shaping or charge shape tuning. In one photon excitation from the ground state one starts from a spherically symmetric charge (zero dipole moment), and tries to mold it by a single operation into a giant dipole whose charge is highly focussed along the field. On the other hand in multi-stage shaping one uses one photon to create from a ground state a not too large dipole of charge distribution that is focussed along the field at an intermediate state followed by another photon absorption from this intermediate state that produces larger dipole whose charge is even more focussed along and so on till one excites the giant dipole in a highly focussed distribution along the field.

We will briefly describe the experimental set up used in these studies.[4] Simultaneous absorption of two photons from a single tunable pulsed laser beam at 243 nm results in excitation from 1s to n=2, and some photoionization of the resulting n=2 population. A second pulsed beam excites states near the continuum from the n=2 state. An atomic beam is formed by effusion from a Wood discharge tube through a multicollimator assembly composed of 25 small glass capillaries. The beam is loosely collimated, but produces a density of about 10^{11} H°/cm^3; the background gas density is on the order of $10^{12}/cm^3$. One of the Stark plates has a 3 mm x 10 mm slot cut into it and covered by a fine mesh to allow the passage of ions. This gives a limit of 1 μs on the detection time. Ions travel through a 100 cm long, field-free drift tube which provides mass analysis and are detected using an 18 stage venetian blind electron multiplier capable of single ion detection. A fraction of the second harmonic of a YAG laser at 532 nm is used to pump one dye laser producing a beam at 630 nm, which is frequency doubled to 315 nm by a KDP crystal and then summed with the residual YAG fundamental by a KDP crystal resuling in a beam at 243 nm of pulse length of about 10 ns, a bandwidth of about 1.5 cm^{-1} and pulse energies on the order of 10 microjoules. A second dye laser produces a beam at about 555 nm which is summed with part of the YAG fundamental to produce a beam with pulse length near 10 ns, bandwidth of .8 cm^{-1}, pulse energies of a few tenths of a millijoule and a wavelength near 365 nm. The data are collected and analyzed using an LSI-11 computer system . A slightly different scheme where the n=2 state is excited using one photon at 1215 Å has been recently reported.[5]

The ability to create focussed dipoles as intermediates is the key to the success of the multi-stage shaping operation.[4] This is explained in Fig. 1 for a two-stage process using n=2 of hydrogen as an intermediate. These are labelled by their parabolic quantum numbers (n_1, n_2, m_ℓ) as 100, 001 and 010, where $n = n_1 + n_2 + |m_\ell| + 1$. Because the fine structure level splittings in n=2 of hydrogen are small enough (.3 cm^{-1}) such that an electric field imposed on the atom which is larger than 5 kV/cm will be able to mix all of these sublevels and hence their charge distributions (each has a zero dipole), which are shown to the left of the figure, to produce distinct dipole distributions, shown to the right of the figure, needed for the shaping process. Our calculations using a field of 16.8 kV/cm show that by utilizing the upfield extended dipole of n=2 as an intermediate (shown as a solid line in Fig. 1), the efficiency can be increased from 10 percent to 30 percent, where as by utilizing the down field extended dipole (shown as a dashed line in Fig. 1) the efficiency is reduced to 1 percent. These were confirmed in our hydrogen experiment as shown in Figs. 2a and 2b respectively. Our further calculations using the

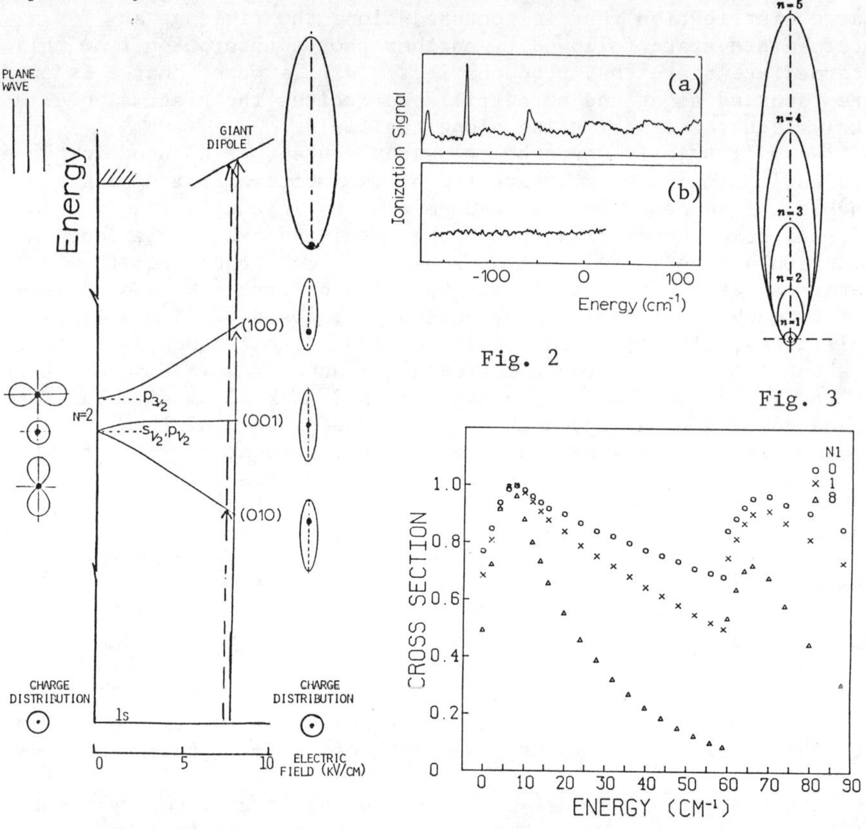

Fig. 1

Fig. 2

Fig. 3

Fig. 4

bluest components of higher n states (n_1 = n-1) whose charge can be focussed along the field more easily as shown in Fig. 3 and hence can be matched or tuned more closely to the charge of the giant dipoles showed dramatic effects on the efficiencies.[3] Figure 4, which gives normalized lineshapes for excitation via the 000, 100, and 800 components of the n=1, 2, and 9 manifolds respectively, shows that excitation via n as high as 9 rejects almost completely the excitation of the continuum. This rejection is accompanied by an enhancement of the excitation cross section of the giant dipole; in specific we find the peak in the case of n=9 is about a factor of 300 larger than that of the case of excitation from the ground state. Moreover, the visibility of the giant dipole states, defined here as $V = (S_{max}-S_{min})/(S_{max}+S_{min})$, reach near 85 percent for n=9.

The enhanced efficiency is very nice, but it is found that it is not possible to increase the lifetimes in this positive energy region. Such inability is related to the fact that the bound motion of all of these giant dipoles in this region are driven by nearly the same charge, most of the nuclear charge $Z_1 \sim 1$, which also dictates very similar orbits where the nucleus is located at the lower tip of the cigar. However, it is found that such enhanced efficiency can be extended to the negative energy region where it is also possible to produce giant dipoles that live quite long. Therefore we will now discuss[3] such promising negative energy regions between E = 0 and $E = -2\sqrt{F}$. In this region the giant dipole atoms take on different properties than the one in the positive energy region. Firstly, the fraction of the charge that drives the found motion can be varied from 0 to 1 by varying the energy of the system, and consequently the position of the nucleus inside the cigar can also be controlled. In Fig. 5 we plotted an indicator of the location of the nucleus inside the cigar as a function of Z_1 along the sketches of some possible orbits (the nucleus is shown as a white dot). This indicator which is the ratio of the excitation cross section from 010, and 100 to the giant dipoles which corresponds to the two processes labelled by dashed and solid lines in Fig. 1 respectively. This ratio is related to

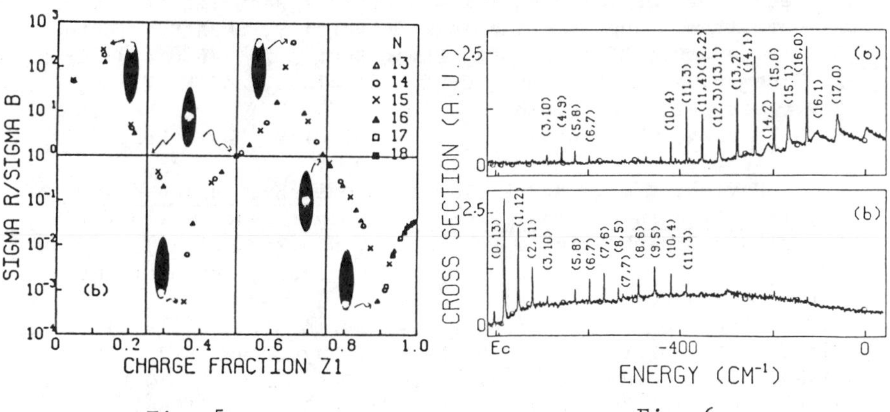

Fig. 5

Fig. 6

the ratio of the area of the orbit below the nucleus to that above the nucleus. The figure shows a remarkable property: for the fractional charges $Z_1 = 1/4$, $1/2$, and $3/4$ the nucleus is located at the center of the cigar (the atom has zero dipole moment). These fractional charges thus constitute demarkation lines across which the direction of the giant dipole reverses. In the first and third quarters the dipole is along the imposed field whereas in the other two quarters it is opposite to it. Given the size of the orbit (can be calculated), one can use the indicator to determine the moment. Moreoever, we have recipes that give us the energy of giant dipoles of given Z_1 values. These features and others have been recently confirmed by our experiments.[3] The giant dipole spectra for the solid and dashed line schemes of Fig. 1 are shown in Fig. 6a and 6b respectively.

Examination of the spectra indeed shows a variety of widths (lifetimes) that range from quite short to quite long. In fact there are giant dipoles that do not show up in our spectrum because they live longer than the time of the measurement which is 100 ns, or because they radiatively decay before they ionize. Moreover some of the states are actually narrower than they appear, because their widths are dominated by the 1 cm^{-1} bandwidth of the laser. Also there are systematics to the ionization lifetime as a function of Z_1 and hence as a function of energy that makes the selection of a giant dipole of given specification easy.

In conclusion, this work showed that as far as strong external field effects, highly excited atomic hydrogen is unique, and highly excited complex atoms become less and less hydrogenic as the external field strength rises. The hydrogen system allows design and preparation of giant dipole one-dimensional atoms with well defined dipole moments, ionization lifetimes and orientation.

This work was supported in part by NSF grant number PHY 81-09305, and in part by the physics department.

REFERENCES

* On leave from the Graduate School of the University of Science and Technology of China, Beijing, People's Republic of China.

† On leave from Shanghai, Jiao Tong University, Shanghai, China.

1. See for example D. A. Harmin, Phys. Rev. A 26, 2656 (1982); R. R. Freeman, N. P. Economou, G. C. Bjorklund, and K. T. Lu, Phys. Rev. Lett. 41, 1463 (1978).
2. M. H. Nayfeh, G. B. Hillard, and W. L. Glab, Phys. Rev. A 32, 3324 (1985).
3. M. H. Nayfeh, K. Ng, and D. Yao in Atomic Excitation and Recombination in External Fields, M. Nayfeh and C. Clark eds., Gordon and Breach Science Publishers, New York (1985); in Laser Spectroscopy VII, T. Hänsch and Y. Shen eds., Springer-Verlag, Berlin (1985).
4. W. L. Glab and M. H. Nayfeh, Phys. Rev. A 31, 530 (1985); W. L. Glab, K. Ng, D. Yao and M. H. Nayfeh, Phys. Rev. A 31, 3677 (1985); W. L. Glab and M. H. Nayfeh, Opt. Lett. 8, 30 (1983).
5. H. Rottke and K. H. Welge, Phys. Rev. A 33, 301 (1986).

TIME REVERSAL INVARIANCE AND ELECTRIC DIPOLE MOMENTS OF ATOMS

F.J. Raab
University of Washington, Seattle, WA 98195

ABSTRACT

This paper discusses experimental searches for permanent electric dipole moments of atoms using optical pumping techniques to achieve high precision frequency shift measurements. Experiments sensitive to either nuclear or electron spin dependent manifestations of broken time reversal symmetry are outlined. Limits for various models, obtained from an experiment on ^{129}Xe are given.

CP non-conservation in the kaon system[1] is the only known evidence of time reversal (T) noninvariance. The existence of a permanent electric dipole moment (EDM) of a neutron, atom, or structurally non-degenerate molecule is a clean manifestation of broken T symmetry. Such an EDM exists in the absence of any applied field and is permanently fixed relative to the angular momentum of the system which defines the symmetry axis for vector observables. The interaction of the EDM with an electric field (E), given by $(-d\vec{\sigma}\cdot\vec{E})$, is odd under T and spatial inversion (P), and so is readily distinguished from other electric dipole moments such as an induced dipole or the static dipole moment which exists in the molecular frame of a polar molecule. An EDM can arise in an atom if the elementary particles which constitute the system possess EDM's or if their interaction is P and T noninvariant.

An atomic EDM will precess in an applied electric field exactly as a magnetic moment precesses in a magnetic field. One can detect the EDM by looking for a change in precession frequency upon reversal of an electric field applied parallel to a magnetic field or by looking for a reversal of the direction of precession when an electric field is reversed in zero magnetic field. Such experiments yield information on electron spin or nuclear spin dependent effects of T noninvariance depending on which spin precession is monitored.

Consequently, our group (E. N. Fortson, B. Heckel, F. Raab, T. Vold, F. Hellinger, and S. Lamoreaux) is pursuing measurements of electron spin precession in heavy alkalis and nuclear spin precession in the $^1S_0$, nuclear spin 1/2, atoms ^{129}Xe and ^{199}Hg. Heavy atoms are used because T noninvariant short range forces or elementary particle EDM's induce an atomic EDM which increases rapidly with nuclear charge and atomic number.

The motion of the electrically charged constituents shields any applied electric field from the interior of the atom and in the limit of exact shielding no information on the EDM's of the constituents is available. This shielding breaks down when some other non-electrostatic force (e.g. relativistic effects for electrons and strong force for nucleons) partially balances the purely electrostatic force on the particles. In an extreme case, the valence electron in heavy atoms has an electric field in its rest frame which is larger than the electric field applied to the atom and a significant enhancement of the atomic EDM results. However shielding is a very severe constraint for constituents of the atomic nucleus.

Two strategies have evolved for overcoming shielding in nuclear EDM experiments. Sandars[2] suggested exploiting the large internal electric field in a polar molecule (TlF) which can be oriented by a much smaller laboratory electric field. In TlF it is estimated that an electric field of a few times 10 kV/cm exists at the valence proton of Tl when a comparable laboratory field is applied, although calculations of this effect have considerable uncertainty. Our strategy is to choose experimental conditions and atoms such that enhanced precision can overcome losses from shielding.

To achieve high precision measurements of frequency shifts we use optical pumping techniques which allow polarization of a large number of spins which precess coherently for a long time in a vapor cell or "bottle." The transmission of the pumping light through the vapor provides a high efficiency, low noise readout of the spin precession, and the random translational motion of the atoms during precession averages velocity dependent systematic effects to zero.

The bottle consists of two platinum coated quartz electrodes which cap a 1 cm high by 2.5 cm diameter quartz cylinder. About 200 Torr of N_2 background gas is used to prevent electrical discharge. For electron EDM experiments Rb (and eventually Cs) metal is distilled into the bottle. These alkalis are readily pumped by about 3 mW of circularly polarized D_1 or D_2 light from a diode laser. For the ^{199}Hg experiment we substitute Hg for Rb and pump with a 253.7 nm resonance lamp. In the ^{129}Xe experiment Rb is used to couple angular momentum between the light and 2 Torr of Xe via spin exchange collisions. A wall coating (paraffin or silane) prevents the metal vapor from forming an electrically conducting monolayer on the cylinder wall.

To reduce the effects of magnetic field fluctuations, two or three identical cells are stacked inside a three layer magnetic shield. Electric fields of about 4 kV/cm are applied, always in opposite directions in adjacent cells. All electric fields are reversed from run to run.

The different EDM experiments can be described by considering a xyz triad with pumping light along z. To measure the Rb EDM (and eventually Cs) we apply electric fields along y. We apply an oscillatory magnetic field along y to wobble the Rb polarization at some nonresonant frequency (300 Hz) and detect the modulation of the transmitted light at this frequency. This signal, proportional to the average precession rate ($\vec{\omega}_y$), is present when there is a static magnetic field along y or if there is an atomic EDM interacting with the electric field. The signal is relatively insensitive to fields along x and z. In the ^{129}Xe experiment the coherence time is long (>500 s), so a pump/probe technique is more practical. Electric fields and a small (0.1 mG) magnetic precession field are applied along x. A large (10 mG) field is applied along z to allow the Xe nuclei to accumulate polarization through spin exchange with polarized Rb. When this field is suddenly switched off, the Xe nuclei freely precess in the y-z plane while the short-lived Rb polarization retains a large z component due to continuous pumping by the light. We use the wobble technique described above to monitor the Xe magnetization along y which acts on the Rb like a 20 μG magnetic field. We measure the Xe precession frequency for 512 s, reverse electric field and repeat the cycle. The Hg experiment will use Hg as a driven atomic absorption oscillator with light along z, electric field along x, driving magnetic field along y and about 20 mG magnetic field at 45 degrees to z and x.

The electron (e) EDM experiment on Rb presently exceeds the current limit on d(e) in precision but further improvements and systematic checks are being pursued. For Hg we have obtained coherence times of several hundred seconds in an applied electric field and a frequency measurement system is under construction.

In Table I, recently derived limits on possible sources of T noninvariance based on our Xe results[3] and theoretical modeling by other groups,[4,5,6,7] are compared with previous atomic experiments (Cs,[8] Tl,[9] and metastable Xe[10]) and the most recently published neutron[11] and TlF[12] experiments. Because of shielding the Xe result, although it exceeds all other EDM determinations in precision, does not provide a better limit on the neutron EDM. The Xe limit for the proton EDM is comparable to the TlF result. Since the hyperfine interaction couples electron EDM information into the nuclear spin precession, Xe and TlF do provide limits on electron EDM but these are not competitive with previous atomic measurements.

Although sensitivity to quark EDM's should scale like nucleon EDM sensitivity, both nucleons and atoms are composed of elementary

Table I Limits on sources of T noninvariance.

Source	Limit from ^{129}Xe	Limits from other work
^{129}Xe EDM (e-cm)	$=(-0.3\pm1.1)\times10^{-26}$	
neutron EDM (e-cm)	$\|d(n)\| < 1\times10^{-21}$	$\|d(n)\|<4\times10^{-25}$ (n)
proton EDM (e-cm)	$\|d(p)\| < 4\times10^{-21}$	$\|d(p)\|<4\times10^{-21}$ (TlF)
electron EDM (e-cm)	$d(e)=(0.4\pm1.4)\times10^{-23}$	$\|d(e)\|<2\times10^{-24}$ (Cs,Tl,Xe*)
$iC_T(\bar{N}\gamma_5\sigma_{\mu\nu}N)(\bar{e}\sigma^{\mu\nu}e)$	$(3/4)C_{Tn}+(1/4)C_{Tp}$ $=(-0.7\pm2.9)\times10^{-6}$	$(1/4)C_{Tn}+(3/4)C_{Tp}$ $=(6\pm9)\times10^{-6}$ (TlF)
$iC_S(\bar{N}N)(\bar{e}\gamma_5 e)$	$C_S=(0.8\pm2.9)\times10^{-4}$	$C_S=(3\pm5)\times10^{-4}$ (TlF) $\|C_S\| \lesssim 3\times10^{-4}$ (Cs,Xe*)
$iK_S(\bar{N}\gamma_5 N)(\bar{e}e)$	$K_{Sn}=(-0.3\pm1.1)\times10^{-3}$	$K_{Sp}=(2.5\pm3.8)\times10^{-3}$ (TlF)
QCD θ parameter	$\|\theta\| < 10^{-7}$	$\|\theta\| < 10^{-9}$ (n)
Kobayashi-Maskawa: $i\eta\,(\bar{n}\gamma_5 n)(\bar{p}p)$	$\eta_{np} = -0.7\pm.24$	$\eta_0 < 0.1$ (n)

particles whose T-odd interactions are perhaps of more fundamental interest. These are expressed as effective phenomenological current-current interactions whose coupling strengths are given as a fraction of the strength of allowed weak interactions ($G_F/\sqrt{2}$). The particle operators N, n, p, e refer to nucleus, neutron, proton and electron, respectively. Our Xe experiment probes T-odd electron-nucleon interactions at the milliweak to microweak scale. T-odd quark-quark interactions induce an EDM in heavy nuclei which is regularly enhanced by several orders of magnitude over the induced neutron or proton EDM. The neutron and Xe results have comparable sensitivity to such interactions when the effective nucleon-nucleon force reduces to kaon exchange (e.g. Kobayashi-Maskawa model), whereas for models like CP nonconservation in quantum chromodynamics (QCD θ parameter) the neutron limit is better.

Future improvements in EDM limits by many orders of magnitude appear feasible based on shot noise estimates. These will ultimately require comparisons of different isotopes in the same bottle to ratio magnetic field fluctuations. Higher density polarized samples should allow yet further improvements. Improved versions of the neutron experiment at ILL-Grenoble, TlF at Yale, and Cs at Amherst should provide interesting new results. We are presently completing a new test of Local Lorentz Invariance using similar optical pumping techniques. Future improvements could lead to a search for a new, intermediate range, P and T odd force carried by axions as well as experiments to study the coupling of gravity to quantum mechanical spin. Clearly there is much promise for atomic physics in the study of the fundamental forces of nature.

This work was supported by NSF grant PHY-8403976.

REFERENCES

1. J. H. Christenson, J. Cronin, V. L. Fitch, and R. Turlay, Phys. Rev. Lett. 13, 138, (1964).
2. P. G. H. Sandars, Phys. Rev. Lett. 19, 1396, (1967).
3. T. G. Vold, F. J. Raab, B. Heckel, and E. N. Fortson, Phys. Rev. Lett. 52, 2229, (1984).
4. A. M. Martensson-Pendrill, Phys. Rev. Lett. 54, 1153, (1985).
5. V. A. Dzuba, V. V. Flambaum, and P. G. Silvestrov, Phys. Lett. 154B, 93, (1985).
6. V. V. Flambaum and I. B. Khriplovich, to be published in Zh. Eksp. Teor. Fiz.
7. V. V. Flambaum, I. B. Khriplovich, and O. P. Sushkov, Phys. Lett. 162B, 213, (1985).
8. M. C. Weisskopf, J. P. Carrico, H. Gould, E. Lipworth, and T. S. Stein, Phys. Rev. Lett. 21, 1645, (1968).
9. H. Gould, Phys. Rev. Lett. 24, 1091, (1970).
10. M. A. Player and P. G. H. Sandars, J. Phys. B 3, 1620, (1970).
11. J. M. Pendlebury, et al., Phys. Lett. 136B, 327, (1984).
12. D. A. Wilkening, N. F. Ramsey, and D. J. Larson, Phys. Rev. A 29, 425, (1984).

AUTODETACHING STATES OF NEGATIVE IONS

Charles W. Clark
National Bureau of Standards, Gaithersburg, MD 20899

ABSTRACT

Autodetaching states (or resonances) of negative ions are very sensitive to electron correlation effects, and the general rules which determine their energetics and stability are not well understood. The experimental techniques used to study them depend strongly on elemental species. For instance, laser photodetachment spectroscopy is one of the preferred ways of investigating alkali negative ions, whereas negative ion resonances of noble gases can only be produced by particle impact. This tends to obscure possible correspondences between the resonance spectra of different elements. However, evidence of common properties of certain classes of autodetaching states is accumulating. This paper describes a few simple examples.

INTRODUCTION

The independent electron model is the backbone of modern atomic physics. Apart from its obvious conceptual advantages, the independent particle picture (i.e. the Hartree-Fock approximation and its surrogates) does reproduce the gross features of atomic and molecular structure (e.g. total energies, ionic radii) with remarkable fidelity, although quantitative discussion at a level approaching spectroscopic accuracy usually requires some account to be made of the effects of electron correlation. However, in some systems the independent particle description gives results which are qualitatively incorrect. This has long been known to be the case for most of the stable atomic negative ions, for which the Hartree-Fock approximation does not yield bound states. Thus, systematic study of negative ions may shed light on some of the fundamental defects of the independent electron picture, and point the way to useful alternative descriptions.

RESONANCE STATES ON THE WANNIER RIDGE

This old, vague hope has been given new life by the recent discovery of the Wannier ridge resonances.[1] These are manifested as long series of sharp resonances in electron - atom scattering, which converge to the ionization limit(s) of the atomic target. Such long series can be produced in atomic systems only by excitation of one or more electrons to ever larger orbits. Since these resonances are found not to be associated with particular excited states of the target, they must consist of <u>two</u> electrons orbiting a singly-charged positive ion core. Thus they can be viewed as doubly-excited states of negative ions.

Elementary considerations suggest that the most stable of such states should be those in which both electrons are at comparable distances from the ion: in a configuration where the electron - ion distances are too unequal, the inner electron neutralizes the ionic Coulomb field acting upon the outer electron, allowing it to escape. Configurations in which the two electrons maintain nearly equal separations from the ion were first identified by Wannier as dominating the process of double electron escape at threshold; such motions take place along a ridge of the six-dimensional potential surface defined by the net Coulomb interaction among electrons and ion.[2]

This simple description is cast in terms of the long-range interaction between charged particles, so it is a statement about universal properties of doubly-excited states. An optimistic analogy can be drawn with the quantum defect theory, which describes all singly-excited states in terms of electron motion in a spherical Coulomb potential with a parametrizable short-range interaction. Substantial theoretical efforts are underway[3] with a view towards similar generalization of the analysis of motion on the Wannier ridge. Success along these lines would probably provide a substantial contribution to understanding the basic physics of multiple excitation. The approach of this paper is much more limited in scope. It is an attempt to identify quantitative similarities between Wannier-type states that lie in the energy range where calculations done by standard methods are still reliable. The examples are taken from R-matrix and variational calculations of scattering amplitudes, and from multiconfiguration Hartree-Fock calculations of resonance energies.

Experimental work on a related class of problems, namely the spectroscopy of double Rydberg states of neutral atoms, is reported elsewhere in this volume.[4]

EIGENCHANNEL ANALYSIS OF SCATTERING AMPLITUDES

A comparison of negative ion resonances of neon and of sodium is made in Fig. 1. The solid lines are calculated[5,6] eigenphases for electron scattering by neon in ^2S symmetry, as a function of incident electron energy in Rydbergs. Markers on the energy axis show

Fig. 1. Eigenphases for electron scattering by Ne and Na.

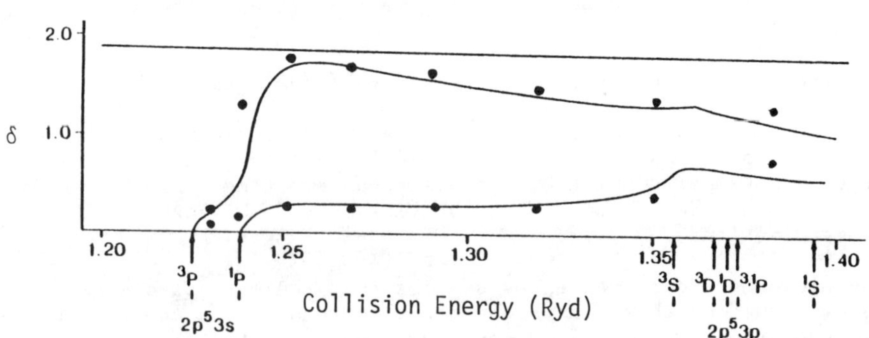

the positions of states in the first two excited configurations of Ne, as obtained in the LS coupling approximation. In the region between the $2p^53s$ and $2p^53p$ configurations there are three open collision channels of ^2S symmetry, viz. $2p^6\varepsilon s$, $2p^53s(^3P^\circ)\varepsilon' p$, and $2p^53s(^1P^\circ)\varepsilon'' p$. The eigenphases are the phase shifts of the three orthogonal eigenchannels in which the scattering matrix is diagonal. The eigenchannels are linear combinations of the collision channels; it is evident that the eigenchannel whose phase shows little energy variation, must be essentially identical to the $2p^6\varepsilon s$ collision channel, which is the only channel that is open at energies below the $2p^53s$ threshold. Compositional analysis of the other two eigenchannels shows[5] that they are well described by the coupling scheme $3s\varepsilon p(^3P^\circ,^1P^\circ)2p^5(^2S)$, the triplet channel being that with the resonance evident near the excitation threshold. Thus the eigenchannel description portrays the resonant state as one in which the coupling between two excited electrons (3s and εp) is of primary importance, and the coupling of the excited pair to the $2p^5$ ion core is secondary.

This view is supported by a comparative analysis of electron scattering by sodium, the eigenphases (from ref. 7) for which are shown by the dots in Fig. 1. At collision energies below the first excitation threshold of Na, the eigen- and collision channels are identical and are described by $2p^63s\varepsilon\ell(^3\ell,^1\ell)$, because the residual ion core is a closed shell. The dots depict the eigenphases for $3s\varepsilon p(^3P^\circ)$ and $3s\varepsilon p(^1P^\circ)$ symmetries; the zero of energy has been placed, not entirely arbitrarily, at the center of gravity of the Ne $2p^53s$ configuration. It is seen that the eigenphases of sodium and of neon are quite similar, even though the residual ion cores are somewhat different.

This raises the question as to what sort of variations in the ion core can be accomodated without changing the character of the external electron pair; or, on the other hand, as to which doubly-excited negative ion states show the least sensitivity to ion core structure. At present, there does not exist sufficient data from full scattering calculations to permit such an investigation over a wide range of negative ions. However, for long-lived resonances it is possible to obtain a fairly realistic description of the wavefunction by standard bound-state techniques, which neglect the interaction of the resonance with the detachment continuum. We now discuss some such results on the Na^- isoionic sequence.

MCHF CALCULATIONS OF LOW-LYING WANNIER RIDGE RESONANCES

The lowest members of the Wannier ridge resonance series are states which have been known for some time and which have been given names in conventional spectroscopic terminology (see e.g. ref. 10 for a catalogue of the noble gas resonances). It is possible to study them fairly economically by standard configuration interaction techniques. There are some features of these low-lying states which are not characteristic of the higher members of the series; for example, they do not have any appreciable admixture of high-ℓ components. On the other hand, the low-lying states should be most sensitive to variations in the positive ion core.

Table I Properties of the lowest Wannier ridge states in the Na^- isoionic sequence

State	Attachment Energy (eV) Calculated	Experimental	MCHF Composition $3s^2$	$3p^2$	$3d^2$	$4s^2$
Na^- $2p^6(^1S)3s^2$	0.541	0.546	0.943	0.309	-0.019	-0.124
Ne^- $2p^5(^2P^\circ)3s^2$	0.486	0.504	0.940	0.319	-0.018	-0.120
F^- $2p^4(^3P)3s^2$	0.425	-	0.936	0.332	-0.016	-0.115
O^- $2p^3(^4S^\circ)3s^2$	0.360	0.366	0.931	0.349	-0.014	-0.109
N^- $2p^2(^3P)3s^2$	0.398	-	0.925	0.365	-0.012	-0.104
C^- $2p(^2P^\circ)3s^2$	0.432	-	0.917	0.388	-0.008	-0.097

Attachment energy is the binding energy with respect to the neutral atom parent state $2p^n3s$ of maximum spin.
Experimental data is from references 9 - 11.

Table I shows the results of some calculations on an isoionic sequence of the Na^- ground state, which is generated by simultaneously removing a 2p electron and decreasing the nuclear charge. The calculations were performed in the multiconfiguration Hartree-Fock approximation, using a modified version of the code of Froese Fischer.[8] An advantage of the MCHF approach is that it determines from a variational principle both the orbitals and the weights that occur in a multiconfiguration expansion of the wavefunction. Thus comparison of results for different ions is not obscured by ambiguities in the choice of orbital bases. The ground state of Na^- is nominally $1s^22s^22p^63s^2$; the calculation reported here assumed a wavefunction of the form $1s^22s^22p^6(a3s^2+b3p^2+c3d^2+d4s^2)$ and solved (self-consistently) for all orbitals and the weights a,b,c,d. Because of the variational nature of the calculation, configurations of the type 3s4s do not appear explicitly, though their effects are present via linear combinations of the configurations $3s^2$ and $4s^2$. Inclusion of higher configurations such as $5s^2$, $4p^2$ does not substantively alter the results shown here. The attachment energy (in this case, the same as the electron affinity of sodium) is computed from the difference of the MCHF total energy and the Hartree-Fock energy of Na $1s^22s^22p^63s$. By this means the pair correlation energy of the two excited electrons is fully accounted for, but the additional intershell correlation energy associated with attachment of an electron is neglected. As it happens, this introduces only a small error.

Calculations for the other states in the sequence have been carried out after the same fashion. It can be seen that although the positive ion core undergoes significant changes, the composition of the external pair wavefunction (in terms of the relative weights of the different MCHF configurations) remains stable. The open 2p shell allows additional configurations to be present which represent transfer of angular momentum between the pair and the core, such as $2p^n3s3d$, but these do not change the basic picture. The composition is also insensitive to the LS term of the positive ion core: for example, one finds essentially the same external pair configuration bound to O^+ $2p^3$ $^4S^\circ$, $^2D^\circ$, and $^2P^\circ$ (there exists some experimental data, not described here, on resonances associated with higher terms of the positive ion cores).[11,12] Finally, the introduction of configuration interaction within the core, e.g. $2s^2 2p$ interacting with $2p^3$, does not much effect the external pair; this seems an obvious extension of the observation of its insensitivity to occupation number of the 2p shell, and it suggests that resonances formed by excitation of an electron out of the 2s shell will also be similar.

The actual spatial character of the wavefunction is also found to change only gradually along the sequence. Figure 2 shows a plot of the external pair wavefunction for the Ne^- state in hyperspherical coordinates θ, the angle between the two electrons, and α, the mock angle which is the inverse tangent of the ratio of the electron radii. This plot displays the wavefunction at a fixed value of the hyperradius $R = (r_1^2 + r_2^2)^{1/2} = 6$ a.u., which corresponds to the region of maximum amplitude. The wavefunction shows a pronounced peak at $\alpha = 45°$, $\theta = 180°$, which is the position of the Wannier ridge. This general pattern is present in the other ions in this sequence; the largest variations occur in the "valley" regions $\alpha \approx 0°, 45°$, where one electron is much closer than the other to the core.

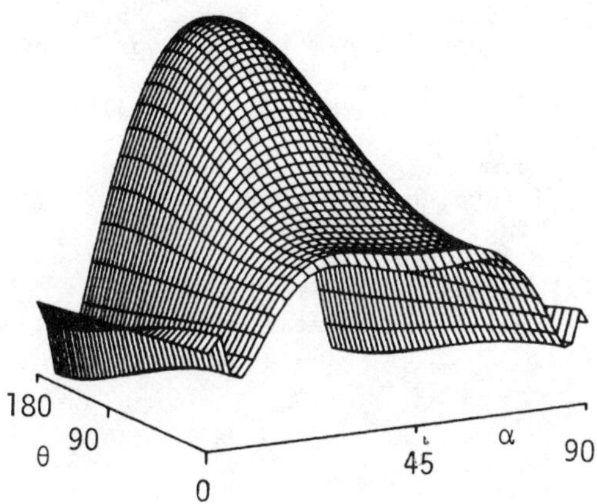

Fig. 2. Pair wavefunction for Ne^- resonance at R = 6 a.u.

The remarks made in this section apply in a general way to the isoionic sequence that proceeds along a column of the periodic table rather than a row. In particular, the noble gas negative ion resonances $np^5 n{+}1s^2$ look quite similar in terms of MCHF composition and probability amplitude upon the Wannier ridge. However, there is not room here to discuss this in detail.

CONCLUSIONS

There is good evidence, from several viewpoints, for "universal" behavior of some low-lying doubly-excited states of negative ions. The extent to which this applies to high members of the Wannier ridge series remains to be investigated in quantitative detail. These regularities should aid the identification of resonance states in atoms with complex core structure, and promote understanding of some aspects of the physics of multiple excitation.

Support of this work by the U.S. Air Force Office of Scientific Research is gratefully acknowledged.

REFERENCES

1. S.J.Buckman, P.Hammond, F.H.Read, and G.C.King, J.Phys.B 16, 4039 (1983)
2. U.Fano, in Atomic Physics 8, ed. I.Lindgren, A.Rosen, and S. Svanberg (Plenum, N.Y. 1983) p.5
3. A.R.P.Rau, J.Phys.B 16, L699 (1983); J.M.Feagin and J.Macek, J.Phys.B 17, L245 (1984); J.Macek and J.M.Feagin, J.Phys.B 18, 2161 (1985)
4. R.R.Freeman, L.A.Bloomfield, J.Bokor, and W.E.Cooke, these proceedings; P.Camus, P,Pillet, and J.Boulmer, these proceedings.
5. C.W.Clark and K.T.Taylor, J.Phys.B 15, L213 (1982)
6. K.T.Taylor, C.W.Clark, and W.C.Fon, J.Phys.B 18, 2967 (1985)
7. A.L.Sinfailam and R.K.Nesbet, Phys.Rev.A 7, 1987 (1973)
8. C.Froese Fischer, Comp.Phys.Commun. 14, 145 (1978)
9. H.Hotop and W.C.Lineberger, J.Phys.Chem.Ref.Data 4, 539 (1975)
10. S.J.Buckman, P.Hammond, G.C.King, and F.H.Read, J.Phys.B 16, 4219 (1983)
11. D.Spence, Phys.Rev.A 12, 721 (1975)
12. J.P.Grouard, V.A.Esaulov, R.I.Hall, J.L.Montmagnon, and Vu Ngoc Tuan, to be published

LASER PHOTODETACHMENT SPECTROSCOPY OF THE NEGATIVE CHLORINE ION

Rusty Trainham, G.D. Fletcher, and D.J. Larson
Department of Physics, University of Virginia
Charlottesville, VA 22901

ABSTRACT

This paper describes recent experiments on photodetachment of negative chlorine ions. The ions are created and stored in a Penning ion trap. Photodetachment is done with a high powered pulsed tunable dye laser. Preliminary results are presented for one and two photon detachment and for a threshold shift in the presence of a strong laser field.

INTRODUCTION

Negative ions permit the study of atomic and molecular properties in systems which are not dominated by the long range Coulomb interaction between a core and valence electrons. For negative ions, electron correlations play a primary role in determining structure. Because of the short range force on the extra electron, negative ions have few bound states, usually one for atomic ions. In addition, the photoexcitation continuum is, to a rather good approximation, simply that of a free electron.

Photodetachment spectroscopy has become the primary technique for high resolution study of the structure of negative ions. In such experiments, ions are produced in a source, the extra electron is detached with light of an appropriate wavelength, and the loss of ions or production of neutrals or free electrons is monitored. Photodetachment spectroscopy has been used to study or measure a large number of properties including electron affinities, structure in the negative ion or neutral such as fine structure or rotational structure, excited states of the ion below and above the detachment threshold, sizes and shapes of detachment cross sections in a variety of circumstances, and the effects of interaction with external fields.

In the present experiments, photodetachment of trapped Cl^- ions is used to measure the threshold and energy dependence of the one and two photon photodetachment cross sections. In addition, the clearly defined continuum threshold allowed observation of a shift in that threshold due to the presence of a strong laser field. Such a shift in the photoionization threshold has been assumed to play an important role in experiments on high power photoionization of neutrals[1] and has been anticipated theoretically in negative ions by a number of authors.[2]

EXPERIMENTAL TECHNIQUE

The basic apparatus in these experiments is a Penning ion trap. The use of traps can offer a number of advantages in ion spectroscopy, including, notably, long storage times, which lead to the possibility of ultra high resolution spectroscopy. Other features which are significant in various experiments include high sensitivity measurement of ion number, straightforward use with a magnetic field, the possibility of having a sample of cold ions, and provision of a relatively benign and well characterized environment. In the present experiments, the long storage times (on the order of many seconds) and the high sensitivity allow straightforward measurement of photodetachment even when the fraction of detachment in a single laser pulse is very small.

The Cl^- ions are created in the Penning trap by dissociative attachment. An electron beam with a current of about 100 nA at an energy of 1 eV incident upon a gas of CCl_4 at about 10 nanotorr produces many thousands of Cl^- ions in a few seconds. The number of ions in the trap is determined by driving the ions' motion and measuring the current induced on a trap electrode. The ions are then subjected to many pulses of light from a Nd:YAG pumped dye laser. A substantial fraction of the ions are destroyed due to photodetachment by the laser light. Finally, the number of ions which survive through the photodetachment period is measured. All of the data are presented by plotting the fraction of ions which survive as a function of laser photon energy.

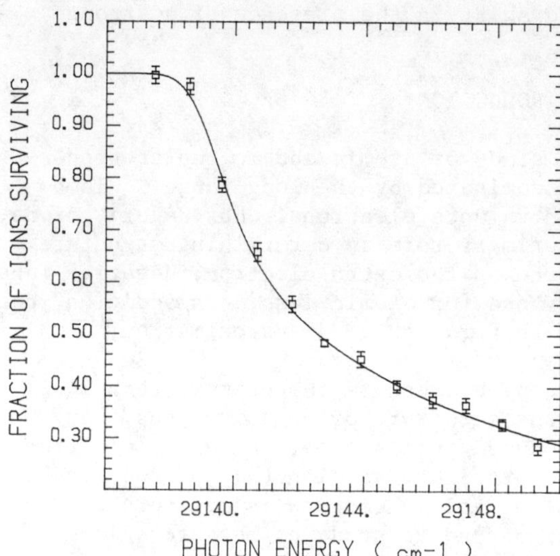

Fig. 1. One photon detachment.

OBSERVATIONS

Figure 1 shows the results for single photon detachment of the trapped Cl^- ions. The data are fit to a curve which includes a cross section which rises from threshold according to the Wigner law[3] for an s-wave continuum ($\sigma \propto (\nu-\nu_o)^{1/2}$) as well as an allowance for the laser linewidth ($\sim$1 cm^{-1}). Oscillations in the cross section due to the presence of a 1 T magnetic field are barely visible because of the relatively large linewidth. The data are in very good agreement with the model and the fit indicates an electron affinity of 29,140(1) cm^{-1}, in agreement with but substantially more precise than the previous measurements of this quantity.[4] Absolute wavelength calibration of the laser is provided by measuring the opto-galvanic resonances in a hollow cathode tube containing argon.

The results for one measurement of the two photon detachment cross section are presented in Figure 2. These data are fit to a curve which includes the Wigner law for transitions to a p-wave continuum ($\sigma \propto (\nu-\nu_o)^{3/2}$). Because of the slow rise in the cross section, data are taken over a much wider wavelength range than in Fig. 1. This is the first measurement of the energy dependence of a nonresonant two photon photodetachment cross section. The data are in good agreement with the model but measurements of the detachment fraction as a function of laser power at a fixed wavelength show a small first order dependence on laser power in addition to the expected second order dependence.

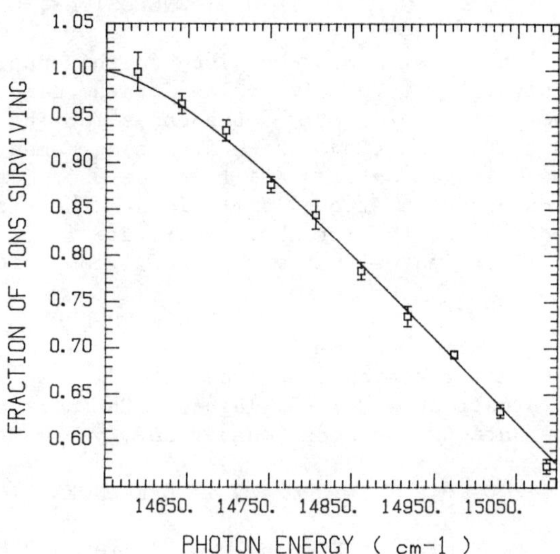

Fig. 2. Two photon detachment

In order to investigate the effects of a strong laser field on the photodetachment process, a fraction of the Nd:YAG laser beam at 1.06 micron wavelength was directed through the trap at the same time as the single photon photodetachment pulse. Data were taken alternately with and without the infrared light. An example of the results is shown in Fig. 3. A shift of the threshold toward higher energy is clearly evident. This shift is approximately 1 cm^{-1} at an infrared intensity on the order of 10^{10} W/cm^2. Theoretically, this shift has been anticipated to be of this order, with the largest contribution arising from the increased energy of the electron in the oscillating optical electric field.

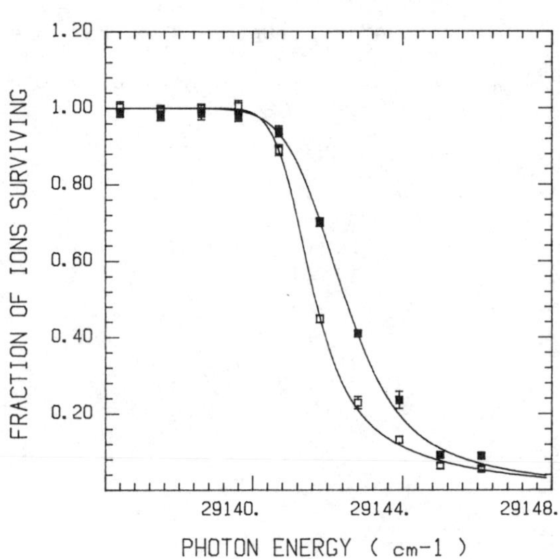

Fig. 3. Threshold shift due to a strong infrared laser pulse.

CONCLUSIONS

The experiments described in this paper have led to the measurement of a more precise value for the electron affinity of Cl, the observation of energy dependences for the one and two photon detachment cross sections which are in agreement with the Wigner law, and the observation of a shift in the photodetachment threshold in the presence of a strong laser field. The sizes of the two photon cross section and the threshold shift are in rough agreement with theoretical estimates, but more careful comparisons should be made.

ACKNOWLEDGEMENTS

This research benefited from discussions with R. Freeman, L. Armstrong and T. Gallagher. The work was supported in part by the National Science Foundation.

REFERENCES

1. P. Kruit, J. Kimman, H.G. Muller and M.J. van der Weil, Phys. Rev. A 28, 248 (1983); L.-A. Lompre, A. L'Huillier, G. Mainfray, and J.Y. Fan, J. Phys. B 17, L817 (1984); M.H. Mittleman, J. Phys. B 17, L351 (1984); S.-I. Chu and J. Cooper, Phys. Rev. A 32, 2769 (1985); H.G. Muller and A. Tip, Phys. Rev. A 30, 3039 (1984).
2. K. Rzazewski, M. Lewenstein, and J.H. Eberly, J. Phys. B 15, L661 (1982); M.V. Federov and A.E. Kazakov, J. Phys. B 16, 3641 and 3653 (1983); S.L. Haan and J. Cooper, J. Phys. B 17, 3481 (1984).
3. E.P. Wigner, Phys. Rev. 73, 1002 (1948); H. Hotop, T.A. Patterson, and W.C. Lineberger, Phys. Rev. A 8, 762 (1973).
4. G. Muck and H.-P. Popp, Z. Naturf. 23a, 1213 (1968); I.S. McDermid and C.R. Webster, J. de Phys. 44, C7-461 (1983).

LASER SPECTROSCOPY WITH RELATIVISTIC BEAMS

H. C. Bryant
Department of Physics and Astronomy
The University of New Mexico
Albuquerque, N. M. 87131

APPLIED RELATIVITY

The man on the street might say, "Relativity is interesting and all, but what good is it?" In this paper I describe how relativistic kinematics makes possible spectroscopy that otherwise would be beyond our reach. The two key ingredients for our experiments with H^o and H^- are the relativistic Doppler shift and the Lorentz transformation of the magnetic field.

THE DOPPLER SHIFT

By directing a laser beam at fixed frequency and variable angle across an atomic beam with near luminal velocity, one has in effect a tunable laser whose intensity varies quite smoothly with angle.

HOW TO MAKE LARGE ELECTRIC FIELDS THE EASY WAY

Large electric fields F appear in the rest frame of an atom transversing even a modest transverse magnetic field B at relativistic velocity.

For example, for 650 MeV H^-'s, and B = 1 Tesla, F = 4.1 MV/cm. To produce such a field in the lab frame would require heroic measures, and the detection of reaction products would be a nightmare.

THE H^- ION

These relativistic principles can be applied to one of the simplest atomic systems. H^- has no singly-excited states. Yet, it has a rich structure of doubly excited states. If one electron stays home in the ground state, it shields the proton so much that an excited electron does not see enough attractive force and wanders off. But, if both electrons leave home, the proton is unshielded and they both can be kept in the neighborhood. The two electrons cannot block one another's view of the proton.

H^- IONS IN ELECTRIC FIELDS

In order to keep H^- intact in an electric field, its highly correlated electrons must avoid the leaky part of the potential where they can be pulled off by the field. The survival rate of doubly-excited state of H^- in electric fields seems to depend strongly upon radial correlations[1].

EXPERIMENTAL METHOD

Fig. 1 shows a schematic diagram of our experimental apparatus. The 650 MeV beam is coming from the left in a series of macropulses of 500 sec spaced 8 msec apart. Each macropulse consists of a string of RF micropulses of 1/4 nsec duration spaced 5 nsec apart (200 MHz). The laser typically fires at 10 pps with an 8 nsec pulse.

THE RESONANCE REGION NEAR n = 2

Fig. 2 shows[1] the photodetachment cross section in H^- in a 300 meV range around 11 eV. We note two striking structures; a Feshbach resonance just below the n=2 threshold, which, since its width is estimated to be around 30 microvolts, shows us that our resolution is about 7 meV. Above the n=2 threshold we find the shape resonance, a structure with a width of some 20 meV.

Over the years our group has studied this region[2] in electric fields. We have found that the response of these two doubly-excited states to electric fields is competely different. The Feshbach resonance undergoes a linear Stark shift because of the presence of a broader ^1S state which is degenerate with the ^1P state. Very quickly, relatively speaking, the members of the Stark triplet quench, the lower lying (m=0) state going first, followed by the middle (m=±1) state. The upper (m=0) state may not quench but simply disappears under the avalanche of cross section spilling down from the broadening of the shape resonance.

THE ROCK OF GIBRALTAR

It is not a limestone mass rising 1408 feet above the water, but the shape resonance behaves as one. It refuses to quench in an electric field as far as we can tell[3]. We have turned the electric field up so high that the parent ground state is stripping at a rate which veils the shape; we are overwhelmed by this background. At the highest fields for which our counters are not swamped, the shape, broadened and symmetric, is still there.

THE SHAPE RESONANCE AT 2.4 MV/cm

Fig. 3. shows what happens[1] to the region shown in Fig. 2 under an electric field of 2.4 MV/cm. This data was taken simultaneously with that shown in Fig. 2. Thus, the normalizations are the same for both cross sections. We can see no trace of the Feshbach resonance. There is a sign of a ^1D resonance induced by mixing with the shape, on the low energy side of the structure.

IS A LITTLE POISON GOOD FOR YOU?

In the field region where the Feshbach resonance has broadened and quenched, the shape resonance appears to narrow[4]. We made a study of this effect using the pulsed magnet, which should cancel out systematic problems, and we see a statistically significant, but small effect of the order of 3%. A report on this work is presently in preparation by our group[4].

THE REGION BELOW n=3

The structures seen near n=2 are not unique. There are also resonant structures associated with the higher excited hydrogenic states. In particular, near n=3 we see a resonance manifesting itself as a dip[5]. Were it not for coherent interference with the direct ^1P background, this structure would be a symmetric Breit-Wigner line shape. This object is much more difficult to study than the shape resonance, since its strength is only about 0.3, whereas that of the shape resonance is around 8.0.

THE n=3 DIP IN STRONG FIELDS

Stanley Cohen is completing his dissertation at UNM on the behavior of the n=3 dip in strong electric fields[6]. Like the shape, the dip appears very resistant to electric field, and does not shift significantly with field.

THE EFFECT OF ELECTRIC FIELDS ON THE THRESHOLD REGION

When $\beta \equiv \frac{v}{c} = -\cos\alpha$, $\delta E/E$, to first order, vanishes. We are familiar with the result that for small velocities this angle is 90°.

To study the threshold of the H^- photodetachment continuum, which starts at an energy corresponding to the H^o electron affinity, 0.7542 eV, we apply this kinematic advantage using the 1st YAG harmonic.

Thus, if we set E = 0.754 and E_L = 1.165 eV, we need an H^- beam energy of 500 MeV. The energy resolution in this region is then governed by the angular uncertainty. That is

$$\delta E = \gamma E_L \sin\alpha\ \delta\alpha.$$

Since $\delta\alpha \sim 1$ mr, our energy resolution is less than 1 meV. We are presently analyzing our data on this region.

CONCLUSION

Over the last few years the use of relativistic H^- beams combined with pulsed ultraviolet lasers has shed new light on the structure of this simplest two-electron system. As accelerator and laser technology continue to evolve, resolutions will improve, and the fine details (where all the action is!) will continue to reveal themselves.

ACKNOWLEDGEMENT

This work was performed under the auspices of the D.O.E. and supported in part under contract DE-AS04-77ER03998. The work reviewed here was done with the collaborators named below.

REFERENCES

1. H. C. Bryant, C. J. Harvey, J. E. Stewart, K. B. Butterfield, D. A. Clark, J. B. Donahue, P.A. M. Gram, D. W. MacArthur, V. Yuan, G. Comtet, W. W. Smith, Contributions to the SASP '86, page 86, 1986.
2. H. C. Bryant, D. A. Clark, K. B. Butterfield, C. A. Frost, H. Sharifian, H. Tootoonchi, J. B. Donahue, P. A. M. Gram, M. E. Hamm, R. W. Hamm, J. C. Pratt, M. A. Yates and W. W. Smith, Phys.Rev. A 27, 2889 (1983).
3. K. B. Butterfield, Ph.D. dissertation, Univ. of New Mexico, 1984 (published as report LA-10149-T by Los Alamos National Laboratory.)
4. G. Comtet, C. J. Harvey, J. E. Stewart, H. C. Bryant, K. B. Butterfield, D. A. Clark, J. B. Donahue, P. A. M. Gram, D. W. MacArthur, V. Yuan, W. W. Smith, and Stanley Cohen, to be published.
5. M. E. Hamm, R. W. Hamm, J. B. Donahue, P. A. M. Gram, J. C. Pratt, M. A. Yates, R. D. Bolton, D. A. Clark, H. C. Bryant, C. A. Frost, and W. W. Smith, Phys. Rev. Lett. 43, 1715 (1979).
6. Stanley Cohen, Ph.D. dissertation, Univ. of New Mexico, 1986, and Cohen et al., to be published.

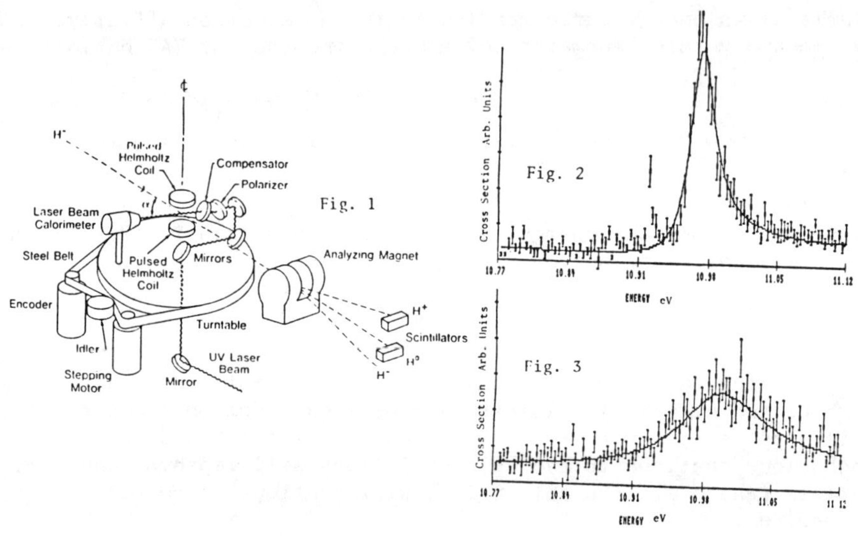

Fig. 1

Fig. 2

Fig. 3

CROSS-SATURATING SIDEBAND ABSORPTION: THE LIMITS OF DETECTIVITY

J. Bialas, R. Blatt, W. Neuhauser, and P.E. Toschek
Universität Hamburg, D-2000 Hamburg, F.R. Germany

INTRODUCTION

Current laser-spectroscopic techniqes are characterized by both high spectral resolution and superior sensitivity. [1,2] Taking advantage of fluorescence detection, the observation of single atomic particles is feasible. [3,4] In *absorption* spectroscopy, hf modulation of the (probe) light plus optical heterodyne detection has dramatically enhanced the sensitivity [5,6], since the frequency of detection is shifted from 0 to a range with negligible technical noise. This method has potential for Doppler-freee operation by the use of counter-running pump and probe beams. However, in order to reach shot-noise-limited signals, additional *modulation of the sample* is required. This can be achieved either by extra modulation of the pump light [7], or by periodic switching the light to a path which bypasses the sample, giving rise to a modulated $\chi^{(3)}$. [8] In this way, the dominant background due to stray light is eliminated.

ULTRA-SENSITIVE TECHNIQUES

Considerable improvement of S/N results from *rf modulation of the pump* and observing, by heterodyne detection, *amplitude modulation transferred*, via $\chi^{(3)}$, to the counter-running probe beam [9,10], s. Fig. 1. The resulting spectra show flat baselines and excel with applications which involve the control of the light frequency by atomic resonances. A variant of this scheme [11] makes use of *two* light frequencies in a single beam, resonant with two coupled transitions. It avoids cross-talk by light scatter and has demonstrated high resolution at μW light power levels.[12] However, the density of the sample, and therefore the number of interacting particles, is large and not well calibrated, in general.

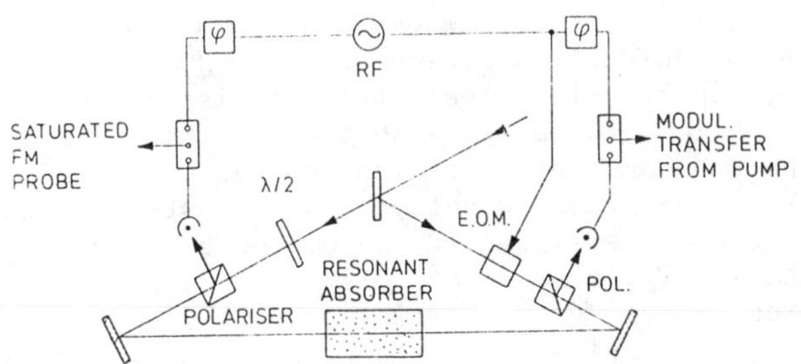

Fig. 1: Scheme of the saturation of a frequency-modulated probe (detection on the left side) and of modulation transfer, from a fm pump beam, to the probe (detection on the right side).

We have applied this technique of cross-saturating sideband absorption (CSA) to small clouds of Ba^+ ions localized in a rf ion trap, and calibrated the signal by observation of single-ion fluoresecence, s. Fig. 2.

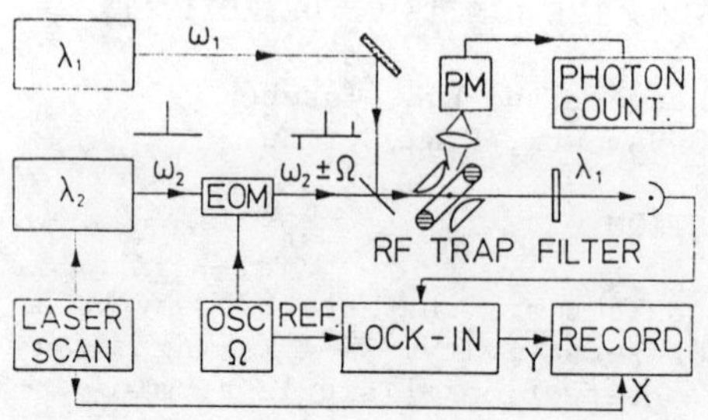

Fig. 2: Experimental scheme of cross-saturating side-band absorption.

FOUR-WAVE MIXING WITH FM LIGHT

A rf-modulated (Ω) light beam (ω_2) interacts with an absorber close to resonance 0-2 such that one sideband ($\omega_2 + \Omega$) and the unmodulated probe light ω_1 (close to resonance 0-1) are two-photon resonant with the electronic Raman transition 1-2. Light of the ω_2 carrier is scattered off the Raman coherence, and ω_1 light is scattered off pulsating population in level 0. Both processes generate the probe sideband $\omega_1 - \Omega$ via the 3rd-order contribution $\chi^{(3)}$ to the susceptibility.[13,14] Optical heterodyne detection of the probe recovers the cross-modulation, Ω. The resonance conditions are

$$\Delta = \begin{cases} \pm (\Omega/2)\ (2k_2/k_1 \pm 1) + \varepsilon \\ \pm (\Omega/2) + \varepsilon \end{cases} ,$$

where $\Delta = (k_2/k_1)\ (\omega_1 - \omega_{10}) - \omega_2 + \omega_{20}$, and $\varepsilon = \pm\ \Omega/2$ for the sidebands of ± 1st order. They are indicated with the recorded spectra.

EXPERIMENTAL

A cloud of up to 10^4 Ba^+ ions has been localized in a rf trap[15] of 3.2 mm Ø, with a 5.8-MHz drive generating a pseudo-potential well of 10 V depth. The Doppler width (FWHM) of fluorescence spectra is 1 GHz corresponding to 500 K which results from parasitic rf heating and weak optical cooling.

Probe light (λ_1 = 650 nm) of .5 MHz rms bandwidth and pump light (λ_2 = 493 nm) of 1 MHz rms bandwidth are generated by cw DCM and Coumarine-102 dye lasers, respectively. The latter one is frequency-modulated (index 1.3) by a $LiTaO_3$ EOM at $2\pi \cdot 36.5$ MHz. The superimposed beams transmit the trap and interact with the ion cloud. After the pump beam blocked, the probe is detected by a fast Si photodiode, and the amplified heterodyne signal (Ω) recovered by a fast lock-in detector. Simultaneously, fluorescence light at ω_2 is collected by a photon counting system.[3] When starting with an empty trap and injecting weak electron pulses into the trap, which is crossed by a weak Ba beam, eventually a Ba^+ is generated whose signature is a step in the time dependence of the fluorescence counting rate. From the size of this step, the ion number in large clouds is calibrated, when also a detectable CSA signal is generated.

RESULTS

A CSA signal from 2000 ions is shown in Fig. 3, where, with fixed tuning of the probe frequency, the pump light was frequency-scanned across $\Delta = 0$. This is the condition for the *coincidence* of two-photon (Raman-) and single-photon resonance in the same ionic velocity class contributing to the Doppler-broadened line. The up and down pointing arrows indicate the location of the two sets of

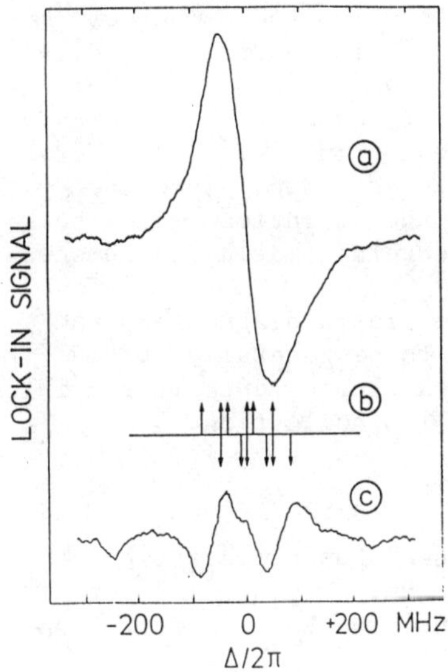

Fig. 3: CSA spectrum: In-phase (a), quadrature (c). Time constant $\tau = 10$ s. Location of individual resonances (b).

six resonances expected from heterodyning the probe carrier with both the lower and upper 1st-order sidebands. This resonance structure is but partially resolved in the quadrature signal. - We have measured the CSA signal from clouds of different size and plotted it

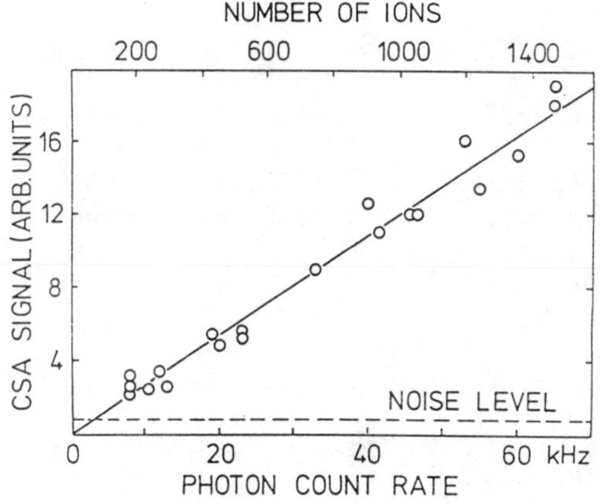

Fig. 4: CSA amplitude *vs.* number of trapped ions. The limit of detectivity, set by quantum shot noise ($\tau = 10$ s), is 65 ions.

vs. the pertaining fluorescence signal (Fig. 4). It equals the level of noise, which is dominated by quantum shot noise, at 65 ions interacting with the light. Since the temperature of the ions in the trap is about 500 K, improvement of this figure seems feasible by better cooling the ions.

CONCLUSIONS

Next to fluorescence detection, CSA has been demonstrated the most sensitive spectroscopic technique in terms of the number of particles interacting with the light. The minimum extinction, determined by the particular absorber, is $5 \cdot 10^{-8}$, which is 8 orders of magnitude lower than the estimated sensitivity of current measurements of two-photon-excited fluorescence from H in flames.[16] Very dilute species may be detected with CSA. In addition, since the signal is contained in *one* spatial mode of the radiation field, background light does not cause problems, and careful imaging is unnecessary.

Applications of this technique include plasma diagnostics and the study of chemical reactions. It seems to be generally suited for the detection of rare species which lack fluorescence intrinsically, or whose environment does not allow its collection.

REFERENCES

1. V.S. Letokhov, V.P. Chebotayev, "Nonlinear Laser Spectroscopy", Springer-Verlag, Heidelberg, New York, 1977
2. See, e.g., various papers in: "Ultrasensitive Laser Spectroscopy", J. Opt. Soc. Am. B2, no 9, Sept. 1985
3. W. Neuhauser, M. Hohenstatt, P.E. Toschek, and H.G. Dehmelt, Phys. Rev. A22, 1137 (1980)
4. D.J. Wineland and W. Itano, Phys. Lett. 82A, 75 (1981)
5. G.C. Bjorklund, Opt. Lett. 5, 15 (1980)
6. J.L. Hall, L. Hollberg, T. Baer, and H.G. Robinson, Appl. Phys. Lett. 39, 680 (1981)
7. M.D. Levenson, W.E. Moerner, and D.E. Horne, Opt. Lett. 8, 108 (1983)
8. M. Gehrtz, G.C. Bjorklund, and E.A. Whittaker, J. Opt. Soc. Am. B2, 1510 (1985)
9. D. Bloch, R.K. Raj, and M. Ducloy, Opt. Communic. 37, 183 (1981)
10. J.L. Hall, L. Hollberg, and Ma Long Sheng, to be published
11. J. Bialas and P.E. Toschek, Verhandl. DPG (VI) 20, 1036 (1985)
12. S. LeBoiteux, D. Bloch, M. Ducloy, A. Pesnelle, S. Runge, and M. Perdrix, Opt. Communic. 52, 274 (1984)
13. J.-L. Oudar and Y.R. Shen, Phys. Rev. 22A, 1141 (1980)
14. S. LeBoiteux, D. Bloch, and M. Ducloy, J. Physique, to be published
15. W. Neuhauser, M. Hohenstatt, P.E. Toschek, and H.G. Dehmelt, Appl. Phys. 17, 123 (1978)
16. R.P. Lucht, J.T. Salmon, G.B. King, D.W. Sweeney, and N.M. Laurendeau, Opt. Lett. 8, 365 (1983)

ATOMIC VAPOR LASER ISOTOPE SEPARATION*

R. C. Stern, J. A. Paisner
Lawrence Livermore National Laboratory, Livermore, CA 94550

Atomic vapor laser isotope separation (AVLIS) is a general process for converting a feed stream into a product stream in which a selected set of isotopes has been enriched or depleted. Laser isotope separation technology for uranium enrichment has been under development at Livermore since 1973 in a joint effort with Oak Ridge. In June 1985, following an intensive year-long review, the Department of Energy selected AVLIS as the technology to meet the nation's future need for competitive production of enriched uranium. Uranium enrichment is sold in separative work units (SWUs). For natural uranium feed, light-water-reactor fuel product, and typical tails assay, one SWU converts 1.24 kg of feed into 0.21 kg of product. The world demand corresponds to processing 40,000 Mg of uranium feed/year. Hence, any uranium-enrichment process involves large-scale hardware. Based on the range of projected prices, we want AVLIS to come in at less than $60/SWU. This corresponds to a process cost of less than $49/kg of feed or less than $286/kg of product.

To evaluate what is required of the laser system to meet these costs, we can estimate the megajoules of laser energy required to process 1 kg of feed (Fig. 1). Putting in values characteristic of the AVLIS process, including the very high process photoselectivity, roughly 100 kJ of laser energy are needed to process 1 kg of uranium feed. Using typical costs for an engineered laser system, we expect the laser cost/kg of feed to be on the order of $10.

The other half of the physics performance is the separative work generated per kg of feed. Figure 2, which is a map of separative work performance vs product and tails assays, shows that high SWU production corresponds to low tails assays, not to high product assays, as one might naively think. Therefore we want a process that strips a very high fraction of the ^{235}U atoms out of the feed.

With this background we consider the physics of the process (Fig. 3). Uranium vapor is generated by an electron beam striking uranium

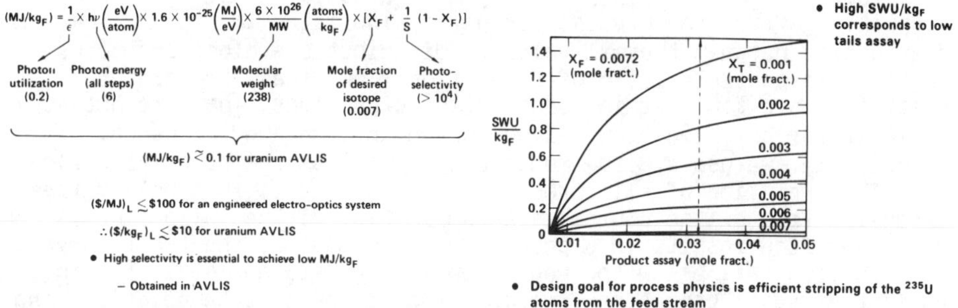

Fig. 1. Laser figures of merit. Fig. 2. Separative work performance.

*Work performed under the auspices of the US Department of Energy by the Lawrence Livermore National Laboratory under Contract W-7405-Eng-48.

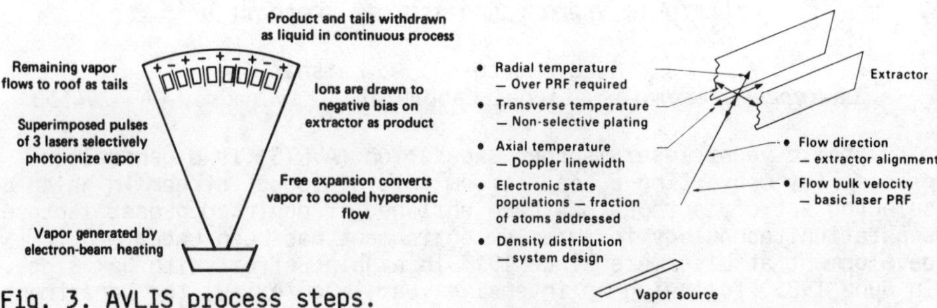

Fig. 3. AVLIS process steps.

Fig. 4. Vapor properties of interest.

metal held in a cooled crucible. This type of vaporization is commonly used. What is unusual about our application is that uranium is highly refractory (the metal boils at 4100 K) and that very high vaporization rates are desired. The vapor then undergoes adiabatic free expansion into vacuum, producing a cooled, hypersonic flow.

As the vapor passes between the extractor plates, it is illuminated by three superimposed laser pulses of the frequencies necessary to drive a three-step photoionization. The initial photoplasma contains, to a very good approximation, only ^{235}U ions. In time, resonant charge exchange between ^{235}U ions and ^{238}U neutrals would restore the ion population to natural abundance. Therefore the photoplasma is rapidly extracted.

Finally, the enriched product is collected from the plates and the depleted tails are collected from the roof. In order to obtain the economic advantages of continuous operation, the streams are collected by liquid flow at high temperature. We are doing precision laser physics inside a foundry.

There is more physics here than one can cover in a brief paper. Therefore we focus on the laser diagnostic measurements of vapor properties and the laser measurements of the spectroscopic parameters that enter the photoionization process.

A major use of lasers in the program has been the measurement of the properties of the vapor stream. A remarkable number of vapor properties affect the process design, as indicated in Fig. 4. The distinction between thermal-velocity distributions along and across the flow is characteristic of transition flow. These temperature splittings are familiar in point source expansions. They are not as familiar for our flow from a line source into an angular wedge. Accordingly, we use a Bird--Monte Carlo model of the transition flow.

The evolution of all three velocity components in a helium free expansion from a line source is shown in Fig. 5. The solid line is the continuum flow result. In a point-source expansion, the two transverse temperatures fall below the continuum line due to geometric cooling, while the radial component freezes out at a constant temperature. The pattern here is essentially different.

The vapor diagnostic system makes simultaneous real-time measurements on a routine basis. Diagnostic laser beams individually modulated at five frequencies are combined into single-mode optical fibers and transported to the separator. There the output from each

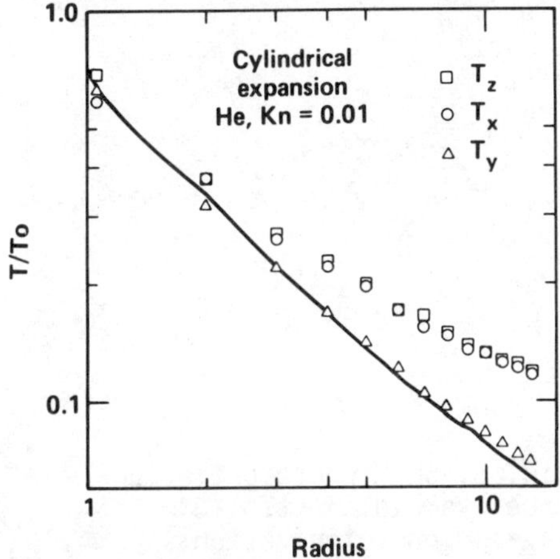

Fig. 5. Temperatures in transition flow.

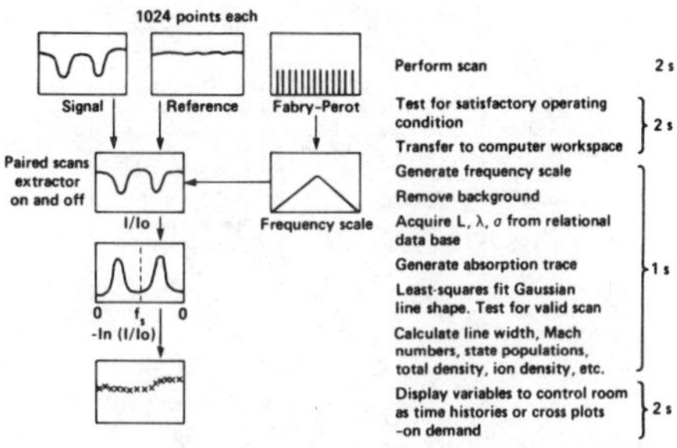

Fig. 6. Real-time reduction of laser measurements.

fiber is again split and sent to one of three types of probes. Up to 12 demodulated signal and reference voltages are sent in parallel to the computer.

The process-control computer captures the signal, reference, and interferometer scans, linearizes the frequency scale, and determines the absorption (Fig. 6). Absolute densities in individual electronic states are determined using vapor path lengths and optical cross sections. Line shapes are fit to determine velocity distributions, temperatures, and Mach numbers. Time histories and cross plots are presented on control-room consoles.

The flow chart in Fig. 7 shows the analysis and experimental data base we apply to the photoionization physics. The first-principles model integrates Schrodinger's equation, using a complete set of experimentally measured spectroscopic parameters for the atomic transitions. The code treats the coherent multiple laser/atom interactions and accounts for every magnetic sublevel and velocity class within each hyperfine level in the excitation sequence. The code also computes the atomic polarizations driven by the light fields and thus facilitates analysis of the back reaction of the atoms on the propagating lasers. Figure 8 shows an example of three-step photoionization of ^{157}Gd (nuclear spin I = 3/2) using simultaneous excitation by single-mode lasers tuned to the center of gravity of levels in a J = 2→2→1→0 ladder.

The simplest measurements in the spectroscopic data base are radiative lifetimes from time-delayed photoionization. Data for atomic gadolinium are illustrated in Fig. 9.

Perhaps the most important spectroscopic parameters are the transition dipole moments. A method we use for accurately measuring these directly observes Rabi flopping of atomic populations (Fig. 10). Here ^{160}Gd was photoionized by three resonant laser pulses. The first- and second-step excitations are derived from pulse-amplified cw dye laser systems, allowing selective excitation and detection of ^{160}Gd without a mass filter. The second and subsequent photoionizing

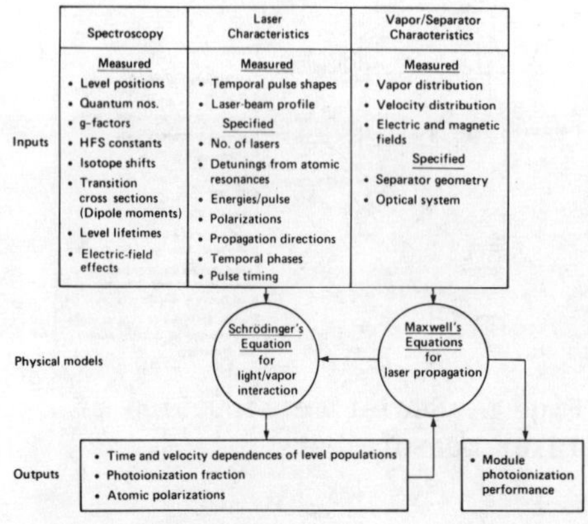

Fig. 7. Modeling of laser/atomic vapor interaction.

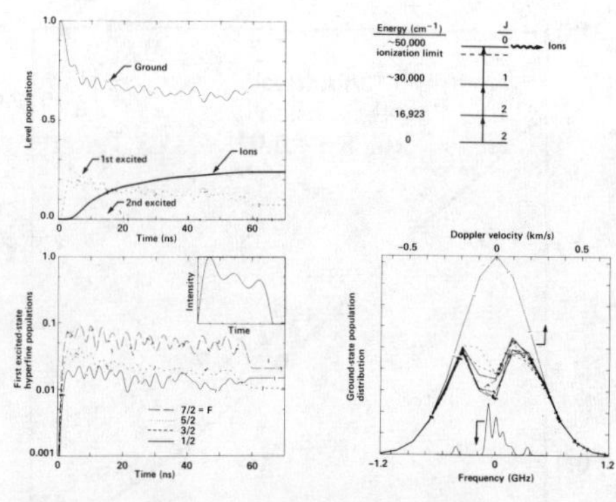

Fig. 8. SHUX photoionization code example: 3-step, J = 2→2→1→0 photoionization sequence in ^{157}Gd.

steps were delayed in time from the first-step excitation pulse. The photoion signal is plotted as a function of the first-step laser fluence. The dramatic oscillations in the photoion signal arise from the coherent evolution of the atom between the manifold of magnetic sublevels in the ground and first excited states. Since each transition in the manifold has its own dipole moment, the signal is not purely sinusoidal. Laser polarizations, fluences, and level J values are known. The only parameter varied to fit the data is the transition oscillator strength. We have used this method to measure transition oscillator strengths for bound-bound transitions originating on both low- and high-J excited states.

The program at Livermore has adopted a fundamental approach to the process physics and developed laser-based techniques to obtain the necessary data base as described above. Once fully developed and deployed for uranium, the AVLIS process can be applied directly to separating many elements economically on an industrial scale.

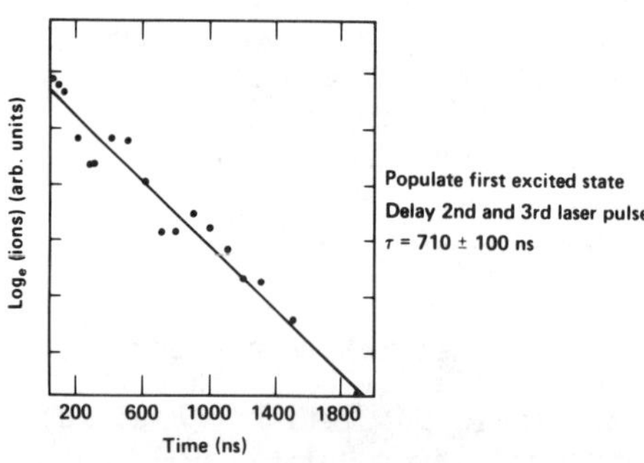

Fig. 9. Lifetime measurement of 17,380.8 cm^{-1} level in atomic Gd using time-delayed photoionization.

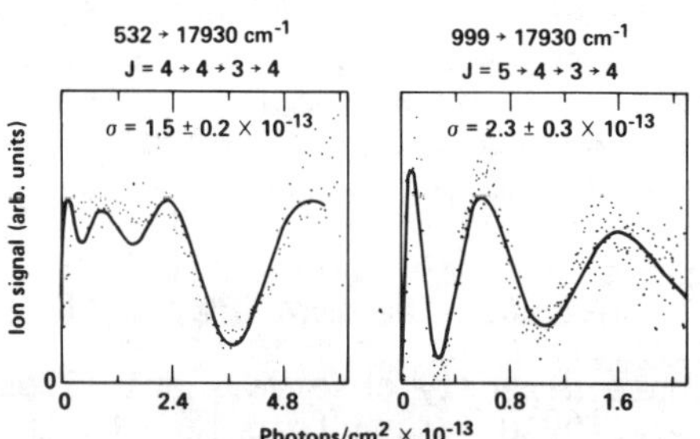

Fig. 10. Cross-section measurements using Rabi oscillations in ^{160}Gd.

OPTICAL LAMB-DICKE CONFINEMENT OF A Ba^+ MONO-ION OSCILLATOR

W. Nagourney, H. Dehmelt and G. Janik
Dept. of Physics, University of Washington, Seattle, Washington 98195

ABSTRACT

We demonstrate the confinement of a single laser cooled Ba^+ ion, held in a radiofrequency trap, to a region of space whose linear dimensions are much less than the wavelength of the interrogating light divided by 2π (Lamb-Dicke regime[1]). This is the condition for the complete freedom from first order Doppler shifts. From the absence of observable Doppler sidebands on a narrow, two-photon feature in the spectrum of Ba^+, we infer an upper bound on the ion orbit radius of 165 nm. This is the first clear demonstration of Lamb-Dicke confinement in the visible.[2]

INTRODUCTION

The suppression of Doppler shifts and broadenings has always been one of the main goals of high resolution spectroscopy. In the recent past, most approaches to Doppler suppression in the optical region have utilized atom-light interactions (such as saturation or two-photon methods) which are insensitive to the motions of rapidly moving atoms. It is only with the recent success of laser cooling of trapped ions that it has been possible to suppress Doppler shifts in a more direct fashion: by essentially bringing the ion to rest. Unlike other methods of Doppler suppression, laser cooling suppresses Doppler shifts to all orders.

EXPERIMENTAL TECHNIQUE

The Ba^+ ion is trapped in a small radiofrequency trap whose endcap separation is 0.5 mm and whose ring diameter is 0.7 mm. The stainless steel trap electrodes are fashioned with circular (not hyperbolic) contours. The trap is enclosed in a highly evacuated (pressure $\approx 4 \times 10^{-11}$ Torr) pyrex envelope with Brewster angle windows for the laser beams.

Two single frequency lasers cause a two photon transition to be made between the $6^2S_{1/2}$ and $5^2D_{3/2}$ states via an almost resonant intermediate $6^2P_{1/2}$ state. The ion is cooled by slightly downtuning the 494 nm laser, which drives the $6^2S_{1/2}$ to $6^2P_{1/2}$ transition. The other laser, at 650 nm, is swept to observe the two photon resonance. The observed signal is due to the fluorescence from the P state to the S state. A small magnetic field reduces the adverse effects of optical pumping in the D level without excessively broadening the $\approx$2 MHz two photon resonances.

The long term stability of the lasers is enhanced by frequency locking both lasers to the same stable, confocal resonator. Before entering the resonator, the

650 nm laser beam is frequency shifted with an acousto-optical modulator driven by a computer-controlled frequency synthesizer. A computer, programmed as a signal averager, simultaneously scans the 650 nm laser (via the synthesizer) over the two photon resonance and integrates the photon counts from the photomultiplier.

RESULTS

It can be easily shown[1] that the first order Doppler broadening of an ion undergoing simple harmonic motion in a trap is manifested by the presence of a series of sidebands spaced from the main feature by the oscillation frequency. For single photon transitions, the intensity of the nth sideband can be shown to be proportional to $J_n^2(\beta)$ where J_n is the Bessel function of order n and β is the ratio of the ion's orbit radius to $\lambda/2\pi$. Thus, the absence of sidebands implies that the ion is confined to a region of space which is much smaller than $\lambda/2\pi$ (Lamb-Dicke regime). This conclusion also holds for two photon transitions, although the dependance of the sideband intensity on β is somewhat different.

By establishing a (noise limited) upper bound on the sideband intensities (see Fig. 1), we estimate that the single laser cooled Ba^+ ion has a maximum orbit radius of 165 nm. Since this is much less than $\lambda/2\pi$, the ion confinement is well into the Lamb-Dicke regime.

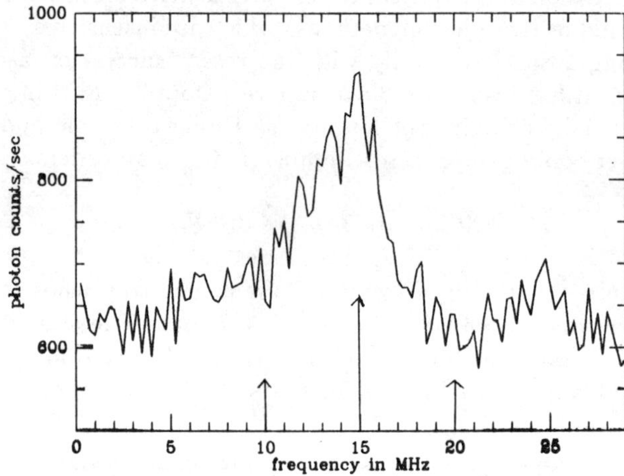

Fig. 1. Single laser cooled Ba^+ 2-photon resonance; arrows show expected Doppler sideband positions.

REFERENCES

1. R. H. Dicke, Phys. Rev. **89**, 472 (1953).
2. G. Janik, W. Nagourney and H. Dehmelt, JOSA B, **2**, 1251 (1985).

HYPERFINE STRUCTURE AND ISOTOPE SHIFTS IN HEAVY ATOMS

Z. Fang, O. Redi, and H.H. Stroke
New York University

R. Neugart
Universität Mainz, Federal Republic of Germany

A. Ahmad, H.T. Duong, and K. Wendt
CERN, Geneva, Switzerland

We are making measurements on the optical spectra of radioactive isotopes of heavy elements to study nuclear structure effects. These experiments are being carried out off-line and on-line with accelerators. A variety of spectroscopic techniques is used including laser spectroscopy, dispersive spectroscopy, and laser optical pumping of an atomic beam detected by Stern-Gerlach analysis.

Isotope shifts in the atomic spectrum of heavy elements reflect directly the variation of the mean square nuclear charge radius as a function of neutron number (mass effects are relatively small). Of particular interest is the irregular, nuclear structure dependent, behavior of these shifts.

Nuclear magnetic moments, obtained from the atomic hyperfine structure are of special value in the region of the doubly-closed-shell nucleus ^{208}Pb because of the relative simplicity of the nuclear wavefunctions. The experimental values can serve to study core excitation and pion exchange contributions to the nuclear magnetic moment.

THALLIUM SPECTROSCOPY

A collinear fast beam experiment on the 4.77-min ^{207}Tl has been performed on-line with the CERN ISOLDE facility in a collaborative effort between Universität Mainz, New York University, Laboratoire Aimé Cotton, and CERN.[1]

The fast atoms are produced in the 6p $^2P_{3/2}$ metastable state from the ion beam by the charge exchange process. The hyperfine spectrum is measured in the 6p $^2P_{3/2}$ -7s $^2S_{1/2}$ 535.0-nm transition. Fluorescence is detected in the 7s $^2S_{1/2}$ -6p $^2P_{1/2}$ 377.6-nm line. The results for ^{207}Tl are: Isotope shift= 1783(3) MHz (relative to ^{205}Tl); nuclear magnetic moment= 1.876(5) μ_N. When combined with the magnetic moment of ^{207m}Tl (experiment now in preparation) the pion exchange contribution can be isolated.[2] The big difference between the ^{207}Tl magnetic moment and the nearly equal values of all the measured odd Tl isotopes to A=195 has been interpreted in a recent paper by Arima and Sagawa on the basis of core excitation.[3]

*Supported in part by the NSF Grant PHY8503971, the Deutsche Forschungsgemeinschaft, and the Bundesministerium für Forschung und Technologie.

In addition to ^{207}Tl and our earlier data by dispersive spectroscopy on 199, 200, 201, 202, ^{204}Tl,[4] we have obtained preliminary results on ^{193}Tl, ^{195}Tl, and ^{199}Tl with the collinear laser method on line at CERN.

BISMUTH SPECTROSCOPY

The long-lived isotopes ^{207}Bi ($\tau_{1/2}$ =38y) and ^{208}Bi ($\tau_{1/2}$= 3.7×10^5y) have been studied in off-line experiments. Nanograms of cyclotron (Princeton) produced, mass separated samples are contained in quartz cells. Spectra at 306.7 nm were obtained by laser fluorescence and by dispersive spectroscopy. The results for ^{207}Bi are[5]: Isotope shift= 2995(60) MHz (relative to ^{209}Bi); HFS interaction constants $A(^4P_{1/2})$= 4887(9) MHz, $B(^4S_{3/2})$= 480(90) MHz; nuclear magnetic moment=4.051(7) μ_N. The behavior of the even neutron isotope shifts for Bi have a regular behavior, IS(205,207) $\simeq$ IS(207,209). This was not indicated in earlier experiments on ^{207}Bi.

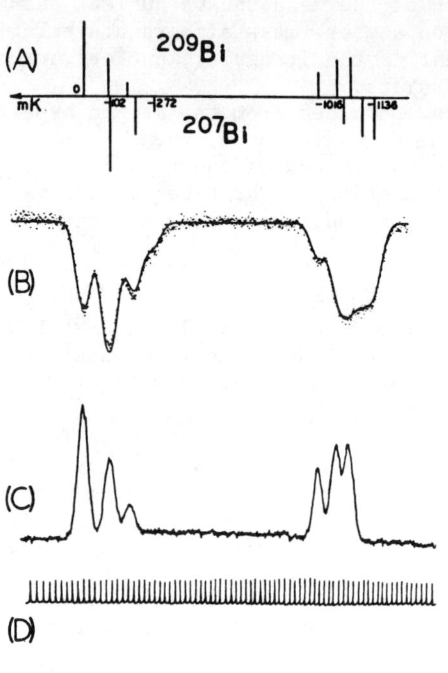

Fig. 1. Hyperfine Spectra of Bismuth at 306.7 nm. (A) Positions of hyperfine components of ^{209}Bi and ^{207}Bi. Units of the scale are in 10^{-3} cm^{-1} = 1mK. (B) Absorption spectrum of a cell containing ^{207}Bi and ^{209}Bi obtained by dispersive spectroscopy. Digitally recorded data (dots) are shown together with a fitted spectrum (solid curve). (C) Fluorescence spectrum obtained by laser spectroscopy of a ^{209}Bi cell. The ring laser output was frequency doubled with an ADA crystal external to the laser cavity. (D) Calibration markers produced by an interferometer with 300MHz free spectral range using the fundamental 613.4-nm laser output.

1. R. Neugart, H.H. Stroke, S.A. Ahmad, H.T. Duong, H.L. Ravn, K. Wendt, Phys. Rev. Lett. 55, 1559 (1985).
2. I. Bergström, A. Kerek, C. Ekström, CERN Report PSCC/80-56/I34.
3. A. Arima and H. Sagawa, to be published.
4. R.J. Hull and H.H. Stroke, J. Opt. Soc. Am. 51, 1203 (1961).
5. M. Barboza-Flores, O. Redi, H.H. Stroke, Z. Phys.A,321,85(1985).

LASER SPECTROSCOPY OF RADIOACTIVE ATOMS USING PHOTON CORRELATION TECHNIQUES

A.G. Martin, S.B. Dutta, W.F. Rogers and D.L. Clark.
Nuclear Structure Research Laboratory
University of Rochester
Rochester, N.Y. 14627

ABSTRACT

Extremely sensitive high resolution laser spectroscopy techniques are required for measurements of the hyperfine splittings and isotope shifts of radioactive atoms. We have developed a computer controlled dye laser system and an on-line atomic beam system for such measurements. Photon time correlation techniques are being applied allowing measurements on atomic beams of short lived nuclear reaction products with fewer than 50 atoms/sec. These measurements are used to extract information on the nuclear size, shape and magnetic moments of nuclei far from stability. We have also demonstrated that by producing a laser beam with a spatially varying polarization, optical pumping effects can be reduced, thus permitting the use of this technique for odd-A isotopes as well as for spin zero even-A isotopes.

EXPERIMENTAL

Figure 1. shows the set-up we have developed to perform laser spectroscopy measurements on radioactive isotopes produced by the University of Rochester MP Tandem accelerator. The heavy

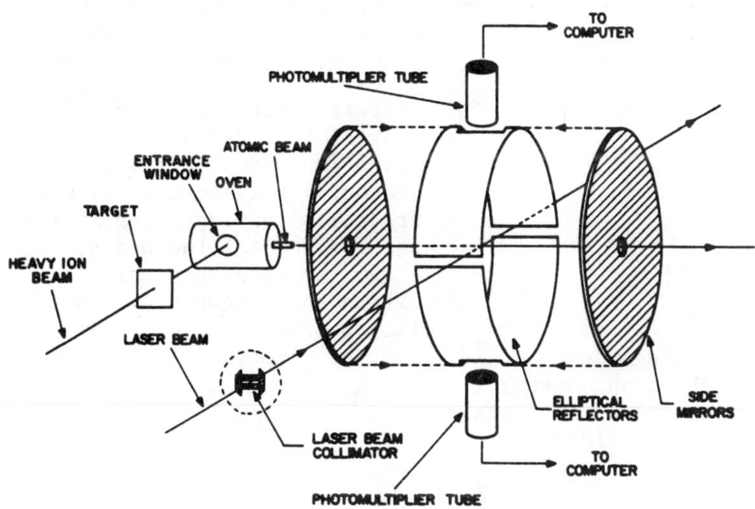

Figure 1. Atomic beam oven and light collection system.

0094-243X/86/1460405-2$3.00 Copyright 1986 American Institute of Physics

ion beam from the accelerator impinges on a thin target, the isotopes under study being produced by fusion evaporation reactions. The nuclear reaction products recoil out of the back of the target and are implanted in a tantalum oven. The oven is heated by e-beam bombardment to temperatures of approximately 1500 °C, and the implanted ions are released from the tantalum and exit the oven forming a thermal atomic beam.

The atomic beam and laser beam intersect at the center of a large solid angle elliptical light collection system which focuses the scattered photons on two high quantum efficiency P.M.T.'s. By studying the time correlation of the photons detected in the phototubes, we discriminate the bursts of photons resonantly scattered by single atoms crossing the laser beam, against the background of randomly scattered oven and laser light[1].

RESULTS

Our initial measurements have been on ^{128}Ba ($T_{1/2}$ = 2.4d) and ^{126}Ba ($T_{1/2}$ = 100min). Figure 2. shows two simultaneously recorded spectra showing the ^{128}Ba peak. The top spectrum is a plot of laser frequency versus P.M.T. counts when all events are included. The bottom spectrum shows only those events in which six photons were detected from each atom crossing the laser beam. As can be seen, using this technique a very large increase in the signal to noise is obtained, which has allowed us to successfully perform laser spectroscopy on fluxes of as few as 10 - 20 atoms per second. Also, by producing a laser beam with a spatially varying polarization using a crystalline quartz wedge, we have been able to extend this technique to odd isotopes by reducing optical pumping effects.

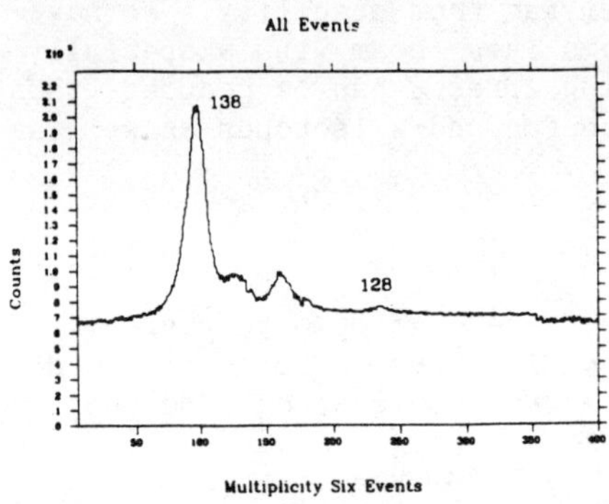

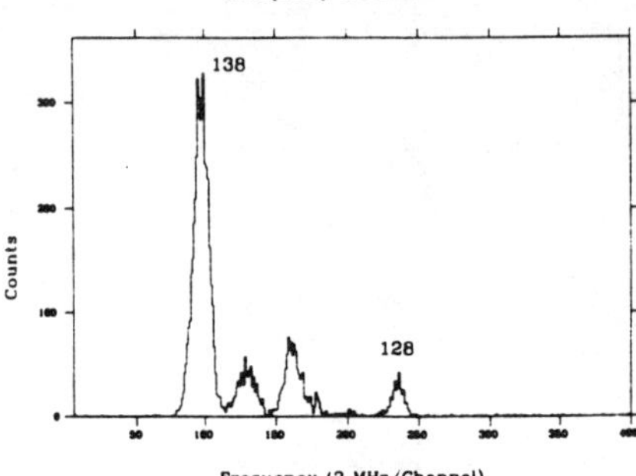

Figure 2. ^{128}Ba scan showing a) all counts, and b) only six fold detection coincidences

REFERENCES

1. G.W. Greenlees et al., Opt. Comm. 23 236 (1977).

* Supported by the National Science Foundation.

A HIGH RESOLUTION COMPUTER CONTROLLED DYE LASER SYSTEM

D.L. Clark, A.G. Martin, S.B. Dutta and W.F. Rogers.
Nuclear Structure Research Laboratory
The University of Rochester
Rochester, N.Y. 14627

ABSTRACT

A high resolution computer controlled dye laser system has been developed for measurements of the hyperfine structure and isotope shifts of radioactive atoms produced by the Rochester heavy-ion accelerator. The laser system is based on an actively stabilized Spectra-Physics 380D ring dye laser. The dye laser is electro-optically locked to a 1 meter Michelson interferometer to enable computer control of the laser wavelength via a CAMAC interface to an LSI-VAX network. This device permits discrete steps of ~1 MHz in the laser output, and sweeps of up to 30 GHz limited only by the laser tuning range. The Sigmameter optical length is locked to a stabilized He-Ne laser, providing a long term dye laser frequency stability of better than 1 MHz per day. The absolute wavelength of the dye laser is measured using a fringe counting system based on a moving mirror Michelson interferometer. Fringe rate multiplication results in an accuracy of 1-2:10^7. Fringe coincidence techniques now being applied show promise of even higher levels of accuracy.

DYE LASER SYSTEM

The dye laser system is based on a Spectra Physics 380D actively stabilized ring dye laser pumped by Spectra Physics 171 Argon ion laser (figure 1.). Short term frequency stability is provided by two external etalons of 500 MHz and 5 GHz respectively. The first provides the laser with a linewidth of less than 1 MHz, and the second is used to return the laser to the correct frequency should a mode hop occur. Long term frequency stability and computer control of the output frequency is provided by the use of a special 1 meter Michelson interferometer, called a Sigmameter[1]. A quarter wave plate is placed in one of the arms of this interferometer, and crossed polarizers are placed in front of two diodes monitoring interferometer output. As the dye laser frequency is varied sine and cosine signals are induced in each of the two diodes respectively. These signal are used to monitor the dye laser frequency, and control the output via special electronics and a CAMAC computer interface. The output from a frequency stabilized He-Ne laser is sent round the interferometer in the opposite direction. By using a similar set of detectors and electronics, the size of the Sigmameter is locked to the He-Ne wavelength, providing the dye laser system with a long term frequency stability of better than 1 MHz per day.

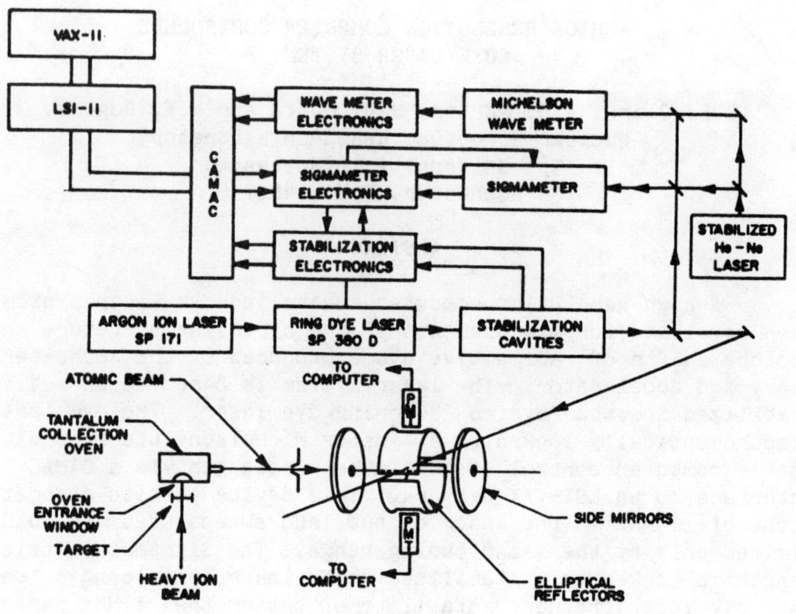

Figure 1. Schematic of computer controlled dye laser system coupled to experiment.

A clock in a custom CAMAC module drives the scanning system, compensating for computer dead time during a laser scan. Laser frequency instabilities are sensed by the stabilization electronics and freeze the scan and data acquisition via this module.

A fringe counting moving mirror Michelson with 30 cm of travel is used to measure the dye laser frequency. Wavelength measurements accurate to 1-2:10^7 are obtained approximately every 5 seconds by using electronic fringe rate multiplication. These wavelength measurements are sent to the data acquisition computer via the CAMAC interface.

REFERENCES

1. P. Juncar and J. Pinard, Opt. Comm. 14, 438 (1975).

* Supported by the National Science Foundation.

RESONANCE IONIZATION MASS SPECTROMETRY OF Mg: The 3pnd Autoionizing Series

R.E.Bonanno, C.W.Clark, J.D.Fassett, T.B.Lucatorto
National Bureau of Standards, Gaithersburg, MD 20899

ABSTRACT

We have utilized stepwise multiphoton excitation to observe autoionizing Rydberg states of magnesium. We observe the 3pnd series which converges to the 3p $^2P^0$ limits of the Mg ion. Our measurements extend up to n=40. A preliminary quantum defect analysis will be presented.

Spectroscopic data on the states of Mg with energies above the first ionization limit is quite sparse. The existing data comes from several techniques but none possess the inherent selectivity and high resolution of laser spectroscopy. This paper reports on a study of Rydberg states converging to the 3p ($^2P_{1/2,3/2}$) limits of Mg^+. Specifically we have investigated the 3pnd series for n ranging from 8 to 40. Until now no member of this series with an effective quantum number greater than 11 has been observed. The need for reliable spectroscopic data in this energy range has been underlined by the continuing uncertainties in the interpretation of experiments on dielectronic recombination of Mg^+, a process to which contributions are made by many different autoionizing Rydberg series of Mg.[1]

Our study of autoionizing Rydberg states of Mg utilizes the resonance ionization mass spectrometry (RIMS) technique which has been described in detail elsewhere.[2] Neutral atoms are produced by resistively heating a magnesium filament. Doubly-excited states are produced by absorption of laser radiation. These states rapidly autoionize and are detected by a magnetic sector mass spectrometer. The Mg^+ 3p states lie 12.1 eV above the Mg $3s^2$ ground state, so that excitation of an autoionizing Rydberg series of the type $3pn\ell$ requires a minimum of three uv photons at wavelengths of $\lambda \approx 300$nm. For states 3pnd with n>10 we first excite the 2-photon transition $3s^2 \rightarrow 3sn\ell$ with ℓ=s or d and n=11-36. A second laser is scanned over the wavelength range which includes the Mg^+ 3s-3p transition. This leads to final states 3pn'd. This scheme is similar to the isolated core excitation (ICE) scheme developed by Gallagher and coworkers.[3] Figure 1 shows a spectra for the $3s^2 \rightarrow 3s17d \rightarrow 3pn'd$ transitions. In addition to strong features at 280.3nm and 279.6nm, associated with

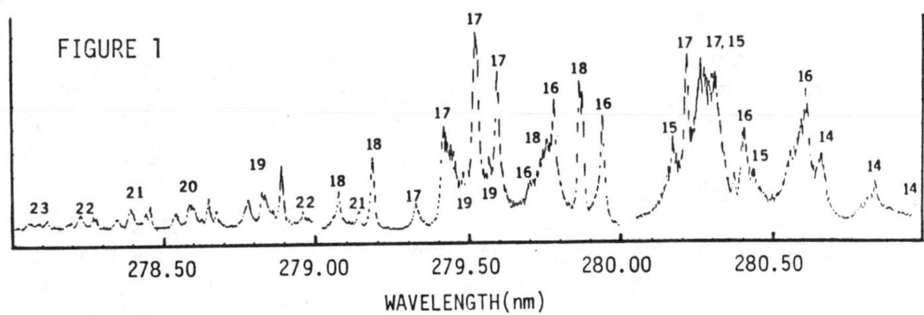

the 3s17d → 3p($^2P_{1/2}$)17d and 3s17d → 3p($^2P_{3/2}$)17d transitions respectively (spectator electron Δn=0 transitions), we observe series members between 3p14d and 3p23d. This relatively wide distribution of Δn for the spectator electron--also referred to as shake-up transitions, is a characteristic of the large difference in quantum defect between the 3snd and 3pnd series. Table I shows a preliminary analysis of the 3pnd levels from the 3s17d spectra. We observe six long series out of ten possible. At this time, we have made identifications of most states in terms of principal quantum number and series limit. Further classification (e.g. total angular momentum) will require improvements in the experimental apparatus which are currently underway.

This work has also included an investigation of the lowest lying autoionizing state in Mg, the $3p^2(^1S)$ state.[4] In addition, the autoionizing states of Be have been studied and reported on elsewhere.[5]

Table I. Preliminary Analysis of the 3pnd series from 3s17d Spectra

n	n* with respect to the limits: $^2P_{1/2}$		$^2P_{3/2}$			
14			13.98	14.27	13.77	
15			15.01	15.19	14.67	
16	16.20	15.82	16.05	16.23	15.67	
17		16.81	16.99	17.28	16.68	
18	18.20	17.81	18.07		17.71	
19	19.20	18.79	19.01	19.19	18.75	
20			20.01	20.27	19.75	19.64
21	21.20	20.84	21.05	21.27	20.76	20.67
22		21.86	22.01	22.24	21.74	21.68
23			22.96	23.17	22.74	
δ	-0.2	0.2	-0.05	-0.2	0.25	0.35

Quantum defect of the 3snd series is δ = 0.4

This work supported in part by the Department of Energy contract number DE-AI01-85ER60302.

REFERENCES

1. G.Dunn, D.S.Belic, B.DePaola, N.Djuric, D.Mueller, A.Muller, and C.Timmer, in Atomic Excitation and Recombination in External Fields, ed. M.H.Nayfeh and C.W.Clark(Gordon and Breach, NY 1985)
2. J.D.Fassett, L.J.Moore, J.C.Travis, and F.E.Lytle, Anal.Chem.55, 765 (1985).
3. T.F.Gallagher, in ref.1.
4. R.E.Bonanno, C.W.Clark, and T.B.Lucatorto, to be published.
5. C.W.Clark, J.D.Fassett, T.B.Lucatorto, L.J.Moore, and W.W.Smith, J.Opt.Soc.Am.B2, 891 (1985).

HIGH-RESOLUTION ABSORPTION SPECTRUM OF THE $6^1S_0 \rightarrow 6^3P_1$ TRANSITION IN MERCURY WITH A CW DYE LASER*

J. K. Crane, G. V. Erbert, S. D. Mostek,
R. C. Kerlin, and J. A. Paisner
Lawrence Livermore National Laboratory
Livermore, CA 94550

ABSTRACT

Using a stabilized single-frequency commercial dye laser and an external cavity doubling crystal, we have measured the isotope shifts of mercury for the $6^1S_0 \rightarrow 6^3P_1$ transition with an accuracy of 4 MHz. We describe the method for generating single-frequency light at 2537 Å and compare the results of our measurements of the isotope shifts with previous work.

DESCRIPTION OF THE EXPERIMENT

In recent years several authors have reported producing single-frequency tunable continuous-UV radiation below 260 nm.[1,2] Typically, these narrow-band UV light sources were applied to spectroscopic measurements using methods that require modest intensities (e.g., Doppler-free saturation spectroscopy or fluorescence spectroscopy). In contrast, absorption spectroscopy requires only enough intensity to provide good signal-to-noise ratios in a linear detection system, which in the case of solar blind photomultipliers may be less than a nanowatt.

We have measured the isotope spectrum of the $6^1S_0 \rightarrow 6^3P_1$ transition in mercury. We use either urea or potassium pentaborate crystals to generate frequency-doubled light from a stabilized ring laser and send this light through an atomic beam. The amount of absorption is measured with a solar blind photomultiplier and recorded with standard phase-sensitive techniques. Based on the specified photomultiplier gain and responsivity at 253.7 nm, we estimate the doubling efficiency for 13 mW in urea to be $\sim 1 \times 10^{-7}$.

RESULTS OF MEASUREMENTS

Figure 1 shows a typical scan over a portion of the isotope manifold along with frequency markers from an oven-stabilized 150-MHz interferometer[3]. The density for this scan was 2×10^{12} cm^{-3}. To enhance the accuracy of our measurement, we divided the entire manifold into four such scans. Each scan included at least two peaks, one of which was a single isotope or hyperfine component. Using this procedure we achieved an accuracy of 4 MHz in measuring the peak separations.

Two pairs of peaks lie too close together to be clearly resolved: 204 and 199A; and 199B and 201c. To resolve the overlapping peaks, we fit the absorption curves to a double Gaussian using a nonlinear least-squares fitting routine.

*Work performed under the auspices of the U.S. Department of Energy by Lawrence Livermore National Laboratory under Contract W-7405-Eng-48.

In Table I we compare the results of our measurement to the values measured by Schweitzer. The values are given in units of 10^{-3} cm^{-1} and are referenced to the ^{198}Hg peak. In all cases, our values agree with those of Schweitzer within the bounds given by our respective error bars.

CONCLUSION

Using absorption spectroscopy we have measured the isotope shifts of the $6^1S_0 \rightarrow 6^3P_1$ transition of Hg in an atomic beam. The technique we have used is very simple yet highly accurate and has wide application for both spectroscopy and laser diagnostics for measuring density or velocity distributions.

Table I. Isotope shifts for the $6^1S_0 \rightarrow 6^3P_1$ transition of Hg with respect to the 198 peak.

Isotope peak	Our values (10^{-3} cm^{-1})	Schweitzer's values[4] (10^{-3} cm^{-1})
199A	-513.59 ± 0.50	-513.99 ± 0.43
204	-510.71 ± 0.50	-510.77 ± 0.43
201A	-489.58 ± 0.44	-488.96 ± 0.33
202	-337.14 ± 0.41	-336.96 ± 0.15
200	-160.45 ± 0.23	-160.29 ± 0.15
201b	-22.52 ± 0.13	-22.56 ± 0.09
198	0.00	0.00
199B	$+224.14 \pm 0.46$	$+224.40 \pm 0.23$
201c	$+228.56 \pm 0.46$	$+229.23 \pm 0.51$

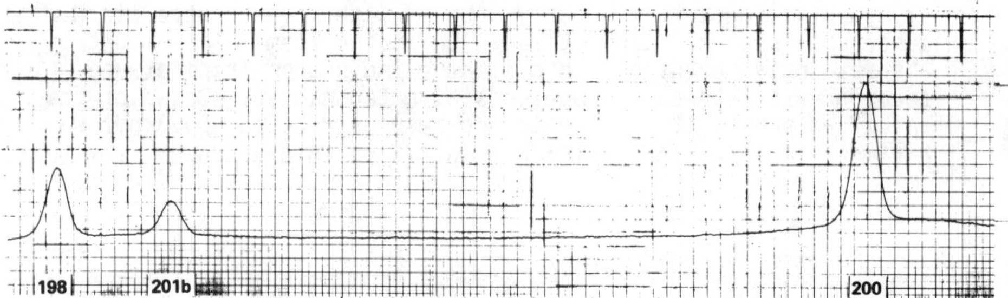

Fig. 1. High-resolution scan across a portion of the $6^1S_0 \rightarrow 6^3P_1$ transition; 150-MHz interferometer markers measure the fundamental from the cw dye laser.

REFERENCES

1. C.R. Webster, L. Woste, and R.N. Zare, Opt. Commun. 35, 435 (1980).
2. L. A. Bloomfield, B. Couillaud, E. A. Hildum, and T.W. Hansch, Opt. Commun. 45, 87 (1983).
3. R. Wyeth, a member of our group, measured 150.280 MHz for the free spectral range of our interferometer using a dual-frequency modulation technique. The essential features of this technique are discussed by R.G. DeVoe and R.G. Brewer, Phys. Rev. A 30, 2827 (1984).
4. W.G. Schweitzer, J. Opt. Soc. Am. 53, 1055 (1963).

*MEASUREMENTS OF THE SPECIFIC MASS ISOTOPE SHIFT IN ^{20}Ne and ^{22}Ne FOR LEVELS OF THE $(2p^5)$3s, 3p, 3d, 4p, 5s AND 4d CONFIGURATIONS USING LASER OPTOGALVANIC SPECTROSCOPY; A PRELIMINARY REPORT.

P.A. Crowell, J.A. Ruff and F.A. Moscatelli
Swarthmore College, Swarthmore, PA 19081

ABSTRACT

We have begun a systematic investigation of the isotope shift and hyperfine structure in refractory elements using both the DC and RF optogalvanic effect as detection. As a preliminary endeavor, we are obtaining Doppler-free measurements of the isotope shifts and ^{20}Ne and ^{22}Ne. Neon is the heaviest element considered to have a negligeable nuclear volume isotope shift. Multiconfiguration Hartree-Fock calculations of the specific mass shift have been most disappointing in the case of neon. A parametric method due to Keller[1] and Bauche and Champeau[2] has achieved much more success. In order for the parameters to be determined accurately, a large number of results for different transitions should be available. Previous results consist of optical emission spectroscopy for the 3s and 3p configurations[3], Doppler free two photon spectroscopy for the 4d and 5s configurations[4], where the signals can be extremely weak and detection based on fluorescence can introduce systematic errors. Shifts for the 3d and 4p configurations were obtained from experiments on infra-red lines[5]. We have obtained signals for over fifty transitions of sufficient intensity to produce results. This is the first experiment in Ne to measure so many lines using a single experimental method. Neon is an interesting case for investigating the specific mass effect since it is the lightest element in which RS coupling is almost completely inappropriate. In addition, it possesses two relatively abundant isotopes both of which have no hyperfine structure.

THE PARAMETRIC METHOD

In an attempt to explain the early results of Odinsov (1965)[3] on the level shifts in the $2p^5$3s and $2p^5$dp configurations, Keller[1,5] has devised an emperical method based on parameters obtained from a restricted set of the measurements. For example, the ten levels of the $2p^5$3p configuration, which result from applying (jl) coupling, $\psi_i(J)$, i=1 - 10, do indeed exhibit different shifts unlike the RS case mentioned earlier. These shifts were interpreted by Keller as follows. The state $\psi_i(J)$ is written in (jl) coupling as

$$\psi_i(J) = \sum_{S,L} a^i_{SLJ}\ \psi(S,L,J)$$

where the summation is over the six LS terms of $2p^5$3p (i.e. ^3P_J, ^3S_J, ^3D_J and ^1P_J ^1S_J ^1D_J). The shift of each level δ_i was assumed to be given by

$$\delta_i = \sum_{L,S} |a^i_{SLJ}|^2\ T(SL) + Z_{2p}\ \zeta_{2p} \qquad \text{eq. 2}$$

where the T(SL) are six isotope shift parameters and Z_{2p} is the parameter added to accommodate a possible J dependence, ζ_{2p} is the spin-orbit integral for a 2p electron. Eq. 2 is one equation with seven unknowns, however, there are ten such eqs. for the shifts of levels so the parameters can be extracted and checked for consistency. Agreement to 3MHz has been achieved.

Keller has suggested, as a physical interpretation for the T(SL) parameters a crossed second order effect of the specific mass shift operator and the electrostatic interaction. The Z_{2p} parameter is assumed to have a purely relativistic origin.

To extend these results to higher configurations required more data and higher accuracy.

EXPERIMENTAL DESCRIPTION

I. Optogalvanic Effect

The optogalvanic effect refers to the observed changes in the electrical impedance of a gaseous discharge caused by irradiation with light resonant to one of the atomic transitions. The actual mechanisms involved are quite complicated and are themselves the subjects of investigations[6]. In any event, the effect is quite strong and is being used as a very sensitive detection method[7]. It possesses several advantages over the traditional techniques of absorption and emission detection, including:

i. Refractory elements can also be studied since an adequate amount of sample, in the vapor state, can be obtained through sputtering in a hollow cathode lamp.

ii. Light intensities (or minute differences in them) need never be measured. Signal is detected simply as a change in voltage drop across a resistor using lock-in detection.

iii. Since the atomic sample is present in a discharge, spectroscopy on excited states, metastable states, and even singly ionized species can be carried out.

The discharge can be achieved by either direct current excitation via electrodes, or by a radio-frequency oscillator coupled to an electrodless discharge cell. In fact, the RF method has the distinct advantage of using lower gas pressures thus minimizing pressure broadening[8].

II. Experimental Setup:

As the frequency of two counterpropagating laser beams directed down the axis of a lamp is scanned, the beams will interact with the same velocity group of atoms only when the frequency is within one homogeneous linewidth of the atomic transition. This group will have longitudinal velocities equal to zero. Now if the intensities of the two beams are chopped at different frequencies (say f_1 and f_2), nonlinearities will allow us to extract this Doppler-free signal by using a lock-in amplifier tuned to the sum $(f_1 + f_2)$ or difference $(f_1 - f_2)$ frequencies. The method is known as intermodulated optogalvanic spectroscopy (IMOGS)[9].

* Supported by the Research Corporation and Swarthmore College.

REFERENCES

1. J.C. Keller, J. Phys. B 6, 1771 (1973).
2. J. Bauche and R.J. Champeau, Adv. Atom. Mol. Phys. 12, 39 (1976)
3. V.I. Odintsov, Optics and Spectrosc. 28, 205 (1963).
4. E. Giacobino, et al. J. Phys. (France) 40, 1139 (1979).
5. J.C. Keller and J.F. Lesprit, Physica 64, 202 (1973).
6. J.E. Lawler, Phys. Rev. A. 22, 1025 (1980).
7. W.E. Bridges, J.O.S.A., 68, 352 (1978).
8. D.R. Lyons, A.L. Schawlow and G.Y. Yan, Optics Comm., 38,35 (1981)
9. J.E. Lawler et al., Phys. Rev. Lett. 42, 1046 (1979).

DOPPLER-FREE COLLINEAR FAST ION BEAM LASER SPECTROSCOPY OF THE NEUTRON-RICH BARIUM ISOTOPES $^{139-142}$Ba AND ^{144}Ba*

H. A. Schuessler, K. Doerschel, U. Trebus, M. Brieger, S. Chen, Y. Wang and J. Zhang
Texas A&M University, College Station, TX 77843

R. L. Gill, A. Piotrowski and C. D. McDonald
Brookhaven National Laboratory, Upton, NY 11573

ABSTRACT

The hyperfine structures and isotope shifts of the Ba II lines at λ=614 nm and λ=585 nm have been investigated by collinear fast ion beam laser spectroscopy. The neutron-rich isotopes $^{139-142}$Ba, ^{144}Ba and the stable reference isotope ^{138}Ba were prepared at the TRISTAN mass separator following fission in a uranium target and were observed on-line. Continuous operation of the surface ionization source was employed to observe the signal from the isotope ^{144}Ba (τ=11 sec). For the longer-lived isotopes a bunched beam release was implemented to increase the signal to noise ratio and to simultaneously reduce the radioactive background count rate produced by isobaric Cs decays.

EXPERIMENTAL METHOD AND RESULTS

In recent years the collinear fast ion beam method has provided both Doppler-free resolution and ultrahigh sensitivity for the minute samples of short-lived isotopes produced on-line. The present experiment is an extension of earlier work on stable isotopes[1] to the short-lived neutron-rich barium isotopes.

Neutron-rich barium isotopes were delivered from a surface ionization source loaded with 5g of ^{235}U and located in a neutron flux of 2x10^{10} neutrons/cm^2 sec. After acceleration to 44 keV and mass separation in a magnetic field the beam is merged collinearly with the beam of a single-mode cw ring dye laser. Typical ion currents produced varied between about 10 pA for the very neutron-rich isotopes to 1 nA for the isotopes closer to stability and consisted to about 0.1% of ions in the metastable 5d $^2D_{3/2,5/2}$ states. The transitions from these metastable states to the 6p $^2P_{3/2}$ state at λ=614.3 nm and λ=585.5 nm were induced by a dye laser using Rhodamine 6G. Resonant excitation was detected by monitoring the fluorescent light from the decay of the 6p $^2P_{3/2}$ state to the 6s $^2S_{1/2}$ ground state at λ=455.5 nm. An interference filter and two dicroic filters blocked the directly scattered laser stray light almost completely. A detection sensitivity of 2x10^{-6} counts/ion including the metastable ion fraction was

*Supported by the U.S. Department of Energy under Contract No.DE-AS05-80 ER 10578, the National Science Foundation through Grant No. PHY 8206960 and the Robert A. Welch Foundation.

achieved. The isotope shifts and hyperfine structure splittings were observed by applying a high voltage ramp to the interaction chamber for Doppler tuning by post acceleration via the kinetic energy of the ions. The fluorescent count rate was stored in a minicomputer as a function of the ramp voltage and both acceleration voltage and the ramp voltage were recorded with an accuracy of 10^{-4} by precision voltage dividers and digital voltmeters The laser frequency was measured to 10^{-6} by a Michelson-type wavemeter.

Fig. 1 displays a resonance curve obtained in a single scan. The observed linewidth of about 80 MHz consists of the natural linewidth of 25 MHz, the residual Doppler width and considerable power broadening. The results of our ongoing evaluation yield nuclear moments and changes in the mean square charge radii and are so far consistent with the results of a different measurement[2] using neutral barium atoms.

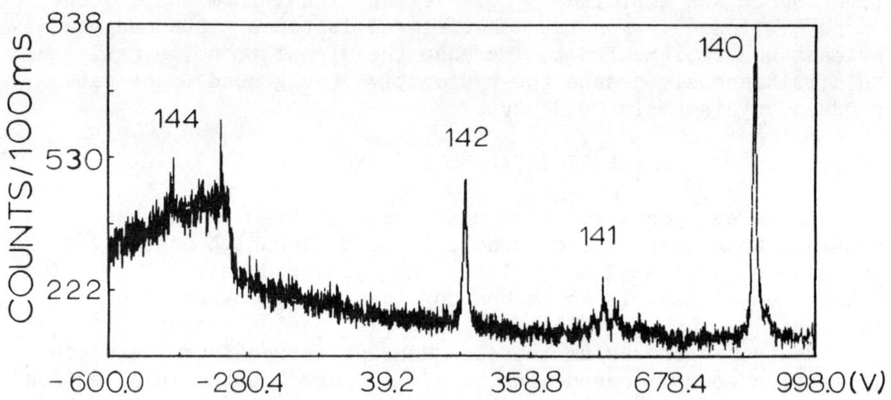

Fig. 1. Fluorescence signals versus the post acceleration voltage. The laser wavelength is fixed at 614.3 nm. The neutron beam initiating the fission process in the target was turned off by a shutter after ^{144}Ba (τ=11 sec) had been observed. The longer-lived isotopes ^{142}Ba (τ=10.7 m), ^{141}Ba (τ=18.3 m) and ^{140}Ba (τ=12.8 d) can then still be detected but without the radioactive background due to the isobaric Cs decays.

REFERENCES

1. M. Van Hove, G. Borghs, P. De Bisschop and R. E. Silverans, J. Phys. B15, 1805 (1982).
2. A. C. Mueller, F. Buchinger, W. Klempt, R. Neugart, E. W. Otten, C. Ekstroem and J. Heinemeier, Nucl. Phys. A403, 234 (1983).

SPECTROSCOPY, RELAXATION, AND LASER ACTION IN Pr^{3+}: LaF_3

R. Kichinski, F. Moshary, and S. R. Hartmann
Physics Department, Columbia Radiation Laboratory
Columbia University New York, New York 10027

ABSTRACT

The photon echo technique was used to investigate the $^3P_0$-$^3H_6$ (5985Å) and the two $^3P_0$ - $^3H_5$(5333.0 Å, 5333.5 Å) transitions in Pr^{3+}:LaF_3 at 4.6 K. On the $^3P_0$-$^3H_6$ transition, the data displayed modulation and spanned a dynamic range of nine decades in intensity. Echo relaxation rates were anomalously fast in view of previous experiments in this sample. We also report observation of 'Satellite Echoes' on the satellite lines of the $^3P_0$-$^3H_4$(4778 Å) transition and present photon echo data obtained on one of the satellite lines. Lasing was observed at 4.6 K and at 77 K during pulse excitation of the $^3P_0$-$^3H_4$ and $^3P_1$ ($^1I_6$)-$^3H_4$ (4633 Å) transitions.

INTRODUCTION

Previous experiments have employed fluorescence line narrowing (FLN),[1-3] and photon echo[4-9] techniques to investigate the $^3P_0$-($^3H_4$),($^3H_6$) and $^1D_2$-$^3H_4$ transitions in Pr^{3+}:LaF_3. FLN has also been used to study the non-resonant $^3P_0$-$^3H_6$ and $^1D_2$-$^3H_5$ transitions.[2,3] However, the linewidth obtained by FLN on a non-resonant transition exhibits a residual Gaussian component inherent to the technique and therefore is not very sensitive to homogeneous broadening mechanisms at very low temperatures. We circumvent this limitation by pulse-exciting the $^3P_0$ state to obtain a prepared excited state which acts as a basis for generation of photon echoes on the $^3P_0$-$(^3H_6)_1$ and -$(^3H_5)_{1,2}$ transitions at LHe temperatures. Echoes were also observed on at least a dozen satellite lines on the longer and shorter wavelength side of the $^3P_0$-$^3H_4$ transition; these lines seem to be symmetric about the $^3P_0$-$^3H_4$ transition.

Kaminskii has reported stimulated emission in a rod of Pr^{3+}:LaF_3 excited by Xe flash lamps on the $^3P_0$-$(^3H_6)_{1,2}$ and -$(^3F_4)_5$ transitions.[10] We observed laser action and stimulated emission in Pr^{3+}:LaF_3 following laser excitation of the $^3P_0$ state at LHe and LN temperatures.

EXPERIMENT

The sample was a 0.1% Pr^{3+}:LaF_3 crystal having a 1.0 cm diameter and a 0.5 cm length. A pair of YAG lasers were used to pump three dye lasers and six transverse amplifiers. The output of one dye laser was used to populate the $^3P_0$ state; the other two dye lasers were used to sequentially excite one of the $^3H_6$ or $^3H_5$ levels from the $^3P_0$ state. Echoes involving $^3H_6$ ($^3H_5$) levels utilized the backward (parallel) wave configuration. The relevant energy level diagram is shown in Fig. 1.

The two pulse echo data taken on the $^3P_0$-$(^3H_6)_1$ transition at 4.6 K is displayed in the upper curve of Fig. 2 where the overall exponential decay has been removed by multiplying the data by the factor exp $(4\,\tau / T_2)$ where T_2=250nsec. The modulation is due to the hyperfine splitting of the terminal $^3H_6$ level. The hyperfine splitting frequencies $\nu_{1,2}$ (and their sum) of the $(^3H_6)_1$ level are obtained by Fourier analyzing the data of Fig. 2: we found ν_1 = 10.0 MHz , ν_2 = 17.4 MHz. and $\nu_1 + \nu_2$ = 27.4 MHz. No modulation was detectable on either of the $^3P_0$- $(^3H_5)_{1,2}$ echoes; the decay time constants deduced from the data taken at 4.6 K are respectively T_2 = 100 (± 10) nsec and 93 (± 5) nsec.

Stimulated echo data were obtained on the $^3P_0$-$(^3H_6)_1$ and -$(^3H_5)_1$ transitions. Both transitions were characterized by a fast initial decay followed by a slower decay with a time constant of the order of the $^3P_0$ lifetime. The initial decay is indicative of a population transfer out of the terminal levels . The data for $^3P_0$-$(^3H_6)_1$ are displayed in the upper curve of Fig 3. The observed modulation is due to the $^3P_0$ hyperfine splitting.

Following Chen et. al.[5], we calculate the modulation pattern of the $^3P_0$-$^3H_6$ echo for various relative orientations of the principal hyperfine axes for comparison with the data. The ground and excited state hyperfine hamiltonians;

$$H_{g(e)} = P_{g(e)} [I_{z(Z)} + \eta_{g(e)} (I_{x(X)} - I_{y(Y)})] \quad (1)$$

are diagonalized by matrices $U_{g(e)}$. If we define $W = U_e U_g^{-1}$, the rephased dipole moment at $t = 2\tau$ may be written as :

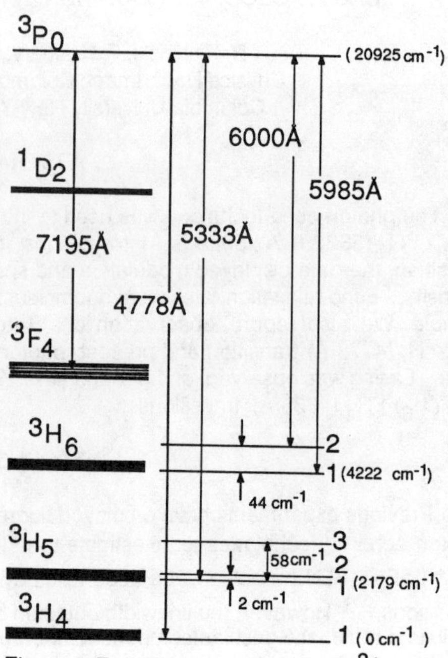

Figure 1. Energy level diagram of Pr^{3+} :LaF_3

$$P(2\tau) = P(0)\,(2I+1)^{-1} \sum_{\alpha,\beta,\gamma,\varepsilon}^{2I+1} (W)_{\alpha\beta}\,(W^{-1})_{\beta\gamma}\,(W)_{\gamma\varepsilon}\,(W^{-1})_{\varepsilon\alpha} \exp[-(\Omega^e_{\alpha\gamma} - \Omega^g_{\beta\varepsilon})\,\tau]$$
$$\cos[(\omega^e_{\alpha\gamma} - \omega^g_{\beta\varepsilon})\,\tau]\,\exp(-2\tau/T_2) \quad (2)$$

where for Pr^{3+} in LaF_3 ; I = 5/2. The ω 's are the hyperfine frequencies of the ground and the excited states, and the Ω's are the nuclear hyperfine transition linewidths. The best fit to the

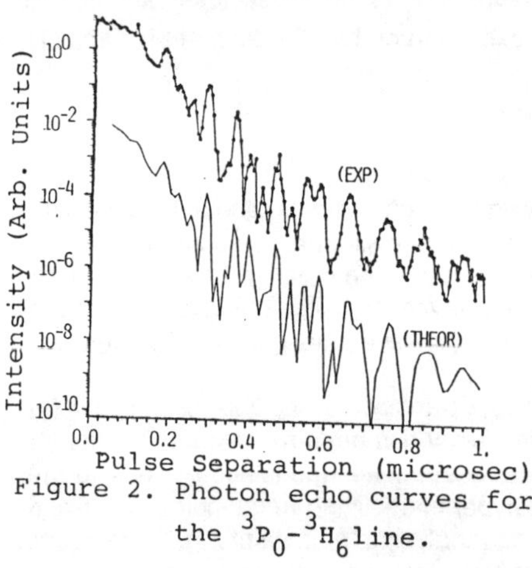

Figure 2. Photon echo curves for the $^3P_0$-$^3H_6$ line.

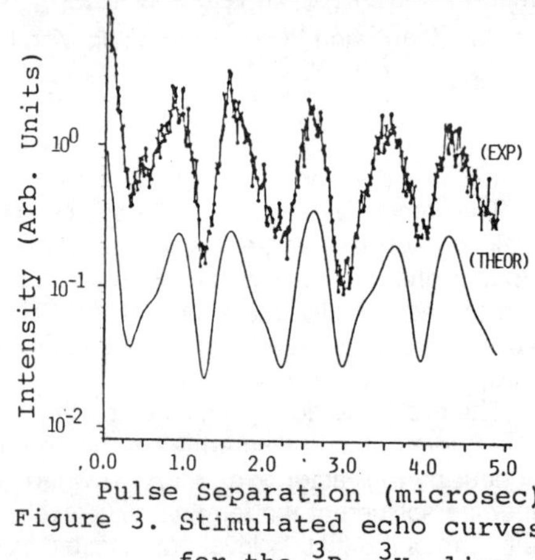

Figure 3. Stimulated echo curves for the $^3P_0$-$^3H_6$ line.

two pulse $^3P_0$-$^3H_6$ echo data is shown in the lower curve of Fig. 2 and represents a rotation of 35° around Y with z || Y. We obtained P_e= 4.4 MHz and η_e= 0.44 from our fourier transform data and P_g= .293 MHz and η_g= .516 from Chen et al.[5] We used $2\Omega/2\pi$= 200 kHz for all three $(^3H_6)_1$ hyperfine nuclear linewidths. This same orientation is corroborated by the calculation for the $^3P_0$-$(^3H_6)_1$ stimulated echo, shown in the lower curve of Fig. 3.

Thermal relaxation data were determined for all three echo transitions. For the $^3P_0$-$(^3H_5)_1$ transition T_2 was obtained by measuring echo intensity vs. pulse separation. In Fig. 4 we plot the homogeneous linewidth $\Delta\nu= [\pi T_2]^{-1}$ as a function of temperature in the range of 2 °K to 8 °K. All transitions we studied indicate the presence of a residual homogeneous width which our stimulated echo studies show to be due to fast population decay out of the terminal levels. This behavior contrasts with that found for echoes produced on the $^3P_0$-$^3H_4$ and the $^1D_2$-$^3H_4$ transitions.[1-8]

We obtained a fit to our temperature relaxation data by invoking phonon relaxation within a manifold[11] and assuming the existence of a temperature independent residual linewidth. The solid curve in Fig. 4 was obtained by considering phonon coupling of the $(^3H_5)_1$ level to the $(^3H_5)_2$ and $(^3H_5)_3$ levels, with spacings $\Delta\epsilon_{12}$=2cm^{-1} and $\Delta\epsilon_1$=58 cm^{-1}, respectively;

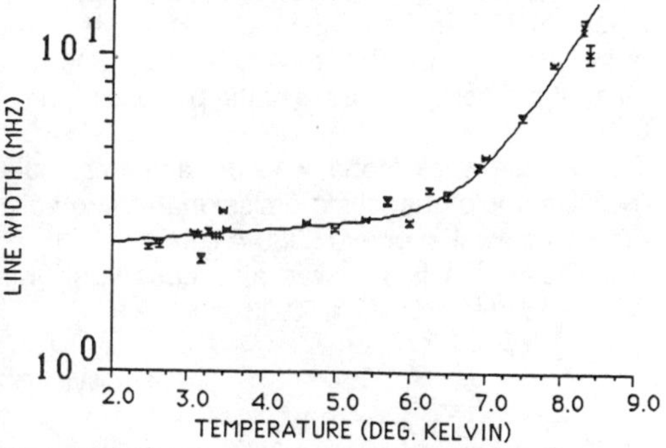

Figure 4. Temperature relaxation data for the $^3P_0$-$(^3H_5)_1$ transition.

$$\Delta\nu[^3P_0 - (^3H_5)_1] = 2.2\times10^{-3} + 4\times10^{-4}\times P(2\ \mathrm{cm}^{-1}) + 260\ P(58\ \mathrm{cm}^{-1})\ \mathrm{GHz} \quad (3)$$

where $P(\Delta\epsilon)=[\exp(\Delta\epsilon/k_BT) - 1]^{-1}$. A similar analysis of the temperature relaxation data for other transitions studied yields:

$$\Delta\nu[^3P_0 - (^3H_5)_2] = 2.2\times10^{-3} + 4\times10^{-4}\times[1 + P(2\mathrm{cm}^{-1})] + 74\ P(56\ \mathrm{cm}^{-1})\ \mathrm{GHz} \quad (4)$$

$$\Delta\nu[^3P_0 - (^3H_6)_1] = 1.2\times10^{-3} + 44\ P(44\ \mathrm{cm}^{-1})\ \mathrm{GHz} \quad (5)$$

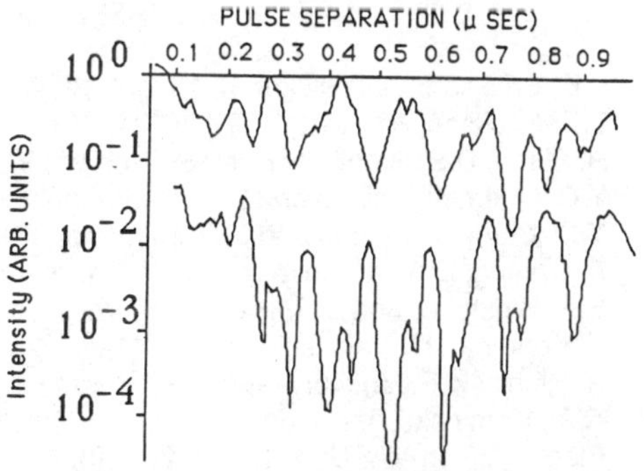

Figure 5. Photon echo data on the $^3P_0$-$^3H_4$ transition (lower curve) and the sattelite line at 4768 Å (upper curve).

In investigating the origins of the residual homogeneous linewidth, we performed a two pulse echo experiment in a concentration c=0.01% sample. We observed a decrease in the value of T_2 when the Pr^{3+} concentration was thus reduced by a factor of ten: the estimated value of T_2 for the c=0.01% sample is 80 nsec, which is three times smaller than for the c=0.1% sample. This suggests that the variation in the relaxation rate observed is due to different amounts of impurities present in the two samples used and the fast relaxation is possibly due to Pr^{3+}-impurity ion interaction.[12,13]

Photon echo data were also taken on the $^3P_0$-$^3H_4$ transition at 4778 Å and on the satellite line at 4768 Å as displayed in Fig. 5 with the upper curve being the satellite echo data. As apparent from the two curves, the character of the modulation of the satellite echo and its relaxation rate are different from those of the echo on the central transition.

Lasing was observed at the $^3P_0$- $(^3H_6)_1$ transition (5985 Å) when a laser pump pulse at 4778 Å ($^3P_0$-$^3H_4$) with a peak power of over 50 kW was focused into the c= 0.1% sample, for temperatures as high as 38 K. Similar lasing was observed in a c= 5% sample at 77 K at three lines: $^3P_0$ - $(^3H_6)_1$ (5895 Å), $^3P_0$ - $(^3H_6)_2$ (6000 Å) , and $^3P_0$-$(^3F_4)_5$ (7195 Å). The 7nsec pump pulse gave rise to a prompt 2 to 3 nsec of intense lasing followed by a longer 300nsec weak lasing exponential tail. Since this laser action is observed regardless of the angle the pump pulse makes with respect to the surface of the crystal we surmise that it is due to amplified spontaneous emission. The lasing efficiency was measured to be 1% in the c= 5% sample. It is interesting to note that the fluorescence decay time of the $^3P_0$ state is ≥ 6 μsec at 77 K in the c = 5 % sample and is observable during the lasing. The time constant of the exponential tail seems to be the same on all the transitions and we estimate that the amount of energy in the tail of the pulse is of the same order as the energy in the initial lasing pulse.

Using a weak probe, we were able to stimulate emission into a single line at 77 K in the c= 5% sample quenching emission to the other two lines. We have observed gain of over 400x on a weak probe at 5985 Å line.

The above lasing was also observed when we pumped the c= 5% sample at the $^3P_1(^1I_6)$ - $^3H_4$ transition (4633 Å) at 77 K.

ACKNOWLEDGEMENT

This work was supported by the National Science Foundation under contract No. DMR 80-06966, and the Joint Services Electronics Program (U.S. Army, U.S. Navy, and U.S. Air Force) under contract No. DAAG29-85-K-0049.

REFERENCES

1. L.E. Erickson, Optics Comm. 15,246(1975).
2. L.E. Erickson, Phys. Rev. B 11,77(1975).
3. R. Flach, D.S. Hamilton, P. M. Seltzer and W. M Yen, Phys. Rev. B 15,1248(1977).
4. Y.C. Chen, K. Chiang and S.R. Hartmann, Opt. Comm. 29,181(1979).
5. Y.C. Chen, K. Chiang and S.R. Hartmann, Phys. Rev. B 21,40 (1980).
6. K. Chiang, E.A. Whittaker and S.R. Hartmann, Phys. Rev. B 23,6142(1981).
7. E.A. Whittaker and S.R. Hartmann, Phys. Rev. B 26,3617(1982).
8. T. Kohmoto, H. Nakatsuka and M. Matsuoka, Jap. J. App. Phys. 22,I571(1983).
9. R. Klchinski,F. Moshary and S. R. Hartmann, Optics Comm. 54,147(1985)
10. A.A. Kaminskii, Int. Conf. on Lasers '80, New Orleans LA, USA. Soc. Opt. Quant. Elec., Mclean VA, USA, 328-342(1980).
11. W.M. Yen, W. C. Scott and A.L. Schawlow, Phys. Rev. 136,271(1964)
12. N. Kraustsky and H.W. Moos, Phys. Rev. B 8,1010(1973).
13. B.R. Reddy, J. Chem. Phys. 77, 2862(1982).

INTERACTION OF TRANSIENT TEMPORALLY MODULATED LASER RADIATION WITH SIMPLE ATOMIC SYSTEMS

Y. S. Bai, W. R. Babbitt, A. G. Yodh, and T. W. Mossberg
Dept. of Physics, Harvard University, Cambridge, Mass. 02138

ABSTRACT

We point out that the frequency-selective nature of the light-matter interaction can be utilized to effect the storage of a spectral image (Fourier transform) of a temporally structured light pulse in the absorption profile of an inhomogeneously broadened atomic ensemble, and that the stored information can be subsequently recalled in the form of a free-induction-decay signal. Applications of this process to optical data storage are discussed. We also consider the interaction of a phase-controlled, amplitude-gated, resonant laser field with the atoms in a collimated atomic beam. We describe experiments wherein suitably prepared atoms are placed in pure stationary-states (dressed states) of the atom + laser-field system, and note that atoms in pure dressed states should have interesting spectral and dynamical properties.

INHOMOGENEOUSLY BROADENED MATERIALS AS SPECTRUM ANALYZERS

Inhomogeneously broadened ensembles of atoms exhibit absorption profiles that are spectrally wider,often by orders of magnitude, than the absorption profiles of their individual constituent atoms. The possibility thus exists, in such materials, of selectively addressing various sets of spatially coincident absorber atoms on the basis of frequency. This fact has been incorporated into proposed optical-data-storage schemes which employ tunable, narrow-band lasers to write a multitude of data bits at each spatial location of a storage material.[1] We point out that the frequency domain can be equivalently exploited by inducing the material to remember the temporal shape of a data pulse and later on to reproduce it.

Consider a temporally structured data pulse of duration, τ_d, which is resonant with the inhomogeneously broadened absorption line of a sample material. Provided that $\tau_d \ll T_2$, where T_2 is the homogeneous dephasing time of the sample material, and that the modulation bandwidth of the pulse is less than the inhomogeneous absorption bandwidth of the excited transition, the data pulse will excite an optical polarization in the material which to first order in perturbation theory is proportional to its own Fourier transform. It follows that the temporal structure of the data pulse can be recreated from the information stored in the polarization. This fact has been demonstrated in various spin and photon echo experiments.[2] On the other hand, the ground- and excited-state populations, which are generally much more stable than the polarization, reflect only the power spectrum of the data pulse; consequently, the shape of the data pulse cannot be deduced from them.

It has been pointed out that the Fourier information available in the polarization can be transferred to populations or Zeeman coherences by means of an interference effect.[3] When two laser pulses (a data pulse and a suitable reference pulse) both occurring within a time T_2 resonantly excite the sample, the ground- and excited-state population densities or Zeeman coherences within a given state contain a modulation in frequency that is proportional to the product of the Fourier spectra of the two pulses.

If a third laser pulse resonantly excites the material in an appropriate fashion, a free-induction-decay signal of the form

$$E_{fid}(t) \propto \int_{-\infty}^{\infty} E_r^*(\omega)\, E_d(\omega)\, E_3(\omega) e^{i\omega t}\, d\omega \qquad (1)$$

will be emitted. Here $E_d(\omega)$, $E_r(\omega)$, and $E_3(\omega)$ represent the Fourier spectra of the data, reference, and third pulses, respectively. In writing Eq. 1, it is assumed that the third pulse occurs delayed from the first two excitation pulses by an interval shorter than the spectral relaxation time of the relevant level population(s) or Zeeman coherences, and for convenience that the reference pulse occurs before the data pulse. If the product $E_r^*E_3$ can be factored outside the integral, $E_{fid}(t)$, has a temporal shape identical to that of the data pulse. $E_r(\omega)$ and $E_3(\omega)$ will be individually constant if the corresponding pulses are sufficiently brief, or their product will be constant if the corresponding pulses are frequency chirped at the same rate over the bandwidth of the data pulse.

To test these results, we performed an experiment on the 555.6 nm $^1S_0$-$^3P_1$ intercombination line of atomic ^{174}Yb vapor. This system does not provide the very long storage times that are characteristic of cryogenic solids, but because of its simplicity allows for detailed comparison of theory and experiment. Pulses were generated by acousto-optically gating the output of a cw ring dye laser. Shown in Figure 1a is a data pulse representing an ASCII encoding of the letters JG. When this data pulse (preceded by a reference pulse and followed by a third pulse, both of approximately the same width as the subpulses shown) is used to excite the Yb sample, the signal shown in Figure 1b is observed. The single-event traces shown in Figure 1 are nearly identical. A slight broadening of the subpeaks in Figure 1b

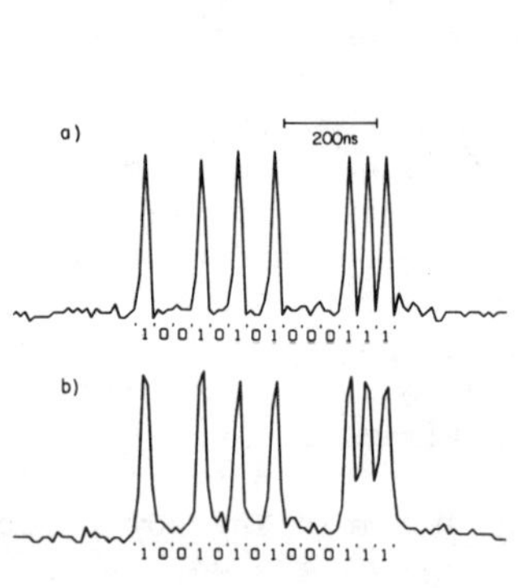

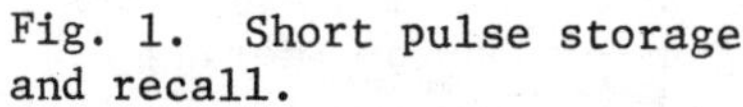

Fig. 1. Short pulse storage and recall.

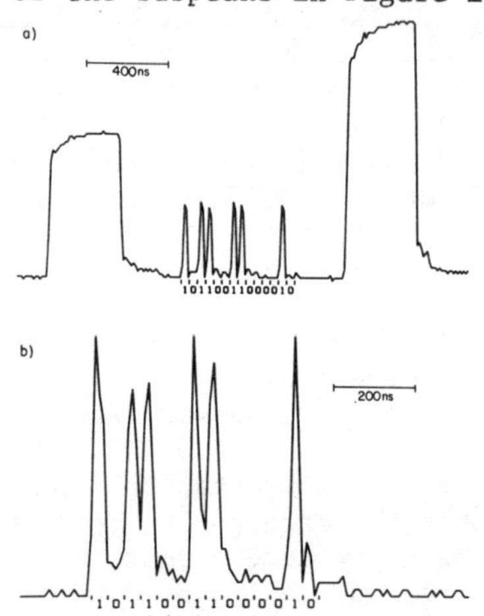

Fig. 2. Chirped pulse storage/recall.

arises from the use of insufficiently brief reference and third pulses. An excitation sequence in which the brief reference and third pulses are replaced by frequency chirped ones is shown in Figure 2a. The single-event putput signal produced, which accurately represents the ASCII encoding of the letters YB, is shown in Figure 2b. Shape reproduction using long chirped pulses is technically important because of difficulties associated with producing short pulses of sufficient power. We note that despite the perturbative nature of Eq. 1, the results presented here were obtained with fairly large area ($\approx \pi/4$) input pulses. As a result, the output signals observed in this strongly relaxing system were roughly 0.05 percent as intense as the original data pulse. In the absence of material relaxation, output signals should be approximately 30 times larger. We note that in certain cases introducing phase noise onto an amplitude encoded data pulse opens the possibility of generating output signals larger in absolute intensity.

Optical memories based on this frequency-selective storage means could operate at multi-gigahertz bit rates (limited by the material's inhomogeneous absorption bandwidth) and achieve ultra-high-storage densities (each bit requires only about 10^6 absorber atoms). In solids, storage times of perhaps years are expected.

INTERACTION OF A TRANSIENT PHASE-CONTROLLED LASER WITH TWO-LEVEL ATOMS

In the semiclassical approximation, the interaction of a two-level atom with an applied laser field is easily visualized using the vector model. In this model, the observables of the atom, i.e. level populations and coherences, are represented in terms of a three-dimensional atomic-state vector which precesses about a driving field vector, where the driving-field vector represents the strength, phase, and detuning of the laser field. The simple atomic behavior predicted by the vector model is quite difficult to observe experimentally because various atoms in the ensembles studied typically possess different detunings and/or have experienced different excitation histories. As a result, elemental two-level atom results must be ensemble averaged to describe actual measurements. We have undertaken an experimental program to study the response of ensembles of essentially identical atoms to transient laser excitation.[4] Such ensembles are realized by right-angle excitation of a collimated ^{174}Yb atomic beam. The ^{174}Yb atom constitutes an ideal two-level atom when excited by circularly polarized light, and its excited $^3P_1$ state decays sufficiently slowly (875 nsec lifetime) to facilitate detailed experimental observation before atom-field phase randomization occurs.

In most cases, an applied laser field causes atomic populations and coherences to undergo amplitude oscillations as the atomic-state vector precesses about the driving-field vector. There are two exceptions to this general result. When the atomic-state vector and the driving-field vector are aligned either parallel or anti-parallel, explicit motion of the atomic-state vector vanishes. These atom-field conditions are special, and in fact, correspond to eigenstates (i.e. dressed-states) of the atom + field system.[4,5] We have resonantly excited atoms into pure dressed states, by suddenly shifting the phase of an excitation field by 90° after the atoms, initially in their

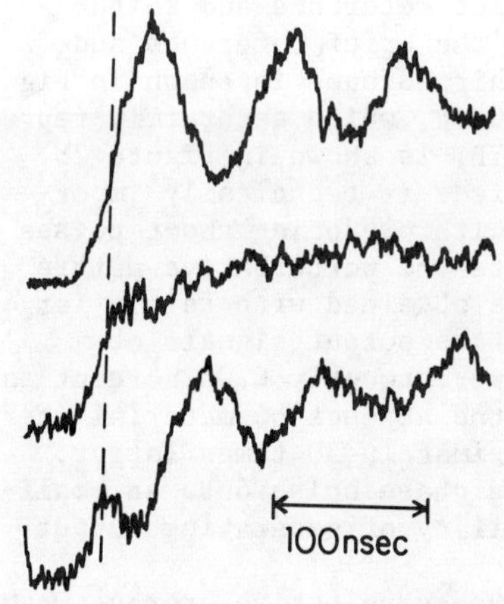

Fig. 3. Fluorescence vs. time. Top down: 0°, 90°, and 180° phase changes at vert. line.

ground state, are excited by a net pulse area of $\pi/2$. The atoms' behavior during this excitation sequence is monitored by detecting (see Figure 3) their resonance fluorescence. When placed in a pure dressed state (middle trace) oscillations in the fluorescence disappear.

One of the most important reasons for studying atoms under controlled excitation conditions is to test existing treatments of natural radiative decay. Steady-state resonance fluorescence from atoms driven by a strong, resonant, driving field is known to display a three-peaked spectrum.[6] Atoms excited to a pure dressed state, however, should initially display a double-peaked spectra. A test of this prediction[7] is planned.

Another aspect of natural decay which has apparently not been subjected to detailed experimental study is its rate in the presence of a strong, resonant, driving field. Despite this fact, it is widely assumed that the rate of natural decay is unaffected by applied optical fields of moderate (Rabi frequency small compared to the optical frequency) strength. Preliminary studies of the decay of atoms prepared in pure dressed states reveal that driving fields with Rabi frequencies up to 10 times the natural decay linewidth do not induce decay rate changes of more than about 25%. Higher precision studies are planned.

We thank the National Science Foundation (PHY-85-04260), the Army Research Office (DAAG29-83-K0040), and the Joint Services Electronics Program (N00014-85-K0724) for financial support of this work.

REFERENCES

1. W. E. Moerner and M. D. Levenson, J. Opt. Soc. Am. B 2, 915 (1985).
2. S. Fernbach and W. G. Proctor, J. Appl. Phys. 26, 170 (1955); V. A. Zuikov and V. V. Samartsev, Phys. Status Solidi A 73, 683 (1982); N. W. Carlson, L. J. Rothberg, A. G. Yodh, W. R. Babbitt, and T. W. Mossberg, Opt. Lett. 8, 483 (1983).
3. N. W. Carlson, W. R. Babbitt, Y. S. Bai, and T. W. Mossberg, J. Opt. Soc. Am. B 2, 909 (1985).
4. Y. S. Bai, A. G. Yodh, and T. W. Mossberg, Phys. Rev. Lett. 55, 1277 (1985.
5. E. Courtens and A. Szoke, Phys. Rev. A 15, 1588 (1977); D. Grischkowsky, Phys. Rev. A 14, 802 (1976).
6. For a review see: P. L. Knight and P. W. Milonni, Phys. Rep. 66, 21 (1980).
7. N. Lu, P. R. Berman, Y. S. Bai, J. E. Golub, and T. W. Mossberg, (submitted to Phys. Rev. A).

TEMPERATURE AND FREQUENCY DEPENDENT OPTICAL DEPHASING OF IMPURITIES IN CRYSTALS

D. Hsu, L. Root and J.L. Skinner
Department of Chemistry
Columbia University
New York, NY 10027
U.S.A.

ABSTRACT

We present a nonperturbative theory of the thermal line broadening of optical transitions due to pure dephasing by phonons. The theory extends previous perturbative results of McCumber and Sturge. The theory is applied to experimental results on mixed organic crystals and ruby. We also present a theory of dephasing at T = 0 K due to interacting impurities. Our results predict a frequency dependence to $1/T_2$. We analyze experiments on $Y_2O_3:Eu^{3+}$ by Macfarlane and Shelby with this theory, and are able to comment on the nature of inhomogeneous broadening in this system.

TEMPERATURE DEPENDENT OPTICAL DEPHASING

The first part of this report discusses the problem of pure dephasing by phonons of dilute impurity optical transitions in crystals. The usual theory of this effect treats the impurity as a simple two-level electronic system, and considers quadratic coupling of this impurity to a harmonic phonon bath. The optical line width is then calculated to second order in the electron-phonon interaction. For a Debye model of acoustic phonons, this approach leads to the well-known McCumber-Sturge result,[1] which displays a T^7 low temperature dependence. For a pseudolocal phonon model, Small[2] has derived an Arrhenius temperature dependence for the line width.

We[3] have performed a nonperturbative calculation of the same model. Our general results are in agreement with a previous treatment by Osad'ko,[4] who used a different method. We have evaluated these results for different phonon models. For Debye acoustic phonons and for moderately small values of the dimensionless quadratic coupling constant, W, we find[5] significant differences between the exact and weak coupling expressions. We[5] have also provided a general analytic expression for the line width that depends on the two parameters W and T_D, the Debye temperature of the crystal. This expression replaces the weak coupling result of McCumber and Sturge, and is useful in the analysis of experimental data.

For pseudolocal phonons we have derived[6] an analytic expression for the line width that depends only on the temperature and the frequencies and lifetimes of the local mode in the ground and excited electronic states. Our result has a bi-Arrhenius form, with the activation energies corresponding to the local mode frequencies. In various limits our result reduces to the weak coupling theory of Small,[2] the exchange theory of Harris,[7] and the uncorrelated phonon scattering theory of deBree and Wiersma.[8] Thus our simple but quite general expression should be useful in the analysis of experimental data.

Finally, we apply[9] the above theory to experiments on ruby[1] and organic mixed crystals.[10-12] In all cases the theory seems to provide a unified and consistent interpretation of the experiments. The theory has also been used by Powell et al[13] to analyze their experiments on alexandrite. Our work in this field is summarized in a forthcoming review article.[14]

FREQUENCY DEPENDENT OPTICAL DEPHASING

In this section we discuss the optical dephasing of <u>interacting</u> impurities in crystals as measured by frequency-dependent photon echoes. The interactions among impurities can lead to a concentration-dependent contribution to $1/T_2$.

Optical transitions of impurities in crystals are inhomogeneously broadened. One would expect that the extent and frequency dependence of optical dephasing would depend on the spatial correlation of the impurity excitation energies. That is, if the spatial correlation length is very large and thus the impurities are in large resonant domains, then the magnitude of $1/T_2$ should be large due to the large number of resonant interactions, but there should be no frequency dependence since the dephasing in domains of different energy should be identical. On the other hand, if the spatial correlation length is small, then the dephasing of a given impurity depends on the number of near-resonant near neighbors, which is clearly greatest at the center of the inhomogeneous line. In this case one would therefore expect a frequency-dependent optical dephasing, with a larger $1/T_2$ at the center of the inhomogeneous line. Thus photon echo experiments can provide us with some information about the spatial correlation length of site energies, which is of very general importance for our understanding of the nature of eigenstates in disordered systems and the Anderson transition.[15]

Recently, frequency-dependent photon echo experiments at 2 K have been performed by Macfarlane and Shelby[16] on the impurity system 2% $Y_2O_3:Eu^{3+}$. They found a frequency-dependent T_2 as described above, and attributed this to Eu^{3+} - Eu^{3+} interactions. We[17,18] have provided a quantitative calculation of this effect, modeling the system with a tight-binding Hamiltonian for the substitutionally disordered Eu^{3+} ions. We find that, indeed, our results for T_2 depend sensitively on the spatial correlation length of the site energies. We find semi-quantitative agreement between theory and experiment if the correlation length is taken to be very short -- only one or two lattice sites.

ACKNOWLEDGMENTS

We thank the National Science Foundation for support from grant No. DMR 83-06429. L.R. acknowledges additional support in the form of an NSF Graduate Fellowship. J.L.S. acknowledges additional support in the form of a Camille and Henry Dreyfus Teacher-Scholar Award, an Alfred P. Sloan Fellowship, and an NSF Presidential Young Investigator Award.

REFERENCES

1. D.E. McCumber and M.D. Sturge, J. Appl. Phys. 34, 1682 (1963).
2. G.J. Small, Chem. Phys. Lett. 55, 501 (1978).
3. D. Hsu and J.L. Skinner, J. Chem. Phys. 81, 1604 (1984).
4. I.S. Osad'ko, in Spectroscopy and Excitation Dynamics of Condensed Molecular Systems, eds. V.M. Agranovich and R.M. Hochstrasser (North-Holland, Amsterdam, 1983) and references therein.
5. D. Hsu and J.L. Skinner, J. Chem. Phys. 81, 5471 (1984).
6. D. Hsu and J.L. Skinner, J. Chem. Phys. 83, 2097 (1985).
7. C.B. Harris, J. Chem. Phys. 67, 5607 (1977).
8. P. deBree and D.A. Wiersma, J. Chem. Phys. 70, 790 (1979).
9. D. Hsu and J.L. Skinner, J. Chem. Phys. 83, 2107 (1985).
10. F.P. Burke and G.J. Small, J. Chem. Phys. 61, 4588 (1974).
11. O.N. Korotaev and M. Yu. Kaliteevskii, Zh. Eksp. Teor. Fiz. 79, 439 (1980) [Sov. Phys. JETP 52, 220 (1980)].
12. L.W. Molenkamp and D.A. Wiersma, J. Chem. Phys. 80, 3054 (1984).
13. R.C. Powell, L. Xi, X. Gang and G.J. Quarles, Phys. Rev. B 32, 2788 (1985).
14. J.L. Skinner and D. Hsu, Adv. Chem. Phys. (in press).
15. P.W. Anderson, Phys. Rev. 109, 1492 (1958).
16. R.M. Macfarlane and R.M. Shelby, Optics Comm. 39, 169 (1981).
17. L. Root and J.L. Skinner, J. Chem. Phys. 81, 5310 (1984).
18. L. Root and J.L. Skinner, Phys. Rev. B. 32, 4111 (1985).

ULTRACOHERENT TRANSIENT SPECTROSCOPY

W. S. Warren and J. L. Bates
Department of Chemistry, Princeton University, Princeton, NJ 08544

ABSTRACT

We have developed sequences of phase and amplitude modulated laser pulses to compensate for inhomogeneities in Rabi frequency and to eliminate off-resonance effects observed with rectangular and Gaussian pulses. We give here two applications for these sequences: highly selective preparation of large populations of vibrationally excited molecules, and generation of one-photon, Doppler-free spectra.

INTRODUCTION

A major and ongoing focus of our research effort has been the development of laser pulse sequences analogous to those used in nuclear magnetic resonance, to study relaxation phenomena in condensed phases. Sequences of multiple rectangular 90° and 180° pulses with various phases have been used extensively in high resolution and solid state NMR for nearly three decades, and have contributed greatly to our understanding of biomolecular structure, electronic distributions in crystals, and cooperative interactions, to name only a few applications. Such complex pulse sequences have been shown to enhance forbidden transitions, selectively refocus some intermolecular interactions, or even increase physiological image contrast for tumor identification. By contrast, the conventional approach to optical coherent transient spectroscopy (splitting a single high power pulse into several parts) generates pulse sequences with well defined delays, but generally totally unknown phases; in addition, the pulse shape is not easily controlled. This approach has proven useful in many cases, particularly if non-collinear pulses are used to exploit propagation effects[1-3]; the functional form of the decay from most commonly used laser pulse sequences, such as photon echoes or stimulated echoes, is independent of the exact flip angle or shape of each pulse. However, growing recognition of the fundamental complexity of collisional and condensed phase relaxation processes (and the defects of a two-level approximation) reveals a fundamental need for a more powerful approach.

We present here two new examples of pulse sequences which are only possible with laser pulse phase and shape control, and which seem likely to be broadly useful. We will call these ultracoherent transient sequences, because they require enormously more control over the radiation field than is needed for such sequences as photon echoes. We have recently shown[4] that we can shape laser pulses arbitrarily with 7 nsec resolution. We have also built an injection locked laser system[5] which gives 150 psec, 10 μJ pulses with partially adjustable

0094-243X/86/1460429-4$3.00

shapes, fully adjustable phases, and delays from 0.2-5 nsec; the decay of a three-pulse photon echo with phase sensitive detection is less sensitive to optical density effects than is a two-pulse echo, and in fact the combination of two-pulse and three-pulse data can determine the real intrinsic T_2[6]. We can use such ultracoherent transient sequences to pump large and selective vibrational population inversions and to generate one-photon Doppler-free spectra.

EXPERIMENTAL APPARATUS

Figure 1 shows the apparatus we use to generate shaped pulses[4]. We produce the desired envelope in digitized form and store it in RAM using ECL circuitry. A fast counter then clocks the data into a video digital-analog converter, the output of which is mixed with an rf pulse and sent to an acousto-optic modulator (AOM). The AOM cuts a shaped piece out of a continuous wave laser beam[4]. Phase and frequency shifts are done at the radiofrequency level; we can easily give frequency shifts of up to 200 Mhz and any of four phases. Some typical selective pulse shapes are also shown.

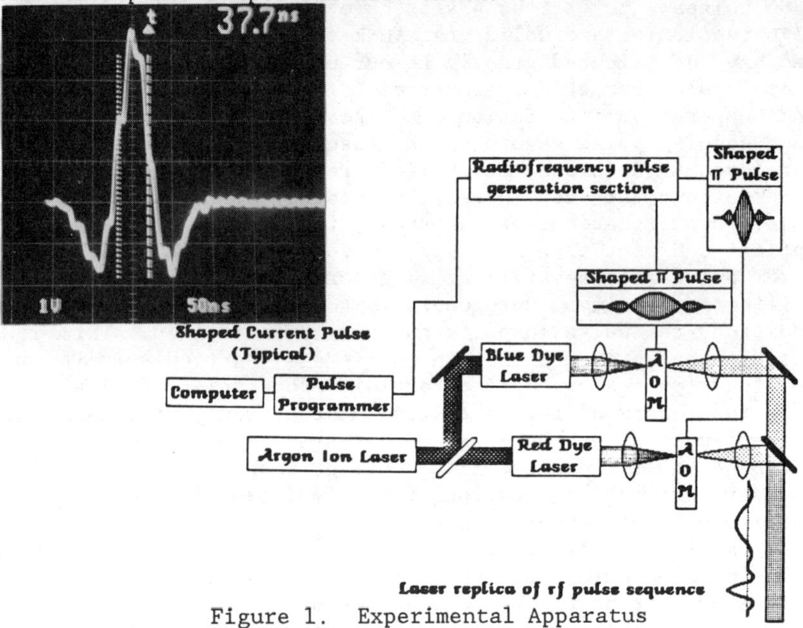

Figure 1. Experimental Apparatus

GENERATION OF SELECTIVE VIBRATIONAL INVERSIONS

Selectively preparing large populations of highly vibrationally excited molecules without simultaneously causing electronic excitation is a helpful tool in the exploration of gas phase molecular dynamics. In some molecules, like I_2 and O_2, electronically excited molecules will predissociate to form radicals which obscure the behavior of the molecules of interest. Unfortunately, the methods by which vibrationally excited populations can be selectively produced leave a substantial number of molecules in electronically excited states. For example, stimulated emission pumping[6] makes use of an electronically excited state and the Franck-Condon principle to populate high vibrational states with two lasers. Using incoherent radiation, no more than 50% of the molecules excited by the first pulse can ever be transferred to the desired level.

Coherent π pulses improve the inversion in principle; however, rectangular or Gaussian pulses have a nonuniform excitation distribution, so off-resonance effects reduce the selectivity. For example, two rectangular π pulses still leave 41% residual electronically excited population (Figure 2a). In addition, eigenstates of any diatomic molecule with $J \neq 0$ is $(2J+1)$-fold degenerate (different values of M_J), with at least $(J+1)$ different Rabi frequencies. Thus, selective inversion requires simultaneous compensation for the off-resonance effects of rectangular pulses and for Rabi frequency deviations.

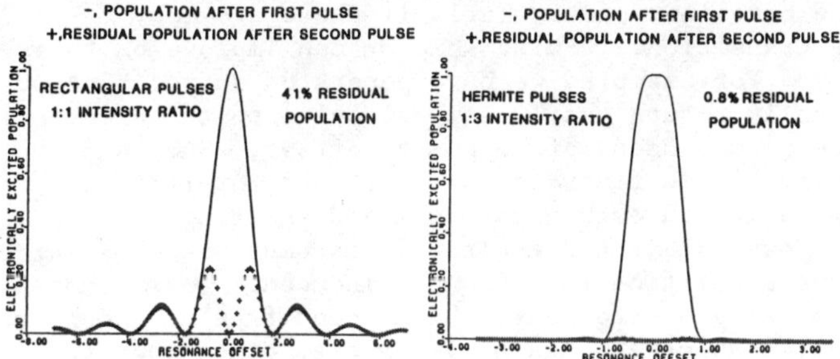

Figure 2. Residual electronic excitation after two π pulses at two different frequencies, used to generate a large population inversion in stimulated emission pumping. Shaping the pulses and adjusting the bandwidth gives a clean inversion.

To correct for off-resonance effects we use pulses with carefully designed shapes. We generated these shapes by extending the formalism known in the NMR literature as average Hamiltonian theory[7] to cover arbitrarily shaped pulses, even with large ($\pi/2$ or π) flip angles. We derived simple, symmetric, single phase pulses which give a uniform and complete inversion; a typical shape is shown in Figure 1. We find that Hermite pulses reduce the 41% residual excited state population expected in a stimulated emission pumping experiment using rectangular pulses to 19%. If the bandwidth of the second pulse is increased, the rectangular pulses leave 17% of the population behind, while Hermite pulses leave only 0.8% (Fig. 2b). In addition, we use "composite pulses" to create a π pulse that is independent of the M_J quantum number. Each is composed of three back-to-back pulses with different phases. The sequence is denoted by $(\pi/2)_x$-π_y-$(\pi/2)_x$. The first and third pulses are identical; the second pulse is four times larger. If all the flip angles are exactly correct, this sequence gives a net rotation of π. If the flip angles deviate from the correct values, it still gives a net rotation of π, to first order in perturbation theory.

TRUE VELOCITY SELECTION WITH OVERLAPPING TRANSITIONS

Crafted pulses can also be used to selectively excite individual velocity components of a completely resolved Doppler broadened transition. In the simplest case, a single Hermite pulse with a narrow inversion profile can select any velocity we choose. However, in I_2, the hyperfine splittings are much smaller than the Doppler broadening so frequency selection and velocity selection are not equivalent.

Counterpropagating continuous lasers can find the v=0 component of any line by locating the Lamb dip, or can find any other velocity component if one of the two beams is frequency shifted. However, the Lamb dips are small decreases on a large fluorescence baseline, particularly if there are many overlapping transitions. Pulse shaping can improve on this considerably. For example, we have generated shaped "2π" pulses which eventually return nearly all molecules to the electronic ground state, independent of resonance offset. Such a pulse generates very little fluorescence by itself, or if it is counterpropagating through a gas cell and exciting a $v \neq 0$ component. However, if the excitation is near $v = 0$, a large standing wave population inversion is induced. Thus, shaped laser pulses can generate nearly background free and Doppler free absorption spectra. Again, velocities other than $v = 0$ can be selected by changing only one of the two counterpropagating pulses.

REFERENCES

1. I. D. Abella, N. Kurnit and S. R. Hartmann, Phys. Rev. 141, 391 (1966).
2. D. E. Cooper, R. W. Olson, and M. D. Fayer, J. Chem. Phys. 72, 2332 (1980).
3. W. H. Hesselink and D. A. Wiersma, J. Chem. Phys. 73, 648 (1980); L W. Molenkamp and D. A. Wiersma, J. Chem. Phys. 80, 3054 (1984).
4. W. S. Warren, J. L.Bates, M. A. McCoy, M. Navratil, and L. Mueller, J. Opt. Soc. Am. B (in press).
5. F. Spano, F. Loaiza, M. Haner, and W. S. Warren, Ultrafast Phenomena IV (Springer, Berlin, 1984) p. 99; Frank Spano and W. S. Warren, Proc. O-E Laser '86 (in press).
6. C. Kittrell, E. Abramson, J. L. Kinsey, S. A. McDonald, D. E. Reisner, and R. W. Field, J. Chem. Phys.,75, 2065 (1981).
7. W. S. Warren, J. Chem. Phys. 81, 5437 (1984).

MODULATED EMISSION IN IODINE VAPOR

A. Z. Genack and P. Mitra
Exxon Research and Engineering Company, Annandale, NJ 08801, and
Queens College of CUNY, Flushing, NY 11367

A. Schenzle
University of Essen, 4300 Essen, West Germany

ABSTRACT

The response of emission in I_2 vapor to intensity modulated light at radial frequency ω is measured and described in terms of the lifetime of the excited state, T_1, the rate of phase interrupting collisions γ, and the inhomogeneously broadened absorption line. When the laser is tuned to resonance, the variation of amplitude of the modulated emission with ω corresponds to a single exponential decay with lifetime T_1. Off resonance, the modulation is described as the sum of the above "delayed" component and a "prompt" in-phase component which is independent of modulation frequency. The ratio of the delayed to prompt component for off resonant excitation is shown to be $2\gamma T_1$. This relation is used to determine the rate of phase interrupting collisions in I_2 vapor.

INTRODUCTION

We have studied the variation in amplitude of the in and out of phase response of emission to intensity modulated excitation as the laser frequency υ is tuned through resonance with an iodine transition. When the laser is tuned off resonance from the inhomogeneously broadened iodine transition, the modulated emission is found to depend upon the rate of phase interrupting collisions γ as well as the excited state lifetime T_1, whereas, when the laser is in resonance, the modulated response depends only upon T_1.

The emission induced by a pulse with well defined laser carrier frequency υ in the linear excitation regime considered here can be given in terms of a convolution of the Fourier components of the transient excitation intensity and the in and out of phase components, $I(\upsilon,\omega)$ and $Q(\upsilon,\omega)$, respectively, of the response to modulated excitation. In earlier studies, Williams et al[1] measured the transient response in emission both in resonance and off resonance from an iodine transition. They found that following the abrupt termination of a resonant exciting pulse the emission decayed with lifetime T_1 from the level reached during the pulse. When the laser is off resonance, they observed a prompt decrease in emission followed by a decay with lifetime T_1.

To better understand the approach to resonance and the influence of dephasing, we have studied the steady state response to intensity modulated light rather than the transient response. The Bloch equations can be readily solved for this simple choice of time varying laser field and the response to a low power pulse can be calculated from these results. Further, $I(\upsilon,\omega)$ and $Q(\upsilon,\omega)$ can be measured as the laser is tuned continuously through resonance.

EXPERIMENT

Fluorescence from the $X\Sigma_g^+(\upsilon''=2,J''=59) \rightarrow B^3\pi_{o^+u}(\upsilon'=15,J'=60)$ transition at 16,956.36 cm^{-1} is studied. A schematic of the experimental appartus is shown in Fig. 1. The output of tunable single frequency Coherent 599-21 dye laser is intensity modulated by a Lasermetrics 3031 Pockels cell and polarization analyzer. This electro-optic modulator is biased at the quarter wave point so that the optical modulation is proportional to the rf input. The modulated light is then passed through an I_2 cell. The laser intensity is in the range 0.05-5 mW/cm^2 and is well below saturation at all pressures. The pressure in the cell is controlled by a side arm immersed in a constant temperature bath. The emission is detected at right angles by a photomultiplier tube through a filter which rejects the laser frequency. The in and out of phase signal is detected in a EG&G PAR 5202 lock-in amplifier in the range of 0.1-20 MHz. A microcomputer collects and averages the lock in outputs and steps the dye laser frequency. Data is averaged for one minute at each point.

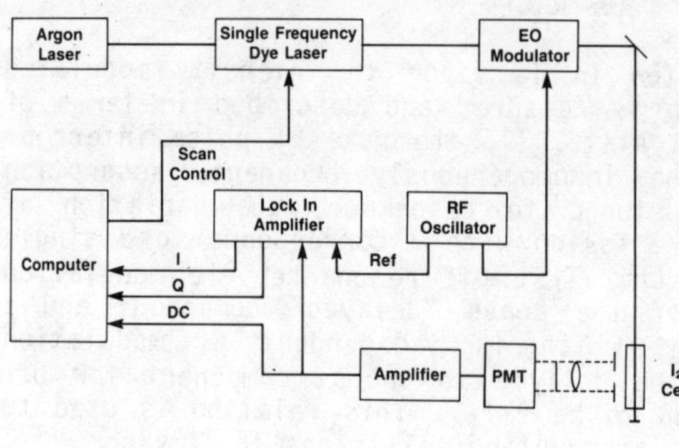

Fig. 1: Experimental Apparatus

THEORY

The modulated laser intensity is represented by

$$I_L(t) = (E_0^2/2)(1 + \cos\omega t)] = E^+(t)E^-(t) \quad (1)$$

where $E^{\pm}(t) = E_0\, e^{\pm i\upsilon t} \cos(\omega/2)t$. The time development of the sample for a two level system with excited state $|2\rangle$ and ground state $|1\rangle$ is described by the Bloch equations for the ensemble average of the polarization operators $P^+ = \overline{\langle|2\rangle\langle 1|\rangle}$, $P^- = \overline{\langle|1\rangle\langle 2|\rangle}$ and the inversion operator $w = \overline{\langle|2\rangle\langle 2|\rangle} - \overline{\langle|1\rangle\langle 1|\rangle}$. If we neglect the counter rotating field, the Bloch equation for the time development of P^+ for a component within the inhomogeneous line which is off resonance by radial frequency Δ from the exciting laser is

$$\dot{P}^+ = \{-(1/T_2') + i\Delta + i\Omega(t)\}P^+ - 2igwE^+. \quad (2)$$

Here $1/T_2' = 1/2T_1$ is the contribution of phase relaxation rate caused by the finite lifetime of the excited state and $g = e\langle 2|r|1|\rangle/h$ is the coupling constant. This expression includes delta correlated phase interrupting collisions with strength $\langle\Omega(t)\Omega(t')\rangle = 2\gamma\delta(t - t')$. In the linear regime we make the approximation that the inversion is undisturbed by the laser and take $w = w_0$, the equilibrium inversion in the absence of the field.

The emission is given by $I(\Delta,t) = \langle P^+(\Delta,t)P^-(\Delta,t)\rangle$ which is the population of the excited state. We consider two important cases. The emission on resonance may be obtained by assuming the inhomogeneous line is independent of frequency, since the emission is from molecules that are within a few homogeneous linewidths of the laser corresponding to a frequency range over which the absorption is nearly constant. The intensity per molecule is the integral of $I(\Delta,t)$ with w_0 assumed independent of frequency,

$$I(t) = 4\pi^2 E_0^2 w_0^2 T_1\{1+(1+\omega^2 T_1{}^2)^{-1}\cos\omega t + \omega T_1(1+\omega^2 T_1^2)^{-1}\sin\omega t\} \quad (3)$$

The amplitude of the modulated emission is just the Fourier transform of a single exponential decay with lifetime T_1. The emission far off resonance at a frequency $\Delta \gg \Delta_{inh},\omega,1/T_1$, where Δ_{inh} is the inhomogeneous linewidth, may be approximated by taking a single detuning frequency Δ which is the difference between the laser frequency and the center of the inhomogeneously broadened line. In this limit the intensity per molecule is

$$\begin{aligned} I(\Delta,t) &= 2g^2 E_0^2 w_0^2/\Delta^2\{(1 + 2\gamma T_1) + \cos\omega t \\ &+ 2\gamma T_1[(1 + \omega^2 T_1^2)^{-1}\cos\omega t + \omega T_1(1 + \omega^2 T_1^2)\sin\omega t]\} \end{aligned} \quad (4)$$

The intensity in eq. 4 is the sum of a constant term, a prompt term, which remains in phase with the excitation and a delayed term which has the same modulation properties as given on resonance in eq. 3. The ratio of the delayed to prompt components far from resonance, D/P, is given by $D/P = 2\gamma T_1$.

RESULTS AND DISCUSSION

The response to modulation $I(\upsilon,\omega)$ and $Q(\upsilon,\omega)$ is measured with a lock-in as the laser frequency υ is scanned through the iodine transition with fixed modulation frequency. The in-phase signal at 5 MHz modulation at an I_2 pressure of 246 mTorr is shown in Fig. 2. Using eq. 4, the response to modulation may be expressed as,

$$I(\upsilon,\omega) = P(\upsilon) + D(\upsilon)(1 + \omega^2 T_1^2)^{-1} \quad (5a)$$

$$Q(\upsilon,\omega) = D(\upsilon)\omega T_1(1 + \omega^2 T_1^2)^{-1}. \quad (5b)$$

These equations are inverted to solve for $D(\upsilon)$ and $P(\upsilon)$ in terms of the experimentally measured quantities $I(\upsilon,\omega)$ and $Q(\upsilon,\omega)$. The ratio $P(\upsilon)/D(\upsilon)$ is shown in Fig. 3. When the laser is in resonance with the I_2 transition $P(\upsilon)$ is zero, but off resonance the ratio of $P(\upsilon)/D(\upsilon)$ reaches a constant value, P/D, in a limited tuning range before dropping again as the laser is tuned into resonance with another I_2 transition. The lifetime, T_1, is obtained by fitting the modulation behavior on resonance to eq. 3 and γ is obtained from the value of P/D. Table 1 gives the result at a number of pressures. We find $\gamma = 0.048\ p \times 10^6\ s^{-1}$, where p is pressure in mTorr.

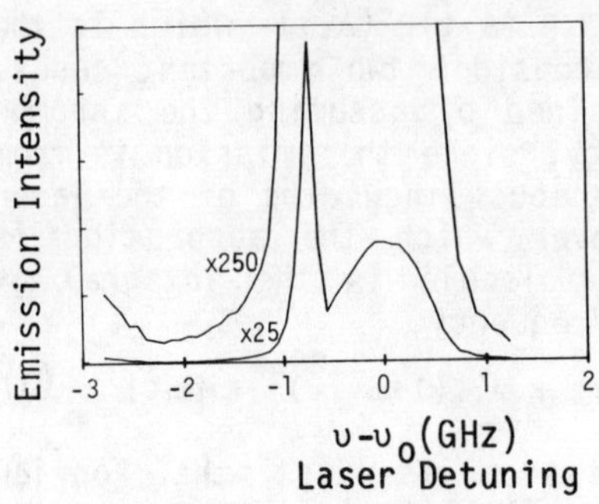

Fig. 2: In phase signal at 5MHz modulation at 246 mTorr.

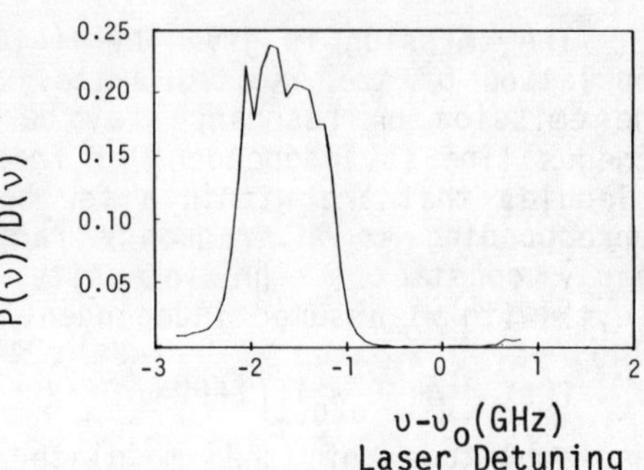

Fig. 3: $P(\upsilon)/D(\upsilon)$ calculated from I and Q at 5 MHz modulation at 246 mTorr.

The model for relaxation we have used in eq. 2 does not include velocity changing collisions. This is appropriate in the limit $\Delta_{inh} << \Delta$ because a change in velocity, which corresponds to a change in magnitude of Δ by at most Δ_{inh} will not appreciably alter the phase of the polarization relative to the effective field in the rotating frame. Thus, the off resonance fluorescence measurements relate specifically to the rate of phase interrupting collisions which are associated with fluctuations in the transition frequency which are at least of order Δ. This is different from on resonance coherent transient experiments in transmission which are sensitive to weak collisions and to velocity changes as well.

Table 1. Pressure dependence of relaxation rates and D/P

Pressure (mTorr)	$1/T_1(\mu sec)^{-1}$	D/P	$\gamma(\mu sec)^{-1}$
31	1.42	2.9	2.04
130	3.13	4.0	6.34
170	3.82	4.35	8.24
200	4.52	4.0	8.96
246	5.15	4.55	11.78

REFERENCES

1. P. F. Williams, D. L. Rousseau and S. H. Dworetsky, Phys. Rev. Lett., 32, 196 (1974).

OBSERVATION AND RELAXATION OF THE TW0-PHOTON-EXCITED-STATE TRILEVEL ECHO IN SODIUM VAPOR

T. J. Chen*, D. DeBeer and S. R. Hartmann
Columbia Radiation Laboratory and Department of Physics
Columbia University, New York, NY 10027

A photon echo from the allowed 3P-3D transition in Na vapor is spontaneously emitted at λ = 8183Å after a single-pulse (λ = 5896Å) one-photon excitation of the $3P_{1/2}$ state is followed by a single-pulse (λ = 6854Å) two-photon excitation of the 3D state, and similarly for $3P_{3/2}$. The collisional relaxation cross section of the excited 3P-3D superposition state of Na with Ar is measured.

Previous studies[1] have measured collisional cross sections of 3P-nD (n = 4, 5, 6 and 7) superposition states of Na perturbed by Ar. Excited state relaxation measurements for the lowest lying D sate with n = 3 were omitted because of the difficulty of generating the long wavelength (λ = 8183Å) excitation pulses which would have been necessary to coherently excite the 3P-3D transition. We have circumvented this difficulty by using a variation of the trilevel echo technique that allows indirect excitation of the 3P-3D superposition state: we first resonantly excite a 3S-3P coherence then time τ later resonantly excite (via a two-photon process) a 3S-3D coherence. This procedure coherently relates the 3P and 3D states so that an echo is subsequently formed; this echo can then be monitored to study 3P-3D superposition relaxation.

The echo formation process can be easily understood using the Billiard Ball Model.[2] The billiard ball recoil diagram for the two-photon-excited-state echo is shown in Fig. 1. At time t_1 the atoms in the ground state, 3S, are excited by a laser pulse which has some probability of exciting the 3P excited state. If an atom absorbs a photon then it must recoil with the momentum of the absorbed photon. The resulting recoil trajectory is shown in Fig. 1 as a dashed line. Similarly, at time $t_2 = t_1 + \tau$ the second pulse generates a 3D state which recoils as shown by the dash-dotted line. At subsequent time $t_e = t_1 + (2k_2 / (2k_2 - k_1))\,\tau$, where k_1 (k_2) is the wavevector amplitude of the first (second) excitation pulse, the 3D excited state overtakes the 3P state and the echo is emitted on the 3D-3P transition with wavevector $\mathbf{k}_e = 2\mathbf{k}_2 - \mathbf{k}_1$.

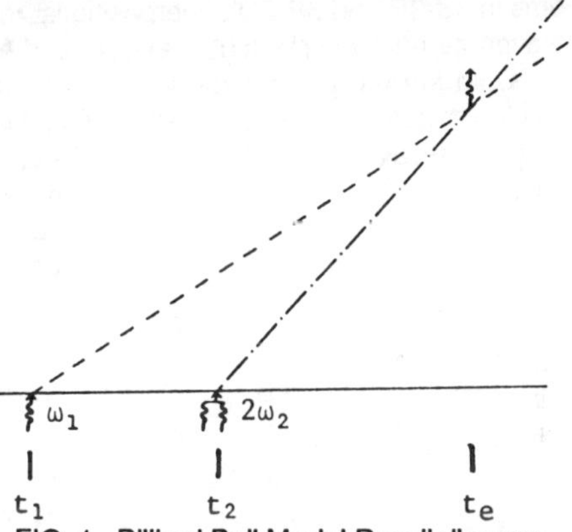

FIG. 1. Billiard Ball Model Recoil diagram for two-photon-excited-state trilevel echo plotting recoil position vs time.

The experimental apparatus is shown in Fig. 2. Dye laser 1 is tuned to be resonant with the 3S-3P transition while dye laser 2 is tuned to be two-photon resonant with the 3S-3D transition. Both dye laser pulses have spectral widths of about 3 GHz and temporal durations of 7 ns. The output of the second laser is delayed optically by 24 ns and then superimposed collinearly with the first pulse. The resultant beam is directed through a Na sample cell maintained at 422 K. The echo is collinear with both excitations; it is separated from

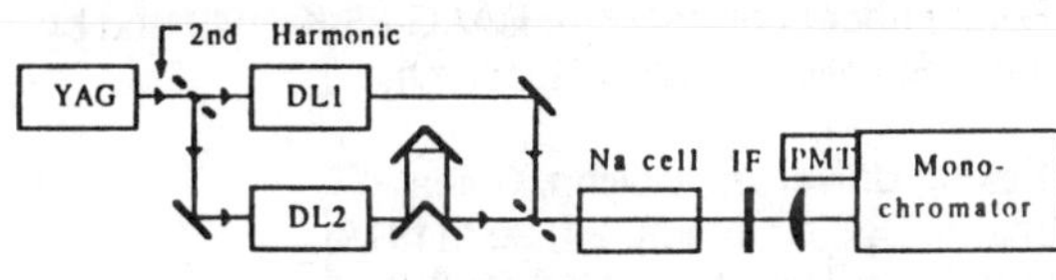

FIG. 2. Schematic diagram of the experimental apparatus.

0094-243X/86/1460437-2$3.00

them by means of an interference filter and a monochromator. The echo is detected with an RCA C31034 photomultiplier tube.

The measurement of the total collisional cross section with Ar of Na in a 3P-3D superposition begins by monitoring the echo intensity as an Ar buffer gas is introduced up to a partial pressure of 1 torr. The echo intensity decays exponentially with buffer gas pressure, P, as[3]

$$I = I_0 \exp(-\beta P)$$

where

$$\beta = 2(N/P)\bar{v}[\sigma_1\tau + \sigma_2(t_e - t_2)],$$

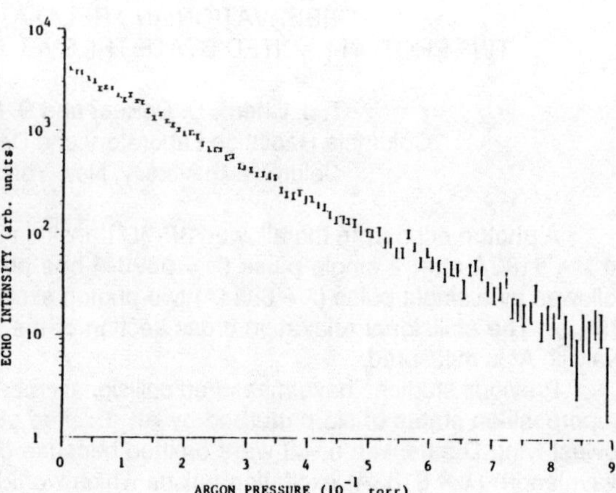

FIG. 3. Typical set of data for decay of echo intensity with Ar gas pressure. Each point represents an average over 100 laser pulses.

σ_1 and σ_2 are the collisional cross sections between Ar atoms and Na atoms in 3S-3P and 3P-3D superpositions, respectively, N is the Ar number density, and $\bar{v}$ is the average relative speed between Na and Ar atoms which is given by $(8k_BT/\pi\mu)^{1/2}$.

Data from a typical experimental run are shown in Fig. 3. The data conform well the expected exponential decay. Performing a least squares fit and averaging over several runs, we obtain decay rates of 7.25(10) and 7.42(10) torr^{-1} for the $3^2P_{1/2}$ and $3^2P_{3/2}$ intermediate states, respectively. The 3S-3P collisional cross sections, σ_1, have been measured previously.[4] Using these data we find

$$\sigma(3^2P_{1/2} - 3D) = 414(30)\ \text{Å}^2$$

$$\sigma(3^2P_{3/2} - 3D) = 429(30)\ \text{Å}^2.$$

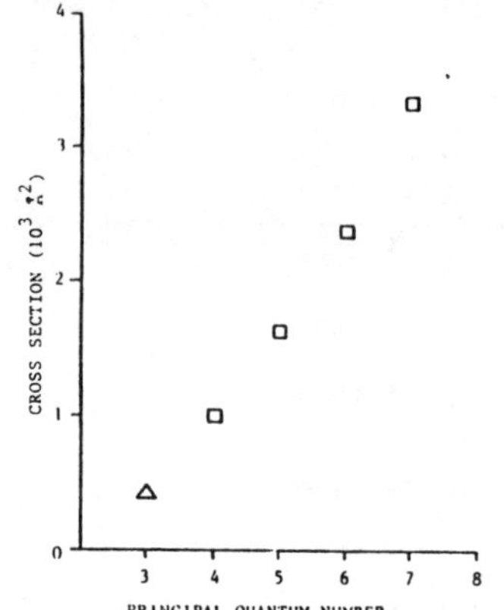

FIG. 4. Collisional cross section of Na in a $3^2P_{1/2} - n^2D_{3/2}$ superposition with Ar, vs principal quantum number n.

This is the first time these 3P-3D cross sections have been determined. In Fig. 4 we show the current determination of the $3^2P_{1/2} - 3^2D_{3/2}$ collisional cross section along with the previous measurements of the $3^2P_{1/2} - n^2D_{3/2}$ (n = 4, 5, 6 and 7) collisional cross sections obtained using the more conventional excited state photon echo technique.[1]

This work was supported by the Joint Services Electronics Program (U.S. Army, U.S. Navy, U.S. Air Force) under contract No. DAAG-85-K-0049 and by the U.S. Office of Naval Research.

*Present address - Physics Department, Peking University, Beijing, China.

1. A. Flusberg, T. Mossberg and S.R. Hartmann, Opt. Commun. 24, 207 (1978).
2. R. Beach, S.R. Hartmann and R. Friedberg, Phys. Rev. A25, 2658 (1982).
3. A. Flusberg, R. Kachru, T. Mossberg and S.R. Hartmann, Phys. Rev. A19, 1607 (1979).
4. R. Kachru, T.W. Mossberg and S.R. Hartmann, J. Phys. B 13, L363 (1980).

NOBLE GAS INDUCED COLLISIONAL LINE BROADENING OF ATOMIC Li RYDBERG STATES nS AND nD (n = 4 to 30) MEASURED BY TRILEVEL ECHOES

Emily Y. Xu, F. Moshary and S. R. Hartmann
Physics Department, Columbia Radiation Laboratory,
Columbia University, New York, N. Y. 10027

ABSTRACT

Collisional relaxation of 2S-nS and 2S-nD superposition states in atomic Li vapor is measured with the trilevel photon echo technique for principal quantum numbers range from n =4 to well into the Rydberg regime (n = 30). With noble gas perturbers (He, Ne, Ar, Kr, and Xe) the variation of collision cross section with n is considerably weaker than previously found for Na vapor.

INTRODUCTION

Collisional relaxation of even-parity superposition states of alkali vapors in noble gas buffer environments has previously been measured by the Doppler-free two photon absorption[1,2] and tri-level photon echo[3] techniques. These measurements have been performed in Na[3] and Rb[2] vapor over a wide range of principal quantum numbers and show the same behavior found by Füchtbauer et al[4] in their study of Na 3S-nP transition broadening. For large principal quantum numbers the theory of Omont[5] accounts for the even-parity nS-mS superposition state relaxation of Na rather well and it was decided to check this behavior for the presumably simpler case of 2S-nS relaxation in Li vapor.

EXPERIMENT

A pair of YAG lasers pump dye lasers to produce three excitation pulses at t_1, t_2, and t_3 and at frequencies ω_1, ω_2, and ω_3 along $\mathbf{k}_1$, $\mathbf{k}_2$, and $\mathbf{k}_3$ respectively. The frequency ω_1 is tuned to the 2S-2P transition at 6708Å while $\omega_2 = \omega_3$ is tuned to one of the 2P-nS or 2P-nD transitions. Pulse 1 is polarized parallel to pulse 2 and normal to pulse 3 while $\mathbf{k}_1$ is antiparallel to $\mathbf{k}_2 = \mathbf{k}_3$. We set $t_1 = t_2 = 0$ and $t_3 = \tau = 29$ nsec. We detect the echo at $t_e = \tau\,(k_2 / k_1)$ along $\mathbf{k}_e = \mathbf{k}_1 + \mathbf{k}_2 - \mathbf{k}_3 = \mathbf{k}_1$ and polarized normal to pulse 1. We measure the echo intensity as a function of the noble gas pressure. The analysis proceeds as in ref. 3 and we extract the collisional cross sections which we plot in fig. 1 as a function of n. Only the 2S-nS cross sections are plotted. The 2S-nD cross sections are generally larger because of L-mixing collisions. The theory of Omont yields the solid lines and does not work as well for Li as for Na and Rb. It does, however, serve to order the magnitudes of the cross sections for the different perturbers. The overlapping dotted lines indicate the geometrical cross sections.

0094-243X/86/1460439-2$3.00

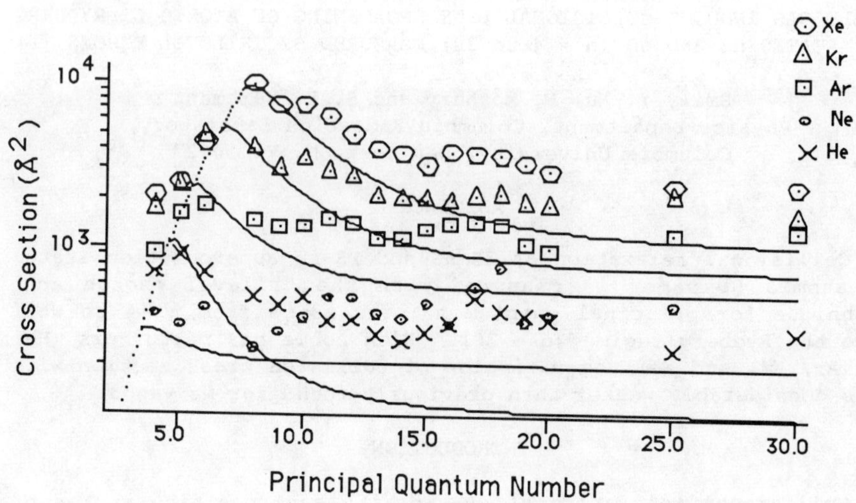

Fig. 1. Li 2S-nS collisional cross section versus principal quantum number n for He, Ne, Ar, Kr, and Xe. The solid curves are Omont's theory and the dotted curves are the extension of Omont's theory to the low n regime.

ACKNOWLEDGEMENT

This work was supported by the U.S. Office of Naval Research and by the Joint Services Electronics Program (U.S. Army, U.S. Navy, U.S. Air Force) under contract No. DAAG29-85-K-0049.

REFERENCES

1. F. Biraben, B. Cagnac, E. Giacobino, and G. Grynberg, J. Physics B 10,m 2369 (1977).
2. K.H. Weber and K. Niemax, Z. Phys. A 307 12 (1982).
3. R. Kachru, T.W. Mossberg, and S.R. Hartmann, Phys. Rev. A 21, 4, 1124 (1980).
4. C. Füchtbauer, P. Schultz and A.P. Brandt, Z. Phys. 90, 403 (1934).
5. A. Omont, J. de Physique, Tome 38, Nov. 1977, 1343.

FEMTOSECOND TIME-RESOLVED SPECTROSCOPY OF COHERENT VIBRATIONAL AND ELECTRONIC EXCITATIONS IN SOLIDS

Leah Ruby Williams* and Keith A. Nelson
Massachusetts Institute of Technology, Cambridge, MA 02139

ABSTRACT

"Impulsive" stimulated scattering (ISS) of femtosecond laser pulses was used to coherently excite and probe a low-lying (61-cm^{-1}) electronic excitation in the cooperative Jahn-Teller crystal, terbium vanadate. Coherent terahertz oscillations and their dephasing were observed in the time domain. ISS is a general aspect of ultrashort-pulse interactions with matter, through which coherent excitations are produced whenever a sufficiently short laser pulse enters a Raman-active medium. Its use for measurement of vibrational and electronic dephasing and lifetimes, and for time-resolved spectroscopy of vibrationally distorted crystals and molecules, is discussed.

Time-resolved observations of coherent vibrational oscillations in solids have recently been carried out using a novel "impulsive" stimulated scattering (ISS) technique.[1, 2] ISS permits direct spectroscopic characterization of vibrationally distorted species, as well as measurement of phonon dynamics. Unlike conventional stimulated scattering schemes, ISS phonon excitation is achieved by crossing two ultrashort laser pulses centered at the same frequency inside a crystal. The excitation pulse duration must be short compared to a single vibrational oscillation period so that the crossed pulses exert a temporally impulsive, spatially periodic force which excites coherent phonons whose wave vector matches the optical interference period. Impulsive stimulated Brillouin[1] and Raman[2] scattering (ISBS and ISRS) have permitted coherent excitation of acoustic and optic phonons using picosecond or femtosecond excitation pulses, respectively. The resulting time-dependent standing-wave vibrational oscillations have been monitored by coherent scattering of variable delayed pulses. ISBS experiments recently permitted characterization of "soft" transverse acoustic phonons near structural phase transitions,[1, 3] while ISRS experiments allowed measurement of rapid optic-phonon dephasing in organic molecular crystals at high temperatures.[2]

Here we discuss recent experimental and theoretical results relating to femtosecond ISRS. First, we note recent experimental results in which ISRS excitation and probing of coherent electronic excitations was carried out.[4] Second, we discuss the potential for unambiguous measurement of both dephasing and population decay times (T_2 and T_1) through ISRS. Finally, we discuss recent theoretical results showing that ISRS coherent excitation occurs not only with crossed femtosecond pulses but whenever even a single sufficiently short pulse enters a Raman-active medium.[5]

*AT&T Bell Laboratories Ph.D. Scholar

ISRS experiments were carried out as described earlier[2] on the cooperative Jahn-Teller crystal, terbium venadate ($TbVO_4$),[5] at room temperature. The 70-femtosecond excitation pulses were vertically (V) polarized relative to the scattering plane, which contained the tetragonal a and c crystallographic axes. The polarization direction was thus coincident with the other equivalent a axis. The scattering wave vector difference was aligned along the c axis. The incident and coherently scattered probe pulses were also V-polarized. The coherent scattering intensity showed 3.65-THz oscillations and a gradual decay corresponding to a 61-cm^{-1} mode with an 8.6-ps dephasing time. The $TbVO_4$ crystal has no 61-cm^{-1} phonon mode,[5] so we believe we are exciting a low-lying electronic level of the Tb^{3+} ion. This has not been observed in earlier V-H polarized Raman-scattering experiments on $TbVO_4$.[5] Given the scattering kinematics, we do not expect that the mode under observation plays a direct role in the cooperative Jahn-Teller phase transition.[5] A temperature-dependent study is under way. These preliminary results demonstrate the utility of ISRS for characterization of electronic as well as vibrational dynamics.

By monitoring the decay of oscillations in ISRS coherent scattering, phonon and electronic dephasing times (T_2) have been determined. We note that since a standing-wave excitation is produced in ISRS, the vibrational or electronic energy density remains spatially periodic even after dephasing occurs. Time-resolved diffraction of light off of this spatially periodic population "grating" should permit determination of population lifetimes (T_1) as well as T_2.

Finally, we note that even a single ultrashort pulse will coherently excite Raman-active vibrational, electronic or other modes through ISRS.[6] The pulse exerts a temporally impulsive, spatially uniform force which excites $k \approx 0$ modes. Thus ISRS is a general feature of the way ultrashort laser pulses interact with matter. Spectroscopic applications of this effect have been pointed out.[6]

We gratefully acknowledge the collaboration of J.M. Huxley, W. Z. Lin, and Profs. J.G. Fujimoto and E.P. Ippen on the experimental work, which was supported by US ARO Contract No. DAAL03-86-K-0002.

REFERENCES

1. M. M. Robinson et al., Chem. Phys. Lett. 112, 491 (1984).
2. S. De Silvestri et al., Chem. Phys. Lett. 116, 146 (1985).
3. M. R. Farrar, L.-T. Cheng, Y.-X. Yan, and K. A. Nelson, IEEE J. Quant Electron., in press.
4. L. R. Williams et al., to be published.
5. R. J. Elliot, R. T. Harley, W. Hayes, and S. R. P. Smith, Proc. R. Soc. Lond. A328, 217 (1972).
6. Y.-X. Yan, E. B. Gamble, Jr., and K. A. Nelson, J. Chem. Phys. 83, 5391 (1985).

LASER AND FOURIER TRANSFORM SPECTROSCOPY: DIATOMICS TO ORGANOMETALLICS

P. F. Bernath
Dept. of Chemistry, University of Arizona
Tucson, AZ. 85721

ABSTRACT

The high-resolution visible spectra of CaOH, SrOH and BaOH were analysed by laser excitation spectroscopy. A large number of new organometallic free radicals were discovered when calcium, strontium and barium vapors were found to react with a wide variety of organic molecules including aldelydes, ketones, alcohols, cyanates, cyanides, thioethers, thiols and carboxylic acids. Laser-induced fluorescence provided electronic and vibrational information for these novel species. The Kitt Peak Fourier transform spectrometer was used to detect the vibration-rotation spectra of CuH and NeH^+, the electronic spectra of MgH and NH and the laser-induced fluorescence spectra of SrOH and C_2N.

ALKALINE EARTH MONOHYDROXIDES

The visible bands of the alkaline earth monohydroxides commonly occur when alkaline earth salts are introduced into flames. They were first observed in 1823 by Herschel, although the carriers of the spectra were not identified until 1955 when James and Sugden[2] recognized the similarity between the isolectronic MOH and MF radicals. The bands are so badly overlapped that they defied analysis until Harris and coworkers[3,4] combined a cool source[5] of CaOH and SrOH with the technique of laser-induced fluorescence.

We have continued the work on CaOH by analysing the $\tilde{B}^2\Sigma^+-\tilde{X}^2\Sigma^+$ transition[6] and reanalysing the $\tilde{A}^2\Pi-\tilde{X}^2\Sigma^+$ transition[7]. For SrOH we have analysed the $\tilde{A}^2\Pi-\tilde{X}^2\Sigma^+$ transition[8] both by laser excitation spectroscopy and Fourier transform detection of laser-induced fluorescence.

The first high-resolution analysis of BaOH and BaOD[9] ($\tilde{B}^2\Sigma^+-\tilde{X}^2\Sigma^+$) proved that BaOH is linear like SrOH and CaOH. The Ba-O bond length was found to be 2.201Å while O-H bond length was 0.923Å in the $\tilde{X}^2\Sigma^+$ state. The excited $\tilde{B}^2\Sigma^+$ state was perturbed by strong interaction with the nearby $\tilde{A}^2\Pi$ state so that e and f parity components had to be fit separately. The $\tilde{X}^2\Sigma^+$ vibrational frequencies for BaOH (BaOD) are 492.4(482.4) cm^{-1} for Ba-O stretch and 341.6(257.6) cm^{-1} for the bend. The electronic transition $\tilde{B}^2\Sigma^+-\tilde{X}^2\Sigma^+$ 000-010 was clearly observed, in violation of the selection rule $\Delta v=2$ for the bending mode of a triatomic[8].

ALKALINE EARTH MONOHYDROSULFIDES

The sulfur analogues of CaOH and SrOH: CaSH and SrSH were discovered by reaction H_2S with Ca and Sr vapors in a Broida oven[5]. The spectra are very complicated but clearly show that CaSH and SrSH are bent molecules. High-resolution analysis of an $^2A''-^2A'$ transition is in progress[10].

ALKALINE EARTH MONOALKOXIDES

The Ca, Sr and Ba vapors react with alcohols, including methanol, ethanol, propanol, isopropanol and t-butanol[11,12] to make free radicals of the form MOR. These molecules strongly resemble the alkaline earth monohydroxides (R=H) in electronic structure. The bonding of these molecules is well described by an M^+ ion perturbed by an ^-OR ligand. (The CaSH and SrSH molecules are bent, presumably because of more covalent character in the bonding). Vibrational frequencies for the M-O stretch ($486 cm^{-1}$ for $CaOCH_3$ to $249 cm^{-1}$ for $BaOCH(CH_3)_2$) and M-O-R bend ($142 cm^{-1}$ for $CaOCH_3$ to $53 cm^{-1}$ for $SrO(CH_2)_3CH_3$) were found in the ground states. Ca (or Sr) reacts with acetone and acetaldehyde to make mainly the alkoxides $CaOCH(CH_3)_2$ and $CaOCH_2CH_3$.

ALKALINE EARTH MONOCYANATES (AND CYANIDES)

The CaOCN and SrOCN molecules were discovered[11] as reaction products of Ca and Sr vapors with HNCO. Note that for this reaction a second laser resonant with the metal atomic $^3P_1$ - $^1S_0$ transition was required. For all of the alkaline earth reactions discussed in this article, the excited metal atom, M*, is always more reactive and provides much more of the product free radical. The CaOCN and SrOCN spectra resemble the corresponding alkoxide spectra, so these molecules are linear and probably oxygen bonded.

A similar reaction of Ca^* vapor with CH_3CN produced the CaCN radical[13,14] but there was no resonance fluorescence or vibrational structure. This suggests that CaCN may be T-shaped like NaCN[13] or very floppy[16].

ALKALINE EARTH MONOFORMATES (AND MONOACETATES)

Formic and acetic acids react with Ca and Sr to produce monoformates and monoacetates[11,17] MO_2CH and MO_2CCH_3 with C_{2v} symmetry. CaH and SrH molecules were also produced by this reaction. In addition to the $^2B_1$, and $^2B_2$ states that correlate with the $\tilde{A}^2\Pi$ state, a new electronic state was found $\sim 1000 cm^{-1}$ below them. The new state probably has $^2A_1$ symmetry and correlates with a $^2\Delta$ state in the linear molecule. Only one vibrational mode was active, the symmetric metal-oxygen stretch ($280 cm^{-1}$ for ground state SrO_2CH).

FOURIER TRANSFORM SPECTROSCOPY

The infrared emission spectrum of a copper hollow cathode lamp was observed with the high resolution Fourier transform spectrometer associated with the McMath Solar Telescope at Kitt Peak. When the lamp was operated with 2 torr of Ne gas and 30 mtorr of H_2 or D_2 the vibration-rotation spectra of NeH^+, NeD^+ and CuH[18] were detected. The CuH work is the first vibration-rotation emission spectrum of a metal hydride to be observed.

The $A^2\Pi$-$X^2\Sigma^+$ transition of MgH was observed with high resolution and high signal-to-noise ratio[19]. The objective was the accurate prediction of the submillimeter wave rotational transitions and the infrared fundamental for astronomical purposes.

The NH molecule is, along with OH and CH, one of the most commonly detected free radicals. The $A^3\Pi-X^3\Sigma^-$ transition and the $c^1\Pi-a^1\Delta$ transition were observed by emission from a hollow cathode lamp[21]. The new analyses are much more extensive and are about a factor of 200 more precise than the best previous work.

Emission spectra of polyatomic free radicals are usually very complicated and difficult to interpret. One powerful method to simplify the spectra is to selectively excite only a few rovibronic energy levels with a laser and then detect the resulting emission with a high-resolution Fourier transform spectrometer. We have successfully applied this technique to the SrOH[8] and C_2N[21] free radicals.

ACKNOWLEDGMENTS

I would like to thank my collaborators, C. R. Brazier, R. S. Ram, L. C. Ellingboe, S. M. Kinsey-Nielsen, A. M. R. P. Bopegedera, J. Black and J. Brault, in this work. The research was supported by funding from NSF, ONR, ACS/PRF and Research Corp.

REFERENCES

1. J. F. W. Herschel, Trans. Roy. Soc. (Edinburgh) 9, 445 (1823).

2. C. G. James and T. M. Sugden, Nature 175, 333 (1955).

3. J. Nakagawa, R. F. Wormsbecher and D. O. Harris, J. Mol. Spectrosc. 97, 37 (1983).

4. R. C. Hilborn, Q. Zhu and D. O. Harris, J. Mol. Spectrosc. 97, 73 (1983)

5. J. B. West, R. S. Bradford, J. D. Eversole and C. R. Jones, Rev. Sci. Instrum. 46, 164 (1975).

6. P. F. Bernath and S. Kinsey-Nielsen, Chem. Phys. Lett. 105, 663 (1984).

7. P. F. Bernath and C. R. Brazier, Astrophys. J. 288, 373 (1985).

8. C. R. Brazier and P. F. Bernath, J. Mol. Spectrosc. 114, 163 (1985).

9. S. Kinsey-Nielsen, C. R. Brazier and P. F. Bernath, J. Chem. Phys. 84, 000 (1986).

10. R. S. Ram, C. R. Brazier and P. F. Bernath, work in progress.

11. C. R. Brazier, P. F. Bernath, S. M. Kinsey-Nielsen and L. C. Ellingboe, J. Chem. Phys. 82, 1043 (1985).

12. C. R. Brazier, L. C. Ellingboe, S. M. Kinsey-Nielsen and P. F. Bernath, J. Am. Chem. Soc. 108, 000 (1986).

13. L. C. Ellingboe, A.M.R.P. Bopegedera, C. R. Brazier and P. F. Bernath, submitted Chem. Phys. Lett.

14. L. Pasternack and P. J. Dagdigian, J. Chem. Phys. 65, 1320 (1976).

15. J. J. van Vaals, W. L. Meerts and A. Dymanus, Chem. Phys. 86, 147 (1984).

16. C. W. Bauschlicher, S. R. Langhoff and H. Partridge, Chem. Phys. Lett. 115, 124 (1985).

17. C. R. Brazier, P. F. Bernath, L. C. Ellingboe, S. M. Kinsey-Nielsen, in preparation.

18. R. S. Ram, P. F. Bernath and J. W. Brault, J. Mol. Spectrosc. 113, 451 and 269 (1985).

19. P. F. Bernath, J. H. Black and J. W. Brault, Astrophys. J. 298, 375 (1985).

20. C. R. Brazier, R. S. Ram and P. F. Bernath, in preparation.

21. C. R. Brazier, L. C. Ellingboe and P. F. Bernath, in preparation.

VIBRATIONAL PREDISSOCIATION SPECTROSCOPY OF MOLECULAR CLUSTERS

James M. Lisy
School of Chemical Sciences, University of Illinois, Urbana, IL 61801

ABSTRACT

The vibrational spectra for a number of van der Waals and hydrogen-bonded clusters have been obtained using a tunable infrared laser in the 2.5 to 4.0 μ range. The spectra were observed via vibrational predissociation in a molecular beam apparatus. The results permitted structural assignments to be made for Ar-C_2H_4 and $(HF)_3$. The dependence of the predissociation dynamics on structure, vibrational mode, molecular complexity and binding energy has been examined. No evidence for rapid ($\lesssim$ 25 psec) predissociation or vibrational relaxation has been observed for binary complexes. However, there appears to be a rapid relaxation process for $(HF)_3$.

INTRODUCTION

The spectroscopy of molecular clusters in molecular beams has been a very active field for the past decade. Advances in laser and detector technologies have enabled researchers to expand the investigations to include binary complexes of large polyatomic molecules and clusters with three or more molecular and/or atomic subunits. One main objective of this research is to enhance our understanding of weak intermolecular forces, i.e. van der Waals and hydrogen bonds. The dynamics of cluster photodissociation, such as final quantum state distributions and fragmentation products has been another area of interest with applications to intermolecular energy transfer.

We have employed vibrational spectroscopy as our probe for molecular clusters for a number of reasons. First, as an alternative to the more established techniques of electronic and microwave spectroscopy, one obtains information on the ground state potential energy surface (PES) from vibrational(-rotational) analysis provided there is a non-zero electric dipole transition moment. Vibrational analysis of electronic spectra[1] usually involves the excited electronic state PES, and microwave spectroscopy,[2] while possessing greater precision, is restricted to systems with non-zero electric dipole moments. Secondly, most polyatomic molecules have more than one infrared-active vibrational mode. This allows for tests of mode-specificity in the predissociation dynamics. Although there has been a recent review[3] of the work in this area, new results[4] and interpretations[5] are challenging a number of earlier conclusions.

EXPERIMENTAL

The vibrational predissociation spectra of the complexes were obtained using the molecular beam depletion method. Briefly, we tune the mass spectrometer to a mass peak corresponding to the complex of

0094-243X/86/1460447-5$3.00

interest. Since the binding energy of the complex is less than the photon energy, the absorption of a single photon by the complex results in the dissociation of the complex, thereby decreasing the signal to the mass spectrometer. This depletion of the beam is plotted versus frequency to obtain the spectrum.

The molecular beam apparatus has been described in detail elsewhere.[6] Our continuous nozzle source operates under relatively gentle expansion conditions - 0.5-1.5 atm backing pressure, dilute mixtures (0.2 to 2.0%) of the species of interest in Ar or "first-run" neon (70% Ne, 30% He). The beam passes through four stages of differential pumping before entering the mass spectrometer chamber, which is pumped to 9×10^{-8} torr. Within this chamber, the beam is ionized (6 mA emission with an electron energy of 60 volts) and then passed through a quadrupole mass filter. The total distance from the nozzle to the ionizer is approximately 80 cm.

The tunable infrared radiation in the 2.5-3.4 μ region was generated by Nd-YAG laser pumping of a $LiNbO_3$ optical parametric oscillator of our own design[7]. This OPO may be operated in either medium-resolution (0.8 cm^{-1} FWHM) or high-resolution (0.06 cm^{-1} FWHM) through the optional inclusion of an intracavity etalon. At 125 mJ/pulse pump energy, we could expect 2-3 mJ/pulse OPO output in medium-resolution and 1 mJ/pulse in high-resolution. The scanning of the laser was controlled by the LSI-11 computer via stepping motor-micrometer drives on the grating, etalon, and crystal. The OPO was aligned in the apparatus to be coaxial to the molecular beam over the entire path from source to ionizer.

STRUCTURAL RESULTS

We have been able to determine the structures for two complexes $Ar-C_2H_4$ and $(HF)_3$ from vibrational predissociation spectra. $Ar-C_2H_4$ was monitored at m/e = 68, the parent mass peak. Two vibrational bands,[8] which originate from the ν_9 and ν_{11} C-H stretching modes in ethylene, are shown in Fig. 1. The rotational contours were

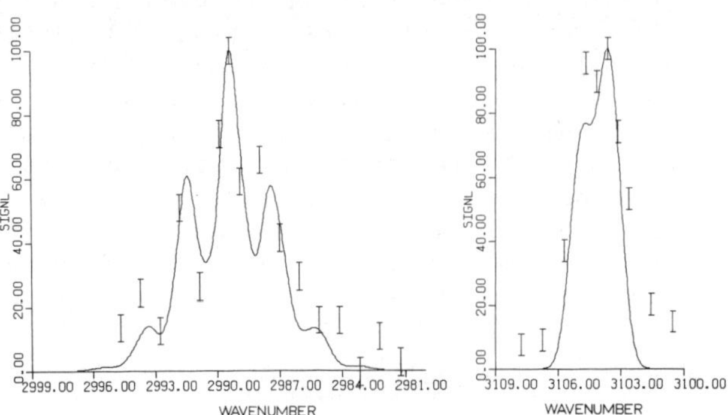

Fig. 1. $Ar-C_2H_4$ ν_{11} (~2989 cm^{-1}) and ν_9 (~3104 cm^{-1}) bands with simulated b and a-type contours respectively. Based on T(rot) = 3 K and a laser linewidth of 0.5 cm^{-1}. Error bars indicate ±σ.

consistent with the argon atom lying in the ethylene plane and along the perpendicular bisector of the double bond 3.6 A away. This structure required the ν_9 transition moment to be parallel to the a-inertial axis and thus gave rise to the observed a-type contour. The rotational temperature was set to the translational temperature of the complex as measured by time-of-flight techniques.[9]

The structure of $(HF)_3$ was determined[10] by the use of deuterium isotopic substitution and a subsequent vibrational analysis of the HF stretching modes in $(HF)_3$, $(HF)_2DF$, $HF(DF)_2$ and $(DF)_3$. HF clusters, following electron impact ionization, fragment extensively. As a result, $(HF)_3$ was only observed at m/e = 21 (H_2F^+). The partially deuterated species could be observed at two mass peaks, for example $(HF)_2DF$ would have mass signals at m/e = 21 (H_2F^+) and m/e = 22 (DHF^+). By varying the amount of DF relative to HF, and monitoring the vibrational bands at the various mass peaks, unambiguous assignments of the vibrational bands to a specific cluster could be readily performed. The vibrational frequencies for all four species are given in Table I.

Table I Trimer stretching frequencies (cm^{-1})

Species	Observed[10] HF	Observed[10] DF	Ne Matrix[11] HF	Ne Matrix[11] DF	Calculated[12] (HF only)
$(HF)_3$	3709	--	3706	--	3710
$(HF)_2DF$	3709	--	3706	--	3710
	3644	--	3645	--	3645
$HF(DF)_2$	3679	--	3672	--	3679
$(DF)_3$	--	2720	--	2721	

Consideration of possible structures for $(HF)_3$ must include the observed spectra for the fully protonated and partially deuterated species. For $(HF)_3$ to possess only one infrared active band, there must be either sufficient symmetry to reduce the number of infrared-active vibrational modes or extremely weak coupling between the HF monomers such that all three HF vibrations are degenerate. The latter possibility can be quickly rejected for the following reasons. In $DF(HF)_2$, the two HF stretching frequencies are split by 65 cm^{-1}, which indicates a strong interaction between the HF stretches. All open $(HF)_3$ structures with reasonable hydrogen-bond interactions, either planar or non-planar, would possess three infrared-active HF stretching modes, and are thus ruled out. The only remaining reasonable geometry is cyclical, i.e. each HF participates in a hydrogen bond. Since in this geometry the HF molecules are identical, a cyclical structure would have C_{3h} symmetry. $(HF)_3$ under this point group would have an infrared active doubly degenerate HF stretch and an infrared inactive totally symmetric HF stretch, which is consistent with the $(HF)_3$ experimental results. Cyclic $DF(HF)_2$ and $(DF)_2HF$ have C_s symmetry with two and one infrared active HF stretches respectively, also consistent with the experimental results.

DYNAMICAL RESULTS

The vibrational predissociation spectra for a series of a binary complexes of HF with molecular nitrogen, acetylene, ethylene and cyclopropane were studied[13] to determine the role of complex size, structure and interaction strength in the predissociation dynamics. These particular complexes were selected in part because the ground vibrational state structures have been previously determined by FT-microwave spectroscopy.[2] This knowledge greatly facilitated the analysis of the vibrational spectra. All of the spectra were obtained at the protonated base mass peak (H-base$^+$). Exciting the HF stretch at medium resolution, the spectra were readily reproduced by a calculated lineshape which used the rotational constants of the complex, the OPO linewidth, the estimated molecular beam temperature and the relative energy of the laser at each frequency.

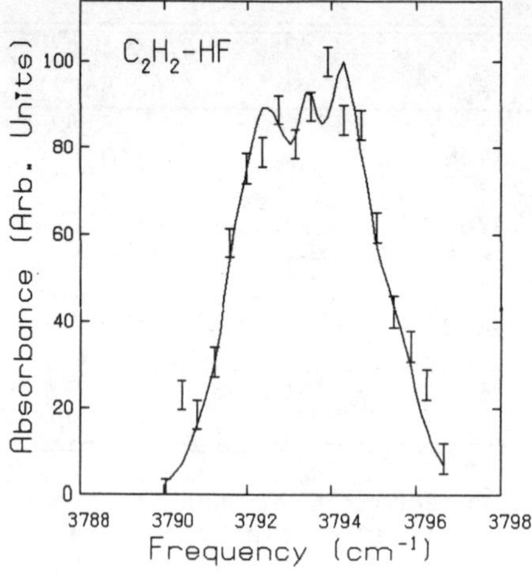

Fig. 2. Rotational contour of C_2H_2-HF. Simulated contour assumes T(rot) = 4 K and laser linewidth of 0.7 cm^{-1}. Error bars indicate $\pm\sigma$.

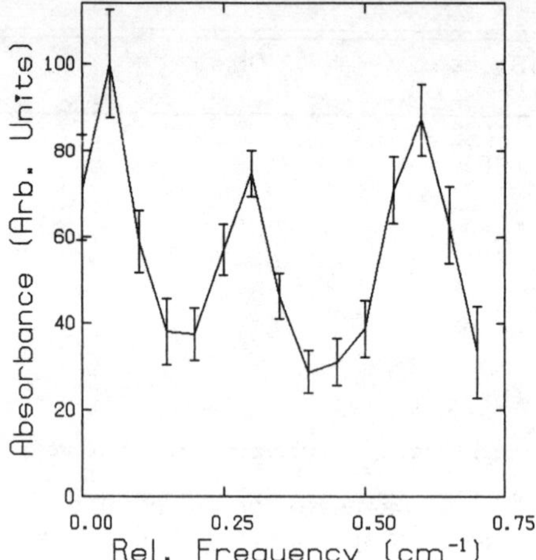

Fig. 3. High resolution spectrum of C_2H_4-HF. Laser linewidth = 0.06 cm^{-1}. Error bars indicate $\pm\sigma$.

Fig. 2 shows a representative observed and simulated spectra for C_2H_2-HF. When examined under high resolution, it was possible to resolve individual rotational lines as shown in Fig. 3.

These results indicate the lifetime of the excited vibrational state is greater than the rotational period of the complex and the linewidth analysis establishes a lower limit of 35 psec. Attempts to resolve rotational structure in the $(HF)_3$ band in high resolution were unsuccessful. Simulations of the perpendicular vibrational band, using rotational constants based on the ab initio $(HF)_3$ equilibrium structure[14] with appropriate vibrational averaging, from the $(HF)_2$ results,[15] would require a vibrational excited state lifetime of <25 psec for consistency with the observed data.

CONCLUSIONS

Vibrational predissociation spectroscopy of molecular clusters using molecular beam mass spectrometric detection has been useful in understanding both the structure of molecular clusters and the dynamics of the predissociation process. When Ar complexed with ethylene, it formed a planar species, as opposed to non-planar C_2H_4-HF and $(C_2H_4)_2$. $(HF)_3$ was shown to have a cyclic-planar structure with C_{3h} symmetry. There was no evidence for a non-cyclic species.

Following excitation of the HF stretch in the binary HF-base complexes, the vibrational relaxation has been shown to be slow with respect to rotational motion. The complexity of the base molecule had no measurable effect on the vibrational lifetime. The behavior of $(HF)_3$ contrasted with the dynamics of the binary complexes. The short (<25 psec) vibrational lifetime of $(HF)_3$ may be due to a relatively facile and rapid ring opening motion.

ACKNOWLEDGEMENTS

We are grateful of the support of this research by the National Science Foundation (CHE 8205811 and CHE 8506698).

REFERENCES

1. D. H. Levy, Ann. Rev. Phys. Chem. 31, 197 (1980).
2. A. C. Legon, Ann. Rev. Phys. Chem. 34, 275 (1983).
3. K. C. Janda, Adv. Chem. Phys. 60, 201 (1985).
4. F. Huisken, H. Meyer, C. Lauenstein, R. Sroka and U. Buck, J. Chem. Phys. 84, 1044 (1986).
5. J. M. Lisy, J. Chem. Phys. 84, xxxx (1986).
6. D. W. Michael and J. M. Lisy, J. Chem. Phys., submitted.
7. D. W. Michael, K. Kolenbrander and J. M. Lisy, Rev. Sci. Inst., submitted.
8. W.-L. Liu, K. Kolenbrander and J. M. Lisy, Chem. Phys. Letts. 112, 585 (1984).
9. "KELVIN-Rare Gas Time-of-Flight Program," Matthew Vernon, LBL Report 12422 (1981).
10. D. W. Michael and J. M. Lisy, J. Chem. Phys., submitted.
11. L. Andrews, V. E. Bondybey and J. H. English, J. Chem. Phys. 81, 3452 (11985). The assignments are ours.
12. Calculated using the FG matrix method in E. B. Wilson, J. C. Decius and P. C. Cross, "Molecular Vibrations" (Dover, New York, 1955), F_{ii} = 762.2 N/m, F_{ij} = -13.8 N/m.
13. K. Kolenbrander and J. M. Lisy, J. Chem. Phys., submitted.
14. S. Y. Liu, D. W. Michael, C. E. Dykstra and J. M. Lisy, J. Chem. Phys., accepted.
15. D. W. Michael, C. E. Dykstra and J. M. Lisy, J. Chem. Phys. 81, 5998 (1984).

THEORETICAL APPROACHES TO UNDERSTANDING MOLECULAR MULTIPLE-PHOTON EXCITATION

C. D. Cantrell
Center for Applied Optics, University of Texas at Dallas
P. O. Box 830688, Richardson, Texas 75083

ABSTRACT

Theories of molecular multiple-photon excitation are reviewed from the point of view of practical implementation. New methods that permit the solution of the time-dependent Schroedinger equation for systems with many energy levels are described.

Molecular physics has experienced a renaissance during the past fifteen years, in part because of the availability of new laser tools that make commonplace hitherto inaccessible regimes of molecular dynamics, and also because of the widespread perception that these new regimes of molecular behavior may find economically important applications. The laser separation of isotopes using molecular working media has been or is actively being pursued in the United States, Canada, the Soviet Union, France, West Germany, Italy, the United Kingdom, Japan, Israel, Brazil, Argentina, and China, to mention only the most prominent nations in this area. Much of what has been learned in the course of research on laser isotope separation may prove to be applicable to the much more general problem of laser chemistry, provided that selective excitation of chemical bonds or moieties proves to be generally possible. The early exuberance of researchers in molecular multiphoton excitation (MPE), which was fueled as much by our own hopes as by heavy funding, has given way to a more cautious attitude in the face of difficulties in measuring, calculating, and above all understanding, the fundamental photophysical processes involved in the excitation of polyatomic molecules by intense infrared laser light.[1,2] A mature scientific field is one that has undergone several such cycles. This field is in its adolescence, a difficult period in which the elevated thoughts of theory are not always in harmony, or even in contact, with the reality represented by the experience of the laboratory, and in which conflicting idealized concepts render a smooth development of the field unlikely.

Several factors have combined to render a comparison between theory and experiment especially difficult in the field of MPE. First, most of the experiments performed so far measure quantities such as the average energy deposited per molecule that are not sensitive to the details of the process of excitation. Second, the highly excited vibrational states of a polyatomic molecule often involve the motion of many of the nuclei that compose the molecular framework, so that indiscriminate excitation of a large number of high vibrational states (which is the usual result of excitation with a single-frequency laser with an uncontrolled pulse shape) may well entail the excitation of the entire molecule. This does not prove that techniques of selective excitation do not exist; indeed, it is the ultimate goal of this research to provide some theoretical guidance in the search for methods of selective excitation. Third, the problem of solving the time-dependent Schroedinger equation for a system with hundreds or thousands of energy levels, as is necessary for a quantitative study of MPE, was considered numerically intractable until recently. The difficulties of this problem do not result from a lack of sufficiently powerful computers. In fact, much useful research can be done with existing computers, at least for small polyatomic molecules. The real obstacle has been a general inability until recently to state the problem in a numerically tractable form. The lack of practical theoretical techniques for a first-principles study of MPE that persisted until very recently may have contributed to the noisy and protracted debate over the nature of intramolecular relaxation (IMR) accompanying MPE.

Approaches to the study of MPE through numerical solution of the time-dependent Schroedinger equation for model systems that have some or many of the features of real molecules fall into two categories: (i) Methods that assume that the laser pulse is turned on suddenly at a given time, and retains a constant amplitude and frequency thereafter; (ii) Methods that are capable of following the temporal profile of a real or model laser pulse that has finite rise and fall times. Approaches of type (i) make extensive use of the dressed states[2] or their analogs, the Floquet eigenstates,[3] which are used when the rotating-wave approximation[2] is not employed. These approaches have the difficulty that the eigenvalue problem may not be numerically feasible for more than a few thousand levels; this difficulty has recently been circumvented in an elegant manner by Nauts and Wyatt, who have adapted a recursive technique that may permit the study of systems with many thousands of energy levels.[4,5] Approaches of type (i) have the more serious disadvantage that laser pulses available in the laboratory do not turn on instantaneously, nor do they remain constant in amplitude thereafter. The photophysics of laser absorption is quite different for slowly rising pulses, for which adiabatic excitation (and even inversion) are commonplace, than for rapidly rising pulses, which tend to induce Rabi oscillations and saturation phenomena that are often well described by the sudden approximation.[6] The approach being followed in this research is of type (ii), and is appropriate no matter what the risetime of the laser pulse. The typical disadvantage of approaches of this type is that it is necessary to solve the time-dependent Schroedinger equation, rather than the possibly simpler eigenvalue-eigenvector problem of finding the dressed states or the Floquet eigenstates.

Brute-force attempts to solve a system of several hundred ordinary differential equations representing the time evolution of the probability amplitudes of the energy levels of a moderately realistic molecular model lead inevitably to very short time-step sizes and to loss of accuracy, as a result of the necessity to follow very high-frequency oscillations resulting from the large detunings from resonance of some states. This is analogous to the problem of stiffness that typically occurs in systems in which the real parts of the eigenvalues of the Jacobian matrix (of the driving functions with respect to the solution functions) are of very different orders of magnitude.[7,8] Unfortunately the numerical methods developed for stiff equations are of little use for systems of equations in which the eigenvalues of the Jacobian matrix have imaginary parts of very different orders of magnitude, because the stiff methods depend for their efficacy on the rapid disappearance of the components of the solution corresponding to the highest decay rates (i.e., the eigenvalues with the largest negative real parts). In a problem where the solutions oscillate, the components of highest frequency do not necessarily disappear. This prevents efficient use of the stiff-equation algorithms, which attempt to increase the step size after the initial transients have died away. Our response to this difficulty has been to restate the problem in such a way that standard numerical methods suffice for its solution[9,10].

We have found that a system with an arbitrary (but smooth) density of states or density of dipole strength as functions of energy can be replaced (for the purposes of solving the time-dependent Schroedinger equation) by an effective system with a far smaller number of energy levels. The similarity transformation that accomplishes this replacement depends on the driving laser field only in a finite (and small) block of the transformation matrix; as a result, the method is useful for laser pulse profiles of arbitrary shape and is valid whether or not the rotating-wave approximation is employed.[9] The system of differential equations into which the time-dependent Schroedinger equation is transformed is less pathological from a numerical point of view than the original system, because each "effective" level interacts only with two others; this tends to reduce the occurrence of very rapid oscillations.[10] An important advantage of this new method is that it provides precise bounds on the errors committed. Since the techniques described in Ref. 10 may be used with an arbitrary set of discrete levels that are coupled (through interaction with the laser) with a quasi-continuum or a continuum, these methods may reasonably be expected to be applicable at all stages of MPE.

ACKNOWLEDGEMENT

This research was supported by the Welch Foundation under Grant No. AT-873, and by the National Science Foundation under Grants CHE-8215245, INT-8116435 and INT-8202413.

BIBLIOGRAPHY

1. For a review of laser chemistry and isotope separation, see (a) C. D. Cantrell, S. M. Freund and J. L. Lyman, pp. 485-576 in The Laser Handbook, Vol. III, edited by M. L. Stitch (Amsterdam, North-Holland, 1979); (b) V. S. Letokhov and C. B. Moore, Sov. J. Quant. Elect. 6, 129 and 259 (1976).

2. For a review of theoretical models of the coherent excitation of multilevel systems by laser light, see C. D. Cantrell, V. S. Letokhov and A. A. Makarov, pp. 165-269 in Coherent Nonlinear Optics: Recent Advances, edited by M. S. Feld and V. S. Letokhov (Berlin, Springer-Verlag, 1980).

3. K. F. Milfield and R. E. Wyatt, Phys. Rev. A 27, 72 (1983).

4. A. Nauts and R. E. Wyatt, Phys. Rev. Lett. 51, 2238 (1983).

5. A. Nauts and R. E. Wyatt, Phys. Rev. 30, 872 (1984).

6. G. L. Peterson and C. D. Cantrell, Phys. Rev. A 31, 807 (1985).

7. C. W. Gear, Numerical Initial Value Problems in Ordinary Differential Equations (Prentice-Hall, Englewood Cliffs, NJ, 1971).

8. L. Lapidus and J. H. Seinfeld, Numerical Solution of Ordinary Differential Equations (Academic Press, New York, 1971), ch. 6.

9. R. S. Burkey and C. D. Cantrell, J. Opt. Soc. Am. B 1, 169 (1984).

10. R. S. Burkey and C. D. Cantrell, J. Opt. Soc. Am. B 1, 451 (1985).

INFRARED SPECTROSCOPY USING OPTOGALVANIC DETECTION OF A RADIO-FREQUENCY DISCHARGE

G. W. Hills
AT&T Bell Laboratories, Allentown Pa. 08103

W. J. McCoy* and R. E. Muenchausen**
University of North Carolina, Chapel Hill, NC. 27514

ABSTRACT

The laser optogalvanic (LOG) technique has been used to observe infrared transitions in the transient molecular species H_2CS and BrO. In addition the dependence of the LOG signals on laser power, discharge gas pressure, and chopping frequency have been characterized. A simple model, which is conceptually consistent with the known LOG effects in a positive column DC discharge, has been developed to explain the observed LOG signal behavior.

SPECTROSCOPIC STUDIES

Laser optogalvanic (LOG) effects in low pressure RF discharges due to infrared absorption have been reported for several molecules.[1] Fig. 1 shows a simplified schematic of the experimental setup; full details have been previously given.[1,2] LOG signals have additionally been observed from C_2H_4, C_3H_4, CF_4, $(CH_3S)_2$ and the short lived molecules H_2CS and BrO using isotopic CO_2 lasers as the excitation source. Using molecular constants provided by other workers, detailed assignments for the C_2H_4 and H_2CS coincidences were possible. Full details of the calculations as well as tabulations of the observations and predictions are reported elsewhere.[3]

BASIC EXPERIMENTAL SET UP

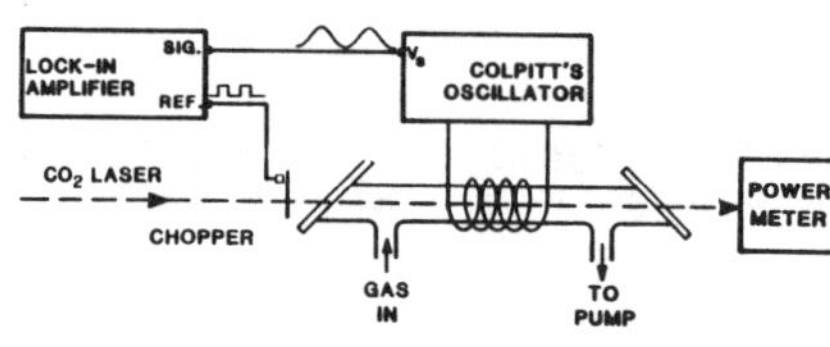

Fig. 1

The observation of a known coincidence[4] of BrO with $^{12}C^{16}O_2$ 10 µm R(8) is the first reported infrared LOG detection of a free radical. The signal was observed both intracavity and extracavity to the CO_2 laser. BrO was produced by the RF discharge in 100 mtorr of O_2 containing a trace of Br_2. From this observation and previous studies,[1,5] the minimum detectable absorption coefficient was estimated to be 5×10^{-7} cm^{-1}.

MECHANISTIC STUDIES

Briefly summarized, it was found that laser absorption produced a decrease in the discharge impedance. The LOG signal was observed to be linearly proportional to laser power over the range of laser power investigated (5 - 100 mW), and inversely proportional to

* St. Andrews College, Laurinburg, NC 28352
** MS G738, Los Alamos National Laboratory, Los Alamos, NM 87545

chopping frequency above 400 Hz. The pressure dependence of the LOG signals in RF discharges of C_2H_4, D_2O and NO_2 is shown plotted in Figure 2. The LOG pressure dependence is seen to be linear over most of the pressure range where the discharge could be sustained except at low pressures (.1 - .2 torr) in C_2H_4 and D_2O discharges, and also in a NO_2 discharge at pressures greater than 1.5 torr.

The observed LOG signal behavior was modeled under the following conditions: 1) The electrical behavior of the RF discharge is well described by a positive column DC discharge. 2) The discharge impedance increases linearly with pressure over the $\simeq$ 1 torr range over which the discharge can be maintained. 3) The LOG signals are essentially due to volume heating of the gas discharge. The net temperature rise due to laser absorption in the gas was calculated for a simple two-level system. 4) The Colpitts circuit response is essentially governed by the impedance mismatch between the discharge "load" and the oscillator "source" impedances. An equation for the LOG signal has been previously derived[5] and can be written as

$$V_s = (1/4\pi r^2)(VpRp/R_s)[\{1-\exp(-1/f\tau)\}/(C_o f+C_1)]h(b,p)P_o\alpha(\nu)p \quad (1)$$

LOG SIGNAL vs GAS PRESSURE

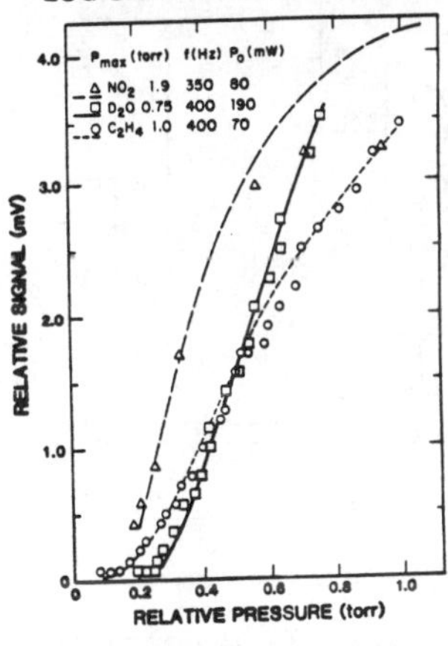

Fig. 2

where r is the cell radius, Vp is the oscillator plate voltage, Rp is the plate resistance, R_s is the effective circuit "source" impedance, P_o is the incident laser power, $\alpha(\nu)$ is the absorption coefficient of the gas (cm^{-1} $torr^{-1}$) and p is the discharge gas pressure. The effect of the laser chopping frequency, f, on the LOG signal is contained within the bracketed term, [], where $\tau(p)$ is the V-T relaxation time. Typically, $f\tau \lesssim 1$ and thus $[] \propto 1/f$. The pressure dependent effect on the oscillator response to a load impedance change is given by

$$h(b,p) = b(bp-1)/(bp+1)^3 \quad (2)$$

where $bp = R_L/R_S$ is the ratio of the load and source impedances. Using estimated discharge parameters, it has been shown that[5] h(b,p) is constant to within ±5% over a $\simeq$ .5 torr pressure range of the discharge, which is consistent with the generally linear pressure dependence of the LOG signals. The fitted curves in Fig. 2 show, however, that the nonlinear pressure effects can be well accounted for using (1) which was scaled by a constant to fit each data set.

REFERENCES

1 R.E. Muenchausen, R.D. May, G.W. Hills, Opt. Comm., 48, 317, (1984)

2 G.W. Hills, R.D. May, W.J. McCoy, R.E. Muenchausen, Proc. Int. Conf. Lasers '83, Ed. V.J. Corcoran

3 W.J. McCoy, Ph.D thesis, Univ. North Carolina, Chapel Hill, 1986

4 G.W. Hills, W.J. McCoy, R.E. Muenchausen, XVIth Int. Symp. on Free Radicals, Lauzelle-Ottignes, Belgium, 1983

5 R.E. Muenchausen, Ph.D thesis, Univ. N. Carolina, Chapel Hill, 1984

COLOR CENTER LASER SPECTROSCOPY OF FREE RADICALS

C. B. Dane, W. B. Yan, D. Zeitz, J. L. Hall,
R. F. Curl, Jr., J. V. V. Kasper, and F. K. Tittel
Rice University, Houston, TX. 77251

An electrical discharge through argon over a coating of polyacetylene has been found to provide an abundant source of free radicals. Using a computer controlled color center laser, magnetic rotation spectroscopy combined with differential detection[1] and a multipass cell was used to obtain high-resolution infrared absorption spectra for both a deuterated and non-deuterated coating.

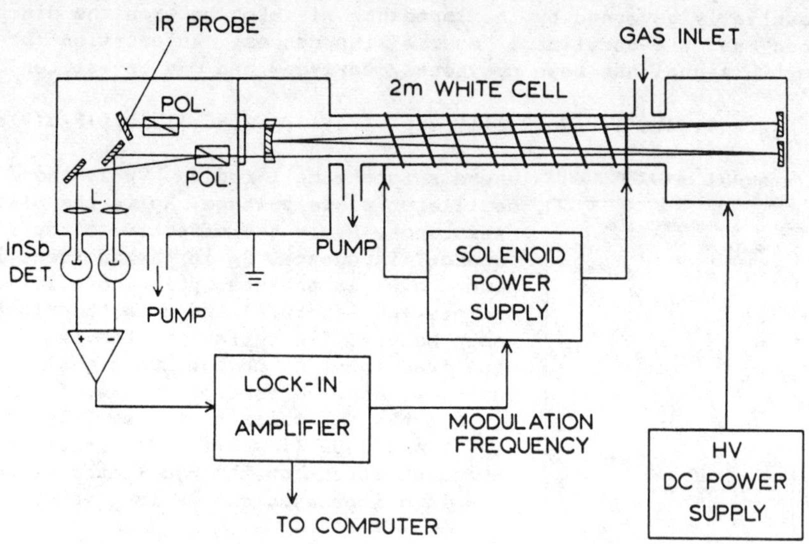

Fig. 1. Experimental apparatus. POL = polarizer, Ge = germanium flat, L=lens.

In previous work, the $A^2\Pi \leftarrow X^2\Sigma^+$ transition of C_2H was observed and assigned[2] and the first observation in absorption of the Ballik-Ramsey $b^3\Sigma_g^- \leftarrow a^3\Pi_u$ system of C_2 was made in this discharge system[3]. Recently the discharge was inadvertently operated with small atmospheric impurities arising from the opening of an undetected sample cell leak. This resulted in the appearance of new bands exhibiting remarkable signal-to-noise. These were found to be due to metastable states of CO and N_2 and were successfully assigned to the $a'^3\Sigma \leftarrow a^3\Pi$ (v=0←2) system in CO[4] and the Wu-Benesch $B^3\Pi_g \leftarrow W^3\Delta_u$ (v=4←2, 5←3, 6←4) system in N_2[5].

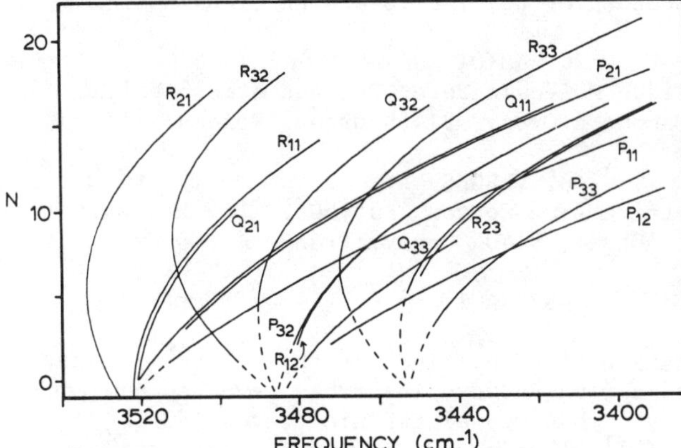

Fig. 2. Partial Fortrat diagram of the $a'^3\Sigma \leftarrow a^3\Pi$ v=0←2 band of CO. The index N is the rotational quantum number K" for large N but the correlation breaks down for N<2.

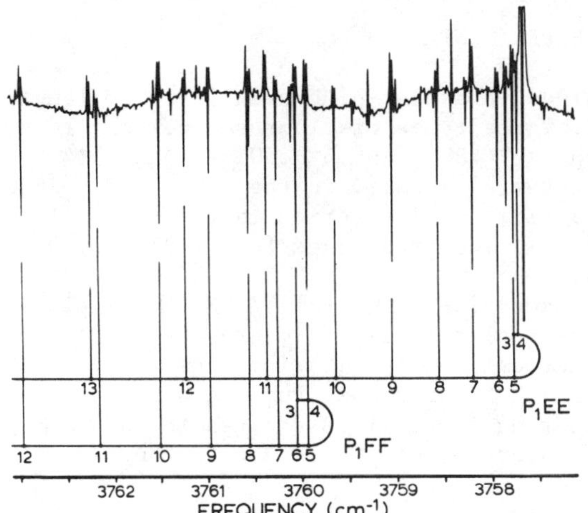

Fig. 3. $B^3\Pi_g \leftarrow W^3\Delta_u$ v=5←3 band of N_2.

Two bands of the spectrum of C_2D have been tentatively identified in the deuterated polyacetylene discharge and their analysis is now in progress. One appears to be a hot band involving the (0,1,0) state.

1. H. Adams, D. Reinert, P. Kalkert, and W. Urban, Appl. Phys B. 34, 179-185 (1984).
2. R. F. Curl, P. G. Carrick, and A. J. Merer, J. Chem. Phys. 82, 3479-3486 (1985).
3. W. B. Yan, R. F. Curl, A. J. Merer, and P. G. Carrick, J. Mol. Spectrosc. 112, 436-446 (1985).
4. C. Effantin, F. Michaud, F. Roux, J. D'Incan, and J. Verges, J. Mol. Spectrosc. 92, 349-362 (1982).
5. D. Cerny, F. Roux, C. Effantin, and J. D'Incan, J. Mol. Spectrosc. 81, 216-226 (1980).

LASER INDUCED FLUORESCENCE OF Cs_2 : FIVE NEW ELECTRONIC STATES

C. AMIOT
Laboratoire de Physique Moléculaire et Optique Atmosphérique,
Bât. 221, Campus d'Orsay, 91405 Orsay, France

J. VERGES
Laboratoire Aimé Cotton, Bât.505,
Campus d'Orsay, 91405 Orsay, France

ABSTRACT

Study of the spectra excited by lines of the Ar^+,Kr^+ ion lasers and by some lines of a ring R6G dye laser has revealed new features on five previously unobserved electronic states ($(2)^1\Sigma_g^+$, $(1)^1\Pi_g$, $(3)^1\Sigma_g^+$, $E\ ^1\Sigma_u^+$, $(3)^1\Pi_u$) of the Cs_2 molecule. Accurate molecular constants and equilibrium internuclear distances are derived.

EXPERIMENTAL

Cesium metal was contained in a heat-pipe oven (~ 10 cm length) which was operated at 300°C. Three types of fluorescence were recognized by excitation with the Ar^+ 488 , 496.5 , 501.7 , and 514.5 nm laser lines : (i) doublet series located in the 4200-4300 , 7600-8500 , and 18000-19600 cm^{-1}; (ii) triplet series in the spectral region 6100-6700 cm^{-1}; and (iii) isolated lines in the 8500-9000 and 18000-21000 cm^{-1} ranges.

Series (i) arise from transitions $E\ ^1\Sigma_u^+-(3)^1\Sigma_g^+$, $E\ ^1\Sigma_g^+-(2)^1\Sigma_g^+$, and $E\ ^1\Sigma_u^+-X\ ^1\Sigma_g^+$ respectively. Triplet series (ii) are due to the $E\ ^1\Sigma_u^+-(1)^1\Pi_g$ transition. Isolated lines (iii) arise from $(3)\ ^1\Pi_u\rightarrow(X)^1\Sigma_g^+$, $(3)\ ^1\Pi_u\rightarrow(2,3)\ ^1\Sigma_g^+$ transitions. Pumping with the Kr^+ 647.1 nm line and by the ring laser lines excites the $D\ ^1\Sigma_u^+$ state. Doublet patterns are observed, which are due to the $D\ ^1\Sigma_u^+\rightarrow(2)\ ^1\Sigma_g^+$ transition.

ANALYSIS

The fluorescence has been analysed by a Fourier Transform spectrometer with an effective resolution ranging from 0.007 cm^{-1} for the spectra recorded in the infrared range to 0.025 cm^{-1} for those in the visible. The accuracy of measured wavenumbers was better than 0.003 cm^{-1}.

Molecular constants were first obtained by a least-squares fit of the vibrational and rotational spacings[1,2,3]. A new analysis has been performed, in which wavenumbers of all the observed transitions are simultaneously used.

RESULTS

The results obtained so far include the molecular constants

for five electronic states. Main constants are listed in Table I below :

TABLE I

Cs_2 MOLECULAR CONSTANTS* (in cm^{-1})

	Te	ω_e	$B_e \times 10^3$
$(X)^1\Sigma_g^+$	0	42.01945(5)	11.7439(3)
$(2)^1\Sigma_g^+$	12114.066(2)	23.3519(5)	7.4591(4)
$(1)^1\Pi_g$	13913.422(2)	18.4393(7)	7.8154(4)
$(3)^1\Sigma_g^+$	15975.347(2)	22.4243(6)	8.2304(4)
$E\ ^1\Sigma_u^+$	20195.321(2)	28.989(2)	8.9022(4)

Accurate potential energy curves have been obtained using these molecular constants. The internuclear equilibrium R_e values are given in Table II.

TABLE II

EQUILIBRIUM DISTANCES R_e

	$(2)^1\Sigma_g^+$	$(1)^1\Pi_g$	$(3)^1\Sigma_g^+$	$E\ ^1\Sigma_g^+$
R_e(Å)	5.8316	5.6977	5.5569	5.3407

Analysis is currently in progress to obtain the molecular constants of the $(3)^1\Pi_u$ state, pumped simultaneously with the $E\ ^1\Sigma_u^+$ state by the Ar^+ laser lines.

REFERENCES

[1] C.Amiot, C. Crépin and J. Vergès, Chem. Phys. Lett. 98, 608 (1983)

[2] C. Amiot, C. Crépin and J. Vergès, J. Mol. Spectrosc. 107, 28 (1984)

[3] C. Amiot, C. Crépin and J. Vergès, Chem. Phys. Lett. 106, 162 (1984)

[4] C. Amiot, J. Vergès, W. Demtröder and C.R. Vidal, J. Chem. Phys. (to be published)

* Quoted uncertainty is equal to one standard deviation.

PROMPT AND DELAYED PHOTOLYSIS OF CS_2 EXCITED AT BLUE WAVELENGTHS

F. Davanloo and C. B. Collins
Center for Quantum Electronics, University of Texas at Dallas
Richardson, Texas 75083-0688

ABSTRACT

In this work a time-delayed, two photon technique was used for the state selective photolysis of Cs_2 with particular attention being placed on the production of the fine structure components of the 5D states of cesium. A quantitative model was constructed to fit the data obtained at seven wavelengths of photolysis over the blue region of the visible spectrum. A delayed source for the production of $Cs(5^2D_{5/2})$ atoms was observed to have a lifetime <100 nsec.

INTRODUCTION

A double resonance method for the correlation of photolysis bands observed in the spectra of simple molecules with their dissociation products had been used in studies[1,2] of the photolysis resulting from electronic excitation of Cs_2. These studies suggested that a highly mixed state of the radiation fields could effectively populate a very selected state of the particles. However, a more general application of such techniques required detailed knowledge of the dissociating state to be excited and of the dynamics of dissociation process which would determine the extent to which the product channels might become mixed. Usually, two types of photolysis have been distinguished that can be conveniently described as prompt and delayed.[3] Prompt photolysis has happened when the time between photolysis and detection was too short to permit collisional mixing or radiative cascading of the populations. An example leading to broad band dispersion curves is reproduced in Fig. 1. The processes leading to the prompt photolysis included dissociation and predissociation of the electronically excited state of Cs_2 that is produced directly from the ground state. In contrast delayed photolysis included predissociation of the state Cs_2 directly excited

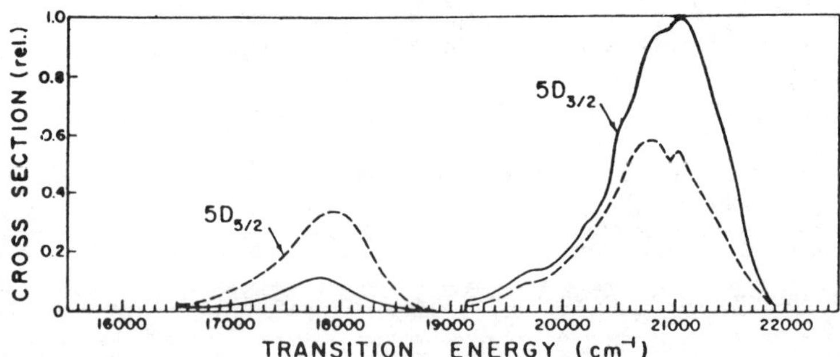

Fig. 1: Relative cross sections as a function of transition energies for the prompt photolysis of Cs_2 into 5D products.

and of other states of Cs_2 populated from the original state by collisional and radiative processes of relaxation.

When the time between photolysis and detection was varied, data produced for the delayed photolysis of Cs_2 at yellow wavelengths could be fit to a quantitative model. A delayed process for the selective production of Cs(5D) atoms was found. Analysis in three wavelengths gave the delayed spectrum through this process.[3] Reported here are measurements of the time dependent evolution of cesium 5D populations resulting from the photolysis of Cs_2 excited at wavelengths of the blue portion of the visible range.

RESULTS

In order to describe the temporal evolution of populations of species excited at seven blue wavelengths of photolysis, a kinetic model containing the minimal number of reactive species sufficient to explain the phenomenology was developed. Data for the population of Cs(5D) atoms were fitted to the populations calculated from the differential equations of this model in a statistical manner.[3] It was found that a delayed source A with lifetime <100 nsec contributing to the production of Cs($5D_{5/2}$) was necessary to fit the data. Delayed contributions to the production of Cs($5D_{3/2}$) was absent in agreement with previous observations and expectations.[4]

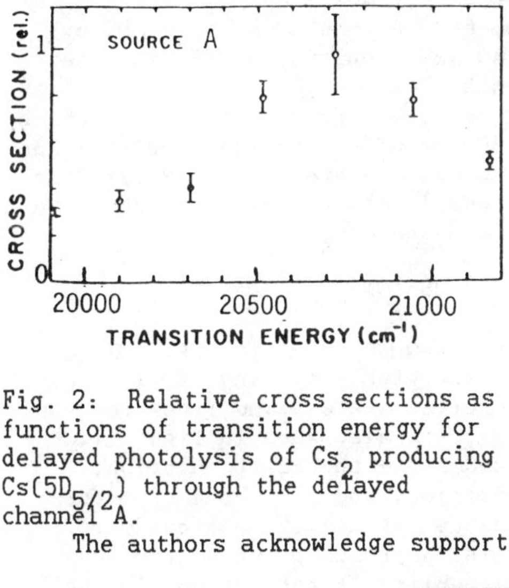

Fig. 2: Relative cross sections as functions of transition energy for delayed photolysis of Cs_2 producing Cs($5D_{5/2}$) through the delayed channel A.

In Fig. 1 relative cross sections for the selective photolysis through the prompt channel were obtained on a single common scale for the prodution of Cs(5D) atoms. It is useful to place the strength of the delayed source contributed by species A at blue wavelengths on the same scale so the relative importance of prompt and delayed source can be readily appreciated. This was done in a similar manner as Ref. 3 and is presented in Fig. 2.

The authors acknowledge support of NSF Grant No. PHY8214273.

REFERENCES

1. C. B. Collins, J. A. Anderson, D. Popescu and I. Popescu, J. Chem. Phys. 74 1053 (1981).
2. C. B. Collins, F. W. Lee, J. A. Anderson, P. A. Vicharelli, D. Popescu and I. Popescu, J. Chem. Phys. 74 1067 (1981).
3. F. Davanloo, C. B. Collins, A. S. Inamdar, N. Y. Mehendale and A. S. Naqvi, J. Chem. Phys. 82 4965 (1985).
4. Z. Wu and J. Huennekens, J. Chem. Phys. 81 4433 (1984).

LIFETIMES OF Na_2 IN THE $b^3\Pi_u$ STATE

B. E. Miller and P. A. Schulz*
Georgia Institute of Technology, Atlanta, GA 30332

INTRODUCTION

Studies of the $b^3\Pi_u$ state of Na_2 have involved narrowband lasers[1-4] or magnetic rotation spectra[5,6]. Here we show that the $b^3\Pi_u$ state is easily observed by delaying the detection of the laser induced fluorescence (LIF). Using this technique we measure the $b^3\Pi_u$-$X^1\Sigma_g^+$ absorption spectrum and lifetimes of the $b^3\Pi_u$ states. These data are fit using previously measured molecular constants for the $A^1\Sigma_u^+$ and $b^3\Pi_u$ states.

EXPERIMENTAL

The investigation was carried out using a crossed pulsed laser-molecular beam apparatus. Sodium metal is heated in an oven to a temperature near 450°C. The metal vapor then expands through a 3.0 mm diameter nozzle into a vacuum chamber operating at about 10^{-1} torr. The pulsed dye laser perpendicularly intersects the sodium beam 2 cm above the nozzle and the subsequent laser induced fluorescence is monitored in a mutually perpendicular direction.

The dye laser power is 2 mJ from 6670 A to 6930 A using LDS 698 dye. It is pumped by a doubled Nd:YAG laser pulsing at 10 Hz. The dye laser output has a 1 cm^{-1} linewidth.

The excitation spectrum is collected by a boxcar integrator which measures the fluorescence 200 ns to 800 ns after the laser pulse. The lifetime data are gathered by an ORTEC Quad Counter with a dwell time of 250 ns/channel. The Quad Counter was interfaced with an IBM-PC in order to facilitate data acquisition and storage.

RESULTS AND DISCUSSION

The Na_2 $b^3\Pi_u$-$X^1\Sigma_g^+$ excitation spectrum is shown in figure 1. Four vibrational progressions are evident with a spacing of ∿150 cm characteristic of this sodium dimer excited state. The linestrengths, which were predicted by a simple model, are also seen in figure 1. The model takes into account the ground state thermal population, Frank-Condon factors for the A-X transition, and $A^1\Sigma_u^+$ -$b^3\Pi_{uo}$ perturbation strength. Some of the discrepancy between the observed spectrum and the predicted linestrengths is due to coupling of the $A^1\Sigma_u^+$ and $^3\Pi_{u1}$ and $^3\Pi_{u2}$ states. This coupling is not present in the Hund's case (a) limit of low rotational quantum number, but appears to be important for J = 30 or higher. Such states were not observed in the supersonic jet experiments.[7] The sharp peak near 14,880 cm^{-1} is the Na 3D-3P fluorescence following the Na 3S-3D two-photon absorption

Some characteristic lifetime curves are given in figure 2. The first plot shows a lifetime indicative of a single decay. The decay seen in figure 2b is for a more complicated process and is therefore

*Support by Cottrell Research Grant from Research Corporation.

less easily quantified. The lifetimes are due to the mixing of the $b^3\Pi_u$ and the $A^1\Sigma_u^+$ states and are directly proportional to the singlet characteristic of the perturbed state. The decay constants displayed in table I were based on a single decay scheme. Our experiment revealed radiative lifetimes ranging from less than 200 ns to near 2 us.

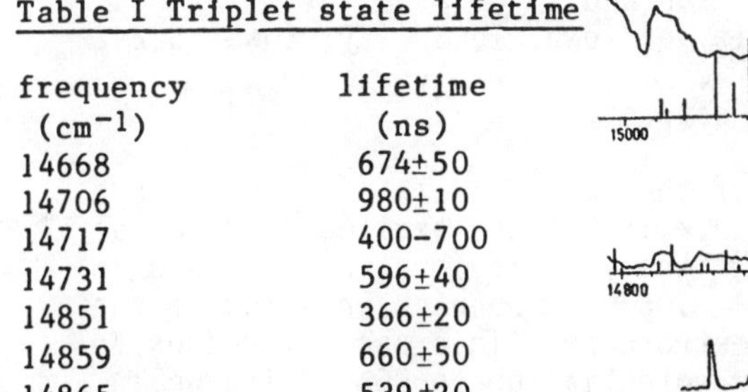

Table I Triplet state lifetime

frequency (cm^{-1})	lifetime (ns)
14668	674±50
14706	980±10
14717	400-700
14731	596±40
14851	366±20
14859	660±50
14865	539±20
14872	397±10
14878	966±50

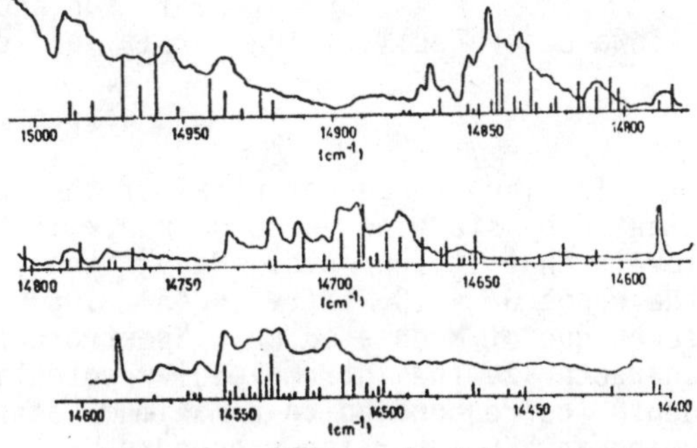

Fig. 1. LIF excitation spectrum.

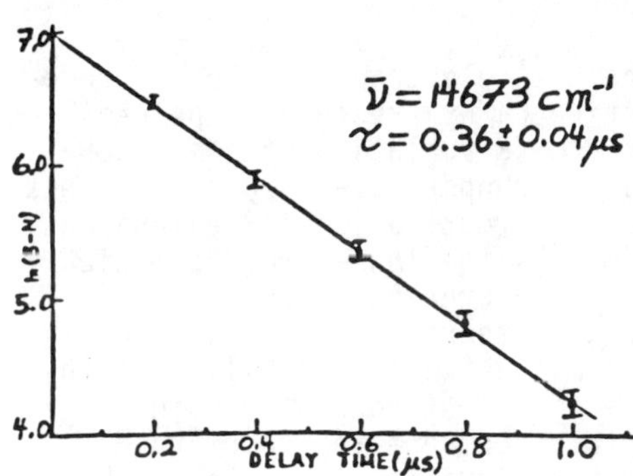

Fig. 2a. Single exponential decay

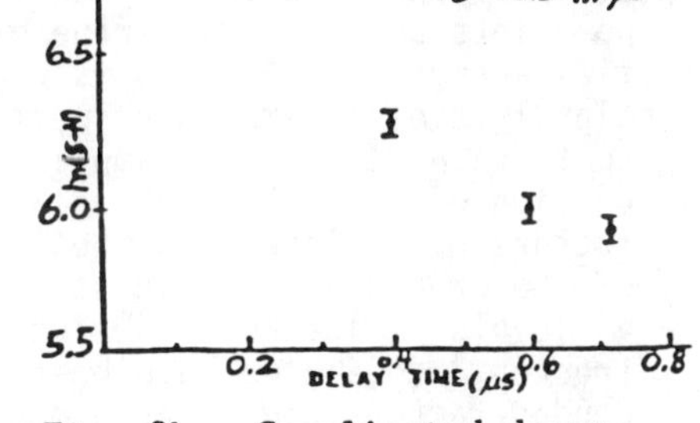

Fig. 2b. Complicated decay.

REFERENCES

1. L. Li and R. W. Field, J. Phys. Chem. 87, 3020 (1983).
2. K. Shimizu and F. Shimizu, J. Chem. Phys. 78, 1126 (1983); F. Shimizu and K. Shimizu, J. Chem. Phys. 78, 2798 (1983).
3. F. Engelke, H. Hage, and C. D. Caldwell, Chem. Phys. 64, 221 (1982).
4. J. B. Atkinson, J. Becker, and W. Demtröder, Chem. Phys. Lett. 87, 92 (1982); ibid 87, 128 (1982).
5. R. W. Wood, Phil. Mag. 10, 408 (1905).
6. W. R. Fredrickson and C. R. Stannard, Phys. Rev. 44, 632 (1933).
7. K. P. Huber and G. Herzberg, Constants of Diatomic Molecules (Van Nostrand Reinhold Company, N. Y., 1979), p. 432.

Acknowledgements: Greg Kintz, Bryan Morris and Ron Jacobsen assisted in the set up of this apparatus. Dr. James Gole kindly loaned us equipment.

Long Range Potential of the $A^1\Sigma_u^+$ State of Na_2 Using Modulated Gain Spectroscopy

G. Chawla, H. S. Schweda, H. J. Vedder, R. W. Field, S. Churassy
Massachusetts Institute of Technology, Cambridge, MA 02139

A. M. Lyyra, W. T. Luh and W. C. Stwalley
Iowa Laser Facility, University of Iowa, Iowa City, Iowa 52242

Abstract

The long range potential of the Na_2 $A^1\Sigma_u^+$ state has been investigated by studying various properties of the excited vibrational levels that it supports. Many high lying rovibrational levels in the range J' = 10, 12, v' = 43 - 105 have been observed using the technique of Modulated Gain Spectroscopy. This has allowed us to characterize the intramolecular potential up to 96% of its well depth (corresponding to a maximum intranuclear separation of 10.7 A) using an RKR inversion procedure.

Introduction

The long range potential of the $A^1\Sigma_u^+$ electronic state is predominantly characterized by an attractive C_3/R^3 resonant dipole-dipole interaction. By virtue of the slow R variation of the potential energy, the A state is expected to support over 200 bound vibrational levels. Previous spectroscopic investigations[1-3] extend only up to v' = 44 thereby sampling primarily the low energy harmonic portion of the potential well. Rapidly decreasing Franck Condon factors of v' levels in the A state with thermally populated levels of the ground electronic state had limited the accessibility of the v' levels of interest. Using the technique of Modulated Gain Spectroscopy[4] (MGS), we have been able to circumvent the poor Franck Condon factors and observe many excited vibrational levels in the range v' = 44 - 105. An RKR inversion of these rovibrational energies yields the full intramolecular potential.

$A^1\Sigma_u^+$ RKR Potential

The MGS spectrum consists of a simple vibrational progression of P(11), R(11) doublets, with a single known v" for each vibrational band. The measured A-X frequencies are reported[4] to an accuracy of 0.006 cm^{-1}. We describe below the concatenation and inversion of the MGS data with that previously observed by Kusch and Hessel[2] for v' = 0 - 20 and Kaminsky *et al.*[1] for v' = 19 - 44. Initial merging attempts of our B_v and G_v values with those of reference 1 were difficult because at two points of overlap, v' = 43 and 44, there existed a 1.5 cm^{-1} discrepancy between G_v values. This prompted a refit of the raw data of reference 1, which once done, removed the above mentioned inconsistencies. The resulting B_v and G_v values for the v' = 0 - 44 levels were combined with our data in the following manner. The vibrational and rotational contributions in the observed E(v, J) levels were separated by first estimating B_v values from P-R separations by assuming a value for the Dunham coefficient Y_{12} and second, generating a trial RKR curve for the v' = 0 - 105 range. From this trial potential energy curve, the centrifugal distortion constants D_v were calculated using the program CDIST by J. Hutson[5]. These D_v values were subsequently used to correct the G_v and B_v values initially used in generating the trial potential energy curve. In an iterative fashion the G_v and B_v values were optimized and used to generate the final RKR curve shown in Figure 1. We note that the potential does not contain more than one inflection point. The potential was tested by using the Numerov integration method to solve the radial Schroedinger equation for the eigenvalues. The calculated energies agree satisfactorily with our observed energies to 0.03 cm^{-1} using a homogeneous integration grid (8000 steps). We note that our calculated energies are consistent with the recently observed A-X bands of Gerber et al.[6]. Extracting van der Waals coefficients C_3, C_6 and C_8 from the long range part of the A state potential will be discussed in a forthcoming paper[7].

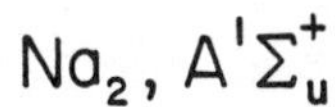

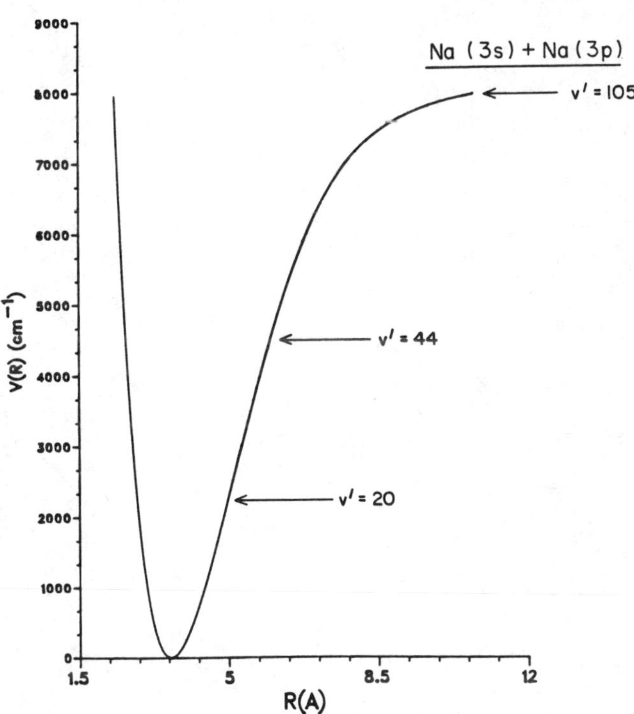

Figure 1. Na_2 $A^1\Sigma_u^+$ potential energy curve.

Acknowledgments

Support from AFOSR and NSF is gratefully acknowledged.

References

1. M. E. Kaminsky, Ph.D. Thesis, Stanford University (1976).
2. P. Kusch and M. M. Hessel, J. Chem. Phys. 63, 4087 (1975).
3. M. E. Kaminsky, J. Chem. Phys. 66, 4951 (1977) and 73, 3520 (1980).
4. G. Chawla, Ph.D. Thesis, Massachusetts Institute of Technology (1985).
5. We thank R. Le Roy for making the program available to us.
6. G. Gerber and R. Moller, Chem. Phys. Lett. 113, 546 (1985).
7. G. Chawla, R. W. Field, A. M. Lyyra, W. T. Luh and W. C. Stwalley, to be published.

High Resolution Laser Induced Fluorescence Spectroscopy of Highly Excited Vibrational Levels in NaAr: $A^2\Pi_r$ and $B^2\Sigma^+$

A. M. Lyyra
Iowa Laser Facility, University of Iowa, Iowa City, Iowa 52242

W. P. Lapatovich
GTE Laboratories, Inc., Waltham, Massachusetts 02254

P. E. Moskowitz
GTE Lighting Products, Danvers, Massachusetts 01923

M. D. Havey
Old Dominion University, Norfolk, Virginia 23508

R. Ahmad-Bitar
University of Jordan, Amman, Jordan

D. E. Pritchard
Massachusetts Institute of Technology, Cambridge, MA 02139

ABSTRACT

This laser induced fluorescence study of NaAr reports the direct observation of vibrational and rotational structure of the $B^2\Sigma^+$ state through spectroscopic means of Reference 1. Crossed laser and supersonic molecular beam technique is used. We study highly excited vibrational levels of both $A^2\Pi_r$ and $B^2\Sigma^+$ states. Several types of rotation induced perturbations are encountered and analyzed. Preliminary intensity calculations for the B-X system lead us to believe that the adiabatic $B^2\Sigma^+$ well minimum must be at smaller internuclear distance than has been estimated before. We estimate that 5.5 A < R_e^B < 6 A and that the $B^2\Sigma^+$ state is at least 30 cm^{-1} deep. This is in agreement with the latest high resolution scattering data[2].

INTRODUCTION

The alkali-rare gas systems are the prototype systems for a variety of studies broadly classified as interatom physics: interatomic forces, atom-atom collisions, line broadening, etc. Among these systems, NaAr is the most studied particularly in its first excited states[3]. The $B^2\Sigma^+$ state, whose well depth is 10% of the $A^2\Pi_r$ state, has not been previously observed directly. The present work is, by choice of the spectral region, sensitive to the $B^2\Sigma^+$ levels.

DISCUSSION

Figure 1 shows the tentative $\Omega = 1/2$ potential energy curves in the observed energy range. RKR potentials were calculated for the ground state and the $A^2\Pi_{3/2}$ state using available data[4-6]. The adiabatic $A^2\Pi_{1/2}$ curve and the $B^2\Sigma^+$ curve (I) are obtained from diagonalization of the off-diagonal spin-orbit interaction for the Ω = 1/2 states. For this we use the *ab initio* $B^2\Sigma^+$ curve[7] in the long range part and the diabatic $A^2\Pi_{1/2}$ curve, which we obtain from the $A^2\Pi_{3/2}$ RKR curve by subtraction of the diagonal part of spin-orbit interaction. The strongest observed transition in the $B^2\Sigma^+$ - $X^2\Sigma^+$ manifold is the (a, 0) band. In our intensity calculations the inner wall of the *ab initio* potential[7] is translated towards smaller internuclear distance in an effort to roughly simulate observed intensities. The vibrational assignment v' = a = 1 for $B^2\Sigma^+$ curve (II) with a minimum at 5.5 A gives the right ratio for intensities of bands from level a to v" = 0, 1, 2 assuming a vibrational temperature of 70 K. Levels v' = b and c are observed to predissociate above J' $\geq$ 3.5.

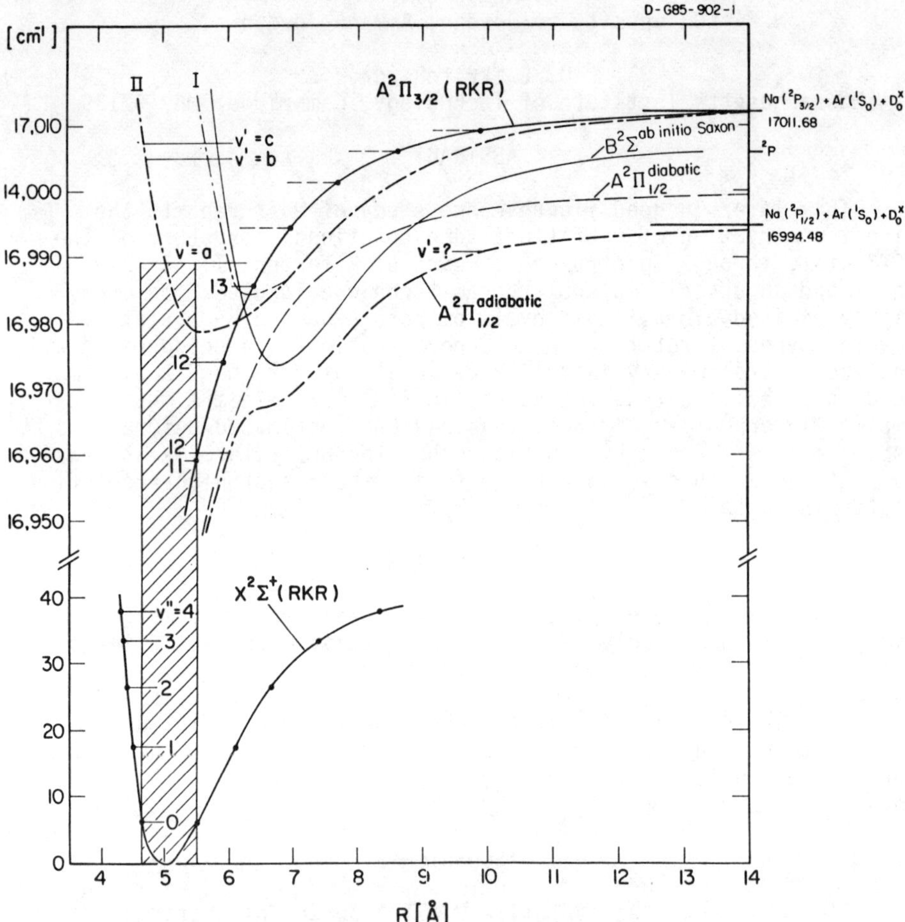

Figure 1. Tentative NaAr $B^2\Sigma^+$ and $A^2\Pi_{1/2}$ potential energy curves.

It is not possible to determine the secondary minimum in the $A^2\Pi_{1/2}$ state based on available data. Too few vibrational levels have been observed to deperturb the data and to accurately determine the Ω = 1/2 potential energy curves. However, it is obvious based on intensity considerations, that the $B^2\Sigma^+$ minimum has to be at smaller internuclear distance than earlier estimates (~6.2 A) based on differential scattering data[8] and theoretical calculations. Recent high resolution differential scattering data[2] yield R_e^B = 5.5 (5) A and D_e^B = 30.7 (20) cm^{-1}, in agreement with our conclusion.

We acknowledge support from the National Science Foundation Grant PHY 83-07172-A01 and the Petroleum Research Fund.

REFERENCES

1. W. P. Lapatovich, Ph.D. thesis, Massachusetts Institute of Technology (1980).
2. F. van den Berg, R. Morgenstern and C. Th. J. Alkemade, Chem. Phys. 93, 171 (1985).
3. A. M. Lyyra, W. P. Lapatovich, P. E. Moskowitz, M. D. Havey, R. Ahmad-Bitar and D. E. Pritchard, J. Chem. Phys. XX, xxxx (1986).
4. R. E. Smalley, D. A. Auerbach, P. S. H. Fitch, D. H. Levy and L. Wharton, J. Chem. Phys. 66, 3778 (1977).
5. J. Tellinghuisen, A. Regone, M. Soo Kim, D. J. Auerbach, R. E. Smalley, L. Wharton and D. H. Levy, J. Chem. Phys. 71, 1283 (1979).
6. G. Aepfelbach, A. Nunnemann and D. Zimmermann, Chem. Phys. Lett. 96, 311 (1983).
7. R. B. Saxon, R. E. Olson and B. Liu, J. Chem. Phys. 67, 2692 (1977).
8. R. Dueren, E. Hasselbrink and G. Moritz, Z. Phys. A 307, 1 (1982).

TWO PHOTON SEQUENTIAL ABSORPTION SPECTROSCOPY OF THE $E(0^+)$ STATE OF IODINE MONOFLUORIDE

B. K. Clark and I. M. Littlewood
University of Missouri-Rolla, Rolla, Missouri, 65401, U.S.A.

ABSTRACT

The energy levels of a new state of iodine monofluoride lying 5eV above the ground state have been measured using the Two Photon Sequential Absorption (TPSA) technique. The energies of eleven vibrational levels of the state were measured. The state has 0^+ character, and dissociates to $I^+ + F^-$ in the diabatic approximation.

INTRODUCTION

We report a detailed spectroscopic study of an ion-pair state of iodine monofluoride. When IF molecules were excited using laser radiation, strong absorption from the B state to an electronic state lying 5eV above the ground state was detected. The spectroscopic constants for this state were determined.

EXPERIMENTAL

The TPSA technique has been described previously [1]. A 1MW nitrogen laser was used to pump simultaneously two dye lasers. One dye laser, refered to as the pump laser, was used to excite IF molecules to specific rotational levels of the v=6 or v=8 vibrational levels of the B state. A second laser, refered to as the probe laser further excited the molecules to the upper state. Broadband fluorescence signals were monitored at right angles to the laser beam axis.

Since the iodine monofluoride molecule is labile, it was created in situ in a slow flow of fluorine established in a stainless steel flow tube, into which iodine vapour was introduced. The iodine vapour was obtained from a small cell containing iodine crystals at room temperature. The gas mixture was allowed to drift for approximately 20 cm before the laser beams were introduced.

RESULTS AND DISCUSSION

All lines which were observed when the probe laser was scanned across its dye profile could be assigned to transitions from the B state to a single electronic state of iodine monofluoride lying around 5eV above the ground state. The rotational energies of eleven vibrational levels of the upper state were identified. The energies can be reproduced using the molecular constants given in Table I. The vibrational numbering was made assuming that the lowest observed level is v=0. Since both iodine and fluorine have only one stable,

0094-243X/86/1460472-2$3.00

naturally occurring isotope, the validity of this assumption could not be confirmed.

Table I Vibrational and rotational constants (cm^{-1}) for the E state of iodine monofluoride

T_e	ω_e	$\omega_e x_e$	B_e	α_e	r_e
41291.74	249.09	0.5037	0.12920	4.4×10^{-4}	2.802Å

Diegelman et al [2] have investigated the fluorescence spectroscopy of a number of diatomic halogens, including iodine monofluoride. They observed a series of bands in the wavelength region 440nm to 500nm, and assigned them to an electronic transition between the D′ ion-pair state and the A′ covalent state. The vibrational constants for the D′ state are similar to those found for the E state.

The intermediate B state has, in case (c) coupling, character 0^+. The E B transition shows no evidence of Q branches, even at the band origin of the system. We therefore assign the E state as 0^+ in case (c) designation. This is the same designation as those of the E states of I_2 [3], and IBr [1].

LeRoy and Lam [4] have detailed a method of estimating the dissociation energy of electronic states of diatomic molecules. King et al [5] have further investigated the case of ionic dissociation. Using the constants given in table I for IF we obtain D = 78434 cm^{-1}, relative to the minimum of the molecular ground state. This value might be compared with the energy of the products $I^+(^3P_2)$ + $F^-(^1S_0)$. This energy is 80032 cm^{-1}, in good agreement with that obtained from the experimental data. Dissociation to the reversed ionic partners would require much higher energies.

If, on the other hand, we assume that the E state of IF dissociates to covalent partners, the predicted covalent dissociation limits all lie in the range 44000 - 48000 cm^{-1}, far from the energy of any possible neutral atomic partners. We conclude, therefore, that the E state of IF, like those of I_2, IBr and ICl, cannot dissociate covalently in the diabatic approximation, but must correlate with I^++F^-.

REFERENCES

[1] G. W. King, I. M. Littlewood and J. R. Robins, Chem. Phys. 62 (1981) 359.

[2] M. Diegelmann, K. Hohla, F. Rebentrost and K. L. Kompa, J. Chem. Phys. 76 (1982) 1233.

[3] G. W. King, I. M. Littlewood and J. R. Robins, Chem. Phys. 56 (1981) 145.

[4] R. J. LeRoy and W-H. Lam, Chem. Phys. Lett. 71 (1980) 544.

[5] G. W. King, N. T. Littlewood and I. M. Littlewood, Chem. Phys. 81 (1983) 13.

DOPPLER-LIMITED COLOR CENTER LASER SPECTROSCOPY OF HYDROGEN-BONDED COMPLEXES: ν_2 HCN---HF AND ITS HOT BANDS

E. K. Kyrö, M. Eliades, A. M. Gallegos,
P. Shoja-Chaghervand and J. W. Bevan
Chemistry Department, Texas A&M University, College Station,
Texas 77843

ABSTRACT

A continuously tunable single frequency color center laser spectrometer has been constructed for Doppler-limited spectroscopic analysis of hydrogen-bonded complexes. Gas phase analyses are reported for the fundamental ν_2 (C-H stretching vibration) and its hot bands $\nu_2 + \nu_4 - \nu_4$, $\nu_2 + \nu_7{}^1 - \nu_7{}^1$ and $\nu_2 + \nu_7{}^2 - \nu_7{}^2$ in the linear heterodimer HCN---HF.

INTRODUCTION

A computer-controlled continuously tunable single frequency color center laser spectrometer has been constructed for broadband Doppler-limited spectroscopy of gaseous phase hydrogen-bonded complexes. The spectrometer has a free running instrumental resolution of $\leqslant$ 3MHz and is capable of tuning over frequency segments of $\geqslant$ 30 cm^{-1} in the range 2.2-3.3 μm by synchronous scanning of its three tuning elements, the cavity folding mirror, 21.7 GHz internal etalon and grating. Absolute rovibrational transition frequencies of the complex are determined by simultaneously recording precisely known frequency standards and etalon markers in a temperature controlled evacuated 150 MHz marker cavity. The absorption spectrum of a gaseous phase spectrum of HCN---HF was recorded using a liquid N_2 InSb detector with temperature controlled White cell at a temperature of 230 ± 0.5K, total pressure of 1 torr and a pathlength of 96 meters.

Rovibrational analysis of the spectra included a multiple least squares regression fitting of the transition frequencies using the familiar energy relationship

$$F(J) = \nu_o + B' [J'(J'+1)-l'^2] \pm q'/4 (v'+1)J'(J'+1) - D'[J'(J'+1) - l'^2]^2 - B'' [J''(J''+1) - l''^2] \pm q''/4 (v''+1)J''(J''+1) + D''[J''(J''+1) - l''^2]^2$$

where q" = 0, l" = 0 for Σ states.

RESULTS

Rovibrational analyses of the ν_2, $\nu_2+\nu_7{}^1-\nu_7{}^1$, $\nu_2+2\nu_7{}^2-2\nu_7{}^2$ and $\nu_2+\nu_4-\nu_4$ bands are given in Table 1.

Table 1

	ν_2	$\nu_2+\nu_4-\nu_4$	$\nu_2+\nu_7^1-\nu_7^1$	$\nu_2+2\nu_7^2-2\nu_7^2$
ν_o/cm^{-1}	3310.32836(31)	3310.1628(14)	3310.04184(34)	3309.87625(58)
B''/cm^{-1}	0.119784(6) (0.119788(1))	0.11773(4)	0.120824(8)	0.121873(52)
B'/cm^{-1}	0.119618(6)	0.11754(5)	0.120658(8)	0.121783(52)
D_J''/cm^{-1}	$2.33(2)x10^{-7}$ $(2.31(1)x\ 10^{-7})$	$2.07(8)x10^{-7}$	$2.702(5)x10^{-7}$	$1.41(5)x^{-7}$
D_J'/cm^{-1}	$2.33(2)x10^{-7}$	$2.57(1)x10^{-7}$	$2.682(5)x10^{-7}$	$1.38(5)x10^{-7}$

Corresponding lower state B" and D_J" constants determined from microwave spectroscopy are given in parenthesis. The latter information was used to confirm spectral assignment in the infrared. The l-type doubling constants $q_{v_7^1}$ and $q_{v_2+\nu_7^1}$ are both determined to be 12.4(8) MHz. It is pertinent to note that the values of $B''_{2v_7^{\circ}}$ and B''_{v_4} are indicative of Fermi resonance between the $2\nu_7^{\circ}$ and ν_4 states of the complex.

The pressure broadening parameter under the experimental conditions used in these experiments has been determined to be 27(6) MHz torr $^{-1}$. A convolved Voigt lineprofile simulation following correction for the 3 MHz laser instrumental linewidth and 161.2 MHz calculated Doppler contribution results in a residual 45 MHz Lorentzian contribution. A 20 MHz correction for pressure broadening leaves a 25(8) MHz residual contribution which is attributed to vibrational predissociation. This permits evaluation of an excited state amplitude lifetime using the equation $\tau = 1/\pi\Gamma$ = 1.3(4) x 10^{-8}s. This excited state amplitude lifetime compares with corresponding values of $1.7(5)x10^{-10}$s and $5.6(4)x10^{-10}$s determined in the ν_1 (H-F stretching) and ν_3 (C-N stretching vibrations of the complex) demonstrating a vibrational state specificity. No rotational dependence was observed transitions specifically within ν_2 or its hot bands. In particular there was no dependence associated with l-type doublets. Furthermore there was no observed vibrational dependence within these bands despite an increase of 76, 148 and 176 cm^{-1} in vibrational metastable energy for the $\nu_2 + \nu_7^1$, $\nu_2 + 2\nu_7^2$ and $\nu_2 + \nu_4$ states relative to the dissociation of the complex.

The X°_{24} and X°_{27} anharmonic cross terms can be estimated as -0.1656(17) and -0.2865(27)cm^{-1} respectively using the available information.

MULTIPHOTON IONIZATION SPECTROSCOPY OF TRIPLET STATES AND RADICALS CREATED IN A SUPERSONIC BEAM

P. M. Johnson and S. W. Sharpe
Department of Chemistry
State University of New York at Stony Brook
Stony Brook, N. Y. 11794

ABSTRACT

Substantial quantities of triplet metastables and radicals are produced in a pulsed supersonic expansion by a pulsed electric discharge. Multiphoton ionization transitions of nitrogen molecule are seen originating from various metastable states throughout the visible and ultraviolet regions. Among these are the np←$E^3\Sigma_g^+$ Rydberg transition. This constitutes the first identification of a triplet Rydberg series in nitrogen. Triplet states have also been studied in benzene. Radicals produced in the supersonic discharge include CCl, where the vibrational and rotational constants of the B state have been established by the resolution of higher vibrational levels for the first time.

INTRODUCTION

Multiphoton ionization (MPI) has found great success in elucidating the structure and dynamics of atomic and molecular systems[1]. The development of an efficient supersonic beam source for producing metastables has enabled us to study these systems using MPI. We have successfully observed and studied the CCl radical, triplet benzene and more recently several metastable states of molecular nitrogen, including the discovery of a new Rydberg series originating from the E $^3\Sigma_g^+$ (3s) state.

EXPERIMENT

A typical experiment[2] proceeds as follows. Samples are mixed with argon and the mixture (2 atm) is expanded into a high vacuum via a pulsed valve. The resulting supersonic jet is subjected to a pulsed electric discharge (~ 30 kV, 1 μs) which, in turn, causes a cw potential(~ 5 kV) to discharge. Both neutral and charged species are skimmed and passed through an electrostatic potential (1 kV). The electrostatic potential deflects all ions out of the molecular beam before reaching the TOF spectrometer where they would cause spurious signals. When the jet reaches the TOF region (100 μs for Ar) the laser is triggered. MPI occurs if the laser is tuned to a suitable frequency and the cations are mass analyzed via a TOF spectrometer.

RESULTS

The CCl Radical

By passing argon over CCl_4, expanding the mixture in a supersonic beam and subjecting the beam to a discharge, large amounts

0094-243X/86/1460476-4$3.00

of CCl ($^2\Pi_{1/2}$) are produced. Rotationally resolved 1+1 MPI spectra for the 0-0, 1-0 and 2-0 $^2\Delta_i \longleftarrow {}^2\Pi_{1/2}$ transitions have been recorded. Based on these results, ω_e' (876.2 cm^{-1}) and $\omega_e x_e$' (12.0 cm^{-1}) have been directly calculated for the first time[3].

Triplet Benzene

Until recently, precise spectroscopic measurement of the energy of $^3B_{1u}$ benzene in the dilute gas phase has evaded experimental determination. By subjecting an Ar/benzene jet to a pulsed discharge we have produced large amounts of vibrationally cold $^3B_{1u}$ benzene. From the single photon ionization threshold, an accurate measure of the energy of the $^3B_{1u}$ state has been determined for both C_6H_6 (29,640 cm^{-1}) and C_6D_6 (29,838 cm^{-1})[4].

Nitrogen Molecule

At least a dozen valence states and a multitude of Rydberg states have been spectroscopicaly observed for N_2[5]. This is not surprising as N_2 is the most abundant diatomic molecule. Although there appears to already be a wealth of data on N_2, we have found new and challenging spectroscopic phenomena related to metastable N_2 created in an electric discharge. Optical transitions of metastable N_2 have been seen throughout the visible and the UV regions. The spectrum of metastable N_2 from 400-300 nm appears in figure 1 and contains several distinct features which deserve individual attention.

Two identical sets of four bands appear at 345 and 321 nm. From vibrational analysis of these eight bands, the lower and upper electronic states are found to be similar to the X $^2\Sigma_g^+$ and B $^2\Sigma_u^+$ states of N_2^+. In addition, the lower electronic state must live long enough to make it into the probe region (>100 μs). From the above observations, experimental data[6] and the theoretical calculations of P. Cremaschi et al[7] we believe that these eight bands are due to a transition between the E $^3\Sigma_g^+$ state and a new Rydberg state. These transitions correspond to the excitation of a valence $2s\sigma_u^*$ electron into the partially filled $2p\sigma_g$ valence orbital while another electron resides in the 3s Rydberg orbital. The final or upper electronic state is actually the lowest member of the Hopfield Rydberg series which converges to the B $^2\Sigma_u^+$ state of N_2^+ and has gone unobserved until now. Hot bands are seen at 372 nm but are weak because of low dye laser power. The vibrational analysis provides ω_e for these two states (ω_e'= 2306 cm^{-1}, ω_e''=2180 cm^{-1}). These numbers may be slightly modified by a more thorough rotational analysis.

A new Rydberg series appearing between 355-335 is attributed to a np(v'=1) $\longleftarrow$ E $^3\Sigma_g^+$ (v"=1) progression, converging to the v'=1 ground state of the ion. Figure 2 is a more detailed spectrum of the higher Rydberg terms (n=15-26) and is summarized in table 1. Because the series appears with an unfocussed laser, while retaining a relatively high signal, this Rydberg series is probably composed of one photon transitions to vibrationally autoionizing levels of the np Rydbergs containing one vibrational quantum of energy. This

METASTABLE NITROGEN MOLECULE
np-15
np-10
np-9
310
320
330
340
350
NANOMETERS
360
370
380
390
400

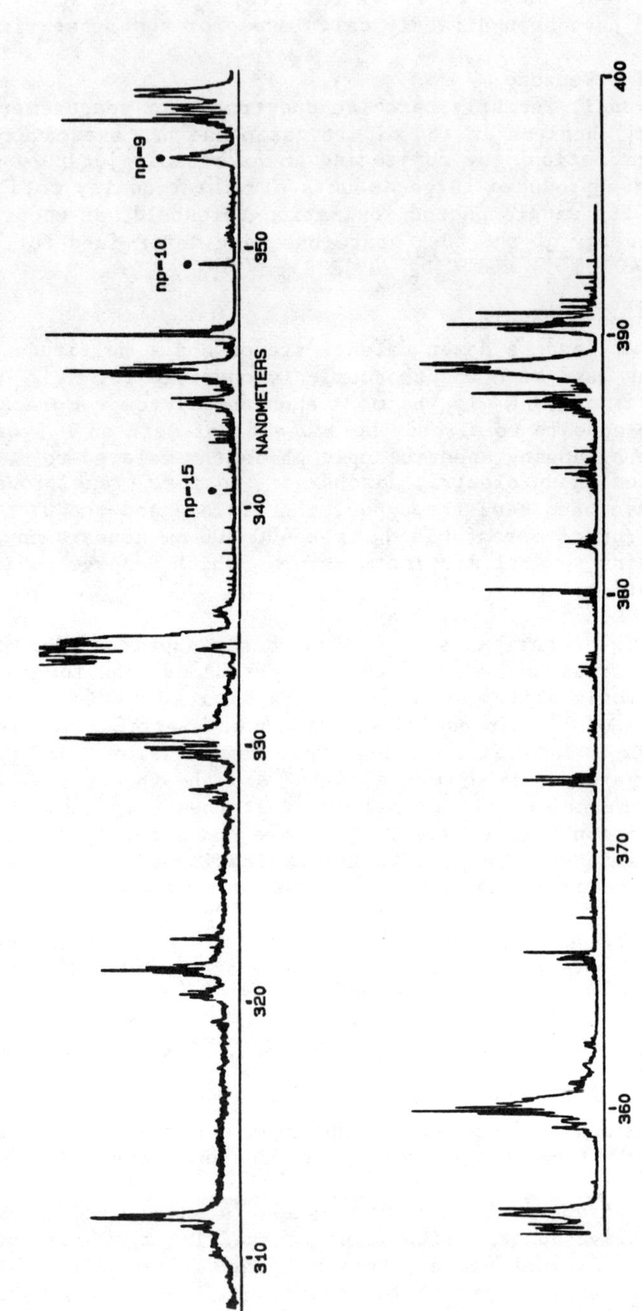

Figure I

assignment is confirmed by using our new vibrational data for the E state along with that for the ground state ion[8]. We then find that the convergence of this Rydberg series comes within one cm^{-1} of the predicted value. Further evidence for autoionization is seen by the lack of n<8 Rydberg terms in the unfocussed spectrum.

In addition to the transitions described above there are many other bands in figure 1. Some of these are likely due to core transitions with the Rydberg electron in various orbitals. For example the bands around 390 nm are at the energy of the B ← X transition of the ion. Those at 330 and 312 nm are vibrationally associated with the A state of the ion. The general density of bands seen in this ultraviolet spectrum continues throughout the visible region when the laser is focussed. These bands will be the subject of future analysis.

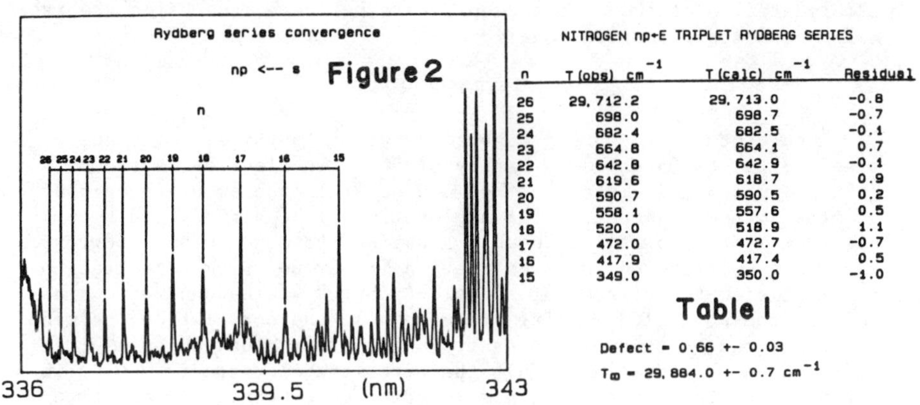

Figure 2

NITROGEN np←E TRIPLET RYDBERG SERIES

n	T (obs) cm^{-1}	T (calc) cm^{-1}	Residual
26	29,712.2	29,713.0	-0.8
25	698.0	698.7	-0.7
24	682.4	682.5	-0.1
23	664.8	664.1	0.7
22	642.8	642.9	-0.1
21	619.6	618.7	0.9
20	590.7	590.5	0.2
19	558.1	557.6	0.5
18	520.0	518.9	1.1
17	472.0	472.7	-0.7
16	417.9	417.4	0.5
15	349.0	350.0	-1.0

Table I

Defect = 0.66 +- 0.03

T_∞ = 29,884.0 +- 0.7 cm^{-1}

Supported by the Department of Energy under contract number DE-AC02-80ER10660.

REFERENCES

1. P. M. Johnson and C. E. Otis, Ann. Rev. Phys. Chem. 32, 139 (1981).
2. S. Sharpe and P. Johnson, Chem. Phys. Lett. 107, 35 (1984).
3. S. Sharpe and P. Johnson, J. Molec. Spect. 116, in press.
4. S. Sharpe and P. Johnson, J. Chem. Phys. 81, 4176 (1984).
5. A. Lofthus and P. H. Krupenie, J. Phys. Chem. Ref. Data 6, 113 (1977).
6. P. K. Carroll and A. P. Doheny, J. Molec. Spect. 50, 257 (1974).
7. P. Cremaschi, A. Chattopadhyay, P. Madhavan and J. Whitten, to be published.
8. K. Huber and G. Herzberg, Constants of Diatomic Molecules (Van Nostrand Reinhold, N. Y., 1979), p. 428.

DOUBLE RYDBERG SPECTROSCOPY

P. Camus, P. Pillet
Laboratoire Aimé Cotton CNRS II, Bât. 505
Campus d'Orsay, 91405 Orsay

J. Boulmer
Institut d'Electronique Fondamentale, Bât. 220
Université Paris-Sud, 91405 Orsay

ABSTRACT

Properties of the neutral doubly excited states, mainly 9dn'd double-Rydberg states have been studied using a two-step two-photon laser excitation. Ionization of neutral Ba atoms in an atomic beam has been investigated by means of a time of flight mass spectrometer. Interpretation of the spectroscopic data has been made using the isolated core approximation model.

INTRODUCTION

One of the interesting problems of atomic physics is to understand better the dynamics of highly excited two-electrons systems ($n\ell n'\ell'$) which is related to the more general problem of three-body quantal systems. Recently, the exploration of highly excited two-electron states so called "double-Rydberg" states has been performed using multiphoton laser excitation in barium. From a theoretical point of view two classes of "double-Rydberg" states can be considered. At the so-called Wannier ridge ($n \sim n'$), the correlation between the electrons becomes predominant. On the other hand, on both sides of the Wannier ridge ($n' >> n$), the outermost electron is moving in the field of the ionic core, partly screened by the inner electron

Properties of such "double-Rydberg" states called "valley states" or planetary atoms by Percival[1] could be described in terms of quantum defect theory.

Freeman et al.[2] have shown the possibility of producing such ($n\ell,n'\ell'$) double-Rydberg states in neutral barium and recent results for $nsn'\ell'$ (n = 7 to 11, n' = 13 to 20 with ℓ' = 0 or 2) states have been reported[3]. In this paper, we give a spectroscopic study of the 9dn'd double-Rydberg states[4] with n' ranging from 18 to 31.

EXPERIMENTAL SET-UP AND PROCEDURE FOR EXCITATION OF THE 9dn'd DOUBLE-RYDBERG STATES

Figure 1 shows the relevant energy levels of Ba and Ba^+, and the two-step two-photon resonant laser excitation. The 6snd $^1D_2$ single-Rydberg state (n' = 17 to 29) is populated from the $6s^2$ ground level by a two-photon absorption process using a nitrogen-pumped dye laser ($\lambda_B \sim 470$ nm). After a fifty nanosecond delay, the single-Rydberg atom is excited to a double-Rydberg state using a frequency doubled Nd-YAG pumped dye laser ($\lambda_{UV} \sim 283$ nm). These states autoionise to $Ba^+(m\ell)$ ($m < n$) and then the ions produced are photoionized to Ba^{2+} ions by the UV laser.

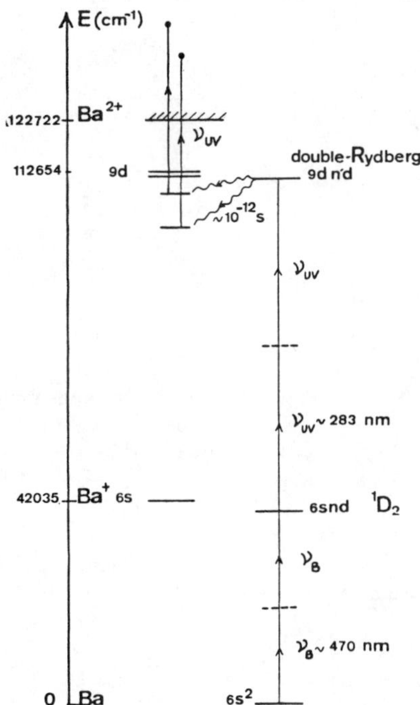

Fig. 1. Ba I and Ba II relevant energy levels.

A time of flight mass spectrometer is used to discriminate and detect the Ba^{2+} ions.

Figure 2 shows a typical record of the Ba^{2+} ion signal recorded out of the 6s28d $^1D_2$ single-Rydberg level by scanning the UV laser around the Ba^+ 6s → 9d ionic transitions. With circularly polarised light only J = 4 levels are excited and two series of double-Rydberg states, $9d_{3/2}n'd$ and $9d_{5/2}n'd$, are observed

ANALYSIS AND INTERPRETATION OF THE SPECTROSCOPIC DATA

The four main peaks of the figure 2 are interpreted in the simple model of the isolated core approximation[5] (two channel quantum defect theory). Calculated ion signal (trace c) corresponding to $9d_{3/2}n'd$ is the product of Lorentzien curves (trace a)(Rydberg series in a continuum) by the square of the overlapping integral (trace b) between the initial 6snd single-Rydberg state and the final 9dn'd double-Rydberg states. Comparison between the observed and calculated spectra show that most of the features as profile and intensity of the lines are explained and understood. This method leads to the determination of the quantum defect 4.17 for the two series $9d_{3/2}n'd$ and $9d_{5/2}n'd$.

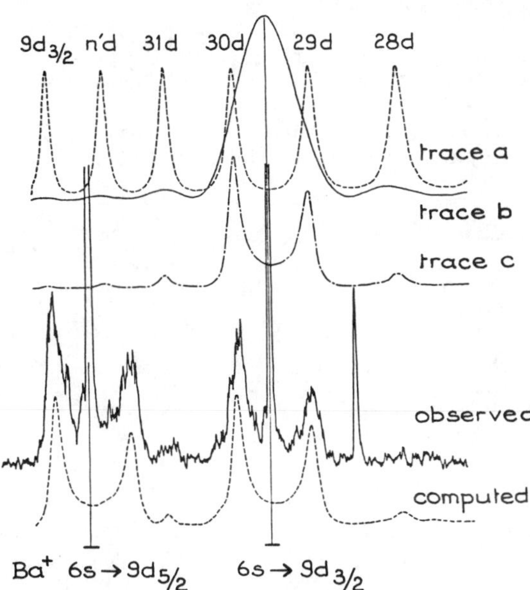

Fig. 2. Ba^{2+} ion spectrum from the 6s28d $^1D_2$.

In this simple approximation, some discrepancy between experimental and theoretical curves appears as shown on figure 3 and are due to coincidence and interference of two members of the series $9d_{3/2}29d$ and $9d_{5/2}27d$. Narrowing of the autoionizing width corresponds to a "stabilization" of the resonance.

A six channel quantum defect analysis is in progress.

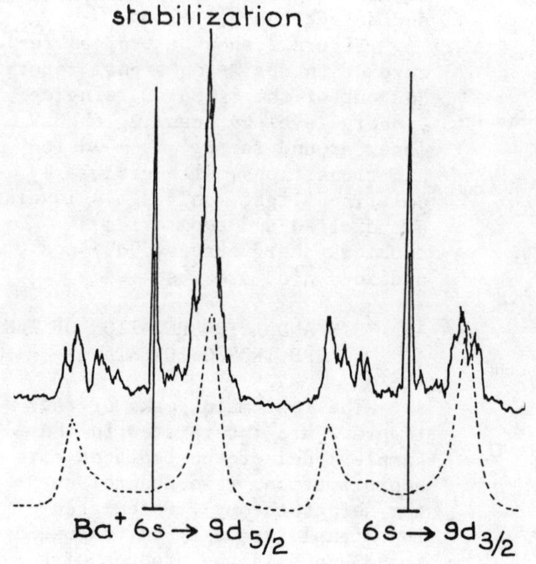

Fig. 3. Ba^{2+} ion spectrum from the 6s26d $^1D_2$.

REFERENCES

1. I.C. Percival, Proc. R. Soc. A 353, 289 (1977).
2. R.R. Freeman, R.M. Jopson, J. Bokor and W.E. Cooke, Laser Spectroscopy VI (H.P. Weber and W. Lüthy, Springer, 1983), p. 220.
3. R.M. Jopson, R.R. Freeman, W.E. Cooke and J. Bokor, Phys. Rev. Lett. 51, 1640 (1983).
4. P. Camus, P. Pillet and J. Boulmer, J. Phys. B. 18, L481 (1985).
5. W.E. Cooke, T.F. Gallagher, S.A. Edelstein and R.M. Hill, Phys. Rev. Lett. 40, 178 (1978).

DYNAMICS OF CLUSTER DISSOCIATION FOLLOWING MULTIPHOTON IONIZATION

A. W. Castleman, Jr.
The Pennsylvania State University, University Park, PA 16802

ABSTRACT

A major advance in the study of unimolecular dissociation and the spectroscopy of clusters has become available through the use of multiphoton ionization coupled with a reflectron introduced into the drift region of a time-of-flight mass spectrometer. Using single and two-color tunable pulsed lasers, the excess energy introduced into a cluster can be well controlled. The power of this method is demonstrated by the results of recent investigations of hydrogen bonded clusters which, following ionization, lead to an internal ion molecule reaction, and cluster fragmentation. The role of dissociation and the influence of the thermochemical stability of cluster ions in effecting the appearance of magic numbers in certain cluster distributions is discussed. The application of this method in determining ionization potentials of probe molecules following successive clustering with a solvent species is also presented.

INTRODUCTION

Studies of the dynamics of formation and dissociation, and the changing properties of clusters at successively higher degrees of aggregation, enable an investigation of the basic mechanisms of nucleation to be probed at the molecular level. Work on systems of increasingly higher degrees of aggregation also has a direct bearing on the development of surfaces, and ultimately solvation phenomena and the formation of the condensed state. The progressive clustering of a molecule involves energy transfer and redistribution within the molecular system, with attendant processes of unimolecular dissociation taking place between growth steps. Related processes of energy transfer and dissociation are operative during the reorientation of molecules about ions following the primary ionization event employed in detecting clusters via mass spectrometry.

Recent advances in the field of molecular beam research, coupled with lasers and time-of-flight mass spectrometry, enable the details of these various processes to be investigated. A major advance in the study of these processes has become available through the use of a reflectron technique introduced into the drift region of a time-of-flight mass spectrometer. Using single and two-color tunable pulsed lasers, the excess energy introduced into a cluster can often be well controlled. The power of this technique is demonstrated by a recent investigation of hydrogen-bonded clusters $(NH_3)_n$ and $(CH_3OH)_n$ which, following ionization, lead to an internal ion-molecule reaction followed by proton transfer and reorientation of the molecular dipoles.

A second example of the exciting prospects provided by an investigation of clusters using pulsed lasers comes from recent work on progressive shifts in the electronic state of a probe molecule due to clustering with a solvent. As an example of the power of this method, findings are presented on the spectroscopic shift of phenyl acetylene and paraxylene in co-clustering with rare gases. Results are interpreted in terms of the relative stability of the excited-state compared to the ground state. Variations in ionization potential with cluster size are also discussed.

EXPERIMENTAL

Clusters of the desired molecules are formed via adiabatic cooling from a pulsed nozzle system, with detection of the products of MPI being made in a time-of-flight mass spectrometer located beyond the ionization region.[1] The combination of ionization and mass selection has the advantage of direct mass determination of the probed van der Waals molecules. Because of the weak bonding in complexes, dissociative ionization may possibly complicate the spectroscopic assignments. In the present work both single and two-color multiphoton ionization is utilized. This is accomplished using a Q-switched Nd:YAG laser to simultaneously pump two tunable dye lasers whose output is frequency doubled in KDP crystals. The arrangement allows continuous tunability from the visible down to 216.5 nm, with a maximum output of 1 to 29 millijoules per pulse (6 ns duration) in the UV. The results show that fragmentation can be greatly suppressed through the use of two-color resonance enhanced photoionization whereby the ionzation process is accomplished with the use of little excess energy in contrast to one-color experiments.

In an alternate method of operation the ions are steered to a reflectron which is positioned off-axis at the end of the drift tube. This reflecting electric field decelerates and reflects the ions in a homogeneous electric field to the entrance of the second particle detector that is positioned above the TOF lens. In this case, ions which do not dissociate between the ion lens and the reflectron give rise to a normal TOF mass spectrum. However, product ions which arise due to dissociation processes occurring between the ion lens and the reflectron, form an additional spectrum that becomes superimposed on the spectrum of the original parent ions with a time difference between corresponding mass peaks.

RESULTS AND DISCUSSION

1. Multiphoton Ionization Studies

Ammonia System

Protonated cluster ions of the form $(NH_3)_n \cdot H^+$ are formed rapidly (in less than 10 ns) in an internal ion molecule reaction. NH_3 monomer units are lost on a longer time scale due to the excess energy created as the ammonia molecules in the cluster reorient from their original configuration to incorporate the structure of the newly formed NH_4^+ entity. A discontinuity in the intensity distributions of clusters at the protonated pentamer was found to be independent of expansion conditions.[1] But, it could be enhanced or diminished via a change in laser fluence and wavelength, providing clear evidence of the role of excess energy in cluster fragmentation processes. The NH_4^+ readily bonds ammonia to each of the hydrogen atoms leading to the enhanced pentamer intensity which corresponds to well-known structures in cluster ions.

An investigation of the cluster intensities as a function of pressure revealed substantial metastable and collision induced dissociation contributions. A finding of some import was the enormous increase of metastable fraction with increasing cluster size. The reflectron technique enabled a determination of reaction channels to be made, and the clear identity of the parents from which the daughter ions were originating. Interestingly, the loss of up to at least five monomer units following cluster ion formation was observed.

Subsequent to the ionization event, ammonium ion is formed via the reaction of NH_3^+ with NH_3, which is is exothermic by 0.74 eV and undoubtedly occurs via an "internal" ion-molecule reaction in a cluster. The reaction and subsequent molecular rearrangement is expected to lead to a reasonably large degree of excess internal energy, the magnitude of which depends on the number of monomer units contained in the species in question. Studies were undertaken to determine the relative importance of collision induced dissociation (CID) compared to unimolecular decay in the various dissociation channels. The dissociation process occurring via loss of one NH_3 unit

$$(NH_3)_n \cdot H^+ \rightarrow (NH_3)_{n-1} \cdot H^+ + NH_3$$

was investigated[1] for various ion cluster sizes from $n = 4$ to 20. The relative intensities were found to increase approximately linearly with background pressure as expected for the thin-target condition, and extrapolation to zero pressure yields a finite intercept where the ordinate is observed to increase with cluster size. The dissociation rates were found to be of order 10^5 s^{-1}. Simple RRK considerations indicate the possible enhancement of the dissociation of larger clusters as observed in the present work.[1]

Alcohol System

Very similar findings were observed in the case of alcohol clusters. Proton transfer reactions ensue following the ionization event, leading to the production of and detection of species of the form $H^+(CH_3OH)_n$. Both collision induced and unimolecular dissociation are observed in the case of alcohol, but no more than two monomer-loss channels have been found.

In addition to the foregoing, one additional reaction channel was observed which has a direct analogy to a well-known ion-molecule reaction.[2] In the case of the protonated dimer ion only, the loss of water was found as follows:

$$CH_3OH_2^+ + CH_3OH \rightarrow [(CH_3)_2O]H^+ + H_2O$$

2. Spectroscopic Shifts and Ionization Potential Variations with Degree of Aggregation

van der Waals molecules of the type $A \cdot M_n$ (A=phenyl acetylene, paraxylene; M=R (rare gas atom), N_2, O_2, N_2O, NH_3, H_2O, CCl_4, and CH_4), were investigated using both one- and two-color resonance enhanced MPI techniques.[3,4] Studies of the perturbed $L_b(^1B_2)$ states reveal a wide range of red spectral shifts as well as small blue shifts with respect to the pure aromatic molecule, depending on the nature of the clustering partner. In cases where M=Ne, Ar, Kr, and Xe, the shift is red; the magnitude of the shift increases linearly with the polarizability of the rare gas atom.

The vibrational excitation of different vdW modes were also measured for M=R. Typical values of the stretching mode range from 35.5 cm^{-1} (PA·Xe) to 42.5 cm^{-1} (PA·Ar). Studies of the spectroscopic shifts show that the magnitude of the shift for ($PA \cdot Ar_2$) is approximately twice the shift for the single argon containing species. Red shifts were also seen for clusters with N_2, O_2, N_2O, CCl_4, NH_3, and CH_4 while blue shifts were observed in the case of CO_2 and H_2O. The spectroscopic studies reveal the expected van der Waals modes of each cluster, shifted to the blue of the band origin. Comparing the spectra of successively larger clusters, and studying variations in intensity with pressure in the stagnation chamber, enables an identification of cluster fragmentation to be readily made.

The appearance potentials of $(PX \cdot Ar_n)^+$ ions were determined through studies in which the energy of one photon was fixed at the L_b resonance and the wavelength of the other was scanned. The observed appearance potentials are found to vary with the square-root of the electric field present in the region of ionization in accordance with expectations and the findings of others. Extrapolation to zero field enables a determination of the ionization potentials of paraxylene and its clusters with rare gas atoms.

Interestingly, although the spectroscopic shifts of large aggregates show features resembling those of the condensed phase, the total shift in the ionization potential is far less than observed in the case of similar molecules trapped in rare gas matrices. For example, the ionization potential of benzene in argon is shifted by approximately 6000 cm^{-1} from the gas-phase value. Evidently, alterations in the work function of a dielectric depend significantly on long-range interactions. It is also interesting to note that delayed ionization has also been found in the case of ionization through the Rydberg states of paraxylene where the product ion comes from a bound partner having a lower ionization potential, namely trimethylamine.

ACKNOWLEDGMENTS

The author thanks Dr. R. G. Keesee, Dr. P. Dao, Dr. O. Echt and Ms. S. Morgan for their assistance in performing the work and for helpful discussions during the course of this investigation. Support by the Department of Energy, Grant No. DE-AC02-ER-60055 and the Department of the Army, Grant No. DAAG29-82-K-0160 is gratefully acknowledged.

REFERENCES

1. O. Echt, P. D. Dao, S. Morgan, and A. W. Castleman, Jr. J. Chem. Phys. 82, 4076 (1985). See also O. Echt, S. Morgan, P. D. Dao, R. J. Stanley, and A. W. Castleman, Jr. Ber. Bunsenges. Phys. Chem. 88, 217 (1984).
2. L. M. Bass, R. D. Cates, M. F. Jarrold, N. J. Kirchner, and M. T. Bowers, J. Am. Chem. Soc. 105, 7024 (1983).
3. P. D. Dao, S. Morgan, and A. W. Castleman, Jr., Chem. Phys. Lett. 111, 38 (1984).
4. P. D. Dao, S. Morgan, and A. W. Castleman, Jr., Chem. Phys. Lett. 113, 219 (1985).
5. P. D. Dao and A. W. Castleman, Jr., "Delayed Ionization Following Resonant Photon Absorption and Intra-Cluster Electron Transfer," J. Chem. Phys., submitted.

PHOTODISSOCIATION DYNAMICS OF SMALL IONIC CLUSTERS: TRIMERS

Hyun-Sook Kim, Martin Jarrold[a], Andreas Illies[b] and
Michael T. Bowers
University of California, Santa Barbara, California 93106

ABSTRACT

Recent results on the photodissociation dynamics of ion-neutral trimer species are presented. Emphasis is placed on the nature of the moiety absorbing the photon, the mechanism of dissociation, and the product energy disposal. Most comments are limited to $(CO_2)_3^+$ and $(NO)_3^+$ but preliminary data on $O_2^+ \cdot (CO_2)_2$ are also briefly discussed.

INTRODUCTION

Interest in clusters has grown rapidly in recent years. van der Waals clusters,[1] ion-neutral clusters[2] and metal clusters[3] are all areas of intense research activity. Our group has concentrated on ion-neutral clusters. In a recent series of papers we have reported the results of studies of photodissociation of a number of ion-neutral clusters. For the most part these are "dimeric" species of either the symmetric,[4] A_2^+, or non-symmetric,[5] AB^+, type. Our motivation is to learn something about the physical and reactive properties of these interesting species. The long term objective is to investigate larger species with the ultimate goal of probing the interface between the vapor and liquid phases. In this paper we will summarize some of what we have learned by extending our studies to trimers.[6-8]

EXPERIMENTAL

The details of the experimental techniques are given elsewhere.[4-8] Briefly, cluster ions are formed by three body association reactions in an ion source, thermalized by collisions, extracted, accelerated, mass selected by a magnet and brought to a spatial focus where they are crossed by the polarized light of an Argon Ion Laser/Dye Laser. Photoproducts are mass and energy analysed by an electrostatic sector and detected by ion counting techniques. Variation of the laser polarization angle allows determination of product kinetic energy distributions and determination of product angular distributions (asymmetry parameters).

a. Present address, AT&T Bell Labs, Murray Hill, NJ 07974
b. Present address, Auburn University, Auburn, Alabama 36849

RESULTS

Due to the space requirements of this paper only selected results can be given. Greater detail is given elsewhere.[6-8] Comments will mainly be restricted to $(CO_2)_3^+$ and $(NO)_3^+$.

Photodissociation Cross Sections:

The variation of the photodissociation cross sections for $(CO_2)_3^+$ and $(NO)_3^+$ are given in Figs. 1a and 1b. For $(CO_2)_3^+$ the cross section almost exactly mimics the $(CO_2)_2^+$ cross section although its magnitude is somewhat larger. This implies that the $(CO_2)_2^+$ moiety acts as the chromophore, a result consistent with conclusions of Johnson, et. al.[9] for $(CO_2)_n^+$ clusters with n=3-10. On the other hand, the $(NO)_3^+$ cross section is strongly blue shifted from $(NO)_2^+$ and has a much larger value at the peak. This result implies $(NO)_3^+$ is a unique molecule and has its own photochemical properties.

Branching Ratios:

There are two energetically allowed product channels for both $(CO_2)_3^+$ and $(NO)_3^+$; A_2^+/A and $A^+/2A$. The fraction of $(CO_2)_2^+/CO_2$ observed is ∿0.04 and increases slightly with wavelength increases from 488 to 650 nm. Over the same range the fraction of $(NO)_2^+/NO$ is ∿0.28. A detailed analysis indicates that ∿50% of the observed

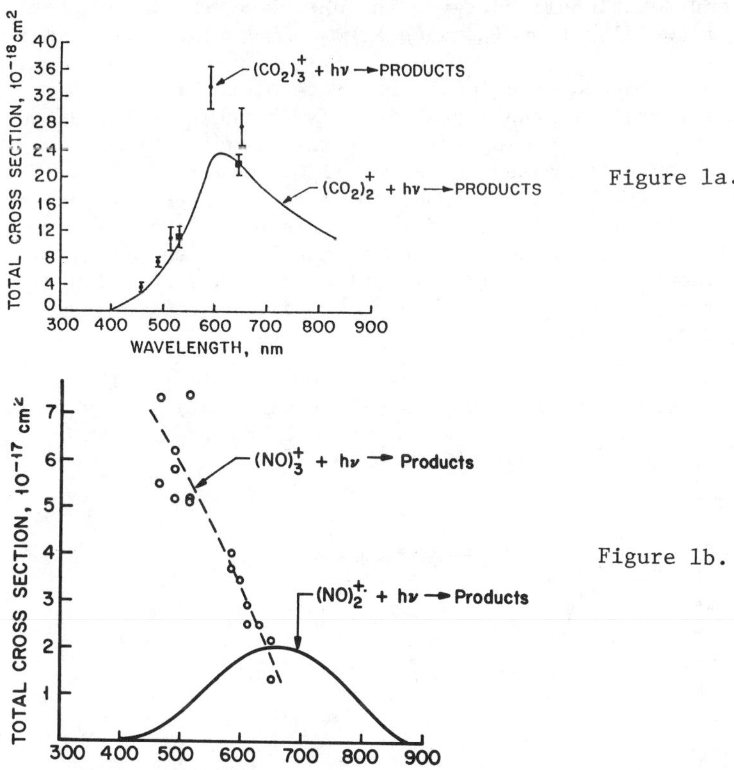

Figure 1a.

Figure 1b.

$NO^+/2NO$ channel comes from a sequential process; $A_3^+ + h\nu \rightarrow A_2^+/A \rightarrow A^+/2A$ while for $(CO_2)_3^+$ essentially 100% of $CO_2^+/2CO_2$ comes from the sequential process. Hence the nascent fractions are ∿1.0 for $(CO_2)_2^+/CO$ and ∿0.67 for $(NO)_2^+/NO$.

Asymmetry Parameter Analysis:

The presence of sequential processes in both systems makes unambiguous interpretations of asymmetry parameters difficult. The details are discussed elsewhere.[6,7] The results indicate that the first step in the sequential process, $A_3^+ + h\nu \rightarrow A_2^+/A$ almost certainly goes through a very short lived A_3^{+*} excited state while the second step $A_2^+/A \rightarrow A^+/2A$ appears to occur by statistical vibrational predissociation of internally excited A_2^+. The direct process forming $NO^+/2NO$ (accounting for ∿33% of the total photochemistry) goes via an excited state with lifetime short relative to a rotational period.

Product Energy Disposal:

The most dramatic feature in the $(NO)_3^+$ system is the large amount of vibrational energy that must be carried off by the NO fragment in the process $(NO)_3^+ + h\nu \rightarrow (NO)_2^+/NO$. An example is given in Fig. 2. Using an $(NO)_2^+$ binding energy of 0.6 eV[10], energy conservation indicates that at least 0.8 eV must be carried off as NO vibrational (and rotational) energy. The narrow width of the peak suggests one or two vibrational states are favored; almost certainly v=3 and 4.

The low photoinduced fractional abundance of the $(CO_2)_2^+/CO_2$ channel and the large metastable peak in this channel precluded investigation of the energy disposal. However, studies of the average kinetic energy released in the $CO_2^+/2CO_2$ were revealing. The data are given in Fig. 3. It is apparant there is agreement between the data and a "transient dimer" model. This model assumes an impulsive ejection of a CO_2 moiety by the initially excited $(CO_2)_2^+$ chromophore followed by statistical vibrational predissociation of a transient dimer formed by the remaining CO_2^+ ion and CO_2 neutral.

The $O_2^+ \cdot (CO_2)_2$ System:

We have preliminary results on the $O_2^+ \cdot (CO_2)_2$ trimer. The CO_2^+/O_2+CO_2 photoproducts strongly predominate with only a very small

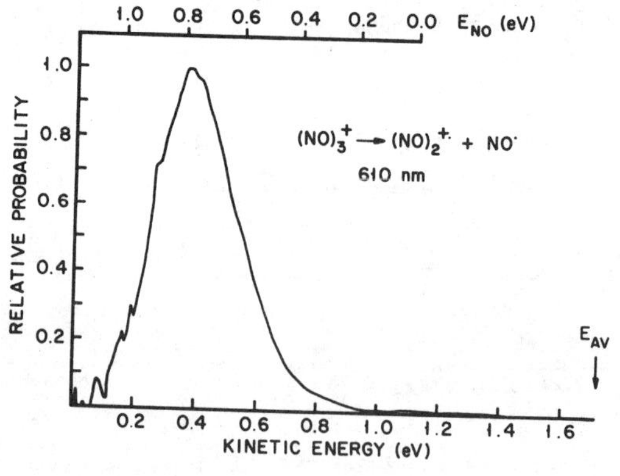

Figure 2.

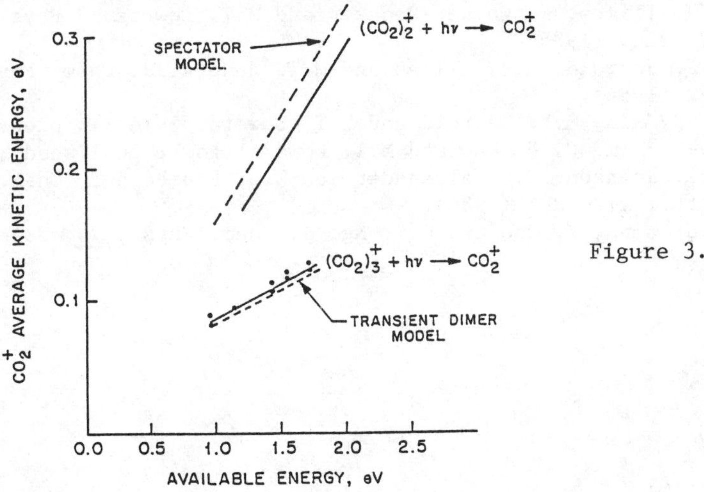

Figure 3.

(∿1%) $O_2^+/2CO_2$ product observed. The $O_2^+/2CO_2$ products are thermodynamically more stable by ∿1.7 eV. This data, along with preliminary cross section data, are very suggestive the photon is initally absorbed by a $O_2^+ \cdot CO_2$ chromophore yielding a repulsive $O_2 \cdot CO_2^{+*}$ charge transfer state that ejects the O_2 molecule. A transient $(CO_2)_2^+$ dimer may be formed before dissociation but we do not have sufficient data to say one way or the other. In the $O_2^+ \cdot CO_2$ dimer system no O_2^+/CO_2 product is observed. Hence, presence of the additional CO_2 moiety does allow some charge transfer to occur from the initial $(O_2 \cdot CO_2^{+*}) \cdot CO_2$ photoexcited state yielding small amount of $O_2^+/2CO_2$ products.

ACKNOWLEDGEMENTS

The support of the Air Force Office of Scientific Research under Grant AFOSR-86-0059 and the National Science Foundation under Grant CHE85-12711 is gratefully acknowledged.

REFERENCES

1. See, for example, D.H. Levy, Adv. Chem. Phys., 47, 323 (1981).
2. T.D. Mark and A.W. Castleman, Avd. At. Mole. Phys., 20, 65, (1984).
3. See, for example, S.C. Richtsmeier, E.K. Parks, K. Liu, L.G. Pobo and S.J. Reily, J. Chem. Phys., 82, 3659 (1985).
4. M.F. Jarrold, A.J. Illies and M.T. Bowers, J. Chem. Phys., 79, 6086 (1983); ibid 81, 214 (1984); ibid 82, 1832 (1985); A.J. Illies, M.F. Jarrold, W. Wagner-Redeker and M.T. Bowers, J. Phys. Chem., 88, 5204 (1984).
5. M.F. Jarrold, L. Misev and M.T. Bowers, J. Chem. Phys., 81, 4369 (1984); A.J. Illies, M.F. Jarrold, W. Wagner-Redeker and M.T. Bowers, J. Am. Chem. Soc., 107, 2842 (1985); M.F. Jarrold,

A.J. Illies, W. Wagner-Redeker and M.T. Bowers, J.Phys. Chem., 89, 3269 (1985).
6. M.F. Jarrold, A.J. Illies and M.T. Bowers, J. Chem. Phys., 81, 222 (1984).
7. H.-S. Kim, M.F. Jarrold and M.T. Bowers, ibid (in press).
8. H.-S. Kim, C.-H. Kuo and M.T. Bowers (to be published).
9. M.A. Johnson, M.L. Alexander and W.C. Lineberger, Chem. Phys. Lett., 112, 285 (1984).
10. S.H. Linn, Y. Ono and C.Y. Ng, J. Chem. Phys., 74, 3342 (1981).

PHOTOIONIZATION AND PREDISSOCIATION OF EXCITED STATES OF NeXe, ArXe, KrXe, AND Xe_2

S. T. Pratt, P. M. Dehmer, and J. L. Dehmer
Argonne National Laboratory, Argonne, Illinois 60439

The rare gas dimers NeXe, ArXe, KrXe, and Xe_2 were studied in the wavelength region of the Xe^* 6p and 5d atomic states using two photon resonant, three photon (2+1) ionization. The apparatus consists of a time of flight mass spectrometer, an electrostatic electron energy analyzer, and a Nd:YAG pumped, frequency doubled dye laser system; and the rare gas dimers were produced in an unskimmed supersonic expansion. New molecular band systems were identified in all four dimers corresponding to bound-bound transitions from the ground electronic state to the resonant intermediate states.[1-3] Figure 1 shows the band system observed in the (2+1) ionization spectrum of Xe_2 in the region to the red of the atomic Xe $5d[5/2]_2$ level. The rotational structure is not resolved and the peaks correspond to the rotational envelope of each vibronic band. The vibrational spacings can be used to determine parameters for the excited state potential curves.

The selectivity of (2+1) ionization allows us to record the rare gas dimer photoelectron spectra (PES) without interference from the corresponding atomic photoelectron peaks. For example, Figure 2 shows the PES obtained at the (0,0) band of Figure 1, and there is no evidence for non-resonant ionization of atomic Xe.

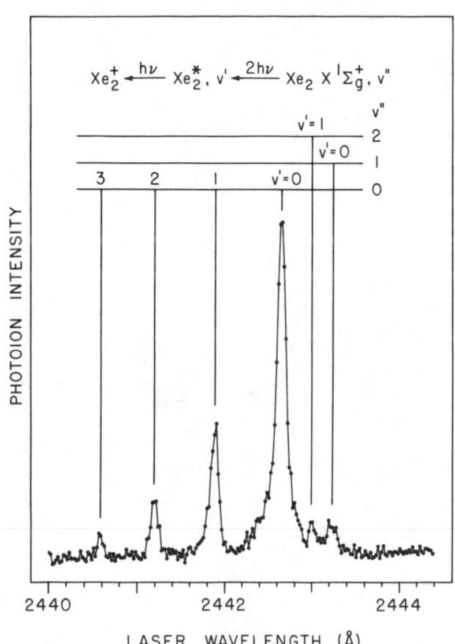

Fig. 1. The (2+1) ionization spectrum of Xe_2 to the red of the atomic Xe $5d[5/2]_2$ level.

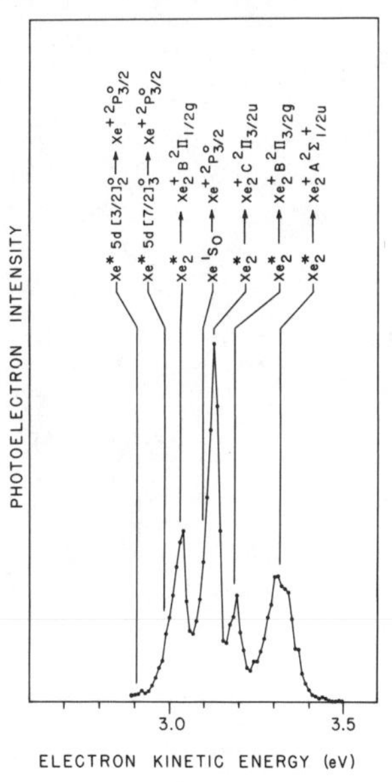

Fig. 2. The PES of Xe_2 obtained at the (0,0) band of Figure 1.

Using this technique we have recorded the first PES of ArXe[1] and KrXe.[2]

While ionization via some of the band systems leads primarily to the production of the corresponding molecular ion, ionization via others leads almost exclusively to the production of Xe^+. The PES show that Xe^+ formation results from predissociation of the resonant intermediate level followed by ionization of the excited atomic fragment, e.g., $Xe_2^* \rightarrow Xe + Xe^* \rightarrow Xe + Xe^+ + e^-$. Thus, these PES provide information on the dissociation dynamics of the resonant intermediate states.

This work was supported by the U.S. Department of Energy under Contract No. W-31-109-Eng-38 and by the Office of Naval Research.

REFERENCES

1. S. T. Pratt, P. M. Dehmer, and J. L. Dehmer, J. Chem. Phys. 82, 5758 (1985); ibid, in press.
2. S. T. Pratt, P. M. Dehmer, and J. L. Dehmer, Chem. Phys. Lett. 116, 245 (1985).
3. P. M. Dehmer, S. T. Pratt, and J. L. Dehmer, to be published.

SPECTROSCOPIC STUDIES OF SUPERSONIC SEMICONDUCTOR CLUSTERS

Y. Liu, J. Heath, S. O'Brien, Q.L. Zhang, R.E. Smalley,
R.F. Curl, and F.K. Tittel
Departments of Chemistry and of Electrical and Computer Engineering
Rice University, Houston, TX 77251-1892

ABSTRACT

Observation of silicon, germanium, and GaAs clusters produced by laser vaporization followed by supersonic expansion in a helium carrier is reported. Evidence that semiconductor clusters differ dramatically from similarly prepared metal clusters is discussed.

DISCUSSION

A detailed description of the apparatus used in this study has been given previously.[1] Since semiconductor elements are more readily available in the form of wafers than in the form of rods as required previously, a rotating disc laser-vaporization cluster beam source was developed. The 2nd harmonic output of a Nd:YaG laser (30 mj/pulse) was focused onto a 0.1 cm spot on the disc, producing a super-heated plasma entrained in a near-sonic flow of pulsed helium carrier gas. Free expansion of this cluster/helium mixture into a large vacuum chamber produced an intense supersonic beam, which was collimated by two skimmers. Subsequently, the cluster beams were characterized by F_2 (7.9 eV) and ArF (6.4 eV) excimer laser ionization accompanied by time-of-flight analysis.

An interesting observation of this study is that Si and Ge cluster beams produce almost identical mass spectra at moderate ArF excimer laser fluence (~0.5 mJ/cm^2), as shown in Fig. 1. Cluster ion signals in the size range of 6 and 11 atoms are particularly strong and show predominant contributions from fragments of higher clusters. This similarity means that the nascent cluster size distributions in the beam must be essentially the same for both elements. Furthermore, large Si and Ge clusters have almost the same fragmentation pattern and fragment primarily by fission into daughter ions in the 6-11 atom size range. This fragmentation pattern is quite different from that observed for metal clusters in the same apparatus. Time delay studies in two-color-experiments involving initial excitation with the SHG (2.4 eV) and THG (3.5 eV) of the NdYag laser followed by excimer laser ionization established that both Si and Ge clusters have excited electronic states with lifetimes of ~100 ns. The existence of such long-lived excited states indicates that there is probably an energy gap between the band of electronic states being excited and the ground state.

A pronounced even/odd alternation in the photoionization cross-section of Ga_xAs_y clusters, was observed depending only on the total number of atoms in the cluster.[2] Clusters in the 5-21 atom range with an odd number of atoms were one-photon ionized by low energy ArF excimer laser pulses (6.4 ev) [Fig. 2a]. The spectrum in Fig. 2b was taken with an F_2 excimer laser at low fluence, showing that all

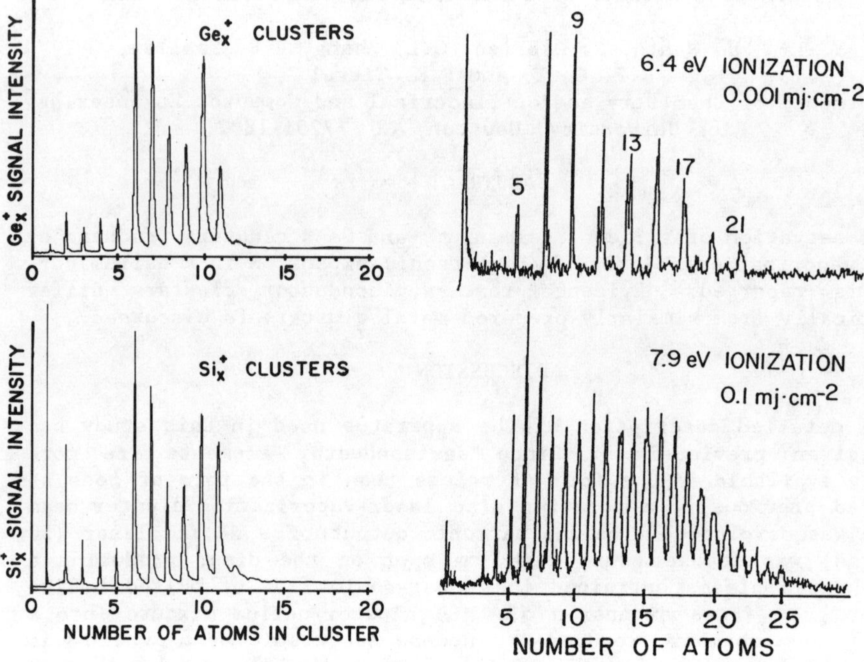

Fig. 1. Time-of-flight mass spectrum of Ge and Si clusters.

Fig. 2. Time-of-flight mass spectrum of Ga_xAs_y clusters ionized with ArF and F_2 excimer lasers.

clusters can be readily one-photon ionized at 7.9 ev. This even/odd alternation in ionization properties of clusters supports the view that the even clusters have fully paired singlet ground states with no dangling bonds, while the odd clusters have unpaired electrons which are more weakly bound.

ACKNOWLEDGMENTS

This work was supported by AROD Contract DAAG 29-85-K-0029 and grants C-071, C-586, and C-689 of the Robert A. Welch Foundation using supersonic beam equipment and lasers supported by the Department of Energy, Division of Chemical Sciences, and the National Science Foundation.

REFERENCES

1. J.R. Heath, Y. Liu, S.C. O'Brien, Q.L. Zhang, R.F. Curl, R.E. Smalley, and F.K. Tittel, J. Chem. Phys., 83, 5520 (1985).
2. S.C. O'Brien, Y. Liu, Q.L. Zhang, J.R. Heath, F.K. Tittel, R.F. Curl, and R.E. Smalley, accepted J. Chem. Phys., 1986.

STATE-RESOLVED PHOTOFRAGMENTATION OF OCS AND CS_2

J.W. Hepburn and I.M. Waller,
Center for Molecular Beams and Laser Chemistry,
University of Waterloo, Waterloo, Ontario N2L 3G1 Canada.

G.E. Hall, N. Sivakumar, I. Burak*, and P.L. Houston,
Department of Chemistry, Cornell University, Ithaca, N.Y. 14853

ABSTRACT

The detailed photodissociation dynamics of OCS and CS_2 have been studied by molecular beam photofragment spectroscopy in which the products of OCS and CS_2 photodissociation are probed by vacuum ultraviolet laser induced fluorescence.

INTRODUCTION

State-resolved photofragmentation of triatomic molecules is currently receiving a lot of attention[1-3], as very detailed experimental and theoretical work has become possible for these systems. Experimentally, one of the significant limitations has been the difficulty of detecting in a state-sensitive manner small densities of the products from these dissociations. The use of tunable coherent vacuum ultraviolet[4] (VUV) for laser-induced fluorescence (LIF) product detection represents a significant advance in the technology of these experiments, as the use of VUV makes it possible to detect such important photofragments as CO, H_2 and a wide range of atoms[4]. As two examples of the experimental power of coherent VUV, we wish to report here some recent results on photofragment spectroscopy of OCS and CS_2, in which the CO fragments from OCS and the S product from both OCS and CS_2 were detected by VUV LIF.

Both OCS and CS_2 are related in that they are 16 electron molecules, and have a similar orbital structure[5]. The excitation of OCS at 222 nm has been assigned to a weak $^1\Delta \leftarrow {}^1\Sigma^+$ transition from the linear ground state to a bent excited state[5]. In CS_2, excitation at 193 nm excites the strong $^1\Sigma_u \leftarrow {}^1\Sigma_g$ transition to a bent excited state[5]. In both cases there is uncertainty about the $S(^1D) : S(^3P)$ branching ratio. Secondary photochemistry studies on OCS suggest a 0.75 : 0.25 branching[6], and experiments on CS_2 are in complete disagreement about the branching ratio[7].

In this paper, because of space limitations, only a portion of our data on OCS and CS_2 will be discussed, and more extensive papers on both studies will be published separately. As well, preliminary results on OCS have recently been published[8], and the discussion here is an extension of these published results.

*Permanent Address: Dept. of Chemistry, University of Tel Aviv, Israel.

EXPERIMENTAL

In order to ensure collision free conditions, pulsed molecular beams of OCS or CS_2 in He were photodissociated by 222 nm (KrCl) or 193 nm (ArF) light respectively from an excimer laser, and the product CO or S atoms probed by LIF, using coherent light tunable in the 145 nm to 150 nm range. The experimental technique is described in more detail in earlier papers on OCS[8] and glyoxal[9]. In the case of CS_2, a much lower density beam was used: 1% CS_2 in 1200 torr He, expanded through a 150 micron nozzle, with the laser crossing 4 cm downstream. To avoid problems with photochemistry of the CS product[10], the excimer laser was attenuated to $\sim$ 200μJ/pulse. The S atom signal was found to be linear in 193 nm pulse energy over the range 5 to 800 μJ/pulse.

The experiments on OCS were carried out at Cornell, using the apparatus described previously[8,9], while the CS_2 experiments were done at Waterloo on a similar machine, with the only major difference being that at Waterloo, the VUV laser system used two Lambda Physik 2002E dye lasers pumped by a Quanta-Ray DCR2A Nd:YAG laser, in place of the excimer pumped dye lasers used at Cornell. The VUV output and linewidth ($0.2\ cm^{-1}$ FWHM) of the two different laser systems is essentially the same and the same detection system is used. The linewidth of the VUV probe was narrow enough to do doppler spectroscopy on both CO and S atom products.

RESULTS AND DISCUSSION

The relative yield of $S(^3P):S(^1D)$ photofragments was measured by comparing the relative intensities of the resonance lines corresponding to the two S atom products, the $^1P_1 - {}^1D_2$ line at 1448 Å and the $^3D_J - {}^3P_J$ lines around 1474 Å. The relative line strengths for these transitions were measured by us in separate experiments[11] and were in agreement with the (fairly uncertain) literature values. For the OCS dissociation at 222 nm, the relative yields of $S(^1D):S(^3P)$ were 0.85 (±0.05):0.15. In the case of CS_2 fragmentation at 193 nm, the relative yields of $S(^1D):S(^3P)$ were 0.30:0.70 (±0.05).

The most interesting findings we wish to report here are the result of doppler spectroscopy on the products of both photodissociations. In the case of OCS, when the CO product from the 222 nm dissociation was probed by VUV LIF, (excitation of the $A^1\Pi \leftarrow X^1\Sigma^+$ bands near 1470 Å) it was found to be 95% in v = 0 and highly rotationally excited[8]. The rotational distribution of the CO product is shown in figure 1 and it is striking in that it shows a very high degree of rotational excitation in a very narrow bimodal distribution. In light of the observed branching between the singlet and triplet channels, this bimodal P(J) distribution could be attributed to the fact that the J = 55 peak was due to the $S(^1D)$ channel and the J = 66 peak to the $S(^3P)$ channel.

In order to test this hypothesis, doppler spectroscopy on the CO product was carried out. Because of the increased available energy in the $S(^3P)$+CO channel, the translational energy for $S(^3P)$ + CO(v=0, J=67) products would actually be significantly larger than

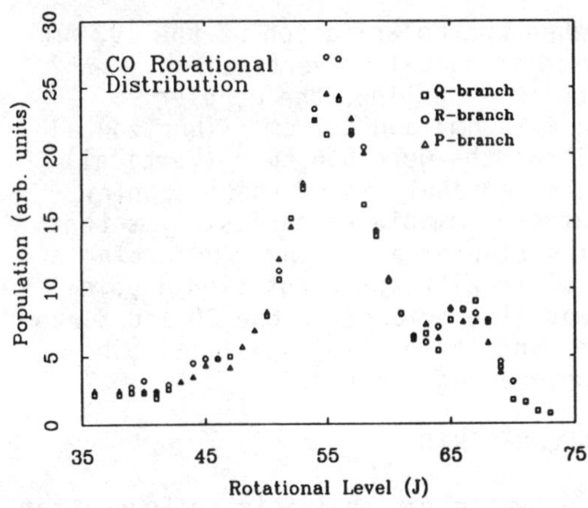

Figure 1
Measured rotational distribution for CO products from 222 nm dissociation of OCS

for S(^{1}D) + CO (v=0, J=56) products: 31.7 kcal/mole (J=67) vs. 12.8 kcal/mole (J=56). In fact, when the linewidths for the Q(56) and and Q(67) lines of the (3,0) band were measured, the Q(67) line was found to be significantly narrower than the Q(56) line. These results are shown in figure 2. The linewidths correspond to both products being from the S(^{1}D) + CO channel, with E_{trans} = 5.3 kcal/mole for J = 67. This means that peaks in the P(J) distribution shown in figure 1 are from the same channel, S(^{1}D) + CO, and thus the bimodal distribution is a manifestation of a microscopic branching in the S(^{1}D)+CO product channel.

In the case of CS_2 dissociation, there has been a lot of work done on photofragment translational spectroscopy of the products[7,12], with the problem always being that it is not clear how to partition the observed product time of flight spectra between the CS(v,J) + S(^{3}P) and CS(v,J) + S(^{1}D) channels. We have provided the data necessary for this partitioning by directly measuring the S(^{1}D):S(^{3}P) branching ratio, and also by measuring the doppler lineshape for both channels. These results will be reported in more detail separately[13]. We also found a significant difference in both

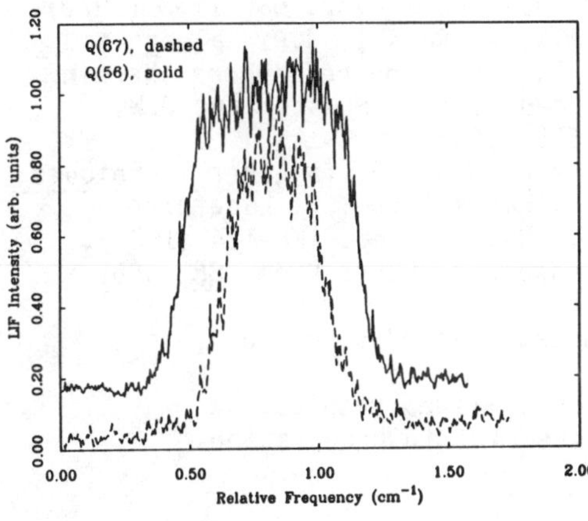

Figure 2
Doppler lineshape for the Q(67) and Q(56) lines of the CO product from OCS.

the $S(^1D)$ and $S(^3P)$ linewidths when the polarization of the 193 nm photolysis laser was rotated from horizontal to vertical polarization. In the case of the ^1P - ^1D line, the doppler linewidths were 0.4 cm^{-1} (vertical 193nm) and 0.6 cm^{-1} (horizontal 193nm). For the $^3D_3$ - $^3P_2$ the linewidths were 0.6 cm^{-1} (vertical) and 0.8 cm^{-1} (horizontal). A detailed analysis of these doppler profiles is in preparation[13], but the immediate conclusion is that there is an anisotropy in the dissociation at 193 nm, indicating a short-lived ($\lesssim$1 psec) excited state with the transition dipole along the CS_2 axis. Anisotropy was also evident in the CO and S atom lines from OCS photodissociation, and these results will also be discussed in more detail in a forthcoming paper[11].

ACKNOWLEDGEMENTS

This work has been supported by the National Science Foundation (CHE-8314146), the Environmental Protection Agency (R811010-01-1) and NSERC (Canada) PLH and JWH are grateful for a NATO award (637/83) which facilitated their collaboration. IMW wishes to acknowledge the support of NSERC in the form of a graduate scholarship. Much of this research was performed in the Facility for Laser Spectroscopy of the Department of Chemistry at Cornell University using lasers provided, in part, by a Department of Defense Instrumentation Grant (AFOSR-83-0279). The Facility gratefully acknowledges support from the Dow Chemical Company Foundation. We are grateful to J. Frey for communicating the results on CS_2 prior to publication.

REFERENCES

1. R. Bersohn, J. Phys. Chem., 88, 5145 (1984).
2. S.R. Leone, Adv. Chem. Phys., 50, 255 (1982).
3. M. Shapiro and R. Bersohn, Ann. Rev. Phys. Chem., 33, 409 (1982).
4. J.W. Hepburn, Israel J. Chem., 24, 273 (1984).
5. J.W. Rabalais, J.M. McDonald, V. Scherr and S.P. McGlynn, Chem. Rev., 71, 73 (1971).
6. W.H. Breckenridge and H. Taube, J. Chem. Phys., 53, 1750 (1970).
7. V.R. McCrary, R. Lu, D. Zakheim, J.A. Russel, J.P. Halpern and W.M. Jackson, J. Chem. Phys., 83, 3481, and references therein.
8. N. Sivakumar, I. Burak, W.-Y. Cheung, P.L. Houston and J.W. Hepburn, J. Phys. Chem., 89, 3609 (1985).
9. J.W. Hepburn, N. Sivakumar and P.L. Houston, in Laser Techniques in the Extreme Ultraviolet, S.E. Harris and T.B. Lucatorto, eds., AIP Conference Proceedings No. 119, pp. 126-134 (1984).
10. G. Dornhofer, W. Hack and W. Langel, J. Phys. Chem. 88, 3060 (1984).
11. N. Sivakumar, I. Burak, P.L. Houston and J.W. Hepburn, in preparation.
12. J. Frey, A. Wodtke, Y.T. Lee, unpublished results.
13. I. Waller and J.W. Hepburn, manuscript in preparation.

OPTICAL-SELECTION IN THE DOUBLE RESONANT TWO-PHOTON PHOTODISSOCIATION OF NO_2

Laurence Bigio and Edward R. Grant
Department of Chemistry, Baker Laboratory, Cornell University, Ithaca, New York 14853

ABSTRACT

Optical-selection is shown to be an effective means to isolate specific initial states for subsequent bound-continuum photodissociation. States in NO_2's dense visible system are selected and then dissociated near the threshold for production of NO + O(^{1}D) by means of variable wavelength two-photon photolysis. Product state distributions confirm the excited oxygen pathway as the dominant route for production of NO(v=0). Rotational and Λ-doublet state populations oscillate in a manner characteristic of the selected intermediate state rotational angular momentum as labeled by separate optical-optical double resonance absorption spectroscopy.

INTRODUCTION

Quantum theories of photodissociation, particularly triatomic ABC → AB + C fragmentation, compute product (rovibrational) state distributions as the optical projection of a specified initial state on the manifold of scattering eigenstates.[1] In a Franck-Condon limit, this projection effectively expands the initial ABC triatomic state in a basis of AB + C atomic and molecular final-state wavefunctions. Interferences can be expected in this projection, and quantum-mechanical models have generally predicted oscillating rotational state distributions for near-threshold photodissociation.[2-4]

The results of conventional photodissociation experiments fail to conform to such predictions.[5-8] For photon energies substantially in excess of the thermochemical threshold for fragmentation, obscured interferences can be attributed to the exit-channel contribution of a highly oscillatory wavefunction in the fragment recoil coordinate, which tends to smear Franck-Condon biases. However, a more substantial source of averaging comes from the nature of the photoexcitation step itself. Bound-continuum transitions promote all members of an initial state distribution to dissociation with almost equal probability. Thus for a triatomic, even under pulsed-jet conditions, the measured state distribution represents the incoherent sum of contributions from numerous initial states.

One thus clearly needs better state specificity in bound-continuum optical preparation. As demonstrated by results in our laboratory[9] and elsewhere,[10] this is readily achieved by inserting an intervening step of bound-bound optical selection in what then becomes an overall process of two-photon photodissociation.

Elegant experiments of Andresen and coworkers[10] use an infra-

0094-243X/86/1460501-6$3.00

red transition to isolate initial rotational quantum states in the 193 nm dissociation of H_2O. We use transitions in the visible to promote the optical-optical two-photon photodissociation of NO_2. As observed for H_2O, we find the final state dynamics for NO + $O(^1D)$ production to be characterized near threshold by oscillating rotational and Λ-doublet state distributions that depend sensitively on the rotational quantum state selected by the first-photon bound-bound preparation step. The present work shows a sample of these dynamics for intermediate states whose |J> assignments are confirmed by separate optical-optical double resonance absorption spectroscopy.

EXPERIMENTAL

NO_2 in a pulsed jet is two-photon photodissociated by the focussed blue-green output of a Lambda Physik EMG-150ES excimer pumped FL2002E dye laser system. A second FL2002 dye laser, frequency doubled to 225 nm and focussed, probes product NO by one-photon resonant two-photon ionization spectroscopy. NO^+ ions so formed are collected and mass analyzed by an Extranuclear quadrupole mass spectrometer and registered by a Galileo channeltron.

Excited states of NO_2 populated by absorption of the first visible photon are characterized by ionization-detected optical-optical double resonance spectroscopy. The same dual dye laser/mass spectrometer arrangement is used. The first laser is set unfocussed to a frequency in the visible matching that for which data on two-photon photodissociation has been obtained. The second dye laser, also unfocussed, is tuned to resonance between the optically-prepared excited state and the first excited bending level of the 3pσ $^2\Sigma^+_u$ Rydberg state of NO_2. Scanned across this band system, the UV laser samples a pattern of double-resonant rotational structure which appears (at m/e = 46) as the one-photon resonant two-photon ionization spectrum of the optically prepared excited state. This structure labels the angular momentum of the intermediate state. To confirm these assignments, the UV frequency can be locked on a transition from an optical level to the Rydberg state. A scan of the visible laser then produces a pattern of lines with the rotational spacing of the ground state levels that connect with that intermediate state.

RESULTS AND DISCUSSION

Two-photon dissociation of NO_2 in the region of 487 nm produces NO predominantly in v=0, from the reaction

$$NO_2 \xrightarrow{2h\nu} NO(\tilde{X}\ ^2\Pi) + O(^1D).$$

This reaction pathway, particularly the generation of electronically excited atomic oxygen, is confirmed by the behavior of the cross section for NO v,J-state production near the thermochemical threshold for this reaction.

A sequence of photolysis laser scans for various monitored NO product states shows a threshold for reaction, evident for each rotational level, that moves up appropriately with increasing energy demand of the NO final state.[9]

The structure exhibited is that of the first photon absorption spectrum. This is confirmed by comparison with the laser induced fluorescence excitation spectrum. Strong features in particular can be identified with specific rotational transitions, as first noted by Levy and coworkers.[11] Their assignments are confirmed for these transitions by Rydberg optical-optical double resonance spectroscopy.

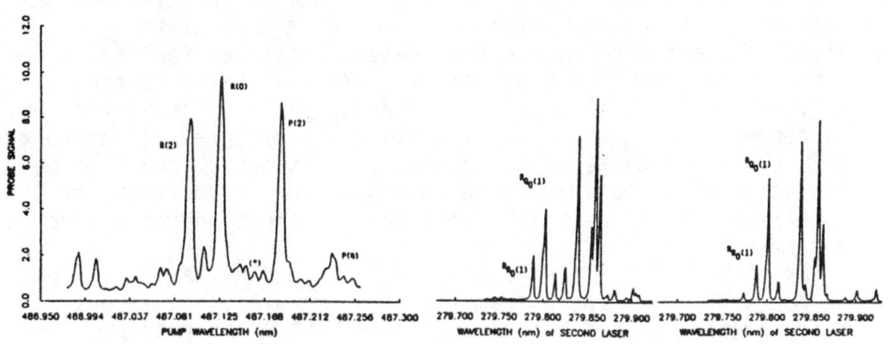

Figure 1 Figure 2

Figure 1 shows a scan of the photolysis spectrum for production of NO(J = 1 1/2) in the region just above threshold. Figure 2 shows OODR spectra of the features labeled in Figure 1 as P(2) and R(0). Portions of these OODR spectra are readily assigned on the basis of the known rotational structure of the 3pσ Rydberg state. These assignable portions are identical, indicating that the resonant contribution to the intermediate state in each case is the same, N=1. Similar analysis establishes that the two outer-lying transitions, labeled P(4) and R(2) also share a common intermediate state N=3.

Figure 3 shows distributions over NO rotational states populated by two-photon photodissociation of NO_2 at frequencies corresponding to the peaks labeled R(0) and P(2) in Figure 1. Note first that these distributions are extraordinarily cold. Photon energies here are just 50 cm^{-1} above threshold for the $O(^1D)$ channel, but 16000 cm^{-1} above that for production of $O(^3P)$. Fragments observed are vibrationally cold, and no rotational levels above J=5 1/2 are detected. This trend persists with increasing photolysis energy, confirming the photolysis pathway as that producing electronically excited atomic oxygen.

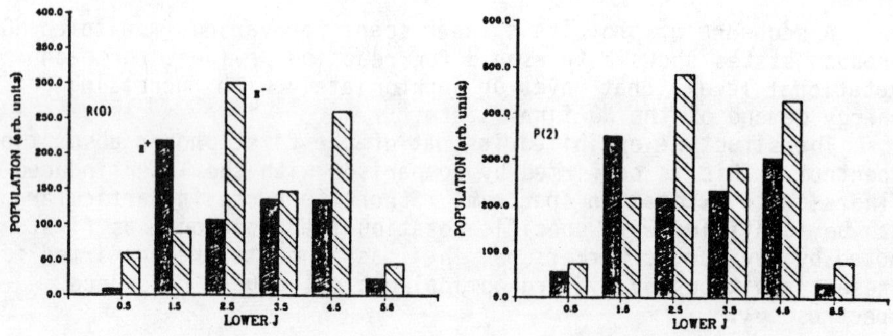

Figure 3

Also interesting is the fact that rotational populations oscillate from J to J, and within the same J, Λ-doublet to Λ-doublet. These trends mirror the behavior observed for state-selected photolysis of H_2O by Andresen and coworkers. Especially significant is the fact that two different photolysis wavelengths, which happened to be tuned through two different transitions to the same (N=1) intermediate state, produce, in essence, identical detailed state distributions. The same behavior is observed for the R(2) P(4) pair of transitions which for the first photon absorption both terminate on N=3.

Distributions for intermediate N=1 differ only a little from those for N=3. Patterns are similar and the average J is only slightly higher for the higher angular momentum intermediate state. What sets these results apart, and lends particular significance to the process of optical-selection evidently at work here, is the fact that at all other wavelengths sampled in the narrow region around these features, the dynamics are very different. Typical are results obtained for the photolysis wavelength marked by the (*) in Figure 1, as shown below in Figure 4. Characteristically for this off-resonance photolysis, product J oscillations are less severe and a more uniform Λ=doublet preference emerges. These trends become more firmly established and even less sensitive to wavelength as the photolysis laser is tuned to the blue.

Figure 4

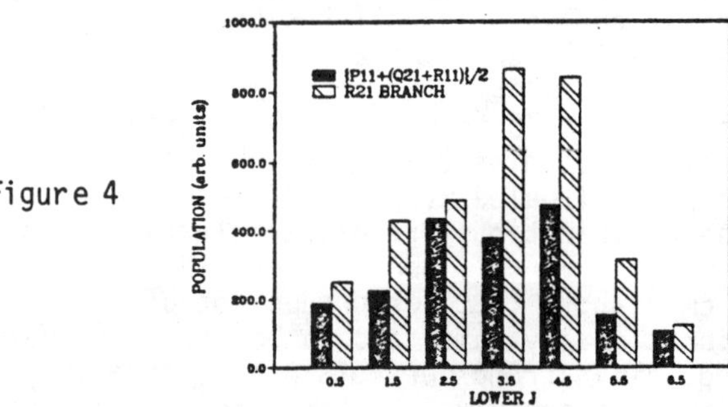

Two characteristics are likely to differentiate on- and off-resonant two-photon photolysis. Sharp strong structure on resonance isolates an ensemble of rotationally cold molecules for dissociation very near threshold. Off-resonance, the structure is congested and must include higher angular momentum states. This both superimposes state-detailed distributions and adds energy: Higher angular momentum intermediate states must be reached from higher rotational ground states. Added to the photon energy this places prepared molecules at higher total energy in the continuum above threshold. These factors are that much more pronounced at higher photon energies off-resonance.

Thus we conclude that optical-selection in two-photon bound-continuum excitation offers a means to increase resolution in state-to-state photodissociation. For NO_2 states of low angular momentum dissociated near threshold, quantum interference effects are evident in distributions over final rotational electronic degrees of freedom. These effects become less distinctive as superpositions of initial states are prepared and/or the total photolysis energy increasingly exceeds threshold.

ACKNOWLEDGMENT

This work was supported by the U. S. Army Research Office. Acknowledgment is gratefully made to the Department of Defense for a DOD-University Research Instrumentation Grant.

REFERENCES

1. M. Shapiro and R. Bersohn, Ann. Rev. Phys. Chem. 33, 409 (1982); G. G. Balint-Kurti and M. Shapiro, Chem. Phys. 61, 137 (1981); G. G. Balint-Kurti and M. Shapiro, Adv. Chem. Phys. 60, 403 (1985).

2. K. F. Freed and Y. B. Band, in Excited States, E. C. Lim, ed., Academic:New York, 1977; Vol. 3, p. 109; Y. B. Band and K. F. Freed, J. Chem. Phys. 63, 3382 (1975); 67, 1462 (1977); 68, 1292 (1978); M. D. Morse, K. F. Freed and Y. B. Band, ibid. 70, 3604, 3620 (1979); M. D. Morse and K. F. Freed, ibid. 74, 4395 (1981); 78, 6045 (1983).

3. R. W. Heather and J. C. Light, J. Chem. Phys. 79, 147 (1983); 78, 5513 (1983).

4. R. Schinke, V. Engel, P. Andresen, D. Häusler and G. G. Balint-Kurti, Phys. Rev. Lett. 55, 1180 (1985).

5. See for example: W. G. Hawkins and P. L. Houston, J. Phys. Chem. 76, 729 (1982).

6. H. Zacharias, M. Geilhaupt, K. Meier, and K. H. Welge, J. Chem. Phys. 74, 218 (1981).

7. R. Vasudeo, R. N. Zare and R. N. Dixon, J. Chem. Phys. 80, 4863 (1984).

8. I. Nadler, H. Reisler, M. Noble and C. Wittig, Chem. Phys. Lett. 108, 115 (1984); I. Nadler, M. Noble, H. Reisler and C. Wittig, J. Chem. Phys. 82, 2608 (1985).

9. L. Bigio and E. R. Grant, J. Phys. Chem. 90, xxx (1986).

10. P. Andresen, V. Beushausen, D. Häusler, H. W. Lülf and E. W. Rothe, J. Chem. Phys. 83, 1429 (1985).

11. R. E. Smalley, L. Wharton and D. H. Levy, J. Chem. Phys. 63, 4977 (1975).

Multiphoton Ionization of Metal β-Diketonates and Metal Tetraphenylporphyrin Complexes: Ligand Dissociation vs. Molecular Ionization

J. B. Morris and M. V. Johnston
Department of Chemistry and Cooperative Institute for Research in Environmental Sciences, University of Colorado, Boulder, CO. 80309

ABSTRACT

We have studied multiphoton ionization mass spectra of various metal β-diketonate and metal tetraphenylporphyrin complexes. The excitation source for this work was the fourth harmonic of a Nd:YAG laser at 266 nm (5 ns pulses) which is directed into the ionization region of a time-of-flight mass spectrometer. β-diketonate complexes are found to exhibit dissociation prior to ionization while the tetraphenylporphyrin complexes undergo ionization prior to fragmentation. The observed behavior of these complexes is discussed in terms of their ligand properties.

INTRODUCTION

The non-linear photochemistry of organometallic compounds has generated much interest in the past few years.[1] In this work, we consider two classes of organometallic complexes which have not been studied previously in multiphoton ionization (MPI), the metal β-diketonates and the metal tetraphenylporphyrins (tpp). The goal of this study is to determine how the dissociation dynamics of organometallic compounds in MPI are influenced by their ligand properties.

EXPERIMENTAL

In these experiments, the fourth harmonic of a pulsed Nd:YAG laser is focused into the ionization region of a time-of-flight mass spectrometer. The samples are introduced via a 1/2" direct insertion probe. About 10 mg of sample is desorbed into the ionization region. The signal obtained is then stored in a computer via a 100 Mhz transient digitizer. The experimental arrangement is described in greater detail in reference 2.

RESULTS AND DISCUSSION

The metal β-diketonate complexes studied were complexes of acetylacetonate, trifluoroacetylacetonate, hexafluoroacetylacetonate, 2,2,6,6-tetramethyl-3,5-heptanedionate, and 6,6,7,7,8,8,8-heptafluoro-2,2-dimethyl-3,5-octanedionate with the Lanthanides and first row transition metals. The MPI mass spectra of these complexes show that ligand dissociation preceeds ionization. In most cases, the bare metal ion is the base peak in the mass spectrum. No large organometallic ions are observed. Along with the bare

0094-243X/86/1460507-2$3.00

metal ion, over half of the compounds studied produce metal oxide and/or metal fluoride ions. A summary of the ions produced is given in reference 2.

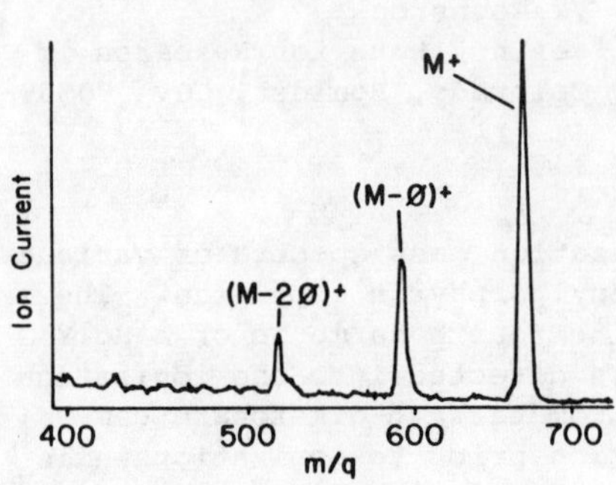

Fig. 1 MPI Mass Spectrum of Cu(tpp). Laser Intensity = 30 MW/cm^2.

The metal tetraphenylporphyrin complexes studied were Co(tpp), Ni(tpp) and Cu(tpp). These compounds are efficiently ionized without ligand dissociation. Figure 1 is a MPI mass spectrum of Cu(tpp) in the high mass region. Both the molecular ion and fragment ions which result from the loss of one and two phenyl rings are observed. As the laser intensity is decreased, the fragment ions disappear, leaving only the molecular ion at laser intensities of less than ca. 5 MW/cm^2. Similar behavior is observed for Co(tpp) and Ni(tpp). A power dependence of two was measured for the molecular ion of Cu(tpp) in the low laser intensity regime, indicating a two photon ionization.

This change in ionization mechanism can be attributed to an increase in the ligand dissociation potential in going from a bidentate β-diketonate ligand to the tetradentate tpp ligand. With the β-diketonates, the ligand dissociation potential is less than the energy of one 266 nm photon, allowing a dissociative charge transfer reaction to occur. With the tetradentate ligand, the energy required to break four metal-nitrogen bonds is greater than the energy of one photon, permitting the absorption of a second photon by the intact complex.

ACKNOWLEDGMENTS

The authors would like to thank Professor R.E. Sievers and coworkers for providing many of the metal β-diketonate complexes used in this study. This work was supported by the National Science Foundation under grant number CHE-8308049.

REFERENCES

1. A. Gedanken, M.B. Robin, N.A. Kuebler, J. Phys. Chem. 86, 4096 (1982).
2. J.B. Morris, M.V. Johnston, J. Phys. Chem. 89, 5399 (1985).

An Overview of Laser Effects on Collision Dynamics

Paul L. DeVries
Department of Physics, Miami University, Oxford, OH 45056

The presence of laser radiation can affect collision dynamics in many ways,[1] not the least of which are initial state selection of the colliding atoms and state analysis of the products of the collision. However, we restrict our attention to those processes in which the radiation is interacting with the quasimolecule formed as the collision unfolds. Such processes were first studied by Gudzenko and Yakovlenko[2]; subsequently, various effects have been investigated by numerous investigators.

Semiclassical methods have always been popular in this field, due in part to the relative ease with which they can be implemented.[3] The Schrodinger equation is expanded in a suitable (molecular) basis set, with time dependent coefficients describing the amplitude of each state. The method becomes semiclassical when a relationship between space and time is established: the specification of a classical path. The radiative interaction is easily described in terms of an oscillating electric field. The evolution of the system is then determined by numerically integrating the equations along the trajectory, often taken to be a rectilinear path.

The quantum theory of these processes[4] is considerably more tedious, due to the presence of many interacting channels. In particular, the total angular momentum J of the field-free system is not a good quantum number, so that a basis set expansion in J will necessarily be rather large and the size of the set will be dependent upon the intensity of the laser![5] However, the theory has its advantages. For example, the spatial and radiative degrees of freedom are treated on an equal footing. This casts the present problem into the same mold as field-free ones, and hence solvable by the same techniques and subject to the same interpretations. Thus expertise and intuition developed in studying field-free collisions is immediately applicable to the problem of laser effected collision processes.

An example of the latter is the role that nonadiabatic coupling plays in laser-induced processes. In describing the process by which radiation is absorbed by two colliding atoms, the following classical picture is often invoked: as the colliding atoms approach on a ground electronic surface, they arrive at an internuclear separation such that they can resonantly absorb the incident radiation and make the transition to the excited electronic state. This description indicates that the absorption process is localized. A more refined description resorts to a "shifting" of the field-free potential energy curves by the energy of a photon; if the Hamiltonian describing the quasimolecule, its interaction with the radiation field, and the radiation itself is then diagonalized, the "dressed states" picture emerges. Where the shifted curves exhibited an actual curve crossing, corresponding to the absorption of a photon, the dressed states manifest an avoided crossing. This (correctly) conveys the idea that the absorption of the photon is not necessarily localized, but introduces the possibility of a severe misinterpretation. Naively, one might assume that all motion of the atoms is along the adiabatic curves, but this is not so. The nonadiabatic coupling, which couples different adiabatic states, must be included in any such analysis; this is, of course, well understood by dynamicists, although sometimes overlooked by others. The magnitude of the interaction, and hence the degree of "avoidedness", increases as the intensity of the radiation increases. Thus, in the strong field limit, nonadiabatic coupling diminishes and the atoms do tend to follow the adiabatic curves. But as lesser fields are considered, the nonadiabatic coupling increases, and the atoms depart from this strict adiabatic following. Certainly, in the limit of zero field, the nonadiabatic coupling is sufficiently large to ensure that <u>every</u> quasimolecule makes the transition from one adiabatic state to the other; they <u>must</u>, since in this limit there is no radiative coupling and the system remains in its diabatic state! This complexity involving nonadiabatic coupling has not fully been appreciated in much of the work to date -- a notable exception is in the recent work of Chu.[6] This work indicates that strongly nonadiabatic behavior persists for fields as high as 4 TW/cm^2.

There are several general classifications of laser effected collision processes: the first is radiative collisions (RC). RC processes occur when both atoms involved in the encounter change internal state. Clearly, the coupling for such transitions must vanish asymptotically. Falcone <u>et al</u>[7] made the first observation of any laser-induced collision process in the Sr-Ca RC system. In optical collisions (OC), only one atom need change its internal state. This might seem to be a trivial distinction, but it is quite significant since the radiative coupling need not vanish asymptotically. Thus optical collisions can occur at large internuclear separations, with correspondingly large collision cross sections.

A significant finding was made in the semiclassical modeling of the optical collision of Sr with Ar.[8] The ground state is coupled to one of the excited states, which asymptotically separates to a P-state of Sr. When all three excited states correlating to that atomic limit were included in a four state calculation, qualitative differences with the simpler two state model (having one excited state) were discovered. In particular, the four state model exhibits significant total excitation cross sections for a wide range of Rabi frequencies, whereas the two state model exhibits significant excitation for only a narrow band of frequencies. This places the commonly used two state model in considerable question and renders the semiclassical approach less attractive than before.

Since the radiative coupling doesn't vanish asymptotically in optical collisions, it is possible that the light which is absorbed during the collision can be subsequently emitted by the separated atom. This "collisional redistribution" of radiation has been of much theoretical and experimental interest, as it represents a new handle on collision dynamics. It is also related to other topics of more traditional interest to collision dynamicists, such as coherence, correlation, relaxation, and collisional depolarization, and so tends to unify the field. The work of Cooper and coworkers,[9] primarily involving collisions of alkaline earths with rare gases, is particularly noteworthy, while Havey <u>et al</u>[10] (experimental) and Kulander <u>et al</u>[11] (theory) have significantly contributed to our understanding of processes involving the alkaline metal - rare gas systems.

Another process that has received considerable attention of late, due in part to its possible application towards an x-ray laser, is laser-assisted charge transfer. Experiments in Ca collisions with Sr have been performed,[12] and helium[6] and lithium[13] ions colliding with hydrogen have been investigated theoretically. A general result of these studies is that reasonable cross sections require very large laser intensities -- but given the potential importance of the process, further work in this area is warranted.

A brief overview such as this does not allow for discussion of all interesting and important processes. In particular, the possibility of free electrons colliding with atoms or molecules in the presence of radiation allows for a wide variety of processes that are deserving of considerable attention in their own right, but are beyond the scope of this article.

Typical collision processes have a duration of a few picoseconds, so that pulsed laser radiation of nanosecond duration can be thought of as being constant during the course of the collision. This is the operating assumption of all theoretical investigations to date. However, experimentalists have succeeded in producing laser pulses as short as a few femtoseconds -- in this amount of time, the atoms have not moved appreciably, and the assumption that the radiation intensity is constant for the entire collision is clearly invalid. Recently, Sizer and Raymer[14] have experimentally verified that the duration of the laser pulse can have a significant effect of the observed excited state population in optical collisions. Perhaps more importantly, simple calculations support the contention that potentially dramatic effects are being "washed out" by their experimental setup, leaving open the possibility that other arrangements might lead to even more interesting results. Theoretical investigations are being initiated in our laboratory to consider such duration effects.

The study of laser effects on collision dynamics, as a field, has matured substantially in the decade and a half of its existence. No longer are we interested in "will there be an effect?" -- by now we have accumulated enough information that a reliable intuition has been developed, and we know under what circumstances effects are likely to occur. Rather, we want to know how to use these effects -- to build an X-ray laser, for example -- or to investigate effects under new conditions, such as ultra short pulses. Let's hope that the next decade and a half will be as exciting as the last.

ACKNOWLEDGMENTS

This work is supported in part by the National Science Foundation under grant number PHY-85-006679. Acknowledgment is also made to the Donors of the Petroluem Research Fund, Administered by the American Chemical Society, for partial support of this research.

REFERENCES

1. T.F. George, J. Phys. Chem. 86, 10 (1982).
2. L.I. Gudzenko and S.I. Yakovlenko, Sov. Phys. JETP 35, 877 (1972).
3. See, for example, reference 8.
4. An excellent review of the quantum mechanical theory is presented by F.H. Mies in Theoretical Chemistry: Advances and Perspectives, Vol. 6B, ed. D. Henderson (Academic Press, N.Y., 1981), p. 127.
5. P.L. DeVries and T.F. George, Mol. Phys. 36, 151 (1978); P.S. Julienne and F.H. Mies, Phys. Rev. A 25, 3399 (1982).
6. T.S. Ho, S.I. Chu and C. Laughlin, J. Chem. Phys. 81, 788 (1984).
7. R.W. Falcone, W.R. Green, J.C. White, J.F. Young and S.E. Harris, Phys. Rev. A 15, 1333 (1977).
8. J. Light and A. Szoke, Phys. Rev. A 18, 1363 (1978).
9. K. Burnett, J. Cooper, P.D. Kleiber and A. Ben-Reuven, Phys. Rev. A 25, 1345 (1982).
10. M.D. Havey, G.E. Copeland and W.J. Wang, Phys. Rev. Lett. 50, 1767 (1983).
11. K.C. Kulander and F. Rebentrost, Phys. Rev. Lett. 51, 1262 (1983).
12. M.D. Wright, D.M. O'Brien, J.F. Young and S.E. Harris, Phys. Rev. A 24, 1750 (1981).
13. T.S. Ho, C. Laughlin and S.I. Chu, Phys. Rev. A 32, 122 (1985).
14. See the article by Sizer and Raymer in these Proceedings.

OBSERVATION OF CURVE CROSSINGS IN RADIATIVE COLLISIONS OF XENON ATOMS

J. W. Keto and N. Böwering
Department of Physics, University of Texas at Austin, Austin, Tx. 78712

ABSTRACT

In experiments described here, xenon atoms are excited to $5p^5 6p$ during a collision by two-photon laser excitation. We measure the probability for collision pairs to dissociate to specific product channels following the half-collision on the excited potential curve. Within the quasistatic range of the absorption line, the initial internuclear separation for this half-collision can be controlled by the laser detuning. As we excite pairs with smaller internuclear separation, we observe a large decrease in the probability for formation of the diabatic dissociative channel. A Landau-Zener model for the first half-collision is in agreement with the data.

INTRODUCTION

In recent years lasers have stimulated interest in photochemistry because of the greater variety and selectivity of reactions which can be studied. An example of simple atomic excitation transfer is given by

$$A + B + h\nu \rightarrow A^* + B \rightarrow A + B^*$$

When using the greater intensities of a laser, less probable reactions can be induced in systems where dipole transitions occur only during the collision and neither A nor B resonate with the laser independently. Cases where both reactants A and B change state have been termed "radiative collisions". Such reactions were first described by Gudzenko and Yakovlenko[1] and studied experimentally by the group of Harris[2]. Gallagher and Holstein[3] have shown how these reactions can be described as dipole-induced transitions between $(AB)^*$ molecular potential surfaces.

In our experiments, $Xe^*(5p^5 6p)$ is excited at various positions on the excited potential curve by two-photon transitions

$$Xe(5p^6) + Xe(5p^6) + 2h\nu \rightarrow Xe^*(5p^5 6p) + Xe(5p^6)$$

The inset of Fig. 1 shows schematically the relevant molecular potentials and the laser-induced transitions observed. The pressure broadening was studied previously to extract bound excited state potentials at intermediate and long range using both unified phase-shift and quasistatic calculations[4,5]. In the quasistatic theory, the shape of the line wing is mapped into the difference potential (difference between ground and excited state potential) by applying the Franck-Condon principle to atom-perturber pairs and taking into account the distribution of nearest neighbors. Thus different frequencies in the line wing correspond to excitation of the excited atom and its perturber at different internuclear separations.

We are studing quenching[6] of excited states of Xe 6p as a function of laser detuning. Avoided crossings of these states occur at long range with the repulsive $0g^+$ and 1g symmetries of Xe 6s' $[1/2]_1$ as is shown in Fig. 1 for the 6p $[1/2]_0$ $0g^+$ state. According to the two-photon selection rules, excitation of this state is allowed while excitation of 6s' $[1/2]_1$ $0g^+$ is forbidden. The line profile is thus expected to be primarily described by the diabatic potential of the 6p $[1/2]_0$ $0g^+$ state. Upon excitation, the 6p $[1/2]_0$ state either radiates in an infrared transition to 6s $[3/2]_1$ at 828 nm or is collisionally deactivated to lower levels. Since at the high pressures ($p > 1000$ Torr) of these experiments the resonant states are radiation trapped and lost by excimer formation[7], laser-induced fluorescence observed at 173 nm represents an accurate measure of the number of excited states produced by the laser.

Excitation transfer can be described using the Landau-Zener model[8,9]. For sufficiently slow collisions, the atoms remain on the adiabatic curves. As the excited pair approaches the crossing radius R_c during the inward half of the collision, there is a probability P that the state will jump to the lower adiabatic curve. In the outward half of the collision the excited pair can jump back to the upper curve or remain on the lower. The probability of dissociating to the upper curve is thus $(1 - P)^2 + P^2$. On the other hand, a pair excited on the outward half of a collision with $R > R_c$ will always dissociate. If, however, the laser is tuned so as to create the excited pair at separations <u>shorter</u> than the crossing radius R_c, the crossing is encountered only once during the collision and the resulting probability for dissociation to the upper curve is only P, thus giving

$$P_{rad}(R) = 1 - P + P^2 , \quad R > R_c \; ; = P \; , \quad R < R_c . \tag{1}$$

According to the Landau-Zener formula, the probability P depends on the interaction energy ε_{12}, the relative velocities of the collision partners v_r and the difference of the slopes of the potentials crossing each other:

$$P = \exp \left\{ - \frac{2\pi}{\hbar v_r} \frac{\varepsilon_{12}{}^2}{|d/dR\,(V_1 - V_2)|} \right\} . \tag{2}$$

RESULTS

Since the vuv fluorescence is a measure of the probability of excitation at each laser wavelength, the <u>ratio</u> of ir to vuv intensity measures the <u>probability</u> for the excited pair to dissociate on the upper adiabatic curve and radiate in the ir. Plotted in Fig. 2 is the dependence of the ratio of the ir to vuv signals on excitation frequency for the 6p $[1/2]_0$ state at 2800 Torr and 296 K. The data shown are an average of seven scans in the line wing for a total of 3750 laser shots per spectral sample[6]. The figure shows that the probability for radiation decreases strongly with increased detuning into the red wing of the line (corresponding to absorption at shorter internuclear separations) and that the decrease has an approximate staircase structure with a dip in the center caused by an increased two-photon absorption cross section in the vicinity of a crossing[6].

We have also considered the effect of bound-bound and free-bound transitions in the data of Fig 2. In the absence of collisions, bound states would be unable to dissociate and would radiate as dimers;

hence they would not fluoresce within the bandpass (~ 2 nm) of the ir monochromator . Since the collision frequency is much larger than the radiative rate[4], it is reasonable to assume that the bound excited pairs are completely thermalized and radiate in a broadened excimer spectrum. The fluorescence line shape for the 6p $[1/2]_0 \rightarrow$ 6s $[3/2]_1$ transition was measured with the laser kept fixed on resonance and when detuned to various positions as far as 70-cm^{-1} into the red line wing. No difference was found in the line shape.

LANDAU-ZENER MODEL

Using Eq. (1), the fact that the ratio data decrease by a factor of ten, and integrating over impact parameter while taking mean thermal energies,

$$10 \int_0^{b_{max}} Pb\ db\ \Big]_{R=4Å} = \int_0^{b_{max}} (1-P+P^2)\ b\ db\ \Big]_{R=10Å}, \qquad (3)$$

where b_{max} is determined by the condition $v_r(b_{max}, E, V_u(R)) = 0$. For the potentials of Fig. 1, $d/dR\ |(V_1-V_2)| \approx 1000\ cm^{-1}Å^{-1}$, an interaction energy of $\varepsilon_{12} = 60\ cm^{-1}$ was calculated.

In the Landau-Zener model described by eqns. 1 and 2, P describes the probability for a pair hopping from one adiabatic curve to another as it moves through the crossing region from a large distance. As the pair is excited closer to the crossing in our experiment, it is created in a linear combination of adiabatic states which describe the $6p[1/2]_0$ diabatic state which has transition rate. This mixing can be described for a two level system, where ψ_i and χ_i are the adiabatic and diabatic states,

$$\begin{aligned} \langle\chi_1|H|\chi_1\rangle &= E_c + F_1(R-R_c) \\ \langle\chi_2|H|\chi_2\rangle &= E_c + F_2(R-R_c) \\ \langle\chi_1|H|\chi_2\rangle &= \varepsilon_{12} = \varepsilon_{21} \end{aligned} \qquad (4)$$

Diagonalizing at $\Delta R = R - R_c$

$$\psi_1 = \cos\Theta\ \chi_1 + \sin\Theta\ \chi_2, \qquad \psi_2 = -\sin\Theta\ \chi_1 + \cos\Theta\ \chi_2$$

$$\Theta = \tan^{-1} \left\{\frac{\Delta F \Delta R}{\varepsilon_{12}} \pm \left(\frac{\Delta F \Delta R}{\varepsilon_{12}} + 1\right)^{1/2}\right\}$$

From the initial set of states the system evolves as the pair move through the crossing arriving at t_1:

$$\begin{aligned} i\hbar|\dot{\psi}_1\rangle &= F_1\ v_r\ t\ |\psi_1\rangle + \varepsilon_{12}|\psi_2\rangle \\ i\hbar|\dot{\psi}_2\rangle &= F_2\ v_r\ t\ |\psi_2\rangle + \varepsilon_{12}|\psi_1\rangle. \end{aligned} \qquad (5)$$

At the beginning of the experiment, where R is defined by the laser, only the state χ_2 with transition rate is excited. Then $\chi_2(t = 0) = 1$ and $\chi_1(t = 0) = 0$; hence $\psi_1 = \sin\Theta$ and $\psi_2 = \cos\Theta$. If we ignor the energy spread caused by the finite $\Delta t = t_1 - t_o$ and treat the initial amplitudes incoherently:

$$R < R_c: \qquad P|\psi\rangle(\infty) \cong P\cos^2\Theta + (1-P)\sin^2\Theta$$

$$R > R_c: \qquad P|\psi\rangle(\infty) \cong P(1-P)\cos^2\Theta + (1-P+P^2)\sin^2\Theta.$$

Recently, A. Gallagher and coworkers [10] solved for the absorption and emission of radiation in the vicinity of a crossing by fourier transforming Eq. 5. Their single crossing spectrum applies directly to our experiment. For the potentials described earlier, their calculations give the dashed-dot line of Fig. 2; this result is similar to our static model although a satellite is clearly visible due to the zero in the difference potential of the upper adiabatic curve. This satellite is not visible in the data; and better agreement is obtained for Gallagher's parameter, $s = (F_{20}-F_{10})/(F_{20}+F_{10}) = -1.25$. For this value the crossing diabatic potentials must both have positive slope in the vicinity of the crossing. This suggests that the crossing observed is with an attractive potential from more highly excited 5d states rather than with the 0_g^+ configuration of $6p[1/2]_0$ as suggested in Fig. 1.

REFERENCES

*Supported by the DOE Div. of Chem. Sci. and the Welch Foundation.

1. L.T. Gudzenko and S.I. Vakovlenko, Sov. Phys. JETP 35, 877 (1972).
2. R.W. Falcone, W.R. Green, J.C. White, J.F. Young, and S.E. Harris, Phys. Rev. A15, 1333 (1977).
3. A. Gallagher and T. Holstein, Phys. Rev. A16, 2413 (1977).
4. T.D. Raymond, N. Böwering, C.Y. Kuo, and J.W. Keto, Phys. Rev. A29, 721 (1984).
5. For a recent review, see: Nicole Allard and John Kielkopf, Rev. Mod. Phys. 54, 1003 (1982).
6. N. Böwering, T.D. Raymond and J.W. Keto, Phys. Rev. Lett. 52, 1880 (1984).
7. T.D. Bonifield, F.H.K. Rambow, G.K. Walters, M.V. Mc Cusker, D.C. Lorents and R.A. Gutscheck, J. Chem. Phys. 72, 2914 (1980).
8. L. Landau, Phys. Z. Sowjet Union 2, 46 (1932).
9. C. Zener, Proc. Roy. Soc. A137, 696 (1932).
10. M. O'Callaghan, T. Holstein, and A. Gallagher, Phys. Rev. 32A, 2754(1985).

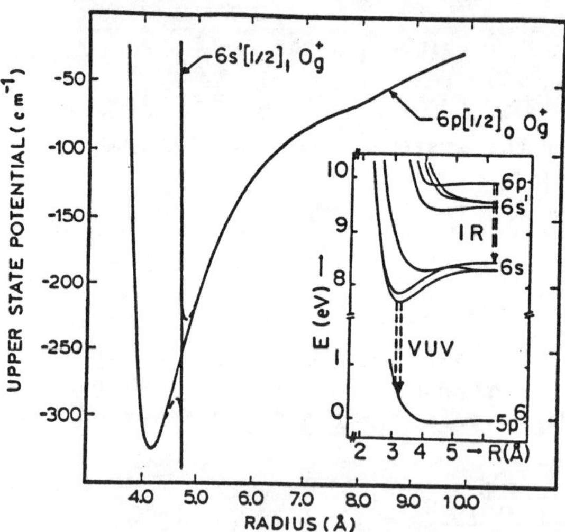

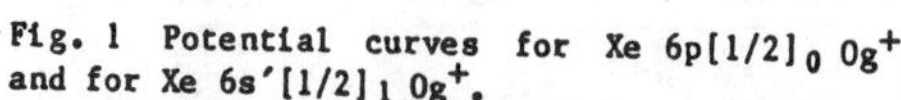
Fig. 1 Potential curves for Xe $6p[1/2]_0\ 0g^+$ and for Xe $6s'[1/2]_1\ 0g^+$.

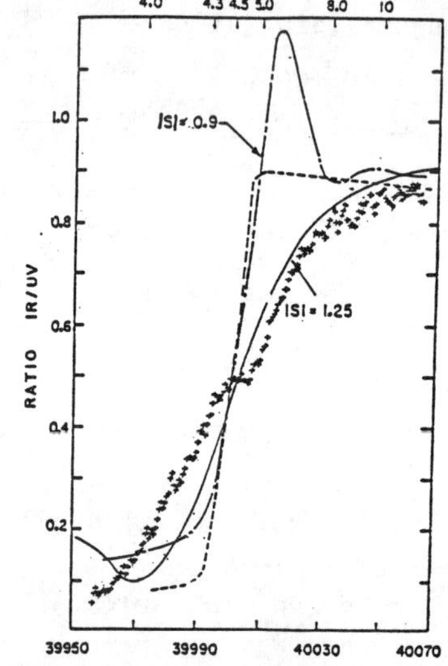

Fig. 2 Ratio of ir/vuv as a function of laser frequency for Xe $6p[1/2]_0$. Also shown as a short dashed line is the static model, dashed-dot line is a model by Gallagher with s = .9 while the solid line is for s = 1.25.

Photodissociation of Na_2 studied by Doppler Photofragment Spectroscopy

G. Gerber, R. Möller
Fakultät für Physik, Universität Freiburg, 7800 Freiburg, Germany

ABSTRACT

The photodissociation of Na_2 has been investigated in a molecular beam experiment. The velocity and angular distributions and internal states of the fragments were determined by Doppler spectroscopy. Three different photodissociation processes were observed and identified: direct photodissociation by excitation from the $X^1\Sigma_g^+$ ground state to the dissociation continuum of the $B^1\Pi_u$ state, dissociation of quasibound levels of the B state, and induced two-photon dissociation of Na_2 leading to Na*(nl) fragments. Direct dissociation of the $B^1\Pi_u$ state yields only $Na^*3^2P_{3/2}$ fragments, which does not support the interpretation of reported $^2P_{3/2}$ and $^2P_{1/2}$ fragment emission.

The photodissociation of alkali-metal dimers, particularly of Na_2, has recently been studied by several groups[1]. However, there are contradictory results concerning the population of fragment fine structure states in the dissociation process. We have applied the technique in Doppler Spectroscopy to the study of photodissociation of Na_2. The fine structure states and the angular and energy distributions of the neutral photofragments were accurately determined.

A molecular beam of Na_2 generated by a stainless steel oven at a temperature of 850 K with a nozzle of 0.2 mm diameter ($p \cdot d \sim 0.12$) and collimated by a heated skimmer is intersected at right angles by two laser beams at a distance of 75 mm from the nozzle. Photodissociation is induced with various Ar^+-laser lines,

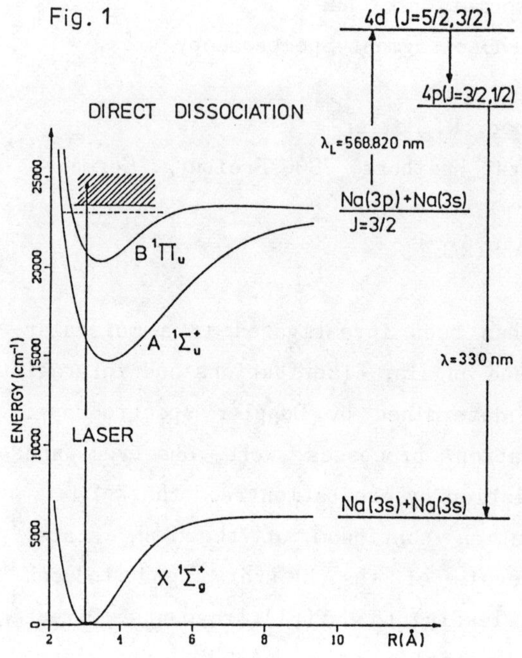

both multi-mode and single-mode. The dissociation into atomic fragments is monitored by the atomic fluorescence at 330 nm (see Fig.1). The Na* 3p atoms produced in the dissociation are excited to the 4d level using a single mode dye laser. Since the excitation wavelengths for the fine structure states $^2P_{3/2}$ and $^2P_{1/2}$ differ by 17 cm^{-1} both are easily distinguished with our detection scheme.

The energy and angular distribution of the fragments produced by a dipole transition can be written in the form:

$$f(\vartheta, v) = g(v)\ (\ 1 + \beta(v) P_2(\cos\vartheta)\ ) \tag{1}$$

ϑ is the angle between the polarization and the axis of the dissociating molecule, v the velocity of the fragment, g(v) the velocity distribution and $\beta(v)$ is the anisotropy parameter, which in general depends on v. By projecting this distribution onto the axis of detection z the following velocity distributions are found:

$$\rho_{\parallel}(v_z) = \frac{1}{2v}\left(1 + \beta(v) P_2\left(\frac{v_z}{v}\right)\right) \qquad E \parallel z \tag{2a}$$

$$\rho_{\perp}(v_z) = \frac{1}{2v}\left(1 - \frac{1}{2}\beta(v) P_2\left(\frac{v_z}{v}\right)\right) \qquad E \perp z \tag{2b}$$

v_z being the z-component of the velocity

In order to determine g(v) and $\beta(v)$ from the observed Doppler spectra, $\rho_{\parallel}(v_z)$ & $\rho_{\perp}(v_z)$, the inversion of these formulae (2) can be

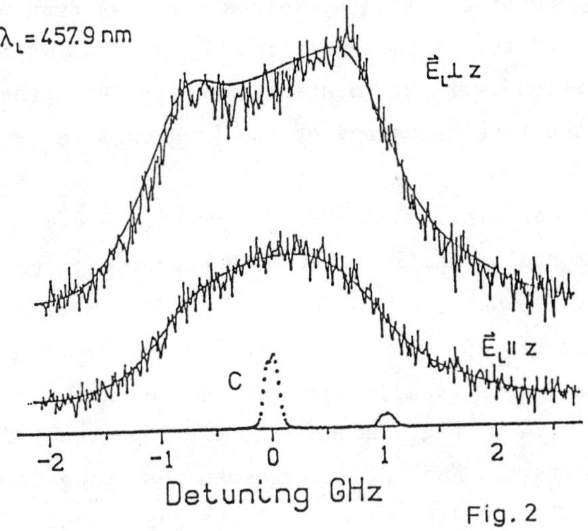

Fig. 2

obtained numerically by a stepwise recursion. The procedure makes use of the fact that for a given point of $\rho(v_z)$ only $v \leq v_z$ can contribute to the velocity distribution. This allows us to construct $\rho(v_z)$ starting with high v_z.

Fig. 2 shows the observed Doppler spectra of the Na $3^2P_{3/2}$-fragments for E∥z and for E⊥z at λ_L=457.9 nm. Both spectra consist of two overlapping Doppler spectra due to the two transitions $3^2P_{3/2}-{}^2D_{5/2}$ and $3^2P_{3/2}-{}^2D_{3/2}$, which are separated by only 1.02 GHz. Trace c shows a Doppler-free atomic excitation spectrum for the two fine structure components.

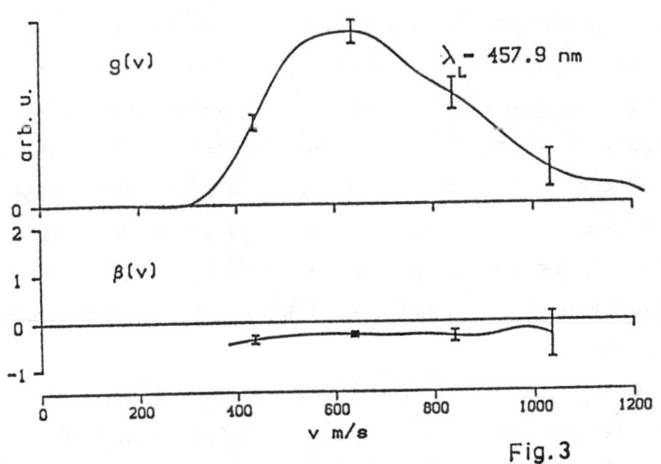

Fig.3

Fig. 3 shows the velocity distribution g(v) and the function β(v) derived from the Doppler spectra obtained with λ_L=457.9 nm. The velocity distribution shows a threshold of about 450m/sec≙E≈385 cm^{-1} , which is approximately the height of the B-state potential barrier (379.3cm^{-1}) .

The angular distributions are characterized by values for $\beta(v)$ ranging from -1 to ≈ 0; this is expected for the dissociation of a rotating molecule following a $\Sigma-\Pi$ transition[3], if the rotational energy is in the same order of magnitude as the kinetic energy of the fragments released in the dissociation.

Within the experimental uncertainties no $3^2P_{1/2}$-fragments are observed at comparable experimental conditions. At much higher oven temperatures $3^2P_{\frac{1}{2}}$-fragments are observed with $\lambda = 476.5$ and $\lambda = 488.0$ nm excitation, but not for excitation at $\lambda = 457.9$ nm. As shown in reference 4 the appearance of $^2P_{\frac{1}{2}}$ fragments can be explained by a two-photon process.

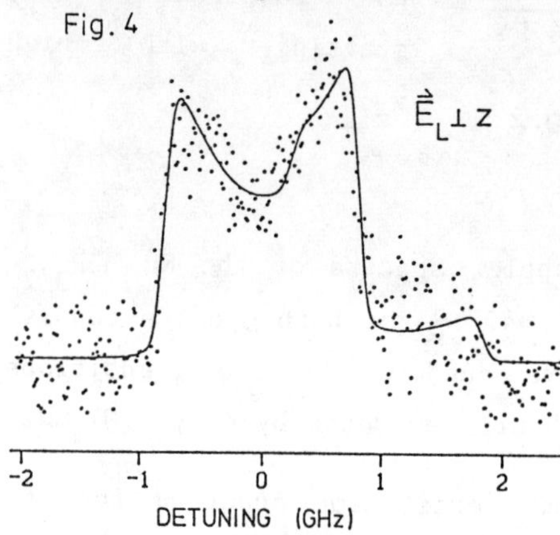

Fig.4 shows a Doppler Spectrum of Na $3^2P_{3/2}$-fragments resulting from dissociation of the quasibound level v'=31, J'=42 of the B $^1\Pi_u$ state. The solid curve represents a fragment distribution with $W=400\text{cm}^{-1}$ kinetic energy and a $\beta=-1$ anisotropy parameter. The obtained value $W=400 \pm 20$ cm^{-1} agrees well with the energy of the v'=31, J'=42 level above the B-state dissociation limit, E=397.7 cm^{-1}. The quantum numbers of this quasibound state v'=31, J'=42, excited by a Q-transition from v"=8,J"=42 by a single mode Ar^+ laser at 457.9 nm, were determined by double-resonance spectroscopy. The linewidth of this state was found to be $\approx$100 MHz corresponding to a lifetime of 1.6 ns. But even in such a "slow predissociation" the initial anisotropy $\beta=-1$ of our excitation process is completely preserved, since for a perpendicular Q-transition, where the transition moment $\vec{\mu}$ is parallel to $\vec{J}$ and to $\vec{E}_L$, the rotation cannot change an angular distribution having rotational symmetry around the $\vec{E}_L$ axis.

The dissociation of the quasibound level v'=31, J'=42 represents one of the very rare examples where, for a neutral molecule, all the quantities governing the process are well determined.

Direct dissociation of Na_2 with wavelengths between 450 and 510nm proceeds via the $B^1\Pi_u$ state and yields Na^* $3^2P_{3/2}$ as the only excited fragment state, indicating that there are no (or very weak) nonadiabatic interactions of the B-state with other electronic states correlated with the $3^2P_{1/2}$ + $3^2S_{1/2}$ dissociation limit. We also have found no evidence for a dissociation via the A-state continuum.

A two-photon absorption process leading to a dissociation and to a population of the $3^2P_{1/2}$ level has also been observed and can clearly be distinguished by its different angular and energy distributions as compared to the one-photon dissociation resulting only in $3^2P_{3/2}$ excited fragments.

References:

1. E.W. Rothe, U. Krause, R. Düren, J. Chem. Phys. 72,5145(1980) V.B. Grushevskii, S.M. Papernov, M.L. Yanson, Opt. Spectrosc. Opt. Spectrosc. 44,475(1978) and J.Phys.B.15,4175 (1982)
2. H.J. Vedder, G.K. Chawla, R.W. Field, Chem.Phys.Lett.111,303(1984)
3. C. Jonah, J.Chem.Phys. 55.1915 (1971)
4. G. Gerber, R. Möller, Phys.Rev.Lett. 55.814 (1985)

BOUND-FREE TRANSITIONS IN MOLECULAR METAL TRIMERS

James L. Gole
Georgia Institute of Technology, Atlanta, Ga. 30332

In an effort which is complimentary to the study of bound-bound transitions via TPI spectroscopy, we have developed techniques for the study of photodissociation and hence bound-free transitions in the alkali and Group IB metal trimers. Our emphasis has been on the sodium[1], potassium[2], and copper trimer[3] molecules. Using bound-free spectroscopy, it has been possible to elucidate features of the bonding and energy levels in sodium and potassium trimer, determine upper and lower bounds for the difficult to measure M_2-M metal cluster bond energies and assess the behavior of these clusters under a variety of perturbing conditions.

In order to study the photodissociation process, single photon optical pumping is used to excite from the ground electronic and low-lying states of the trimer to repulsive excited state levels with an open channel to dissociation. We monitor the electronically excited atomic states which are the products of this dissociation. Hence, we observe laser induced atomic fluorescence, LIAF, which in the case of sodium trimer corresponds to the production of ^2P sodium atoms monitored using the well known Na D-line. As the dye laser which induces photodissociation is frequency scanned, we record the intensity of emission from the $^2P_{1/2}$ and $^2P_{3/2}$ components of the Na D-line. A typical LIAF spectrum obtained for the Na_3 produced cold in high purity expansion is shown in Figure 1. As the dye laser frequency is scanned, one produces a series of "fluctuation bands" in the intensity for both the $^2P_{1/2}$ and $^2P_{3/2}$ atomic emissions dominated by what appears to be a progression of features separated by $\sim$130-135 cm^{-1}. These bands are thought to reflect, in large part, the vibronic structure of the ground electronic state of sodium trimer, however, more detailed studies involving Li_3 will be needed to remove some ambiguity in this interpretation. Preliminary analysis indicates that the sodium trimer LIAF bands correspond to a vibrationally heated yet rotationally cold distribution.

Using the spectrum in Figure 1, we have the potential for modeling the vibronic structure in the ground electronic state of the sodium trimer molecule as well as obtaining information on the repulsive states which dissociate to excited state atoms. The measured high frequency limits of the Na_3 (21200) and K_3 (16600) LIAF spectra allow the determination of an upper bound for the trimer bond energies. Further, the onsets for the Na_3 (20000 cm^{-1}) and K_3 (15100 cm^{-1}) LIAF spectra appear to provide lower bounds for these bond energies.[2] Using E(Na*) = 16950 cm^{-1} and E(K*) = 12990 cm^{-1} we deduce upper and lower bounds, $3000 \leq D_o \leq 4250$ cm^{-1} and $2020 \leq D_o \leq 3610$ cm^{-1} respectively for the sodium and potassium trimer bond energies. These results are in good agreement with the theoretically determined bond energies for Na_3 (3000 cm^{-1})[4] and K_3 (2200 cm^{-1})[5] and with mass spectrometric measurements.[6] For potassium trimer, this represents the first experimental measurement of this

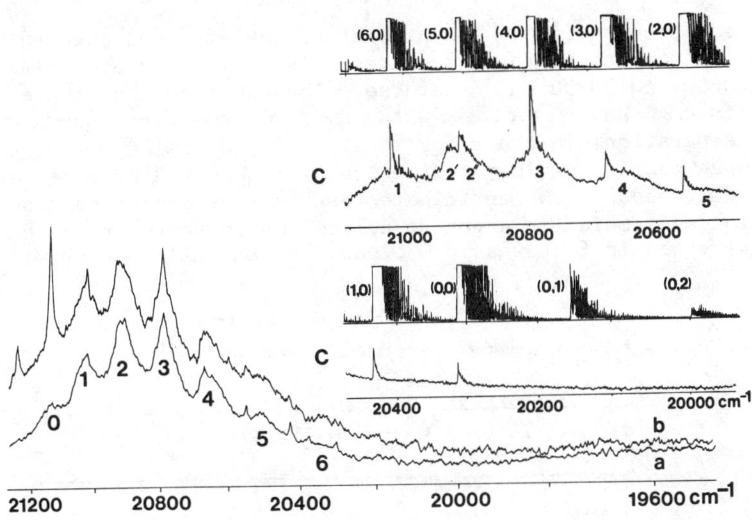

Fig. 1. (a) Laser Induced Atomic Fluorescence (LIAF) from the photodissociation of "cold" Na_3 with an Ar ion pumped (coumarin 102) dye laser, a = $^2P_{1/2}$, b = $^2P_{3/2}$ LIAF. (c) Close-up of LIAF corresponding to the $^2P_{3/2}$ component of the Na D-line vs. the cold Na_2 $B^1\Pi_u$-$X^1\Sigma_g^+$ excitation spectrum with bands denoted (v', v").

quantity. For sodium trimer, the accuracy obtained exceeds that of mass spectrometry (D_o (Na_2 - Na) = 2608 + 1337[6] cm^{-1}).

There is a strong correlation between the sodium trimer LIAF spectrum and the bound-bound spectra observed by Schumacher[7], Wöste[8], and coworkers. The Na_3 LIAF spectrum is thought to arise primarily from a strongly allowed ($1a_2'' \leftarrow 1a_1'$) transition which theory predicts at ∿ 4800 Å. Dominant LIAF signals as those for Na_3 which correspond to efficient dissociation are countered by a complete absence of features in the bound-bound spectrum. In addition, several features which are weak in Na_3 two-photon spectroscopy are also weak in the Na_3 LIAF spectrum (features extending from ∿ 20000 cm^{-1} to lower frequency) indicating the presence of selective predissociation. In other words the two techniques produce spectra which are virtually mirror images of each other. The interpretation of the nature of the potentials giving rise to the strong bound-free LIAF spectra would appear to involve a consideration of Slonczewski resonances.[9]

We have also been concerned with the finer details of certain features in the LIAF spectrum of sodium trimer again with the impetus of correlating these observations with the two photon bound-bound spectroscopy of the trimer.[7,8] The Na_3 LIAF spectrum is dominated by a progression of features (2-6 Fig. 1) separated by ∿ 130-135 cm^{-1}. Two of these features at 20800 and 20900 cm^{-1} have now been studied at much higher alkali flux in several pure sodium supersonic expansions. Comparisons of the attained spectra have revealed several vibronic features (Table I) associated with both the 20800 and 20900 cm^{-1} features. Thus we find that these are regions which have associated with them a vibronic fine structure with peak separations in the range 15-30 cm^{-1}. We believe that these features bear an analogy to the features observed by Wöste et al. in TPI spectroscopy. We anticipate that further seeded supersonic expansions of sodium in argon or helium should reveal these and additional vibronic features more clearly in the LIAF spectrum.

Sodium Trimer LIAF Spectrum

Fine Structure of "20800cm^{-1}" and "20900cm^{-1}" Regions

Region	Peak Position (cm^{-1})	Δ (cm^{-1})
	20738	
		28
"20800"	20766	
		13
	20779	
		16
	20795	
		16
	20811	
		15
	20826	
	20875	
		24
	20899	
		23
"20900"	20922	
		24
	20946	

Finally, we note that the Na_3 LIAF spectrum demonstrates both a selective and extremely efficient energy transfer process.[1] In contrast to the $^2P_{1/2}$ LIAF spectrum, the $^2P_{3/2}$ LIAF spectrum is accompanied by sharply (vs. the broad LIAF features) peaking regions in close frequency coincidence with the Na_2 $B^1\Pi_u - X^1\Sigma_g^+$ excitation spectrum also shown in Fig. 1(c). These features are thought to result from the combination of a surprisingly efficient energy transfer process and a subsequent dissociation[1] viz.

$$Na_2^* + Na_3 \rightarrow Na_3^* + Na_2 \quad (1)$$

$$Na_3^* \rightarrow Na_2 + Na^* \quad (2)$$

$$Na^* \rightarrow Na + h\nu \text{ (D-line)} \quad (3)$$

strongly favoring the $^2P_{3/2}$ component.

Detailed quantum chemistry indicates that the lowest energy electron configuration of sodium trimer, of symmetry $^2E'$, Jahn-Teller distorts to a $^2B_2$ obtuse angled and $^2A_1$ acute angled configuration. The $^2B_2$ configuration corresponds to the ground state (minimum on the Na_3 surface) and lies ∿ 0.6 kcal (200 cm^{-1}) below the $^2A_1$ acute angled configuration which corresponds to a saddle point on the Na_3 potential energy surface. Another saddle point, a $^2\Sigma_u^+$ linear conformation, is thought to lie some 3 kcal/mole (1050 cm^{-1}) above the $^2B_2$ conformation. Based on the location of the $^2B_2$ state and the $^2A_1$ saddle point, a ∿ 200 cm^{-1} pseudorotation barrier characterizes the system.

The pseudorotation interaction in the ground state of Na_3 leads to a vibronic coupling of the bending and assymmetric stretching modes which will necessarily produce complicated spectral effects. However, the manifold of levels corresponding to the symmetric stretch should be only mildly affected. If several trimer ground state vibronic levels are populated, one might anticipate a complicated spectral pattern from which emerge peaks associated with the symmetric stretching mode. Figure 1 would appear to provide support for this suggestion. The readily observed peak separations for the highest energy features (lowest levels in the ground state) are on the order of 100-110 cm^{-1} although 10-30 cm^{-1} satellite structure accompanies features 1-3 (Fig. 1 - Table I). It would not be unreasonable to associate the dominant energy separations for the higher energy features with significant vibronic coupling. Higher resolution scans will be needed for definitive identification of this rather overlapped region, however, the 130-136 cm^{-1} separations for bands "2"-"6" bear a correlation with the TPI spectrum and are consistent with a ground state symmetric stretch frequency ∿ 134 ± 3 cm^{-1}. This is in good agreement with a calculated value of 147 ± 20[4] cm^{-1} and well conceived estimates of 137[7,10] and 138 cm^{-1} [5,11] based on (1) the extrapolation of detailed vibronic calculations on Li_3[10] and (2) the determination of the symmetric stretch frequency for the sodium trimer ion.[11] If the symmetric stretch frequency for the ground electronic state of sodium trimer is ∿ 135 ± 3 cm^{-1} and the anharmonicity of the symmetric stretch manifold is reasonably small, the features 3-6 in Fig. 1 may correspond, at least in part, to vibrationally hot levels of the symmetric stretch manifold.

REFERENCES

1. J. L. Gole, G. J. Green, S. A. Pace, and D. R. Preuss, J. Chem. Phys. 76, 2247 (1982).
2. J. S. Hayden, R. Woodward, and J. L. Gole, J. Phys. Chem. xxxx (1986).
3. W. H. Crumley, J. S. Hayden, and J. L. Gole, J. Chem. Phys., xxxx (1986).
4. R. L. Martin and E. R. Davidson, Mol. Phys. 35, 1713 (1978).
5. S. C. Richtsmeier, R. A. Eades, D. A. Dixon, and J. L. Gole, A.C.S. Symposium Series No. 179 - Metal Bonding and Interactions in High Temperature Systems with Emphasis on Alkali Metals (Ed. J. L. Gole and W. C. Stwalley) Am. Chem. Soc. Wash. D.C. (1982) pg. 177. T. C. Thompson, G. Izmirlian Jr., S. J. Lemon, D. G. Trular, and C. A. Mead, J. Chem. Phys. 82, 5597 (1985).
6. K. Hilbert, Ber. Bunsenges Phys. Chem. 88, 260 (1984).
7. A. Hermann, M. Hofmann, S. Leutwyler, E. Schumacher, and L. Wöste, Chem. Phys. Lett. 62, 216 (1979). Also W. H. Gerber and E. Schumacher, J. Chem. Phys. 69, 1692 (1978). E. Schumacher, W. H. Gerber, H. P. Harri, M. Hofmann, and E. Scholl, "Preparation, Electronic Spectra, and Ionization of Metal Clusters", A.C.S. Symposium Series No. 179 - Metal Bonding and Interactions with Emphasis on the Alkali Metals (J. L. Gole and W. C. Stwalley, eds.) pg. 83.
8. G. Delacretaz and L. Wöste, Surf. Science 156, 770 (1985).
9. J. C. Slonczewski, V. I. Moruzzi, Physics 3, 237 (1967).
10. W. H. Gerber "Theorie des dynamischen Jahn-Teller Effekts in Li_3 and Untersuchung von. Lithium-Molekularstrahlen, Ph.D. Thesis, Bern University (1980).
11. R. A. Eades, M. L. Hendewerk, R. Frey, D. A. Dixon, and J. L. Gole, J. Chem. Phys. 76, 3075 (1982).

IONIZATION THRESHOLD ENERGIES FOR METAL CLUSTERS

D. M. Cox, R. L. Whetten, M. R. Zakin
D. J. Trevor, K. C. Reichmann and A. Kaldor
Exxon Research and Engineering Company
Route 22 East
Annandale, NJ 08801

ABSTRACT

We have measured the ionization threshold energies as a function of cluster size for V, Nb, and Fe clusters. The metal clusters are produced by laser vaporization of a metal substrate inside the throat of a pulsed nozzle. The clusters are detected by photoionization TOF mass spectrometry. Using tunable UV lasers, the ionization thresholds are measured as a function of cluster size. In addition, ionization thresholds for clusters with dissociatively chemisorbed hydrogen are found to increase significantly over that of the corresponding bare cluster.

INTRODUCTION

The chemistry and physics of clusters have long been recognized as being of technological importance in such diverse areas as metallurgy, catalysis, and combustion. Considerable effort has been expended to characterize supported clusters and to study the spectroscopy of small clusters in matrices. But experimental information regarding the intrinsic properties of unsupported clusters containing 3 to a few hundred atoms has been particularly elusive since production and non-destructive probing of such unsupported clusters has been very difficult. Recent advances in experimental techniques now make it possible to generate and detect gas phase clusters in an environment free from the perturbing influence of supports. Metal clusters are synthesized through sudden condensation of laser produced vapor. An intense pulsed laser beam focussed onto a target rod evaporates off metal atoms. If these atoms are injected into the throat of a high pressure pulsed nozzle, nucleation occurs and clusters are produced. This particular technique is important in that clusters of virtually any material can now be produced and studied. We have made clusters of even the most refractory transition metals, numerous oxides, carbides and intermetallic alloys of these elements as well as clusters of carbon and semiconductor materials. The clusters are ionized by tunable UV lasers and mass resolved by time of flight mass spectroscopy. Measurement of the ionization threshold behavior of these clusters and their chemical reactivity are two powerful techniques we have used to probe how the electronic structure and chemical behavior of these novel materials varies as a function of cluster size, and when and how they approach bulk properties. In this paper we will show how the ionization thresholds for clusters of iron, vanadium and niobium exhibit strong non-monotonic behavior

0094-243X/86/1460527-4$3.00

as a function of cluster size and will present evidence demonstrating a strong anticorrelation of ionization threshold and hydrogen chemisorption reactivity for clusters containing eight or more metal atoms.

RESULTS AND DISCUSSION

From our previous measurements[1] on the ionization thresholds of iron clusters, we find that as the iron cluster size increases, a highly non-monotonic decrease in the ionization threshold energies occurs. In particular, oscillations in the ionization threshold energies are observed in which local maxima centered around Fe_4 and Fe_{15} encompass a local minimum centered around Fe_{10}. The oscillations are not predicted by any current model of metal clusters: in particular they represent a significant deviation from the classical metal droplet model[2] used to explain trends in the ionization threshold energies of alkali metal clusters[3]. For Fe as well as all other transition metals, the ionization threshold drops very rapidly toward the bulk work function.

In addition to probing the electronic structure of clusters directly via ionization threshold measurements, we have found that certain chemical reactions (hydrogen chemisorption in particular) appear to be a sensitive chemical probe of electronic structure. In these measurements the metal clusters are injected into a chemical reactor which is attached to the end of the cluster source's condensation channel[4]. A helium/reactant mixture is injected into the reactor at a time which coincides with the time at which the metal clusters pass through. The extent of reaction is indicated by the extent of reduction of the bare cluster signal and is corroborated by the apppearance of new mass peaks which correspond to metal clusters with chemisorbed molecules. (For H_2 on metal clusters the H_2 is dissociatively chemisorbed). Using hydrogen (deuterium) as the reactant, we uncovered a strong anti-correlation between the reactivity of the bare iron clusters and the ionization threshold energy. For clusters containing 8 or more atoms, those with low ionization thresholds react much more rapidly with H_2 than those which have high ionization thresholds. The relative rates for H_2 chemisorption closely follow variations in cluster ionization threshold energies[5]. Such a correspondence can be qualitatively understood simply in terms of a requirement to overcome a relatively small activation barrier ($\lesssim$ 5kcal/mol) in the initial chemisorption step. The magnitude of the activation barrier will be sensitive to the symmetry and energies of the metal cluster orbitals near the Fermi level[6]. Thus sensitivity to ionization threshold energy is rationalized.

Our results for the ionization threshold energies and relative reactivities toward deuterium chemisorption for vanadium and niobium are summarized in Table I. From Table I we see that niobium clusters exhibit a dramatic oscillation in ionization thresholds between Nb_7 and Nb_{11}. Nb_8 and Nb_{10} have considerably larger ionization threshold energies than do the adjacent Nb_7, Nb_9, or Nb_{11}. In addition local maxima in ionization thresholds are observed at Nb_{16} and in the region of Nb_{24}, and Nb_{26}. The high ionization thresholds for Nb_8, Nb_{10}, and Nb_{16} are of particular interest since they are the specific clusters which were found by Smalley[7] to be least reactive toward hydrogen. Overall the local maxima (minima) in ionization threshold correlate reasonably well with local minima (maxima) in niobium reactivity toward chemisorption of H_2.

Table I. Photoionization thresholds and reactivity toward D_2.

Cluster Size (X)	Photoionization Threshold[a] (ev) Vanadium	Niobium	Reactivity (R_x)[b] Vanadium	Niobium
1	6.74	6.88	--	
2	6.10	--	-0.01	
3	5.49	--	0.15	
4	5.63	5.58	-0.03	1.24
5	5.47	5.43	0.77	1.75
6	5.37	5.34	0.03	1.19
7	5.24	5.32	0.28	0.42
8	5.36	5.46	0.14	0.07
9	5.20	4.93	0.39	0.30
10	5.17	5.33	0.25	0.03
11	4.99	4.72	0.50	0.59
12	4.95	4.91	0.35	0.24
13	4.93	4.84	0.78	0.77
14	4.96	4.65	0.80	0.69
15	≤4.70	≤4.60	0.99	0.91
16	4.99	4.76	1.07	0.19
17	5.00	≤4.60	1.44	0.32
18	4.91	4.60	1.58	0.38
19	≤4.70	≤4.60	1.72	0.36
20	≤4.70	≤4.60	1.71	0.85
21	≤4.70	≤4.60	1.81	1.33
22	≤4.70	≤4.60	1.83	1.78
23	--	4.67	1.84	0.69
24	--	4.73	1.83	0.45
25	--	4.68	1.96	0.77
26	--	4.77	1.81	0.48
27	--	4.67	1.75	1.62
28	--	~4.60	1.83	1.14

(a) Ionization thresholds have estimated uncertainties of ± 0.05 eV.
(b) Reactivity is defined as $R_x = -\ln S_x$, where S_x is the experimentally determined bare cluster survival fraction after exposure to D_2.

For vanadium the ionization thresholds do not show the dramatic oscillations observed for niobium, but again a quite good correlation is obtained between ionization threshold energy and hydrogen chemisorption reactivity.

Finally we have also bracketed the ionization threshold energy for clusters with chemisorbed hydrogen. For all cases examined thus far (Fe, Nb, and V) we find the ionization thresholds of clusters increase significantly upon chemisorption of hydrogen. For instance V_3 has an ionization threshold energy of 5.49 eV but the V_3H_4 cluster has an ionization threshold energy of >6.42 eV, an increase greater than 0.9 eV. Similar increases are found for other small vanadium clusters which have chemisorbed 3 to 5 hydrogen molecules. Iron clusters, Fe_{22}-Fe_{25} upon chemisorption of 8 to 10 hydrogen molecules exhibit an increase greater than 0.4 eV.

SUMMARY

The ionization threshold energies of small transition metal clusters are found to vary dramatically and non-monotonically with cluster size. The variation in ionization threshold energy is found to correlate with variation in reactivity of the clusters toward hydrogen chemisorption. The ionization threshold energies for clusters with chemisorbed hydrogen are found to be significantly higher than corresponding bare clusters.

REFERENCES

1. E. A. Rohlfing, D. M. Cox, A. Kaldor, and K. H. Johnson, J. Chem. Phys. 81 3846 (1984).
2a. D. M. Wood, Phys. Rev. Lett. 46, 749 (1981).
2b. K. I. Peterson, P. D. Dao, R. W. Farley, and A. W. Castleman, J. Chem. Phys. 80, 1780 (1984).
3. E. Schumacher, M. Kappes, K. Marti, P. Radi, M. Schar, and B. Schmidhalter, Ber. Bunsenges. Physik. Chem. 88, 220 (1984).
4a. R. L. Whetten, D. M. Cox, D. J. Trevor, and A. Kaldor, J. Phys. Chem. 89 566 (1985).
4b. S. C. Richtsmeier, E. K. Parks, K. Liu, L. G. Pobo, and S. J. Riley J. Chem. Phys. 82, 3659 (1985).
4c. M. E. Geusic, M. D. Morse, S. C. O'Brien, R. E. Smalley Rev. Sci. Instrum. 56 2123 (1985).
5. R. L. Whetten, D. M. Cox, D. J. Trevor, and A. Kaldor, Phys. Rev. Lett. 54 1494 (1985).
6. J-Y. Saillard and R. Hoffman, J. Am. Chem. Soc. 106, 2006-(1984).
7. M. D. Morse, M. E. Geusic, J. R. Heath, and R. E. Smalley J. Chem. Phys. 83, 2293 (1985).

POSSIBLE OBSERVATION OF EMISSION FROM THE K + NaCl REACTION COMPLEX

Sydney J. Ulvick, Philip R. Brooks, R. F. Curl, and James H. Spence
Chemistry Department and Rice Quantum Institute
Rice University
Houston, TX 77251

ABSTRACT

Molecular beams of K and NaCl are crossed in the cavity of a cw dye laser operated in the near IR. In preliminary experiments, blue shifted emission is observed in the 650-750 nm range, and we tentatively suggest this may result from emission of a quasi-bound excited state of the reaction complex.

INTRODUCTION

In an attempt to directly probe a chemical system during the reaction process, we have recently(1) initiated a study of the reaction between K and NaCl in the cavity of a CW dye laser. The laser is tuned to various wavelengths in the red (600-750nm) where neither the reagents (K and NaCl) nor the products (Na and KCl) absorb. Nevertheless, we observe emission of a reaction product, Na*. We believe that our observations, summarized below and in refs. 1-3, are most easily explained by assuming that a "reaction complex" or "transition state", $[KNaCl]^{\dagger}$, is formed which absorbs a photon in the red and which consequently falls apart to yield KCl and Na* according to the following mechanism:

FORMATION	$K + NaCl \rightarrow [KNaCl]^{\dagger}$	(1)
EXCITATION	$[KNaCl]^{\dagger} + h\nu \rightarrow [KNaCl]^{\dagger *}$	(2)
DECOMPOSITION	$[KNaCl]^{\dagger *} \rightarrow KCl + Na^*$	(3)
RADIATION	$Na^* \rightarrow Na + h\nu'$	(4)

The chemically evolving system passes thru a series of reaction complexes, so $[KNaCl]^{\dagger}$ denotes a set of states whose properties must change on a picosecond time scale. As the reaction progresses, it is thus possible that a complex becomes chemically tuned into resonance with the laser and absorption occurs as in (2) above.

The experiments have so far been restricted to the crossing of quasi-effusive beams in the cavity of a cw dye laser tuned over the range 600-730 nm with bandwidth ~2 cm^{-1}. Emission at 589.0 nm is observed with a cooled PMT and the details of the apparatus and data acquisition system are contained in refs. 2 and 3. Briefly, we observe a signal at the Na D line which arises when all three beams are on, which is called the 3-beam signal (3BS). The 3BS is blue shifted from the exciting laser and persists from ~600 nm to a threshold of ~730 nm, consistent with the exoergicity of the chemical reaction supplying the energy deficit. The 3BS is linear with respect to each reagent, K, NaCl and $h\nu$, and since single-photon transitions occur for neither the isolated reagents nor products, we conclude that we have observed the laser-assisted reaction

$$K + NaCl + h\nu \rightarrow KCl + Na + h\nu' \qquad (5)$$

which is probably best viewed as the sequence of reactions, (1)-(4).

Light absorption by the reaction complex can be visualized by referring to Fig. 1, where we show collinear potential energy surfaces from pseudopotential calculations of Roach & Child (4).

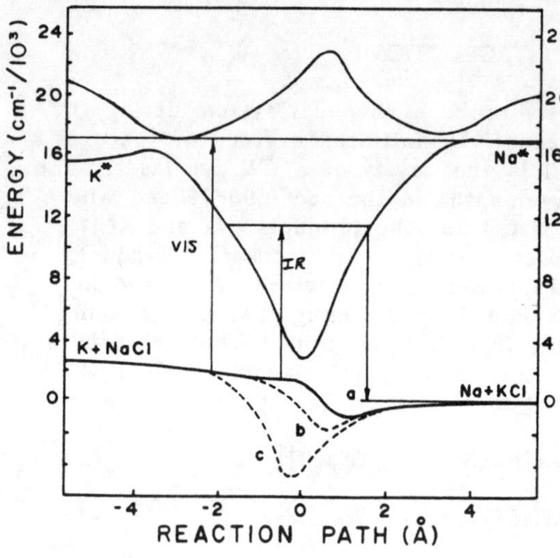

Fig. 1. Potential energy variation along the reaction path for the three lowest σ states of the collinear K + NaCl reaction. Curves b and c are for bent geometries. Arrows schematically represent possible absorptions of the reaction complex in the visible and IR.

The thermally excited X state can be excited to either the B state or to a predissociative level of the A state, and this is symbolized by the transition labeled "VIS". The experiments so far discussed and reported (refs. 1-3) have been restricted to wavelengths such that only states lying above the Na* asymptote could be accessed. The preliminary excitation spectrum (Na* fluorescence vs. excitation wavelength) for this process is featureless, which might be expected since absorption represents transition between an unbound upper state and an unbound ground state which occurs during reaction.

Although careful measurement of a truly continuous spectrum may help elucidate the molecular potentials sampled during the course of reaction, we believe a structured spectrum would be more informative. If the excitation wavelength were to lie in the IR, the possibility of evolution to either K* or Na* would be closed, and a transition to a quasi-bound level of the A state would result. This is illustrated in Fig. 1 by the transition labelled "IR." The system might return to the X state by some internal conversion process, but it might also fluoresce back to the X state. Since the calculations give small potential wells for non-collinear orientations, the A state could radiate to a bound level of the X state. Because of the possibility of observing structured spectra, both the absorption and subsequent emission are of great interest to us, and we have conducted some preliminary experiments in the IR to explore these possibilities.

EXPERIMENTAL

The apparatus is basically that described in ref. 1-3, except for major changes in the optical cavity. The laser operates in the region 820-850 nm using LDS 820 pumped by all lines of the Ar^+ laser. Blue-shifted fluorescence is viewed by a PMT through a filter passing the window 650-750 nm. This window is much larger than the 0.5 nm of the Na D line filter and extraneous photons are much more of a problem. A very serious background contribution arises from photoluminescence of K_2 (apparently from the A state) which is excited by the fluorescence of the dye in the laser. In order to minimize this interference, a prism was inserted into the cavity to disperse away the dye fluorescence, and the dye laser was mounted above the table on stilts in what was known as the "Kamakazi Configuration". This configuration allowed us to maintain vertical laser polarization without catastrophic loss of intracavity power. Despite the apparent lack of stability, this configuration did lase and dispersion of the dye fluorescence significantly reduced the K_2 photoluminescence. Further reduction of K_2 photoluminescence was obtained by inserting a K heat pipe in the cavity to absorb light resonant with K_2.

RESULTS AND DISCUSSION

Preliminary results for a typical IR experiment are shown in Table I for a counting period of about 5 minutes. The count rates are shown for various elementary signal processes Rijk where i, j, k are 0 or 1 to indicate K, NaCl or laser, off or on, respectively. These elementary rates are assumed to be additive so that the signal observed with all beams on, S_{111}, is given by

$$S_{111} = R_{111} + R_{110} + R_{101} + R_{011} + R_{100} + R_{010} + R_{001} + R_{000}.$$

("Origin" denotes the apparent source of signal. "Dark current" is observed when the PMT views the inside of the machine with all beams flagged, "K photoluminescence" arises from photoluminescence of some species in the K beam, and so on.)

As in the visible experiments the 3BS is positive and highly significant, but unlike the visible experiments, we have not yet completed auxilliary experiments to rule out various possible artifacts. Thus, the large K photoluminescence signal shows that K_2 is being excited, and this may play some role in the 3-beam observations. Replicate runs without the $K(K_2)$ heatpipe in the cavity shows much more K photoluminescence without affecting the 3BS very much. The K_2 present in the heatpipe thus removes some of the spontaneous dye fluorescence which is responsible for the "K photoluminescence" but has no effect on the 3BS. This suggests that excited K_2 is not responsible for the 3BS, but more definitive experiments need to be performed.

It is clear that IR excitation of the system produces emission in detectable amounts, and observation of the process illustrated in Fig. 1 seems possible, although we have not yet ruled out a number of artifacts. We are currently working to do that, and hope that these observations will provide us with information about the X and A state of the complex in the region where bonds are making and breaking.

We gratefully acknowledge support of this research by the National Science Foundation Grant CHE-837764 and by Robert A. Welch Foundation Grant C-919.

Table I

Count rates (± 1 σ) for observation of emission passed by a 650-750 nm filter with various combinations of K, NaCl, and light beams for irradiation at λ = 840 nm.

Signal	Origin	count rate s^{-1}
R_{000}	Dark Current	444 ± 6
R_{001}	Scattered Light	330 ± 9
R_{100}	K Background	226 ± 9
R_{010}	NaCl Background	4068 ± 19
R_{101}	K Photoluminescence	8021 ± 29
R_{010}	NaCl Photoluminescence	-37 ± 27
R_{110}	Chemiluminescence	178 ± 27
R_{111}	3-Beam Signal	1479 ± 55

REFERENCES

1. T. C. Maguire, P. R. Brooks, R. F. Curl, *Phys. Rev. Lett.*, **50**, *1918*, (1983).
2. T. C. Maguire, Ph.D. Thesis, Rice University (1984).
3. T. C. Maguire, P. R. Brooks, R. F. Curl, J. H. Spence, S. J. Ulvick, *J. Chem. Phys.*, 1986 (to be published).
4. A. C. Roach and M. S. Child, *Mol. Phys.*, **14**, (1968)

DYNAMICS OF REACTIVE COLLISIONS BY FAR WING LASER LIGHT SCATTERING*

P. D. Kleiber, A. M. Lyyra, K. M. Sando,
V. Zafiropulos and W. C. Stwalley
University of Iowa, Iowa City, Iowa 52242

ABSTRACT

We report on our studies of the far wing absorption of laser light into the MgH_2 reactive collision complex. We have measured the absorption profile leading to both the non-reactive channel (Mg* formation) and into a specific ro-vibrational state of the reactive channel (MgH (v" = 0, J" = 23) and have developed a simple theoretical model to explain some of the general features of the results.

In a recent paper[1] we presented preliminary results of our experiments on the far wing laser absorption into the MgH_2 collision complex. The process may be written symbollically as

$$Mg + H_2 + h\nu_L \rightarrow (MgH_2) + h\nu_L \rightarrow (MgH_2)^* \tag{1}$$

followed by

$$(MgH_2)^* \rightarrow Mg^* + H_2$$

or

$$ MgH(v'', J'') + H. \tag{2}$$

Here we discuss qualitatively the implications of these results for our understanding of the dynamics of this excited state reactive collision, and present a semi-quantitative comparison of these experimental data with a very simple theoretical model based on approximate potential curves and a modified orbiting model of the collision dynamics.

The major results of the work are given in Figure 1. The far wing profiles show the relative populations in (a) the $Mg^*(3^1P_1^0)$ level (absorption into the non-reactive channel) and (b) the MgH(v" = 0, J" = 23) to corresponding absorption into a particular final state-specific reactive channel) as a function of pump laser detuning from the Mg resonance line ($\Delta \equiv \omega_L - \omega_0$). The absorption profiles have been multiplied by a factor Δ^2 to expand the scale and enhance the structure. The collisional redistribution profile of Figure 1(a) was obtained from the spectrally and temporally integrated emission on the Mg* resonance line, while the reactive collision profiles of Figure 1(b) are obtained from the integrated probe LIF signal. Also shown are the results (solid lines) of a simple semi-quantitative theoretical model. Our theoretical model is based on the ab initio SCF-CI potential curves of Chaquin et al.[2] for C_{2v} and $C_{\infty v}$ geometries (where the H-H distance is fixed at equilibrium). Following the dynamical arguments of Breckenridge and Umemoto[3] we

* Research supported by the National Science Foundation under grant CHE 83 13352 and by the Petroleum Research Fund under grant PRF 15244G.

correlate the high rotation product state MgH (v" = 0, J" = 23) with side-on (C_{2v}) attack of Mg onto H_2. We then apply the classical Franck-Condon principle to calculate the spectral absorption profile in the quasistatic limit. Complete potential energy surfaces would be required to accurately evaluate the subsequent excited state dynamics including branching into a distribution of exit channels (non-reactive and reactive). Lacking the complete surfaces we instead use simple classical arguments to describe the dynamical evolution of the system through the reaction. Specifically, we assume that all free-bound and free-quasibound absorption will lead to reaction. This is probably reasonable because of the observed high reaction probability in the excited state accompanied by the long lifetimes of the bound and quasibound states. Free-free transitions to states below the rotational barrier are assumed not to react while free states above the rotational barrier are assumed to react on incoming trajectories only.

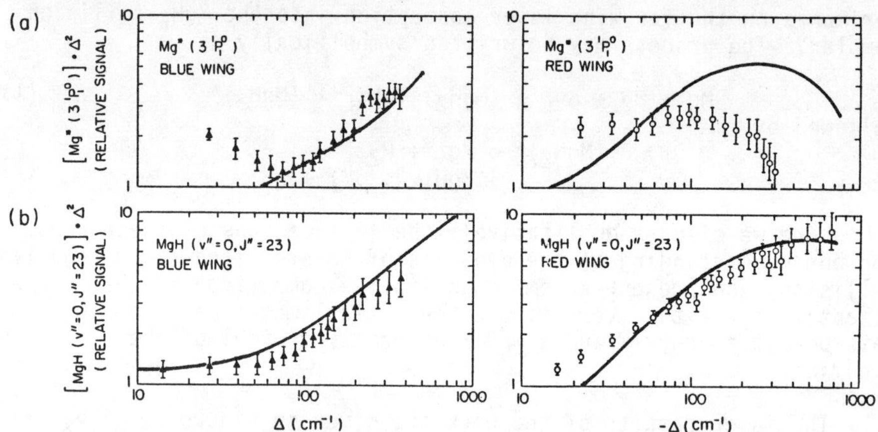

Figure 1. (a) Collisional redistribution profile showing $Mg^*(3^1P_1^0)$ emission, and (b) reactive absorption profile showing laser-induced fluorescence signal from MgH (v" = 0, J" = 23), as functions of pump-laser detuning from the magnesium resonance ($\Delta \equiv \omega_L - \omega_0$). The open circles correspond to red detunings and the solid triangles to blue detunings.

The overall agreement between experiment and theory is quite reasonable. In particular note that the theoretical profiles generally demonstrate the correct qualitative behavior, particularly at shorter internuclear range (larger detunings). Our experimental results indicate reasonable agreement with the potential surface calculations of Chaquin et al. to laser detunings of $\lesssim 500\ cm^{-1}$ (corresponding to $Mg\text{-}H_2$ internuclear separations $\gtrsim 3.0$ A estimated from the theoretical difference potentials).

We believe the rapid decrease in the branching ratio of non-reactive to reactive absorption in the red wing can be intuitively explained in terms of channel competition effects. As the laser is tuned deeper into the attractive excited potential well, the non-reactive channel becomes less favorable energetically; the very sharp drop at $\Delta \sim 200\ cm^{-1}$ (red) is probably due to the onset of bound or quasibound absorption in the MgH_2 complex. This qualitative picture is roughly born out in our theoretical model.

The red/blue wing asymmetry for the reactive profiles of Figure 1(b) (MgH(v" = 0, J" = 23)) is correctly predicted by our simple theoretical model. This agreement is perhaps somewhat unexpected since the model assumes a unit reaction probability for absorption on incoming trajectories into the highly repulsive 2^1A_1 state. This large reaction probability for trajectories on the repulsive curve is in contrast to the usual theoretical arguments suggesting that the excited state reaction occurs preferentially via the attractive 1^1B_2 level. It is, however, consistent with the observation of the large (essentially gas kinetic) overall reactive quenching cross section measured by Breckenridge and Umemoto[3]. Of course, our picture of the repulsive nature of the 2^1A_1 curve must be modified when the H-H distance is allowed to vary; this observation can perhaps be understood in terms of an insertive attack of Mg stretching the H-H bond following blue wing absorption to the repulsive 2^1A_1 curve at the equilibrium H-H separation.

REFERENCES

1. P. D. Kleiber, A. M. Lyyra, K. M. Sando, S. P. Heneghan and W. C. Stwalley, Phys. Rev. Lett. 54, 2003 (1985).
2. P. Chaquin, A. Sevin and H. Yu, J. Phys. Chem. 89, 2813 (1985).
3. W. H. Breckenridge and H. Umemoto, J. Chem. Phys. 80, 4168 (1984).

POLARIZATION AND ALIGNMENT EFFECTS IN COLLISIONS INVOLVING OPEN-SHELL MOLECULES

Millard H. Alexander and Gregory C. Corey
University of Maryland, College Park, MD 20742

Stephen L. Davis
George Mason University, Fairfax, VA 22030

Paul J. Dagdigian
The Johns Hopkins University, Baltimore, MD 21218

ABSTRACT

Two collaborative experimental-theoretical projects aimed at the elucidation of polarization and alignment effects in molecular collisions are discussed. In both studies general theoretical propensity rules have been derived, independent of a particular choice of potential energy surface, and then verified experimentally.

I. INTRODUCTION

Most past work on inelastic molecular collisions has been devoted to the determination of degeneracy averaged differential or integral cross sections, which represent sums and averages over all relative orientations of the molecular angular momentum vector J before and after the collision. The investigation of the dependence of the cross sections on the projection quantum number can provide additional information on the detailed nature of the dynamics and on the way in which the anisotropy in the intermolecular potential allows the orbital angular momentum of the collision partners to couple with and ultimately reorient J.[1-5]

The orientation of this angular momentum vector can be defined relative (a) to the separation vector of the two collision partners (often designated the collision frame), (b) to some laboratory fixed axis system, or, finally, (c) to a molecule-fixed coordinate system. For example, a differential scattering experiment would naturally fall into category (a), since the initial and final relative velocity vectors are well defined. In contrast a cell experiment would fall into category (b), because the axis of quantization, which is imposed by an external electric or magnetic field, is fixed in the laboratory. Since the relative velocity vector of the collision partners is not uniquely oriented with respect to a laboratory frame, the observed M-dependent cross sections will represent an average over different orientations of J in the collision frame.[6-8]

An additional degree of complexity occurs in collisions involving open-shell molecules. Here the orbital angular momentum of the collision partners can couple not only with the nuclear rotational angular momentum of the molecule but also with the electronic orbital and spin angular momenta, as well as with the spin of the nuclei. A study of collision-induced transitions between atomic or molecular fine- and hyperfine-structure multiplet levels would fall into category (c) of the preceding paragraph and would probe the degree to which the electrostatic interaction potential (or potentials) can decouple and recouple the various angular momenta in a molecule-fixed coordinate system.

This article reviews two recent exemplary studies: (1) transitions between molecular hyperfine levels in collisions of $CaBr(X^2\Sigma^+)$ with Ar[9] and (2) transitions between individual M levels in collisions of $CaF(A^2\Pi)$ with Ar.[10,11]

II. COLLISIONAL TRANSITIONS BETWEEN MOLECULAR HYPERFINE LEVELS

In contrast to the body of earlier work devoted to collisions between atomic hyperfine levels,[12] there has been little similar work on molecules. Recently, Dagdigian and Bullman[13] demonstrated that an electric quadrupole will selectively refocus different hyperfine levels of CaBr in its ground $X^2\Sigma^+$ electronic state. This can be used to investigate the variation of rotationally inelastic CaBr+Ar cross sections with hyperfine level.[9]

Since the atom-molecule interaction potential is purely electrostatic in origin,[14] both the electronic spin **S** and nuclear spin **I** remain spectators during a collision. The recent angular momentum decoupling scheme introduced by Corey and McCourt[15] can then be used to write the integral cross section for a transition between two hyperfine-state-resolved levels as

$$\sigma_{NJF\to N'J'F'} = \frac{\pi}{k^2}\,[JJ'F']\sum_K \begin{Bmatrix} J & J' & K \\ F' & F & I \end{Bmatrix}^2 \begin{Bmatrix} N & N' & K \\ J' & J & S \end{Bmatrix}^2 P^K(N,N') \quad . \qquad (1)$$

Here $\{:::\}$ is a 6j symbol,[16] $[x_1 \ldots x_n]$ is a product of $2x_i+1$ degeneracy factors, k is the wavevector in the initial channel, and J, N, and F designate, respectively, the total molecular (excluding nuclear spin), the nuclear rotational and the total molecular (including nuclear spin) angular momenta. For $^2\Sigma^+$ molecules the levels with $J=N+1/2$ are designated **e** and those with $J=N-1/2$ are designated **f**.[17] The quantity $P^K(N,N')$ is a tensor opacity, which can be interpreted as the probability that a collision will reorient the nuclear rotational angular momentum vector **N** by K units.[18,19] The details of the collision dynamics are contained in these tensor opacities. The dependence of the cross sections on the spectator angular momenta I and S appears solely in the vector coupling coefficients in Eq. (1).

Collision-induced transitions between atomic hyperfine levels have been interpreted within a M-randomizing model,[12] in which collisions are assumed not to affect the orientation of I while at the same time leading to a total randomization of the orientation of **J**. One can show[9] that this model corresponds to a purely statistical limit for the tensor opacities, namely that the probability of collisional reorientation of **N** is proportional to the statistical weight associated with the degree of reorientation. The hyperfine resolved $F\to F'$ cross sections in this M-randomizing limit are then predicted[9] to be purely statistical, proportional to the degeneracy of the final hyperfine level and independent of the initial quantum number.

Quantum mechanical calculations[18,20] as well as many experiments confirm the idea that tensor opacities for rotationally inelastic collisions with closed-shell targets will be large only for small values of the tensor order K. This is because most molecular collisions are glancing encounters at large impact parameter which give rise to only small changes in both the magnitude and direction of **N**. For small K, the 6j symbols which appear in Eq. (1) are largest when $\Delta F=\Delta J=\Delta N$.[9,18] Thus we predict, without the introduction of dynamical approximations, that in contrast to the M-randomizing model, for a given initial hyperfine level the only facile transitions will be those to final hyperfine levels which satisfy this $\Delta F=\Delta J=\Delta N$ selection rule.

The experimental technique used to study hyperfine resolved transitions in the ground $X^2\Sigma^+$ electronic state of CaBr involves electric quadrupole state selection of a supersonic beam of CaBr, produced by the vaporization of a mixture of Ca and $CaBr_2$.[9,13,21] For example, in the case of the N=3,e level, only the hyperfine levels with the two highest values of the total angular momentum F (F=5 and F=4) are focussed. After passage through the quadrupole the state selected beam enters a scattering chamber, where the incident and inelastically scattered molecules are interrogated by cw dye laser fluorescence excitation on the $B^2\Sigma^+-X^2\Sigma^+$ band system. Experimental scans were carried out to probe collision-induced N=3,**e**→N=5,**e** and N=2,**e**→N=1,**e** transitions. In both cases the spectra obtained were in **close agreement** with spectra simulated from Eq. (1) using values of the relevant tensor opacities taken from our previous study[20] of $CaCl(X^2\Sigma^+)$+Ar collisions, but **differed significantly** from the predictions of the M-randomizing (statistical) model.[9]

III. TRANSITIONS BETWEEN INDIVIDUAL M LEVELS IN COLLISIONS OF $CaF(A^2\Pi)$ WITH Ar

In collision experiments involving laser-excited atoms or molecules, the linear or circular polarization of the excitation laser results in a nonuniform distribution of M levels in the excited state.[5,6,22] This implies that the observed cross sections will carry information on the dependence of the inelastic process on the initial and final projection quantum numbers of the excited molecule (or atom). In cell experiments involving M state preparation the observed cross sections represent an average of the collision frame M-dependent scattering amplitudes over all possible orientations of the relative velocity vector with respect to the laboratory fixed axis of quantization. The expression for the JM→J'M' integral cross section, where the projection quantum numbers refer to the laboratory fixed axis of quantization, can be written in the particularly simple form[8,11,23]

$$\sigma_{JM\to J'M'} = \frac{\pi}{k^2} \sum_{KQ} \begin{pmatrix} J & J' & K \\ -M & M' & -Q \end{pmatrix}^2 P^K(J,J') \qquad (2)$$

Here $(\vdots\vdots\vdots)$ is a 3j symbol[16] and the quantity $P^K(J,J')$ is a tensor opacity. As discussed in Sec. II above, these tensor opacities can be interpreted as the probability that a collision will reorient the angular momentum **J** through K units. The reader should not confuse the tensor orders which appear in Eq. (2) with those which appear in the expansion of the density operator, for either the initial or final states, in spherical tensor components.[6,8,22] In fact, the relative efficiency of collisional transfer of either the first moment (polarization) or the second moment (alignment) of the initial density matrix can be expressed, in a manner similar to Eq. (2), as a sum over tensor opacities.[24]

Equation (2) is reminiscent of the Wigner-Eckart theorem,[16] except that the 3j symbol is squared. Since the projection quantum numbers in the denominator of a 3j symbol can never exceed the corresponding angular momenta in the numerator,[16] a given tensor opacity will contribute only to transitions for which $|M-M'|\leq K$. Since most molecular collisions are grazing encounters, we expect, as discussed in Sec. II, that the largest tensor opacities will correspond to the smallest allowed values of K.

In molecules in a $^2\Pi$ state the rotational levels are split into two Λ-doublet levels of opposite parity, designated **e** and **f.**[25] We have shown that for transitions within a given spin-orbit manifold of a $^2\Pi$ molecule the tensor opacities which appear in Eq. (2) are vanishingly small unless $(-1)^{J+J'+K} = \pm 1$, where the plus sign refers to **e/f** changing transitions and the minus sign, to **e/f** conserving transitions. This selection rule arises from the rotational invariance of the scattering operator.[11,18,24]

Let us specialize to transitions between the J=1/2 Λ-doublet levels. The triangular relations[16] contained in the 3j symbol in Eq. (2) and the selection rule introduced in the preceding paragraph imply that only the K=0 tensor opacity will contribute to the cross section for M changing transitions within one Λ-doublet level (**e/f** conserving) and only the K=1 tensor opacity will contribute to the cross section for transitions between the Λ-doublet levels (**e/f** changing). A similar argument implies that for $J=1/2\to J'=3/2$ transitions only the K=1 tensor opacity will contribute to the **e/f** conserving cross sections while only the K=2 tensor opacity will contribute to the **e/f** changing cross sections.

The first implication of this discussion is that $M=1/2\to M'=-1/2$ transitions within either of the J=1/2 Λ-doublet levels will be **forbidden.** This is because the properties of the 3j symbol in Eq. (2) imply that the K=0 term, which makes the only contribution to these transitions, can couple only levels with the same projection quantum numbers. In other words **collisional depolarization of a J=1/2 Λ-doublet level will never occur.** This prediction, made several years ago by Alexander and Davis,[11] is independent of the potential surface for the collision in question. A simple physical analogy is provided by the metastability of the 1s2s ^2S level of the He atom. Because this level has the same parity as the ground state, only even multipole terms in the field-matter interaction can cause the transition. Yet, because of the low angular momentum in these two states (J=0) the quadrupole (or higher even multipole) coupling vanishes, as it similarly does in the present case between two J=1/2 states of identical parity.

The cross sections for the **e/f** changing transitions between the J=1/2 Λ-doublets involve only the K=1 tensor opacity. Numerical evaluation of the 3j symbol in Eq. (2) implies that the cross section for the M=1/2→M'=-1/2 transition will be twice as large as the cross section for the M=1/2→M'=+1/2 transition. A further scaling relation emerges[9] within the energy sudden limit,[26] where the tensor opacities will be independent of the initial and final rotational quantum numbers. In this case the J=1/2,M=1/2→J'=3/2,M' **e/f conserving** cross sections, which also involve only the K=1 tensor opacity, can be related directly to the J=1/2→J'=1/2 **e/f changing** cross sections. The relative cross sections are given in Table I.

Table I. Relative values of **e/f conserving** J=1/2,M=1/2→J'=3/2,M' cross sections, normalized to the **e/f changing** J=1/2,M=1/2→J'=1/2,M'=1/2 cross section.

M' =	3/2	1/2	-1/2	-3/2
relative cross section	3	2	1	0

Both the selection rule forbidding M=1/2→M'=-1/2 transitions within one J=1/2 Λ-doublet level as well as the scaling relations summarized in Table I are now being verified by Norman and Field[10] in a series of double resonance experiments involving the $A^2\Pi$ state of CaF. Preliminary results confirm two of our theoretical predictions: (1) There is no evidence of M=1/2→M'=-1/2 transitions within the pumped Λ-doublet, and (2) the four J=1/2,M=1/2→J'=3/2,M' cross sections obey the scaling relation shown in Table I. However, for the two possible transitions across the J=1/2 Λ-doublet (M=1/2,**e/f**→M'=±1/2,**f/e**) the experimental ratio differs significantly from the theoretical prediction of 2:1. A possible explanation involves the role of hyperfine coupling, which was ignored in our theoretical analysis.[11]

IV. DISCUSSION

In addition to the polarization effects discussed above, it is also of fundamental interest to investigate the effect of collisions on the orientation of the molecular angular momentum in a frame defined by the initial and/or final relative velocity vectors of the collision partners. As mentioned in the Introduction, this is the natural coordinate system for the description of a beam experiment. In this context we mention some beautiful recent experiments,[4,5] most notably by Bergmann and co-workers,[5] on collisions involving Na_2 molecules with optical pumping state selection. Clear evidence was seen of the conservation of the orientation of J along the kinematic apse (the bisector of the initial and final relative velocity vectors), which confirms an earlier prediction of Khare, Kouri, and Hoffman[27] based on a simple impulsive model.

ACKNOWLEDGMENT

The authors are grateful to Jeffrey Norman and Robert Field (MIT) for useful discussions about the details of their experiment. The authors' work has been supported by the National Science Foundation, Grants CHE81-08464 and CHE84-08528, and by the U. S. Army Research Office, Grants DAAG-29-81-G-0102, DAAG-29-84-G-0078, and DAAG-29-85-G-0018. Gregory Corey is grateful to NATO for a postdoctoral fellowship.

REFERENCES

1. M. H. Alexander, P. J. Dagdigian, and A. E. DePristo, J. Chem. Phys. 66, 59 (1977).
2. D. A. Case, G. M. McClelland, and D. R. Herschbach, Mol. Phys. 35, 541 (1978).
3. M. D. Rowe and A. J. McCaffery, Chem. Phys. 43, 35 (1979).
4. M. A. Treffers and J. Korving, Chem. Phys. Lett. 97, 342 (1983).
5. A. Mattheus, A. Fischer, G. Ziegler, E. Gottwald, and K. Bergmann, Phys. Rev. Lett., submitted.
6. A. Omont, Progr. Quantum Electron. 51, 69 (1977).
7. Ph. Bréchignac, A. Picard-Bersellini, R. Charneau, and J. M. Launay, Chem. Phys. 53, 185 (1981).
8. M. H. Alexander and S. L. Davis, J. Chem. Phys. 78, 6754 (1983).
9. M. H. Alexander and P. J. Dagdigian, J. Chem. Phys. 83, 2191 (1985).
10. J. Norman and R. W. Field, unpublished.
11. M. H. Alexander and S. L. Davis, J. Chem. Phys. 79, 227 (1983).
12. A. Omont, J. Phys. 26, 26 (1965); E. A. Franz and J. R. Franz, Phys. Rev. 148, 82 (1966).
13. P. J. Dagdigian and S. J. Bullman, Chem. Phys. 88, 479 (1984).
14. M. H. Alexander, J. Chem. Phys. 76, 3637 (1982).
15. G.C. Corey and F. R. McCourt, J. Phys. Chem. 87, 2723 (1983).
16. D. M. Brink and G. R. Satchler, **Angular Momentum**, 2nd. ed. (Oxford University, Oxford, England, 1975).
17. J. M. Brown, J. T. Hougen, K.-P. Huber, J. W. C. Johns, I. Kopp, H. Lefebvre-Brion, A. J. Merer, D. A. Ramsay, J. Rostas, and R. N. Zare, J. Mol. Spectrosc. 55, 500 (1975).
18. G. C. Corey and A. D. Smith, J. Chem. Phys. 83, 5663 (1985); M. H. Alexander, J. E. Smedley, and G. C. Corey, ibid., in press.
19. J. Derouard, Chem. Phys. 84, 181 (1984).

20. M. H. Alexander, S. L. Davis, and P. J. Dagdigian, J. Chem. Phys. 83, 556 (1985).
21. See also S. J. Bullman and P. J. Dagdigian, J. Chem. Phys. 81, 3347 (1984); P. J. Dagdigian and S. J. Bullman, ibid. 82, 1341 (1985).
22. K. Blum, **Density Matrix Theory and Applications** (Plenum, NY, 1981).
23. See, however, L. Monchick, J. Chem. Phys. 75, 3377 (1981).
24. M. H. Alexander, J. Chem. Phys. 71, 5212 (1979); M. H. Alexander and T. Orlikowski, ibid. 80, 1506 (1984).
25. M. H. Alexander and P. J. Dagdigian, J. Chem. Phys. 80, 4325 (1984).
26. V. Khare, J. Chem. Phys. 68, 4631 (1978).
27. V. Khare, D. J. Kouri, and D. K. Hoffman, J. Chem. Phys. 74, 2275 (1981).

STATE-SPECIFIC COLLISION DYNAMICS OF OH RADICALS AND N ATOMS

Richard A. Copeland, David R. Crosley, and Jay B. Jeffries
Molecular Physics Department, SRI International,
Menlo Park, CA 94025

ABSTRACT

Open shell species, with a variety of spectroscopically accessible quantum states, offer opportunities for laser generation of nonequilibrium spatial distributions in both the laboratory (m_J) and molecular (Λ-doublet) frames. In addition, laser excitation can also select the relative orientation of the spin and orbital angular momentum in these systems. The quantum state distributions can be interrogated by a delayed second laser pulse or by resolving the polarization and/or wavelength of the resulting fluorescence. In experiments on the OH radical, we have observed propensities for retention of electronic parity and spatial orientation during rotationally inelastic collisions of, respectively, the $X^2\Pi_i$ and $A^2\Sigma^+$ electronic states. In nitrogen atoms, we have qualitatively determined the magnitude of fine structure, m_J and electronic state changing collisions in the $2s^22p^23p\ ^4D^o$ electronic state.

INTRODUCTION

Combining initial quantum state selection and final quantum state resolution in an experimental technique permits a detailed examination of the state specific processes which occur during the collision of two species. Often these interactions are extremely state specific and show surprising mechanistic selectivity for complex molecular and atomic collision events. Here, we describe three different experiments recently performed in our laboratory which examine the collision dynamics in the ground ($X^2\Pi_i$) and first excited ($A^2\Sigma^+$) state of OH and the $^4D^o$ state ($2s^22p^23p$) of N atoms.

Λ-DOUBLET PROPENSITIES IN OH($X^2\Pi_i$) COLLISIONS WITH H_2O

Doublet pi states of diatomic radicals possess nearly degenerate but spatially distinct Λ-doublet electronic states having opposite parity, in addition to the angular momenta resulting from electron orbital motion, spin and nuclear rotation. By observing changes in the electronic parity (e/f states) following rotationally elastic and inelastic collisions, we gain insight into the potential surfaces involved in the interaction.

0094-243X/86/1460545-4$3.00 Copyright 1986 American Institute of Physics

Using a two-laser infrared pump/time-delayed ultraviolet probe technique, we examined collisions of ground state OH(v=2) with water vapor.[1] Although the e/f pair differs in energy by less than 0.25 cm^{-1}, the excitation laser bandwidth and the characteristics of the A-X electronic transition permits selection of the initial e/f state and interrogation of the final e/f state. These experiments were performed at 0.14 Torr of H_2O in a room temperature discharge flow cell. For the lowest rotational level in the $^2\Pi_{3/2}$ state, elastic f→e transfer is one third to one half of the total collision removal rate. For inelastic collisions to the next higher rotational level f→f is favored over f→e by a factor of 2.7.

The results for Λ-doublet propensities in rotationally inelastic collisions are perhaps the most interesting and surprising in that the OH-H_2O system will most likely form a hydrogen bonded collision complex. Some indirect experimental evidence for the complex is the rapid electronic quenching of the $A^2\Sigma^+$ state of OH by H_2O.[2] Hence, one might expect little or no memory of the initial state in this strongly interacting collision system. This experiment clearly shows that even in systems where long range dipole-dipole forces dominate, collision processes can be controlled by the orientation of the electronic orbitals.

POLARIZED FLUORESCENCE FROM OH IN ATMOSPHERIC PRESSURE FLAMES

Observation of alignment effects in molecular collisions is not restricted to the well controlled environment of a low pressure cell or a molecular or atomic beam. When polarized lasers are used to examine systems at atmospheric pressure, nonequilibrium spatial distributions can be manifested in fluorescence which is significantly polarized. In this experiment we measured the polarization and relative magnitude of laser-induced fluorescence signals generated by a polarized laser which excites molecules to the $A^2\Sigma^+$ state of OH in the burnt gases of an atmospheric pressure methane/oxygen flame.[3] The extent of the polarization can have a significant effect on the quantitative measurement of OH in systems where collision processes dominate (i.e. flames and the atmosphere).

In systems at atmospheric pressure collisions occur at a rate of ~10 ns^{-1}, suggesting that the elastic depolarizing collisions (J conserving, m_J changing) might totally destroy any polarization effects generated by the nonequilibrium m_J distribution created by the laser. However the important quantity is not the depolarizing rate D but the relative probability that a given collision will lead to depolarization. The total removal rate R, which includes electronic quenching plus vibrational and rotational energy transfer, is very rapid for collisions with the primary combustion products H_2O and CO_2. If $R>D$ polarized emission will be observed; however, if $D>R$ the emission will be unpolarized. For the initially prepared rotational level significant polarization is seen by wavelength resolved fluorescence scans in the different relative laser and detection polarizations.

While quantitative results are difficult to obtain for this combustion system several important and new qualitative features emerged from this investigation. The first is the similarity in the magnitude of the rates R and D for $OH(A^2\Sigma^+, v=1)$ resulting in polarized fluorescence from the initially excited rotational level. A surprising result was the polarization of the fluorescence from neighboring rotational levels with the same spin component ($F_1 \rightarrow F_1$, $F_2 \rightarrow F_2$). This shows that rotationally inelastic collisions at these relative velocities (~1500K) prefer a retention of the m_J component. Collisions which reorient the spin ($F_1 \rightarrow F_2$) and rotational angular momentum, however, generate no observable polarization.

These experimental observations clearly show polarization of the fluorescence must be considered when making absolute concentration measurements in atmospheric pressure systems, and even in the complex environment of the flame the state specific nature of the collision dynamics is important.

COLLISION DYNAMICS OF NITROGEN ATOMS

In atomic systems the angular momentum J is the vector sum of the orbital angular momentum and the electron spin. Nitrogen in the $^4D^o$ state has four distinct fine structure components corresponding to total angular momentum J=1/2, 3/2, 5/2 and 7/2. Using the two photon transition at ~210 nm we excited nitrogen atoms from the ^4S ground state to a particular J component of the $^4D^o$ excited state.[4] Using the wavelength, polarization and time dependence of the fluorescence we can unravel the rates and pathways of collisional processes in this highly excited atomic species.[5]

From the time dependence of the fluorescence we obtained the radiative lifetime of the excited state and collisional quenching rates. He and N_2 were the collision partners examined. From the pressure dependence of the observed fluorescence decay we obtain a radiative lifetime for the $^4D^o$ state of N of 45±5 ns, and a collisional quenching rate for N_2 of $5 \pm 1 \times 10^{-10}$ $cm^3 s^{-1}$ independent of the fine structure component. He does not significantly remove the excited state over the pressure range of 1 to 10 Torr.

The two-photon excitation process creates non-equilibrium distributions in the magnetic sublevels of the excited state, thus generating spatially anisotropic polarized fluorescence. The emission from the initially excited state was polarized. Hence m_J changing collisions are competitive with quenching in the case of N_2 and with fine structure changing collisions for He. To remove any ambiguities in the fine structure changing results due to non-equilibrium m_J distributions, a magnetic field was applied to randomize the magnetic sublevels and destroy the initial laboratory frame orientation produced by the laser.

Figure 1 shows resolved fluorescence scans after excitation of each of the four fine structure components of the $^4D^o$ state at a fixed pressure of N_2 and He. The peaks correspond to transitions between different fine structure components of the $^4D^o$ and ^4P states; the solid line is the best fit for the four $^4D^o$ populations contributing to the fluorescence. Similar data at different pressures of N_2 and He plus values of the radiative lifetime and quenching rates yield state-to-state fine structure changing collision rates. This phase of the investigation is currently underway; however, several qualitative features are already apparent. Most importantly, $\Delta J=1$ transitions occur significantly faster than those with $\Delta J=2$ and $\Delta J=3$ for all J. The $1/2 \rightarrow 3/2$ rate is the largest of all.

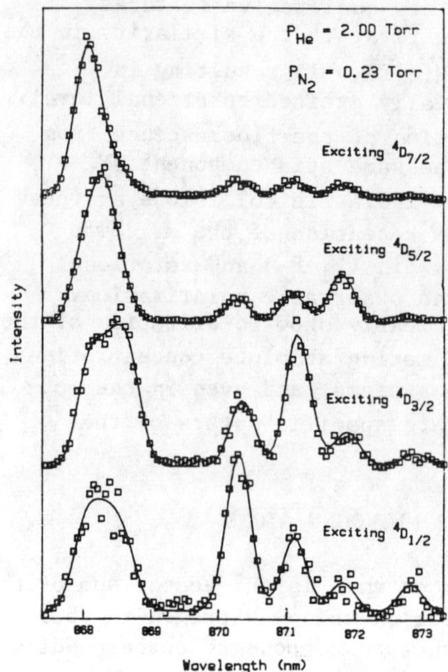

Fig. 1 N atom fluorescence scans.

These preliminary results on N atoms demonstrate how combining initial state selection and final state resolution can reveal the details of state-specific collisions.

ACKNOWLEDGMENTS

Research in this area has been supported by NASA Contract NAS1-16956, NSF Grant CPE-8319610 and AFOSR Contract F49620-85-K-0010. We gratefully acknowledge the experimental contributions of Paul Doherty on the flame study and Mark J. Dyer on the Λ-doublets.

REFERENCES

1. R. A. Copeland and D. R. Crosley, J. Chem. Phys. 81, 6400 (1984).
2. R. A. Copeland, M. J. Dyer, and D. R. Crosley, J. Chem. Phys. 82, 4022 (1985).
3. P. M. Doherty and D. R. Crosley, Appl. Opt. 23, 713 (1984).
4. W. K. Bischel, B. E. Perry, and D. R. Crosley, Appl. Optics 21, 1419 (1982).
5. J. B. Jeffries, R. A. Copeland, and D. R. Crosley, to be published.

ANISOTROPIES IN LASER-INDUCED CHEMICAL PROCESSES: EJECTION OF OH BY THE HONO $\tilde{A}$ STATE

R. Vasudev
Wright-Rieman Chemistry Laboratories
Rutgers, The State University of New Jersey
New Brunswick, New Jersey 08903

ABSTRACT

Aligned trans HONO is optically prepared in specific N=O stretching vibrational levels of the quasi-stable $\tilde{A}$ state. The ejected OH fragment is characterized by laser polarization and Doppler spectroscopy. The rotational and translational anisotropies, and the quantum state populations are probed through the $A^2\Sigma^+-X^2\Pi$ transition. The fragment is found to be rotationally aligned, and its unpaired π electron orbital lies preferentially in the plane of rotation. These effects become more pronounced at higher rotational energies. The OH translational velocity is anisotropic, resulting in characteristic Doppler-split line shapes. These observations, together with the measured internal state populations, permit the construction of a detailed mechanism for the fragmentation process.

INTRODUCTION

The dependence of the outcome of chemical events on the initial conditions is of considerable interest. State-selected photofragmentation offers a convenient means of addressing this issue. In the absence of exit-channel effects, the details of the dynamics and the fate of the products of such a process are determined by the lifetime, geometry, force field and intramolecular interactions of the parent molecule. It has been recognized[1] that for obtaining incisive information on reaction dynamics one should measure not only the vectorial and scalar properties of the products, but also the correlations among them. We describe here such a study on the translational and rotational anisotropies in laser-induced dissociation, and their correlation with the product quantum state populations. The system chosen is HONO, for the reasons given below.

In the near ultraviolet region ($\bar{\nu}$=25000-34000 cm^{-1}), HONO exhibits an extensive system of bands that are very diffuse and show no signs of fine-structure or local undulations at moderately high resolution.[2] If the widths of the absorption features are assumed to be entirely due to the dissociation of the excited state, the estimated fragmentation time is about 90 femtoseconds. According to King and Moule,[2] the absorption spectrum is due to the excitation of a nonbonding electron on the oxygen atom in the N=O chromophore to a π^* orbital, so that the overall symmetry change is $^1A''\leftarrow{}^1A'$ with the transition moment perpendicular to the molecular plane. In the trans isomer of HONO, the only geometrical parameter that undergoes a significant change upon electronic excitation is the N=O bond length and, consequently, the absorption spectrum shows a long prominent progression in the predominantly N=O stretching vibration ν_2. The near UV absorption ($\tilde{A}^1A''\leftarrow\tilde{X}^1A'$) thus offers a convenient means of optically selecting one of the initial conditions in the half-collision (fragmentation) process, namely the number of quanta of the N=O stretching vibration. This system is therefore well-suited for probing how the quantum states of reactants control the outcome of chemical events. In addition, since the $\tilde{A}$ state lifetime is shorter than the rotational period, HONO photofragmentation is ideal for observing the product rotational and translational anisotropies. We describe below (a) the measurement of these anisotropies in the photoejected OH fragment and their correlation with the quantum-state populations, and (b) the state-to-state dynamics inferred from these measurements. We begin with a survey of our published results on the dissociation of the HONO $\tilde{A}$ state with 1,2 and 3 quanta of the ν_2 vibration.[3,4] We then describe some recent unpublished results, and finally mention some of our future plans.

RESULTS

In our experiments, <u>trans</u> HONO is excited to specific N=O stretching vibrational levels of the $\tilde{A}$ state by polarized laser photoselection. The ejected OH is probed through the $A^2\Sigma^+$-$X^2\Pi$ transition by laser polarization and Doppler spectroscopy. Since the HONO $\tilde{A}$-$\tilde{X}$ transition moment is perpendicular to the molecular frame, the ensemble of the photoexcited molecules are preferentially aligned with the molecular plane perpendicular to the electric vector ($\hat{\epsilon}_p$) of the photolysis laser (see Fig 1). In the limit of fast fragmentation, the product is therefore expected to be highly aligned. We describe the fragment rotational alignment in terms of a parameter $\mathcal{A}_0^{(2)}$ which is positive for $\vec{J}_{OH} \parallel \hat{\epsilon}_p$, negative for $\vec{J}_{OH} \perp \hat{\epsilon}_p$, and zero for an isotropic sample.[4] We measure $\mathcal{A}_0^{(2)}$ through excitation spectra with (a) the electric vector $\hat{\epsilon}_p$ parallel to the electric vector $\hat{\epsilon}_a$ of the analyzing probe laser, as shown in Figure 1; and (b) $\hat{\epsilon}_p \perp \hat{\epsilon}_a$. We find that the OH Q branch transitions are stronger and the P,R lines are weaker in case (a) than in case (b), implying that the OH fragment is preferentially aligned in the initial HONO plane. We also find that $\mathcal{A}_0^{(2)}$ varies with the quantum number N, as shown in Figure 2.

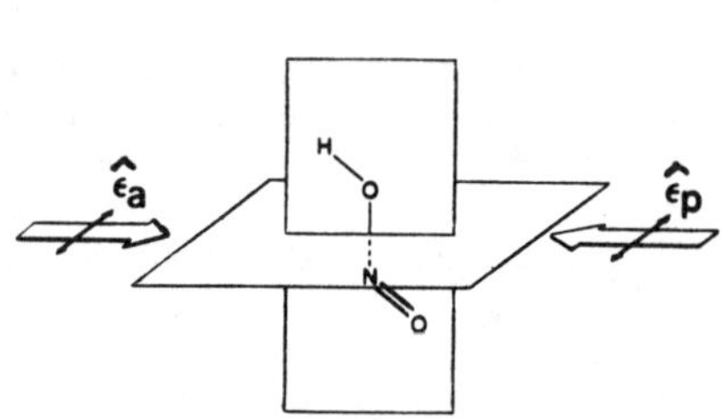

Fig 1. One of the experimental geometries used for the measurement of $\mathcal{A}_0^{(2)}$.

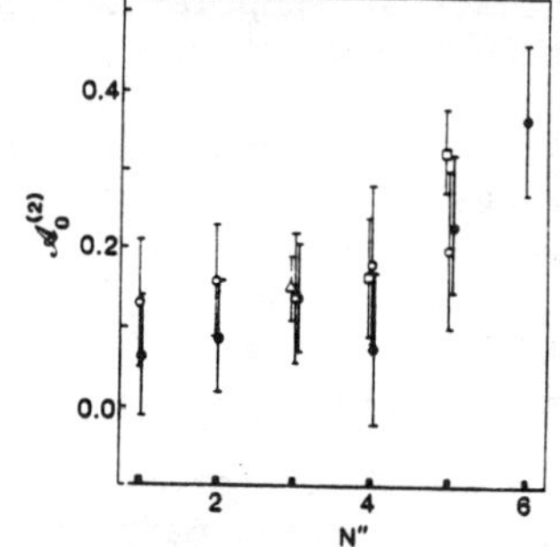

Fig 2. $\mathcal{A}_0^{(2)}$(J) vs. N". ●: $X^2\Pi^+_{3/2}$, 355 nm photolysis (v_2=2); Δ: $X^2\Pi^-_{3/2}$, v_2=2; □: $X^2\Pi^+_{1/2}$, v_2=2; o: $X^2\Pi^+_{3/2}$, v_2=3.

From the fragment excitation spectra with $\epsilon_p \parallel \epsilon_a$ and $\epsilon_p \perp \epsilon_a$, it is possible to deconvolute the polarization effects and hence accurately deduce the rotational population distributions. We find that, although the OH $X^2\Pi$ spin-orbit components F_1 and F_2 are not in equilibrium, the rotational distribution within each component can be approximately described by a "rotational temperature".[4]

Since the HONO $\tilde{A}$-$\tilde{X}$ transition moment is perpendicular to the central O-N bond, the center-of-mass angular distribution of the OH product (for a single recoil velocity), namely $W(\theta)=[1+\beta.P_2(\cos\theta)]/4\pi$, is characterized by the anisotropy parameter $\beta=-1$. We probe the corresponding Doppler profile by measuring the line-shapes with a narrow-band probe laser.[3,4] We find that the experimental profiles can be satisfactorily simulated by a small fragment velocity distribution with a sharp cutoff on the high energy side, and a mean recoil velocity of 2.50 km/s. We also find that the Doppler profiles are independent of OH rotation and, in addition, their widths are the same for the fragmentation of the 2^1, 2^2, and 2^3 vibrational levels of the HONO $\tilde{A}$ state.[4]

DISCUSSION

We now describe the mechanisms for the generation of OH fragment internal states and translation. Among the possible sources of OH rotation,[4] we first consider the impulse model. Here, the impulse associated with the O-N bond rupture exerts a torque on the central O atom and, if the O-H bond is rigid enough, this generates a rotation of the OH fragment. This model implies a sharply-peaked rotational distribution and also a partitioning of the available energy into OH rotation and translation. We do not observe either of these predictions, implying that the O-H bond is "floppy" and, therefore, the H-end of the molecule does not respond significantly to the impulse on the central O atom. All the available energy must then be channelled into OH translation.

The most satisfactory model for the fragment rotational excitation appears to be the one due to Freed et al.[5] Here, the OH rotation is derived from the angle-bending vibrations in the parent. Two such vibrations exist in HONO: the in-plane HON bend ν_3 and the out-of-plane torsion ν_6. Since we do not excite these vibrations in our experiments, it is the associated zero-point motions that are carried over into the fragment rotation. Since ν_3 has a higher frequency than the ν_6 vibration, this model predicts that the OH rotation will be predominantly in the initial HONO plane and, further, that the alignment should increase with rotation, as observed. We note, however, that part of the observed variation arises from the fact that $\mathcal{A}_0^{(2)}$ shown in Figure 2 refers to the alignment of $\vec{J}_{OH}$ whose direction coincides with that of the nuclear rotational angular momentum $\vec{R}$ only in the classical (high J) limit.[4,6]

The validity of the Freed mechanism, described above, can be verified through an analysis of the fragment populations in the Λ-doublet fine-structure states. We recall that the $p\pi$ orbital with the lone unpaired electron in the OH fragment is initially associated with the central O-N bond in HONO. The Freed mechanism therefore implies that the OH pπ orbital should be aligned with respect to the rotational angular momentum. In particular, the orbital should be preferentially in the plane of rotation, and this preference should increase with rotational energy. In other words, the Π^+ Λ-doublets should be populated in preference to the Π^- components and this effect should be more pronounced at higher rotations, which is what we observe experimentally.[4]

We now consider the OH fragment vibrational and translational energy content. The experimental results on these two degrees of freedom can be explained in terms of a simple Franck-Condon (FC) model. For convenience, we make the simplifying assumption that the HONO $\tilde{A}$ state lifetime is long enough that, just before fragmentation, the molecule does not retain much memory of its $\tilde{X}$ state history. Consequently, our model considers the wavefunctions of the HONO oscillators only in the $\tilde{A}$ state, and those in the fragments. The quantity of interest here is the probablilty of the process HONO $(\tilde{A}, r_{ON}, v_1, v_2) \rightarrow$ OH (X, v_{OH}) + NO (X, v_{NO}) + E_t, where E_t is the final interfragment recoil energy, ν_1= the O-H stretch, and ν_2= the predominantly N=O stretching vibration in HONO. Since the coupling between the vibrations in trans HONO is not strong,[2] the probablity of a particular vibration-translation partitioning in the fragments may be written as

$$P = \langle\psi_{v_1}^{HONO}|\psi_{v_{OH}}\rangle^2 \ \langle\psi_{r_{ON}}^{HONO}|\psi_t\rangle^2 \ \langle\psi_{v_2}^{HONO}|\psi_{v_{NO}}\rangle^2 \qquad (1),$$

where $\psi_{r_{ON}}^{HONO}$ is the HONO (continuum) wavefunction along the unstable coordinare r_{ON}, and ψ_t refers to the fragment recoil and determines the final relative linear momentum through the the deBroglie equation. The rest of the symbols in the above equation have their obvious significance.

The OH fragment vibrational populations are determined by the first term in eq (1). Since the O-H bond length in the parent[2] (0.98 Å) is very close to the value (0.97 Å) in the fragment, the first FC factor in eq (1) is diagonal. In our experiments, v_1=0, and thus the fragment is predicted to be generated exclusively in v_{OH}=0, as observed. The N=O bond lengths in the parent and the NO fragment are, however, are quite different (1.28 and 1.15 Å, respectively). Consequently, the third FC factor in eq (1) is nondiagonal, so that the NO fragment vibration is predicted to have a distribution. Therefore, by energy conservation, we expect a distribution of energy along the O---N translational coordinate [given by the second term in eq (1)]. Thus, if the NO fragment vibrational distribution peaks on the high energy side, then the recoil energy should peak on the low energy side, and vice versa.

We estimate the NO vibrational distribution through crude (harmonic oscillator-based) calculations of the FC factors $\langle\psi_{v_2}^{HONO}|\psi_{v_{NO}}\rangle^2$. Figure 3 shows the results as a function of v_{NO} for various values of v_2. The only point we wish to make through this approximate treatment is that, for v_2=1, 2 and 3, the NO fragment vibrational distributions are predicted to be similar and peaked somewhat sharply on the low energy side. This implies that the fragment recoil energy distribution should peak on the high energy side and should be similar for v_2=1,2 and 3. This is what we concluded earlier, independently of the present considerations.

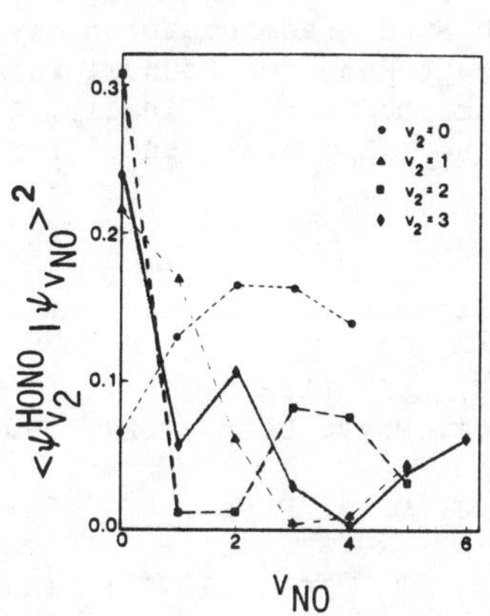

Fig 3. The calculated NO vibrational populations.

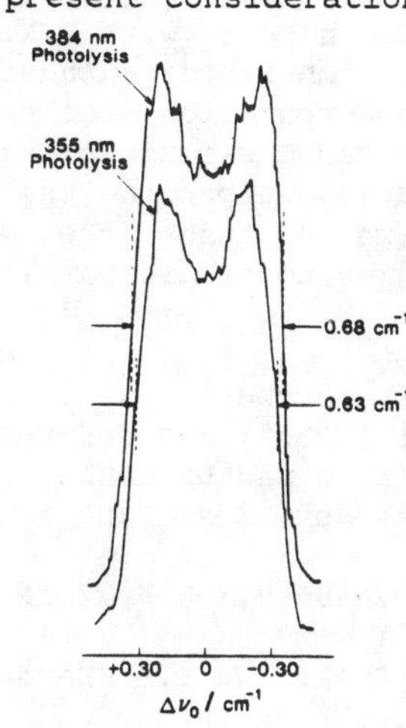

Fig 4. The OH photoproduct Doppler profiles.

The calculations summarized in Figure 3 also predict that, for the excitation of the HONO $\tilde{A}$-$\tilde{X}$ origin band (v_2=0), the NO vibrational distribution should be much wider than for v_2=1, 2 and 3. This implies that the corresponding fragment recoil energy distribution should also be wider for v_2=0. We have recently performed experiments to test this prediction and the results are presented in Figure 4, where we compare the Doppler profiles of the OH photofragment generated at 384 nm (v_2=0) with that at 355 nm (v_2=2). We note that the former is decidedly wider and the width of the latter, as mentioned previously, is the same as in the case of v_2= 1 and 3. These observations are in agreement with the qualitative predictions of the FC model.

A detailed comparison with better theory must, of course, await measurements on the NO fragment. In our earlier experiments,[4] we did not probe the NO product because HONO coexists with NO_2, which also fragments to produce NO. However, for the fragmentation at 384 nm (v_2=0), the interference from NO_2 should be minor because it generates only vibrationally cold NO. This experiment is planned for the near future. It should be interesting to correlate the measurements on the NO fragment with that on the OH product. We are also interested in experiments on (a) DONO, mainly to probe the effects of spectroscopically observed[2] large ν_2-ν_3 anharmonic coupling on the state-to-state dynamics; and (b) cis HONO, in order to probe the dynamical consequences of intramolecular hydrogen bonding[2] in this isomer.[7]

ACKNOWLEDGEMENTS

The work described here was performed in collaboration with R.N. Zare and, in part, with R.N. Dixon. I would like to thank R.N. Zare for his continued enthusiastic support. I am grateful to the Research Corporation, the Petroleum Research Fund (administered by the American Chemical Society), and the Rutgers Research Council for financially supporting the type of work described here. Finally, I would like to thank the organizers of the ILS conference for inviting me to participate.

REFERENCES

1. J.D. Barnwell, J.G. Loeser and D.R.Herschbach, J. Phys. Chem., 87, 2781 (1983); and references therein.
2. G.W. King and D. Moule, Can. J. Chem., 40, 2057 (1962).
3. R. Vasudev, R.N. Zare and R.N. Dixon, Chem. Phys. Lett., 96, 399 (1983).
4. R. Vasudev, R.N. Zare and R.N. Dixon, J. Chem. Phys., 80, 4863 (1984).
5. K.F. Freed and Y.B. Band, Excited States, 3, 109 (1977); and references therein.
6. P. Andresen, G.S. Ondrey, B. Titze and E.W. Roth, J. Chem. Phys., 80, 2548 (1984).
7. R. Vasudev and J.H. Shan, work in progress.

ELECTRONIC SELF QUENCHING AND ENERGY TRANSFER RATE CONSTANTS FOR Br_2 B $^3\Pi$ (0_u^+)

M.C. Heaven and L.J. van de Burgt
Illinois Institute of Technology, Chicago, Il. 60616

ABSTRACT

The self quenching behavior of selected v′ and J′ states of Br_2 (B) has been investigated in the 0.005–3 torr pressure range. Direct decay lifetime measurements were used to construct Stern-Volmer plots, and for low J′ values a pronounced curvature of these plots was noted at low pressures (< 0.5 torr). Br_2 (B) exhibits a strong rotationally dependent predissociation, and the curved Stern-Volmer plots have been interpreted in terms of highly efficient rotational energy transfer to predissociated levels. This process has been simulated using a master equation approach, and approximate rates for electronic quenching ($4.2 \pm 0.5 \times 10^{-10}$ cm^3 molecule^{-1} s^{-1}), near-resonant vibrational energy transfer $k_{v-r,t} \sim 3.5 \times 10^{-10}$ cm^3 molecule^{-1} s^{-1}) and rotational energy transfer ($\sum_f k_f < 8 \times 10^{-10}$ cm^3 molecule^{-1} s^{-1}) have been extracted from the model.

INTRODUCTION

The B $^3\Pi$ (0_u^+) state of Br_2 is subject to both spontaneous[1] and collision induced[2] predissociation. The spontaneous process involves the interaction of the B state with a repulsive $^1\Pi$ (1_u) state, and the predissociation occurs at a rate given by the expression Γ (s^{-1}) = C_v^2 J(J + 1), where J is the rotational quantum number and C_v^2 is a constant for a given vibrational level.[1] Consequently, the B state fluorescence quantum yield is strongly dependent on the rotational state. Electronic quenching of the B state occurs via collision induced predissociation, but direct measurement of the rate constant for this channel is hampered by the presence of the spontaneous channel. Rotationally inelastic collisions can promote the excited molecules to rotational levels which have very low fluorescence quantum yields, and this can lead to an overestimation of the quenching rate constant.

In the present work we have examined the deactivation of Br_2 (B) by collisions with Br_2 (X). A detailed kinetic analysis of the real time fluorescence decay behavior has provided an accurate determination of the quenching rate constant, and estimates for the rotational and vibrational energy transfer rate constants.

EXPERIMENTAL

The apparatus and techniques used for these measurements have been described elsewhere.[3] Br_2 (B) state lifetimes were measured by observing the fluorescence decay following pulsed laser excitation. Decay curves were recorded for initial excitation to individual rotational levels of the v′ = 7, 11, and 14 vibrational states. The fluorescence was not spectrally resolved, so that emission from both optically and collisionally populated levels was observed. Signal averaged decay curves were fitted to a single exponential decay expression by a non-linear least squares fitting program. For each level examined, the Br_2 pressure was varied from 0.005–3 torr, and the data was used to construct a Stern-Volmer (S-V) plot.

RESULTS AND DISCUSSION

Fig. 1 shows the results for excitation to the v′ = 11, J′ = 11 level. Typical of the S-V plots for levels with J′ < 15, Fig. 1 is essentially linear at pressures above 1 torr, but there is significant negative curvature in the 0–1 torr region. The degree of curvature is dependent on the value of J′ excited; the plots became progressively less curved with increasing J′, and for J′ > 25 they were effectively linear at all pressures. At pressures above 1.5 torr all of the S-V plots reached a limiting slope of 4.2×10^{-10} cm^3 molecule^{-1} s^{-1}.

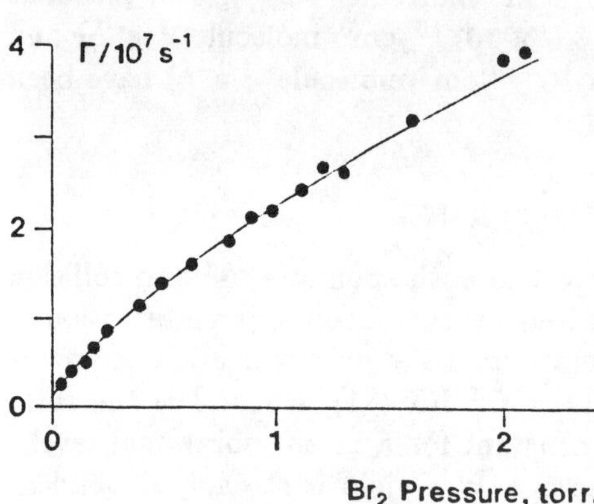

Fig. 1. Self quenching S-V plot for Br_2(B) v′ = 11, J′ = 11. The curve is generated by computer simulation.

The curvature of the S-V plots is a consequence of rotational energy transfer. The predissociation rate increases with rotational energy so that upward energy transfer enhances the spontaneous decay rate. Competition between spontaneous decay and quenching at low pressures results in curvature of the S-V plots for levels appreciably below the maximum in the rotational population distribution (J′ = 44 at 298 K). Excitation to levels near the maximum of the population distribution is followed by both upward and downward energy transfer. In this case, the net effect on the observed deactivation rate is small, and this is consistent with the linearity of the S-V plots for the high J levels. At high pressures, electronic quenching is much faster than predissociation, regardless of the level excited, and the limiting slope of the S-V plots defines the electronic quenching rate constant.

The role of energy transfer in the deactivation kinetics was quantitatively investigated by computer modeling the S-V plots. As rotational energy transfer is generally more efficient than vibrational energy transfer, the model was initially designed to simulate the fluorescence decay from a single vibrational manifold. Starting with the excited state population confined to a single rotational level, the model followed the evolution of the rotational state population distribution by numerical integration of the coupled rate equations for levels in the range $0 \leq J' \leq 120$. The master equation employed radiative lifetimes and predissociation rates taken from the literature.[4] A matrix of microscopic rate constants for transitions between pairs of rotational states was generated using a modified version of the Polanyi-Woodall model.[5] The validity of this matrix was evaluated by using these rates to simulate a collisionally relaxed resolved fluorescence spectrum,[6] and in this process the rate constant for rotational energy transfer was found to be $6 \pm 2 \times 10^{-10}$ cm^3 molecule^{-1} s^{-1}. A more complete description of the calculations is given in reference 3.

Model S-V plots were generated from the set of rate constants described above, and they were compared with the experimental results. The computed plots could not adequately reproduce the curvature seen in the data for the low J levels because the transfer to strongly predissociated levels was not fast enough. Rapid access to predissociated levels was incorporated into the model by adding a near-resonant vibrational energy relaxation channel. Each v', J' state was allowed to relax to the energetically closest $v' - 1$, $J' + \Delta J'$ state and a single transfer rate constant k_v was used to represent this process. The curvature of the S-V plots was readily simulated using this model, and a value of $k_v = 3.5 \times 10^{-10}$ cm^3 molecule^{-1} s^{-1} gave good fits to the data for the full range of rotational states investigated.

The high efficiency of the near resonant vibrational energy relaxation is somewhat surprising. The rate constant for vibrational relaxation of I_2 (B) by I_2 (X) is an order of magnitude smaller, and the transfer occurs with very little exchange of angular momentum. Further experiments will be undertaken to examine the apparently anomalous relaxation mechanism for Br_2 (B) in more detail.

This work was supported by the AFOSR under grant 83–0173.

REFERENCES

1. M.A.A. Clyne, M.C. Heaven and J. Tellinghuisen, J. Chem. Phys. **76**, 5431 (1982).
2. M.A.A. Clyne, M.C. Heaven and S.J. Davis, J. Chem. Soc. Faraday Trans. II **76**, 961 (1980).
3. L.J. van de Burgt and M.C. Heaven, Chem. Phys. accepted December 1985.
4. M.A.A. Clyne and M.C. Heaven, J. Chem. Soc. Faraday Trans. II **74**, 1992 (1978).
5. J.C. Polanyi and K.B. Woodall, J. Chem. Phys. **56**, 1563 (1972).
6. G. Perram and S.J. Davis, private communication.

MODELING OF ION MOLECULE REACTIONS AT HIGH PRESSURES

C. D. Eberhard and C. B. Collins,
University of Texas at Dallas, Richardson, Tx 75083-0688

J. Stevefelt
GREMI, University of Orleans, Orleans, France

ABSTRACT

A model of the charge transfer reactions of He_2^+ based on the formation of a reaction complex was formulated. Both coherent and incoherent collisional processes of excitation and deexcitation of the complex were included. The overall reaction rates for the charge transfer reactions with N_2, CO_2 and Ne at atmospheric pressures were predicted. Model parameters were fixed by matching the experimentally measured reaction rate coefficients at 300^oK^1, and the temperature dependences of the termolecular rate coefficients were predicted for these species.

INTRODUCTION

Reaction rates exceeding the Langevin rate have been observed in high pressure laser plasmas. It has been shown[1] that the observed reaction rate of molecular ions, such as He_2^+, must include a rate at which they are destroyed in 3-body reactions, as $He_2^+ + X + He \rightarrow$ Products. The destruction rate may be generally expressed as

$$R_d = R_o + K_2\ [X] + K_3\ [X]\ [He]$$

where R_o is the loss rate of He_2^+ in pure helium, K_2 is the bimolecular rate coefficient and K_3 is the termolecular rate coefficient. Rates derived using only the bimolecular rate coefficients may be satisfactory at low pressures but are inadequate to describe the processes at atmospheric pressures and often lead to erroneous predictions.

The presence of a high pressure background gas must enhance the reaction rates through collisions with the reacting ion-molecule system. When the ion-molecule system is struck by a structureless projectile (a helium atom), the collision can increase the overall reaction rate by 1) changing otherwise glancing collisions between the ion and molecule into capture collisions, or 2) raising or lowering the internal energy of the complex.

MODEL OF THE REACTION COMPLEX

In the specific examples considered here, the reaction complex consisted of the He_2^+ ion and the reactant, weakly bound by the attractive polarization potential. The polarization potential was closed at the binding energy, $-E_D$, a principal model parameter. The energy levels of the complex were those of a hard spherical well whose

0094-243X/86/1460558-2$3.00

radius depended on the energy level, which was measured from the binding energy, $-E_D$. A typical pattern of energy levels is shown in Figure 1. As the energy of the complex approached the binding energy, the spacing between the levels decreased, until just below the dissociation limit, the energy levels were approximated by a quasi-continuum. Each level had a degeneracy computed from the density of states in a multi-dimensional spherical potential well. Additional degrees of freedom were accomodated by increasing the degeneracy at each of those levels. Semiclassical rate coefficients were calculated from the rate of energy transfer resulting from collisions with both components of the complex. These rate coefficients were used for individual steps up and down the ladder of energy levels. Coherent two-body capture with collisional and nonlinear stabilization was also included. Dissociation levels of the complex at which charge transfer occurred due to the crossing of potential surfaces were a principal model parameter.

RESULTS

Rate matrices were obtained from which the termolecular and bimolecular rate coefficients were derived. All reactions considered could be modeled with a binding energy of 0.052 ev or 2.0 KT at T = 300°K and from 1 to 16 dissociating levels. Reasonable temperature dependences were obtained. The choices of model pararmeters were not unique. Figure 2 shows the temperature depedences of three possible solutions for the reaction of He_2^+ with N_2 for a dumbbell complex. These solutions differ only in the number and location of the dissociating states. Since these solutions have different behaviors in temperature regions accessible to future experiments, it may be possible to determine which solutions is appropriate.

1. F.W. Lee, C.B. Collins, R.A. Waller, J. Chem. Phys.65, 1605(1976).

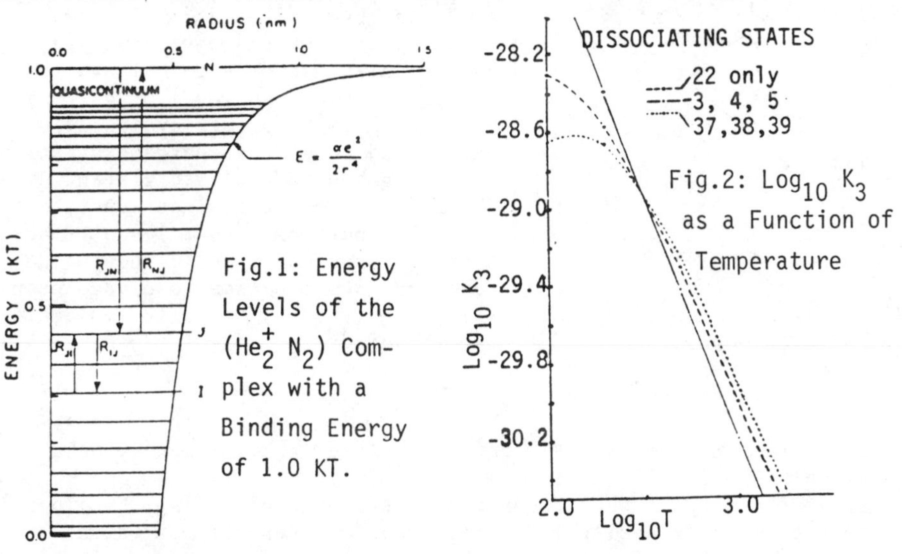

Fig.1: Energy Levels of the ($He_2^+ N_2$) Complex with a Binding Energy of 1.0 KT.

Fig.2: $Log_{10} K_3$ as a Function of Temperature

THE IMPORTANCE OF THREE-BODY PROCESSES TO REACTION KINETICS AT ATMOSPHERIC PRESSURES *

V.T. Gylys, H.R. Jahani, Z. Chen +, and C.B. Collins,
Center for Quantum Electronics, University of Texas at Dallas,
P.O. Box 830688, Richardson, TX 75083-0688.

J.M. Pouvesle and J. Stevefelt,
G.R.E.M.I., CNRS/Université d'Orléans, 45067 Orléans Cédex 2, France.

The purpose of this work was to reexamine the significance of three-body effects that enhance the reaction rates for non associative energy transfer and ion-molecule reactions in atmospheric pressures of inert gas diluent.[1-4] The widest possible range of background helium pressures, varying from 1 to 6 atmospheres, and of reactant concentrations were examined.

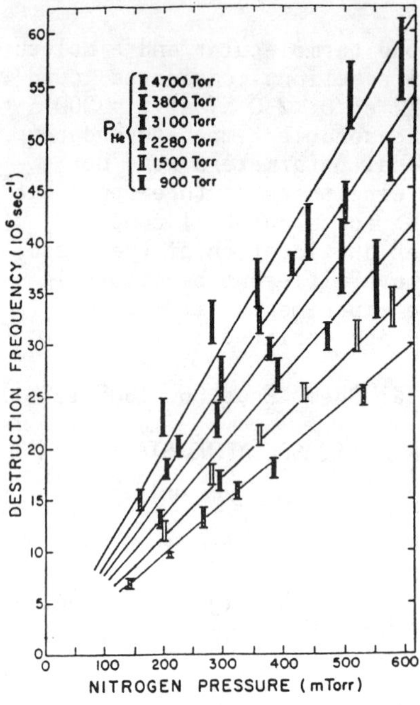

Fig.1. Destruction frequency of the He_2^+ population as function of N_2 pressure for different diluent pressures.

The experimental method exploited the intense fluorescence emitted from the product state $N_2^+(B)$ of the reactions of the energy storing species He_2^+ and $He(2^3S)$ with nitrogen, added as a convenient monitor.[1,2,4] Following the termination of an external source of ionization, the energy decays in the early afterglow primarily by the ion-molecule reactions of He_2^+. Subsequently, there is a time during which the energy is mainly carried by the metastables $He(2^3S)$ which generally have longer lifetimes.

In order to resolve the exponential character of each domain, two experimental systems were employed, one in Dallas and one in Orléans, each exploiting the high level of gas purity and system integrity available with preionized pulsed discharge excitation, but using substantially different discharge energies to enhance separate afterglow periods. The relevant reaction frequencies, ν could then be determined from the time resolved intensity, I emitted by the $N_2^+(B)$ molecules, $\nu = I^{-1}\, dI/dt$.

* Supported in part by NSF Grant ECS 8314633 and in part by NATO Grant 655/84.

+ Present address : Department of Electrical Engineering, Nanjing Institute of Technology, Nanjing, People's Republic of China.

Table I. Summary of the bimolecular and termolecular rate coefficients, in units of 10^{-10} cm^3 s^{-1} and 10^{-30} cm^6 s^{-1}, respectively, for the reactions of He_2^+ and $He(2^3S)$ measured in this work. Also shown are the classical Langevin limits for the ion-molecule reactions in units of 10^{-10} cm^3 s^{-1}.

Reactant	Langevin	He_2^+ k_2	k_3	$He(2^3S)$ k_2	k_3
Ar	11.6	2.0±0.7	7.2±1.2	0.76±0.01	0.20±0.04
Xe	17.1	2.5±1.0	10.0±3.0		
N_2	12.4	11.5±1.2	13.1±0.2	0.68±0.03	0.15±0.04
O_2	11.6	10.0±4.0*	24.0±8.0*	2.1 ±0.05	0.17±0.17
CO_2	14.5	work in process			

*Preliminary results

Figure 1 presents ν obtained from the early-time portions of the UTD data as functions of reactant N_2 concentration. The results clearly depend on the diluent pressure, and the linear approximations producing the best fit to the data define a reaction rate coefficient, $k = k_2 + k_3[He]$, as would be expected from independent bimolecular and termolecular reaction channels.[2] No evidence of saturation can be found, and at six atmospheres of helium pressure the charge transfer from He_2^+ to N_2 is proceeding at a rate 2.5 times in excess of the Langevin rate that has been traditionally considered to be an upper limit on the overall rate of reaction.

The reactions with other gases were studied by adding these as a third component to the helium-nitrogen mixture. The results, summarized in Table I, generally confirm the earlier measurements of ion-molecule reactions at atmospheric pressures,[1,2,4] although the new values for the termolecular reactions with Ar and O_2 are about one third of the e-beam results.[2] The reactions of He_2^+ with Xe are, to our knowledge, reported here for the first time.

The termolecular reactions of the metastable atom $He(2^3S)$ proceed at rates that are one order of magnitude slower than previously reported results.[3,4] However, the latter have been found to reflect the bimolecular reaction of vibrationally excited He_2 $(^3\Sigma_u)$, the population of which depended upon both $He(2^3S)$ concentration and helium pressure. In the present studies, this population was consumed by an excess of reactant N_2, and the reactions of $He(2^3S)$ could be studied without being masked by those of $He_2(^3\Sigma_u)$,

Of the thirty-nine reactions of the ions He_2^+, Ne_2^+, and Ar_2^+ that have been examined to date, only those of Ar_2^+ with N_2 and H_2 seem to be exclusively bimolecular. The former is endothermic and generally does not proceed by either channel so there is actually but a single system that is persistently bimolecular at all pressures examined.

1. F.W.Lee, C.B.Collins & R.A.Waller, J. Chem. Phys., 65, 1605 (1976).
2. C.B. Collins & F.W. Lee, J. Chem. Phys., 68, 1391 (1978).
3. C.B. Collins & F.W. Lee, J. Chem. Phys., 70, 1275 (1979).
4. J.M. Pouvesle, A. Bouchoule, & J. Stevefelt, J. Chem. Phys., 77, 817 (1982).

GAS-PHASE OXIDATION OF ATOMIC BORON AND BORON MONOXIDE

Richard C. Oldenborg and Steven L. Baughcum
Chemistry Division
Los Alamos National Laboratory
Los Alamos, New Mexico 87545

ABSTRACT

Rate constants for the reactions of B + O_2 and BO + O_2 have been measured over the temperature range 298-1180 K using the laser photolysis/laser-induced fluorescence technique. The rate of the B + O_2 reaction increases slightly with increasing temperature. In contrast, the BO + O_2 reaction has a negative temperature dependence and is believed to proceed via a stable intermediate complex.

INTRODUCTION

The emphasis in much of the previous boron slurry rocket fuel research has been on particle ignition, yet roughly half of the potential energy content of boron fuels is released in the gas-phase oxidation of BO(g) to B_2O_3(g). We are investigating the kinetics of the key gas-phase boron oxidation reactions over a wide temperature range so that a comprehensive reaction scheme can be assembled and evaluated. Our initial effort has concentrated on the BO + O_2 → BO_2 + O reaction, which is considered to be the key BO oxidation step in dry atmospheres.

EXPERIMENTAL

Volatile boron compounds such as BCl_3 are photolyzed with an excimer laser to provide an essentially instantaneous source of boron atoms. These quickly react with O_2 to yield BO molecules,

$$B + O_2 \rightarrow BO + O$$

which in turn react with additional O_2 to produce BO_2.

$$BO + O_2 \rightarrow BO_2 + O$$

The time histories of the B, BO, and BO_2 are monitored using laser-induced fluorescence (LIF) and chemiluminescent techniques.

Systematic variation of O_2 pressure yields the desired rate constants. Experiments are conducted in a high-temperature cell capable of 298-1500 K operation. All experiments are in an argon diluent at 25 torr total pressure.

RESULTS AND DISCUSSION

The $B + O_2$ reaction rate constant was measured to be 7×10^{-11} cc/molecule-s at 298 K and increases slightly with increasing temperature. In contrast, the $BO + O_2$ reaction rate constant is somewhat slower at 298 K, i.e., 2×10^{-11} cc/molecule-s, and decreases with increasing temperature in a non-Arrhenius fashion. The reaction may proceed via a stable BO_3 intermediate

$$BO + O_2 \rightleftarrows BO_3 \longrightarrow BO_2 + O \quad ,$$

which may help account for this interesting temperature dependence. Experiments are in progress to better establish the functional form of this temperature dependence and to test the role of third-body collisions at both high and low gas pressures. Other key boron oxidation reactions are also presently under investigation.

ACKNOWLEDGMENTS

The technical assistance of Kenneth Winn of Los Alamos National Laboratory is gratefully acknowledged. The work is supported by the Air Force Office of Scientific Research (AFOSR).

PHOTOCHEMISTRY OF TRIMETHYLGALLIUM WITH APPLICATIONS TO ATOMIC GALLIUM REACTION KINETICS

Steven L. Baughcum and Richard C. Oldenborg
Chemistry Division
Los Alamos National Laboratory
Los Alamos, NM 87545

ABSTRACT

The photodissociation of gas-phase trimethylgallium at 193 nm has been investigated using the laser photolysis/laser induced fluorescence technique. Gallium atoms are produced by multiple-photon processes during the time of the excimer laser pulse and continue to be produced at long times after the photolysis pulse. The observed dependences on photolysis laser fluence, trimethylgallium pressure, and buffer gas pressure are consistent with a mechanism in which highly excited gallium methyl radicals undergo unimolecular decomposition to produce gallium atoms. Since this process can happen on time scales long compared to the photolysis laser pulse, these results may have important implications for studies of laser deposition of thin metal films.

INTRODUCTION

Volatile organometallic compounds provide a potentially attractive source of refractory metal atoms for deposition in such applications as chemical vapor deposition of Group III-V compound semiconductors and laser deposition processes of thin films. The Group III metals (gallium, indium, and thallium) are also interesting as potential energy transfer laser candidates because of the spin-orbit splitting of the ground ^2P state.

Trimethylgallium is photolyzed at 193 nm using a rare gas halide excimer laser operating on ArF. Atomic gallium is detected in the ground state ($4^2P^\circ_{1/2}$) and low-lying metastable state ($4^2P^\circ_{3/2}$) by laser induced fluorescence (LIF). The experiments are carried out at a variety of excimer laser fluences, trimethylgallium pressures, and argon buffer gas pressures.

RESULTS AND CONCLUSIONS

Both ground state and excited state gallium are produced directly in the multiple-photon photolysis of trimethylgallium at 193 nm. At high photolysis laser fluence (20 mJ/cm^2), the amplitude and time behavior of the ground state was observed to be sensitive to 193 nm energy, trimethylgallium pressure, buffer gas pressure, and even probe laser energy. Under these conditions, higher-order processes, including multiple-photon dissociation effects and radical-radical reactions, appear to be important.

Under conditions of low 193 nm fluence (1 mJ/cm^2), low trimethylgallium pressure (0.6-5 mtorr) and 25 torr of argon buffer gas, ground state gallium is produced promptly followed by a rise which continues for approximately 500 μs. Experiments indicate that this rise is not due to collisional quenching of metastable Ga($4^2P^\circ_{3/2}$), which is much faster than the observed rise under these conditions. The rise was observed to slow with increasing trimethylgallium or argon pressure and was not sensitive to excimer laser energy at low fluences. This rules out secondary radical-molecule or radical-radical reactions as the source of the delayed atomic gallium production. Rather, these results are consistent with the production of gallium atoms by the unimolecular decomposition of excited gallium methyl radicals produced directly in the photolysis.

ACKNOWLEDGMENTS

The technical assistance of Kenneth Winn and Douglas Hof of Los Alamos National Laboratory is gratefully acknowledged. This work is supported by the Air Force Office of Scientific Research (AFOSR).

LASER-INDUCED CHEMISTRY OF THE ALLYL HALIDES

J. K. McDonald and J. A. Merritt
U.S. Army Missile Command, Redstone Arsenal, AL 35898-5248

S. P. McManus
University of Alabama in Huntsville, Huntsville, AL 35899

ABSTRACT

The continuous wave CO_2 laser-induced chemistry of several allyl halides has been studied at various pressures and laser powers. The majority of products observed in these reactions are also reported in gas phase pyrolysis. From the kinetic studies of allyl chloride and bromide, overall reaction orders were determined. New reaction mechanisms for formation of cyclic products based upon results obtained in these reactions have been proposed.

INTRODUCTION

The absorption of infrared laser radiation by molecules promotes the molecules into excited vibrational states, thus making the molecules reactive. Energy relaxation can then occur with a redistribution of energy among the different vibrational modes as well as translational and rotational levels. Furthermore, at pressures greater than a few torr, intermolecular redistribution accompanies collisions. Consequently, after irradiation by an infrared laser, only microseconds are required for the sample to reach thermal equilibrium. Normal thermolytic reactions are expected when the laser energy has been redistributed throughout a molecule. However, laser driven reactions can differ from typical pyrolysis processes[1].

We have studied the laser-induced reaction of the allyl halides and compared them where possible with reported pryolytic processes.

RESULTS AND DISCUSSION

The CO_2 laser irradiation of the allyl halides was performed in a single-line mode over a range of 25 to 150 W/cm^2 of laser power and times of .1 to 60 seconds. The reactions were carried out in stainless steel cells fitted with ZnSe or KCl windows. Product analyses were determiend by gas chromotography and fourier transform infrared spectroscopy.

In addition to the hydrogen halides, the laser-induced reaction (LIR) of allyl bromide produced mainly benzene, propene, ethylene, acetylene and methane. Trace components of allene and propyne were also observed. The same products were observed in the LIR of allyl chloride although concentrations were different. Allene and propyne were not in trace quantities and 1,3 butadiene, 1,3 cyclopentadiene (trace), 1,3 cyclohexadiene (trace) were also observed. The LIR of allyl fluoride produced mainly propyne and allene. Smaller amounts of acetylene and ethylene were formed with traces of benzene, propene, methyl fluoride and vinyl fluoride. The main products observed for allyl bromide and chloride agreed with those observed for the literature reported products from pyrolytic processes[2,3]. Allene, propyne and acetylene had not been previously reported.

Overall reaction orders were determined for allyl bromide and chloride by monitoring their concentrations for irradiation times from .1 to 2 seconds. Allyl bromide was found to follow a first order reaction while allyl chloride was determined to follow a three halves order reaction. The overall orders reported in the literature for pyrolytic processes is first for both compounds[2,3]. The three halves order is in better agreement with a radical chain reaction that is observed for the allyl chloride.

From the LIR of allyl chloride and some other useful molecules the mechanistic pathway for cyclization to benzene has been proposed[4,5].

CONCLUSIONS

From the products obtained the laser-induced reactions of the allyl halides follow basically the same processes that are involved in pyrolysis processes, where comparisons are possible. Although pyrolysis can produce larger quantities of materials, the LIR process is much more convenient for carrying out small scale reactions and studying the products.

REFERENCES

1. J. I. Steinfeld, Ed, Laser-Induced Chemical Processes, (Plenum Press, New York, 1981).
2. S. Kunichika, Y. Sakakibara, and M. Taniuchi, Bull. Chem. Soc. Japan, 42, 1082 (1969); L. M. Porter and F. F. Rust, J. Am. Chem. Soc., 78, 5571 (1956); A. M. Goodall and K. G. Howlett, J. Chem. Soc., 2599 (1954).
3. A. Maccoll, J. Chem. Soc., 965 (1955).
4. J. K. McDonald, J. A. Merritt, B. J. Alley, and S. P. McManus, J. Am. Chem. Soc., 107, 3008 (1985).
5. J. K. McDonald, J. A. Merritt, J. R. Durig, V. F. Kalasinsky, and S. P. McManus, to be published in Ind. Eng. Chem. Prod. Res. Dev.

PHOTOFRAGMENT VIBRATIONAL EXCITATION AND THERMAL CHEMISTRY FOLLOWING 351 nm PHOTOLYSIS OF $Mn_2(CO)_{10}$

R. G. Bray, P. F. Seidler, J. S. Gethner and R. L. Woodin
Exxon Research and Engineering Company, Annandale NJ 08801

ABSTRACT

We have demonstrated the application of infrared emission to probe the structure and dynamics of organometallic intermediates.[1] We identified the species produced by photolysis of gas phase $Mn_2(CO)_{10}$, measured the vibrational relaxation rates of $Mn(CO)_5$ for seven gases and have begun to characterize the secondary chemistry.

INTRODUCTION

Ultraviolet photolysis of gas phase $Mn_2(CO)_{10}$ may produce vibrationally excited species by two primary pathways (eq 1,2).[2]

$$Mn_2(CO)_{10} \rightarrow 2Mn(CO)_5 \quad (1); \qquad Mn_2(CO)_{10} \rightarrow Mn_2(CO)_9 + CO \quad (2).$$

We used standard techniques[3] for detecting time-resolved infrared emission from species produced by pulsed laser photolysis (~351 nm, 15 ns pulse width, 2 Hz repetition rate) of $Mn_2(CO)_{10}$ in a static cell (P=7-60 Mtorr, T=79°C). The observed emission near 2000 cm^{-1} is bi-exponential: the fast component decayed within 0.5-10 μs while the weaker, slower component decayed within 2-10 ms.

IDENTIFICATION OF EMITTING SPECIES

The infrared emission could arise from the primary photoproducts (reactions (1), (2)), or from chemiluminescent molecules, e.g., $Mn_2(CO)_{10}$, produced by secondary reactions. We crudely characterized the emitting species by measuring the intensities of each component for a series of bandpass filters relative to a broad band filter (2500-1800 cm^{-1}). No emission from free CO was detected (2090-2210 cm^{-1}). Comparison with the intensities calculated from the known solution or gas phase spectra[2b] distinguished between three possible emitting species (Table 1). We attribute the fast emission to $Mn(CO)_5$ and the slow emission to either $Mn_2(CO)_9$ or $Mn_2(CO)_{10}$. Mn-Mn bond cleavage produces $Mn(CO)_5$ fragments having sufficient excitation for infrared emission, consistent with recently reported Mn-Mn bond dissociation energy of 32-41 kcal mol^{-1},[4] but insufficient excitation for further dissociation, assuming the same partitioning of the available energy into the degrees of freedom as for $Re_2(CO)_{10}$.[2a] Mn-CO bond scission either is a minor process at 351 nm or produces fragments having little vibrational excitation relative to $Mn(CO)_5$.

TABLE I. Comparison of Observed and Calculated Emission Intensities for $Mn_x(CO)_y$ Species

Filter	Broad Band	>1980 cm^{-1}	2040-1940 cm^{-1}	<1980 cm^{-1}
Observed				
Fast Decay	1.00	0.41	0.84	0.30
Slow Decay	1.00	0.98	0.41	0.06
Calculated				
$Mn(CO)_5$	1.00	0.57	0.95	0.13
$Mn_2(CO)_9$	1.00	1.11	0.59	0.10
$Mn_2(CO)_{10}$	1.00	1.31	0.33	0.02

INTERPRETATION OF FAST DECAY KINETICS

A linear pressure dependence is observed for the pseudo-first-order decay rates of the fast component for seven buffer gases indicating that vibrationally excited $Mn(CO)_5$ decays by collisional deactivation. Table II compares the second order rate constants for collisional deactivation of $Mn(CO)_5$ with published data obtained with the same technique for highly vibrationally excited azulene.[5] The deactivation rates for $Mn(CO)_5$ and azulene are remarkedly similar given the disparity in structure, the mass of the constituent atoms and the reactivity of these species and confirm that collisional deactivation of vibrationally excited $Mn(CO)_5$ and not a chemical transformation causes the emission. Average energies transferred per collision were calculated from the observed rate constants for $Mn(CO)_5$.[5] Even though the values are 1.5-3 times smaller than for azulene the approximations involved do not allow a definitive statement concerning the dependence of the average energy transferred per collision on the internal energy.

TABLE II. Phenomenological Second-Order Rate Constants[a] and Average Energy Transferred per Collision[b] for Vibrational Collisional Deactivation

	$Mn(CO)_5$		Azulene			
	(15 kcal mol^{-1})[c]		(48 kcal mol^{-1})[c]		(85 kcal mol^{-1})[d]	
Collider Gas	$10^{11}k$	$\langle \Delta E \rangle$	$10^{11}k$	$\langle \Delta E \rangle$	$10^{11}k$	$\langle \Delta E \rangle$
$Mn_2(CO)_{10}$	27.4	325	—	—	—	—
Azulene	—	—	25.2	1217	13.6	1425
$CH_3CH_2CH_2CH_3$	9.26	216	11.7	707	9.41	1234
NO	4.99	110	—	—	—	—
CH_4	4.92	78	3.10	173	3.70	448
CO	3.75	77	1.92	151	2.20	375
Ar	3.24	80	1.29	119	1.07	215
He	1.69	17	0.663	37	0.69	83

[a] From linear least-squares fits of data; units are cm^3s^{-1}.
[b] Units are cm^{-1}. [c] average excess vibrational energy

INTERPRETATION OF SLOW DECAY KINETICS

We interpret the slow emission as infrared chemiluminescence from either $Mn_2(CO)_{10}$ or $Mn_2(CO)_9$ formed via exothermic reactions since thermal equilibration of the primary fragments is complete in << 1 ms. The decay rate is independent of partial pressure of non-reactive buffer gases so diffusion of emitting species from the detection volume cannot account for the observed decay. Similarly thermal conduction effects are negligible since the calculated temperature jump (<20°K) is insufficient to observe infrared emission. We are continuing to investigate the chemistry leading to the slow decay.

REFERENCES

1. R. G. Bray, P. F. Seidler, J. S. Gethner and R. L. Woodin, J. Am. Chem. Soc., accepted for publication.
2. (a) A. Freedman and R. Bersohn, J. Am. Chem. Soc., 100, 4116 (1978); (b) S. P. Church, H. Hermann, F. -W. Grevels and K. Schaffner, J. Chem. Soc., Chem. Commun. 785 (1984).
3. S. R. Leone, Acc. Chem. Res., 16, 88 (1983).
4. J. A. Martinho Simoes, M. C. Schultz and J. L. Beauchamp, Organometallics 4, 1238 (1985).
5. J. A. Barker, J. Phys. Chem. 88, 11 (1984).

KINETIC SPECTROSCOPY USING A COLOR CENTER LASER

J. W. Stephens, J. L. Hall, D. Zeitz,
R. F. Curl, Jr., G. P. Glass, J. V. V. Kasper, and F. K. Tittel
Rice University, Houston, TX. 77251

The powerful method of infrared kinetic spectroscopy[1] has been applied to the investigation of the reaction between NH_2 and NO. The amidogen radical, NH_2, was produced in high concentrations by photolysis of NH_3 with the 193 nm ArF line of an excimer laser. Probing with a high resolution color center laser spectrometer combined with fast infrared detectors allowed the NH_3, H_2O, NO and NH_2 and OH radicals to be monitored on microsecond time scales.

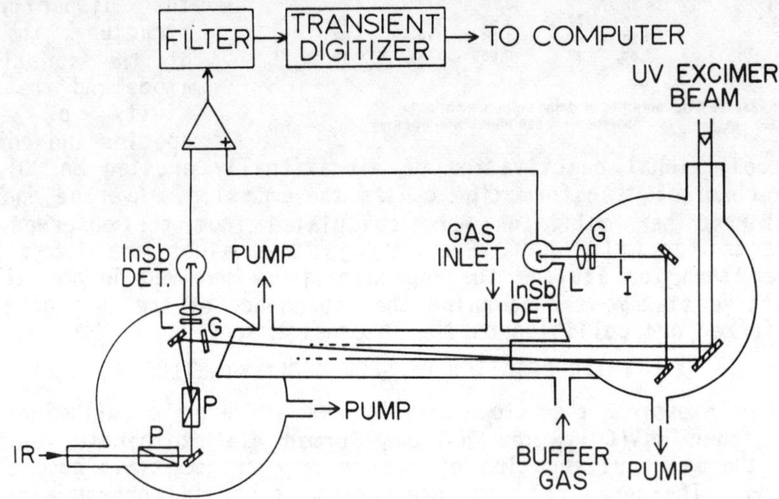

Fig. 1. Experimental layout. P = polarizer, G = germanium flat, L = lens, I = iris.

The NH_2 + NO reaction, proposed as a key step in the industrially important "Thermal DeNOX" process, has been studied extensively by others. The rate constant (~1.6×10^{-11} cm^3 s^{-1}) has been measured by several workers[2-4] and the two major product channels are generally agreed to be

$$NH_2 + NO \rightarrow N_2 + H_2O \quad (\Delta H^\circ = -127 \text{ kcal mol}^{-1})$$
$$\rightarrow N_2H(?) + OH \; (\Delta H^\circ = -3 \text{ kcal mol}^{-1}).$$

Determination of the branching ratios of these two channels, however, has been somewhat more elusive.

The current work indicates that the OH channel accounts for 13±2% of the reaction. Measurements of OH lines were made with and without NO_2 added to the reaction mixture. The reaction of NO_2 with the H atoms formed in the NH_3 photolysis forms OH radical, and thus gives a measure of NH_2 produced. Comparison of the OH signal with and without NO_2 enables the calculation of the percentage of NH_2 which reacts to form OH.

Two different pairs of NH_3 and H_2O lines were measured with 20 Torr of buffer gas to minimize the temperature rise of the system. Values of 0.85±0.09 and 0.66±0.03 for the ratio of H_2O formed to NH_3 photolyzed come close to representing a "materials balance" with the contribution of the OH channel. All of the H_2O signals exhibited a pronounced induction period suggesting that H_2O is formed in high vibrational states. The behavior of the signals were consistent with simulations in which a stepwise sequential loss of vibrational energy occurs with equal quenching cross-sections for each step.

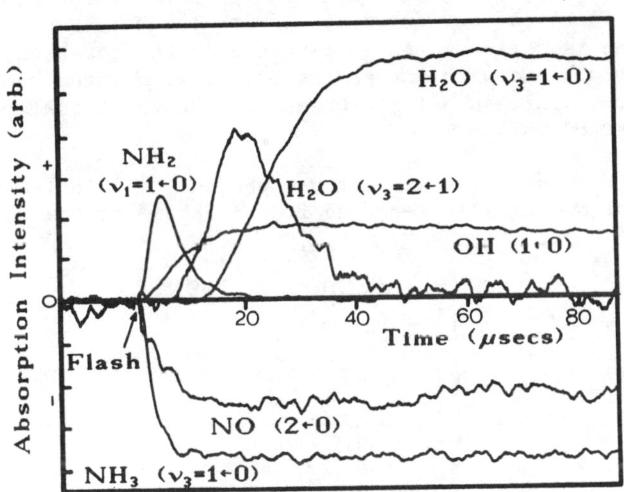

Fig. 2. The time behavior of a variety of species can be monitored. (Initial conditions, P_{NH_3} = 70 mTorr, P_{NO} = 1 Torr, and P_{He} = 8 Torr.)

1. H. Adams, J. Hall, L. A. Russell, J. V. V. Kasper, F. K. Tittel, and R. F. Curl, J. Opt. Soc. Am. B 2, 776 (1985).
2. P. Andresen, A. Jacobs, C. Kleinermanns, and J. Wolfrum, Nineteenth Symposium (International) on Combustion, p.11, The Combustion Institute, 1982.
3. J. A. Silver and C. E. Kolb, J. Phys. Chem. 86, 3240 (1982).
4. L. J. Stief, W. D. Brobst, D. F. Nava, R. P. Borkowski, and J. V. Michael, J. Chem. Soc. Faraday Trans. 2, 78, 1391 (1982).

TWO PHOTON SPECTROSCOPY OF AUTOIONIZING LEVELS OF NITRIC OXIDE

P. M. Dehmer, J. L. Dehmer, and S. T. Pratt
Argonne National Laboratory, Argonne, Illinois 60439

We have obtained the two photon ionization spectrum of supersonically cooled nitric oxide (NO) from the NO^+ X $^1\Sigma^+$, v = 0 ionization threshold (~ 2677 Å) to just above the NO^+ X $^1\Sigma^+$, v = 2 threshold. Although no strict selection rule applies to the electronic state symmetry for two photon absorption in NO, the two photon ionization spectrum is completely different from the corresponding single photon ionization spectrum,[1] and in fact, all of the observed bands are new. Of particular interest are the previously unobserved ndσ,π v = 1, 2 Rydberg series, which are observed in the two photon spectra for n = 7 - 12. As an example, Figure 1 shows the spectrum in the region of the NO^+ X $^1\Sigma^+$, v = 0 threshold, and Figure 2 shows the spectrum of the 5dσ,π v = 2 band. The beam temperature is low enough to considerably simplify the spectrum, but it is also high enough for the rotational structure to reveal the interaction with the nsσ, v series.[2] The interaction is reflected in an increase in the rotational constants of the ndσ,π bands with increasing principal quantum number, in contrast to a corresponding decrease in the rotational constants of the (n+1)sσ bands.[2]

The two photon resonant, three photon (2+1) ionization spectrum of NO was also recorded from ~ 2865 Å to the NO^+ X $^1\Sigma^+$,

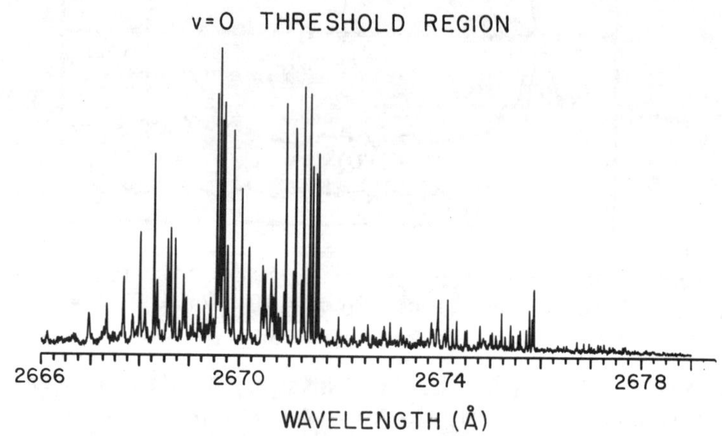

Figure 1. The two photon ionization spectrum of supersonically cooled nitric oxide in the region of the NO^+ X $^1\Sigma^+$ v = 0 threshold. The ionization threshold from the NO X $^2\Pi_{1/2}$ ground state corresponds to 2676.61 Å. However, ionization may occur below that wavelength due to the non-negligible population of the NO X $^2\Pi_{3/2}$ state in the molecular beam.

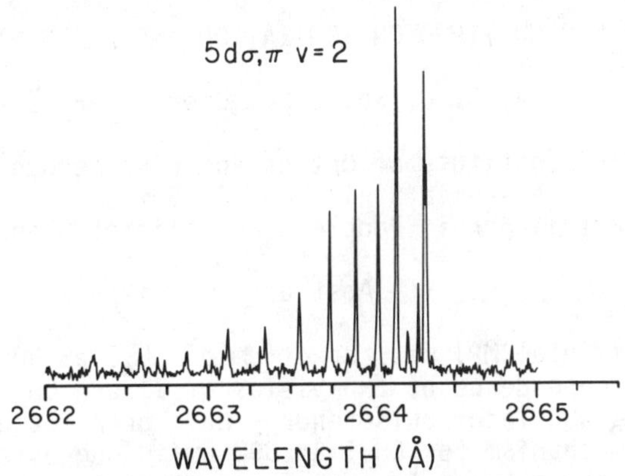

Figure 2. The two photon ionization spectrum of the 5dσ,π v = 2 ← X $^2\Pi_{1/2}$ band of nitric oxide.

v = 0 threshold, in an effort to observe the corresponding ndσ, π v = 0 bands. We have thus far identified the 5dσ,π v = 1 band below threshold. A number of bands observed previously in single photon absorption studies[3] are also observed in the (2+1) ionization spectrum. One particular advantage of the (2+1) ionizaton spectrum over the absorption spectrum is that the strong continuous absorption is not observed in the former, which helps to bring out the weaker discrete structure. We are currently analyzing the spectra in more detail.

This work was supported by the U.S. Department of Energy, Environmental Research Division, under Contract W-31-109-Eng-38 and by the Office of Naval Research.

REFERENCES

1. Y. Ono, S. H. Linn, H. F. Prest, C. Y. Ng, and E. Miescher, J. Chem. Phys. 73, 4855 (1980).
2. See, for example, E. Miescher, Can. J. Phys. 54, 2074 (1976).
3. See, for example, E. Miescher, J. Mol. Spectrosc. 20, 130 (1966); E. Miescher and F. Alberti, J. Phys. Chem. Ref. Data 5, 309 (1976).

THE MECHANISM FOR MULTIPHOTON IONIZATION AND FRAGMENTATION OF CH_3I

J. C. Han, Y. F. Gua, S. J. Su, J. F. Chen, Y. F. Ji and S. H. Liu

Anhui Institute of Optics and Fine Mechanics
Academia Sinica, P. O. Box 25
Hefei, Anhui, People's Republic of China

Abstract

The experimental MPI mass spectrum of CH_3I at 308 nm has been reported. The dependence of the yields of I^+ and CH_3I^+ ions on the sample pressure and laser pulse energy have been measured. The fragmentation mechanism for CH_3I at 308 nm is suggested, which involves competition between two MPI-fragmentation pathways.

Introduction

Methyl iodide is a hybrid system in the sense that it is clearly class A, ionization followed by fragmentation, in one spectral region, but class B, fragmentation followed by ionization, in another[1]. The class A-to-class B shift of nonlinear photochemistry displayed by methyl iodide is due to a change of resonance state on scanning through the UV region, i.e. from (5P, σ^*) valence shell to (5Pπ, 6P) Rydberg[2].

In the present work we have studied MPI of CH_3I at 308 nm in order to determine the mechanism of MPI-fragmentation of CH_3I in the transitive region, where a shift from class A to B behavior has taken place.

Experimental

The MPI mass spectrum experimental setup for CH_3I consists of three parts: a XeCl excimer laser, a molecular beam device, and a modified quadrupole mass spectrometer and signal processing system.

Results and Discussion

At 308 nm, three mass peaks were observed, which belong to CH_3^+, I^+ and CH_3I^+. The ion signal for CH_3^+ was always small, while the ion signal of I^+ was always large for various laser powers.

The dependences of yields of I^+, CH_3I^+ ions on the parent molecule pressure shown in Figure 1 were found to be linear, as is expected. This indicates that formation of I^+ and CH_3I^+ is a unimolecular process.

Power data are shown in Figure 2 as log-log plots for CH_3I^+ and I^+. A plot of log ionization current versus log optical power for I^+ has a slope of 3.19, whereas the log-log plot for CH_3I^+ has a

0094-243X/86/1460574-2$3.00

slope of 1.7 at low power and a slope of 3.22 at high power.

We analyzed our results in terms of population rate equations for various mechanisms. The following mechanisms can be used to explain our data:

When a XeCl excimer laser at 308 nm is applied, single-photon energy is between 32500 cm^{-1} and 32000 cm^{-1}, and the CH_3I^+ production is probably determined by the following mechanism:

$$CH_3I + 2\ hw\ (308\ nm) \rightarrow CH_3I\ (5P\pi,\ 6P) \quad (1)$$

$$CH_3I\ (5P\pi,\ 6P) + hw\ (308\ nm) \rightarrow CH_3I^+ + e^- \quad (2)$$

The first step is a two-photon near resonance process, which belongs to (5Pπ, 6P) Rydberg excitation.

Production of the I^+ ion is obviously predominantly determined by the following process:

$$CH_3I + hw\ (308\ nm) \rightarrow CH_3 + I\ (^2P^0_{1/2},\ ^2P^0_{3/2}) \quad (3)$$

$$I(^2P^0_{3/2}) + 2hw\ (308\ nm) \rightarrow I(5s^2\ 5p^4\ 6p\ ^4P^0_{3/2}) \quad (4)$$

$$I(5s^2\ 5p^4\ 6p\ ^4P^0_{3/2}) + hw\ (308\ nm) \rightarrow I^+ + e^- \quad (5)$$

The I^+ ion is also produced by dissociation of the parent ion at high power, which can be shown by the deviation of CH_3I^+ ion yield from second order behavior at high intensities. Two competing channels seem to exist simultaneously.

The ion yield of CH_3^+ is very small at 308 nm. The CH_3^+ ion is probably produced by excitation and dissociation of the parent ion.

References

1. N. A. Kuebler, J. Phys. Chem. 86, 4096 (1982).
2. A. Gedanken, J. Chem. Phys. 76, 4798 (1982).

PROBE INVESTIGATION OF LASER PRODUCED PLASMAS

A. M. Abdel Nasser, L. Zaki and L. El Nadi

Physics Department
Cairo University
Cairo, Egypt

Abstract

The plasma charge separation voltage generated when a CO_2 laser pulse was focused on metal targets of Al or Bi was measured for an irradiance ϕ of 0.1 - 4.5 GW cm^{-2} at a pressure of 5×10^{-2} - 10^{-5} Torr. The current flow between target and grounded wire probe 1 cm ahead was found to be proportional to $\phi^{1/3}$. The electric field potential detected by a transmitting wire probe was found to scale as $\phi^{0.8}$. The obtained results are interpreted in terms of a model for the expansion of an incompletely ionized laser produced plasma.

Introduction

It was reported by several authors[1-5] that currents and electric fields appear in the plasma formed by optical breakdown in air or low pressure gas near conducting or insulating targets. However unambiguous conclusions from the observed data are difficult to obtain. The present work is an attempt to clarify some aspects for laser produced plasmas.

Experiment

A multimode TEA CO_2 laser provided 50 ns, 60 MW single pulses which were focused on targets centrally mounted in a vacuum chamber. Laser powers were monitored by a photon drag detector on a dual beam 7844 Tektronix oscilloscope together with: (i) target voltage signals from the isolated target through a 1/20 fast response attenuator, or (ii) axial current signals between the isolated target and a grounded bare tip wire probe set at 1 cm facing the target and 0.5 cm above the beam axis, or (iii) electric field potential signals obtained through the attenuator from a floating wire probe (which connected the target to the chamber to form a transmitting probe assembly of RC = 70 μs).

Results and Conclusion

As shown in Figure 1, the target voltage signal is a single spike in time following the laser pulse (a time delay is imposed). Threshold and saturation behavior of V vs ϕ is consistant with the data[4] for Cu. The lower threshold at low pressure proves that the plasma initiated in metal vapor is assisted by that initiated in the residual gas. The saturation voltage is dependent on the chamber geometry; the conical geometry for the Cu experiment provided[4] a

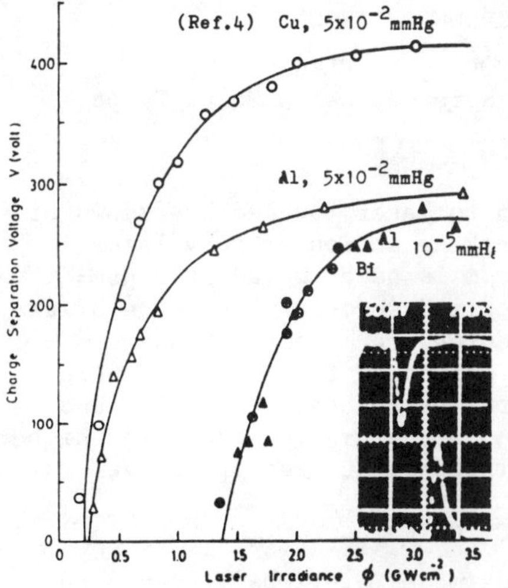

Figure 1. Target voltage versus irradiance and typical signals.

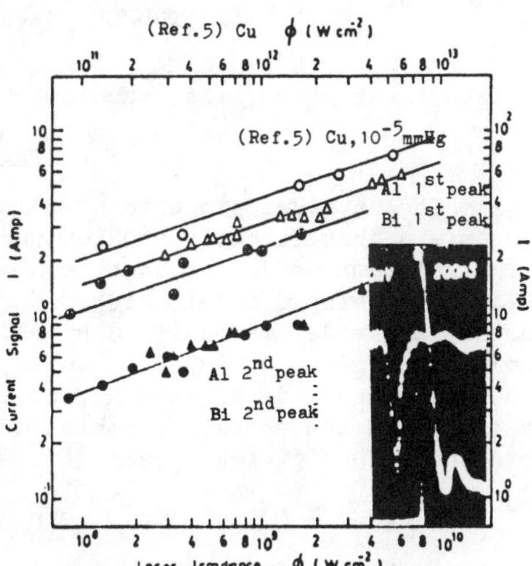

Figure 2. Axial current irradiance and typical signals.

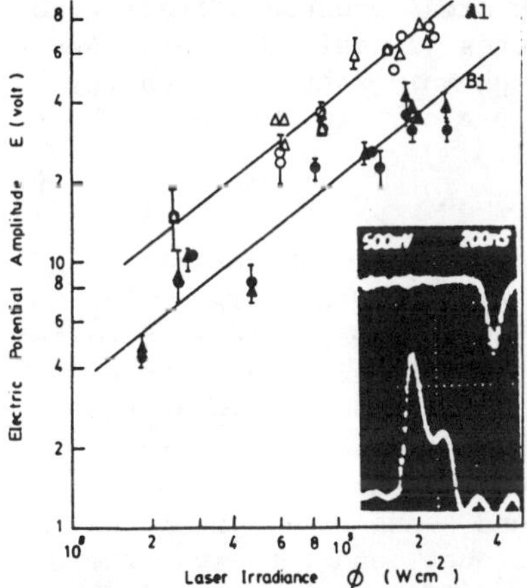

Figure 3. Electric field potential vs. irradiance.

higher efficiency of charge collection. In Figure 2 the axial current is formed of two 50 ns peaks. The first is due to the prompt laser produced plasma and the second may be initiated by the plasma UV radiation occuring within 60 ns of its production[5]. The amplitude of the current I scales as $I \sim \phi^{1/3}$, as for Cu. The electric field potential signal in Figure 3 is double peaked with peaks at 50 and 120 ns. The first is correlated with the prompt plasma. The later one is relevant to the expanding plasma. A scaling of $E \sim \phi^{0.8}$ is explained by assuming self similar expansion[6] of an incompletely ionized plasma.

References

1. A. I. Barchukov et al., Sov. Phys. JETP 51, 482 (1980).
2. H. M. Thompson et al., J. Appl. Phys. 48, 3307 (1977).
3. S. I. Andreev et al., Sov. Phys. Tech. Phys. 15, 1109 (1971).
4. W. T. Silfvast et al., Appl. Phys. Lett. 31, 726 (1977).
5. M. G. Drouet et al., Appl. Phys. Lett. 28, 426 (1976).
6. V. N. Timoshchenko et al., Sov. Phys. Tech. Phys. 25, 588 (1980).

RUBY LASER INDUCED BREAKDOWN OF ARGON

Yosr E E-D GAMAL, I.M. AZZOUZ , M. EL NADY
Department of Physics, Faculty of Science, Cairo Univ. Egypt.

ABSTRACT

The physical processes leading to laser induced breakdown of argon are theoretically studied under the action of ruby laser light. A simple model of the argon atom is used. We adopt a previously developed model which was applied successfully to He breakdown. This model is modified by allowing the momentum transfer collision frequency (MTCF) $\nu_m(\varepsilon)$ to depend on the electron energy. Good agreement is obtained between the calculated thresholds and the experimental ones. Recombination losses are introduced at higher pressures thus giving agreement over the whole pressure range.

INTRODUCTION

In the present work the cascade model previously developed by Evans and Gamal[1] is modified to include the exact correlation between the MTCF and the electron energy. This model is used to investigate the pressure dependence on the threshold intensity for breakdown. We attempt to analyze the physical processes leading to laser induced breakdown. At high pressures recombination has been included to determine the channel through which the laser energy deposited in the gas. The computations are carried out under the experimental conditions of Morgan et al.[2]

THEORY

In allowing for the variation of the MTCF with the electron energy a curve fit expression based on experimental data giving in ref.3 for the momentum transfer collision cross-section is obtained (See Fig.1 Curve 1) then $\nu_m(\varepsilon)$ is calculated (Fig. 1 Curve 2) and introduced into equation (3) ref.1. The model includes the impact ionization of ground state atoms and impact excitation followed by either photoionization or collisional ionization by low energy electrons. The depletion of ground state atoms due to excitation and ionization is taken into acount. Under the experimental conditions used in this interpretations diffusion losses and photoionization of ground state atoms are neglected.

RESULTS AND COMPARISON WITH EXPERIMENTS

The relevant excitation and ionization cross-sections $\sigma_x(\varepsilon)$, $\sigma_i(\varepsilon)$ used in the calculations are taken from ref. 4. The two step ionization cross section for excited argon atom is assumed as a lower limit to have the same value as $\sigma_i(\varepsilon)$ for $\varepsilon>\varepsilon_i-\varepsilon_x$ The three photon photoionization coefficient for agron is calculated using the results of Grey Morgan[5] to be C=1. 146 x 10^{-34}.

Assuming a breakdown criterion as the degree of fractional δ 0.01% of the neutral atoms in the focal region, the calculated values of the threshold intensities were ploted versus the pressure as shown in Fig. 2. Curve A, shows a good agreement over pressure range $5 \times 10^2 - 3 \times 10^3$ torr. The discrepancy observed above this pressure was attributed due to the omission of recombination losses from computations. Curve B in the figure 2 is calculated using constant value of $\nu_m = 3.9 \times 10^9$ P sec^{-1} (P is the pressure in torr).

The threshold intensities at higher pressures have been calculated when recombination losses control the breakdown using a formula obtained by Morgan et al[2] (Curve C fig.2). Good agreement has been obtained over the whole pressure range.

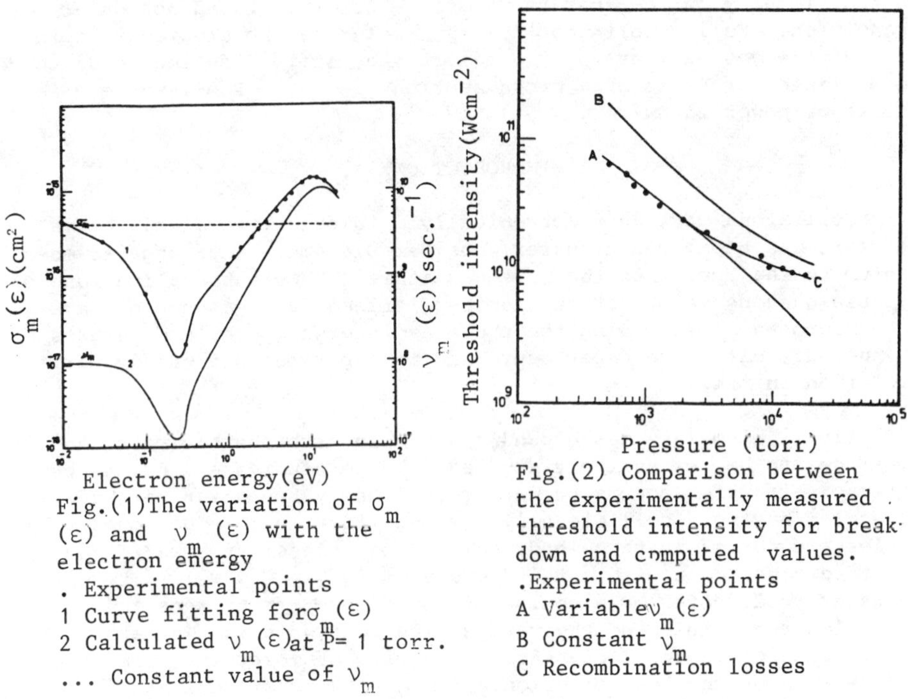

Fig.(1) The variation of $\sigma_m(\varepsilon)$ and $\nu_m(\varepsilon)$ with the electron energy
. Experimental points
1 Curve fitting for $\sigma_m(\varepsilon)$
2 Calculated $\nu_m(\varepsilon)$ at P= 1 torr.
... Constant value of ν_m

Fig.(2) Comparison between the experimentally measured threshold intensity for breakdown and computed values.
. Experimental points
A Variable $\nu_m(\varepsilon)$
B Constant ν_m
C Recombination losses

REFERENCES

1. C.J.Evans and Y.E.E. D Gamal, J.Phys.D;Appl.Phys. 13,1447 (1980).
2. F.Morgan, L.R. Evand and C. Grey Morgan: J.Phys.D : Appl. Phys. 4, 225 (1971).
3. M.Hayashi, Report of At.data, IPP/Nogoya University (1981).
4. Ya.B. Zel'dovish and Yu. p.Raizer, Sov.Phys.JETP. 20,3,772 (1965).
5. C.Grey Morgan, Rep.Prog. Phys., 38, 628 (1975).

LASER-ENHANCED SPIN-EXCHANGE COLLISIONS

L. L. Vahala and M. D. Havey
Old Dominion University, Norfolk, Va. 23508

ABSTRACT

Spin-exchange collisions between optically pumped alkali atoms and rare gas atoms can polarize the nuclei of rare gas atoms. We present calculations which indicate that the process should be readily modified by relatively weak (<1KW/cm^2) CW laser radiation tuned in the vicinity of the Rb-rare gas 5sΣ - 6sΣ free-bound molecular transition. Full coupling of the spin-polarized Rb electronic spin to the rare gas nuclear spin in the excited state lifetime leads to an estimate for Rb-Xe of a cross section $\sigma \simeq 10^{-22}$ Pcm^2, where P is the laser power in KW/cm^2.

INTRODUCTION

Collisions between electronically spin-polarized alkali atoms and rare gas atoms can transfer considerable amounts of angular momentum to the nuclei of the rare gas atoms. Numerous applications of a dense gaseous sample of such spin-polarized rare gas nuclei have been proposed, [1] including their use as internal polarized targets [2] in nuclear scattering experiments and in experiments to study parity violation in ^{21}Ne. [3]

Alkali atom-rare gas ground state spin exchange has been, and continues to be, extensively studied. These studies [1,4] have shown that spin-transfer occurs primarily in loosely bound van der Waals molecules having a collisionally-limited lifetime τ. The bound molecules are formed in three-body collisions between an alkali atom and a rare gas atom with a third gas atom (e.g., H_2, N_2). The spin transfer is effected by a contact interaction $\alpha\vec{S}\cdot\vec{K}$ between the alkali electron spin $\vec{S}$ and the rare gas nuclear spin $\vec{K}$. The efficiency of transfer is low because of the rarity of molecular formation and because of the loss of orientation to the rotational angular momentum $\vec{N}$ of the alkali rare gas pair via a spin-rotation interaction $\gamma\vec{S}\cdot\vec{N}$.

The Hamiltonian [1] for the spin dynamics is

$$H = A\vec{I}\cdot\vec{S} + \alpha\vec{S}\cdot\vec{K} + \gamma\vec{S}\cdot\vec{N}$$

where $\vec{I}$ is the alkali nuclear spin, A the alkali hyperfine coupling constant. Theoretical expressions exist [5,6] for the coupling constants $\alpha(R)$ and $\gamma(R)$, where R is the alkali-rare gas separation, and yield good agreement with experimental numbers.

DISCUSSION

We propose here [7] a means by which the spin-transfer may be modified by relatively weak radiation. The basic idea is to laser-excite a colliding spin-polarized alkali atom and a rare gas atom to the lowest bound (excited) molecular Σ state. In this state the spins evolve for a radiative lifetime τ, allowing for effective spin-transfer to occur via H . . . just as in the ground state. A potential advantage of the method is that it is externally controlled via the laser intensity and polarization (circular polarization of the exciting laser will transfer orientation to $\vec{N}$ and hence, to some degree, to $\vec{K}$). The method should also work for systems with little or no ground state binding where spin-transfer is very small and via binary encounters ($\sigma = 6.6 \times 10^{-24} cm^2$ for $^{87}Rb^{21}Ne$). The coupling constants $\alpha(R)$ and $\gamma(R)$ depend strongly on R and values for excited states may be more favorable than for the ground state. Finally, the full spectrum of values for $\vec{N}$ is limited by the optical collision dynamics to $\mu v b_{max}/\hbar$, where μ is the reduced mass and b_{max} the maximum impact parameter for the collision.

We make an estimate of σ for $^{87}Rb^{129}Xe$ by taking $\sigma = \sigma_{abs}\, Q$ where σ_{abs} is the optical absorption cross section for the RbXe $5s\Sigma$ - $6s\Sigma$ transition and Q is the probability of spin-exchange in the molecular lifetime τ. σ_{abs} has been measured [8] to be 5.9×10^{-21} Pcm^2 where P is the laser power in KW/cm^2. The spin transfer probability Q is given as a spin transfer coefficient in Happer, et. al. [1] For a range of values of α and γ (taken in the vicinity of the ground state values), we obtain $Q \simeq 0.015$ and $\sigma \simeq 10^{-22}$ Pcm^2 for $^{87}Rb^{129}Xe$. Comparable numbers are obtained for Rb-Ne, Rb-Ar and Rb-Kr.

This work is supported by the NSF under grant number PHY-8509881.

REFERENCES

1. W. Happer, E. Miron, S. Schaefer, D. Schreiber, W.A. van Wijngaarden and X. Zeng, Phys. Rev. A29, 3092 (1984).
2. T.E. Chupp, W. Happer and A.B. McDonald, in Proceedings of the Workshop on Polarized Targets in Storage Rings, Argonne Ill. (Argonne Nat. Lab., Argonne Ill.) ANL-84-50, p. 177 and other papers therein.
3. T.E. Chupp and K.P. Coulter, Phys. Rev. Lett. 55, 1074 (1985).
4. C.C. Bouchiat, M.A. Bouchiat and L.C.L. Pottier, Phys. Rev. 181, 144 (1969).
5. R.M. Herman, Phys. Rev. 137A, 1062 (1965).
6. Z. Wu, T.G. Walker and W. Happer, Phys Rev. Lett. 54, 1921 (1985).
7. M.D. Havey and L.L. Vahala, submitted to Phys. Lett. A.
8. G. Moe, A. C. Tam and W. Happer, Phys. Rev. A14, 349 (1976).

EXPERIMENTAL FINE-STRUCTURE BRANCHING RATIOS FOR Na-RARE GAS OPTICAL COLLISIONS

M. D. Havey, F. T. Delahanty, L. L. Vahala, and G. E. Copeland
Old Dominion University, Norfolk, Va. 23508

ABSTRACT

Experimental ratios for branching into the fine-structure levels of the Na 3p multiplet, as a consequence of an optical collision with He, Ne, Ar, Kr, or Xe, are reported. The process studied is $Na(3s^2S_{1/2})+RG+nh\nu \rightarrow Na(3p^2P_j)+RG+(n-1)h\nu$, where RG represents a rare gas atom and where the laser frequency ν is tuned in the wings of the Na resonance transitions. The branching ratios are defined as I(D1)/I(D2) where I(D1) and I(D2) are measured intensities of the atomic Na D1 and D2 lines. The ratios are determined for detunings ranging from about 650 cm^{-1} blue to 170 cm^{-1} red of the Na 3p multiplet. The branching is found to be strongly detuning dependent in the vicinity of the NaKr and NaXe near red-wing satellites. The blue wing branching ratios show a detuning-dependent approach to a recoil or sudden statistical limit of 0.5, irrespective of the rare gas.

INTRODUCTION

Final state distributions produced as a result of absorption of light by atoms in collision can depend strongly upon nonadiabatic mixing among the molecular states of the system. For diatomic systems with fine-structure, off-diagonal spin-orbit and rotational terms in the molecular Hamiltonian, along with the interatomic potentials, constitute the primary determinants of the fine-structure branching ratios in free-free scattering.

The process considered here is the optical collision $Na(3s^2S_{1/2})+RG+nh\nu \rightarrow Na(3p^2P_j)+RG+(n-1)h\nu$, where RG represents a rare gas and where the laser frequency ν is tuned into the wings of the Na resonance transitions. The branching ratios $B(\Delta)$ are defined as I(D1)/I(D2) where I(D1) and I(D2) are the measured intensities of the atomic Na D1 and D2 lines. These ratios provide a measure of nonadiabatic mixing among the molecular terms of the NaRG molecules. We have measured the ratios for detunings Δ ranging from about 650 cm^{-1} blue to 170 cm^{-1} red of the Na 3p multiplet.[1] Because the mixture of states produced by optical excitation is detuning dependent, $B(\Delta)$ is also.

RESULTS

To illustrate the range of behavior of $B(\Delta)$, we present some of our data in Figs. 1-4. All data presented is independent of laser power, rare gas pressure, and Na density. Alignment of the $3p^2P_{3/2}$ state as a result of the optical collision produces a maximum decrease of $B(\Delta)$ of 7%; theoretical calculations [2] of the alignment show the effect to be less than 2%. No corrections for this effect have been made. The blue wing $B(\Delta)$ for NaHe, NaNe, and NaKr in Figs.

1-3 display a detuning-dependent approach to a sudden statistical limit of 0.5.[3] This limit is expected to obtain whenever the dissociation is fast compared to an inverse fine-structure coupling frequency. The approach to this limit from above 0.5 (as for NaKr) as opposed to below 0.5 (as for NaHe and NaNe) is due to the considerably stronger binding of the NaKr BΣ state. This greater binding produces an extremum in the XΣ-AΠ difference potential and hence a satellite in the total absorption coefficient. The resulting satellite in the branching ratio, presumably due to the adiabatic correlation of the $A^2\Pi_{\frac{1}{2}}$ state to the Na $3^2P_{\frac{1}{2}}$ state, is shown in Fig. 4. Details of our experiments on all the NaRg molecules are presented elsewhere.[1]

This work is supported by the NSF under Grant PHY-8509881.

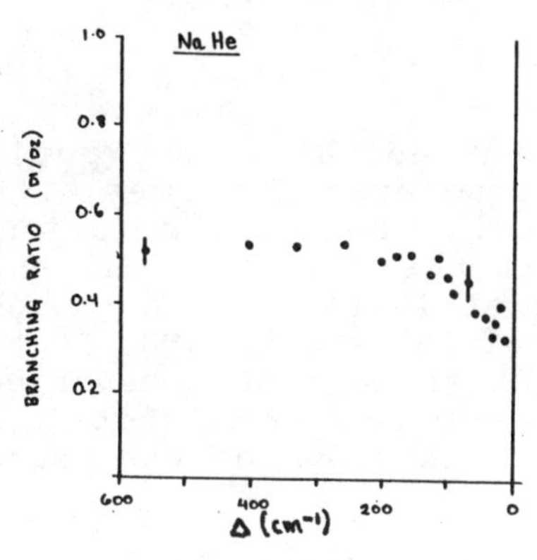

Fig. 1. NaHe blue wing.

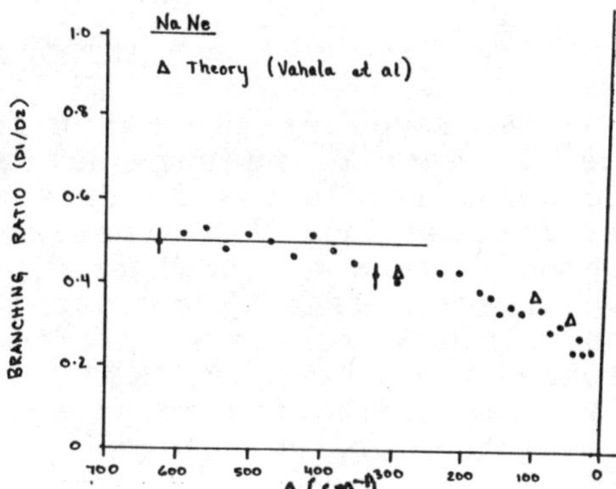

Fig. 2. NaNe blue wing.

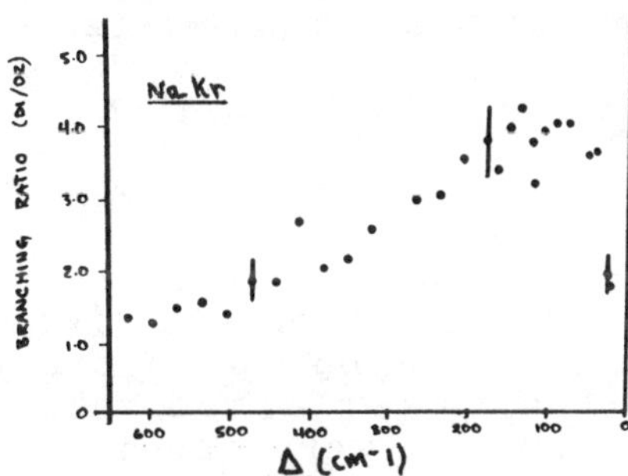

Fig. 3. NaKr blue wing.

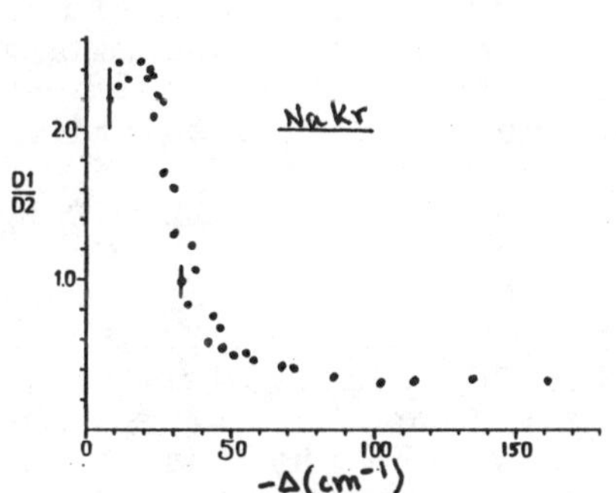

Fig. 4. NaKr red wing.

REFERENCES

1. M.D. Havey, F.T. Delahanty, L.L. Vahala and G.E. Copeland, submitted to Phys. Rev. A.
2. L.L. Vahala, P.S. Julienne and M.D. Havey, submitted to Phys. Rev. A.
3. S.J. Singer, K.F. Freed and Y.B. Band, J. Chem. Phys. 81, 3064 (1984).

SPECTROSCOPIC TECHNIQUES IN THE FEMTOSECOND RANGE

J.-C. Diels, I. C. McMichael*

Center for Applied Quantum Electronics, P.O. Box 5368
North Texas State University, Denton, Texas 76203

ABSTRACT

A temporal resolution beyond the pulse duration can be achieved with an accurate diagnostic of the source. The use of pulse sequences results in a spectral resolution better than the bandwidth of each individual pulse.

INTRODUCTION

It is generally taken for granted that, in experiments meant to measure ultrafast relaxation times, the temporal resolution is determined by the pulse duration. We discuss first an experiment in the time domain, and demonstrate a temporal resolution of 10 fsec, even though the laser pulses have a duration of 75 fsec (FWHM). It is shown that, for this particular experiment of Degenerate Four Wave Mixing (DFWM), the temporal resolution is directly correlated to an accurate knowledge of the phase modulation of the pulses.

In the frequency domain, the resolution is not always limited to the bandwidth. Calculations of multiphoton excitation of an anharmonic ladder of vibrational-rotational transitions by sequences of picosecond pulses, demonstrate spectral resolution within the pulse bandwidth.

SUB-PULSEWIDTH TEMPORAL RESOLUTION

As an example of a determination of a time constant shorter than the pulse duration, we present an experiment of DFWM mixing leading to the measurement of a phase relaxation time. The experiment being performed intracavity, the physical conditions are set very precisely, and are accurately reproducible. A ring mode-locked dye laser is used, with an intracavity prism which enables us to control the phase modulation and duration of the output pulses [1]. The saturable absorber (DODCI) is the nonlinear medium for this experiment. The two counterpropagating pulses inside the cavity which meet simultaneously in the DODCI form the pump signals for the DFWM. The probe pulse is taken from one of the outputs of the laser, after being sent through an adjustable delay line and having its polarization rotated by 90 $^{\circ}$. A 22 mm focal distance lens

* Present address: Rockwell International Science Center, P.O. Box 1085, Thousands Oaks, CA 91360.

is used to focalize the probe in the interaction volume, at right angle to the pump beams. This particular geometry minimizes the propagation effects that can broaden the signal [2]. Indeed, tight focusing of the pump pulse with the 3 cm curvature cavity mirrors results in a beam waist of 4 μm. The interaction length for the probe pulse (at right angle to the pump) is that for which the **squared** pump intensity is within a factor 2 of its peak value. Hence it is only a 2 μm thick slice of medium that is being probed, and the signal broadening introduced by propagation effects is less than 5 fsec.

Two distinct contributions to the signal can be identified: (a) scattering of one of the pump signals off the grating formed by the other pump with the polarization created by the probe pulse and (b) creation of two photon excitation by the two counterpropagating pumps, and subsequent two photon emission stimulated by the probe pulse. In the first case (a), the **leading edge** of the signal versus delay is a measure of the phase relaxation of the single photon excitation in the absorber, since a grating will only be formed if the probe pulse precedes the pumps by no more than the dephasing time of the induced dipoles. In the second case (b), the **trailing edge** of the signal versus delay is a measure of the phase relaxation time of the two-photon excitation, since the off-diagonal matrix of the two-photon excitation will only subside for that time lapse after the intense pump pulses. Since the phase relaxation times are expected - and confirmed by these experiments - to be shorter than the pulse duration, a simple steady state approximation of the density matrix equations can be used to estimate the signal intensities from both mechanisms. In the case of a single photon interaction, the signal intensity is proportional to [3]:

$$|E_4|^2 = |E_1E_2E_3|^2 \frac{1}{\left[\Delta\omega^2 + \left(\frac{1}{{}^1T_2}\right)^2\right]^2} \tag{1}$$

where $E_i = \mu_{01}\,\mathcal{E}/2h$ (μ_{01} being the dipole moment) is the transition rate between levels 0 and 1 induced by the field i (pumps: i =1, 2; probe: i = 3; and signal: i = 4), $\Delta\omega$ is the difference between the transition and the light frequency, and ${}^1T_2$ is the phase relaxation time for the induced polarization. In the case of the two-photon interaction, similar approximations [3] lead to a signal proportional to:

$$|E_4|^2 = |E_1E_2E_3|^2 \left\{\frac{{}^2T_2}{\Delta\omega}\right\}^2 \tag{2}$$

where ${}^2T_2$ is the phase relaxation time for the element of the density matrix connecting the lower and upper states of the resonant two-photon transition. Consistently with the experimental data and measurements in the frequency domain, the single photon phase relaxation rate ${}^1T_2^{-1}$ is of the order of the pulse bandwidth, which is in this case also equal to the detuning $\Delta\omega$. The ratio of the two photon to the single photon contribution to the DFWM is thus roughly $4[\Delta\omega \cdot {}^2T_2]^2$, which is much larger than unity.

Computer simulation indicate that the shape of the DFWM signal versus delay is a sensitive function of the pulse shape and phase modulation. Accurate knowledge of these parameters is therefore essential to extract a correct phase relaxation time from the measurements. The amplitude and phase of the laser pulses is determined by simultaneous iterative fitting of the spectrum, intensity and interferometric autocorrelation of the laser output [4]. The envelope of the laser field is found [3] to have the asymmetric shape $\mathcal{E}_o/[\exp\{-t/\tau_r\}+\exp\{t/\tau_f\}]$ with a risetime $\tau_r = 57$ fsec and a fall time $\tau_f = 95$ fsec. The phase modulation is a linear down frequency sweep of only $4\ 10^{-4}$ $fsec^{-1}$/fsec. Fitting of the computed DFWM signal versus delay for various values of the two-photon phase relaxation time ${}^2T_2$ to the experimental data (Fig. 1) results in a determination of ${}^2T_2 = 50$ fsec ($\pm$ 10 fsec).

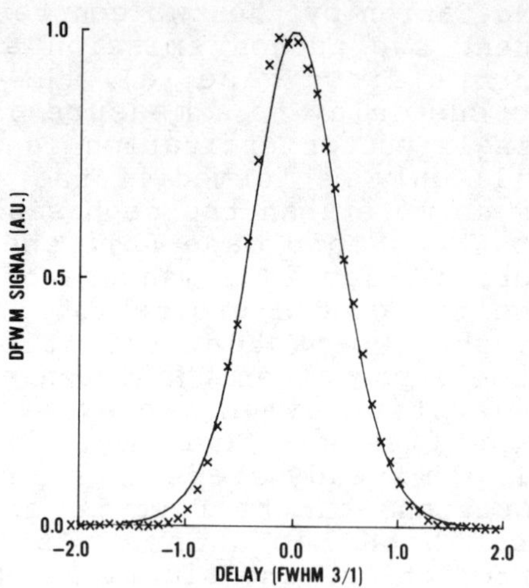

Figure 1: Theoretical (solid line) and experimental (crosses) DFWM signal.

FREQUENCY SELECTIVITY WITH ULTRASHORT PULSE SEQUENCES

We present calculations of multiphoton excitation of an anharmonic ladder of vibrational-rotational levels, to demonstrate spectral resolution within the pulse bandwidth. The excitation is assumed to occur in a time short compared to the dephasing times. To solve the Schrödinger equation for the system interacting with the optical field, the wave function is expanded in a series of energy eigenfunctions corresponding to the quantum numbers v, J and K [5]. For a system of N vibrational levels, the time evolution of the coefficients of this expansion is determined by a set of N x N_R differential equations, where N_R is the number of of rotational lines being considered. In each equation, the dipole matrix elements is also a function of the quantum numbers J and K. The dimension of the problem is however greatly reduced if the transverse moment of inertia of the molecule is neglected. Within this approximation, we solve numerically the system of equations applied to the molecule CH_3F. The rotational line distribution has a regular structure that could be matched by the spectrum of a regular sequence of pulses spaced by a delay of $2n\pi/B$ (n being an integer). This type of excitation maximizing the spectral overlap provides maximum pumping in incoherent interaction. Systematic calculations of the excitation to the fifth level for sequences of 1 psec pulses of various relative phases and delays leads to an optimal five pulse sequence only slightly departing from that condition. The final population distribution among rotational lines in the upper level is plotted in Fig. 2 (heavy line). The final rotational line distribution of the lower level (thin line) and the initial Boltzman distribution (shaded area) are also shown. The particular pulse sequence selected results indeed in a **cooling** of the rotational temperature in the excited state. The pulse sequence has resulted in transferring most of the populations in the J=2 and J=3 rotational lines of the upper level. These calculations demonstrate thus clearly the possibility of vibrationally as well as rotationally selective multiphoton excitation with sequences of pulses.

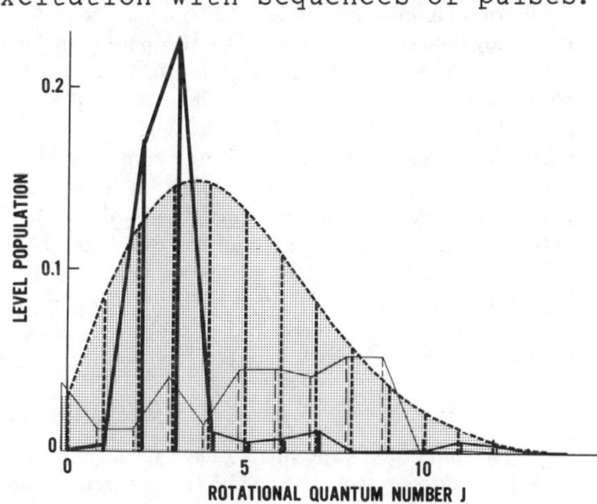

Figure 2. Population distribution among the rotational lines in the fifth level of an anharmonic ladder, following excitation by an optimized five pulse sequence. The initial distribution in the ground state (shaded area) corresponds to a temperature of 250 ^{o}K.

This work was supported by the NSF grant ECS-8406985.

REFERENCES

1. W. Dietel,J. Fontaine,and J.-C. Diels, Optics Letters **8**, 4(1983).
2. J.-C. Diels, W. Wang, and H. Winful, Appl. Phys. B26, 105 (1981).
3. J.-C. Diels and I. C. McMichael, JOSA B, **3** (April 1986).
4. J.-C. Diels,et al.,Applied Optics **24**, 1270-1282 (1985).
5. Townes & Schawlow, "Microwave Spectroscopy", McGraw-Hill, (1955).
6. "Multiphoton Coherent Excitation of Molecules", J.-C. Diels and S. Besnainou, submitted to J. Chem. Phys. (1986).

CHARACTERIZATION OF ULTRASHORT LASER PULSES BY THE METHOD OF SELF-DIFFRACTION

P.M. Fauchet, W.L. Nighan, Jr.
Princeton University, Princeton, NJ 08544

R. Trebino
Stanford University, Stanford, CA 94305

ABSTRACT

The pulsewidth and coherence time of a picosecond Nd:YAG laser has been measured by the method of self-diffraction, either in the two pulses or in the three pulses configuration. Experimental results have been obtained with thin dye cells or thin semiconductor films. A theoretical model has been developed that explains our results.

INTRODUCTION

Non-transform-limited laser pulses can be characterized by two parameters, the pulsewidth and the coherence time. In nonlinear optics, both parameters are important. Standard second-harmonic autocorrelation measurements do not provide both parameters conveniently. We demonstrate that self-diffraction is a convenient method for measuring these parameters. Furthermore, since it relies on absorption in a material, self-diffraction can be used at short wavelengths.

EXPERIMENTAL

Frequency-doubled pulses from an actively-passively modelocked Nd:YAG oscillator/amplifier system are split into two equal-intensity components. A variable delay line is introduced in one arm. The two parallel-polarization pulses are incident on a sample with a small angle between them and form an interference pattern when they overlap temporally. The samples are a one micron thick film of silicon deposited on fused quartz or a one millimeter thick cell filled with Rh6G in methanol. In the two pulses configuration, the diffraction of one of the two writing beams is recorded as a function of delay between the two writing beams. This can be done either in transmission or in reflection. In the three pulses configuration, the transmitted part of one of the two writing beams is retroreflected onto the interference pattern and the diffraction of that beam is recorded as a function of time delay between the two writing beams. The delay of the third beam is equal to 1 ns. Figure 1 shows results obtained with both configurations on Rh6G. Results obtained with SOI are very similar, except that in the three pulses configuration it was difficult to observe the signal.

DISCUSSION

Self-diffraction experiments have recently been performed by others [1,2]. The results have been interpreted in terms of the coherence time of the pulse only. One of us [3] has developed a theory that includes in an exact way the effects of pulsewidth τ_p and coherence τ_c time. If $\tau_p >> \tau_c$, then the grating efficiency is given by

$$<\eta> \cong \frac{\tau_c}{\tau_p} \exp\left[-2\ln 2\left(\frac{\tau_d}{\tau_p}\right)^2\right] + \exp\left[-\Pi\left(\frac{\tau_d}{\tau_c}\right)^2\right] \qquad (1)$$

where τ_d is the time delay, and a gaussian pulse envelope and a gaussian coherence function have been assumed. For the three pulses configuration, a fit to our data yields $\tau_p \simeq$ 23ps, $\tau_C \simeq$ 6ps. We presently have no definitive explanation for the dips observed on both sides of the coherence peak. The theory for the two pulses case is more complex. Considering the pulsewidth effect only and assuming that the grating lifetime exceeds the pulsewidth, we find that the FWHM of the pedestal is $\simeq 1.1\text{x}\tau_p$ and should peak around $\tau_p/6$. The data yield $\tau_p \simeq$ 23ps and $\tau_C \simeq$ 7ps, in excellent agreement with the other configuration. Noncollinear second harmonic autocorrelation measurements performed at the fundamental output of the laser showed good agreement with τ_p obtained by self-diffraction. Note that the maximum of the pedestal was observed to shift to $-\tau_p/6$ in the two pulses configuration if the self-diffracted signal was produced by the fixed pulse, not by the delayed pulse. This effect is due to the simultaneity of the creation and the probing of the grating, and is predicted by theory.

CONCLUSIONS

Self-diffraction has been demonstrated to be a powerful method for characterizing ultrashort pulses. Its usefulness stems in part from the fact that the amplitude of the coherence peak is magnified by a factor of τ_p/τ_C compared to the pedestal. We have shown that widely different materials can be used to perform the measurement, which can easily be scaled to shorter wavelengths (where nonlinear crystals absorb and thus are not used) and to shorter pulses. We acknowledge the support of a Cottrell Research grant from Research Corporation (Princeton) and of AFOSR (Stanford).

REFERENCES

1. H.J. Eichler, U. Klein, and D. Langhans, Appl. Phys. **21**, 215 (1980)

2. A.L. Smirl, T.F. Boggess, B.S. Wherrett, G.P. Perryman, and A. Miller, IEEE J. Quant. Electron. **QE-19**, 690 (1983)

3. R. Trebino, Ph.D. Thesis, Stanford University, 1983

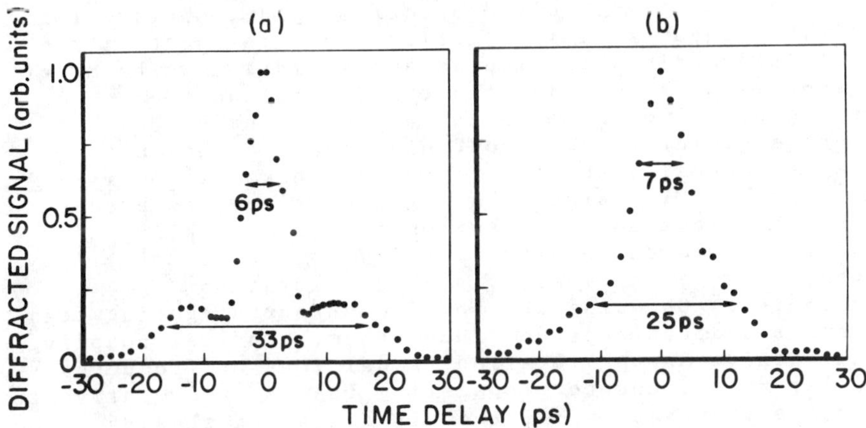

Fig. 1. Diffracted signal vs. time delay between the two writing beams
(a) three pulses configuration
(b) two pulses configuration (self-diffraction)

Time Resolved Picosecond Raman Induced Phase Conjugation In Liquids and Solids

P.J. Delfyett, R. Dorsinville, and R.R. Alfano
Institute for Ultrafast Spectroscopy and Lasers
and
Photonic Engineering
The City College of New York, NY 10031

ABSTRACT

The Raman induced phase conjugation spectroscopy technique has been extended to the picosecond regime. Using a broadband picosecond supercontinuum beam generated from D_2O, Raman spectra spanning 3000 cm^{-1} were obtained in CS_2, Nitrobenzene and calcite using a single 50 mJ, 35 ps pulse from a Quantel Nd:Yag laser system. Delaying one of the counterpropagating pump beams and monitoring the intensity of the Stokes wave in the phase conjugate direction, the dephasing times of molecular vibrations and optical phonons were measured.

Introduction

Raman induced phase conjugation (RIPC)[(1)] has been used as a new time-resolved spectroscopic technique for obtaining fundamental information about resonant vibrational interactions in liquids and solids. Two beams at frequencies w_L and w_L and the third continuum[2] beam spanning over 4000 cm^{-1}, enter a sample in the phase conjugate geometry. The continuum beam is chosen to propagate opposite to one of the monochromatic beams. A fourth beam, (RIPC signal) at frequency w-v, is produced in the phase conjugate direction due to a coupling of the input beams via the nonlinear third order susceptibility X^3(w-v; w, w-v, -w) where v are the vibration frequencies. Choosing the third counterpropagating beam to be a broadband supercontinuum beam, provides coupling to all active Raman vibration transitions at frequency v. By delaying the counterpropagating beam at w_L relative to the beams at w_L and w_L-v, the vibrational dephasing times can be measured from the intensity of the RIPC signal.

Experiment

In the experiment, a frequency doubled, 35 ps laser pulse is generated from an Nd:YAG laser system. The fundamental and second harmonic are separated using a harmonic beam splitter. The second harmonic is then divided into two beams using a beam splitter and then recombined in the sample at a small angle. The fundamen-

tal is used to pump the continuum generator cell. The RIPC signal is generated via the four wave mixing mechanism and examined using a spectrometer and an optical multi-channel analyzer.
The ultimate time resolution in the experiment is less than 5 ps. This has been determined by measuring the temporal duration of the stokes frequency component in the continuum pulse using a narrow band filter and a 2 ps resolution streak camera.

Results

Using this technique, RIPC spectra in CS_2, nitrobenzene and calcite, spanning 3000 cm^{-1} have been obtained. A typical RIPC spectra from a 1 cm cell of nitrobenzene is given in Fig. 1.

By delaying the counterpropagating second harmonic beam and monitoring the intensity of a desired Stokes frequency component, information about the vibrational dephasing times can be obtained. Performing the time resolved experiment in CS_2 and calcite, a vibrational dephasing time of 24 ps for the 656 cm^{-1} vibration in CS_2, and 8 ps for the 1086 cm^{-1} phonon mode in calcite were measured. A typical picosecond time resolved curve for the 1086 cm^{-1} vibration in calcite is shown in Fig. 2. The lifetimes are in good agreement with the work of Fischer and Lauberau[3] (22 ps for CS_2) and Alfano and Shapiro[4] (8.5 ps for calcite) using conventional time resolved excite and probe Raman techniques.[5] A salient feature in the time resolved curve of Fig.2 is the asymmetry. This is due to the short temporal duration of the stokes frequency in the continuum pulse.

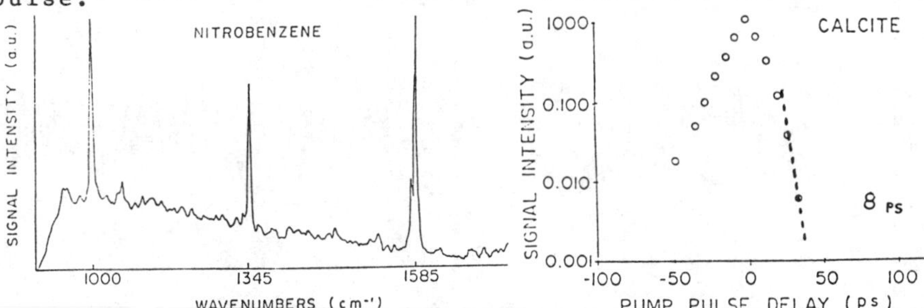

References

1) S.Saha and R.Hellwarth, Phys.Rev. A, 27, 919 (1983).
2) Alfano,R. and S.Shapiro,Phys.Rev.Lett. 24, 584 (1970).
3) Fisher,S. and Laubereau,A,Chem.Phys.Lett. 35, 6(1975)
4) Alfano,R. and S.Shapiro,Phys.Rev.Lett. 26, 1247(1971)
5) Alfano, R.R. & S.L. Shapiro, Scien. Amer. 228,(1972)
Supported by AFOSR.

TIME-DELAYED FOUR-WAVE MIXING IN SODIUM VAPOR

D. DeBeer, L. G. Van Wagenen, R. Beach and S. R. Hartmann
Columbia Radiation Laboratory and Department of Physics
Columbia University, New York, NY 10027

Two separate but related experiments are discussed. Time Delayed Four Wave Mixing (TDFWM) on the Na D doublet is reported using intense 7 nsec incoherent pulses which spectrally cover both transitions,[1] and using a pair of lasers, each tuned to one of the transitions. In both cases a 1.9 psec beat is observed as a function of relative pulse delay. For the latter case the beats occur throughout the pulse overlap region and well into the normal photon echo regime where the excitation pulses no longer overlap. This phenomenon leads to the possibility of doing high resolution spectroscopy.

The experimental apparatus for the first TDFWM experiment is shown in Fig. 1. The spontaneous emission from a YAG pumped Rhodamine 6G cell provided 7ns pulses 12 Å wide which covered the Na D line transitions at 5890Å and 5896Å. These pulses were split and then recombined with a variable relative delay τ to provide two pulses directed along $\mathbf{n}_1$ and $\mathbf{n}_2$ and angled 2 mrad with respect to each other. These pulses overlapped spatially throughout a 10 cm long active region Na cell. A pinhole in the focal plane of a 45 cm lens passed the TDFWM signal in the phase matched $2\mathbf{n}_2 - \mathbf{n}_1$ direction while blocking the excitation pulses. The signal was detected with either a photodiode or a PMT. In all cases the signal was integrated over its full 7 nsec duration.

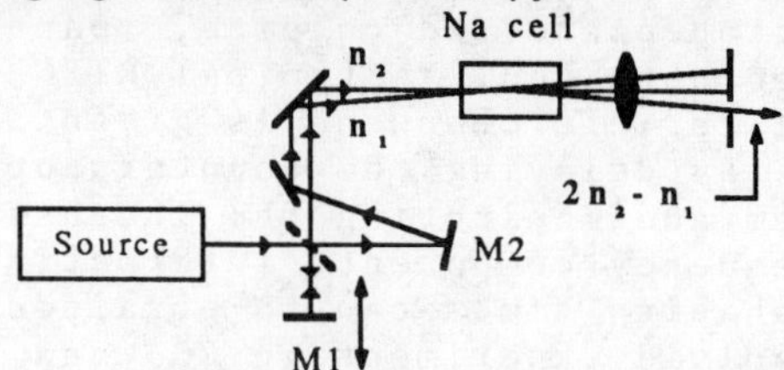

FIG. 1. Schematic diagram of the experimental apparatus used to generate TDFWM. Mirror M1 is mounted on a precision translation stage which provides variable relative delays of up to 1 ns in increments of <100 fs. Mirror M2 can be repositioned to obtain large offset delays.

Working at an oven temperature of 425 K we obtain the results shown in Fig. 2. A nonperturbative theory gives the result expressed by the solid line.[1] The signal falls off at large τ due to noise induced relaxation and at small τ due to inhomogeneous dephasing.

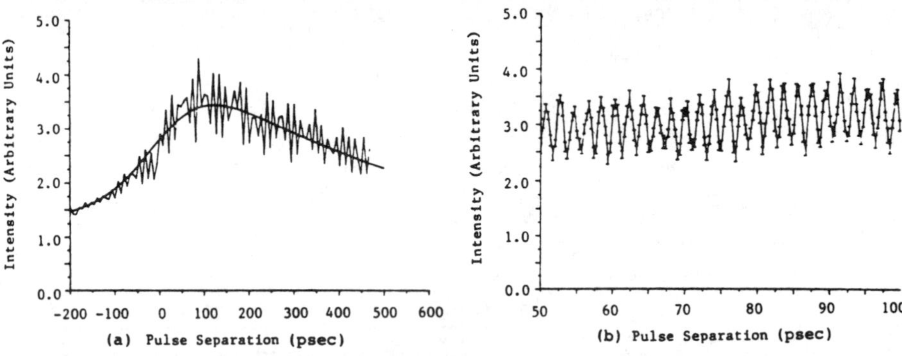

FIG. 2. (a) Experimentally measured signal intensity vs excitation pulse separation using a broadband source. The pulse separation was stepped in approximately 5 ps increments.
(b) Data taken under the same experimental conditions as in (a) but with a finer incremental change in the excitation pulse separation so that the 1.9 ps modulation is clearly visible.

The 1.9 ps modulation corresponds to the inverse of the sodium 3P fine structure splitting of about 6Å. This beating is due to a modulation of the ground state population with τ and not an interference of two radiation fields. The possibility of using this beating to do high resolution spectroscopy leads us to the second experiment in which the incoherent source above was replaced by the combined outputs of two narrow-band lasers tuned to the two Na D-line transitions. Figure 3 shows a typical run in this second configuration in which the TDFWM signal was monitored as the relative pulse delay τ was varied. In this run a spectrometer was used to look at just the 5890Å component of the signal; similar results were obtained for just the 5896Å component and when no spectrometer was used and both transitions were viewed simultaneously. We have taken data in the range of delays from minus a few hundred psec to +30 nsec. All traces exhibit 1.9 psec modulation. Thus the modulation persists well into the photon echo regime where the pulses no longer overlap. Previous work has shown that echoes may be observed over twenty fluorescence lifetimes after the initial excitation pulse.[2] Since the beating can be made Doppler free, this leads to the possibility of high resolution spectroscopic studies.

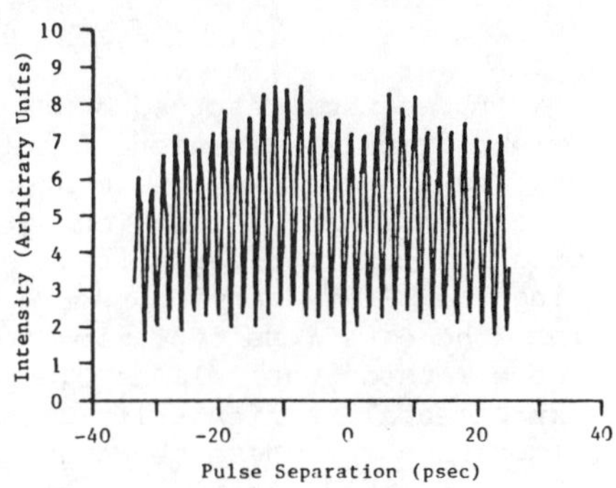

FIG. 3. Typical set of data plotting signal intensity vs relative pulse delay using narrow band sources.

The theoretical analysis of the observed signal will be published elsewhere.[3] We present here only the result for the signal intensity in the photon echo regime. Assuming the lasers to be single mode and transform limited the integrated echo intensity is:

$$\int dt\, I_{echo} \sim 1 + \exp\{(-1/8)[\tau v_{eff}(\Omega'-\Omega)/c]^2\} \cos[(\Omega'-\Omega)\tau(1+\bar{v}/c)]$$

where Ω and Ω' are the two transition frequencies, $v_{eff} = 2\pi c/\Omega\tau_p$ corresponds to the linewidth of the interacting atoms, τ_p is the excitation pulse duration which is assumed to be much longer than the Doppler dephasing time, and $\bar{v}$ is the average velocity of the interacting atoms. The modulation degradation factor has a time constant of about 3 µsec and is thus negligible in our experiment. The modulation term is not Doppler free in that it depends on $\bar{v}$. However, the error is only $(\Omega-\Omega')\bar{v}/c$ and not $\Omega\bar{v}/c$. Furthermore, the error can be eliminated entirely if the laser pulses excite an atomic sample that has an average velocity $\bar{v}=0$. Several possibilities exist for exciting such a distribution. One can use a multimode laser that effectively interacts with the entire Doppler line due to a shot to shot frequency jitter. One can use short laser excitation pulses having durations less than the Doppler dephasing time. Finally, one can use spectrally broad incoherent light that covers the entire Doppler line. In these cases this technique becomes Doppler free and the accuracy to which an energy level splitting can be measured depends on how many beats can be observed. As mentioned previously echoes can be observed many lifetimes after the initial excitation and thus the energy level splitting should be measurable to very high precision.

This work was supported by the U.S. Office of Naval Research and by the Joint Services Electronics Program (U.S. Army, U.S. Navy, U.S. Air Force) under contract No. DAAG29-85-K-0049.

[1] R. Beach, D. DeBeer, and S.R. Hartmann, Phys. Rev. A32, 3467 (1985).
[2] R. Beach, B. Brody, and S.R. Hartmann, J. Opt. Soc. Am. B1, 189 (1984).
[3] D. DeBeer, L.G. Van Wagenen, R. Beach, and S.R. Hartmann, to be published.

COMPARISON OF PICOSECOND FOUR-WAVE MIXING TECHNIQUES FOR MEASURING ORIENTATIONAL RELAXATION TIMES

Anne B. Myers and Robin M. Hochstrasser
Department of Chemistry, University of Pennsylvania, Philadelphia, PA 19104

ABSTRACT

Three time-resolved four-wave mixing techniques for measuring orientational relaxation of molecules in solution are compared. A unified theoretical treatment of two-pulse polarization spectroscopy and three-pulse transient grating diffraction from both population and purely orientational gratings is developed by incorporating the slowly varying reorientation of the transition dipoles into the equations of motion for the density matrix elements. All three methods, as well as fluorescence polarization, yield the same experimental reorientation times. The practical advantages and limitations of these techniques and their applicability to the study of other dynamical processes are discussed.

INTRODUCTION

The rotational motion of molecules in solution has been studied extensively in order to probe the nature and dynamics of solvent-solute interactions. Rotational reorientation of molecules in both ground and excited electronic states can be measured using nonlinear four-wave mixing techniques, which yield essentially background-free signals and pulsewidth-limited time resolution.

THEORY AND RESULTS

In two-pulse time-resolved polarization spectroscopy (Fig. 1a), a linearly polarized excitation pulse creates an anisotropic distribution of excited states that acts to rotate the polarization of a delayed probe, causing partial transmission of the probe through crossed polarizers. In a transient grating experiment (Fig. 1b), two time-coincident pump pulses having either parallel or perpendicular polarizations cross inside the sample to form an anisotropic population or orientation grating, which can diffract a delayed probe into a new direction. Both techniques can be treated as parametric four-wave mixing processes in which three fields interact through the third-order susceptibility to generate a fourth field whose intensity is detected. We have derived the fully resonant

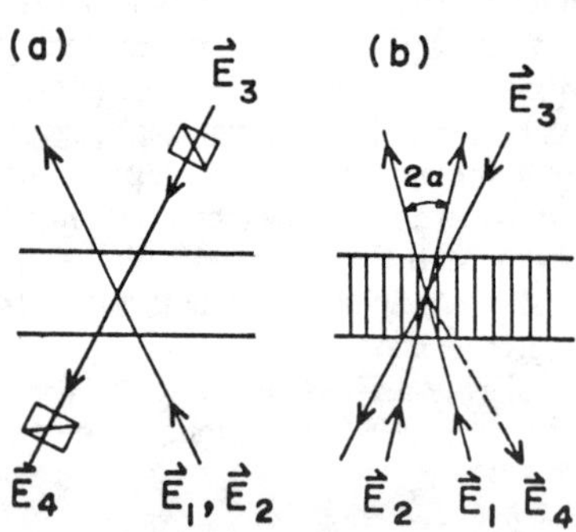

Figure 1. Four-wave mixing beam configurations.

third order response function (the polarization induced by δ-function pulses) for a general four-level system in which the transition dipoles reorient slowly compared with the pulse envelopes and dephasing times.[1] In the limits appropriate to many condensed phase experiments, the signal as a function of pump-probe delay time T becomes

$$S(T) \propto e^{-2T/\tau_{ex}} \left|\langle \vec{\mu}_{ab}(0)\cdot\hat{e}_1\, \vec{\mu}_{ba}(0)\cdot\hat{e}_2^*\, \vec{\mu}_{\alpha\beta}(T)\cdot\hat{e}_3\, \vec{\mu}_{\beta\alpha}(T)\cdot\hat{e}_4^*\rangle\right|^2$$

$$= \mu_{ab}^4\, \mu_{\alpha\beta}^4\, [A + B\cdot r(T)]^2\, e^{-2T/\tau_{ex}}$$

where

A = 1/3, B = 2/3 for population grating probed parallel
A = 1/3, B = -1/3 for population grating probed perpendicular
A = 0, B = 1/2 for orientation grating or polarization spectroscopy

and for isotropic rotational diffusion,

$$r(T) = \frac{2}{5} P_2(\cos\lambda)\, e^{-T/\tau_R} \quad .$$

τ_{ex} and τ_R are the excited state lifetime and orientational relaxation time, $\vec{\mu}_{ab}$ and $\vec{\mu}_{\alpha\beta}$ are the transition moments for the absorptions on resonance with the pump and probe frequencies, respectively, and λ is the angle between these transition moments in the molecule-fixed frame.

These theoretical results have been tested by comparing the τ_R values obtained using the polarization spectroscopy, population grating, orientation grating, and, where applicable, fluorescence polarization methods. All methods yield the same experimental reorientation times. In addition, the predicted 9:1:1 intensity ratio in the parallel:perpendicular:orientational grating experiments near time zero has been verified.[1]

DISCUSSION

The orientational grating configuration may be the method of choice for these measurements, as it does not exhibit significant interference from either thermal gratings or "external" birefringence. These four-wave mixing techniques can be used to compare ground and excited state rotational times and to study internal motions and electronic state dynamics as well as overall rotation. They can also be extended to examine rotation-vibration coupling in gas phase molecules.[2]

REFERENCES

1. A. B. Myers and R. M. Hochstrasser, IEEE J. Quantum Electron., submitted.
2. A. J. Bain, P. McCarthy, and R. M. Hochstrasser, Chem. Phys. Lett., submitted.

RECOMBINATION AND RELAXATION DYNAMICS OF DIATOMIC MOLECULES IN CONDENSED PHASES

N. A. Abul-Haj and D. F. Kelley[a-c]

University of California, Los Angeles
Los Angeles, Ca. 90024

and

Colorado State University[d]
Fort Collins, Co. 80523

ABSTRACT

Picosecond absorption kinetics of I_2 and Br_2 in room temperature CCl_4 are presented. Several probe wavelengths were used following excitation with 30ps pulses of 532nm light. The results indicate that population of the electronically excited $A(^3\pi_{1u})$ and $A'(^3\pi_{2u})$ states occur rapidly (<30ps) following dissociation. Analysis of transient absorption intensities indicates that repopulation of the ground electronic state of I_2 requires about 30-50ps. The difference in population times may be understood in terms of trapping in weakly bound electronic states (perhaps $^3\pi_{o}-_{u}$) which relax to populate the ground electronic state, or in terms of slow atom recombination into the ground state. Vibrational relaxation rates through the ground state manifold were interpreted in terms of vibration to translation energy transfer as given be the Schwartz, Slawsky, Herzfeld theory.

INTRODUCTION

The photodissociation and geminate recombination of diatomic molecules in solution are among the simplest and most extensively studied of all chemical processes. Photodissociation of the diatomic molecules produces separated atoms which are held in close proximity to each other by the surrounding solvent "cage", greatly enhancing the probability of geminate recombination. Atom recombination results in a vibrational and/or electronically excited diatomic molecule. Relaxation of the nacent diatomic molecule occurs as this excess vibrational and/or electronic energy is given up to the solvent. The first picosecond studies of the geminate recombination were performed on I_2 in CCl_4 by Eisenthal *et. al.*[1] These studies showed that significant repopulation of the ground vibronic state

a) Alfred P. Sloan Fellow
b) IBM Faculty Development Award Recipient
c) Author to whom correspondence should be addressed
d) Present address of D. F. Kelley

occurs on the 150ps timescale. This was interpreted in terms of slow atom recombination.

Subsequent theoretical studies have suggested that all recombination occurs very rapidly and that vibrational relaxation limits the rate of ground vibronic state repopulation.[2] These calculations make one rather questionable assumption: Level hopping of the separated atoms to the ground electronic states occurs very rapidly. Recent theoretical studies by Miller _et. al._[3] suggest that this may not always be the case.

Further questions have been raised about the mechanisms of vibrational relaxation. Specifically, the roles of vibration to translation (V-T) and vibration to vibrational (V-V) energy transfer have been addressed both theoretically[2,4] and experimentally.[4]

In this paper the results of picosecond absorption studies and dynamic simulations are presented. These studies determine the rates of geminate recombination, and elucidate the mechanisms of vibrational relaxation. The experimental apparatus used in these studies has been described in detail elsewhere.[5]

RESULTS AND DISCUSSION

Transient spectral difference kinetics (transient minus static absorption intensities) at various wavelengths for I_2/CCl_4 and Br_2/CCl_4 are shown in Figures 1 and 2, respectively. In all cases, the kinetics are biphasic. Previous studies have shown that a significant fraction of both I_2 and Br_2 recombine into the electronically excited $A(^3\pi_{1u})$ and $A'(^3\pi_{2u})$ states.[6] Population in these states gives rise to the long lived red absorbances seen in Figures 1 and 2. Data taken at longer times indicates A' state lifetimes of 2.7 and 5.5ns for I_2/CCl_4 and Br_2/CCl_4, respectively. The depopulation of these states results in the corresponding slow component of ground state recovery observed in the 500nm kinetics. Curves corresponding to 2.7 and 5.5ns decays are also shown in Figures 1 and 2, respectively.

Comparison of the experimental kinetic curves with the calculated exponential decay curves reveals the presence of additional red absorbances and ground state depletion at short times. The Franck-Condon factors indicate that the upper vibrational levels of ground electronic state will absorb in the red and near IR regions. We conclude that the short lived (100-200ps) red transient absorptions are due to population in the excited vibrational levels produced from recombination directly into the X state. Specifically, Franck-Condon calculations on I_2 indicate that the 580nm transient is due to population in the v=1-4 levels and the 640nm transient is due to population in the v=4-10 levels. As vibrational relaxation procedes, these populations are reduced, and the ground vibrational state (500nm kinetics) is repopulated.

We have performed numerical simulations of the geminate recombination/vibrational relaxation process. These calculations assume that vibrational relaxation may be described by a simple isolated

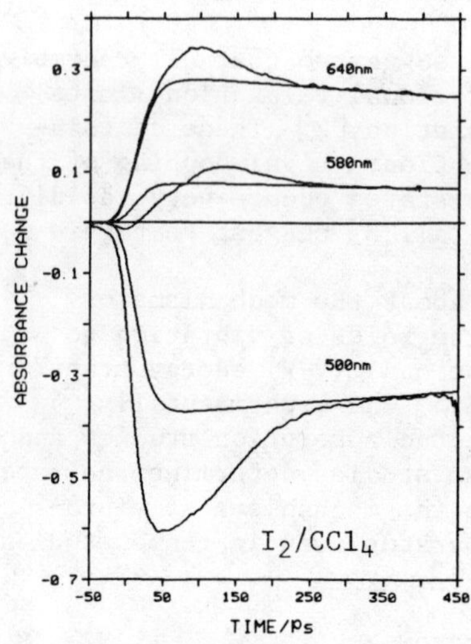

Fig. 1. Experimental transient absorption difference (transient minus static absorption intensities) kinetics of I_2 in CCl_4 at 500, 580 and 640 nm. Also shown with the 580 and 640nm kinetics are curves calculated from the convolution of 2.7ns exponential decays with the instrument response function. The curve shown with the 500nm kinetics was calculated from the convolution of the instrument response function with a 2.7ns exponential bleach recovery and a small (.17 of the total bleach) constant bleach. The instrument response function was 35ps full width at half maximum.

Fig. 2. Experimental absorbance changes at 640nm, 560nm and 500nm versus time. Also shown are curves calculated from the convolution of a 15ps rise and a 5.5ns decay with the instrument response function (35ps full width at half maximum). The calculated curves have amplitudes corresponding to the total maximum absorption and the the long lived (5.5ns) component. The arrow by the 560nm curve indicate 65 and 105ps following excitation.

binary collision model (SSH theory[7]) and described in detail elsewhere.[8] These calculations, when coupled with the Franck-Condon spectral analysis, simulate the experimental absorption kinetics. Some of the major conclusions resulting from the comparison of the calculated and experimental results are summarized below.

1) A significant fraction (40% for I_2/CCl_4 and ~60% for Br_2/CCl_4 of the molecules rapidly (<30ps) recombine into electronically excited $A(^3\pi_{1u})$ and $A'(^3\pi_{2u})$ states.

2) Repopulation of the ground electronic state of I_2 requires ~50ps. This may be understood in terms of slow geminate recombination or in terms of trapping in shallow electronically excited states.

3) The I_2/CCl_4 system vibrationally relaxes primarily via a vibration to translation (V-T) rather than a vibration to vibration (V-V) mechanism, despite a vibrational frequency resonance. This is probably a result of very poor coupling due to the symmetry of the resonant CCl_4 normal mode.

4) Vibrational relaxation of I_2 in CCl_4 may be described in terms of Schwartz, Shawsky, Herfeld theory.

5) The Br_2/CCl_4 system relaxes faster than the I_2/CCl_4 system suggesting that Br_2/CCl_4 relaxes by a V-V mechanism.

ACKNOWLEDGEMENT

Acknowledgement is made to the National Science Foundation, the Research Corporation, IBM Corporation, and the donors of the Petroleum Research Fund administered by the American Chemical Society, for support of this research.

REFERENCES

1) T.J. Chuang, G.W. Hoffman, and K.B. Eisenthal, Chem. Phys. Lett., 25, 201 (1974).

2) D.J. Nesbitt and J.T. Hynes, J. Chem. Phys., 77, 2130 (1982).

3) D.P. Ali and W.H. Miller, Chem. Phys. Lett., 105, 501 (1984).

4) P. Bado, C. Dupuy, D. Magde, K.R. Wilson, and M. Malloy, J. Chem. Phys., 80, 5531 (1984), and references therein.

5) D.J. Jang and D.F. Kelley, Rev. Sci. Inst., 56, 2205 (1985).

6) D.F. Kelley, N.A. Abul-Haj, and D.J. Jang, J. Chem. Phys., 80, 4105 (1984).

7) J.D. Lambert, "Vibrational and Rotational Relaxation in Gases", Oxford (1977), and references therein.

8) N.A. Abul-Haj and D.F. Kelley, J. Chem. Phys., in press.

PICOSECOND REORIENTATIONAL DYNAMICS OF OPPOSITELY CHARGED DYE MOLECULES. CORRELATION WITH THE DIELECTRIC FRICTION

Eva F. Gudgin Templeton and G.A. Kenney-Wallace
Department of Chemistry
University of Toronto
Toronto, Ontario
M5S 1A1 Canada

ABSTRACT

The rotational reorientation times (τ_{rot}) of charged dyes have been measured in polar protic and aprotic solvents, using difference frequency modulation picosecond pump-probe spectroscopy. The dielectric friction (DF) model has been shown to explain nonhydrodynamic behaviour in alcohols and alcohol/water mixtures but is unsuccessful in predicting observed differences between the dye molecules in other solvent systems. These deviations are correlated with specific solvation interactions for cations in some solvents.

INTRODUCTION

The dynamics of reorientational motion are fundamental to chemical reactions in which relative motion of reactants will influence reaction pathways and dynamics. We are concerned with the role the solvent plays in the reorientational motion of charged probe molecules (see Figure 1).

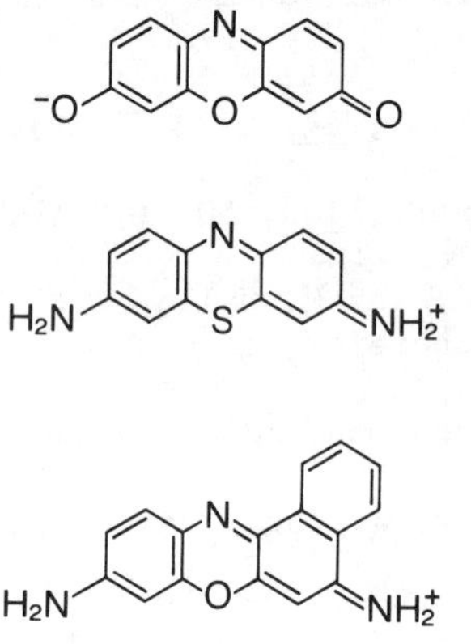

The most common model applied to reorientational motion has been the continuum hydrodynamic model[1] in which τ_{rot} is described by the Debye-Stokes-Einstein (DSE) equation:

$$\tau_{rot} = V_{hyd}\ \eta\kappa/k_B T \qquad [1]$$

and thus is dependent on the bulk shear viscosity, η, the hydrodynamic volume of the probe V_{hyd}, the temperature T, and the coupling parameter κ which for stick or slip boundary condition is equal to 1 or 0, respectively. This model does not allow for specific or short-range probe-solvent interactions and has limited predictive capability due to uncertainties in values for V_{hyd} and κ. We have attempted to correlate τ_{rot} with the microscopic properties of the system by using pure and mixed solvents capable of differing interactions with probe molecules of opposite charge.

Figure 1 Probe molecules resorufin (R^-), thionine (Th^+) and cresyl violet (CV^+).

0094-243X/86/1460600-4$3.00

EXPERIMENTAL

We have measured τ_{rot} using a variation of the picosecond pump-probe technique. The probe dye molecule is excited by pump and probe pulses of ~1 ps duration from an Ar-ion synch-pumped Rhodamine 6G dye laser. The probe pulse is delayed and relative pump-probe polarizations varied so that the change in transmission $\Delta T(t)$ for $\parallel$, $\perp$ and 54.7° is measured as function of delay. Then from the polarization anisotropy R(t) we can obtain the τ_{rot} as given by equation [2] (the ground state recovery time τ_{gsr} due to $S_1 \rightarrow S_0$ relaxation processes is obtained from $\Delta T_{54.7}$).[3]

$$R(t) = \Delta T_{\parallel} - \Delta T_{\perp} \propto e^{-t/\tau_{rot}} e^{-t/\tau_{gsr}} \qquad [2]$$

Solutions of resorufin (sodium salt, Aldrich), thionine (acetate salt, Aldrich) or cresyl violet (perchlorate salt, Exciton) were $\sim 10^{-5}$ M at 298 K. The usual excitation wavelengths were 580 nm (R^-), 590 nm (CV^+) and 600 nm (Th^+).

RESULTS

The results are presented as hydrodynamic plots in Figure 2.[4] The observed linear correlation between τ_{rot} and η for pure alcohol solvents can be used to obtain V_{hyd} from [1] for the three molecules assuming $\kappa = 1$. These can be compared with predicted hydrodynamic volumes V^*, obtained from the estimated van der Waal's volume of the

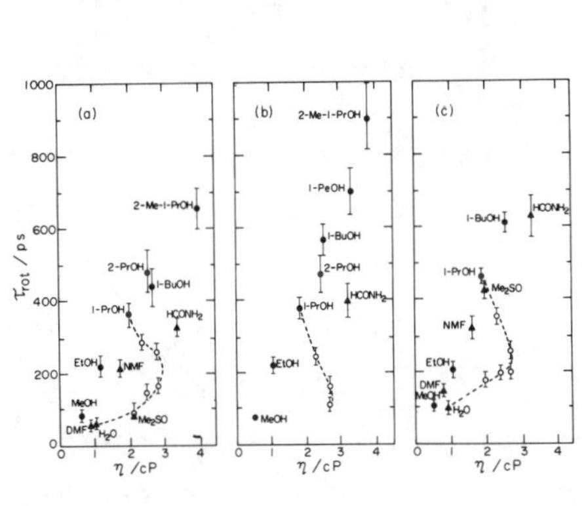

Figure 2 Hydrodynamic plots for a) R^- b) Th^+ c) CV^+ in ● alcohols, ▲ other pure solvents, ○ 1-PrOH/H_2O mixtures (χ_{ROH} = 0.65, 0.35, 0.25, 0.15, 0.10 from top to bottom).

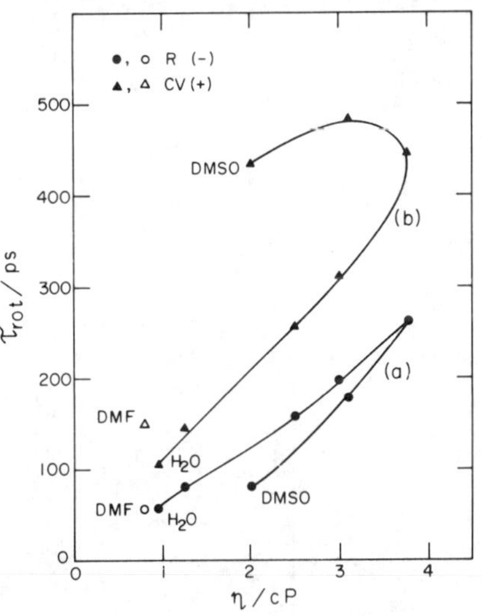

Figure 3 Hydrodynamic plots for a) R^- and b) CV^+ in dimethylsulphoxide/H_2O mixtures. χ_{Me_2SO} = 1.00, 0.96, 0.88, 0.80, 0.68, 0.41, 0.00, reading from Me_2SO to H_2O. Data for DMF are also given.

ellipsoidal probe with a known axial ratio[5], and we observe that in all cases, V_{hyd} exceeds V^* by a factor of about 1.5 to 2.5, e.g. for R^-, V_{hyd} = 690 Å while V^* = 270-360 Å. This is a nonhydrodynamic result (called "superstick" behaviour).[6] Also, severe deviations from linearity in the mixed solvents are observed in all cases except R^- in Me_2SO/H_2O mixtures. Finally, significant differences are seen between τ_{rot} values for R^- and CV^+ in the amides dimethylformamide, N-methyl-formamide and formamide, as well as Me_2SO.

DISCUSSION

We have previously observed a correlation between τ_{rot} for R^- in alcohols and 1-PrOH/H_2O mixtures, and τ_d, the dielectric relaxation time of the H-bonding network of the solvent.[3] This is consistent with the dielectric friction (DF) model[7] in which the probe is pictured as a rotating dipole μ in a cavity within a solvent continuum of viscosity η and dielectric constant ε. The solvent is polarized around the dipole and exerts a retarding torque on the probe's motion, whose relaxation is characterized by the solvent Debye dielectric relaxation time τ_D. Then the calculated τ_{rot} is the sum of τ_{DSE} obtained from eq. [1] and τ_{DF} using [3][4,7]:

$$\tau_{DF} = A'\left[\frac{4}{\tau_D^{-1} + 12D_r/\varepsilon} + \frac{2}{\tau_D^{-1} + 12D_r} + \frac{6}{\tau_D^{-1} + 2D_r/\varepsilon} + \frac{3}{\tau_D^{-1} + 2D_r}\right] \qquad [3]$$

where $A' = \frac{4}{45}\pi \frac{\mu}{V_c k_B T} \frac{(\varepsilon - 1)}{(2\varepsilon + 1)^2}$

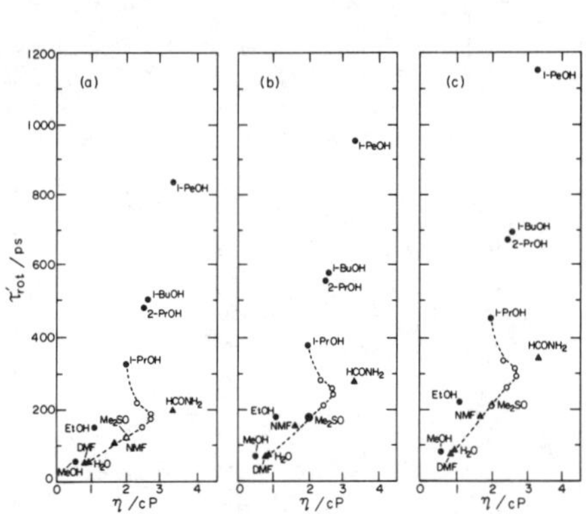

Figure 4 Calculated τ'_{rot} using DF model for a) R^- b) Th^+ c) CV^+ Symbols as in Figure 2.

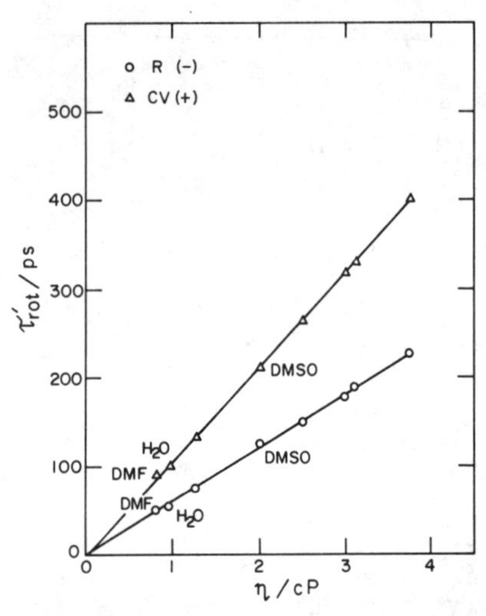

Figure 5 Calculated τ'_{rot} using DF model for a) R^- and b) CV^+ in Me_2SO/H_2O mixtures. Mole fractions as in Figure 3.

and D_r is the rotational diffusion constant of the probe, obtained from τ_{DSE}. For our calculations we must estimate some of our parameters such as μ and τ_{DSE} based on the experimental data which means that the relative magnitudes of τ'_{rot} are more significant than absolute magnitudes. Bearing this in mind, we can compare the results shown in Figures 4 and 5 with those of Figures 2 and 3. The DF model can successfully predict super-stick behaviour in alcohols and non-linear hydrodynamic plots for alcohol/water mixtures at least qualitatively. However, it predicts smaller τ_{rot} values in amides than observed and cannot predict the differences between R^- and CV^+. The CV^+ in Me_2SO/H_2O data are inconsistent with the essentially linear DF prediction in the Me_2SO-rich region.

The success of the DF model for the alcohol systems is consistent with a picture in which the probe molecules do not interact specifically with the solvent but are governed by solvent medium and long-range dynamics (characterized by τ_d and η). On the other hand, it is known that cations are capable of strong solvent binding in amides and Me_2SO.[8] This would invalidate the DF model in these solvents, since μ, τ_D and V_{hyd} might all vary. The DF model should then be more successful for anionic probes in the solvents studied, while in order to understand reorientational motion of cations a model involving molecular parameters must be formulated.

REFERENCES

1. A review of hydrodynamic models is presented in J.L. Dote, D. Kivelson and R.N. Schwartz, J. Phys. Chem., 85, 2169 (1981).
2. E.L. Quitevis, E.F. Gudgin Templeton and G.A. Kenney-Wallace, Appl. Optics, 24, 318 (1985).
3. E.F. Gudgin Templeton, E.L. Quitevis and G.A. Kenney-Wallace, J. Phys. Chem., 89, 3238 (1985).
4. E.F. Gudgin Templeton and G.A. Kenney-Wallace, submitted to J. Phys. Chem.
5. J.L. Dote and D. Kivelson, J. Phys. Chem., 87, 3889 (1983).
6. D. Kivelson and K.G. Spears, J. Phys. Chem., 89, 1999 (1985).
7. P. Madden and D. Kivelson, J. Phys. Chem., 86, 4244 (1982).
8. D. Martin and H. Hauthal, Dimethylsulphoxide (van Nostrand Reinhold, Berkshire U.K., 1975).

UNIMOLECULAR REACTIONS OF ISOLATED, COLLISIONAL GAS PHASE AND SOLVATED MOLECULES: CONNECTIONS BETWEEN STILBENE ISOMERIZATION RATES UNDER SUPERSONIC JET AND THERMAL GAS PHASE CONDITIONS

M.W. Balk, S.H. Courtney and G.R. Fleming
Dept. of Chemistry, The University of Chicago, Chicago, IL 60637

ABSTRACT

Photochemical isomerization provides a practical testing ground for activated barrier crossing theories. Stilbene is a prototype, because (1) reaction may be initiated by a short optical pulse and the decay of the excited trans isomer followed optically, and (2) the isomerization rate is slow compared to vibrational or solvent relaxation processes in solution. The goal is to understand the effects of the medium upon the reaction kinetics proceeding from the isolated molecule to the low and high pressure gas, and liquid solution. Towards this end time correlated single photon counting experiments of stilbene isomerization are discussed from isolated molecule conditions up to 90 atm of methane gas. In statistical theories of barrier crossing the transition state rate at temperature T, $k_{TST}(T)$, represents an upper limit for the reaction rate. However, contrary to expectation, for pressures greater than about 2 atm the isomerization rate was found to exceed $k_{TST}(T)$ by over a factor of 2 at 5 atm, and by as much as 10 times at 90 atm. Connections between the various types of experimentally observed rates will be discussed, taking into account their different statistical properties.

INTRODUCTION

The dependence of chemical reaction dynamics on the surrounding solvent medium is a topic of much current interest. As discussed elsewhere[1], the photochemical isomerization of stilbene is an excellent model system for such studies. By combining the techniques of ultrafast spectroscopy with supersonic jet spectroscopy it has been possible to study the isomerization process progressively from isolated molecules prepared with a well defined initial energy, through thermal but still isolated (on the isomerization time scale) molecules, to collisional gas phase conditions and finally into the liquid phase. The solvent friction is varied over about 6 orders of magnitude in these studies, which can therefore provide stringent tests of current theories of activated barrier crossing.

In this paper we discuss the relationships between the various types of experimentally observed rate constants, using as a basis for our comparisons the isomerization rates k(E), measured in the jet experiments[1,2], for the microcannonical ensemble as a function of rovibrational energy. Since it is well-established from extensive liquid state studies that no process other than radiative loss from S_1 competes with isomerization, the isomerization rate may be determined by measuring the total rate of fluorescence decay.[1-3]

0094-243X/86/1460604-4$3.00

RATES UNDER ISOLATED THERMAL GAS PHASE CONDITIONS

The vapor pressure of stilbene at room temperature is less than 1 mTorr. Thus, no collisions occur during the fluorescence lifetime. However, in contrast to the single exponential decays observed in jet cooled stilbene[2], non-exponential decay is observed in the thermal isolated molecule experiments (Fig. 1). This may be attributed to the initial Boltzmann distribution over rovibrational levels in the ground electronic state. A simple model of the non-exponential decay assumes that the initial Boltzmann distribution is taken up into the first excited singlet (a good approximation for stilbene[1]), and that the fluorescence decay out of each level is a single exponential as suggested by the jet experiments. The normalized transient fluorescence is given by[1,3]

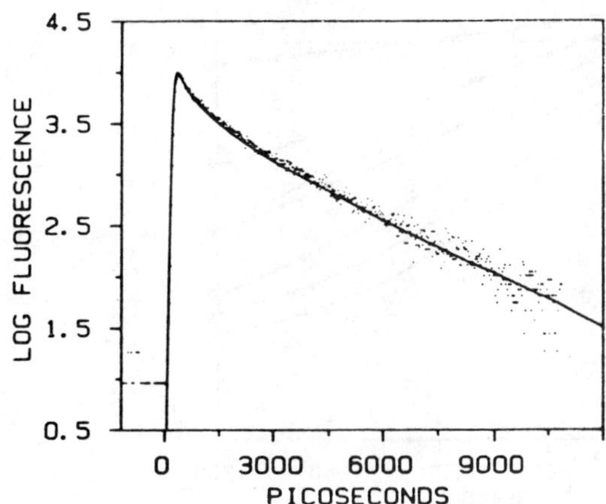

Fig. 1. Observed (···) thermal isolated molecule fluorescence of stilbene at 296K, and model decay (——) (Eq. 1).

$$G_F(t) = 1/Q(T) \int_0^{\infty} dE\,\rho(E)\,\exp[-E/k_BT]\,\exp[-k_F(E)\,t] \qquad (1)$$

where $\rho(E)$ is the rovibrational density of states[1] and $k_F(E)$ is the fluorescence rate measured in the jet experiments. The decay curve calculated via Eq.(1) (and convoluted with the experimental instrument function) is also shown in Fig. 1. It is important to note that this is not a fit to the data; there are no adjustable parameters.

As shown elsewhere[3], $G_F(t)$ decays initially as a single exponential with rate k_{TST}. This is expected since the reactant population is initially at thermal equilibrium and k_{TST} is the rate of decay of an equilibrium distribution.

LOW PRESSURE COLLISIONAL THERMAL GAS PHASE RATES

When buffer gas is added the isomerization rate initially increases, since collisions try to maintain the Boltzmann distribution in the upper levels [larger k(E)]. As the collision rate is increased further, so is the friction exerted on the stilbene molecules by the buffer gas. Eventually the friction becomes large enough so that the isomerization rate begins to decrease as the collision rate is increased still further. This turnover in the rate with friction is predicted by Kramers' barrier crossing theory[4] as well as more modern treatments.[5] These theories also predict that the observed rates will be less than $k_{TST}(T)$. While the turnover has apparently been observed,[1,6] the rates are not bounded by $k_{TST}(T)$.

The deconvoluted decays[3] for 0 to 5 atm of added methane are shown in Fig. 2.

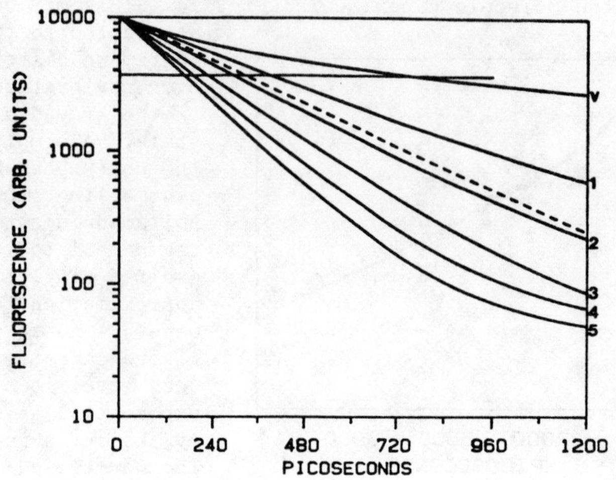

Fig. 2. Deconvoluted experimental decays of stilbene-methane mixtures from pure vapor (v) to 5 atm methane at 296K. Dashed curve is $\exp[-k_{TST}t]$. The horizontal line marks the 1/e points.

The inverse 1/e time of the decay exceeds k_{TST} at about 2 atm, and is about $2k_{TST}$ at 5 atm. The low, high pressure and liquid state rates are plotted in Fig. 3 versus the inverse diffusion coefficient (a convenient friction scale[1,7]). While Fig. 3 seems to indicate that the turnover exists in the gas phase at about 90 atm, the rate at this pressure exceeds k_{TST} by about a factor of 10. All the liquid state rates shown also exceed k_{TST}. Possible explanations for this behavior include:

(1) The isomerization is non-adiabatic in the isolated molecule, but becomes more adiabatic as the collision rate increases.[8]

(2) Intramolecular vibrational redistribution is incomplete in the isolated molecule and energy flow from non-reactive to reactive modes is enhanced by collisions.

(3) The intramolecular potential surface is substantially perturbed during collision, lowering the barrier height.

Our current state of knowledge does not permit us to confirm or rule out any of the above mechanisms. However, they should provide a guide for future work on this problem since we believe that our data clearly demonstrate that purely statistical models of barrier crossing dynamics cannot fully describe stilbene isomerization.

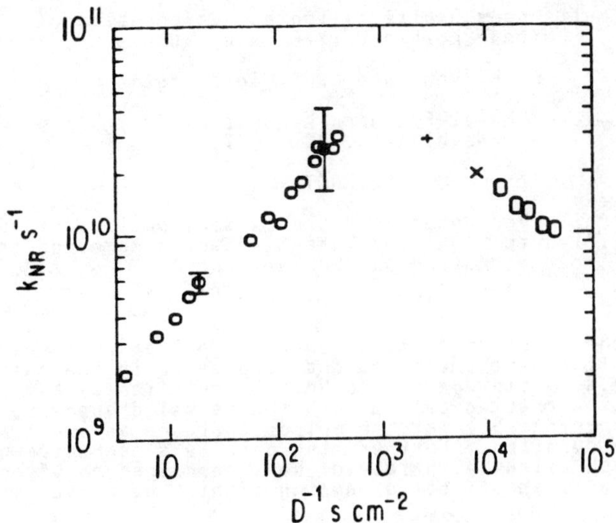

Fig. 3. Stilbene isomerization rates[1] (296K) versus inverse diffusion coefficients (see text). o, in gaseous methane varying the pressure; +, in liquid ethane; x, in liquid propane; 0, in higher alkanes[9].

REFERENCES

1. G.R. Fleming, S.H. Courtney and M.W. Balk, J. Stat. Phys., 42, 83(1986).
2. J.A. Syage, P.M. Felker and A.H. Zewail, J. Chem. Phys., 81, 4706(1984).
3. M.W. Balk, and G.R. Fleming, J. Phys. Chem., submitted.
4. H.A. Kramers, Physica, 7, 284(1940); P. Hanggi, J. Stat. Phys., in press.
5. See for example, R.F. Grote and J.T. Hynes, J. Chem. Phys., 73, 2715(1980); B. Carmeli and A. Nitzan, Phys. Rev. Lett., 51, 233(1983); J.L. Skinner and P.G. Wolynes, J. Chem. Phys., 72, 4913(1983).
6. M. Lee, G.R. Holtom and R.M. Hochstrasser, Chem. Phys. Lett., 118, 359(1985).
7. G. Maneke, J. Schroeder, J. Troe and F. Voss, Ber. Bunsenges. Phys. Chem., 89, 896(1985).
8. P.M. Felker and A.H. Zewail, J. Phys. Chem., in press.
9. G. Rothenberger, D.K. Negus and R.M. Hochstrasser, J. Chem. Phys., 79, 5360(1983).

Simultaneous Amplification and Compression of CW Mode Locked Nd:YAG Laser Pulses

D. F. Voss and L. S. Goldberg

Naval Research Laboratory
Washington, DC 20375

ABSTRACT

Pulses spectrally broadened in a single-mode optical fiber have been amplified to high peak intensity and compressed in the gain medium without a grating pair delay line.

INTRODUCTION

We report the first experiments in which pulses broadened in an optical fiber are amplified and compressed in the same gain medium. Using a multistage pulsed Nd:YAG amplifier system we obtain, without a grating pair or other external dispersive delay line, highly reproducible intense pulses that are as short as 12 picoseconds in duration. Further, the precise synchronism afforded by this approach allows a variety of nonlinear frequency conversion processes, as well as effective pumping of high gain dye amplifier stages.

EXPERIMENTAL PROCEDURE

The experimental configuration is shown in Fig. 1. The output of a commercial CW mode locked Nd:YAG laser is coupled into 150 meters of single-mode optical fiber. The circular core fiber has a diameter of 5.6 microns and exhibits low loss at 1064 nm. The fiber output is recollimated, and the polarization adjusted with a half-wave plate. The input power is attenuated to yield less than 1 watt from the fiber to avoid stimulated Raman scattering. A burst of 8 pulses is selected from the self-phase modulated CW train at a repetition rate of 10 Hz by an electro-optic shutter. This burst is amplified in two passes through the first stage before a single pulse is switched out. A second double pass amplifier stage and a final single pass stage increase the overall gain to more than 10^6, giving an output pulse energy of ~10 mJ. The amplified pulse is frequency doubled and examined with a 2 picosecond resolution S-20 response streak camera; the spectrum of the pulse at the fundamental frequency is recorded with a 3/4 meter spectrograph and diode array.

RESULTS

Fig. 2A shows a streak camera record at 532 nm of a pulse amplified directly from the CW mode locked oscillator (calibrated

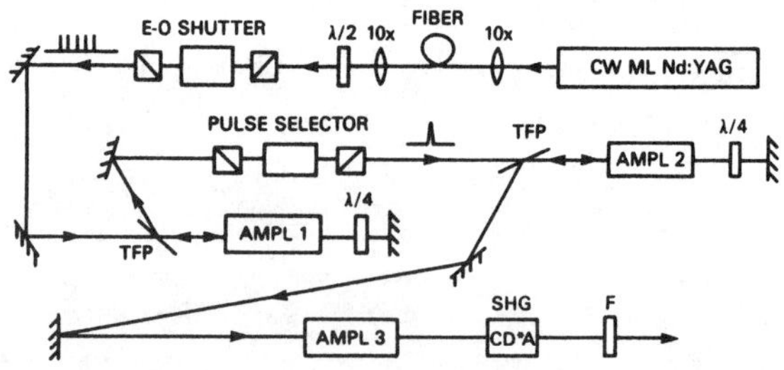

Fig. 1: Schematic of the experimental configuration.

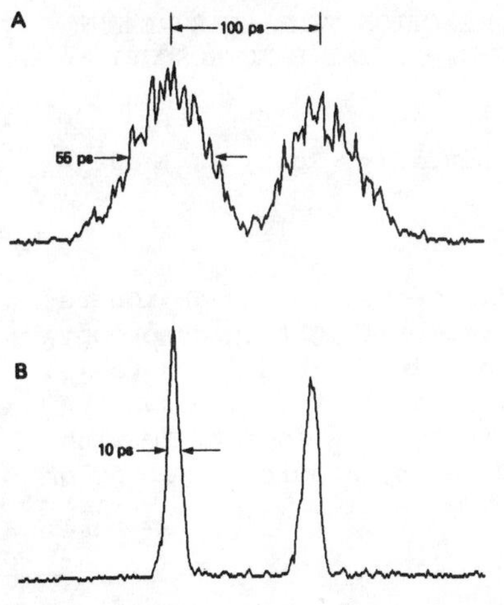

Fig. 2: Streak camera measurements A. oscillator pulse; B. amplified pulse.

Fig. 3: Duration of amplified pulse vs. spectral broadening.

by reflection from a 100 ps etalon). The corresponding IR pulse is approximately 80 picoseconds in duration, depending on RF power to the acousto-optic modelocker. When the pulses are passed through a 150 meter length of 5.6 micron diameter fiber and amplified, the dramatic pulse shortening shown in Fig. 2B is obtained. The compressed IR pulses are consistently under 15 picoseconds, with the shortest being 12 picoseconds in duration. This compression has been achieved <u>without the use of gratings</u>.

The degree of spectral broadening can be controlled by carefully adjusting the power coupling at the input objective. Figure 3 shows streak camera measurements of the amplified pulse duration as a function of input spectral width for a 7 micron diameter, 100 meter fiber. The shortest pulses are obtained when the input spectral width approaches the amplifier bandwidth (6 cm^{-1}).

The mechanism giving rise to the pulse shortening is considered to be the selective amplification of frequencies within the gain bandwidth occurring near the peak of the pulse [1,2]. At the high peak power levels encountered in the fiber, the intensity dependent refractive index causes spectral broadening via self-phase modulation. The slowly varying portions of the pulse (namely, the peak and the wings) will undergo the smallest frequency shift, and thus remain within the amplifier gain bandwidth.

CONCLUSION

In summary, we have for the first time directly amplified a CW mode locked Nd:YAG pulse train, spectrally broadened by self-phase modulation in an optical fiber, and have observed substantial pulse compression in the gain medium. This approach provides a very reproducible repetitive source of high peak power pulses 12 picoseconds in duration.

REFERENCES

1. R. A. Fisher and W. K. Bischel, J. Appl. Phys., <u>46</u>, 4921(1975).

2. A. J. Duerinckx, H. A. Vanherzeele, J-L. van Eck, and A. E. Siegman, IEEE J. Quant. Elect., <u>QE-14</u>, 983(1978).

FEMTOSECOND RESOLUTION ULTRAFAST ION REACTION TIME MEASUREMENT A PULSED-LASER FIELD DESORPTION TIME-OF-FLIGHT SPECTROSCOPY*

T. T. Tsong
The Pennsylvania State University, University Park, PA 16802

ABSTRACT

Using a pulsed-laser field desorption technique and an ion reaction time amplication scheme for the time-of-flight spectroscopy, dissociation by atomic tunneling of $^4HeRh^{2+}$ in a field of ~3 to 4.5 V/Å has been measured to be 790 ± 21 fs. This is the time for the ion to rotate 180°. A strong isotope effect by replacing ^4He with ^3He confirms the ion reaction to be produced by atomic tunneling of a mass dependent potential barrier.

INTRODUCTION

Compound ions can dissociate in electric field by an atomic tunneling effect,[1] similar to field ionization of atoms by tunneling of atomic electrons. We have studied field dissociation of $^4HeRh^{2+}$ in a field of 3 to 4.5 V/Å using a pulsed-laser stimulated field desorption time-of-flight spectroscopy.[2] In some respects the effect is similar to α-decay which occurs by tunneling of a He nucleus. Field dissociation, however, involves with a potential barrier which depends on mass, molecular orientation and the applied field strength. Thus, the effect is theoretically most interesting. The resolution of our reaction time measurements is ~20 fs.

RESULTS AND DISCUSSIONS

$^4HeRh^{2+}$ can be produced by low temperature (<100K) field evaporation of a Rh tip in He of 1 x 10^{-8} Torr or higher and in a field of 4.5 to 5.0 V/Å. Under such field, each surface Rh atom in the more protruding position is field adsorbed with a He atom. A ToF spectrum shown in Fig. 1 is obtained if the field evaporation is done with a weak stimulation of laser pulses of 300 ps width. It contains a Rh^{2+} line with a double-peak structure and a $HeRh^{2+}$ line. The main peak of Rh^{2+} is identical to those Rh^{2+} produced by field evaporation in vacuum. A flight time difference of 30 ns of the main and secondary Rh^{2+} peaks corresponds to an energy difference of 51 eV. The secondary peak is produced by the reaction:

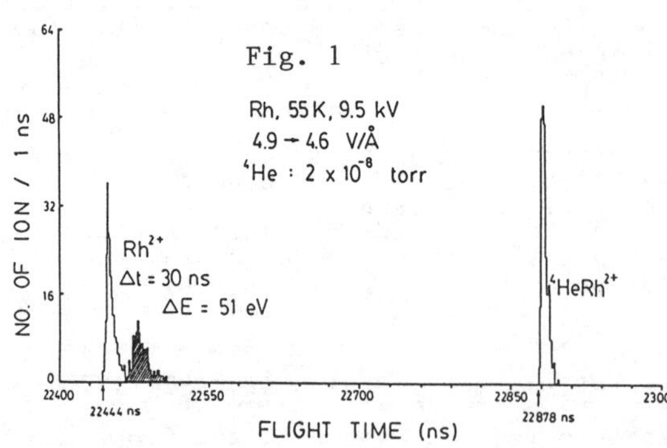

$^4HeRh^{2+} \xrightarrow[\text{tunneling}]{\text{atomic}} {}^4He + Rh^{2+}$.

Using the field distribution near the tip one finds that Rh^{2+} in the secondary peak is formed in a spatial zone of ~150Å width which is centered at 220Å above the emitter surface. It takes a $^4HeRh^{2+}$ 790 fs to travel that distance. Thus the dissociation time of $^4HeRh^{2+}$ is 790 ± 21 fs.

Fig. 2

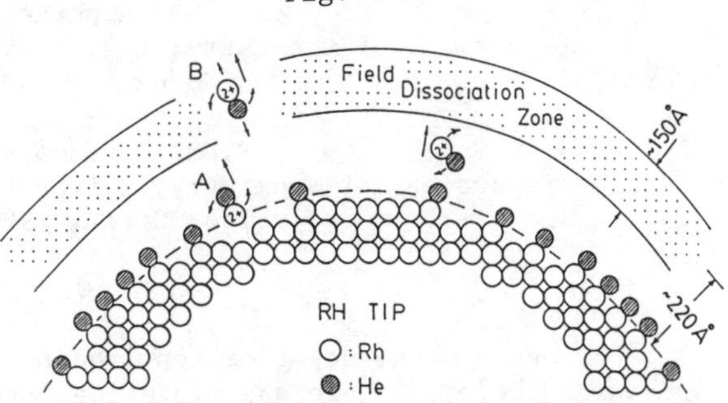

A model based on a Q.M. theory of field dissociation has been constructed, as shown in Fig. 2. A $^4HeRh^{2+}$ cannot field dissociate as desorbed since the orientation is wrong. It has to rotate 180° to come to the correct orientation for the tunneling to be possible. Thus 790 fs is the rotation time of 180° of $^4HeRh^{2+}$ in a field of 4.5 to 3.0 V/Å. The question of how the rotational quantum number changes discretely by the applied electric torque, and the rotation time are fundamental problems of interest to quantum theorists.

That the dissociation is achieved by atomic tunneling is supported by a very strong isotope effect, and the lack of a noticeable effect of the surface temperature. When ^4He is replaced with ^3He the secondary Rh^{2+} peak disappears completely, indicating no field dissociation of $^3HeRh^{2+}$. This observation can be explained by a WKB calculation of the particle tunneling probability using an effective, mass dependent potential $[U(\vec{r}_n) - \frac{2eFz_n}{1+m/M}]$ where $U(\vec{r}_n)$ is the interparticle potential energy, $\vec{r}_n$ is the nuclear distance, z_n is the component of $\vec{r}_n$ along the field direction, M and m are the mass of Rh^{2+} and He, respectively. From the potential diagram we can also derive the dissociation energy of $HeRh^{2+}$ to be 0.7 ± 0.3 eV.

* This work was supported by NSF under DMR-8217119.

REFERENCES

1. J. R. Hiskes, Phys. Rev. 122, 1207 (1961).

2. T. T. Tsong and Y. Liou, Phys. Rev. Lett. 55, 2180 (1985); Phys. Rev. B32, 4340 (1985); T. T. Tsong, Phys. Rev. Lett. (Dec. 1985).

DIRECT OBSERVATION OF HIGHLY EXCITED SINGLE VIBRONIC LEVEL RELAXATION IN CONDENSED MEDIA

J. B. Hopkins
Department of Chemistry, Louisiana State University,
Baton Rouge, LA 70803

P. M. Rentzepis
Department of Chemistry, University of California,
Irvine, Calif. 92717

ABSTRACT

By means of direct time resolved picosecond techniques we have measured the electronic and vibrational relaxation process occuring in large molecules in low temperature matrices. The fluorescence spectra originating from upper vibrational levels of the excited state exhibit sharp well resolved vibronic structure. Vibrational relaxation rates have been time resolved and several "intermediate bottlenecks" in the decay process have been identified. Solvent effects on decay rates have been investigated to elucidate the primary mechanisms responsible for the decay process.

We shall present in this paper a discussion on the direct observation of picosecond kinetics and spectra of S_2 Azulene in rare gas matrixes. Several consequences of the molecular size and structure will be noted with regard to their effect on the rate and mechanism of the vibrational relaxation process. First due to the large rigid molecular structure of Azulene the molecule is incapable of significant rotation, therefore we shall only consider the vibrational degrees of freedom. Second the density of vibrational states reduces the vibrational spacing and therefore the amount of energy which must be exchanged, at a selected period of time, with the matrix is reduced. This has the effect of increasing the overall decay rate from $\sim 10^6$ sec^{-1} for small diatomic and triatomic molecules to $\sim 10^{11}$ sec^{-1} for large molecules such as Azulene.

The dynamics of vibrational energy relaxation were studied by exciting specific upper vibrational levels of S_2 and monitoring directly with the streak camera the emission lifetime of the individual vibronic level.[1,2] However, it was found that the time integrated emission spectrum is not effected by the vibrational level excited which indicates that the decay lifetime of the upper vibronic levels is very fast compared to the fluorescence lifetime. The vibrational decay rate was measured, by monitoring the emission risetime of the $S_2(v=0)$ state subsequent to excitation at higher vibrational levels. We find that, the decay rate can be fitted to a single process which represents the summation of all decay pathways which populate the $S_2(v=0)$ state from the nth excited vibronic level. In principle this could lead to a complicated multiexponential decay. It is found experimentally however, that in this case the risetime of the relaxed fluorescence can fit a rather smooth single exponential when deconvolved from the apparatus response function.

In a previous publication[1] it was shown that the relaxation process involved very fast, 2-5ps, vibrational decay to low lying bottleneck states followed by a slower decay ~ 14ps to the S_2 vibrationless level. Matrix media effects on vibrational decay rates were also investigated and the results[1] shown to give credence to statistical theories of condensed phase vibrational relaxation.[3,4]

One of the consequences of such statistical models is that molecular matrixes will have much higher intermolecular coupling terms for exchange of vibrational energy compared to atomic matrices. In a molecular matrix, vibrational modes exist with large vibrational frequencies which greatly reduce the energy gap between the vibrational levels of the guest and the host molecules of the matrix. It is therefore expected that under these conditions intermolecular energy exchange should be exceedingly fast.

Experimentally, this prediction can be investigated by measuring the rates of intermolecular energy exchange between a quest molecule such as Azulene in a N_2 matrix at 4K. In the N_2 matrix the Nitrogen intra-molecular vibrational frequency is ~2200 cm^{-1}. The vibrational decay within the S_2 electronic manifold of Azulene was measured as a function of excitation energy above the S_2 vibronic level. It was found that the rate of repopulation of $S_2(v=0)$ was essentially constant over the region of 1500-2600 cm^{-1} of initial excitation in S_2. The implication of this result is the opposite of what was predicted using statistical theories[3,4] in that the decay pathway is the same for Azulene in a N_2 matrix as well as atomic matrixes.

Large molecules such as Azulene posess very high densities of vibrational states in the vacinity of 2200 cm^{-1} (the N_2 molecular frequency). Apparently, the decay pathway which involves stepwise relaxation within the Azulene vibrational manifold dominates the overall decay process due to the small amount of energy which must be exchanged with the matrix for each step down the strongly coupled vibrational ladder in the S_2 electronic manifold. For smaller molecules especially in diatomics where the vibrational level density is low, the decay characteristics may be altered and the overall decay process may possibly be dominated by an intermolecular energy exchange to the N_2 stretching vibration.

REFERENCES

[1] J.B.Hopkins and P.M.Rentzepis, Chem.Phys.Lett.117, 414 (1985).

[2] D.Huppert and P.M.Rentzepis, J.Phys.Chem. in press.

[3] D.J.Diestler, J.Chem.Phys.60, 2692 (1974).

[4] A.Nitzan, S.Mukamel, and J.Jortner, J.Chem.Phys.60, 3929 (1974).

COLLIDING PULSE MODELOCKING OF SOLID STATE LASERS

G.P. SHENG and LIN LIHUANG
Shanghai Institute of Optics and Fine Mechanics, Academia Sinica, Shanghai, P.R.C.

ABSTRACT

Colliding pulse modelocking of solid state lasers with an antiresonant ring structure has been studied experimentally. By using pentamethylidyne dye as a saturable absorber, we have obtained pulses close to Fourier-transform limit from colliding pulse mode-locked Nd:YAG and Nd:phosphate glass lasers. The average pulse duration throughout the pulse train is 8 psec and 3.7 psec, respectively.

After Fork et al. used the colliding pulse modelocking (CPM) technique in generating optical pulses shorter than 0.1 psec in a cw ring dye laser, Vanherzeele et al.[1] demonstrated CPM in a Nd:YAG laser using Kodak 9740 dye as the saturable absorber, and Buchert et al.[2] also demonstrated CPM in a Nd:silicate glass laser using Kodak 9860 dye.

We have experimentally studied CPM of Nd:YAG and Nd: phosphate glass lasers by using an antiresonant ring cavity configuration also. The saturable absorber used is pentamethylidyne dye possessing a recovery time of 9 ± 1 psec dissolved in 1,2-dichloroethane.

The experimental setup is shown in Fig.1. The laser cavity consists of a partially reflecting concave mirror M and a nearly 100% reflecting antiresonant ring; the cavity roundtrip time is about 8.5 nsec. The thickness of the saturable absorber is 1 mm.

The pulse duration averaged over a whole pulse train is measured by using a conventional two-photon fluorescence (TPF) arrangement, assuming pulses of Gaussian shape. By converting the fundamental beam to its second harmonic in a type-I KDP crystal, the second harmonic spectral width of pulses is measured with a 1-m grating spectrograph, then the fundamental spectral width is deduced from it.

(A) CPM Nd:YAG laser

The mirror M had a reflectivity of 20% at 1064 nm and a radius of curvature of 2.6 m. The Nd:YAG rod was 4 mm in diameter and 67.5 mm in length. When the low-intensity transmissivity T_0 of the saturable absorber was about 25%, the averaged pulse duration t_p and total output energy E of a pulse train as functions of the displacement |x| of the dye cell from the position of equal optical path in the antiresonant ring are shown in Fig.2. Mode-locked pulses with duration of 11 psec can routinely be obtained in this laser, and pulses as short as 8 psec have been recorded under optimum conditions. The corresponding fundamental spectral width deduced is 2.3 cm-1, and the time-bandwidth product is 0.56, close to the transform-limited value of 0.441.

(B) CPM Nd:phosphate glass laser

The mirror M had a reflectivity of 50% at 1054 nm and a radius of curvature of 10 m. The Nd:phosphate glass rod was 5.5 mm in diameter and 86 mm in length and contained Nd_2O_3 of 2.2% in weight. When T_0 was about 34%, pulses as short as 3.7 psec have been recorded. The corresponding fundamental spectral width is 4.6 cm^{-1}, and the time-bandwidth product is 0.51, close to the transform-limited value too.

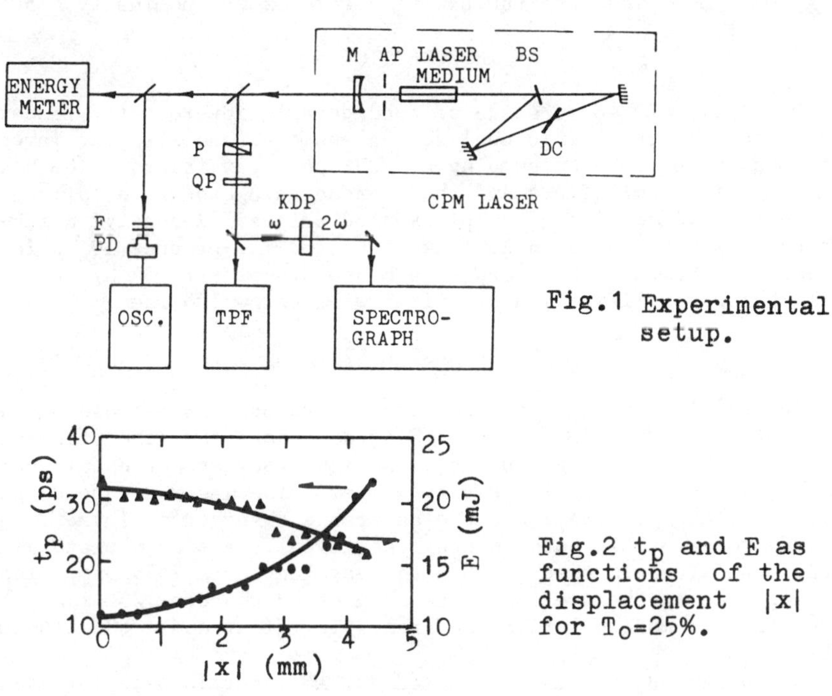

Fig.1 Experimental setup.

Fig.2 t_p and E as functions of the displacement |x| for T_0=25%.

REFERENCES

1. H. Vanherzeele, J. L. Van Eck and A. E. Siegman, Appl. Opt. 20, 3484 (1981).
2. J. M. Buchert, C. Tzu and R. R. Alfano, J. Appl. Phys. 55, 683 (1984).

SUBPICOSECOND HIGH POWER UV - LASER SYSTEM

A. P. Schwarzenbach, T. S. Luk, U. Johann, I. McIntyre,
A. McPherson, K. Boyer, and C. K. Rhodes
Department of Physics, University of Illinois at Chicago
P. O. Box 4348, Chicago, Illinois 60680

ABSTRACT

A synchronously pumped dye laser with saturable absorber jet and cavity dumper is used as the source for producing a seed beam for excimer amplifiers. An optical fiber after the dye laser and a grating pair are used to compress the dye laser pulse to 250 fsec. A two stage dye amplifier brings the pulse to about 0.1 mJ at 745.2 nm. Frequency doubling followed by summing the second harmonic with the fundamental in two KDP crystals produces radiation at 248.4 nm to be amplified in two KrF amplifiers. The UV pulse duration was measured after the first amplifier to be 450 ± 150 fsec. The pulse energy was 23 ± 2 mJ, and the power therefore, was nominally ∿ 50 GW.

INTRODUCTION

High power UV sources are of considerable interest for research in surface science, atomic nonlinear phenomena, and also for investigations in multiphoton pumping of VUV and x-ray lasers. The high gain of excimer amplifiers and their broad gain bandwidth, allow efficient amplification of subpicosecond pulses. Recently, a subpicosecond system based on XeCl was reported[1] to produce 10 mJ in 350 fsec at 308 nm. The broad gain bandwidth of KrF (34 Å)[2] potentially allows amplification of pulses well below 100 fsec.

EXPERIMENTAL SETUP

Our preliminary results achieved in producing subpicosecond high power KrF pulses[3] are presented. Figure 1 shows the schematic setup. A mode-locked, frequency doubled CW Nd:YAG laser synchronously pumps a double jet dye laser with cavity dumper. The laser dye used is 4×10^{-2} molar Pyridine 2 and the saturable absorber employed is 2×10^{-5} molar HITCI, both in ethylene glycol. A single plate birefringent filter is used to tune the wavelength to 745.2 nm. The 1 psec pulses of the dye laser are compressed to below 250 fsec using a single mode polarization preserving fiber and a 600 line/mm grating.

A Q-switched, frequency doubled Nd:YAG laser is used for transverse pumping of the subsequent two stage dye amplifier. The dye cells contain 10^{-3} molar Oxazine 750 in propylene carbonate/methanol. A solid saturable absorber (S.A.) between the cells reduces the amplified spontaneous emission (ASE). The energy of the amplified pulses is ∿ 130 μJ and the pulse duration of ∿ 270 fsec is measured by autocorrelation using conventional background free noncollinear second harmonic generation in a KDP crystal.

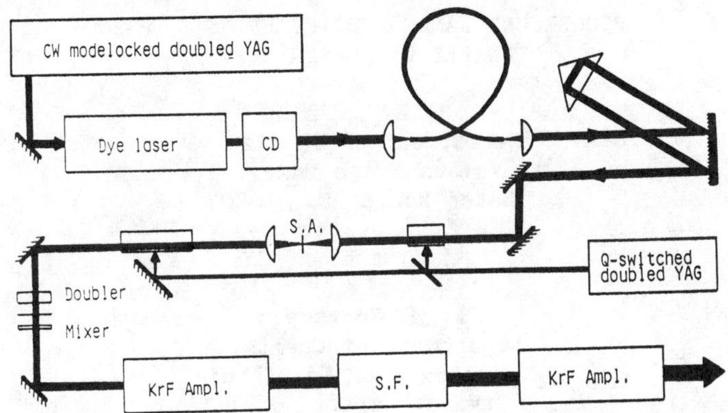

Fig. 1: Experimental setup for KrF laser system.

The KrF wavelength is achieved by first doubling the 745 nm radiation in a 3 mm KDP crystal cut for Type I collinear SHG, and then by mixing the fundamental with the doubled light resulting in 248 nm radiation. The mixing crystal is a 0.5 mm length of KDP cut for Type I collinear phase matching. The crystal thicknesses are chosen to produce good conversion efficiency but, conversely, also to achieve phase matching for the whole bandwidth and limited pulse broadening due to the different group velocities for the interacting wavelengths in the crystals.

The 248 nm seed beam is passed through the first KrF excimer amplifier followed by a vacuum spatial filter (S.F.), is expanded and then passed through the second excimer amplifier. The pulse energy was 23 ± 2 mJ with less than 20% ASE. The spectral width of the amplified pulse is 5.8 Å. The pulse duration after the first excimer amplifier was measured by cross-correlation, using difference frequency mixing of 745 nm and 248 nm in a crystal, to be 450 ± 150 fsec. The gain saturation seen in the second excimer amplifier demonstrates efficient energy extraction from the excited KrF, even with subpicosecond pulses, and shows that scaling to large aperture amplifiers will allow terawatt power levels.

ACKNOWLEDGEMENTS

The authors wish to acknowledge R. Slagle, J. Wright, R. Bernico, and T. Pack for technical assistance. This work was supported by the ONR, the AFOSR, the SDIO(ISTO), the DOE, the LLL, the NSF, the DARPA, and the LANL.

REFERENCES

1. J. H. Glownia, G. Arjavalingham, P. P. Sorokin, and J. E. Rothenburg, (to be published).
2. Ch. A. Brau, Excimer Lasers, C. K. Rhodes ed. (Springer-Verlag, Berlin, 1979) p.91.
3. A. P. Schwarzenbach, T. S. Luk, I. A. McIntyre, U. Johann, A. McPherson, K. Boyer, and C. K. Rhodes (submitted to Opt. Lett.).

PICOSECOND RAMAN STUDIES USING A 5 KHz TUNABLE DYE LASER SYSTEM

J. B. Hopkins
Department of Chemistry
Louisiana State University
Baton Rouge, LA 70803

and

P. M. Rentzepis
Department of Chemistry
University of California
Irvine, Calif. 92717

ABSTRACT

We have developed a very high sensitivity picosecond spectroscopic technique which eliminates many of the deliterions effects found in high power low repetition rate experimental systems. We find that the commonly occuring multiphoton, saturation, and decomposition problems which often occur in chemical and biological systems studied with low repetition rate high power lasers are practically eliminated. In addition, this system enablels us to utilize photon counting technics allowing for the detection of extremely weak signals.

Detailed information regarding the role of vibrational motions in condensed phase chemical dynamics is difficult to obtain due to the lack of vibrational structure frequently observed in the conventional picosecond spectroscopic techniques such as absorption and fluorescence spectroscopy. Vibrational spectroscopic techniques such as infrared, Raman, and coherent antistokers Raman are well documented in their capacity to provide well resolved vibrational structure even in condensed phases. To a large extent however, they have not been exploited on the picosecond time scale largley because of experimental difficulties in detecting the weak signals generated by Raman scattering and infrared absorption processes. Recent technological advances however have made it possible to generate observable picosecond transient Raman spectra for some species.[1,2] In these, picosecond Raman experiments a high repetition rate (~800 KHz) amplified sync pumped dye laser was used.

In our case we utilized a recently developed high sensitivity picosecond Raman apparatus[3] which consists of a dye laser synchronously pumped by a cw Nd/YAG mode locked laser and amplified by a high repetition rate (5 KHz) high energy laser source.[4] The 5 KHz picosecond Raman system described here and in detail in reference 3 is ideal because it provides a rather high energy per pulse which is essential for the low cross section Raman process. The average energy is two orders of magnitude higher than that used in the previous picosecond Raman experiments[1,2] and therefore results in the generation of large concentrations of the transient intermediates which consequently allow for their detection. It is important also to note that the peak powers are not sufficiently high to induce multiphoton absorption and other nonlinear processes such as stimulated Raman scattering. The application of this new experimental technique is discussed here with respect to the study of excited state isomerization in substituted diphenylethylenes.

Detailed studies of the ground and excited states of isotopically substituted trans stilbenes have been published previously using Raman spectroscopy.[6,7] The 1100 cm^{-1} to 1600 cm^{-1} region of the spectrum is dominated by three bands which are believed to be the symmetric C-phenyl stretch (1148 cm^{-1}), the phenyl-CH in plane bend and the 1566 cm^{-1} band due to either a C-C stretch or phenyl ring stretch $\nu 8a$. These assignments are based on extrapolation from similar frequencies observed in the ground state where the assignments are more accurately known.

The true normal mode character of the 1566 cm^{-1} band has been investigated in the experiment described here in order to develop an in depth knowledge of the S_1 electronic potential surface which is pivotal to the isomerization process. Substitution of heavy groups in one of the rings of t-Stilbene should provide the necessary spectral information for the determination for the assignment frequency of the 1566 cm^{-1} vibrational mode. If this vibration is a C-C stretch, the 1566 cm^{-1} vibration would be expected to shift to lower frequencies with substitution. However, the spectrum of the phenyl ring stretch $\nu 8a$ would be expected to split into two vibrations for the substituted and unsubstituted ring.

Consider the 1566 cm^{-1} vibration of t-Stilbene; The substituted compound spectrum was found to have only a single vibration which is slightly shifted to lower frequencies by ~25 cm^{-1} with no apparent broadening. This data strongly suggest that this vibration belongs to the C-C stretch, and agrees with the previous tentative assignment of this band.[6,7]

REFERENCES

1. T.L.Gustafson, D.M.Roberts, and D.A.Chernoff, J.Chem.Phys.79 1559 (1983).
2. T.L.Gustafson, D.M.Roberts, and D.A.Chernoff, J.Chem.Phys.81, 3438 (1984).
3. J.B.Hopkins and P.M.Rentzepis, Chem.Phys.Lett. in press.
4. J.B.Hopkins and P.M.Rentzepis, Appl.Phys.Lett.47(8), 776 (1985).
5. J.A.Syage, W.R.Lambert, P.M.Lambert, P.M.Felker, A.H.Zewail, and R.M.Hochstrasser, Chem.Phys.Lett.88, 266 (1982).
6. H.Hamaguchi, C.Kato and M.Tasumi, Chem.Phys.Lett.100, 3 (1983).
7. H.Humaguchi, C.Kato and M.Tasumi, Chem.Phys.Lett.106, 153, 1984.

APPLICABILITY OF RESONANT TWO PHOTON IONIZATION IN SUPERSONIC BEAMS TO HALOGENATED AROMATIC HYDROCARBONS

R. Tembreull and D. M. Lubman
University of Michigan, Ann Arbor, MI 48109

ABSTRACT

This work investigates problems encountered in the application of one color resonant two-photon ionization as an ionization source in supersonic beam mass spectroscopy. Of particular interest is the fact that photons of one color may not provide sufficient energy to cause ionization even when the laser source is tuned to an excited vibronic molecular state. We have therefore correlated trends in ionization potential with molecular structure for simple systems, specifically, halogenated aniline, phenol and toluene derivatives and mono- and di-substituted benzenes. In the case of para-substituted compounds where there is little substituent group interaction ionization occurs efficiently at the $S_0 \rightarrow S_1$ origin, provided ultrafast processes are not active as in the case of iodo substituted benzenes. Many ortho compounds, however, are found not to ionize efficiently. This is probably due to a combination of coulombic and steric interactions which result in an increase in ionization potential. These types of effects have been qualitatively related to the electron releasing and withdrawing properties of the substituent groups thereby allowing reasonable predictions to be made regarding those types of substituted benzenes which can be probed with R2PI.

INTRODUCTION

With the advent of coherent lasers it is now possible to routinely study processes which involve multiple photon excitation. One such process, resonant two photon ionization (R2PI) has significant analytical potential as a wavelength selective ionization source in mass spectrometry. We have therefore addressed the question of the general applicability of R2PI as an analytical tool in this work.

R2PI involves the absorption of two photons by a molecule and results in the production of a positive ion and concurrent release of an electron. Absorption of the first photon is strongly enhanced when the frequency of the ionizing laser is tuned to an excited vibronic state. As a result a large ionization signal is observed when the laser is in resonance with such an intermediate state, e.g. a vibronic π^* state in an aromatic molecule. Absorption of the second photon, on the other hand, is into an ionization continuum where the cross section for absorption is not a strong function of wavelength. Consequently, R2PI spectra are in effect absorption spectra. R2PI therefore provides wavelength selectivity thereby adding an extra dimension to conventional analytical mass spectrometry. R2PI is also characterized by excellent sensitivity allowing detection below ppb levels. Further, R2PI provides an extremely efficient means of "softly" ionizing molecules for mass spectroscopic detection. Thus it is possible to obtain the molecular weight of fragile biological or inorganic molecules which tend to fragment in electron impact ionization.

In order to fully exploit the wavelength dependence and spectroscopic information available from R2PI we have coupled our time-of-flight mass spectrometer (TOFMS) with a supersonic jet source in order to provide ultracold molecules. This is significant because at room temperature ultraviolet absorption spectra are broad and often featureless in contour due to the population of a large manifold of rovibrational states in the ground electronic state. Hence, wavelength selectivity is difficult or impossible to achieve. However, with the use of supersonic beam methodology, sharp rovibronic bands are observed which are typically ~2 cm^{-1} in width. As a result individual components can be identified in complex mixtures even when the analyte is present at a very low concentration.

Analytically a major drawback of R2PI supersonic beam mass spectrometry is the meager and often conflicting data base available. For example, many molecular ionization potentials are unavailable in the literature and those which are available are frequently in error. Such knowledge is crucial in this technique because in order to observe a resonant ionization signal, the sum of the energy of the two photons absorbed by a molecule must exceed its ionization potential. Therefore in this study the ability to make reasonable predictions regarding the effect various substituents have on the energy states involved in the resonant ionization of simple benzenoid systems is investigated. Trends in ionization potential are correlated with the type and position of substituent groups on a benzene ring.

EXPERIMENTAL

Our experimental apparatus consists of a six port vacuum cross in which a laser beam, the supersonic jet and a time of flight mass spectrometer are all mutually perpendicular. The supersonic molecular beam is generated by a pulsed valve which reduces the duty cycle of the gas flow compared to a continuous flow so that a small 6" pumping station can be used to maintain pressures at 5 x 10^{-6} torr. This molecular beam modulator allows the use of a large expansion orifice (0.5 mm) which can provide high (~10^{15} molecules/cc) on-axis intensity in the free jet. In turn, high sensitivity can be obtained in our measurements relative to present continuous jet expansions which require an orifice 100x smaller to maintain similar pumping conditions. The ions are created in the jet by tunable UV radiation (~1-3 mJ) generated by frequency doubling the output of a tunable Nd:YAG pumped dye laser system. The ions are then mass analyzed in a time-of-flight mass spectrometer in which resolution >1000 can be obtained and the spectroscopy of a particular mass peak is monitored by a gated integrator in order to obtain "mass selected" ionization spectra.

RESULTS AND DISCUSSION

Three states are involved in R2PI. For the substituted benzenes these states are the π electronic ground state, S_o, the intermediate π* state, S_1, and the ionic ground electronic state, S_o*. As pointed out by[1] previous investigators the effect of a substituent on the intermediate π* state of benzene is similar to its effect on the ground ionic state. This is reasonable since in the intermediate state an electron has been promoted from the stable 4n+2 aromatic π bonding core structure to a π* antibonding orbital creating a "photohole", whereas in the ground ionic state an electron has been removed altogether from the benzene ring. Release of electron density to or removal of electron density from the ring system therefore affects the energy of these states in a similar manner. Electron releasing groups such as $-CH_3$, -OH, and $-NH_2$ stabilize these upper states. This is reflected in the absorption spectra and PES of these compounds, both of which exhibit a red shift dependent on the ability of the substituent to release electron density to the ring system. The converse is true for electron withdrawing groups. Significantly, substituents which interact strongly with the aromatic π system do not exhibit linearly additive effects and are difficult to characterize theoretically. We have therefore looked for experimental trends upon which a simple model may be based.

The first class of molecules studied were the monosubstituted benzenes. The shift to lower energy of both the $S_o \rightarrow S_1$ absorption and the ionization potential of aniline, phenol and toluene are easily rationalized in terms of the overall dipole properties of each species. A simple electrostatic dipole model predicts that both the "permanent dipole" associated with each substituent as well as the dipole induced in the substituent by the presence of charge on the ring center favor release of electron density to the ring. For the monohalogenated benzenes however, the magnitude of the induced dipole is comparable to that of the permanent dipole and opposite in direction. Thus, halogen substituents may be either electron releasing or electron withdrawing depending on what other substituents are also present on the ring. In all the cases investigated herein the halogens were found to act as electron releasing substituents towards the aromatic π system.

The disubstituted benzenes proved to be quite interesting. When a strong electron releasing group, such as -OH or $-NH_2$, is "ortho" to a halogen substituent two photons of one color are either insufficient to cause ionization or resonant ionization is very weak.[1] Several ortho dihalogenated compounds were also found not to exhibit any detectable spectrum well to the blue of their electronic origins, e.g. o-dibromobenzene and o-bromochlorobenzene. These effects are believed to be due to coulombic interaction between adjacent groups which are particularly effective in stabilizing the S_o and S_1 states relative to the ionic ground electronic state S_o*. This results in an increase in the energy gap between the S_1 and S_o* states. Thus, not only is ionization usually not observed at the 0→0 origin, but furthermore, when two photons of one color are sufficient to cause resonant ionization they must proceed through a highly excited vibrational level in the S_1 state. Poor Franck-Condon factors therefore result in signals of small intensity. These types of effects were not observed for halogenated benzenes containing a moderately electron releasing group as Fig. 1 illustrates. It is also noteworthy that electronic transitions other than ππ* transitions may be probed for some compounds as has been demonstrated for nπ* transitions in several azabenzenes.[2]

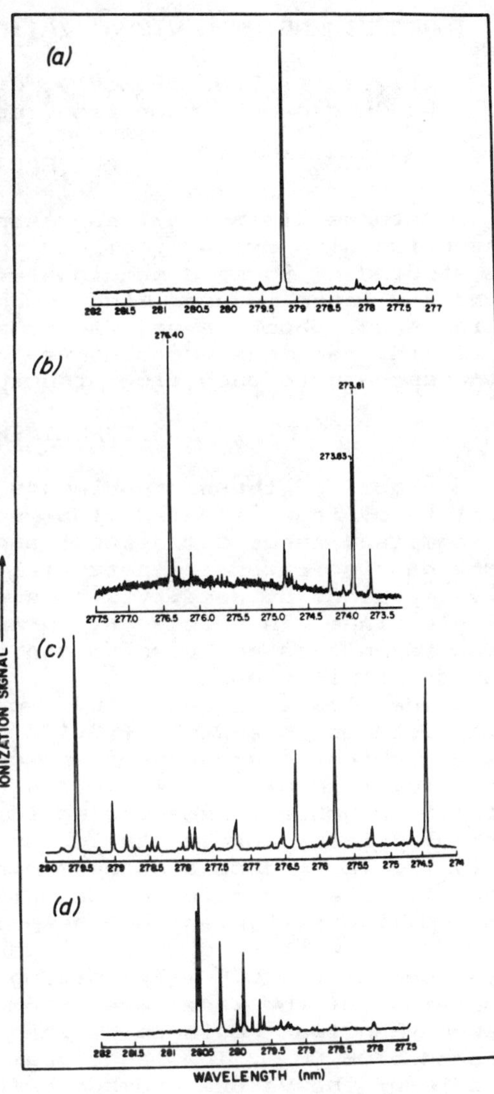

Fig. 1. Resonant two-photon ionization spectra taken in an expansion of 1 atm back pressure of Ar of (a) 2,4-dichlorotoluene, (b) 2,6-dichlorotoluene, (c) 2,5-dichlorotoluene, and (d) 3,4-dichlorotoluene. Only the molecular ion is monitored in these spectra using a TOF mass spectrometer.

1. R. Tembreull & D.M. Lubman, Anal. Chem., 56, 1962 (1984).

2. R. Tembreull, C.H. Sin, H.M. Pang & D.M. Lubman, Anal. Chem., 57, 2911 (1985).

THREE DIMENSIONAL VISUALIZATION OF SUPERSONIC FLOWS

G.I. Segal, G.D. Wiemokly, C.E. Gardner, M.A. Kwok
The Aerospace Corporation, Los Angeles, Ca., 90009

ABSTRACT

A three dimensional supersonic flow, generated by insertion of a cylindrical rod into an axisymmetric flow, is studied by three dimensional flow visualization techniques and laser induced fluorescence using I_2 as the molecular seed. Shock waves, wakes, and boundary layers and their interactions are observed and the relation of I_2 fluorescence to such flow structures is discussed.

EXPERIMENT

A goal of these experiments has been to assess the utility of laser induced fluorescence (LIF) in the study of complex, three dimensional supersonic flows. Previous work has indicated the potential of LIF to give quantitative mappings of density, temperature and velocity in a single plane. In this work, three dimensional visualization techniques are used to map the entire volume of a non-symmetric flow.

The flow is produced by a converging-diverging nozzle with a throat diameter of 0.125". A plenum pressure of 12 psia, ambient chamber pressure of 2 torr and a constant mole fraction of I_2 was used in all experiments. The high plenum to ambient pressure ratio developed a nitrogen free jet of primary wavelength to throat diameter of approximately 15. To produce a non-symmetric flow a 1.6 mm cylindrical rod was placed in the uniform region of the flow 1 cm from the exit plane of the nozzle.

A 2.5 watt krypton ion laser run TEM_{oo} and broadband was used as the LIF illuminating source. The effective bandwidth of the laser was 10 GHz FWHM which pumped no fewer than four rovibronic lines in the B-X manifold of I_2 around 5308.68 A. The laser beam is focused into the middle of the vacuum chamber by a long focal length mode matching lens giving an effective Rayleigh length of 4 cm. An illuminating sheet approximately 0.2 mm thick and 4 cm wide is formed by a cylindrical lens and mirror system parallel to the flow (yz plane). The fluorescence produced in the flow is imaged on a silicon intensified target video camera.

Initially the camera optics are tightly focused on a single fluorescent sheet in the flow. After establishing the desired flow conditions, the illuminating laser sheet and video camera are stepped in unison accross the flow volume. At each position, an averaged image of the induced fluorescence is digitized and stored by the mini-computer. The geometry of the volume of scanned images is preserved by assuring that the separation between planes is equal to the separation between video lines within each plane. A scan consisting of 220 images is required to map the entire flow volume. Typical data image planes, corrected for scattered light and the pump beam profile, with and without the rod present in the flow are shown in Figures 1 and 2 respectively.

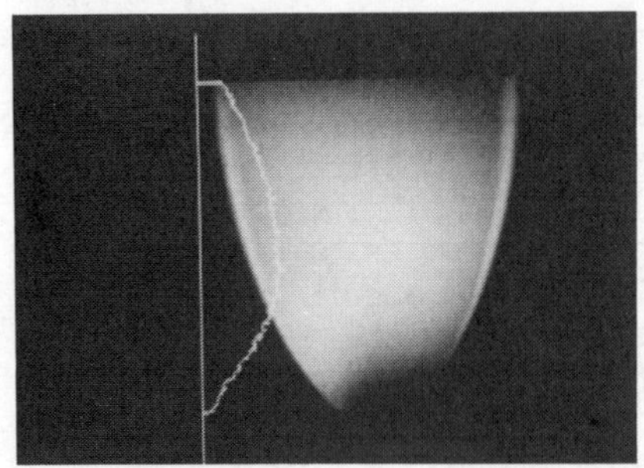

Fig. 1. Data yz image plane of baseline flow. Flow is in the z direction.

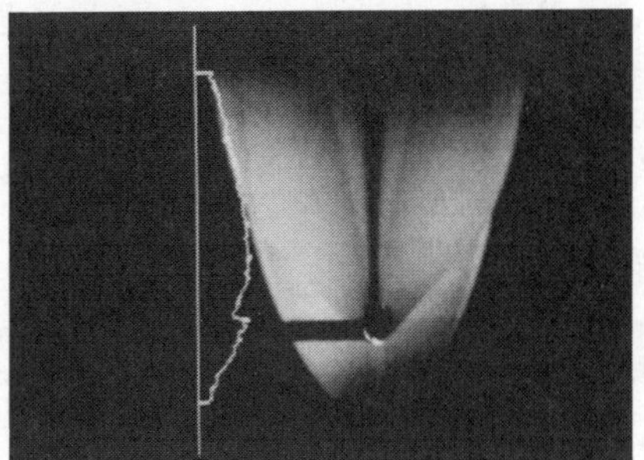

Fig. 2. Corresponding yz plane with rod in flow. Plot taken to right of the rod.

Images of planes in any orientation with respect to the flow axis can be generated from the collected image data. In particular, planes transverse to the flow axis (xy planes) can be reconstructed (Fig.3) showing a view of the flow normally unattainable.

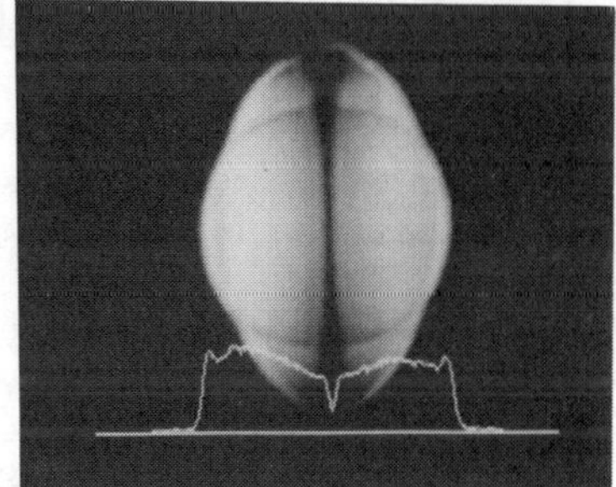

Fig. 3. Reconstructed xy plane with rod. 7mm above rod centerline.

ANALYSIS

It has been shown that using broadband illumination in the limit of low intensity, the induced fluorescence signal is dependent on the pressure through the quenching term in the Stern Volmer factor and on the temperature through the Boltzmann distribution. The quenching complication is an unavoidable consequence of the long lifetime of the upper states of the I_2 molecule. The temperature dependence is due to the broadband laser used to pump the iodine. Since the temperature changes continuously due to flow expansion and abruptly at shock discontinuities, the fraction of molecules in each state will also change. One can see the result of these effects in crossing the oblique shock in Figure 2 where analysis would suggest that the temperature and density should increase but where the fluorescence observed decreases through a combination of increased quenching and shifting of the population out of the pumped states.

GAS DYNAMIC FOCUSING IN SUPERSONIC JETS: APPLICATIONS TO CHEMICAL ANALYSIS

S.W. Stiller, B.D. Anderson, and M.V. Johnston
Univ. of Colo., Dept. of Chemistry/CIRES, Boulder, CO 80309

ABSTRACT

Supersonic jet spectroscopy is useful as an analytical technique for selectively discriminating between compounds that have similar electronic absorption spectra when used with laser induced fluorescence. In applying supersonic jet spectroscopy to analytical problems, sensitivity is limited by the small region of laser-sample interaction. A solution to this problem is gas dynamic focusing. The sample stream is enveloped in a sheath gas which restricts the dispersion of the sample stream in the jet, allowing the laser to interact with a larger fraction of the analyte molecules. Both cooling and sample enrichment is seen in the jet. Detection limits are in the mid-picogram range.

INTRODUCTION

The selectivity of supersonic jet spectroscopy as a potential analytical tool is now well established. Early attempts at interfacing laser spectroscopic detection to gas chromatography have either shown no cooling[1], or poor detection limits[2]. Our approach uses a gas dynamic focusing system similar to one used in a non-supersonic environment by Keller and Nogar[3].

EXPERIMENTAL

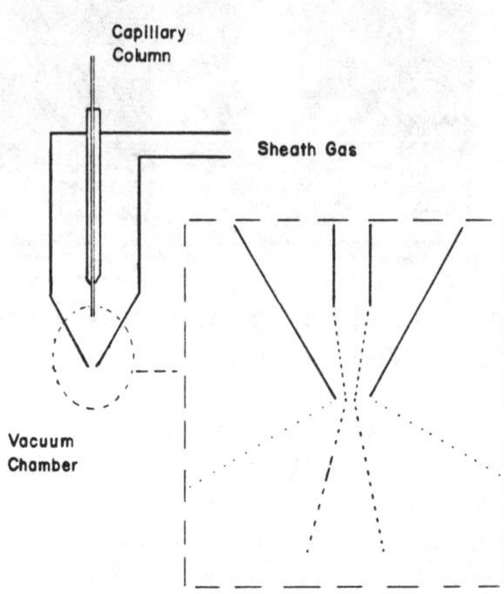

Fig. 1 Diagram of gas dynamic focusing nozzle.

The focusing nozzle consists of a capillary column which ends in a cone through which a He and/or Ar sheath gas flows. A diagram of the nozzle is shown in fig. 1. Supersonic jet conditions are met for the sheath gas and the capillary effluent is injected into the sheath gas near the 200 μm nozzle orifice. The vacuum chamber is pumped by a 6" diffusion pump. Either the frequency doubled output of a Nd:YAG pumped dye laser or the 266nm, fourth harmonic of the Nd:YAG cross the sample stream 50 nozzle diameters down from the nozzle orifice. Fluorescence is focused through a bandpass filter into a PMT. Signals are recorded as the output from a box-car integrator.

RESULTS

The sample density profile of an unfocused free supersonic jet falls off as $\cos^2\theta$ of the angle off of the center of the beam. In the focused jet, the sample density is at least an order of magnitude greater in the center of the beam and falls off much faster than in the analogous free jet.

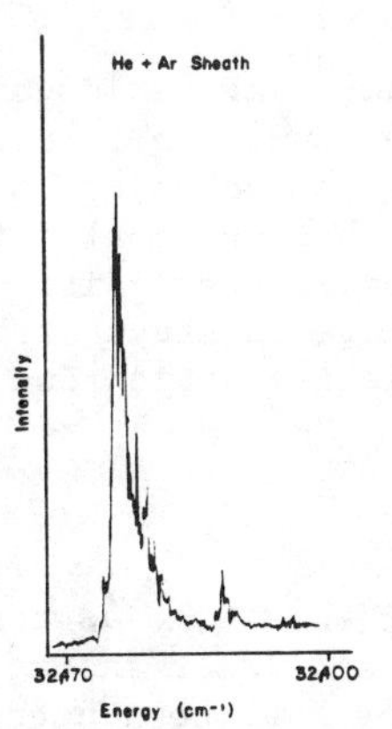

Fig. 2 Excitation spectrum of napthalene $\bar{8}^1_0$ vibronic.

The best cooling has been seen with a mixture of He and Ar as the sheath gas. The carrier gas does not have very much effect on cooling in that it comprises less than 0.5% of the total gas flow into the vacuum chamber. Fig. 2 shows the excitation fluorescence spectrum of the $\bar{8}^1_0$ vibronic of napthalene. The 1:1 mixture of He and Ar in the 250 torr sheath gas with He carrier gas give a FWHM of 3 cm^{-1} (0.3 Å) for the jet cooled napthalene vibronic. The spectrum shown is not limited by the laser bandwidth and therefore reflects the rotational contour. This corresponds to a rotational temperature of <10°K. This is on the same order of magnitude to rotational temperatures seen in a free jet.

In contrast to a previous attempt to interface a chromatographic column to a supersonic jet technique, the chromatography is not degraded by the interface. In the focusing nozzle, the effluent freely flows into the vortex of the sheath gas near the nozzle orifice. The chromatography can be optimized, giving narrow peaks, and therefore better peak intensities. Fluorescence intensity is linear from in the range of sub-nanogram to microgram. Detection limits (S/N = 3) have been found to be approximately 250 picograms. The noise is limited by laser scatter that is transmitted through the bandpass filter. A monochromator is being installed and detection limits at least an order of magnitude better are expected.

The gas dynamic focusing interface represents a significant advance in quantitative jet spectroscopy. The coupling of capillary gas chromatography with focused jet laser induced fluorescence can lead to sensitive spectroscopic analysis of samples.

REFERENCES

[1] G.Rhodes,R.B.Opsal,J.T.Meek,and J.P.Reilly, Anal. Chem., 1983,55,280.
[2] J.M.Hayes and G.J.Small, Anal. Chem., 1982, 54, 1202.
[3] R.A.Keller and N.S.Nogar, Appl. Optics, 1984, 23, 2146.

ULTRAVIOLET LASER APPLICATIONS TO COMBUSTION DIAGNOSTICS

Andrzej W. Miziolek, Brad E. Forch*, Rosario C. Sausa*,
Mark A. DeWilde
US Army Ballistic Research Laboratory
Aberdeen Proving Ground, MD 21005-5066

ABSTRACT

Ultraviolet lasers have been used for flame diagnostics of oxygen atoms by two-photon fluorescence at 225.6 nm. The focussed uv laser has been observed, however, to induce photochemical perturbation in certain parts of the flame. This photochemical interaction, in turn, has been explored with respect to the ignition characteristics of reactive gas mixtures. We have observed strong wavelength resonance effects and highly efficient ignition due to multiphoton photochemistry.

INTRODUCTION

Ultraviolet lasers have been utilized for some time as combustion diagnostic tools for Raman scattering and laser induced fluorescence (LIF) studies. More recently, they have been used for imaging of O_2 in combustion flows[1,2] as well as in nonlinear spectral probing[3], which typically requires rather tight focussing of the laser excitation beam.

Short wavelength lasers are also capable of inducing single or multiphoton photochemistry and, thus, it is reasonable to expect that photochemical perturbation may occur during certain types of laser combustion diagnostic experiments. Such a perturbation was first reported in a laser flame study where C_2 emissions were observed during excitation of OH at 282 nm.[4] In a related experiment C_2 emissions were observed during 2-photon excitation of CO at 230 nm.[5]

Oxygen atom emission, which resulted from multiphoton photolysis of the oxidizer (N_2O) with subsequent two-photon O-atom excitation (Fig. 1), was first reported in the preheat region of a $CH_4/N_2O/N_2$ flame.[3] Similarly, O-atom production from O_2 was recently observed during oxygen atom fluorescence imaging in a C_2H_2/O_2 flame.[6] Subsequent to the initial probe laser-induced perturbation studies in our laboratory, experiments were undertaken to investigate other aspects of laser-induced photochemistry, particularly with respect to activating reactive gaseous mixtures. This paper represents a summary of our work in this area, i.e. in multiphoton photochemical ignition.

* NRC Postdoctoral Research Associates

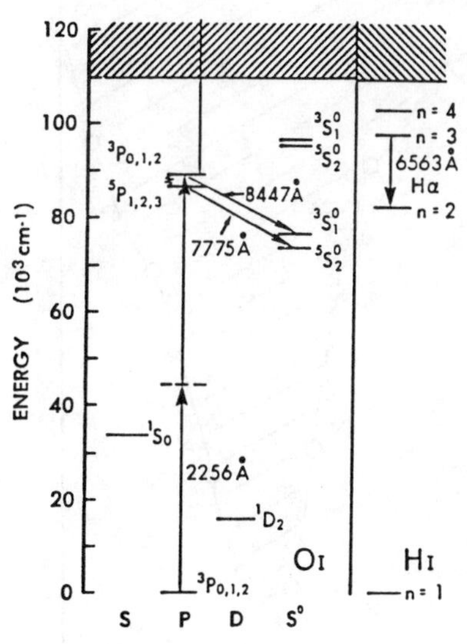

Figure 1. Partial Energy Level Diagram For O Atoms

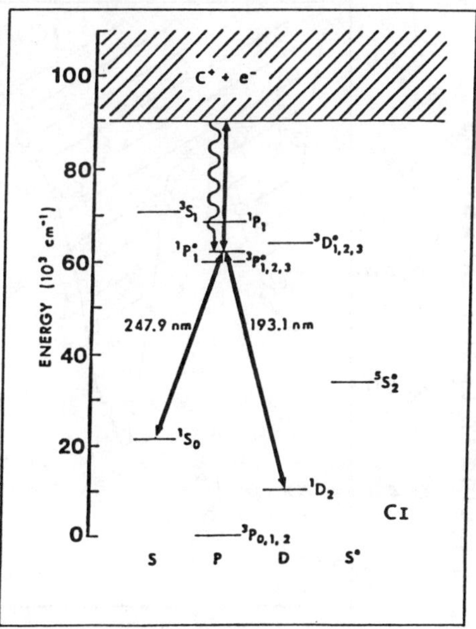

Figure 2. Partial Energy Level Diagram for C Atoms

RESULTS AND DISCUSSION

Multiphoton photochemical ignition was first demonstrated on reactive mixtures of small hydrocarbons with N_2O or air as the oxidizer.[7] One of the most interesting results of this initial study was the observation of laser ignition of C_2H_2/air with as little as 0.25 mJ of incident ArF laser (193 nm) energy. The fact that the ArF laser is particularly efficient in activating a reactive mixture containing C_2H_2 is not too surprising since previous work[8] has described a strong interaction between the ArF laser and the C_2H_2 molecule leading to extensive photofragmentation. Furthermore, there is a coincident overlap between the ArF laser and a strong C-atom transition at 193.1 nm (see Fig. 2) leading to the excitation of a high-level atomic electronic state as well as to the ionization of C-atoms due to the absorption of a second photon. A determination of the relative importance of these different factors (i.e. photochemistry, atomic excitation, ionization) to the ignition mechanism required further work on a different reactive system, namely H_2/O_2 and H_2/N_2O.

Figure 3 shows the dependence of the amount of incident laser energy required to ignite a premixed flow of H_2/N_2O on the laser wavelength in the 225.6 nm region using a 50 mm focal length lens. Clearly, there are sharp laser wavelength resonances[9] involved

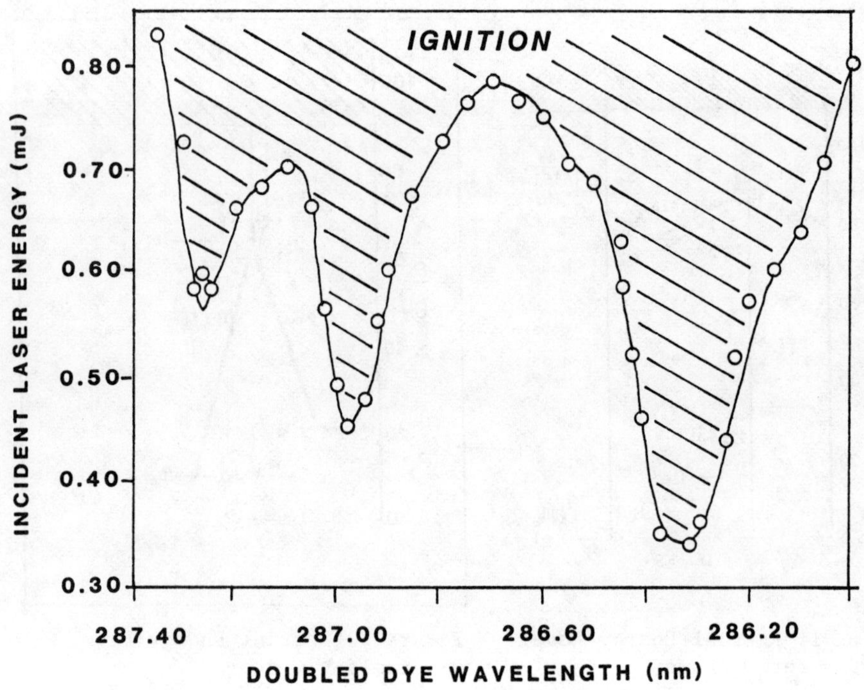

Figure 3. Incident Laser Energy Necessary to Ignite a Premixed Flow of H_2/N_2O as a Function of Laser Wavelength in the 225.6 nm (Doubled Dye + 1.06 Micron) Region.

which correspond to the peaks of oxygen atom two-photon excitation of the ground electronic spin-orbit split states.[10] In other words, the same focussed laser induces photodissociation of N_2O into O-atoms which are then further excited by a two-photon resonant process (Fig. 1). A similar behavior as shown in Fig. 3 is found for premixed H_2/O_2 flows as well. Further experiments[10] on these reactive systems have shown that the photoionization channel is a key step in the ignition process since it supplies the seed electrons which initiate the formation of a laser-produced microplasma. A microplasma of sufficient strength leads to the creation of a viable ignition kernel. Multiphoton photochemical ignition, therefore, is a new laser ignition phenomenon which combines multiphoton photochemistry and resonance ionization with the laser gas breakdown mechanism to yield an efficient and highly-controllable ignition tool.

CONCLUSION

Ultraviolet lasers are expanding laser applications to combustion studies in significant new ways since they can be used as diagnostic tools as well as a means for activating and/or perturbing reactive gas mixtures through single and multiphoton photochemistry. It is expected that the increased use of these lasers will impact this field significantly in areas of fundamental understanding of combustion chemistry and in practical applications where efficient and highly controllable activation of reactive systems is a requirement.

ACKNOWLEDGEMENT

This research is supported in part by the US Air Force Office of Scientific Research, Contract #86-0008.

REFERENCES

1. M. P. Lee, P. H. Paul and R. K. Hanson, Opt. Lett. 11, 7 (1986).
2. J. E. M. Goldsmith and R. J. M. Anderson, Opt. Lett. 11, 67 (1986).
3. A. W. Miziolek and M. A. DeWilde, Opt. Lett. 9, 390 (1984) (and references therein).
4. M. Alden, H. Edner and S. Svanberg, Appl. Phys. B 29, 93 (1982).
5. M. Alden, S. Wallin and W. Wendt, Appl. Phys. B 33, 205 (1984).
6. M. Alden, H. M. Hertz, S. Svanberg and S. Wallin, Appl. Opt. 23, 3255 (1984).
7. R. C. Sausa and A. W. Miziolek, Comb. and Flame (submitted).
8. J. R. McDonald, A. P. Baranovski and V. M. Donnely, Chem. Phys. 33, 161 (1978).
9. B. E. Forch and A. W. Miziolek, Opt. Lett. 11, (1986).
10. B. E. Forch and A. W. Miziolek, submitted to 21st Symposium (International) on Combustion (1986).

PHOTOFRAGMENT FLUORESCENCE AS AN ANALYTICAL TECHNIQUE: APPLICATION TO GAS-PHASE ALKALI COMPOUNDS

Richard C. Oldenborg and Steven L. Baughcum
Chemistry Division
Los Alamos National Laboratory
Los Alamos, New Mexico 87545

ABSTRACT

Photodissociation of gas-phase compounds using a laser at suitably short ultraviolet wavelengths can produce electronically excited photofragments. Detection of fluorescence from these excited fragments, particularly atomic fragments, allows sensitive and quantitative density measurements while signal strength as a function of dissociation laser wavelength allows differentiation between compounds that yield the same photofragment. Application of the technique to the detection of gas-phase alkali compounds is discussed and the results of experiments to detect sodium and potassium chlorides is presented.

INTRODUCTION

Alkali compounds are corrosive contaminants in many fossil fuel combustion processes. The objective of this research is to develop a laser-based optical diagnostic technique applicable to the direct monitoring of trace levels of alkali compounds within the post-combustion coal gas stream. The diagnostic technique that has been adopted is photofragment fluorescence. An ultraviolet laser excites the parent compound to a repulsive potential, which in turn dissociates to yield an excited alkali atom. This atom then emits at its characteristic wavelength and the intensity of this emission is proportional to the original concentration of the parent compound. The sensitivity of this atomic photofragment fluorescence technique as an analytical method should be high since the atomic alkali emission is strong (~20 ns radiative lifetime) and the absorption cross sections of the alkali compounds in the desired band are large ($\sigma > 10^{-17}$ cm^2). In addition, since practical applications involve measurements in high-temperature environments (600 to 950°C for gas turbine inlet temperatures), the limiting source of noise is due to statistical fluctuations in the blackbody emission background, and the inherently narrow bandwidth of the atomic emission allows for good wavelength discrimination from the blackbody background and correspondingly better sensitivity.

Compounds of the different alkalis (e.g., NaCl, KCl) can readily be distinguished by the emission wavelengths of the excited atoms (589 nm for Na*, 766 nm for K*). Discrimination between compounds of the same alkali but different anions (e.g., KCl, KOH) is more difficult, but can be achieved by examining the dependence of the alkali emission on the excitation laser wavelength. This can best be illustrated by comparison of threshold

0094-243X/86/1460632-2$3.00

wavelengths, i.e., the wavelength needed to dissociate a vibrationally cold parent compound to yield the excited alkali atom. Threshold wavelengths vary among different compounds as the dissociation energies vary. For example, the calculated energy required for the production of excited potassium atoms in the lowest $^2P°$ state and the corresponding threshold wavelengths for KCl and KOH are 138 kcal/mol (207 nm) and 123 kcal/mol (233 nm), respectively. Absorption of vibrationally excited compounds is important at the temperatures of interest and this "hot-band" absorption significantly broadens the bands in the photodissociation spectra to longer wavelengths. From Boltzmann energy distribution considerations and assuming comparable sensitivities, we estimate that if the threshold energies differ by at least 10 kcal/mol, then 100:1 discrimination can be achieved given some knowledge of the temperature. The variation in threshold energies for these and the other important gas-phase molecules expected to be present in fossil fuel environments is large enough that good discrimination between compounds containing the same alkali, but different major anion groups (halides, hydroxides, monoxides, dioxides, and sulfates), should be feasible.

RESULTS AND DISCUSSION

Sodium and potassium chloride vapors were irradiated at 193 nm from an ArF laser and other wavelengths generated by stimulated Raman scattering in H_2 and D_2. Fluorescence from electronically excited alkali atoms in the lowest $^2P°$ level was observed and the dependence of the fluorescence intensity on cell temperature, vapor pressure, and dissociation laser wavelength was investigated.

The atomic alkali emission intensity from the photodissociation of the alkali chlorides at wavelengths near the corresponding alkali hydroxide threshold wavelength was down roughly three orders of magnitude from the peak, implying that reasonable discrimination between species containing different anions should be possible. The emission intensities were found to track linearly with the alkali chloride monomer densities. Sensitivities in the sub-ppb range were observed for single laser shots. Similar experiments on alkali hydroxide photodissociation are now in progress.

ACKNOWLEDGMENTS

The technical assistance of Douglas Hof and Kenneth Winn of Los Alamos National Laboratory is gratefully acknowleged. This work is supported by Morgantown Energy Technology Center (DOE).

LASER DIAGNOSTICS OF SEMICONDUCTOR PROCESSING SYSTEMS

Alan C. Stanton* and Joda Wormhoudt
Aerodyne Research, Inc. Billerica, MA 01821

ABSTRACT

Laser spectroscopic methods for the study of advanced semiconductor processing and fabrication techniques, such as chemical vapor deposition, plasma etching, and plasma deposition, are being increasingly utilized as tools for probing the detailed mechanisms of these processes and may see eventual applications in process control. The current emphasis is on the application of laser techniques to the measurement of gas phase radical species which are believed to be important as etchants or as intermediates in deposition systems. This paper provides an overview of such applications and presents recent results on the in situ measurement of chlorine atoms in an etching plasma.

INTRODUCTION

The etching of silicon and silicon dioxide by halogen-containing rf discharge plasmas is a major semiconductor fabrication process. Deposition of silicon, in plasmas or by pyrolysis and chemical reactions of silanes, is of interest for solar cells and a variety of electronics applications. Chemical vapor deposition is also being extended to the formation of many other materials, such as silicon carbide and silicon nitride. The in situ characterization of these advanced processing systems, by measurement of gas phase species concentrations, temperature, flow patterns, local electric field, etc., can contribute to a better understanding of the controlling mechanisms and may lead to improvements in process design or to eventual on-line monitoring and control.

We have previously reviewed the application of laser induced fluorescence and laser absorption to the detection of stable and transient species of interest in etching or vapor deposition systems.[1-2] These reviews have included detailed discussions of operational considerations for both techniques as well as estimates of the minimum detectable number densities for a large number of stable and radical species. Rather than repeat these discussions here, we will merely mention a few qualitative considerations pertaining to the two techniques. A few examples of their implementation in deposition and etching systems are cited in References 1 and 2.

The techniques of linear, single photon laser induced fluorescence (LIF) using tunable UV-visible dye lasers, and laser absorption using tunable infrared diode lasers provide powerful and

*Present address: Southwest Sciences, Inc., Santa Fe, NM 87501

sensitive methods for measurement of most gas phase species of interest in deposition and etching systems. A qualitative comparison of the two techniques is given in Figure 1. Neither method is universally applicable (for example, many polyatomic molecules cannot be detected by LIF), and the choice of technique often depends on the measurement goals. For relative measurements of spatially resolved species concentrations, LIF is the clear choice when it can be used, while laser absorption may be the better technique if absolute species concentrations are desired. Other factors such as quenching or background chemiluminescence, in the case of LIF, or line broadening or spectral interferences, in the case of laser absorption, may limit sensitivity.

MEASUREMENT OF ATOMIC CHLORINE CONCENTRATIONS

As a specific example of the application of laser absorption to diagnostics of a plasma etching system, we discuss the *in situ*, quantitative measurement of chlorine atoms in a molecular chlorine etching plasma. Detailed discussions of these measurements and analyses of the results will appear in forthcoming publications.

IR ABSORPTION (TUNABLE DIODE LASER)	UV/VISIBLE LIF (TUNABLE DYE LASER)
• WIDE GENERALITY, INCLUDING "FEEDSTOCK" GASES (E.G. SiH_4)	• NOT APPLICABLE TO MANY POLYATOMICS
• LOWER SENSITIVITY THAN LIF ($\sim 10^{11}$ CM^{-3})	• HIGH SENSITIVITY ($\sim 10^{6}$ TO 10^{9} CM^{-3} AT LOW PRESSURE)
• ABSOLUTE CONCENTRATION MEASUREMENTS POSSIBLE	• ABSOLUTE CALIBRATION USUALLY DIFFICULT
• LINE OF SIGHT MEASUREMENT (CAN BE SPATIALLY SCANNED)	• CAN PROVIDE 2-D OR 3-D SPATIAL RESOLUTION

Figure 1. Qualitative Comparison of Laser Fluorescence (LIF) and Laser Absorption Diagnostic Methods

A capability for *in situ* measurement of atomic halogen species, particularly atomic fluorine and chlorine, is of interest in studies of plasma etching because under some conditions these species may be the primary etchants. For the halogen atoms, available measurement techniques in the visible or ultraviolet wavelength regions are limited either to vacuum uv resonance fluorescence/absorption using atomic resonance lamps or to nonlinear laser methods such as multiphoton absorption combined with fluorescence or ionization detection.[3] The resonance lamp techniques, while useful for many laboratory applications, may not provide the desired temporal and spatial resolution for system diagnostics measurements. The nonlinear laser methods can be expensive to implement and apparently result in only modest detection sensitivities compared with the resonance lamp techniques.

An alternative detection method utilizes the $^2P_{1/2}-^2P_{3/2}$ spin orbit transitions in the infrared, as suggested by Schlossberg in the case of atomic fluorine.[4] The F atom transition near 404 cm^{-1} has been fully characterized in terms of absorption line positions[5-6] and cross sections[5] using tunable diode laser techniques. The analogous transition in atomic chlorine near 882 cm^{-1} has been observed in previous high resolution spectroscopy studies.[7-8]

As a preliminary step in developing an in situ diagnostic capability for measurement of absolute concentrations of chlorine atoms, we have measured the absorption line strengths of the Cl $^2P_{1/2} \leftarrow {}^2P_{3/2}$ transition.[9] In these laboratory studies, a low pressure discharge flow apparatus is used to prepare known concentrations of atomic chlorine by reaction of Cl_2 or HCl with excess fluorine atoms. A tunable diode laser combined with a multipass White cell transverse to the Cl atom flow is used to measure absorbance at line center as a function of Cl concentration for the strongest hyperfine component of the transition. The measured line center absorption cross section for this line is $\sigma_o = (1.80 \pm 0.38) \times 10^{-18}$ cm^2, corresponding to an integrated line strength of $S = (4.14 \pm 0.89) \times 10^{-21}$ cm^2 $molecule^{-1}$ cm^{-1}. The estimated Cl atom detection sensitivity using the diode laser absorption technique is $\leq 5 \times 10^{11}$ cm^{-3}, which is one to two orders of magnitude more sensitive than the estimated limits for Cl atom detection by two photon LIF[3] or CARS.[10]

Measurements of chlorine atom concentrations in a plasma etching reactor have been made in collaboration with the research group of Prof. Herbert H. Sawin of the Chemical Engineering Department at M.I.T. The goals of this study were to perform quantitative measurements of atomic chlorine in a molecular chlorine etching plasma in order to aid the development of models of chemical and ion-driven etching. The effects of rf power and frequency, discharge pressure, gas additions, electrode spacing and materials, and presence of silicon substrate were studied.

The plasma etching research apparatus used for this study is designed to accommodate a variety of diagnostics, including optical measurements of emission, fluorescence, or absorption, as well as mass spectrometric sampling of plasma constituents. The Cl concentration measurements were made using a multipass absorption cell formed by mirrors mounted in two opposed windows on the apparatus. The absorption path traverses the chlorine plasma contained between parallel stainless steel electrodes (12.5 cm diameter) nominally spaced 3 cm apart (although the spacing is adjustable). Absorption signals, measured for a wide range of plasma conditions, were digitized and stored for later analysis.

An example of some results obtained from these measurements is shown in Figure 2, which is a plot of measured absolute Cl concentrations as a function of discharge pressure. The rf discharge frequency was 13.5 MHz at a power of 75 watts for these measurements. The absolute number density is obtained from measurement of the line center absorbance, combined with the experimentally determined cross section. A temperature correction is applied, based on the measured Doppler line width. The absorption path length is estimated from visual observations of the spatial extent of the luminous region of the plasma. The overall uncertainty in Cl number density is believed to be approximately ±25%. Present work is aimed at correlation of such results with other observations and measurements of system performance (such as etching rate, directional characteristics of etching, etc.) in conjunction with the development of kinetic models of chlorine containing plasma etching systems.

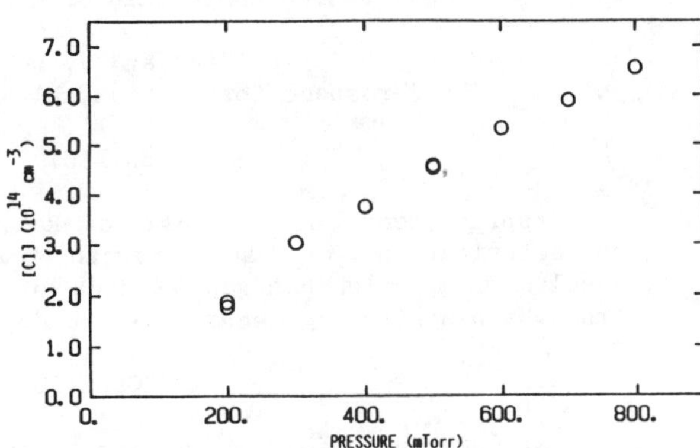

Figure 2. Measured Chlorine Atom Concentrations As a Function of Pressure In a Plasma Etching Reactor

ACKNOWLEDGMENTS

This work was supported by the Air Force Office of Scientific Research (AFSC) under Contract F49620-84-C-0036. The in situ measurements of atomic chlorine were made by A.D. Richards and H.H. Sawin at M.I.T., under sponsorship of the Semiconductor Research Corporation.

REFERENCES

1. J. Wormhoudt, A.C. Stanton, and J. Silver, Proc. SPIE 452, 88 (1984).
2. J. Wormhoudt, A. Stanton, and J. Silver, in Plasma Synthesis and Etching of Electronic Materials, Materials Research Society Symposia Proceedings, Vol. 38 (Materials Research Society, Pittsburgh, PA, 1985), pp. 91-98.
3. M. Heaven, T.A. Miller, R.R. Freeman, J.C. White, and J. Bokor, Chem. Phys. Letters 86, 458 (1982).
4. H. Schlossberg, J. Appl. Phys. 47, 2044 (1976).
5. A.C. Stanton and C.E. Kolb, J. Chem. Phys. 72, 6637 (1980).
6. G.A. Laguna and W.H. Beattie, Chem. Phys. Letters 88, 439 (1982).
7. M. Dagenais, J.W.C. Johns, and A.R.W. McKellar, Can. J. Phys. 54, 1438 (1976).
8. P.B. Davies and D.K. Russell, Chem. Phys. Letters 67, 440 (1979).
9. A.C. Stanton, Chem. Phys. Letters (in press).
10. D.S. Moore, Chem. Phys. Letters 89, 131 (1982).

STIMULATED RAMAN GAIN SPECTROSCOPY OF GaAs

S. Beck and J. Wessel
The Aerospace Corporation, El Segundo, CA 90254

ABSTRACT

Applications of stimulated Raman gain spectroscopy to characterization of semiconductor materials is discussed. Results from below-bandgap studies of bulk GaAs are presented, from which scattering mechanisms are deduced.

INTRODUCTION

Stimulated Raman gain spectroscopy (SRGS) provides an exceedingly sensitive probe[1,2] of phonon structure in solids. The method is promising for both materials analysis applications, in which properties such as impurity content, defects, strain, and carrier concentrations are measured, and for fundamental studies where the sensitivity is used to measure new properties. Selection rules and scattering mechanisms are similar to those appropriate to conventional Raman scattering, however, SRGS can be performed in forward scattering geometry, whereas conventional Raman must ordinarily be performed in backward scattering geometry. Forward scattering probes phonons with near zero phonon momentum (q), whereas conventional backward scattering accesses phonons with appreciable momentum. Thus, Stimulated Raman provides a valuable probe for low q processes in semiconductors. The q dependence of Raman scattering in semiconductors provides a critical test of scattering mechanisms[3]. Currently there is considerable interest in mechanisms because recent above-bandgap studies[4] indicate that impurity induced scattering plays a major role in GaAs. Until now, experimental studies have emphasized above-bandgap scattering. In this paper we describe an application where the q dependence of below-bandgap scattering provides a critical test of mechanisms. The results show that conventional deformation potential and Frohlich mechanisms provide an excellent description of below-bandgap Raman processes. Very near bandgap, an impurity induced contribution becomes observable. In addition, we point out future potential applications of Raman gain spectroscopy to materials analysis.

EXPERIMENTAL TECHNIQUE

The experimental apparatus used for the SRGS experiments is described in Figure 1. Two tunable mode-locked dye lasers synchronously pumped by a mode-locked argon ion laser provide output that is combined, using a beamsplitter, on a sample mounted at Brewster's angle in a liquid helium dewar. The pump laser, denoted ω_1, passes through an electrooptic modulator operated at 14 MHz. The probe laser, denoted ω_2, is separated from the pump by a diffraction grating and is sensed by a photodiode. Output is detected as the component of the probe beam synchronously modulated at the pump beam frequency, at which laser noise is greatly reduced relative to lower frequencies. By this technique, Raman signals are easily recorded to within 20 meV of bandgap for 0.3 mm thick GaAs samples. Spectral scans and data processing including signal averaging are performed by computer.

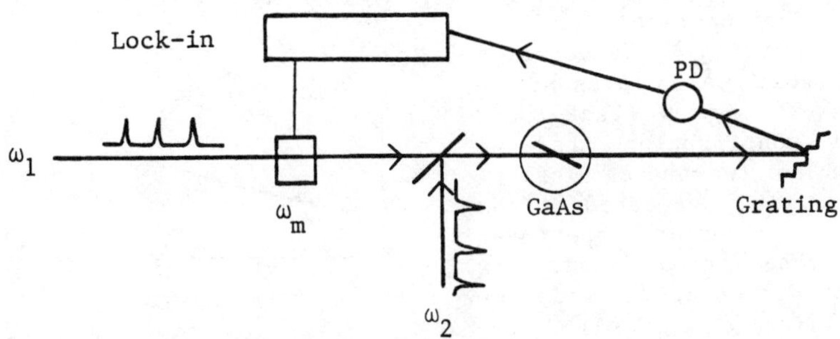

Fig. 1. Apparatus used for SRGS. The electrooptic modulator operates at ω_m and a photodiode (PD) detects the signal.

RESULTS

Spectra recorded for the LO phonon band of intrinsic GaAs are shown in Figure 2, with polarization orientations chosen to reveal scattering selection rules. Deformation potential scattering should occur for $(0\bar{1}1)$ and (011) polarization orientations. Frohlich induced Raman should occur for incident and scattered polarizations oriented along either (010) or (001). However, the Frohlich contribution should be weak for forward scattering. The observed spectra are in accord with these expectations. We have calculated the expected magnitude of Frohlich and deformation

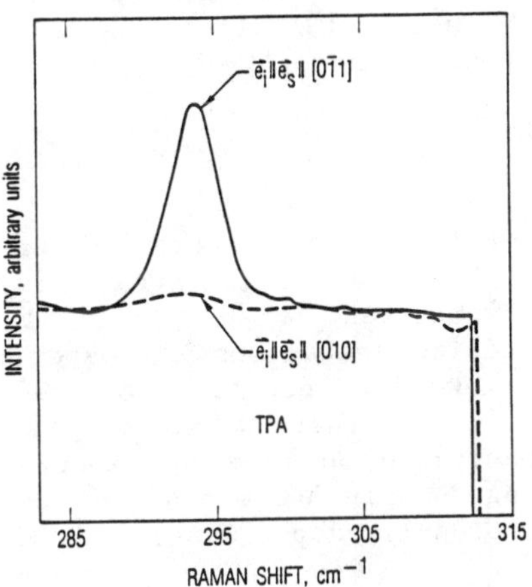

Fig. 2. SRGS of GaAs with two-photon absorption (TPA) background.

potential contributions, as shown in Figure 3. Only deformation potential scattering is expected under the forward scattering conditions. This is consistent with observations, except that a small contribution may be present in the Frohlich orientation. This may be due to imperfect sample orientation, or due to impurity scattering mechanisms. Also present in Figure 2 is a background signal contribution that we assign to two-photon absorption. As expected from theory and prior experimental observation, it is largely orientation independent, it has the same time dependence as the Raman signal and as the pulse cross correlation function, and it is moderately enhanced near bandgap resonance. The signal level is consistent with known two-photon absorption cross sections.

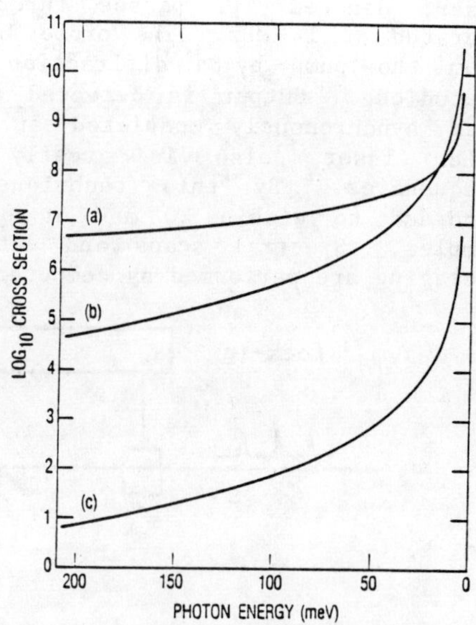

Fig. 3. Raman calculated for the deformation potential mechanism (a), for the Frohlich mechanism with backward scattering (b), and forward scattering (c).

DISCUSSION

Bulk Raman spectra of GaAs were readily recorded by a straightforward stimulated Raman gain technique. The results are consistent with predicted selection rules for intrinsic scattering mechanisms. However, the observations are surprising in view of recently reported above-bandgap measurements on GaAs[4]. In that work, an impurity scattering mechanism was demonstrated to play an important role. This mechanism was not strong in our below-bandgap measurements on either intrinsic or intentionally doped GaAs.

The results demonstrate that stimulated Raman gain spectroscopy provides excellent sensitivity when applied to bulk material. It is characterized by stringent selection rules that make it an ideal candidate for sample orientation and strain measurements. In future applications, we intend to exploit its high sensitivity and rapid response capability, combined with the electric field sensitivity of forbidden scattering configurations, to probe scattering from semiconductor interface regions. For example, in a Schottky barrier structure, an applied voltage generates a large electric field in the small depletion region between the metal and semiconductor. Scattering in this region will be subjected to electric field induced perturbations. Thus signals originating from the interface region can be extracted from the larger bulk signals by using detection synchronous with a modulating field. A typical sample configuration is shown in Figure 4. The electric field will occur between the metal surface and a shallow depletion region (cross-hatched in Figure 4) within the semiconductor. Both pump and probe beams are incident on this region and the probe beam reflected from the junction is detected using a lock-in amplifier.

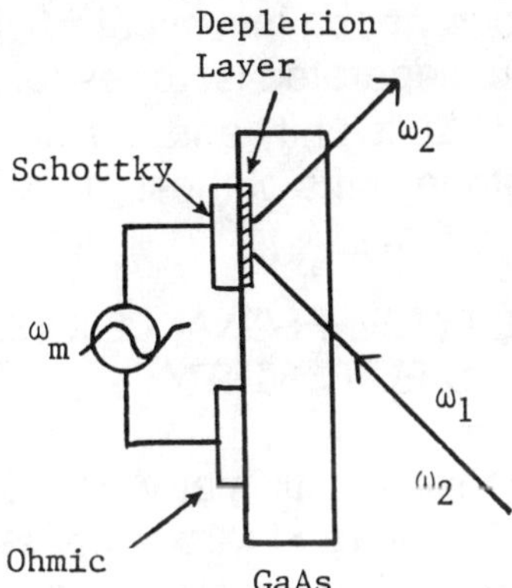

Fig. 4. Sample for Schottky barrier interface studies by electric field modulated SRGS.

Experiments to date on interface modulation demonstrate that electrically modulated signals can be recovered from surface layers on GaAs samples, however, problems associated with near bandgap absorption induced interfering background signals. We intend to implement wavelength modulation procedures in order to minimize this contribution and to achieve the desired high sensitivity for interface layers.

REFERENCES

1. B. G. Levine and C. G. Bethea, IEEE JQE-16, 85 (1980).
2. J. P. Heritage, Appl. Phys. Letters, 34, 470 (1979).
3. R. M. Martin, Phys. Rev. B4, 3676 (1971).
4. J. Menendez and M. Cardona, Phys. Rev. B31, 3696 (1985).

HIGH RESOLUTION, MASS RESOLVED SPECTRA OF RARE ISOTOPES

C. M. Miller, R. Engleman, Jr., and R.A. Keller
Los Alamos National Laboratory, Los Alamos, NM 87545

ABSTRACT

Resonance ionization mass spectrometry is used to acquire high-resolution optical spectra of rare isotopes. Hyperfine spectra of the ${}^2D^o_{3/2}$ ' ${}^2D_{3/2}$ lutetium transition at 22 125 cm^{-1} are presented for ${}^{173-176}$Lu. ^{173}Lu and ^{174}Lu are rare isotopes whose optical spectroscopy has not previously been investigated. We required only tens of picograms of these unseparated isotopes for our studies. We confirmed that the spin of ^{174}Lu is unity and derived values for the hyperfine constants and isotope shifts of these nuclei.

ANALYTICAL CAPABILITY OF RESONANCE IONIZATION MASS SPECTROMETRY

Resonance-ionization mass spectrometry (RIMS) can be used to obtain high resolution optical spectra of rare isotopes without prior chemical or isotopic separation. The sensitivity and selectivity of RIMS is documented by the recent analysis of the isotopic composition of a lutetium sample. A 60 ng sample containing trace amounts of ^{173}Lu (–10^8 atoms) and ^{174}Lu, and a thousandfold excess of the isobaric interfering isotopes ^{173}Yb and ^{174}Yb was analyzed to yield (1):

$$ {}^{173}\mathrm{Lu}/{}^{175}\mathrm{Lu} = (0.44\pm0.07)\times10^{-6} $$
$$ {}^{174}\mathrm{Lu}/{}^{175}\mathrm{Lu} = (0.36\pm0.01)\times10^{-5} $$

The relative abundances of ^{173}Lu and ^{174}Lu was confirmed by radioactive counting. It is noteworthy that the analysis using RIMS was completed in hours while the radioactive counting required several days. The efficiency of the RIMS process is –10^{-4} (ions counted/atoms in the sample).

The photoionization of lutetium, made with a broadband laser, was

0094-243X/86/1460642-4$3.00

element selective but not isotope selective. Isotope ratios were measured by magnetic dispersion in the mass spectrometer. Large dynamic range isotope measurements are often limited by space charge effects or scattering from the predominant isotope. We believe that the range of isotope ratio measurements could be increased by using a single frequency laser to affect isotopically selective photoionization of the minor isotope. In order to accomplish this task, it is necessary to know hyperfine splittings and isotope shifts of rare and sometimes highly radioactive isotopes. We demonstrate that this data can be obtained without prior chemical or isotopic separation, by tuning the mass spectrometer to the isotope of interest and sweeping a single frequency dye laser through the spectral range of interest. In addition to our analytical interests, the high resolution spectra of rare nuclei can lead to interesting information about the structure of the nucleus.

HIGH RESOLUTION SPECTROSCOPY

High resolution spectra of the ${}^2D^o_{3/2}$, ${}^2D_{3/2}$ lutetium transition at 22 125 cm^{-1} for ${}^{173-176}Lu$ are shown in the accompanying figure. One to ten microgram samples containing tens of picograms of ${}^{173}Lu$ and ${}^{174}Lu$ were deposited from solution onto a rhenium filament and then atomized by heating the filament to – 1500 K in the mass spectrometer. A cw laser (50 mW) was focused to a 50 µm waist several mm above the filament. Photoionization was accomplished by scanning the laser across the transition to the resonant intermediate state; excited atoms were then ionized by photons of the same wavelength. Details of the experiment are given in Ref. 2.

A least squares procedure was used to analyze the spectra taking into account the known intensities of the hyperfine components as modified by laser saturation (2,3). The resulting parameters are listed in the table. Parameters determined in this work are underlined. Of particular interest is the confirmation that the nuclear spin of ${}^{174}Lu$ is unity. Previous to our measurements, this value was uncertain (4). Note at the laser power used in these measurements, saturation broadening is appreciable.

APPLICATION TO ISOTOPE SELECTIVE IONIZATION

It is interesting that although isotope shifts are less than the Doppler widths, partially selective ionization can be accomplished by taking advantage of the different hyperfine patterns. Line broadening could be decreased by using less power to saturate the resonant intermediate state and a second, high power laser at a different wavelength for ionization. Doppler free ionization using counter-propagating laser beams is presently under investigation; it is expected that this will improve the isotopic selectivity of photoionization by several orders of magnitude.

PARAMETERS FOR THE LUTETIUM, $^2D^o_{3/2}$ ' $^2D_{3/2}$ TRANSITION

ISOTOPE	176	175	174	173
NUCLEAR SPIN	7	7/2	1	7/2
HALF LIFE (y)	$3.6\text{x}10^{10}$	STABLE	3.3	1.4
DOPPLER WIDTH (cm^{-1})	0.036	0.036	0.036	0.036
LORENTZ WIDTH (cm^{-1})	0.050	0.050	0.050	0.050
μ (nm)[a]	3.139	2.3799	1.94	2.34
Q (barns)	8.0[a]	5.68[a]	0.1+0.5	5.7±0.6
ISOTOPE SHIFT	0.01	0.00	-0.009	-0.037

[a] Ref. 4.

REFERENCES

1. N. S. Nogar, S. W. Downey, and C. M. Miller, Resonance-Ionization Spectroscopy 1984, G. S. Hurst and M. G. Payne, eds. (Institute of Physics, Bristol, England, 1984), p. 91.
2. C. M. Miller, R. Engleman, Jr., and R. A. Keller, J. Opt. Soc. Am. B. 2, 1503 (1985).
3. R. Engleman, Jr., R. A. Keller, and C. M. Miller, J. Opt. Soc. Am. B. 2, 897 (1985).
4. C. M. Lederer and V. S. Shirley, eds., Tables of Isotopes, 7th ed. (Wiley, New York, 1978).

HYPERFINE STRUCTURE OF LUTETIUM ISOTOPES

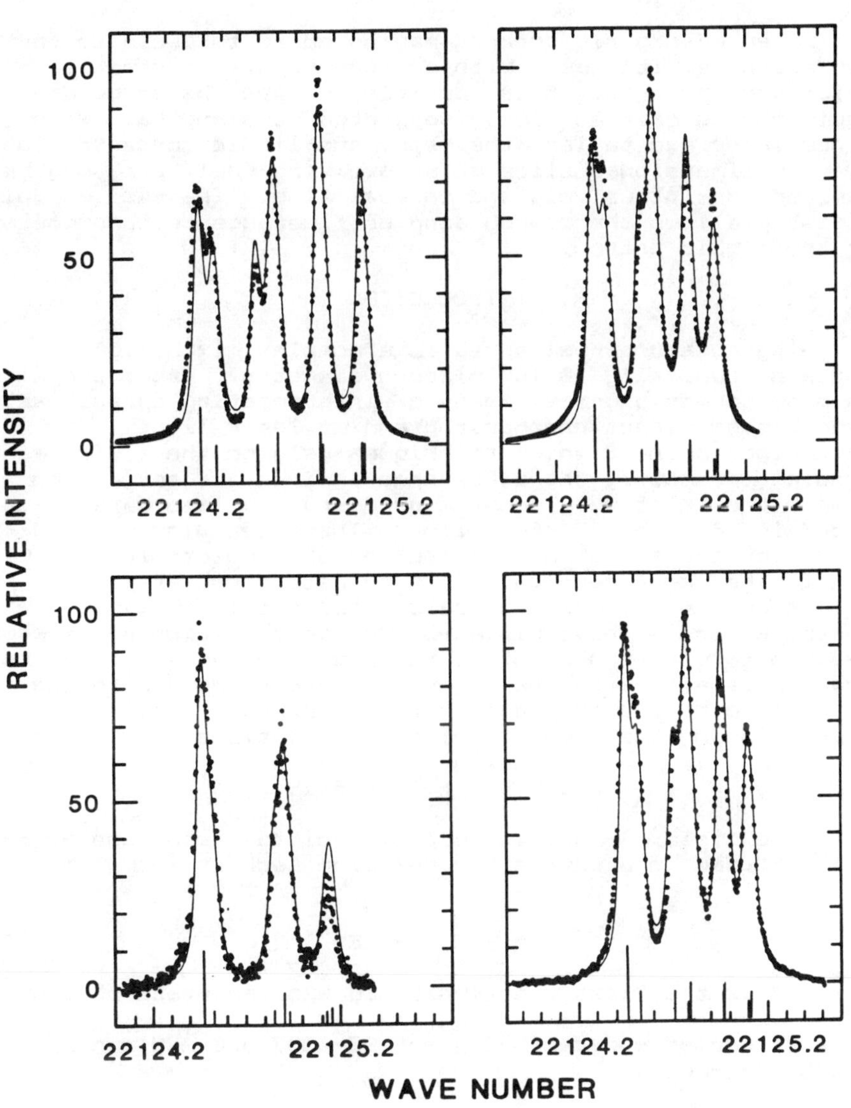

EXCIPLEX-BASED DIAGNOSTICS FOR FUEL SPRAYS*

L. A. Melton and A. M. Murray
University of Texas at Dallas, Richardson, TX. 75080

J. F. Verdieck
United Technologies Research Center, East Hartford, CT. 06108

ABSTRACT

Selected organic dopants may be added to fuels to form fluorescent exciplexes. With an appropriate combination of exciplex components, it is possible to have the vapor and liquid phases emit at widely separated wavelengths. With planar laser excitation, the vapor and liquid concentrations in a two-dimensional slice of an evolving fuel spray may be examined non-intrusively and in real-time. The same exciplex chemistry allows the compounding of fluorescence thermometers for the liquid phase.

INTRODUCTION

A hydrocarbon fuel spray is a complex mixture of droplets (typically 10-100 microns diameter), vapor and air. The most common physical means of interrogating sprays has been through direct photography and/or laser light scattering, both of which techniques rely on the a single physcial parameter, the difference in index of refraction between the relatively dense liquid and the less dense vapor/air, for their information. These techniques yield the number density and size distribution of droplets in the spray. The vapor density, if it is determined at all, results from subtraction of calculated total liquid concentrations. The exciplex-based visualization systems are complementary to the light scattering techniques: they are directly sensitive to the vapor concentrations and to the temperature of the droplets, but they provide little information on the droplet size distributions.

EXCIPLEX PHOTOPHYSICS

An exciplex -- shortened from excited state complex -- may be formed in a reversible reaction such as shown in eqn.(1),

$$M^* + G \longleftrightarrow E^* \tag{1}$$

where M* is the first electronically excited state of the

*Work performed under AFOSR grant 83-0307 and ARO contract DAAG29-84-C-0010.

monomer M and G is an apropriately chosen ground state molecule, and E* is the exciplex. E* is bound with respect to separated M* and G, and there is very little interaction between ground state M and G; thus the E* emission is necessarily red-shifted with respect to the M* emission, a shift which may be 100-200 nnm. In the liquid, the concentration of G may be set sufficiently high that the dominant emission is from the liquid -- by a factor of 100 perhaps -- is from E*. In the vapor, the concentrations are much lower and the relatively polar exciplex is less stable, and hence, the dominant emission from the vapor is from M*, again perhaps by a factor of as much as 100. These two arguments taken together mean that a filter which suppresses the monomer emission and passes the exciplex emission allows one to photograph the liquid only and a filter which suppresses the exciplex emission and passes the monomer emission allows one to photograph the vapor only.

The reaction in equation (1) is temperature dependent, either through the viscosity of the fuel at low temperatures or through the temperature dependence of the equilibrium constant at high temperatures. Thus the ratio of exciplex emission to monomer emission, as seen in the changing shape of the fluoresence spectrum, allows one to measure the temperature in the liquid.

RESULTS AND DISCUSSION

An exciplex-based vapor/liquid visualization system containing 2.5% naphthalene(G)/ 1.0% N,N,N',N'-tetramethyl-p-phenylenediamine(M)/96.5% hexadecane (w/w) was excited with the fourth harmonic of a Nd:YAG laser (266 nm, 50 mJ/pulse) in a fuel spray into nitrogen at approximately 200 C.[1] The photographs showed that the liquid and vapor phases could be separtely measured. With high speed black and white film (ASA 2000) , a single laser pulse gave adequate exposures. Since thee fluorescence lifetimes of M* and E* are less than 100 nanoseconds, the system has an effective shutter speed of less than 1 microsecond, and the droplet motion is frozen.

Fluorescence thermometers based on the naphthalene /N,N,N',N'-tetramethyl-p-phenylenediamine system (145-265 C) and on a 3,10-dicyanophenanthrene/diethyaniline system (25-140 C) have been demonstrated.[2] Within the stated temperature ranges, if the ratio of the exciplex fluorescence intensity to that of the monomer can be measured with 1% accuracy, then the temperature can be determined within 1 C.

REFERENCES

1. L. A. Melton and J. F. Verdieck, Combustion Sci. and Tech. 42, 217 (1984).
2. A. M. Murray and L. A. Melton, Appl. Opt. 24, 2783 (1985).

LASER-PHOTOACOUSTIC SPECTROSCOPY OF HUMID AND POLLUTED AIR

M.W. Sigrist, St. Bernegger, J. Hinderling, and P.L. Meyer
Physics Department, ETH, CH-8093 Zurich, Switzerland

ABSTRACT

The application of laser-photoacoustic spectroscopy (PAS) to the measurement of the weak IR water vapor absorption as well as to air pollution monitoring is discussed. Good agreement between experimental data on the pressure- and the temperature dependence and theoretical predictions of a H_2O dimer model is found for the water vapor continuum absorption. In addition four weak absorption lines could be assigned for the first time as pure rotational H_2O transitions. Both a CO laser-based and a CO_2 laser-based PAS system are applied to the monitoring of gaseous air pollutants. Results on the sensitivity and the dynamic range of the method are presented. Experimental PAS spectra of specific species are compared to literature data. First examples of gas analysis on the basis of single spectra are discussed.

INTRODUCTION

The 8 to 14 μm atmospheric window exhibits a water vapor continuum absorption which still eludes a complete theoretical explanation. The far-wing absorption of present line-shape models does not yield a satisfactory agreement with the measured continuum absorption [1]. Alternative hypotheses for the explanation of the continuum include equilibrium water dimers with a strong librational band centered at 780 cm^{-1}, non-equilibrium water dimers and large aggregates of water vapor.

In addition to the continuum absorption water vapor also exhibits a number of weak absorption lines within the 8 to 14 μm window. Until recently, a strict identification of these lines by comparison with accurate experimental results, has not been published.

Today, the monitoring of various specific air pollutants is of great interest for the understanding of the numerous chemical and physical processes occurring in the terrestrial atmosphere which finally result in the formation of smog, acid rain etc. Apart from conventional methods, more recently also laser techniques have been applied to this type of studies [2-4]. In this report we concentrate on the opto- or photoacoustic detection scheme [2,3,5] which has been demonstrated to permit measurements of minimum absorption coefficients $\alpha_{min} \gtrsim 10^{-10}$ cm^{-1} at a pathlength of only 10 cm.

PHOTOACOUSTIC SPECTROSCOPY (PAS) WITH LASERS

The photoacoustic effect is based on the conversion of the excitation energy into translational energy of the absorbing atoms or molecules. This is manifested as a pressure change Δp in a closed system and detected with a microphone. In general one finds

$$\Delta p(\lambda) \text{ prop. } N\sigma(\lambda)P_0 \tag{1}$$

where N is the number density of absorbing species, $\sigma(\lambda)$ their absorption cross section at the wavelength λ and P_0 the laser power in the medium. Thus absorption spectra are obtained by measuring Δp versus λ of the amplitude-modulated laser beam. If $\sigma(\lambda)$ is known, the concentration N can be deduced directly from the PA spectrum.

In comparison to other absorption techniques, PAS offers several advantages [6]. The high sensitivity allows the measurement of weak absorptions even with short pathlengths which is important for getting reliable data for our water vapor studies. With respect to the pollution monitoring,PAS permits the detection of trace constituents at the ppb or even subppb level. The dynamic range includes up to eight orders of magnitude, i.e. the method is applicable for the detection of pollutants close to and far away from the source of pollution. Specific detection of numerous gases with one system is achieved by consideration of the different absorption spectra of the gases within the given emission range of the laser.

WATER VAPOR ABSORPTION STUDIES

The experimental setup for the studies on the pressure- and temperature dependence of the water vapor absorption has been described elsewhere [7]. A step-tunable cw CO_2 laser is used as excitation source. The mechanically chopped radiation is directed through an acoustically resonant PA cell which contains a water vapor/nitrogen mixture at a well-defined partial pressure and temperature. The total pressure is kept at ca. 1 atm. The controversial continuum absorption has been measured at different CO_2 laser transitions over a wide range of partial pressures and for temperatures down to -20°C. Examples have been presented previously [7,8]. In all cases good agreement is obtained with a theoretical fit which is based on the water dimer model by Kassner et al. [9]. From the measured temperature dependence an average dimer binding energy $\langle E_2 \rangle = -6.2$ kcal/mole is deduced which is in reasonable agreement with ab-initio calculations [10]. The conclusion that dimers have to be taken into account for the explanation of the observed negative temperature dependence of the continuum absorption is supported by our previous measurements in supersaturated vapor which yielded an increased continuum absorption at the maximum of supersaturation [11].

In contrast to the continuum absorption, the line absorptions measured at the near-coincident 9R(36), 9P(38), 10R(20) and 10P(40) laser transitions exhibit a positive temperature dependence. Our detailed studies that have been discussed elsewhere [7,12] permitted the identification of these four absorption lines as pure rotational transitions of the H_2O molecule.

AIR POLLUTION STUDIES

Both a CO- and a CO_2-laser PAS system have been set up for the trace gas detection [13-16]. A special PA cell has been constructed which yields a sensitivity which corresponds to $\alpha_{min} \simeq 10^{-8}$ cm^{-1} or e.g. to a minimum detectable ethylene (C_2H_4) concentration in N_2 of 0.5 ppb. For the case of C_2H_4/N_2 mixtures the PA signal has been found proportional to the C_2H_4 concentration from <1 ppb to >100 ppm. Our cell can be operated with flowing gaseous mixtures with flow rates up to 1 ℓ/min. Thus, on-line measurements can be performed. First PA spectra have been taken from numerous gases with certified trace gas/nitrogen mixtures. In general good agreement is obtained with experimental literature data or with own calculations on the basis of AFGL trace gas compilation data [13]. As a first example of a gas analysis, the PA spectrum of a mixture of HNO_3 and H_2O vapor with N_2 as carrier gas, taken with the CO laser system, has been discussed recently [16]. HNO_3 is known to decompose into NO_2, H_2O and O_2. The complete PA spectrum has been analyzed on the basis of the individually measured spectra of HNO_3, H_2O, NO_2 and NO. A good fit with the complete PA spectrum was obtained with these four gases and the individual gas concentrations could be deduced from this fit. Other examples include studies on air samples taken from a car parking garage as well as from car exhausts.

A PA spectrum of the garage air taken with the CO_2 laser system is presented in Fig. 1.

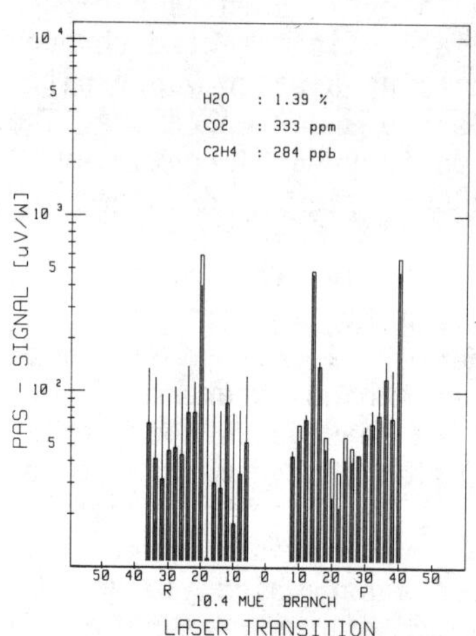

Fig. 1. CO_2 laser PA spectrum (10.4 μm branch) of an air sample from a car parking garage. Experimental conditions: P_{tot} = 942 mb, T = 300 K.
—— experimental data
══ calculated data

The peak at 10P(14) originates from C_2H_4 absorption whereas the two dominant peaks at the 10R(20) and 10P(40) laser transitions are due to water vapor absorption. A first analysis on the basis of single PA spectra including phase considerations yields the concentrations given in the Figure for H_2O vapor, CO_2 and C_2H_4. Although PA spectra of additional trace gases have to be taken into account in order to fit the complete spectrum in this case, the analysis of gas mixtures is feasible with this technique. However, our main aim is to monitor specific pollutants on-line by performing measurements at selected laser transitions which are characteristic for the gas under study and which do not exhibit an uncontrolled interference from other gases. For in situ studies the CO_2 laser-PAS system is now installed in a trailer. First measurements concern the continuous monitoring of C_2H_4 at a main road.

ACKNOWLEDGMENTS

Our studies are supported by the Swiss Natl. Science Foundation (Natl. Research Program 14), the GRD of EMD and ETH Zurich.

REFERENCES

1. M.E. Thomas and R.J. Nordstrom, J. Quant. Spectr. Radiat. Transfer 28, 81 (1982)
2. E.D. Hinkley, ed., "Laser Monitoring of the Atmosphere" (Topics in Appl. Phys., vol. 14, Springer Verlag, Berlin, 1976)
3. C.K.N. Patel, Science 202, 157 (1978)
4. A.I. Carswell, Can. J. Phys. 61, 378 (1983)
5. Yoh-Han Pao, ed., "Optoacoustic Spectroscopy and Detection" (Academic Press, N.Y., 1977)
6. M.W. Sigrist and J. Hinderling, Proc. Int. Conf. LASERS'83, San Francisco; R.C. Powell, ed. (STS press, McLean, VA (USA), 1985), p. 115.
7. J. Hinderling, M.W. Sigrist and F.K. Kneubühl, Proc. Int. Conf. LASERS'84, San Francisco; K.M. Corcoran, D.M. Sullivan and W.C. Stwalley, eds. (STS press, McLean, VA (USA), 1985), p. 634
8. M.W. Sigrist and J. Hinderling, Digest 4th Int. Topical Meeting on Photoacoustic, Thermal and Related Sciences, Ville d'Estérel (Canada), August 1985, paper MC 4
9. S.H. Suck, A.E. Wetmore, T.S. Chen and J.L. Kassner, Jr., Appl. Opt. 21, 1610 (1982)
10. G. Finger and F.K. Kneubühl, in "Infrared and mmWaves"; K.J. Button, ed., Vol. 12, Part II (Academic Press, N.Y. 1984), p. 145
11. J. Hinderling, M.W. Sigrist and F.K. Kneubühl, J. de Phys.-Colloque C6, 559 (1983)
12. J. Hinderling, M.W. Sigrist and F.K. Kneubühl, Infrared Phys. 25, 491 (1985)
13. S. Bernegger, P.L. Meyer and M.W. Sigrist, Helv. Phys. Acta 58, 829 (1985)
14. P.L. Meyer, S. Bernegger and M.W. Sigrist, Helv. Phys. Acta 58, 833 (1985)
15. M.W. Sigrist, S. Bernegger and P.L. Meyer, see Ref. 8, paper MC8
16. M.W. Sigrist, S. Bernegger and P.L. Meyer, Can J. Physics, to be published 1986.

Light Scattering Diagnostics
Applied to High Voltage Spark Discharges*

Alexander Scheeline, M. J. Zoellner, and M. A. Lovik
School of Chemical Sciences, University of Illinois at Urbana-Champaign
Urbana, IL 61801

ABSTRACT

The atmospheric-pressure high-voltage spark, used for elemental analysis of alloys, gives rise to working curves whose shape and precision vary with alloy family and metallurgical history. Raman, Mie, and Thomson scattering diagnostics are of use in understanding these variations by revealing changes in small molecule formation and diffusion, particulate condensation, and free electron behavior. The high background emission levels and radio frequency noise produced by the spark complicate measurement of the desired scattering signals. Work reported includes approaches taken to circumvent the measurement difficulties, and includes spatially - and temporally-resolved data on nitrogen diffusion into the spark channel, particulate condensation, and particulate trajectories.

INTRODUCTION

Although the spark discharge is used for emission spectrochemical analysis, much of its behavior can be understood only when non-emitting species are mapped. Raman and Mie scattering have been used to observe molecular nitrogen and condensed particulates respectively, in an attempt to understand both material transport and energy exchange within the spark and between the spark and its environs.

EXPERIMENTAL

Raman measurements were made using a frequency doubled Nd:YAG laser as primary light source with a Spex 0.5 meter spectrometer and R928 photomultiplier as disperser and detector. Gated photon counting with a Tektronix 7854 oscilloscope acting as amplifier/discriminator was used for data logging. Baffling and aperture stops were sufficient to limit stray light at mid-gap to less than 1 detected photon per 600 laser firings. Mie measurements used a 2 mW HeNe laser as primary light source, a 0.35 m Heath Czerny-Turner spectrometer as disperser, and the Tektronix oscilloscope in single sweep mode to obtain time-resolved measurements. Stigmatic optics allowed for spatial resolution of 0.05 mm for the Mie Scattering experiment; resolution was 0.2 mm for the Raman experiment.

*Supported in part by NSF Grants CHE 81-21809 and CHE 81-16702

RESULTS AND DISCUSSION

Initial Raman measurements on the N_2 vibrational band at 2330 cm^{-1} shift were made for argon flowing through the spark gap but the discharge not being triggered. Stray light from the interaction of the edges of the laser beam with the electrodes was a problem when the beam center was within 1 mm of either electrode. In the center of a 4 mm gap, complete exclusion of N_2 can be obtained for an argon flow of 18 ml s^{-1} for our particular flowjet geometry. Flow of 9 ml s^{-1} resulted in turbulent mixing of atmospheric gases (specifically N_2) with the argon flow as evidenced by significant scattering (~ 20% of ambient signal) in the center of the argon flow, and increased fluctuations in signals from just outside the previous boundaries of the flow. Gas flow which results in laminar flow with the spark not firing corresponds to conditions giving rise to minimal spark wander when current passes through the gap.

Temporally- and spatially-resolved measurements of nitrogen Raman scattering with the spark running reveal unexpectedly simple behavior. Following spark firing, nitrogen scattering decreases within 2 mm of channel center, and increases at 3 and 3.5 mm from channel center. As time passes, all scattering signals from outside the core argon flow return to their quiescent levels with 200 microseconds of current cessation. This appears to be simply a displacement followed by rediffusion. There is no evidence of the toroidal cloud known to form around the discharge channel; indeed nitrogen appears to have returned to it's pre-spark distribution at times when the torus is still fully formed.

Ratios of Mie scattering to line emission during sparking show vague but discernable trends over the course of several thousand sparks. Both emission and scattering peak during the first few hundred discharges, then subside, then grow again. However, the ratio of scattering to emission gradually declines, then levels off. This indicates that the mechanisms giving rise to particulate formation and to free atom formation are correlated, but that the relative importance of various paths of formation vary at different rates. On the spark axis, the mean size of particulate found over a several minute burn is approximately 4.4 microns, as determined from the angular structure of the forward lobe of the scattered light.

CONCLUSION

It has been shown that the spark acts as a piston, displacing atmospheric constituents. These relax to their original spatial distribution after cessation of discharge current. Particulate velocity falls off within a few millimeters of the discharge axis, with super-micron particles forming on a millisecond time scale.

LITHIUM PLASMA GENERATION FOR PBFA-II ION DIODES*

James K. Rice, Robert A. Gerber, and Gary C. Tisone
Sandia National Laboratories, Albuquerque, NM

ABSTRACT

The selection of lithium as the ion for PBFA II has made the development of a suitable lithium ion source one of the highest priority areas in the light-ion fusion program. Several schemes for producing lithium ions in a suitable diode configuration are being actively pursued. One attractive option is to use the LIBORS laser ionization technique on a thin layer of lithium vapor near the anode. A space- and time-dependent model of this process has been developed to guide system design studies, and experiments are being performed to validate model predictions.

Sandia National Laboratories' advanced light-ion accelerator, PBFA II, is designed to deliver a 30-MV, 5-MA lithium ion beam to an inertial confinement fusion target, which may produce the first laboratory demonstration of ignition (self-heating of the fuel by fusion reactions) if theoretical performance can be attained. Lithium ions were selected for PBFA II for several reasons. Their use allows a relatively large anode-cathode spacing in the diode so that less impedance change occurs during the pulse. Since the first ionization potential of lithium is low relative to the second, a singly-charged lithium ion source without significant contamination from higher charge states appears feasible. Also, lithium ions experience less magnetic deflection in the diode than lighter ions so they can be focused for a longer period of time. Ions heavier than lithium would require operating voltages higher than 30 MV or acceleration of multiply charged ions.

The selection of the lithium ion option for PBFA II has made the development of a suitable ion source one of the highest priority areas in Sandia's light ion beam fusion program. Several source concepts are under active investigation. One leading candidate is the BOLVAPS/LIBORS scheme. BOLVAPS (boil-off lithium vapor source)[1] is a way to produce a thin lithium vapor layer near the anode surface in an ion diode by rapid ohmic heating of a thin-film laminate containing lithium.

*This work was supported by the U. S. Department of Energy under contract number DE-AC04-76DP00789.

This vapor layer is then ionized by LIBORS (laser ionization based on resonance saturation), a laser technique for efficiently producing a singly ionized lithium plasma.

More than 15 years ago, Measures[2] suggested that plasmas could be produced efficiently by exciting the 1s-2p resonance transition in the vapor. In 1977, McIlrath and Lucatorto[3] achieved nearly 95% ionization in a column of lithium vapor with a 1-MW dye laser tuned to the 670.8-nm resonance line. The ionization process proceeds as follows. As the resonance transition is saturated, a large population of lithium atoms in their excited 2p state is produced. A few free electrons are also formed through multi-photon ionization of 2p-state atoms and laser-induced Penning ionization (the collision of two 2p-state atoms in the presence of the laser field). These electrons are heated by superelastic collisions with the resonantly excited atoms so that they become effective in direct collisional ionization and in populating intermediate energy levels that can in turn be ionized by a one-photon process. As the electron density increases, electron collisions can lead to a rapidly accelerating and nearly complete ionization of the vapor.

We have developed a mathematical model to describe the LIBORS ionization process in lithium. This model treats the lithium atom as a five-level system. These five levels are the ground (1s) and first excited (2p) states, an intermediate level at 3.85 eV that represents the effect of the n=3 and 4 levels, a level at 5.16 eV to represent high n-value levels near the continuum, and the ionization continuum, which begins at 5.37 eV. The time-dependent collisional coupling of these levels with each other and with electrons and a description of electron, ion, and neutral specie temperature evolution follows the rate-equation formalism developed over the past several years by Measures and his co-workers.[4] In addition, the present model includes a one-dimensional representation of laser radiation transport and motion of the ionization front coupled with the rate-equation description of ionization and recombination processes. For reasons that will be discussed later, we also included the possibility of using a uv laser in addition to the visible one to enhance the ionization process. The uv laser (e.g., KrF at 248 nm) can ionize a 2p-state atom with a single photon.

The purpose for developing this model was to help guide the definition of laser requirements for producing ionization in useful ion-diode configurations. Thus, we have considered lithium densities that are 1×10^{15} cm^{-3} or greater and vapor thicknesses of 1 cm or less. As described by Measures,[4] when a laser tuned to the resonance transition at 670.8 nm enters a column of lithium vapor, the electron density builds slowly until a critical density is reached, at which time runaway collisional ionization occurs, leading to nearly complete ionization of the vapor. Typically, the time required to reach the critical density, which is usually 1% or less of the total atom density, is very much longer than the ionization burnout phase. The model calculations indicate that the time to reach >90% ionization scales with laser intensity I and atom density D approximately as $I^{-.9}D^{-0.5}$. This result implies that lower power lasers are somewhat less energy efficient in ionizing a given density vapor, and, for a fixed laser power, it takes more laser energy to ionize a lower density vapor. Of course, for sufficiently high vapor densities (or low laser power), the above scaling does not apply since electron-ion recombination limits the maximum ion density to less than 90%.

The burnout ionization front begins at the end of the vapor column nearest the laser and propagates along the laser axis. Initially, the velocity of the front is limited by the rate at which the laser supplies energy to the region of gas undergoing ionization. However, as the front progresses, it slows considerably. This behavior is shown in the figure, which contains calculated distance versus time plots for the position of the ionization front for a column whose initial density is constant.

Propagation of the burnout ionization region for a laser normal to the BOLVAPS lithium vapor source is complicated by the density gradient inherent with this source. Model calculations show a complex distance- and time-dependent growth of the ion density. The low-density region furthest from the anode surface takes much longer to reach full ionization than does the higher density interior region. The ionization rate of the low-density vapor can be greatly enhanced by the addition of 248 nm laser radiation, as shown in the curves of the figure, but the burnout rate at high vapor density is not much affected.

Scaling studies of the type discussed above are continuing in an effort to define optimum laser ionization conditions for various diode configurations.

REFERENCES

1. P. L. Drieke and G. C. Tisone, J. Appl. Phys. (in press).
2. R. M. Measures, J. Quant. Spec. Radiat. Trans. 10, 107 (1970), J. Appl. Phys. 39, 5232 (1968).
3. T. J. McIlrath and T. B. Lucatorto, Phys. Rev. Lett. 38, 1390 (1977).

4. R. M. Measures, N. Drewell, and P. G. Cardinal, J. Appl. Phys. 50, 2662 (1979), Appl. Opt. 18, 1824 (1979), R. M. Measures, P. L. Wizinowich, and P. G. Cardinal, J. Appl. Phys. 51, 3622 (1980), R. M. Measures and P. G. Cardinal, Phys. Rev. A23, 804 (1981).

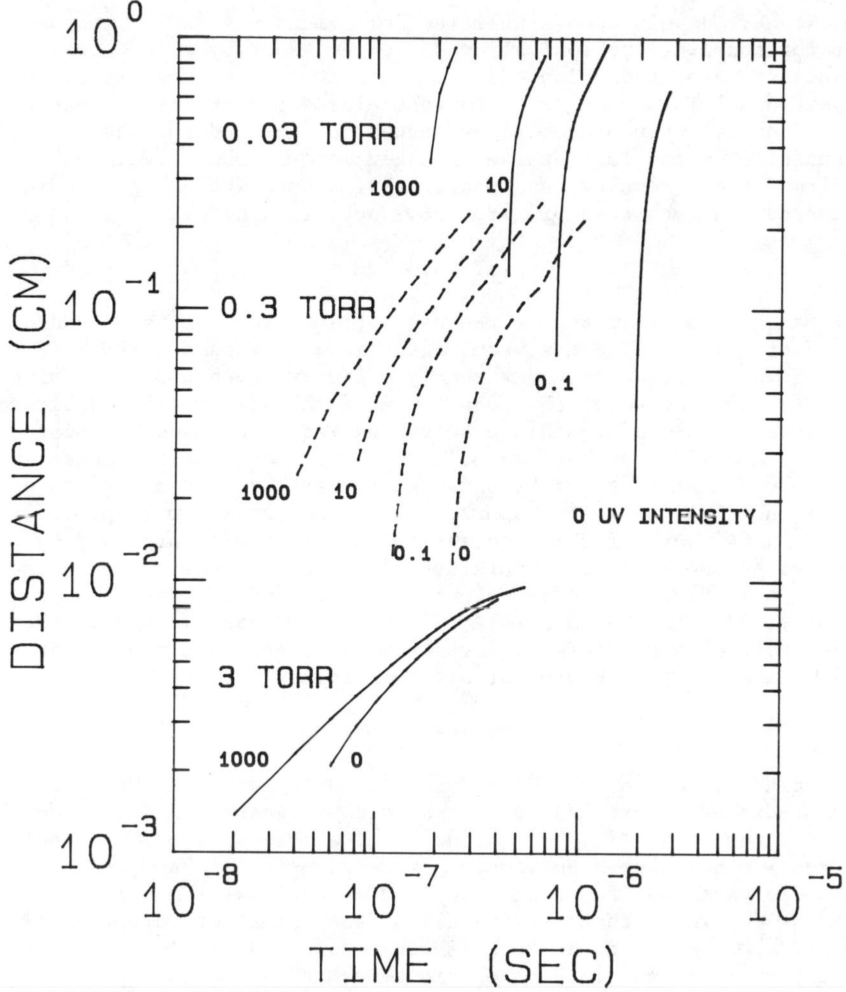

Figure. Propagation distance of the ionization front versus time after the beginning of the laser pulse. Each curve is labelled with the initial lithium vapor pressure and the intensity of the uv laser at 248 nm in kw/cm^2. The intensity of the 670.8-nm laser is 10 kw/cm^2.

LASER RESONANCE SATURATION

R. M. Measures, M. A. Cappelli, R. S. Kissack and S. K. Wong
Institute for Aerospace Studies, University of Toronto
4925 Dufferin Street, Downsview, Ontario, M3H 5T6

ABSTRACT

We have developed a simplified collisional-radiative computer code for modeling the ionization of sodium vapor by a short pulse of intense laser radiation tuned to saturate one of the resonance transitions. This code predicts that plasma temperatures somewhat lower than had been previously expected can be produced when attenuation of the laser pulse is taken into account. We have confirmed these results in a preliminary experiment using a sodium heat sandwich oven that provides 360° optical access.

INTRODUCTION

Saturation of an atomic resonance transition for an extended period of time (at least several lifetimes of resonance level) has been shown to represent an extremely effective method of achieving new total ionization of the constituent involved for atom densities in excess of $10^{15}cm^{-3}$. This observation was first made by Lucatorto and McIlrath [1] in the case of sodium vapor and has subsequently been confirmed many times by numerous researchers for a variety of alkali and alkaline earth vapors. The basic mechanism responsible for the attainment of full ionization was originally proposed by Measures[2] and involves superelastic collisional energy conversion of the laser field into free electron energy and ionization. Recently, Jahreiss and Huber[3] have directly measured the degree of ionization in the case of Strontium vapor and have shown that 100% ionization is achieved at high densities.

THEORETICAL MODELING

Laser ionization based on resonance saturation (LIBORS) has been predicted to proceed in four stages by Measures and Cardinal[4] and Measures et al[5]. The influence of a molecular gas upon this interaction has also been studied by Measures et al[6]. In this work the sodium atom was represented by a 20-energy level model and a code was developed that permits all of the important collisional-radiative processes to be taken into account. Recently, we have shown that it is possible to obtain almost the same results using the simplified 5-energy level model of the sodium atom presented as figure 1. This LIBORS code is used to solve the relevant set of energy and population equations and provides the temporal variation of the densities and temperatures. Representative temporal variations of the free electron density and temperature for two values of laser fluence 189mJ cm^{-2} and 18.9mJ cm^{-2} are presented as figure 2. It is evident that there is a substantial drop in both

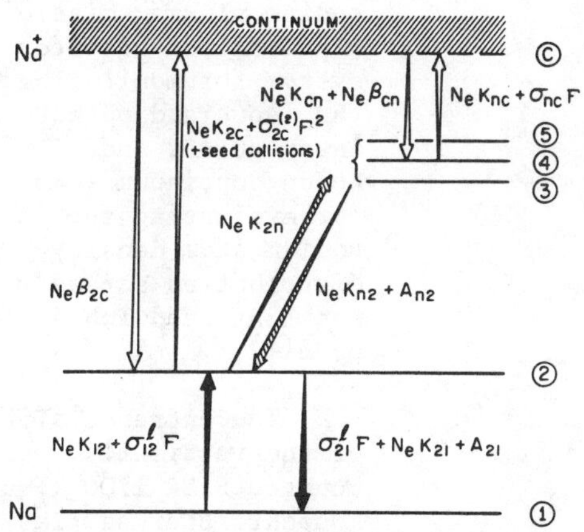

Fig. 1. Five level model of sodium atom used in the multislab LIBORS computer code.

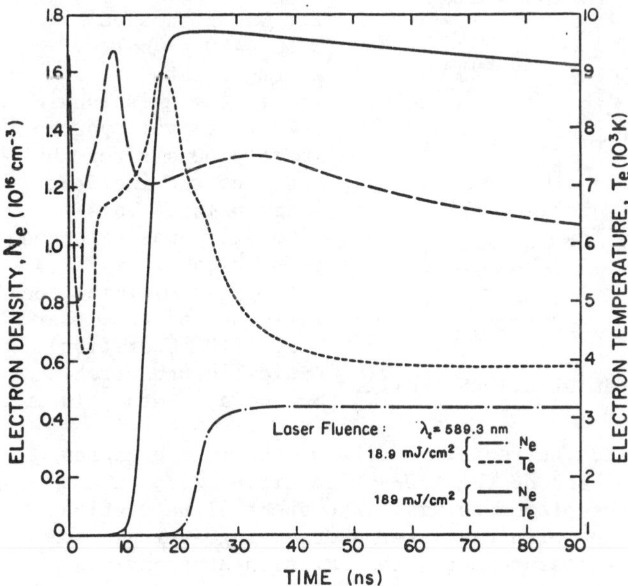

Fig. 2. Temperal variation of free electron density and temperature of sodium vapor of atom density $1.7x10^{16}cm^{-3}$, subject to a 40ns laser pulse of either 189 or $18.9mJcm^{-2}$ fluence tuned to 589.3nm.

the degree of ionization and the free electron temperature associated with the lower laser fluence.

Consequently, we have developed a multislab model of LIBORS that takes account of the energy absorbed from the laser pulse as it propagates through the vapor. This new multislab five level (MSFL) LIBORS code indicates how the degree of ionization and electron temperature can be expected to vary along the path of a given laser pulse allowing for the axial variation of the sodium atom density measured in our experiments(7).

EXPERIMENTAL FACILITY AND RESULTS

Our LIBORS facility is schematically presented as figure 3 and can be seen to comprise: a Nd-YAG laser pumped dye laser, a specially designed sodium heat sandwich oven that provides 360° optical access, a photodetection system for undertaking spectroscopic

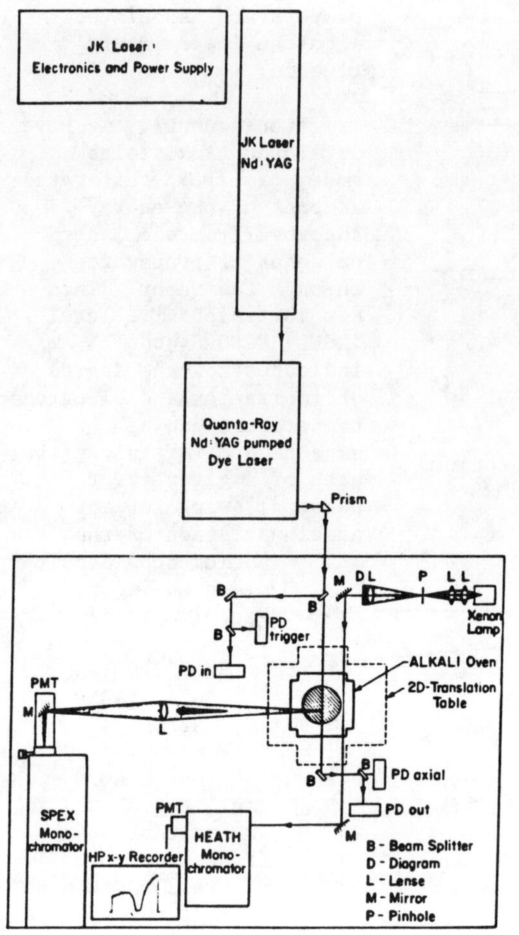

Fig. 3. Schematic diagram of LIBORS facility

measurements on the sodium plasma emission, a photodiode detector system for monitoring the input and output laser pulses and a Xenon continuum lamp for evaluating the sodium atom density distribution across the heat sandwich oven(7).

The entrance slit of the vertically mounted SPEX 1700 II monochrometer samples a small slab of the sodium plasma emission at various heights (y) above the centre line of the laser beam. The signal from the RCA 7265 photomultiplier was processed by an EG&G 4420 signal averager gated to sample the emission in a 2ns interval approximately 60ns after the onset of the incident laser pulse. This time delay was selected to simultaneously attempt to minimize the influence of the laser field (40ns duration), optical depth effects and decay of the plasma.

In order to evaluate the radial profile of the free electron density $N_e(r)$ spectral scans of the 4^2D-3^2P multiplet were taken for 15 heights across the plasma column. The spectral resolution was estimated to be 0.05nm with the laser shots being averaged to constitute one intensity measurement. The emission at such wavelength was deconvoluted to yield the 4^2D-3^2P multiplet spectrum at several radial positions. This allowed the electron density to be determined at each of these positions from a knowledge of the stark broadening and shifting of the lines that comprise the multiplet.

In a preliminary experiment a 37mJ laser pulse was tuned to

the 589.0nm sodium resonance line and the peak sodium atom density was determined to be about $1.7\times10^{16}cm^{-3}$. Four representative 4^2D-3^2P multiplet spectra, corresponding to y=0, 0.5, 1.5 and 2.5mm are displayed by open circles in figure 4. The electron density at

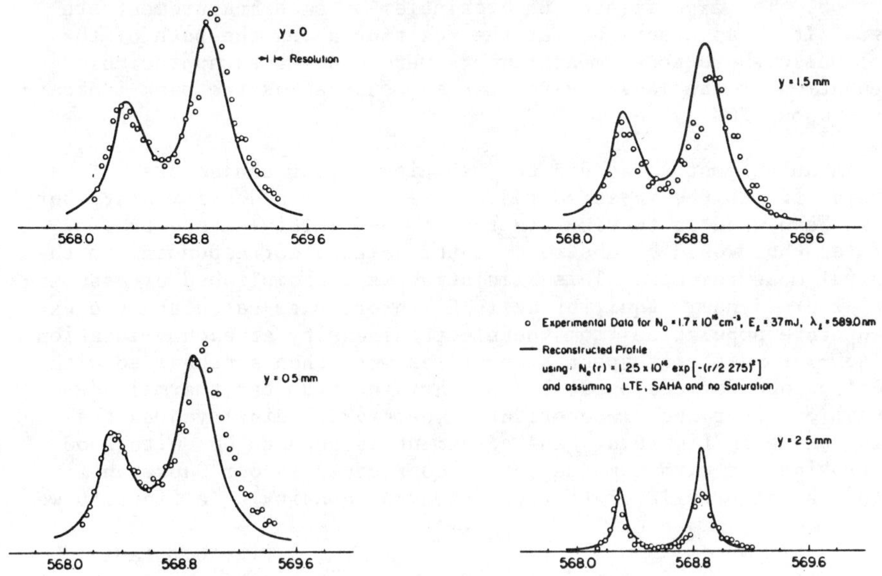

Fig. 4 Comparison of experimentally observed 4^2D-3^2P multiplet spectra (O) at four y-locations with curves predicted on basis of LTE and Gaussian fit to measured free electron density radial profile.

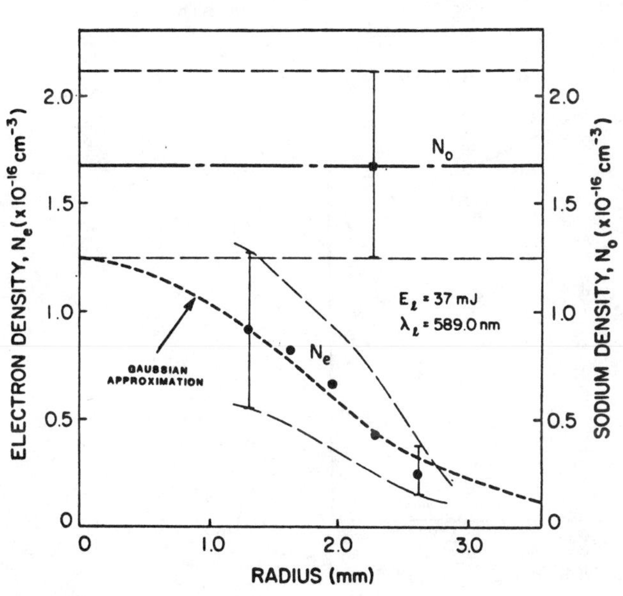

Fig. 5. Radial free electron density N_e profile evaluated from deconvoluted 4^2D-3^2P multiplet spectra. Also shown is the initial sodium atom density N_o and Gaussian fit to the electron density data.

five radial positions, evaluated from the spectral data of this experiment, is presented with the corresponding measured sodium atom density in figure 5. These five electron densities were fitted by the Gaussian radial profile also shown as the broken curve in figure 5. The experimental uncertainties of each measurement are shown. It is apparent that at the position along the path of the laser pulse where these measurements were undertaken appreciable attenuation of the laser pulse must of occurred as the peak ionization is about 70%

In an attempt to assess the validity of our conjectured Gaussian fit to the measured radial free electron density distribution we have used it to simulate the 4^2D-3^2P multiplet spectral profiles that would be observed at the heights corresponding to the original measurements. This simulation was accomplished by assuming Local Thermodynamic Equilibrium (LTE) in order to calculate the excited state population from the electron density at each y-location. The 4^2D-3^2P multiplet spectral profiles were then synthesized with due allowance for the optical depth arising from the thermal 3^2P population. Four such reconstituted spectra are displayed as the solid curves in figure 4. The agreement is seen to be quite good and provides us with some degree of confidence in our Gaussian fit to the radial profile of the free electron density. In figure 6 we

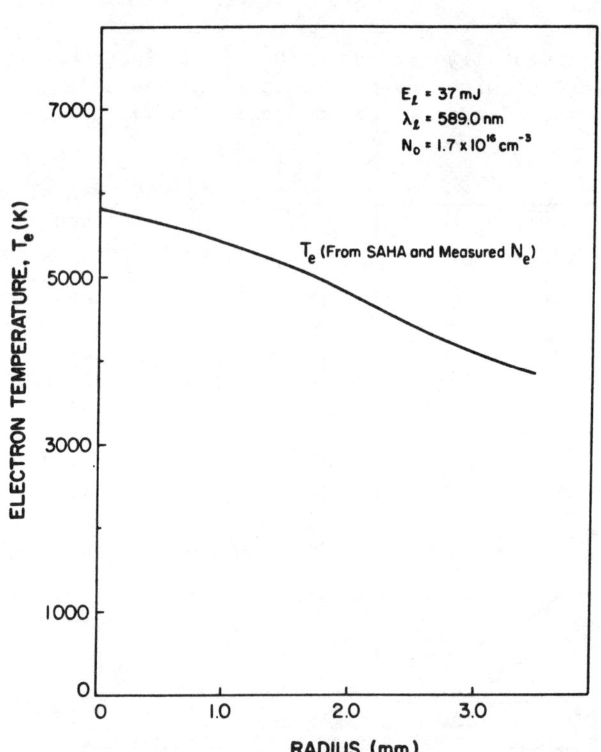

Fig. 6. Radial profile of free electron temperature corresponding to Gaussian fit to free electron density radial distribution and LTE.

have presented the corresponding (LTE) radial profile of the free electron temperature. This temperature is rather lower than expected from earlier computational results but is not too far out of line when allowance is made for attenuation of the laser pulse. Indeed, low electron temperatures are predicted when the laser energy is insufficient to drive the ionization to completion, see figure 2.

ACKNOWLEDGEMENTS

Financial support for this research was provided by the United States Air Force Office of Scientific Research under Grant No. 85-0020 and the Natural Science and Engineering Research Council of Canada.

REFERENCES

1. T. B. Lucatorto and T. J. McIlrath, Phys. Rev. Letters, 37, 428-431, 1976.
2. R. M. Measures, J. Quant. Spectrosc. Radiat. Transfer, 10, 107-125, 1970.
3. L. Jahreiss and M. C. E. Huber, Phys. Rev. A, 28, 3382-3401, 1983.
4. R. M. Measures and P. G. Cardinal, Physical Review A, 23, 804-815, 1981.
5. R. M. Measures, P. G. Cardinal and G. W. Schinn, J. Appl. Phys. 52, 1269-1277, 1981, and 52, 7459, 1981.
6. R. M. Measures, S. K. Wong and P. G. Cardinal, J. Appl. Phys. 53, 5541-5551, 1982.
7. M. A. Cappelli, P. G. Cardinal, H. Herchen and R. M. Measures, Rev. Sci. Instrum. 56, 2030-2037, 1985.

CROSSED-BEAM THERMAL LENS MICROSCOPE

D.S. Burgi, W.A. Weimer, T.G. Nolan, and N.J. Dovichi
Department of Chemistry, University of Wyoming, Laramie, WY 82071

ABSTRACT

The crossed-beam thermal lens technique provides measurements of very weak absorbances with good spatial resolution. By recording the thermal lens signal as a function of position within an inhomogeneous sample, it is possible to form an image based upon the absorbance and thermal diffusivity of the sample.

INTRODUCTION

Thermo-optical techniques are based upon the periodic temperature rise induced within a sample by absorbance of a modulated pump laser beam. Since the index of refraction is a function of temperature, the heated region of the sample will produce a time-varying refractive index perturbation. This refractive index perturbation acts to deflect or defocus a second probe beam which passes through the heated region. Several thermo-optical techniques are based upon a crossed-beam design and produce a measure of sample absorbance within a localized volume[1-3].

CROSSED-BEAM THERMAL LENS MICROSCOPE

The crossed-beam thermal lens provides very sensitive measurements of sample absorbance within a small probed volume[4-11]. In this technique, a modulated pump beam and cw probe beam are crossed at right angles within a sample. The heated sample acts as a cylindrical lens to defocus the probe beam about the plane containing the two beams. The change in the probe beam center intensity is demodulated with a lock-in amplifier and is proportional to sample absorbance. An image is formed by recording the thermal lens signal on a point-by-point basis.

A typical experimental diagram is shown in figure 1. An argon ion laser beam, λ = 514.5 nm and power = 1 mw, is chopped at low audio frequencies with a mechanical chopper. A portion of this pump beam is split to a reference detector while the remainder of the beam is focused into the sample with a 16mm focal length microscope objective which produces a spot size of about 2 micron. The probe beam is produced by a low power helium-neon laser operating at λ = 632.8 nm and focused into the sample with a 18.0 mm focal length microscope objective. After propagating through the sample, the probe beam center intensity is isolated in the plane containing the two beams with a thin, 200 micron, slit. The transmitted light is filtered to remove any scattered pump laser radiation and focused onto a silicon photocell. The output of the reference and signal detectors are sent to a lock-in amplifier for demodulation. The dc signal from the signal detector and the amplitude and phase from the lock-in are digitized and sent to a computer.

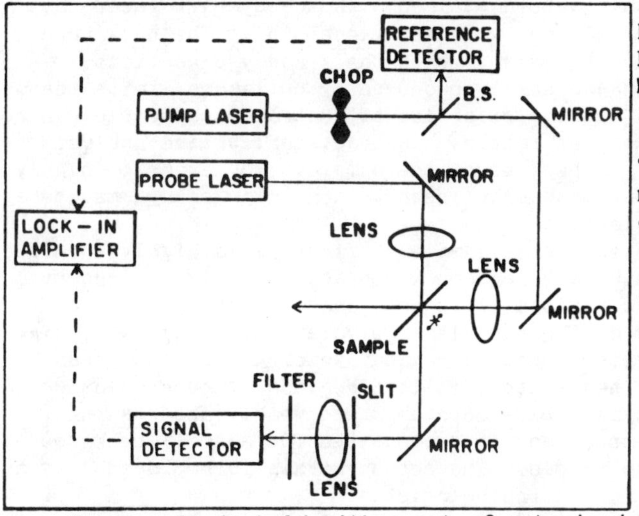

Figure 1. Crossed-Beam Thermal Lens Microscope. Chop is a chopper, B.S. is a beam splitter. The sample is mounted on a pair of motorized translation stages. A computer records the thermal lens signal as a function of position.

The sample is held with a set of motorized, computer driven translation stages. Both the phase and the ratio of the lock-in amplifier voltage with the dc signal detector voltage are recorded point-by-point as the sample is translated in two directions through the intersection region of the beams. By dividing the lock-in signal with the dc voltage, perturbations generated by variations in the sample transmission at the probe beam wavelength are eliminated.

RESULTS AND DISCUSSION

Measurements of small absorbances in small regions of large, inhomogeneous materials will be of interest in a number of disciplines. For example, we have published preliminary data with the scanning laser microscope for a sample of the mineral sphalerite taken from the Creete mine in southern Colorado[6]. Certain regions within the sample were tentatively identified as small fluid filled inclusions with the larger mineral matrix.

Scanning laser microscopy will also be of value to the medical community. Figures 2a and b present the amplitude and phase of the crossed-beam thermal lens signal for a thin, 6 micron, section of

Figure 2 -- AFB Stained Tuberculosis Tissue

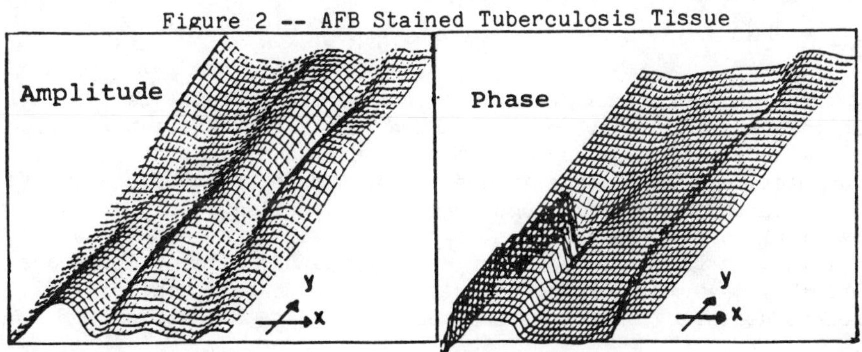

human epidermis which has been stained with the dye AFB (acid fast bacteria) and counter stained with methylene blue. The data were recorded at 5 micron intervals in both the x and y directions.

The wavelength range easily produced by cw lasers limits the spectral information available for identification of specific materials within a complex sample. Instead, qualitative analysis must rely upon other properties of the sample. For example, highly specific chemical reactions which result in a color change may be employed for identification of analyte.

The stain employed on the sample of figure 2 is highly specific for tuberculosis bacteria; regions which are strongly stained by the dye correspond to regions of tuberculosis infection. The dye AFB absorbs strongly at the pump laser wavelength. Regions of high signal in figure 2a correspond to regions which absorb the green pump laser beam and, hence, to localized regions of tuberculosis infection. The methylene blue counter stain weakly absorbs the argon ion pump laser beam and is responsible for a small background signal throughout the sample. The counter stain is employed only as a visualization aid for the pathologist.

The phase of the crossed-beam thermal lens signal is related to the rate of thermal transport within the sample. We have shown that variation in the phase of the signal may be employed to distinguish between regions with different thermal diffusivities in inhomogeneous materials[6,7]. The phase image presented in figure 2b may be due to variations in the thermal diffusivity between normal and tuberculosis infected regions of the sample. For example, the phase shift may be due to either variations in the lipid concentration of normal and infected tissue or differences in the affinity of the tissue regions for the embedding material used to prepare the sample. On the other hand, the variation in phase may be a result of variations in sample thickness caused by a scratch in the microtome which sliced the sample. Presumably, the heat flow in the very thin sample would be highly influenced by sample thickness.

The phase image demonstrates the high spatial resolution of the crossed-beam thermal lens microscope. Phase shifts from $+150^{\circ}$ to -160° are observed over a distance of 5 microns. It appears that the spatial resolution of the phase image is limited by neither the size of the pump beam nor the thermal diffusion distance but instead by the smallest step-size produced by our motorized translation stage system.

CONCLUSION

The primary advantage of the crossed-beam thermal lens for microscopy is the small probed volume, defined by the intersection of the pump and probe beams. On the other hand, conventional transmission measurements integrate the sample absorbance along the optical path. To obtain good spatial information on analyte distribution with conventional techniques, it is necessary to utilize very thin samples. For example, pathological samples are typically 5-10 microns in thickness. Unfortunately, the embedding process necessary to to prepare samples for the microtome is quite tedious. Typically, 24-48 hours are required to prepare the sample. An exciting application of the crossed-beam thermal lens microscope will be for imaging of relatively thick histological samples prepared as frozen sections. These frozen sections are rapidly prepared but difficult to study under the visible microscope. Hopefully, the crossed-beam thermal lens microscope will significantly speed the preparation and study of histological materials for rapid diagnosis of pathological states.

ACKNOWLEDGMENTS

Ms. Paula T. Jaeger of the pathology laboratory, Ivenson Memorial Hospital, Laramie, is gratefully acknowledged for the tuberculosis tissue standard. This work was funded by the donors of the Petroleum Research Fund, administered by the American Chemical Society, and by the National Science Foundation, Grant CHE8415089.

REFERENCES

1. H. Eichler, Optica Acta 24, 631 (1977).

2. A.C. Boccara, D. Fournier, W. Jackson, and N.M. Amer, Opt. Lett. 5, 377 (1980).

3. S.D. Woodruff and E.S. Yeung, Anal. Chem., 54, 1174 (1982).

4. N.J. Dovichi, T.G. Nolan, and W.A. Weimer, Anal. Chem. 56, 1700 (1984).

5. T.G. Nolan, W.A. Weimer, and N.J. Dovichi, Anal. Chem. 56, 1704 (1984).

6. D.S. Burgi, T.G. Nolan, J.A. Risfelt, N.J. Dovichi, Opt. Engin. 23, 756 (1984).

7. W.A. Weimer and N.J. Dovichi, J. Appl. Phys. 59, 225 (1986).

8. W.A. Weimer and N.J. Dovichi, Appl. Opt. 24, 2981 (1985).

9. W.A. Weimer and N.J. Dovichi, Appl. Specto. 39, 1009 (1985).

10. T.G. Nolan, B.K. Hart, and N.J. Dovichi, Anal. Chem. 57, 2703 (1985).

11. T.G. Nolan and N.J. Dovichi, IEEE Circuits and Devices Magazine, 2, 54 (1986).

Linear Imaging of Gas Velocity Using the Photothermal Deflection Effect.*

Robert J. Cattolica and Jeffrey A. Sell**
Combustion Research Facility
Sandia National Laboratories, Livermore, Ca 94550

Photothermal deflection (PD) has been shown to be a viable technique for point measurement of velocity in both liquids[1] and gases.[2,3] There are two different approaches to photothermal deflection velocimetry. One is a transit-time method also called the travelling thermal lens[2-4], where the fluid is heated by a pump laser pulse and a probe laser beam located further downstream is deflected by an index of refraction gradient after a transit-time delay. The velocity is computed from the time delay and the separation of the pump and probe beams. The other method involves measuring the rate at which the probe beam is deflected and returns to the original undeflected position. The velocity can be determined by fitting the rate of deflection and relaxation to either an analytical[2] or numerical[3] expression for the deflection as a function of time.

In this work an experimental technique for the extension of photothermal velocimetry to the simultaneous measurement of a single spatial component of gas velocity along a line is presented. The principle is similar to the single-point method, but a sheet or plane of probe laser radiation is used rather than a beam focussed to a spot. A high-speed, two-dimensional video detector and recording system is used to record the deflection of the probe sheet due to the convection of a temperature perturbation induced by a pump laser pulse. In this experiment a cold laminar jet was studied. The velocity profile across the jet was measured with the photothermal velocimetry technique for a series of volumetric flow rates and compared with the expected theoretical profiles.

EXPERIMENT

The apparatus used in the experiments is shown schematically in Fig. 1. The pump laser was a Lumonics Model 861 excimer laser converted into a grating tuned TEA CO_2 laser operating on the P(14) line at 949.8 cm^{-1}, with a pulse duration of 1 microsecond and energy of 0.03 Joules. The CO_2 beam was focussed with a 250 mm focal length ZnSe lens to a 1mm diameter 2mm above a 6mm diameter nozzle. The gas chosen for absorption of the CO_2 beam was a 3.74% mixture of ethylene in nitrogen. In the experiment the gas flowed through a 6 mm diameter nozzle forming a jet. Four flow conditions were studied: 8.2, 16.3, 32.6, and 40.8 cc/sec.

---

*Research Sponsored by Office of Basic Energy Science, U. S. Department of Energy.

**Visiting Scientist, Physics Department, General Motors Research Laboratories, Warren, MI.

For the velocity imaging technique the planar probe beam was formed by using the 514.5 nm line from a continuous wave Ar^+ laser. After passing through an acoustic-optic modulator to obtain a 1 microsecond light pulse, synchronized with the framing rate of the video detector the beam, was formed into a thin sheet of light by passing it through a spherical lens and a pair of cylindrical lenses (see Fig. 1). The resulting probe sheet was about 7 mm wide and 0.8 mm thick at the nozzle tip. The pump and probe beams crossed at right angles above the nozzle with 1.3 mm separation vertically between them; the pump beam was closer to the jet. After transmitting the experiment the probe sheet passed through an additional 300 mm focal length cylindrical lens onto the high speed video camera.

The detector was a Kodak-Spin Physics SP2000 high-speed video camera and recorder. The camera is a CCD array of Si detector elements, 238 by 192 pixels spaced 30 microns apart. It operated a 12000 frames per second with each image being 1/6 of a full frame (238 x 32 pixels).

RESULTS

The photothermal deflection mesurements shown in Fig. 2, for the 32.6 cc/sec flow, show the centroid position of the probe sheet on the video detector is plotted as a function of the horizontal position across the jet. Each line represents the image of the probe sheet taken every 83.3 microseconds with the video detector. The exposure time is limited to 1 microseconds with the acoustic-optic modulation of the probe laser. The first six images of the probe sheet taken over a 0.5 ms time interval are shown in Fig. 2a. The laser heated gas is just starting to deflect the probe beam in the last image of Figure 2a. The next six subsequent images in Fig. 2b show the completion of the interaction of the line of heated gas as it moves through the plane of the probe beam. The distribution of gas velocity across the jet can be computed directly from the data shown in Fig. 2 using the rate of deflection.

Figure 3 shows velocity distributions computed from the probe laser images for four different gas flow rates ranging from 8.2 cc/s to 40.8 cc/s. In each case the velocity distribution is roughly parabolic as expected for laminar flow. Also, on the vertical scale in Fig. 3 arrows indicate the maximum velocity expected at the jet center computed from the flow rate and the cross sectional area of the jet. The agreement between these values and the peak velocities is good for all four runs. Fig. 3 illustrates the capability of this technique to simultaneously image velocities along a line. Although the present experiment deals with steady flows, this technique would be very useful for imaging in unsteady situations such as those involving pulsed jets.

REFERENCES

1. W. A. Weimer and N. J. Dovichi, Applied Optics, 24, 2981 (1985).
2. H. Sontag and A.C. Tam, Optics Letter, 10, 436 (1985).
3. J. A. Sell, Applied Optics, 24, 3725, (1985).
4. A. Rose and R. Gupta, Optics Letters, 24, 532 (1985).

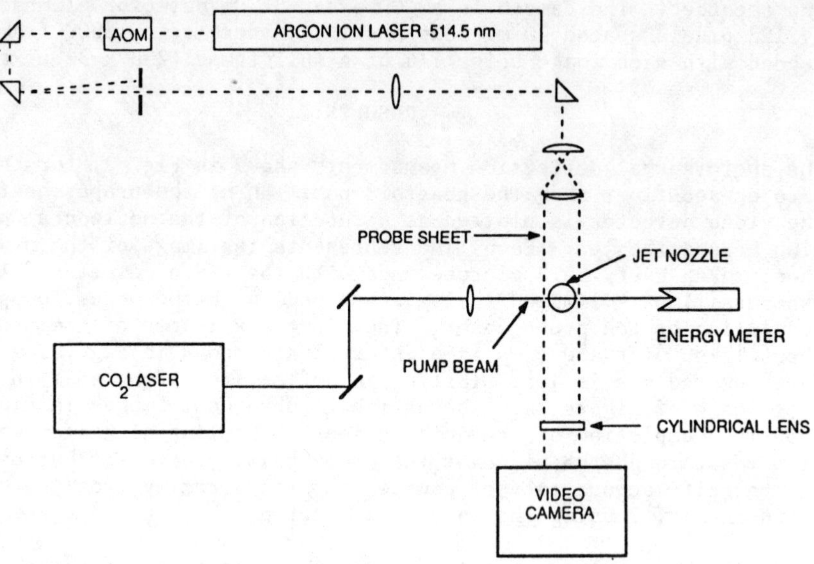

Fig. 1 Schematic diagram of the experiment for linear imaging of velocity using the photothermal deflection technique.

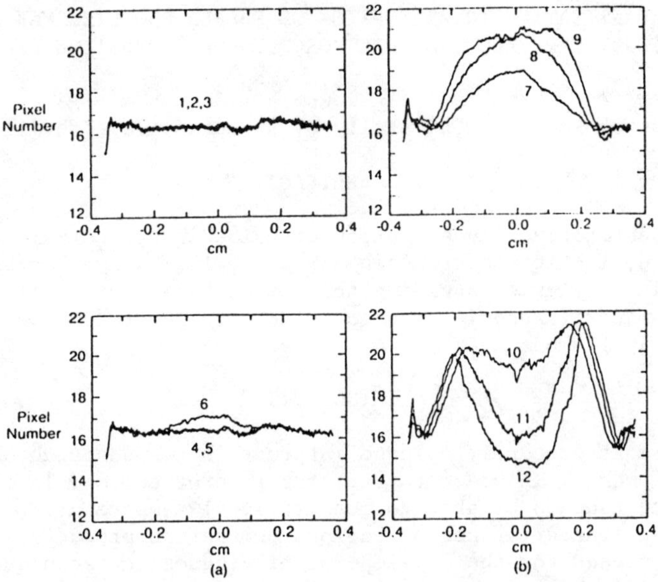

Fig. 2 Vertical pixel distribution of the centroid of the planar probe beam for 12 sequential images after laser excitation of the C_2H_4 - N_2 jet for the 32.6 cc/sec flow condition.

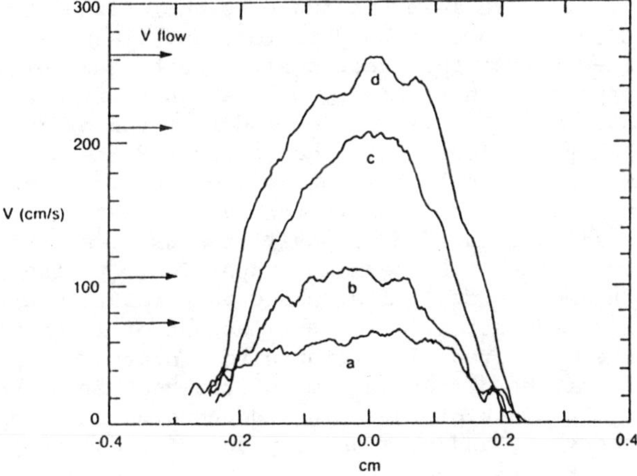

Fig. 3 Gas velocity profiles measured with the planar photothermal-deflection technique for four flow rates: (a) 8.2, (b) 16.3, (c) 32.6, and 40.8 cc/sec. Expected peak velocity denoted by horizontal arrows.

A QUANTITATIVE INVESTIGATION OF PULSED PHOTOTHERMAL AND PHOTOACOUSTIC DEFLECTION SPECTROSCOPY FOR COMBUSTION DIAGNOSTICS

R. Gupta
Department of Physics, University of Arkansas, Fayetteville, AR 72701

ABSTRACT

A quantitative investigation of pulsed photothermal and photoacoustic deflection spectroscopy in a flowing medium has been performed with a view of applying these techniques for absolute measurements of minority species concentration, temperature, and flow velocity in flames.

INTRODUCTION

There is presently a large interest in the research on combustion phenomena. In order to test the theoretical models of combustion, one needs to be able to measure local temperatures, flow velocities, and concentrations of various combustion products. Hence, there is a need for the development of diagnostic techniques capable of measuring these parameters. In the last few years we have made some investigations of photothermal deflection spectroscopy (PTDS) and photoacoustic deflection spectroscopy (PADS) for combustion diagnostics. Both of these techniques are non-perturbing techniques suitable for in situ measurements and have a high degree of spatial and temporal resolution.

The basic idea behind the PTDS technique is quite simple: A dye laser beam (pump beam) passes through the region of interest (flame in this case). The dye laser is tuned to one of the absorption lines of the molecules that are to be detected, and the molecules absorb the optical energy from the laser beam. Due to quenching collisions this energy appears as heating of the dye laser irradiated region, leading to changes in the refractive index of the medium in that region. Now if a probe laser beam, generally a He-Ne beam, overlaps the pump beam, the probe beam is deflected due to the gradient in the refractive index of the medium created by the heating. This deflection can be easily measured by a position sensitive optical detector. The deflection of the probe beam is proportional to the number density of absorbing molecules. If a pulsed pump laser is used, the probe beam gets deflected shortly after the instant of laser firing, as discussed above, and gradually returns to its original position on the time scale of the diffusion time of the heat from the irradiated region. Thus the width of the signal depends on the thermal diffusion time constant, and consequently on the local temperature of the flame. In the presence of a flow, the heat pulse produced by the absorption of the dye laser travels downstream with the medium. The transit time of the heat pulse between two positions of the probe beam a distance Δd apart is measured. The flow velocity is derived simply from $v = \Delta d/\Delta t$, where Δt is the measured transit time.

We have also investigated the closely related PADS technique. The heating of the pump beam irradiated region is accompanied by an

0094-243X/86/1460672-4$3.00

increase in pressure. If the pump beam is pulsed, a pressure pulse travels outward from the pump beam irradiated region. This pressure pulse causes changes in the refractive index of the medium which may be detected by the deflection of a probe beam placed a few mm away from the pump beam. The strength of the PADS signal is proportional to the concentration of the absorbing molecules. Moreover, acoustic velocity in the flame can be measured by measuring the arrival time of the PADS signal after the instant of laser firing, and from a knowledge of the probe-pump-beam separation. Since acoustic velocity depends on the temperature, this information directly yields the local flame temperature. In the presence of a flow, the acoustic velocity is modified. Therefore if two probe beams are used, one upstream and one downstream from the pump beam, then both the flow velocity and the acoustic velocity can be measured simultaneously.

In order to be able to apply these techniques for absolute measurements in a flame, we need to have a quantitative understanding of the size and the shape of the signals in a flowing medium. We have attempted to gain such an understanding by solving the appropriate equations and subjecting the theories to experimental tests. This paper describes some of the theoretical and experimental results.

THEORY

We have derived an approximate analytical expression to predict the shape and strength of the PTDS signals in a flowing medium. Although this expression is approximate, it gives excellent agreement with the exact numerical results.[1] The deflection angle is given by

$$\phi_T(t) = \frac{2\alpha E_o(\partial n/\partial T)}{\sqrt{2\pi} n t_o \rho C_p [4Dx + v_x a^2 + 4Dv_x(t - t_o)]}$$

$$\{[a^2 + 8D(t - t_o)]^{\frac{1}{2}} \exp(-2[x - v_x(t - t_o)]^2/$$

$$[a^2 + 8D(t - t_o)])$$

$$-[a^2 + 8Dt]^{\frac{1}{2}} \exp(-2[x - v_x t]^2/[a^2 + 8Dt])\} \quad (1)$$

In the derivation of Eq.(1) we have assumed that the pump laser propagates along the z-axis and the probe beam propagates along the y-axis. The flow velocity v_x is directed along the x-axis, and the deflection ϕ_T in the x-direction is measured as a function of time t. The separation between the probe and the pump beams is x, and the $(1/e^2)$-radius of the pump beam is a. E_o and t_o are the energy and the duration of the laser pulse. α, n, ρ, C_p, T and D are, respectively, the absorption coefficient, refractive index, density, specific heat, temperature, and the diffusion constant of the medium.

The photoacoustic deflection signal $\phi_A(t)$ has been derived for a static medium since the typical flow velocities in a laboratory flame ($\sim$1 m/s) do not appreciably affect the size and shape of the PADS signal. The deflection angle is given by[2]

$$\phi_A(t) = -\frac{\ell E_o(\partial n/\partial p)}{n(2\pi\varepsilon)^{3/2}}\frac{\alpha\beta}{C_p}\left(\frac{v}{x}\right)^{\frac{1}{2}}\left[\frac{1}{2x}F(\xi) + \frac{1}{v\varepsilon}F'(\xi)\right] \qquad (2)$$

where

$$F(\xi) = \left[M(-\tfrac{1}{4},\tfrac{1}{2},\xi^2)\Gamma(\tfrac{3}{4}) - 2\xi M(\tfrac{1}{4},\tfrac{3}{2},\xi^2)\Gamma(\tfrac{5}{4})\right]e^{-\xi^2} \qquad (3)$$

and

$$\varepsilon = (t_o^2 + a^2/2v^2)^{\frac{1}{2}} \; ; \; \xi = (t - x/v)/\varepsilon \qquad (4)$$

Here ℓ is the interaction length, v is the acoustic velocity, p is the pressure and β is the coefficient of thermal expansion. $F'(\xi)$ is the derivative of $F(\xi)$ with respect to ξ, M is the degenerate hypergeometric function and Γ is the Gamma function.

RESULTS

Our experimental setup is shown in Fig. 1. The pump beam is provided by a flash-lamp-pumped dye-laser (1 μsec duration pulses) and the probe beam is provided by a He-Ne laser. PTDS experiments were performed on an open jet of N_2 seeded with 1000 ppm NO_2. Seeding was required because N_2 does not have optical absorption in the dye-laser spectral range. PADS experiments were performed in a quartz cell filled with N_2 and 1000 ppm NO_2. When the dye laser is fired, a transient deflection of the probe beam is produced

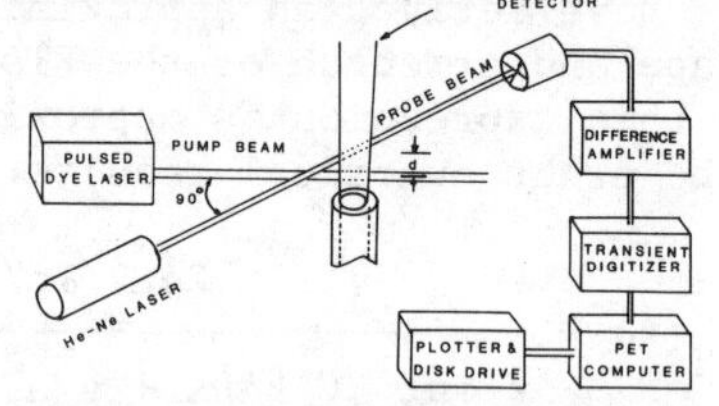

Fig. 1. Experimental setup.

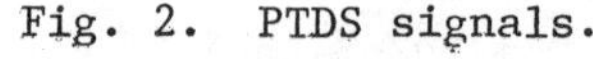

Fig. 2. PTDS signals.

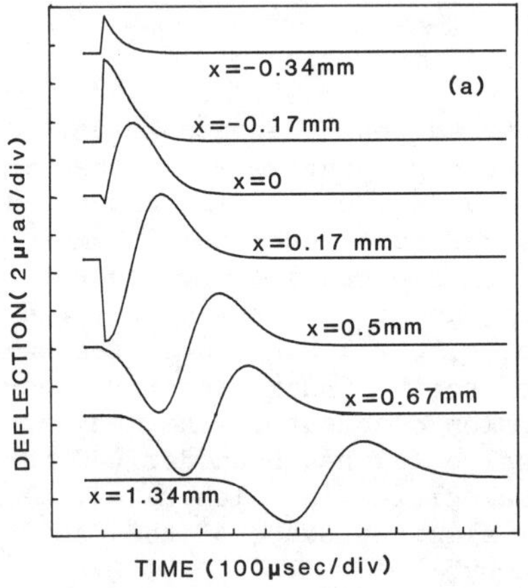

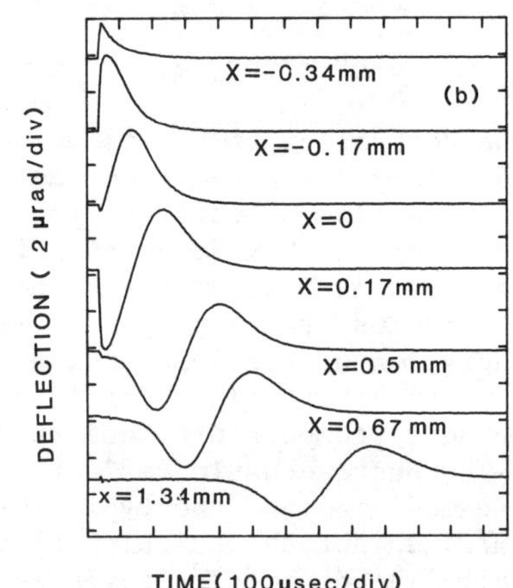

which is detected as a difference signal from a quadrant detector. The distance between the probe and pump beams, x, is variable.

Fig. 2 shows the (a) theoretical and the (b) experimental PTD signals as a function of time for $v_x = 1.96$ m/s with x as a parameter. Different values of x are given on each curve. Negative x represents probe beam upstream from the pump beam and positive x represents probe beam downstream. We note that as the probe beam is moved upstream, the signal becomes smaller and narrower because the heat has to diffuse against the gas flow. As the probe beam is moved downstream, the signal gets stronger at first and then acquires a shape which is essentially the derivative of the spatial profile (assumed Gaussian here) of the pump beam. As x is increased further, the signal becomes smaller and broader due to the thermal diffusion effects. There is an excellent agreement between theoretical prediction and the experimental results. Note that the theoretical results are absolute with <u>no adjustable parameters</u>. Preliminary results of PADS are shown in Fig. 3. The agreement between the theory and the experiment is fair, considering the uncertainties associated with the spatial profile of our pump beam.[2]

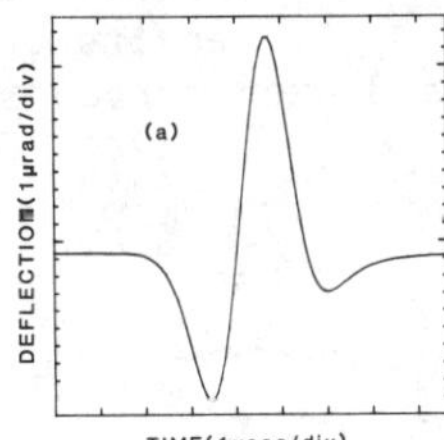

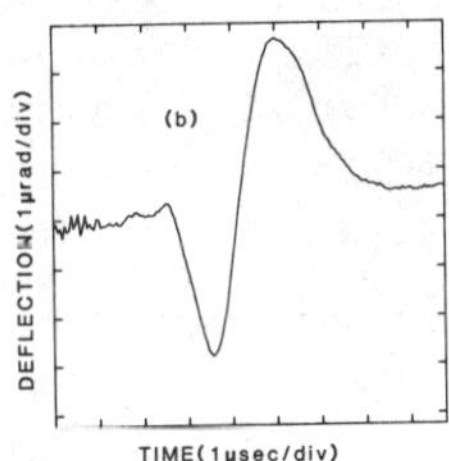

Fig. 3. PADS signals. (a) Theoretically predicted signal. (b) Experimental signal.

This work was performed in collaboration with Allen Rose, Reeta Vyas, and Yu-Xin Nie, and was supported by Air Force Wright Aeronautical Laboratories.

REFERENCES

1. A. Rose, Reeta Vyas, and R. Gupta, Applied Optics (to be published).
2. A. Rose, Y.-X. Nie, and R. Gupta, Applied Optics (to be published).

AN ULTRASENSITIVE FIBER-OPTIC MAGNETOMETER

K. A. Arunkumar
Spectron Development Laboratories
3303 Harbor Blvd., Ste. G-3
Costa Mesa, CA 92626

ABSTRACT

A current carrying conductor undergoes displacement in a magnetic field. Using this concept and an all fiber Mach-Zehnder interferometer, a sensitive magnetometer with theoretical sensitivity on the order of 10^{-18} Tesla/meter of the fiber has been proposed. This magnetometer will be capable of sensing magnetic field direction as well.

MAGNETOMETER CONCEPT

We propose a fiber-optic magnetometer capable of both vector and scalar measurements based on a concept different from those using magnetostriction or Faraday rotation.[1] The magnetic field measurement is to be done using an all-fiber Mach-Zehnder interferometer. One arm of the Mach-Zehnder will have a metallic coating. Field detection and measurement can be done by sensing the force this conductor will experience in the field when current passes through it.

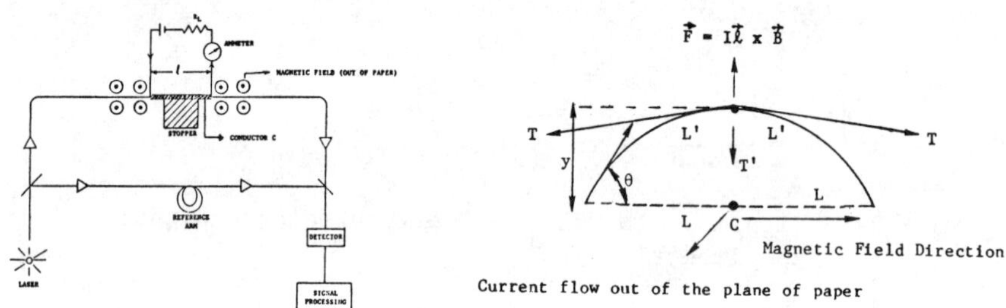

The schematic of the magnetometer is shown in Figure 1. The output of the fibers are superimposed to generate an interference pattern, which is sensitive to any change in the interferometer path lengths. One would exploit this fact in developing the FO magnetometer.

If the magnetometer is in the vicinity of a magnetic field, displacement of the current carrying conductor will cause the fiber to bow. The resulting stretch will increase path length by, δL, for the light beam traversing through it. δL cn be related to the force exerted by the field on the conductor C and hence, to the magnetic field. The direction in which the fiber bows will depend on the magnetic field and current flow directions. Thus, knowing the current flow direction and using a stopper, the magnetic field direction can be inferred.

0094-243X/86/1460676-2$3.00

ANALYSIS

If I is the current passing through C of length ℓ, the force exerted on it by the unknown magnetic field B is $\vec{F} = I\vec{\ell}\times\vec{B} = I\ell B_{\perp}$. The force displaces the conductor in a direction determined by the above relation. For small displacement y, Figure 2 can be approximated to a triangle. If 2L is the length of the sensor arm then $\delta\phi = (2\pi/\lambda)\left[2\sqrt{y^2+L^2}-2L\right]$ and $y = (L\lambda/2\pi)^{1/2}\ \delta\phi^{1/2}$. Under equilibrium condition, $F = T' = 2T\sin\theta$ where T is the tension in the fiber. $\delta\phi$ is also $= (2\pi/\lambda)\ (2L/Y)\ (T/\pi r^2)$ where r is the fiber core radius and Y the Young's modulus for glass. Expessing T in terms of F and combining the above equations we get $IB_{\perp} = (\lambda/2\pi L)^{3/2} \cdot (Y/\ell) \cdot \pi r^2 \cdot (\delta\phi)^{3/2}$, where it is assumed that $L \overset{\perp}{\cong} L'$ for small values of y. Unit in the last equation is Tesla • amp. in the MKS System.

SENSITIVITY

For a $\delta\phi$ detectability of 10^{-6} rad[1], $\ell = 1$ cm, $r = 2\mu m$, length 2L = 1m and $Y = 0.55 \times 10^{11} N/m^2$ and $\lambda = 0.63\ \mu m$, the minimum detectable field will be 6.2×10^{-18} Tesla or 6.2×10^{-14} gauss for one amp. current through the conductor. If the 1 amp current causes any thermal problem, the magnetometer can be operate in pulsed mode or else I can be lowered. Thus, for I = 1mA, the minimum detectable field will be of the order of 10^{-11} gauss/m of the fiber.

DISADVANTAGES

Possible heating of the fiber because of the current through the conductor. Pulsing the current might solve the problem.

ADVANTAGES

1. No hysteresis
2. Can have large dynamic range by controlling conductor current
3. Capable of sensing field direction
4. Sensitivity of the magnetometer controlled by the two parameters viz;

   a. length of sensor arm
   b. magnitude of current through the conductor

REFERENCES

1. T. G. Giallorenzi, J. A. Bucaro, A. Dandridge, G. H. Sigel, J. H. Cole.

LASER-INDUCED BREAKDOWN OF UF_6 AND ITS APPLICATION TO FLOW DIAGNOSTICS

S. W. Allison, M. R. Cates
Oak Ridge Gaseous Diffusion Plant,* Oak Ridge, TN 37831

B. W. Noel
Los Alamos National Laboratory, Los Alamos, NM 87545

ABSTRACT

Breakdown of gaseous UF_6 can be produced with relatively low fluence (approximately 10^6 W/cm^2) near-uv pulsed laser light. A broad spectrum is produced consisting of hundreds of atomic (U_I and U_{II}) lines. Following breakdown, particles are formed from the dissociation/ionization products and the sample volume remains ionized for a long time. This sample volume is elevated in temperature, and a shock wave is produced. Other pertinent details are presented. A schlieren technique is described for observing the motion, hence velocity, of the temperature defect and shock wave. Also given is a flow visualization method based on imaging a long-lived emission component with a gated (≥10-ns) image intensifier system. These and other methods are prospective ways to perform flow diagnostics in gas centrifuges.

INTRODUCTION

Breakdown produces a bright, bluish-white emission, typically 2 mm in length and 3/4 mm in diameter. The emission intensity is exponentially dependent on laser intensity and pressure. An exhaustive description of the breakdown characteristics is presented in reference 1. Photodissociation products, UF_5 polymers, are produced but their scattering cross section and growth rate is small. Ions that are created, however, can be used as velocity tracers.[2]

ACOUSTIC WAVES AND DENSITY DEFECT

The sudden deposition of energy into the gas leads to creation of shock waves and a hot "hole" density defect in the gas. Figure 1 shows a schlieren-type setup to measure these effects. The shock velocity decays very rapidly to the acoustic speed; hence, by observing the wave speed in a flowing system, the velocity may be inferred by subtraction from the measured speed. The density defect is longer lived but physically broader than the acoustic wave.

*Operated by Martin Marietta Energy Systems, Inc., for the U. S. Department of Energy.

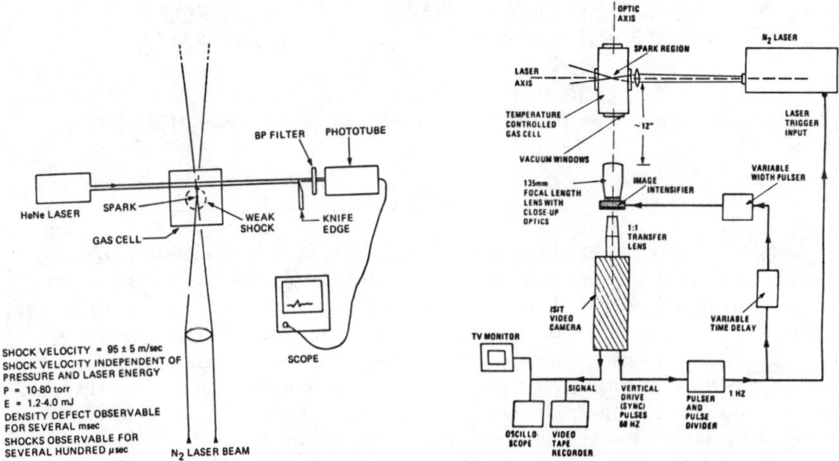

Fig. 1. Schlieren setup.

Fig. 2. Image system.

TIME-RESOLVED VIDEO MEASUREMENTS

Most of the spark emission decays within the first microsecond. This is followed by a long-lived afterglow which can be used to track high-speed flows. Figure 2 shows how a gated image intensifier and video camera can be used to measure this low light level image. Gates varied from 10 to 1000 ns and images were recorded for up to 25 μs after breakdown initiation.

REFERENCES

1. S. W. Allison, University of Virginia dissertation, 1979.

2. R. Krauss, Proceedings Fifth Conference on Gases in Strong Rotation, University of Virginia, Charlottesville, Va., 1983.

IMAGING OF CONTINUUM EMISSION FOR DIAGNOSTICS OF LASER SUSTAINED PLASMAS

R. P. Welle and D. R. Keefer
University of Tennessee Space Institute, Tullahoma, TN 37388

ABSTRACT

Axially symmetric plasmas in flowing argon have been sustained using the focused beam from an unstable oscillator, continuous, carbon dioxide laser having a nominal power of 1.5 kW. Digital images of the plasma within a narrow wavelength region of the continuum emission were acquired using a CID (charge injection device) camera and a digital image processing computer calibrated in absolute spectral radiance. These images were processed using an Abel inversion to obtain spatially resolved plasma emission coefficients which were used to determine the temperature distribution within the plasmas. Using the measured temperatures and a geometric ray trace through the plasma, it was possible to calculate the spatially resolved power absorbed from the laser beam and the thermal radiation lost from the plasmas.

INTRODUCTION

In plasmas sustained in flowing argon, with temperatures in the range of 10,000 to 15,000 K, it is reasonable to assume local thermodynamic equilibrium and a constant pressure. Under these assumptions, detailed spatial characterization of the plasma consists primarily of determining local temperatures. From this, all other significant properties can be determined.

DIAGNOSTIC TECHNIQUE

In order to measure temperatures in the laser sustained plasmas, digital images of a narrow portion of the continuum emission of the plasma were obtained using a CID video camera and a real time computer image acquisition system. The signal-to-noise ratio was improved by averaging 16 consecutive frames into each image acquired. The system was calibrated over the required intensity range by imaging a standard source of known spectral radiance through a succession of neutral density filters in combination with the narrow bandpass filter used for data acquisition. For the argon plasma, the filter was selected to isolate a 1 nm wide portion of the spectrum centered at 626.5 nm, a region free of line radiation in the argon spectrum.

In running the experiments, special care was taken to ensure axial symmetry; by using a cylindrical flow chamber with a coaxial laser beam, and by operating the chamber with its axis oriented in the vertical direction to eliminate non-axisymmetric bouyancy effects. A fold point of minimum variance was found in each radial scan of plasma radiance, and an axis of symmetry of the entire image was defined as the straight line which was the least squares best fit to the minimum variance fold points. The assumption of axial symmetry allowed the use of an Abel inversion technique to obtain a three dimensional spatial distribution of emission coefficient in the plasma.[1] The procedure used for the Abel inversion was to fold each radial scan about the axis of symmetry, and fit the data points, again in a least squares sense, with a series of cubic splines. The resulting curve was then inverted mathematically to obtain the emission coefficient. From the emission coefficient, the temperature was obtained by interpolating in the tabulated data published by Morris.[2]

Knowing the temperature and pressure allowed calculation of two other plasma properties of interest, the radiative power loss, and the absorption coefficient at 10.6μ. This, in turn, allowed the calculation of total energy conversion efficiency in the plasma. Kozlov et al.[3] have published an expression for the radiation losses in an argon plasma which depends only on the temperature and species concentration. In an LTE plasma, species concentration is a function of temperature and pressure, and has been tabulated for argon plasmas by Drellishak et al.[4] Thus, the radiation loss was calculated as a function of position and numerically integrated over the volume to find the total radiation loss of the plasma. Absorption of 10.6μ radiation in an argon plasma occurs only by the inverse Bremsstrahlung mechanism. A method of calculating the coefficients of this process was given by Kemp and Lewis.[5] Using again the thermodynamic data tabulated by Drellishak, and the Gaunt factors published by Karzas and Latter,[6] the absorption coefficient was calculated as a function of position in a grid pattern in the plasma. To calculate total power absorption, the known profile of the incoming laser beam was broken mathematically into a sum of discrete rays, each of which was traced geometrically through the power absorption grid. At each step through the grid, the appropriate fraction of the power was subtracted from the beam and added to the plasma. This procedure gave both the spatial distribution of power absorption and the total power absorbed by the plasma. Subtracting the radiation loss from the total power absorbed gave the total energy conversion efficiency of the plasma.

REFERENCES

1. Shelby, R. T. and Limbaugh, C. C., "Smoothing Technique and Variance Propagation for Abel Inversion of Scattered Data," AEDC-TR-76-163, 1977.

2. Morris, J. C. and Yos, J. M., "Radiation Studies of Arc Heated Plasmas," ARL 71-0317, 1971.

3. Kozlov, G. I., Kuznetsov, V. A. and Masyukov, V. A., "Radiative Losses by Argon Plasma and the Emissive Model of a Continuous Optical Discharge," *Soviet Physics JETP,* Vol. 39, No. 3, Sept. 1974,

4. Drellishak, K. S., Aeschliman, D. P. and Cambel, A. B., "Tables of Thermodynamic Properties of Argon, Nitrogen, and Oxygen Plasmas," AEDC-TR-64-12, 1964.

5. Kemp, N. H. and Lewis, P. F. "Laser-Heated Thruster Interim Report," NASA CR-161665 (PSI TR-205), Feb. 1980

6. Karzas, W. J. and Latter, R., "Electron Radiative Transitions in a Coulomb Field", *The Astrophysical Journal, Suppliment Series, Suppliment Number 55,* Vol. VI, May 1961. pp. 463-468.

DEVELOPMENT OF A HIGH-SENSITIVITY, HIGH-RESOLUTION TRANSIENT ABSORPTION SPECTROPHOTOMETER

Junko Nakamura
The Institute of Physical and Chemical Research, Wako Saitama 351-01

ABSTRACT

We have developed a new system of semi-automatic recording on-line transient absorption spectrophotometer. Characteristic features of a pulsed dye laser used as the absorption monitoring light have been fully utilized to give an excellent performance with high resolution in both time and spectrum and high sensitivity which allows the use of low concentration samples. Among them, the amplifiability of dye laser was the key to make the whole system very simple and easy to operate. As an example, a precise spectrum of $S_n \leftarrow S_1$ absorption of anthracene was measured with this system.

INTRODUCTION

Observation of a transient absorption spectrum is important for the study of photochemical reactions. Conventional system for this purpose, which measures the time profile of absorption at a fixed wavelength, covers well from seconds to nanoseconds, yet the perturbations such as the electric noises and fluorescence from the sample make the measurement at the initial stages of reactions difficult, if not impossible. Difficulty in the optical path alignment is another problem. Aiming to overcome these barriers, we have developed a new system which revises the "laser to laser" method [1] introducing a new idea for the easy and rapid operations.

EXPERIMENTAL SETUP OF NEW SYSTEM

Characteristic features of a pulsed dye laser such as 1) narrow band width, 2) tunability, 3) short pulse, 4) high peak intensity, 5) low total energy, 6) non-divergent beam, 7) instantaneous lasing, and 8) amplifiability were effectively utilized in the new system for a time-resolved spectroscopy, a schematic alignment of which is shown in Figure 1. Here, an exciting pulse laser(e.g. UV 24 nitrogen laser, Molectron) is used to excite the sample and to make the dye oscillate, and the last potion is accepted with a phototransistor (P.T., OS18) and a detector D3. The pulsed dye laser beam is used as the monitoring light, a part of which passes the excited region of the sample cell and the rest from a beam splitter (B.S.) serves as the reference, and their intensities are measured with D1 and D2, respectively. Sample excitation is controlled with a shutter. Optical delay (O.D.) is inserted to measure the delayed-time spectra. In case of monitoring at shorter wavelength, a second harmonic generator (SHG.) was used. Signals from D1, D2, and D3 (photomultipliers grounded with 100k ohm resistors) are processed through a sequence of sample-hold circuit, A/D converter (ANALOG-PRO, Canopus) and personal computer (PC-9801F2, NEC). Stepwise and semi-automatic measurement of transient absorption is achieved by scanning the dye laser, and the spectrum is directly observed on the display.

The alignment of the optical path is much simplified as follows. First, we set a cell with the same dye solution as the oscillating stage in place of the sample and align the cell as to obtain the maximum amplification of laser, and set a few pinholes to fix the path of dye laser. Then the dye cell at amplifying stage is replaced with the sample cell. This procedure is done visually and gives the best condition for the transient absorption measurement.

Thus we need no monochromators, time-resolving detectors nor high speed memories. Construction of this system is easily done by converting a commercial dye laser set with a transverse amplifying stage. Overall time resolution is determined with exciting laser.

RESULTS AND DISCUSSION

The present system well deserves to observe the transient absorption spectrum just after the excitation and is not perturbed with fluorescence of the sample. As an example, we measured the $S_n \leftarrow S_1$ absorption spectrum of anthracene in benzene at room temperature. The experimental conditions are as follows: Sample concentration 1 x 10^{-5} mol· dm^{-3}, cell length 10 mm, time resolution 10 ns, delay time 0 ns, spectral resolution 0.01 nm, and observing step 0.1 nm. The observed transient spectrum is shown in Figure 2 together with the region of fluorescence and $S_1 \leftarrow S_0$ absorption in arbitrary intensity units.

Time-resolved spectra at 0 to 10 ns delays were observed and gave the decay time of 5.8 ns (at 427.5 nm) in good agreement with the fluorescence decay of 5.9 ns.

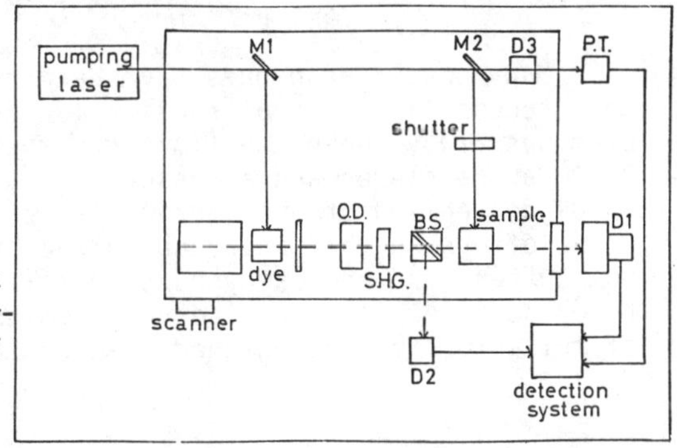

Fig.1 Schematic diagram of optics for the Transient Absorption spectrometer.

REFERENCES

1. E. Sahar and I. Wieder , Chem. Phys. Letters 23, 518 (1973).

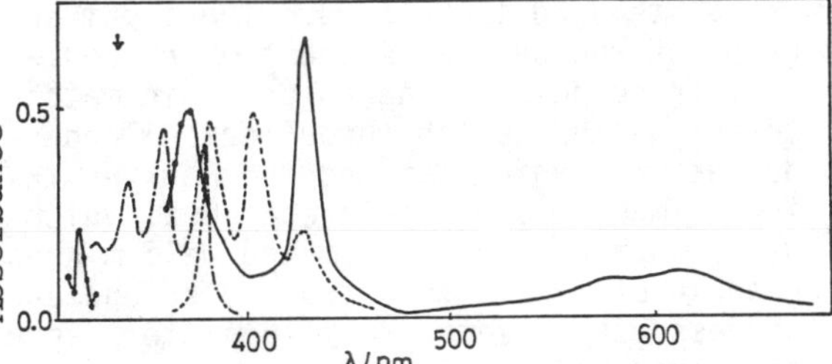

Fig.2 Spectra of anthracene. —— ; $S_n \leftarrow S_1$ absorption. ---; fluorescence. -·-; $S_1 \leftarrow S_0$ absorption. Arrow; excitation = 337.1 nm.

Image Information Transmission, Storage, and Acquisition Using an Acousto-Optic Cell and Optical Matched Filters

Don A. Gregory and Laura L. Huckabee
U.S. Army Missile Command
Redstone Arsenal, AL 35898-5248

James F. Hawk
University of Alabama at Birmingham
Birmingham, AL 35294

ABSTRACT

An acousto-optic beam deflector has been incorporated into a coherent VanderLugt optical correlator in an attempt to increase the volume of stored optical recognition information. The current research is limited to one dimensional deflection, however the extension to two dimensional arrays of matched filters seems possible. The system discussed is real-time; employing a Hughes liquid crystal light valve as an incoherent to coherent image converter. A modified liquid crystal television has also been used successfully for the same purpose.

INTRODUCTION

Many novel techniques have been proposed to make coherent pattern recognition more attractive to applications.[1] A major problem has always been the limited number of memories (matched filters) that can be stored and accessed in near real time. Acousto-optic (A-O) deflection of the Fourier transform of the coherent input scene offers another approach to the problem. This technique does not suffer from the background correlation noise associated with holographic elements or the very limited memory and low efficiency of multiple exposure matched filters.[2]

EXPERIMENTAL RESULTS

Experiments were done using two basic arrangements. The initial technique is shown in Fig. 1. The matched filters were made using standard techniques. The input scene was an aerial photograph of Huntsville, Alabama which contained a good mixture of spatial frequencies. The holographic matched filters were made by interfering the zeroth (undeflected) order from the Isomet LD401-2Y deflector, (which carried the Fourier transform of the coherent input image), and a reference beam which was derived from the original expanded and collimated Helium-Neon laser beam. The holographic plate was then translated and another exposure made. This process was repeated until 6 filters were made. The plate was then developed and replaced in the correlator. The deflector was then used to address the row of filters in near real time, (a few milliseconds). The correlation signals obtained by the CCD television camera shown in Fig. 1 were quite bright and possessed the expected translational invariance. Filters were made using a transparency as well as a Hughes liquid crystal light valve to supply the input scene.

A second arrangement which was implemented involved the use of two A-O deflectors. The process of A-O deflection causes a small frequency shift in the deflected beam thereby making interference with an undeflected beam impossible. This problem can be avoided if both the object and reference beams are frequency shifted an equal amount. This was done by employing a second A-O deflector, located in the reference beam shown in Fig. 1. Four matched filters were successfully made and addressed using this technique.

Lastly, an investigation was made into modifying a commercially available liquid crystal television (LCTV) for use as an addressable transparency. This proved to be quite successful. The low resolution of the LCTV was adequate for making and addressing matched filters using an arrangement similar to Fig. 1 with the addition of a frequency plane aperture to remove some of the LCTV pixel structure. A photograph of the modified LCTV is given in Fig. 2.

CONCLUSIONS

It has been shown that acousto-optic deflection is a realistic technique for addressing optical memories (holographic matched filters) in near real time. These filters may be made using undeflected object and reference beams, then addressed with a deflected object beam or made using deflected object and reference beams. It has also been shown that low cost liquid crystal televisions may provide a viable alternate to the expensive light modulators now in use.

REFERENCES

1. A. VanderLugt, Appl. Opt. 5, 1760 (1966).
2. D. A. Gregory and H. K. Liu, Appl. Opt. 23, 4560 (1984).

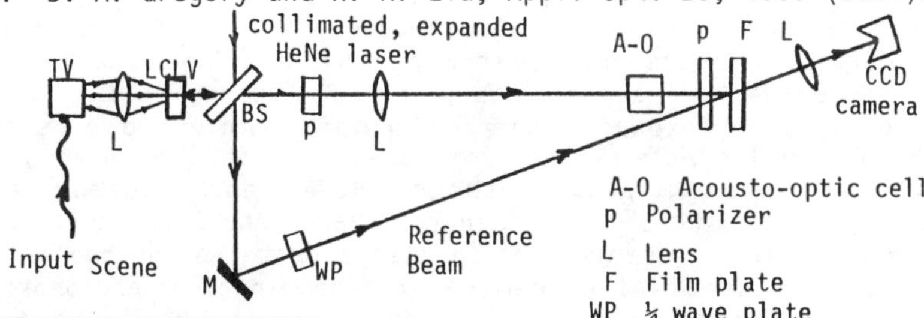

Fig. 1. Basic arrangement for acousto-optic correlator. The input scene was provided by an external camera to the TV monitor.

Fig. 2. A photograph of the modified liquid crystal television. The light diffuser and polarizers were removed and the screen held vertical.

STUDIES OF FAST PROCESSES IN BIOLOGICAL SYSTEMS

Carey K. Johnson
University of Kansas, Lawrence, KS 66045

Robin M. Hochstrasser
University of Pennsylvania, Philadelphia, PA 19104

Abstract

A remarkably wide range of biologically important phenomena are intrinsically fast and require pulsed lasers for study. In these studies, the interaction of intense laser pulses with biological molecules may lead to effects which must be carefully evaluated in the interpretation of results. Among the effects to be noted are multiphoton processes and strong-field effects. Recent results are reported from the first pump-probe time-resolved Raman study of heme relaxation following photodissociation. Relaxation of the porphyrin core-size marker bands occurs between 100 and 400 ps in myoglobin and after 8 ns in hemoglobin following photodissociation of carbonmonoxy-myoglobin or hemoglobin. Possible sources of this relaxation include the heme-protein coupling.

Introduction

Laser methods of optical spectroscopy have radically extended the capabilities of biological research. In some applications, lasers are used to study biological processes which naturally require light, such as vision, purple membrane systems of light-activated bacteria, and photosynthesis. In other applications, lasers are used as a structural probe, for example by Raman spectroscopy. Another class of experiments is an extension of pre-laser relaxation methods whereby a biological system is perturbed away from equilibrium and the recovery of the system followed by laser spectroscopy. Heme-protein photolysis is an extensively studied example. Not only are lasers applied to biological systems, but their application is leading to biologically relevant interpretations which could not have been reached by other experimental techniques. An example is the study of biophysical events that are intrinsically fast and therefore require ultrashort light pulse methods. Ultrafast processes are ubiquitous in biology and include energy transfer and electron transfer in photosynthesis, isomerization in visual proteins and bacterial membranes, photodissociation, geminante recombination, and protein dynamics in heme proteins, and chain dynamics in nucleic acids. Picosecond and subpicosecond studies have shown that important fundamental events such as vision and photosynthesis occur with great efficiency at nearly the maximum conceivable rate for such processes.

Interaction of Intense Laser Pulses with Biological Molecules

Picosecond or subpicosecond laser pulses interacting with biological molecules generate laser intensities as high as 10^9 W cm^{-2}.

0094-243X/86/1460686-4$3.00

High laser intensities may engender unexpected optical and photochemical processes. There is the possibility, for example, of effects produced by the strength of the laser field. A strong Raman excitation intensity can cause broadening and, if the field is strong enough, splitting, of the emission signal.[1] Such an effect could resemble shifts in the Raman spectrum due to vibrational mode changes and, if unrecognized, could be interpreted as structurally significant chemical behavior.

Another possible effect of the interaction of laser pulses with biological systems is multiphoton processes. In a typical experiment, a laser pulse excites a system to an excited state which relaxes to biologically relevant states which are probed with a second laser pulse. If the pumping pulse is intense enough, however, it can induce further absorption from the initially excited state or from intermediate states to produce intermediate species which are not biologically relevant. These multiphoton processes may be difficult to detect. For example, the absorption coefficients of the intermediates at a given wavelength may be such that the change in absorption as a function of laser intensity appears linear.

Another effect of laser pulses is to heat the molecule by depositing energy into vibrational modes. This possibility is illustrated by carboxymyoglobin, which absorbs a 532 nm photon to dissociate rapidly. The ground state of the photodissociated species lies some 7,800 cm^{-1} higher in energy than the ground state of carboxymyoglobin. The remaining 11,000 cm^{-1} is deposited in the protein. If this energy is distributed over the heme vibrational modes, the temperature of the heme increases by some 200°K. Asher and Murtaugh recently published results showing that shifts of several cm^{-1} in the

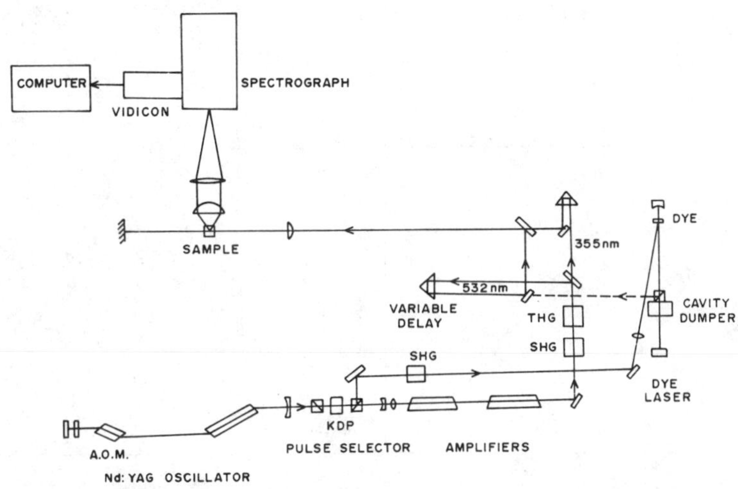

Fig. 1. Picosecond time-resolved Raman spectrometer.

frequency of a porphyrin skeletal stretching mode can be induced by temperature changes of a few hundred °K in model porphyrins.[2] This mode has been probed in hemeprotein photolysis and shifts of a few cm^{-1} in this band have been observed. Are these shifts simply due to heating of the molecule? Our answer is no for several reasons.[3] One is that we were unable to induce shifts or broadening in stable deliganded myoglobin, which also absorbs strongly at 532 nm. Secondly, an estimate with a typical protein thermal diffusion coefficient indicates that the heme cools rapidly, within our 30 ps pulsewidth. Finally, as the results presented herein show, the shift in Raman frequency lasts far longer in hemoglobin than could reasonably be expected for vibrational energy loss. The cooling rate may be significant in some applications of intense laser pulses to biological molecules on short time scales, however.

Picosecond Time-Resolved Raman Studies of Hemeproteins

The study of hemeprotein dynamics by flash photolysis and picosecond probing of the relaxing heme is an example of a laser perturbation and relaxation method. Hemeproteins have been extensively studied on the picosecond time scale by laser flash photolysis and transient absorption measurements.[4] Such measurements of photolyzed carboxyhemoglobin (HbCO) and carboxymyoglobin (MbCO) showed differences in the absorption spectra on the picosecond time scale from the

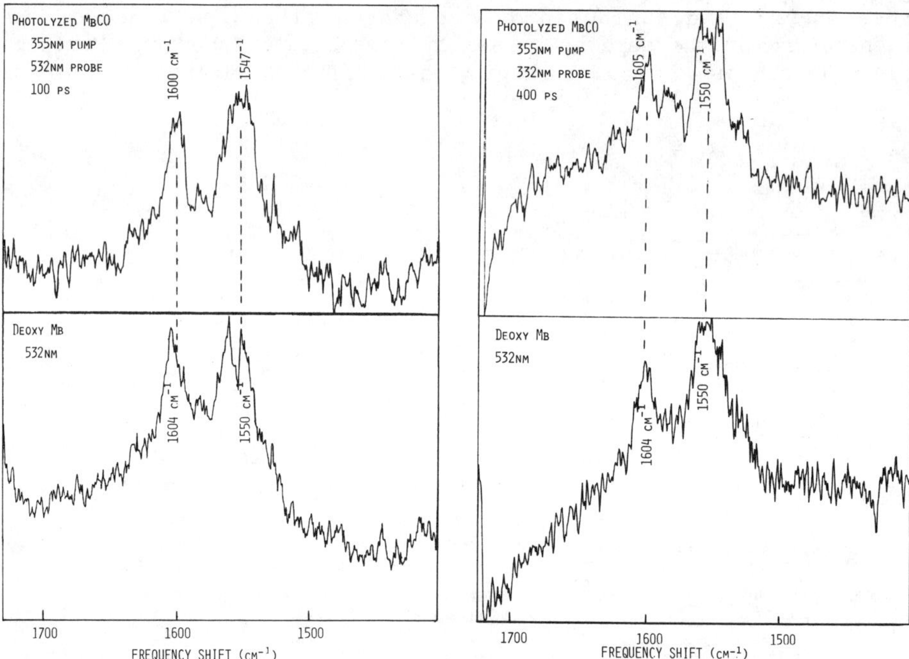

Fig. 2. Picosecond Raman spectrum of MbCO 100 ps and 400 ps after photolysis (top); Raman spectrum of stable Mb (bottom).

equilibrium spectra of deliganded myoglobin (Mb) or hemoglobin (Hb).[5] Raman spectroscopy was first used to probe photolyzed HbCO in experiments where one beam of pulses both photolyzed the molecule and probed its Raman spectrum.[6] These experiments found shifts in Raman band frequencies in photolyzed Hb relative to stable Hb when 30 ps or 20 ns pulses where used. In collaboration with T.G. Spiro and coworkers, we have shown that in myoglobin also, using 30-ps pulses, these Raman bands are shifted relative to stable Mb, but using 7-ns pulses, no shift is observed.[3]

Recently we have employed a pump-probe Raman technique with an actively and passively mode-locked 20-Hz Nd:YAG laser which allows us to observe the evolution of the Raman spectrum on the picosecond time scale (figure 1).[7] The third harmonic of the laser is used as the pump beam, and the second harmonic as the probe. Our most recent results were to obtain the first picosecond pump-probe Raman spectra of hemeproteins. These results show the Raman spectrum of photolyzed MbCO from 100 ps to 1 ns following photolysis. The 100-ps and 400-ps spectra are shown in figure 2. In the 100-ps spectrum the core-size marker Raman bands are shifted 4 cm-1 to lower frequency from stable Mb. These shifts are similar to those observed for Hb.[6] The 400-ps spectrum, however, shows the Raman bands at their stable position. We have also recorded pump-probe picosecond Raman spectra of photolyzed HbCO. We observe that 8 ns after photolysis the Raman spectrum of photolyzed HbCO is still shifted to lower frequency by 4 cm-1 relative to Hb. The pump-probe result verifies that the observation of a shift in HbCO Raman spectrum in the earlier single-pulse experiment was not due to heating of the sample by the ns laser pulse, since by 8 ns the heat deposited upon excitation must have dispersed.

Studies of model compounds have demonstrated a correlation between porphyrin core size and the position of the Raman bands observed in these experiments.[8] Our results imply that the heme core is expanded following photolysis and that this expansion relaxes between about 100 ps and 400 ps in Mb, but after 8 ns in Hb. We suggest that the full relaxation of the iron out of the heme plane is restrained for much longer times in Hb than in Mb following photolysis.

References

1. B. Dick and R.M. Hochstrasser, J. Chem. Phys. 81, 2897 (1984).
2. S.A. Asher and J. Murtaugh, J. Am. Chem. Soc. 105, 7244 (1983).
3. S. Dasgupta, T.G. Spiro, C.K. Johnson, G.A. Dalickas, and R.M. Hochstrasser, Biochem. 24, 5295 (1985).
4. P.A. Cornelius and R.M. Hochstrasser, in Picosecond Phenomena III, ed. K.B. Eisenthal *et al.*, pp. 288-293 and references therein.
5. P.A. Cornelius, A.W. Steele, D.A. Chernoff, and R.M. Hochstrasser, Proc. Natl. Acad. Sci. USA, 78, 7526 (1981).
6. J. Terner, T.G. Spiro, M. Nagumo, M.F. Nicol, and M.A. El-Sayed, J. Am. Chem. Soc. 102, 2328 (1980).
7. C.K. Johnson, G.A. Dalickas, S.A. Payne, and R.M. Hochstrasser, Pure Appl. Chem. 57, 195 (1985).
8. S. Choi, T.G. Spiro, K.C. Langry, K.M. Smith, D.L. Budd, and G.N. La Mar, J. Am. Chem. Soc. 104, 4345 (1982).

ULTRAVIOLET RESONANCE RAMAN SPECTROSCOPY OF BIOPOLYMER COMPONENTS*

Bruce Hudson
University of Oregon, Eugene, Oregon 97403

ABSTRACT

Ultraviolet resonance Raman spectroscopy has recently been demonstrated to be a useful technique for obtaining new information about the vibrational spectra and excited electronic states of large molecules.[1] Advances in the application of ultraviolet resonance Raman spectroscopy to the study of biopolymer components are discussued in this paper. Specifically, Raman excitaion with laser radiation in the 200 - 300 nm region is shown to be a useful method for obtaining spectra without interference from biopolymer fluorescence, with enhanced sensitivity due to resonance, and with new selectivity and spectral features.

INTRODUCTION

The laser technology needed for generation of coherent radiation in the far ultraviolet region is now sufficiently reliable that it can be used in a routine fashion to obtain Raman spectra. The use of radiation in the 200 - 300 nm region of the spectrum makes many "colorless" chemical groups "chromophores" and permits resonance enhancement of their Raman spectra. Resonance enhancement has several distinct advantages for studies of biopolymers including enhanced sensitivity and selectivity for the group or groups with electronic transitions near resonance. The use of far ultraviolet radiation (wavelengths less than 250 nm) avoids the problem of interfering fluorescence because relaxation to lower electronic states moves this emission to a spectral region removed from the vibrational resonances of the Raman spectrum. The particular vibrations that show pronounced resonance enhancement when the laser excitation frequency is tuned to a particular electronic transition of a molecule or component of a molecule provide information about the nature of the geometric change associated with that excited electronic state relative to the ground state. This can be useful in determining the nature of the electronic excitation.

We have recently been exploring the utility of this technique for obtaining basic conformational information about biopolymer components. This conformational information derives from the fact that the vibrational frequencies of large molecules are dependent on their conformation. The experimental method used will be briefly presented and then the results of these studies to date will be outlined.

* This work is supported by the National Institutes of Health and the National Science Foundation

EXPERIMENTAL METHODS

The Raman spectrometer used in these studies is based on a high power, Q-switched, flash lamp pumped Nd:YAG laser. Non-linear optical crystals are used to convert the fundamental radiation at 1064 nm into harmonic frequencies at 532, 355, 266 and 213 nm. The first three of these frequencies can then be converted by stimulated Raman shifting in gasses to produce collimated radiation with either shorter or longer wavelengths separated by an interval corresponding to the vibrational frequency of the gas. Table I shows the values of the wavelengths obtained when hydrogen gas is used as the stimulated Raman medium. Other wavelengths can be generated by using gasses with smaller vibrational intervals.

Table I Values of the wavelengths obtained by stimulated Raman shifting the indicated Nd:YAG harmonic in hydrogen

532:	320	282	253	229	209	192
355:	309	273	246	204	188	
266:	299	266	240	218	200	184
213:	213					

The selected radiation is then directed to the sample which is a circulating solution stream. This type of sample is used because it avoids the requirement for containment in a cuvette. This is needed because, upon prolonged irradiation, the cuvette becomes damaged by the radiation and often becomes fluorescent. Circulation also minimizes the effects of sample damage. Backscattering geometry is used because of the high optical density of the samples at many of the wavelengths used.

The scattered radiation is then dispersed with a large single monochromator and detected with a "solar blind" photomultiplier that is insensitive to longer wavelength radiation such as fluorescence. The photomultiplier output is processed with a box-car amplifier and transferred to a minicomputer. Other details of the experimental arrangement are described in a forthcoming publication.[2]

NUCLEIC ACID BASES

The application of this technique to the nucleic acid bases has been described in two publications.[3, 4] The nucleic acid bases have a very complex electronic spectrum with several broad overlapping transitions in the 280 - 200 nm region. Resonance Raman excitation profiles (the variation of the intensity of a particular vibrational band with wavelength) reveal that these states have distinct geometrical changes relative to the ground state. Conversely, the pattern of the excitation profiles can be used to locate distinct electronic excitations even though they overlap in the absorption spectrum. Some of the vibrational modes observed with far UV excitation are sensitive to the state of ionization of the nucleic acid base and new bands are observed that may be sensitive to the state of hydrogen bonding.

PROTEIN COMPONENTS

The diverse side chain structures of proteins provide several opportunities for selective ultraviolet enhancement. The aromatic amino acids tryptophan, tyrosine and phenylalanine have strong transitions at wavelengths shorter than 280 nm. These groups are intensely fluorescent. This restricts studies of proteins containing these groups to wavelengths shorter than about 260 nm in order to avoid the reduced signal to noise associated with high background fluorescence levels.

Our primary emphasis has been on resonance in the far ultraviolet with the peptide bond itself.[5-7] A simple model peptide, N-methylacetamide,[7] shows a number of interesting features in its resonance Raman spectrum. The first is that the amide II vibration, a mixture of the C-N stretching motion and the C-N-H angle bending motion, becomes very intense under resonance conditions (wavelengths shorter than 230 nm). This vibrational band is essentially absent in Raman spectra obtained with visible excitation. In the isotopic variant in which the amide hydrogen has been replaced by deuterium, the corresponding amide II' vibration is the only strong band under resonance conditions. Isotopic labeling of the peptide bond with nitrogen 15 and carbon 13 in this deuterated N-methylacetamide shifts the amide II' vibration to lower frequency by an amount that is comparable to that expected if the motion associated with this vibration is purely C-N stretch. The strong enhancement of this vibration indicates that the structural change associated with electronic excitation of the peptide bond is almost purely elongation (or contraction) of the C-N bond. In the protonated form of N-methylacetamide the intensity of the amide II' vibration is now roughly equally divided between the amide II and amide III vibrations. This demonstrates that isotopic substitution has a profound influence on the form of the normal mode of vibration. Specifically, these two vibrations must consist of roughly equal amounts of C-N stretch and C-N-H angle bending motions (with opposite phase). This coupling in the hydrogen-containing form is due to near-equality of the intrinsic frequencies of C-N stretching and C-N-H angle bending. Replacement of the amide hydrogen by deuterium shifts the bending motion to lower frequency.

Another interesting feature of the spectra of N-methylacetamide is the observation that the amide I (C=O stretch) mode is much broader in aqueous solution than in acetonitrile.[7] In fact, in aqueous solution this band appears only as a broad background with a width of several hundred wavenumbers. A similar width is observed in D_2O solution indicating that this breadth is not due to coincidence with nearby solvent vibrations. This extreme width in aqueous solutions is presumably due to the presence of a variety of hydrogen bonded forms with distinct vibrational frequencies.

In recent work we have investigated the possibility of detecting cis/trans isomerization of the peptide linkage to proline residues. This isomerization process is important in understanding the kinetics of protein folding. Kinetic measurements have detected a "slow-folding" fraction in this reaction. This has been ascribed to the presence of a population of incorrect isomers at the X-proline linkage. Unlike the normal peptide bond where the trans isomers is favored by over a factor of 100, the X-proline linkage has a trans/cis equilibrium ratio of only about 2. The native globular structure of a protein may have either cis or trans X-proline linkages. The conversion of one conformation to the other for this linkage is very slow because of the partial double bond character of the peptide bond. A formidable spectroscopic problem in determining the conformation of the X-proline bond in a folded or unfolded protein is that there are generally only a few such groups in a typical protein of perhaps 100 residues. The alkyl substitution at the imide nitrogen of the X-proline bond shifts the electronic absorption to longer wavelengths. We have shown [8] that excitation in the 220 - 240 nm region selectively enhances the vibrational features characteristic of this linkage relative to that of the predominant secondary peptide linkages by a factor of about 30. This may permit detection of the conformation of the protein at this connection and therfore permit spectroscopic monitoring of this aspect of protein folding.

REFERENCES

1. B. Hudson, Spectroscopy 1, 22 (1985).

2. B. Hudson and L. C. Mayne, Methods in Enzymology, in press.

3. L. D. Ziegler, B. Hudson, D. P. Strommen and W. L. Peticolas, Biopolymers 23, 2067 (1984).

4. W. L. Kubasek, B. Hudson and W. L. Peticolas, Proc. Natl. Acad. Sci. USA 82, 2369 (1985).

5. L. C. Mayne, T. Ramahi, T. Oas, and B. Hudson, Biophys. J. 45, 322a (1984).

6. L. C. Mayne, G. Harhay, and B. Hudson, Biophys. J. 47, 88a (1985).

7. L. C. Mayne, L. D. Ziegler, and B. Hudson, J. Phys. Chem. 89, 6399 (1985).

8. L. C. Mayne, G. Harhay and B. Hudson, Biophys. J. 49, 330a (1986).

TWO-PHOTON LASER-INDUCED FLUORESCENCE OF THE TUMOR-LOCALIZING PHOTOSENSITIZER HEMATOPORPHYRIN DERIVATIVE

David S. KING
National Bureau of Standards, Molecular Spectroscopy Division, Gaithersburg, MD 20899

Donald F. HELLER and Jerzy KRASINSKI
Allied Corp., 7 Powderhorn Dr., Mt. Bethel, N.J. 07060

Richard S. BODANESS
National Center for Health Services Research, OHTA, Park Building, Room 3-10, Rockville, Md 20857

SUMMARY

The tumor localizing photosensitizer hematoporphyrin derivative (HPD) is shown to undergo simultaneous two-photon excitations upon intense laser irradiation at 750 or 1064 nm, a spectral region where there is no significant HPD one-photon absorbance in aqueous solution. Evidence for the two-photon excitation consists in the observation both of the HPD fluorescence spectrum in the region of 615 nm as a result of 750 or 1064 nm excitations and the quadratic dependence of this fluorescence emission intensity upon the excitation laser intensity. Since the penetration depth of ultraviolet and visible light into tissue varies logarithmically with wavelength (red penetrating more deeply than blue), these studies suggest the possibility that two-photon induced localization of tumor-bound HPD might facilitate the detection of deeper lying tumors than allowed by the current one-photon photolocalization method.

INTRODUCTION

Photosensitized reactions are a topic of considerable interest in biochemistry and clinical medicine. Photosensitizing molecules, including porphyrins, have been utilized to catalyze the oxidation of many biological constituents such as proteins. Many of these oxidations are known to proceed via molecular oxygen in its first excited singlet electronic state. From the viewpoint of clinical medicine, much of the interest in photosensitizers has centered around hematoporphyrin derivative because the active component of this mixture of photosensitizing porphyrins (proposed to be dihematoporphyrin ether) binds with increased avidity to tumor tissue. It has been shown by Profio that 24 hours after injection of HPD into mice with subcutaneously implanted tumors, the concentration of HPD retained by the tumor tissue was more than 10-fold higher than in the surrounding healthy muscle.[1] Subsequent to activation with photons of an appropriate energy, HPD can be used for tumor photo-localization via its visible fluorescence and for cancer photochemotherapy, in which case the singlet molecular oxygen generated via energy transfer from the excited porphyrin is thought to be the distal tumoricidal agent.

A major limitation in conventional tumor photolocalization is the attenuation of the excitation source (i.e., laser beam) by the tissue through which it must pass. The penetration depth of visible and ultraviolet light into tissue has been shown to increase logarithmically with wavelength, i.e., red light penetrating deeper than blue.[2] It is for this reason that excitation sources at 630 nm are used with HPD for photochemotherapy, despite the fact that 630 nm radiation is only weakly absorbed by the HPD (decadic molar extinction coefficient of ≈500 M^{-1} cm^{-1}). In photolocalization experiments with HPD, however, photons in the neighborhood of the strongly absorbing, 400 nm Soret band have been utilized. The rationale for near-UV excitation, rather than excitation at 630 nm is to provide a high level of spectral discrimination between the excitation source and the emitted fluorescence which results following relaxation to the ground vibrational level of the S_1 state and which exhibits maxima at approximately 615 and 675 nm. Unfortunately, this scheme essentially limits photolocalization to the detection of surface layer tumors due to the minimal tissue transmittance in the blue and UV. An approach whereby these problems may be circumvented takes advantage of the two-photon spectroscopy of HPD.

We have demonstrated the two-photon absorption of HPD at 1064 nm[3] and the resonance enhancement of the HPD two-photon excitation cross-section arising from excitation wavelengths closer to the visible (600 nm) S_1 electronic absorption bands, e.g., at 750 nm. The results of our studies suggest the possibility of utilizing the resonance enhanced 750 nm two-photon excitation of HPD in clinical cancer photolocalization and cancer photochemotherapy.

EXPERIMENTAL

HPD was obtained in aqueous solution from Photofrin Medical, Inc.[4] Further dilutions were made in 10 mM potassium phosphate buffer pH 7.4 to a final HPD concentration of 5.4 µg per ml. Two-photon excited fluorescence spectra were taken at an approximate 1 nm resolution using either a Q-switched YAG with a 20 ns output pulse duration and maximum pulse energy of 500 mJ at 1064 nm or a Q-switched alexandrite laser with 55 ns pulse duration and 50 to 150 mJ pulse energy (depending on wavelength) in the 730 to 780 nm range. To assist in the initial optical alignment, the output of the lasers was frequency doubled in KDP crystals. This near uv second harmonic was used to directly excite the sample at 375 or 532 nm. Since the absorption coefficient and fluorescence quantum yield for single-photon excitations are known,[3] this single-photon ultraviolet-excited fluorescence signal can be used to calibrate the excitation efficiency of the two-photon visible-excited fluorescence. For these latter measurements the KDP crystals were removed from the laser beam path.

RESULTS

Fluorescence spectra obtained following two-photon excitations at 1064 and 750 nm are shown in figure 1. Curve 1(A) presents the one-photon fluorescence spectrum obtained with a Spectrolab f1.6 monochromator upon irradiation of HPD at 532 nm with a pulse energy of 6 µJ at the sample cuvette. Curve 1(B) is the two-photon fluorescence spectrum obtained using the same apparatus with a 1064 nm pulses of 50 mJ energy. In the alexandrite experiments a long-wavelength blocking interference filter combination was used to eliminate laser scatter. The transmission curve for these filters is shown in curve 1(C). Curve 1(D) was recorded for an excitation wavelength of 750 nm (using a Jarrell-Ash 0.25 m monochromator) and laser pulse energy of 50 mJ (500 MW/cm^2). The fluorescence peak usually reported at approximately 675 nm (as shown in curves A and B) is totally suppressed by the long wavelength cut-off filter; however curve 1(D) is virtually identical to spectra obtained following single-photon excitations at 375 nm using the same monochromator and filter combination.

The dependence of the HPD two-photon fluorescence intensity at 615 nm on excitation laser intensity was measured for incident laser energies from 1 to 55 mJ per pulse. The dependence of two-photon fluorescence intensity on incident laser power density was quadratic for excitation wavelengths 750 and 1064 nm. The actual power dependence obtained from linear least squares fitting the experimental data were $m=1.9\pm0.1$ and $m=2.1\pm0.1$, respectively. Calibrations to determine the two-photon cross-sections were made using single-photon fluorescence data. The two-photon excitation cross-section at 750 nm (of 1.5×10^{-49} cm^4 s) was approximately 100-fold greater than that at 1064 nm; we attribute this to a resonant enhancement contribution from the HPD S_1 state.

DISCUSSION

Our results demonstrate the direct two-photon laser excitation of HPD in solution. The evidence consists of both the identicalness of the fluorescence spectra recorded following, for example, excitations at 1064 nm (two-photon absorption) and at 532 nm (one-photon absorption), and the quadratic intensity dependence of the fluorescence.

A major limitation to the use of single-photon HPD fluorescence, as presently utilized in photolocalization, lies in the requirement for the use of relatively high energy photons (approximately 400 nm) to achieve the adequate energy separation needed for unambiguous discrimination between scatter from the excitation beam and the fluorescence emission. Tissue transmission at 400 nm is relatively low, and therefore the photolocalization effect is inherently limited to the tissue surface. In the alexandrite laser excitation experiments the HPD two-photon fluorescence was of sufficient intensity to be observed visually in a darkened room through a single dielectric filter (see Fig.1C), implying potential usage for tumor localization at substantial tissue depths since for skin there is an approximately one-thousand fold increase in tissue transmission at 750 nm as compared to 400 nm.[2]

This same approach might apply to photochemotherapy as well. HPD photochemotherapy is customarily driven at 630 nm, utilizing the weak low energy one-photon HPD absorption band. For skin, there is approximately a 10-100 fold increase in transmission at 750 nm as compared to 630 nm. In the 750 nm experiments a relatively long (55 ns) laser pulse was used. Since the two-photon absorption rate is proportional to the square of the laser intensity, the application of picosecond laser pulses should greatly increase the efficiency of the two-photon absorption process, permitting the use of substantially reduced laser pulse energies to achieve equivalent ends. Lower pulse energies (i.e., doses) are desirable to reduce the potential for damage to surrounding tissue. Use of the 50 ps, 1 mJ pulses available, for example, from a pulsed, mode-locked alexandrite laser oscillator operating at 10 pps will only deliver 2% of the energy dose used in the present in vitro two-photon experiments. The two-photon excitation rate should, however, be enhanced a million-fold in relation to excitation by a 50 ns pulse of the same energy. Under the experimenal conditions described herein, this would lead to complete saturation of the HPD two-photon absorption.

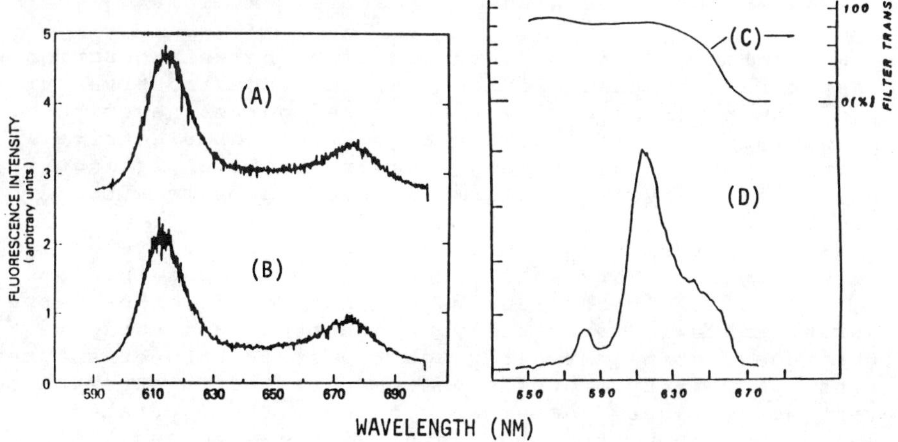

FIG.1 Fluorescence spectra of dilute aqueous solutions of HPD: (A) following direct one-photon excitation at 532 nm; (B) following two-photon excitation at 1064 nm; (C) combined transmission curve of the optical filters used in the 750 nm two-photon fluorescence experiments; (D) following two-photon excitation at 750 nm (with filters of curve C anterior to the detector).

ACKNOWLEDGEMENTS: We are indebted to Mr. Paul Papanestor for technical assistance with the alexandrite laser.

REFERENCES

1. A.E. Profio, IEEE J. Quant. Electron. QE-20, 1502-1507 (1984).

2. S. Wan, J.A. Parrish, R.R. Anderson, and M. Madden, Photochem. Photobiol. 34, 679-681 (1981).

3. R.S. Bodaness and D.S. King, Biochem. Biophys. Res. Commun. 126, 346-351 (1985).

4. Certain commercial equipment, instruments, or materials are identified in this paper in order to adequately specify the experimental procedure. In no case does such identification imply recommendation or endorsement by the National Bureau of Standards, the Department of Health and Human Services, or the U.S. Public Health Service, nor does it imply that the materials or equipment identified are necessarily the best available for the purpose.

LASER REPAIR OF SEVERED NERVES

Edward E. Almquist, M. D.
University of Washington, Seattle, WA 98195

Returning good nerve function after nerve laceration is one of the greatest difficulties in reconstructive surgery of the upper extremity. Lacerated nerves never regain complete function because the pattern of electrical circuitry in the tens of thousands of single cell conduits in each major nerve in the arm is never the same after healing.

Microsurgical techniques have been employed recently in nerve repair. Very small sutures, 25 microns in diameter, are used, but the repairs are not ideal. Each nerve or nerve bundle is imperfectly reconnected, and escapement of the regenerating nerve through the space between sutures results. A variety of cuffs has been tested in attempts to prevent scar tissue from growing between the lacerated nerves and to shunt the growing nerve cells toward their distal counterparts. None has been clinically successful.

The argon laser offers unique advantages for peripheral nerve repairs.[1] It has pinpoint accuracy and tissue specificity. The blue-green wave lengths of laser energy are almost completely absorbed by red-colored substances such as red blood cells, yet are almost completely reflected by white surfaces, such as nerves. Thus when an argon laser beam is applied to a small amount of blood on a nerve, the light is absorbed by the red blood cells and turns them into an adherent coagulant. The laser light is reflected by the white nerve, leaving it undamaged. An adherent minitubule is thus constructed around the lacerated portions of the nerve, binding the ends together, theoretically without damage to the nerve itself.

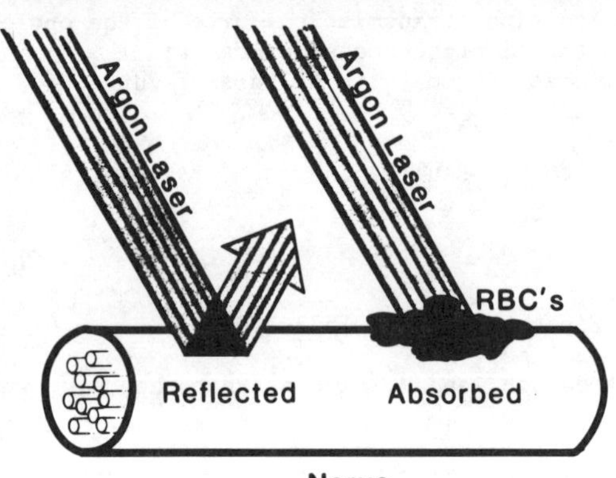

Fig.1. Blue-green argon laser energy is reflected by white nerve but is absorbed by red blood cells.

A fiberoptic laser system was developed in our laboratory consisting of an argon laser fitted with a mirror shutter, a timer, and a foot switch enabling the shutter to open for a pre-set time. The laser energy was applied in

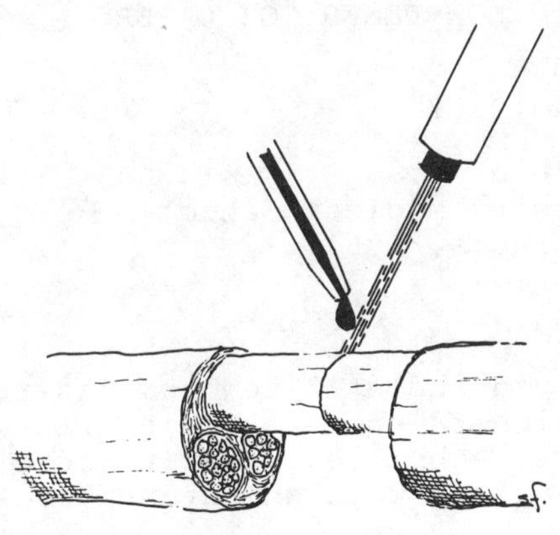

Fig.2. The laser energy is applied to a nerve fascicle with a cuff of blood.

peripheral nerve surgery. Trial and error showed that 0.75 watts of laser energy applied for 0.5 seconds was the ideal power-time relationship for coagulating red blood cells without damaging the nerve through heat build-up. (Fig. 2)

Rat and primate sciatic nerves were repaired with this system, and at intervals up to six months following repair, the nerves were removed and evaluated using scanning electron microscopy, transmission electron microscopy, and traditional light microscopy.[2,3] The system appeared to produce a technically superior nerve repair. The small weld-like cuff of blood appeared to shunt the growing axones toward their appropriate destination, and it seemed to prevent in-growth of scar tissue. (Fig. 3) The problem of foreign body reaction and the relative inaccuracy of suture repair are avoided. Accurate quantitation of function could not be done in animals; this was a qualitative study. The system is presently being used in human clinical trials and awaiting quantification.

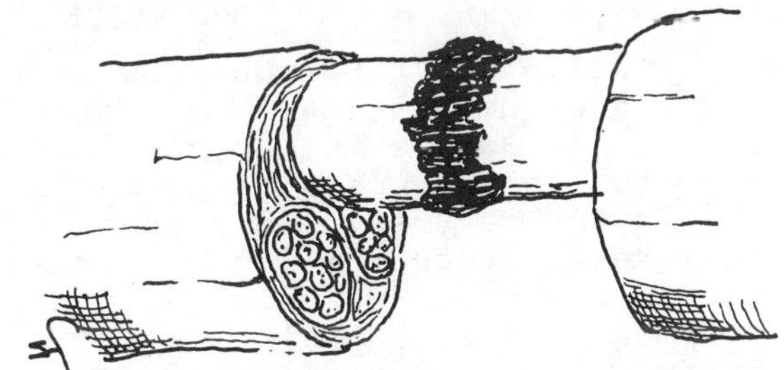

Fig.3. The coagulated blood forms an adherent cuff around the fascicle.

REFERENCES

1. L. Goldman, Biomedical aspects of the laser; the introduction of laser applications into biology and medicine. New York, 1967, Springer-Verlag, Inc.
2. A.L. Van Beek, S.C. Jacobs, E.G. Zook: Examination of peripheral nerves with the scanning electron microscope. Plast.Reconstr.Surg 63:509-19, 1979.
3. M.G. Orgel and J.W. Huser: A comparison of light and scanning electron microscopy in nerve regeneration studies. Plast.Reconstr.Surg 65:628-34, 1980.

NERVE ANASTOMOSIS WITH LOW-POWER CO_2 LASER

J.E. Bailes
M.R. Quigley
V. Sahgal
L.J. Cerullo
Northwestern University Medical School
Chicago, Illinois, 60611

ABSTRACT

A rat model was developed and utilized for laser-assisted anastomoses of peripheral nerves, which was compared to conventional suture techniques. Laser nerve anastomosis is technically feasible with an advantageous histologic healing response.

INTRODUCTION

Advances in surgical technique have allowed for improvement of the repair of peripheral nerves. In the modern era, the two most noteworthy achievements have been the development of microsurgical techniques and the utilization of electrophysiological evaluation. Recently, it has become possible to reconstruct tissues using low-power CO_2 laser energy. A pilot study was undertaken to assess the feasibility of joining nerves using a sutureless laser technique in a rat model.

TECHNIQUE

Twelve adult Sprague-Dawley rats had laser-assisted nerve anastomosis (LAVA) performed on the sciatic nerve of one limb while the opposite limb underwent conventional suture anastomosis using microsurgical technique and 10-0 nylon sutures. In another group of 12 animals, a unilateral procedure was performed wherein an interposition nerve graft was placed in the sciatic nerve using a portion of the contralateral sciatic nerve as a donor graft. Twelve additional animals were operated upon using a suture technique as a control graft. Animals were allowed to recover and were sacrific at 6 and 12 weeks. Pre-operative and post-operative nerve conduction velocities were performed in addition to histological evaluation with light and electron microscopy, myelin stains and horseradish peroxidase permeability studies.

RESULTS

At sacrifice, all suture specimens were found to be anatomically intact. LAVA specimens had dehiscence in 3 of 12 (25%) of end-to-end and 2 of 12 (17%) of interposition grafts. Nerve conduction velocities showed no statistical difference between pre- and post-operative values in either the LAVA or suture anastomo-

0094-243X/86/1460700-2$3.00

sis groups. Histologically, neuroma formation appeared less in LAVA as compared to suture anastomosis.

DISCUSSION

Laser vascular "welding" has been shown to yield results comparable to conventional suture techniques in regard to tissue approximation and patency. Intuitively, it appears plausible that a thermal seal of the epineurial collagen and loose connective tissue may create a union or bond that would restrict abberant axonal sprouting as nerve regeneration occurs. From our early results, it appears that laser-assisted anastomosis of peripheral nerves is technically achievable. It is encouraging that electrical continuity was seen in LAVA across the anastomotic site at 6 and 12 weeks. In addition, neuroma formation was less with LAVA; however, quantitative analysis is pending. Further research in this and higher animal models would be a prerequisite to potential human application.

REFERENCES

1. M.R. Quigley, et al, LASERS SURG MED 5:357 (1985)
2. M.R. Quigley, et al, LANCER ii:334 (1985)
3. J.E. Bailes, et al, MICROSURGERY 6:163 (1985)

LASER-ASSISTED VASCULAR ANASTOMOSIS

M.R. Quigley
J.E. Bailes
H.C. Kwaan
L.J. Cerullo
Northwestern University Medical School and
VA Lakeside Hospital, Chicago, Il. 60611

ABSTRACT

The milliwatt CO_2 laser was used to anastomose the femoral artery in the rat. Aneurysm formation was a significant problem occurring in 18.6% of the 113 vessels studied. The rate was 29.0% in vessels examined one week or more following the procedure. Histologic analysis showed loss of media elastic elements as a probable etiologic factor. A parallel study of the bursting strength of sutured vs. lasered vessels showed a significant depression in the laser group as compared to suture at 1 day through 1 week.

INTRODUCTION

The use of laser energy to join biologic tissues is a new one, employing argon[1], Nd:YAG[2], and CO_2[3] sources. We have recently investigated the histologic changes following low power CO_2 laser application for the welding of small blood vessels in rats[4]. A large series of animals is herein reported

MATERIALS AND METHODS

One hundred and twenty-five rats underwend end-to-end anastomosis of the femoral artery with either standard suture microsurgical technique with 8-10 10-0 nylon sutures or a laser-assisted technique using 3 stay sutures placed 120 degrees apart and low power CO_2 laser energy (multiple applications .07W, 150 u spot). Animals were sacrificed from 1 hour to 12 weeks following surgery. Thirty two animals underwent bursting strength determinations by vessel cannulation and saline infusion.

RESULTS

Histology

Vessel tissue bonding is effected by non-specific coagulation necrosis of the medial proteins. There is widespread loss of elastic elements up to 200 u from the anastomosis and the elastic laminae disappears, never to return. The damaged media is replaced by spindle-shaped cells, probably myofibroblasts[5]. The lumen is initially stripped of endothelium, which returns in approximately 1 week. Subsequently, there is significant intimal myoproliferation which is present to 12 weeks time.

Aneurysms

The rate of anastomotic aneurysm formation was significant ($p<.001$) reaching 18.6% for the entire series, but 19.8⇒ 29,8 in vessels followed one week and beyond. Histologically, aneurysms manifested cessation of elastic laminae at the neck and dome composed of spindle-shaped myofibroblasts and adventitia.

Bursting Strength

There was significant depression of bursting strength in lasered versus sutured vessels at 1 and 3 days ($p<.001$) and 1 week ($p<.01$). Other time points were statistically equivalent (Table I)

TABLE I: BURSTING STRENGTH OF VESSELS IN mmHg

TIME	SUTURE MEAN±SEM (n)	LASER MEAN±SEM (n)	DIFFERENCE (PAIRS)
1 HOUR	691+107.2(6)	606+58.0(6)	N.S. (6)
1 DAY	952± 33.4(5)	456±27.1(5)	P<0.001(4)
3 DAYS	976± 4.0(5)	470±25.1(4)	P<0.001(4)
7 DAYS	1,144± 37.1(5)	745±73.6(4)	P<0.01 (4)
2 WEEKS	1,036± 20.4(5)	904±57.1(5)	N.S. (5)
3 WEEKS	1,058± 20.6(5)	992±18.8(4)	N.S. (5)

SEM, standard error of the mean; N.S., not significant ($P>.05$)

DISCUSSION

The low power CO_2 laser may be used to anastomose small blood vessels with acceptable patency rates[3]. The disadvantages of this technique are widespread medial injury which destroys the supporting elastic structures leading to aneurysm formation. The depressed bursting strength from 1 day-1 week might explain the occurrence of aneurysms during this period.

REFERENCES

1. O.M. Gomez, et al, TEX HEART INST J 10:145 (1983)
2. K.K. Jain, VASC SURG 17:240 (1983)
3. A. Serure, SURG FORUM 34:634 (1983)
4. M.R. Quigley, et al, LASERS SURG MED 5:357 (1985)
5. W.J. McCarthy, et al, J VASC SURG (In press, 1986)

MECHANICAL PROPERTIES OF END-TO-END LASER-ASSISTED AND SUTURED ARTERIAL ANASTOMOSES UNDER AXIAL LOADING

J. LoCicero III, R. S. Hartz, S. R. Shih, W. J. McCarthy
J. S. T. Yao, and L. L. Michaelis
Northwestern University Medical School, Chicago, IL 60611

INTRODUCTION

Laser energy can be used to perform vascular anastomoses with long-term patency rates similar to sutured anastomoses.[1,2] No clear advantage of laser technique has been demonstrated. This study analyzes the mechanical properties of end-to-end anastomoses.

EXPERIMENT

Right and left common carotid arteries of 92 rabbits were dissected microscopically. In 28 rabbits, the neck was closed after two hours (sham group). In 56 animals, the left carotid artery was sectioned and anastomosed with microsuture technique (10-12 10-0 nylon sutures). The right carotid artery was sectioned and anastomosed using low power $CO2$ laser energy ($250W/cm^2$). Four sutures were necessary to coapt the blood vessel for the laser anastomoses. The vessels were removed at 0, 1, 3, 7, 14, 21, and 42 days (eight operated and four sham per group), observed to be patent, and subjected to destructive axial loading in an Instron® testing machine. Eight control arteries were also tested. Tensile strength and elastic modulus were calculated from the data.

RESULTS

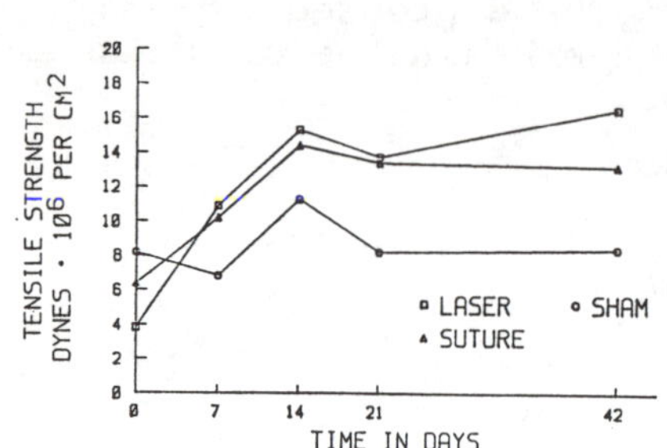

FIGURE 1

Figure 1 shows the results of recalculated tensile strength from the experiment. The laser-assisted (LA) group demonstrated a three-phase healing curve. Phase I (0-7 days): LA ($3.51 \pm 0.44 \times 10^6$ dyne/cm^2) was significantly weaker ($p < 0.01$) than sutured ($6.79 \pm 0.6 \times 10^6$ dyne/cm^2) as well as sham ($8.47 \pm 0.68 \times 10^6$ dyne/cm^2) and normal controls ($8.18 \pm 0.62 \times 10^6$ dyne/cm^2). Phase II (7-21 days): No significant difference between groups ($15.25 \pm 1.3 \times 10^6$ dyne/cm^2). Phase III (21-42 days): LA ($16.48 \pm 0.4 \times 10^6$ dyne/cm^2) was found to be significantly stronger ($p < 0.01$) than suture ($13.14 \pm 0.46 \times 10^6$ dyne/cm^2).

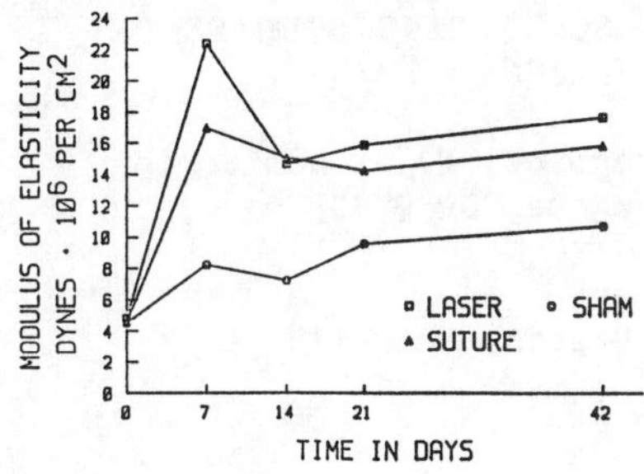

FIGURE 2

Figure 2 shows the modulus of elasticity data. During Phase I, LA ($22.38 \pm 3.99 \times 10^6$ dyne/cm^2) and sutured ($16.94 \pm 4.03 \times 10^6$ dyne/cm^2) were significantly stiffer ($p < .01$) than sham ($8.22 \pm 1.63 \times 10^6$ dyne/cm^2) and normal control ($5.01 \pm 0.58 \times 10^6$ dyne/cm^2), but were not significantly different from one another. During Phase II there were no significant differences noted. LA anastomoses were $15.87 \pm 3.16 \times 10^6$ dyne/cm^2 and sutured $14.25 \pm 1.2 \times 10^6$ dyne/cm^2. During Phase III, a parallel rise in stiffness was noted among all three operated groups: laser, sutured, and sham.

Since sham operated vessels were not stronger than the nonoperated control vessels during healing, perivascular adhesions were not considered to be a major factor in the superior strength of the laser-assisted and sutured anastomoses. In the early stages of healing, the sutures present in both laser and suture groups contributed to the strength of the anastomoses. Since laser-assisted anastomoses gained significant strength with time, however, the changing nature of the laser bond itself must have been responsible. Remodeling of elastin and collagen is most likely the major factor in this process.

CONCLUSION

The healing response demonstrated by the cumulative tensile strength and elastic modulus data implies that the photochemical laser bond not present in the sutured anastomoses does not add significant early strength to the vascular anastomosis. Laser strength does not appear to affect the intermediate phase of wound healing but stimulates the late phase of wound healing by yet undetermined mechanisms.

REFERENCES

1. D.K. Dew, et al, Lasers Surg. and Med. 3, 135-137 (1983).
2. R.S. Hartz, et al, Surg. Forum 36, 457-459 (1985).

ULTRAVIOLET RESONANCE RAMAN SCATTERING STUDIES OF ELECTRONIC EXCITATIONS

B. Hudson, P. B. Kelly, R. R. Chadwick, R. A. Desiderio
University of Oregon, Eugene, OR 97403

L. D. Ziegler
Northeastern University, Boston, MA 02115

ABSTRACT

Ultraviolet resonance Raman scattering spectroscopy has recently been shown to be a useful technique for the study of excited electronic states of molecules.[1-12] This technique can be used to determine the symmetry of upper electronic states in those cases where there is a symmetry forbidden electronic transition from the ground state. The same technique can be used to study large conformational changes associated with electronic excitation. In specific cases the pattern of rotational-vibrational lines observed in the resonance Raman spectrum can be used to measure the homogeneous line width of single rovibronic transitions in congested spectra. Finally, this technique provides new information about highly excited vibrational levels of the ground electronic state of polyatomic molecules.

RESONANCE RAMAN SCATTERING AND GEOMETRY CHANGE

Resonance with allowed electronic excitations that result in large geometry changes produces spectra with long progressions of the vibrations corresponding to this change in the structure. This has been demonstrated for ethylene,[2] ammonia,[4] oxygen,[6] carbondisulfide,[7] N-methylacetamide,[8] and benzene[10]. These results have been reviewed.[11, 12]

VIBRONIC COUPLING IN RAMAN SCATTERING

Symmetry forbidden electronic transitions can be induced by non-totally symmetric vibrations whose excitation results in mixing the excited state with another state giving rise to an allowed transition. If this coupling is linear in the vibrational coordinate then resonance with this electronic excitation will result in the unique enhancement of the vibrational transition corresponding to two quanta of the "promoting" mode for this transition.[1, 11, 12] This can be used to establish the electronic symmetry label of the upper state.[11, 12] This has been applied to the state of benzene near 210 nm,[1, 10] simple methyl substituted benzenes,[3] and butadiene.[9] In the case of benzene this method confirmed the theoretical prediction that the state near 210 nm is indeed one of B_{1u} symmetry. In the case of butadiene the resonance Raman spectra identified the low energy A_g excited state at an energy slightly below that of the allowed electronic transition to the B_u state.

PHOTODISSOCIATION AND ROTATIONAL STRUCTURE: AMMONIA

In the case of ammonia the rapid photodissociation of the upper A state results in broadening of the rovibronic lines so that they overlap. Analysis of the rotational structure associated with the resonance Raman spectrum can be analysed to determine the width of these lines and thus the rate of this dissociation reaction.[4] This circumstance occurs because in this case only a few rovibronic lines contribute significantly to the resonance Raman cross section. The number of such lines that are important is determined by their width, the damping factor in the Kronig-Kramers-Heisenberg expression for the Raman effect. The number of rotational states that make a contribution in turn determines the pattern of rotational side bands in the Raman spectrum.

REFERENCES

1. L. D. Ziegler and B. Hudson, J. Chem. Phys. 74, 982 (1981).

2. L. D. Ziegler and B. Hudson, J. Chem. Phys. 79, 1197 (1983).

3. L. D. Ziegler and B. Hudson, J. Chem. Phys. 79, 1134 (1983).

4. L. D. Ziegler and B. Hudson, J. Phys. Chem. 88, 1110 (1984).

5. L. D. Ziegler, P. B. Kelly and B. Hudson, J. Chem. Phys. 81, 6399 (1984).

6. P. B. Kelly and B. Hudson, Chem. Phys. Lett. 114, 451 (1985).

7. R. A. Desiderio, D. P. Gerrity and B. Hudson, Chem.Phys. Lett. 115, 29 (1985).

8. L. C. Mayne, L. D. Ziegler and B. Hudson, J. Phys. Chem. 89, 3395 (1985).

9. R. R. Chadwick, D. P. Gerrity and B. Hudson, Chem. Phys. Lett. 115, 24 (1985).

10. D. P. Gerrity, L. D. Ziegler, P. B. Kelly, R. A. Desiderio and B. Hudson, J. Chem. Phys. 83, 3209 (1985).

11. B. Hudson, P. B. Kelly, L. D. Ziegler, R. A. Desiderio, D. P. Gerrity, W. Hess and R. Bates, in Advances in Laser Spectroscopy, Vol. 3, edited by B. A. Garetz and J. R. Lombardi, (John Wiley & Sons, New York) in press.

12. B. Hudson, Spectroscopy 1, 22 (1986).

SINGLE-PHOTON LASER FLUORESCENCE STUDIES OF TRYPTOPHAN DYNAMICS IN PROTEINS

B. Hudson, D. A. Harris, R. D. Ludescher and L. McIntosh
University of Oregon, Eugene, OR 97403

ABSTRACT

Picosecond laser excitation of the fluorescence of tryptophan residues of soluble proteins is applied to the problem of the nature of the dynamic stability of proteins. Theoretical simulations of the internal dynamics of proteins indicate considerable flexibility and rapid motion due to thermal fluctuations. Experiments described here will test these simulation calculations using an experimental observable that can be calculated directly from the dynamic simulation.

INTRODUCTION

Recent molecular dynamics simulation calculations of the internal dynamics of proteins results in a picture of protein structure that is quite different from the usual static view deduced on the basis of results from x-ray crystallography. The validity of these simulations is dependent on the accuracy of the numerical methods used to solve Newton's equations of motion and on the inherent reliability of the potential energy function describing atom-atom interactions. The time scale of these computations is limited to times on the order of 100 ps to perhaps a few nanoseconds depending on the size of the protein and the nature of the treatment of solvation.

In order to test the validity of these simulations it is necessary to have a measurement that responds to very short time motions of the atomic interior. The best technique available for this purpose appears to be time resolved fluorescence anisotropy determinations of the reorientation of tryptophan residues that are interior to proteins. This paper will briefly outline the experimental techniques used in such studies and then describe results for two simple globular proteins, bovine pancreatic phospholipase A2 and T4 phage lysozyme.

EXPERIMENTAL METHODS

The experimental methods used in these studies involve excitation with the second harmonic frequency (near 300 nm) of a rhodamine 6G dye laser that is synchronously pumped with the second harmonic of a mode-locked cw Nd:YAG laser. The dye laser is cavity dumped to provide pulse repetition rates on the order of 100 kHz. The pulse width is a few ps. Single-photon counting detection is used. The microchannel-plate photomultiplier used has an intrinsic response time of about 140 ps. Deconvolution procedures result in a limiting temporal resolution on the order of 20 - 50 ps. The details of this method have been described elsewhere.[1, 2]

OBSERVED PROTEIN ANISOTROPY DECAYS

Phospholipase A2 is an enzyme with a single tryptophan residue near its surface. This tryptophan is important in the ability of this enzyme to have enhanced activity for micellar lipid substrates. The decay of the fluorescence of this protein is unusually complex for reasons that are not understood.[1] The decay of the anisotropy ot the tryptophan residue show rapid internal motion that is "quenched" upon binding to micelles.[2] This indicates that the tryptophan residue buries itself in the micelle upon binding in agreement with other non-dynamic techniques. The mobility of this residue is consistent with a static protein structure given the range of angular space available to it. These fluorescence results also show that the zymogen of this enzyme, prophospholipase A2, exhibits increased motional freedom for its tryptophan consistent with the disorder observed in its crystal structure. This may be related to the inability of the proenzyme to recognize micellar lipids as special substrates.

T4 phage lysozyme contains three tryptophans but can be genetically engineered to contain only one. Trp-138 is a buried residue whose motion requires flexibility of the entire protein. Studies to date indicate that this residue is immobile at room temperature but experiences considerable reorientational motion at slighlty elevated temperatures.[2] The ability to modify this protein at a great variety of sites makes it an excellent candidate for testing dynamics simulations. Preliminary results for a mutant form, where a replacement occurs at a position neighboring Trp-138, show that greatly enhanced mobility results. The ability of molecular dynamics calculations to reproduce the effect of this protein modification and the temperature dependence of internal mobility are in progress.

REFERENCES

1. R. D. Ludescher, J. J. Volwerk, G. H. de Haas and B. Hudson, Biochemistry 24, 7240 (1985).

2. B. Hudson, D. L. Harris, R. D. Ludescher, A. Ruggiero, A. Cooney-Freed and S. A. Cavalier, in Fluorescence in the Biological Sciences, edited by D. L. Taylor, A. S. Waggoner, F. Lanni, R. F. Murphy and R. Birge. (Alan R. Liss, New York, 1986), in press.

TISSUE MORPHOLOGIC ANALYSIS AND ABLATION RATES IN THE UV AND VISIBLE FOR LASER ANGIOPLASTY

M. Sartori, P.D. Henry, and R. Roberts
Section of Cardiology, Baylor College of Medicine, Houston, TX 77030

R. Sauerbrey and F.K. Tittel
Department of Electrical and Computer Engineering, Rice University
Houston, TX 77251-1892

ABSTRACT

Ablation rates were determined in human and canine aortas subjected to excimer and visible laser radiation. For UV and pulsed frequency doubled Nd:YAG lasers ablation rates were constant and depended linearly on average laser power, while for cw argon lasers ablation rates depended nonlinearly on laser power.

DISCUSSION

A major difficulty in applying lasers to cardiovascular therapy is arterial perforation. Therefore, it is important to assess in quantitative terms how changes in laser parameters influence the extent of laser induced injury. In this study, ablation rates for different lasers are experimentally determined.

The lasers used in this study were an excimer, argon ion, and a pulsed frequency doubled Nd:YAG laser. The excimer laser was operated at 193, 248, and 351 nm. The argon laser radiation used was continuous wave (cw) and chopped (duration 50 msec). The pulsed frequency doubled Nd:YAG laser was chosen because it generates a wavelength (532 nm) that is similar to the wavelength of the argon laser and pulse width (7 ns) close to the pulse duration of the excimer laser (12-20 ns). Segments of dog and human aortas were mounted perpendicular to the path of the laser beam at the focal point of a CaF_2 lens ($f \cong 50$ cm). Perforation of the sample was detected by placing an energy meter behind the sample.

The relationship between power and tissue penetration of the cw argon laser in the dog aorta is illustrated in Fig. 1. For power densities up to 12 W/mm^2, no ablation occurred in a time period of 60 seconds. However, with only a minor increase in power density to 14 W/mm^2, there was a dramatic increase in crater depth accompanied by wall perforation. A similar phenomenon was observed for argon laser radiation in the chopped mode. Thus, in these experiments the ablation rate has a strongly nonlinear dependence on the laser power density with cw and chopped argon laser radiation.

Effects of UV pulsed lasers on dog aorta are shown in Fig. 2. The crater depth produced by the lasers at a constant power density and repetition rate is plotted as a function of the irradiation time. In contrast to the argon laser, the crater depth (d) was proportional to the duration of exposure to irradiation (t). Thus, the slopes of the straight lines representing the ablation rates ($R = d/t$) were independent of d and t. Similarly, with pulsed Nd:YAG irradiation the crater depth increased linearly with time for constant power density and repetition rate. By varying power density and/or repetition rate of the pulsed UV and Nd:YAG lasers, it was demonstrated that R was a linear function of the product of power density and repetition rate only.

Comparing ablation rates at the same average power density of 1 W/mm^2, the R values for KrF, Nd:YAG, ArF, and XeF were 120, 75, 15, and 12 μm/s. No power density threshold was observed for ArF and KrF within the investigated parameter range.

This study shows the advantage of pulsed lasers as compared to cw laser irradiation for the effective control of tissue ablation. The linear relationships between ablation rate and laser power density for pulsed lasers make the ablation predictable. In the case of cw lasers, ablation rates increased nonlinearly with increasing power, making the selection of appropriate irradiation parameters difficult. The lack of control with cw lasers may explain the high incidence of perforation reported with the use of such lasers [1,2].

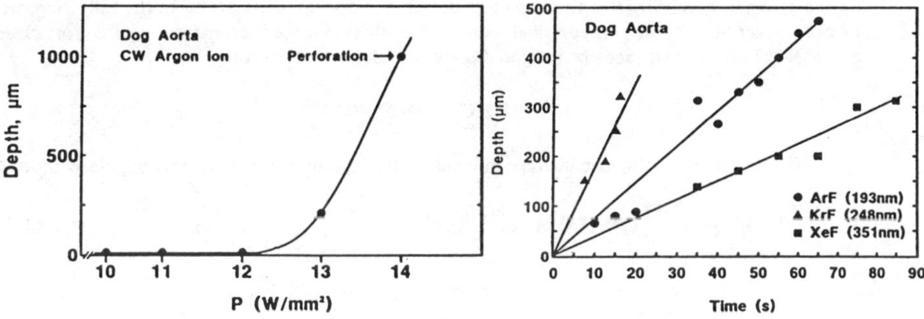

Fig. 1: Ar ion laser penetration depth as a function of power density for dog aorta.

Fig. 2: Excimer laser penetration depth as function of irradiation time for a constant power density of .375 W mm^{-2}.

REFERENCES

1. G.S. Abela, S.J. Norman, D.M. Cohen, D. Frausini, R.L. Feldman, F. Cress, A. Fenech, C. Pepine, and R. Conti, Circulation 71, 403-411 (1985).

2. J.M. Isner, R.F. Donaldson, J.T. Funai, L.I. Deckelbaum, N.G. Pandian, R.H. Clarke, M.A. Konstam, D.N. Salem, and J.S. Bernstein, Circulation 72, 184-191 (1985).

SECOND-HARMONIC GENERATION FROM CRYSTALLINE SURFACES

T.F. Heinz, M.M.T. Loy, and W.A. Thompson
IBM T.J. Watson Research Center, Yorktown Heights, NY 10598

ABSTRACT

Optical second-harmonic generation from crystalline materials has been found to be a sensitive probe of symmetry and order of the atomic structure of the surface layer. The nature of this process is discussed and illustrated with experimental results for Si(111) surfaces held under ultrahigh vacuum.

INTRODUCTION

In this paper, we review briefly our work concerning the application of optical second-harmonic generation (SHG) to the study of well-defined crystalline surfaces.[1–3] The technique, just as for the case of adsorbed monolayers, [4] relies on the fact that SHG is (electric-dipole) forbidden within the bulk of a centrosymmetric material, but is allowed in the surface region. The strength and polarization dependence of the surface-specific SH radiation carry information on the symmetry and order of the atomic structure at a crystalline surface. In the following, we give an overview of the theoretical concepts and present experimental results for clean Si(111)-2x1 and 7x7 surfaces investigated under ultrahigh vacuum conditions.

THEORETICAL DESCRIPTION

SHG from centrosymmetric media can be characterized by a nonlinear source polarization

$$\vec{P}^{NLS}(2\omega) = \vec{P}_s^{NLS}(2\omega)\delta(z) + \vec{P}_b^{NLS}(2\omega), \qquad (1a)$$

with

$$\vec{P}_s^{NLS}(2\omega) = \chi_s^{(2)}:\vec{E}(\omega)\vec{E}(\omega) \qquad (1b)$$

$$\vec{P}_b^{NLS}(2\omega) = \chi_b^{(2)}:\vec{E}(\omega)\nabla\vec{E}(\omega). \qquad (1c)$$

In this expression, $\vec{E}(\omega)$ is the applied electric field at frequency ω. The term $\vec{P}_s^{NLS}$ corresponds to the dipole-allowed polarization at the surface of the medium defined by $z = 0$; the second term represents the weak magnetic-dipole and electric-quadrupole terms allowed within the bulk of the medium. It is the form and magnitude of the tensor elements of of the surface nonlinear susceptibility, $\chi_s^{(2)}$, that provide us with information on the surface properties of the material under study. For the measurements on surfaces prepared in ultrahigh vacuum discussed here, the bulk contribution to the SH signal was found to be negligible. In general, the bulk contribution may be comparable with the surface contribution, in which case its influence must be taken into account. [5]

We summarize some information on the form of the surface nonlinear susceptibility tensor $\chi_s^{(2)}$ for different surface symmetries in Table I. For each symmetry class, the number of nonzero elements of $\chi_s^{(2)}$ and the number of independent elements of the tensor are shown. It should be noted that only for 4mm and 6mm symmetries does the surface SHG process respond as it would for a fully disordered surface (one with arbitrary rotational symmetry and mirror planes). Consequently, for the other cases, we can assess the degree of ordering of the surface by examining certain tensor elements of $\chi_s^{(2)}$.

Surface Symmetry	Nonzero elements of $\chi_s^{(2)}$ for SHG (for SFG or DFG)	Independent elements of $\chi_s^{(2)}$ for SHG (for SFG or DFG)
1	27 (27)	18 (27)
m	14 (14)	10 (14)
2	13 (13)	8 (13)
mm2	7 (7)	5 (7)
3	19 (21)	6 (9)
3m	11 (11)	4 (5)
4	11 (13)	4 (7)
4mm	7 (7)	3 (4)
6	11 (13)	4 (7)
6mm	7 (7)	3 (4)

Table I. Number of nonzero and independent elements of the surface nonlinear susceptibility tensor $\chi_s^{(2)}$ for crystal faces of differing symmetry. Values are given both for SHG and for sum- and difference-frequency generation.

The form of $\chi_s^{(2)}$ is, of course, related to the polarization dependence of the SH radiation. We illustrate this property with results relevant to the Si(111) surfaces studied experimentally. If the surface layer preserves the symmetry of the bulk crystal, the Si(111) surface will exhibit three-fold rotational symmetry with one of three mirror symmetry planes lying perpendicular to the $[01\bar{1}]$ direction (3m symmetry). For this surface symmetry, the independent elements of $\chi_s^{(2)}$ are $[\chi_s^{(2)}]_{zzz}$, $[\chi_s^{(2)}]_{zxx} = [\chi_s^{(2)}]_{zyy}$, $[\chi_s^{(2)}]_{xzx} = [\chi_s^{(2)}]_{yzy}$, and $[\chi_s^{(2)}]_{xxx} = -[\chi_s^{(2)}]_{xyy} = -[\chi_s^{(2)}]_{yxy}$, where x lies along the $[2\bar{1}\bar{1}]$ direction and y lies along the $[01\bar{1}]$ direction. Tensor elements with permuted second and third indices are equal and have not been indicated.

Explicit expressions for the polarization dependence of the surface SHG process can be obtained for a given form of $\chi_s^{(2)}$ by solving Maxwell's equations for the radiation from a sheet of current oscillating at the harmonic frequency.[5] With normally incident pump radiation, the polarization dependence of the SH intensity is given by

$$I(2\omega) \propto | \hat{e}(2\omega)\cdot\chi_s^{(2)}:\hat{e}(\omega)\hat{e}(\omega) |^2, \quad (2)$$

where $\hat{e}(\omega)$ is the polarization vector of the pump field and $\hat{e}(2\omega)$ is the polarization of the detected SH radiation. For the example of the surface layer with 3m symmetry, we then find for SH signals polarized along the x ($[2\bar{1}\bar{1}]$) and y ($[01\bar{1}]$) directions:

$$I_x(2\omega) = A\, | [\chi_s^{(2)}]_{xxx} |^2 \cos^2 2\theta \quad (3a)$$

$$I_y(2\omega) = A\, | [\chi_s^{(2)}]_{xxx} |^2 \sin^2 2\theta. \quad (3b)$$

Here A is a constant and the pump polarization lies at angle θ with respect to the x-axis.

If the surface has only a single plane of mirror symmetry, rather than the full 3m symmetry, the corresponding equations for the SH intensity are

$$I_x(2\omega) = A \mid [\chi_s^{(2)}]_{xxx}\cos^2\theta + [\chi_s^{(2)}]_{xyy}\sin^2\theta \mid^2 \tag{4a}$$

$$I_y(2\omega) = A \mid [\chi_s^{(2)}]_{yxy} \mid^2 \sin^2 2\theta. \tag{4b}$$

In these expressions, the mirror plane is oriented normal to the $[01\bar{1}]$ direction.

EXPERIMENTAL RESULTS

We have applied the SH technique to the study of Si(111) surfaces prepared in ultrahigh vacuum in both the 2x1 and 7x7 reconstructions.[6] Clean surfaces in the metastable 2x1 reconstruction were prepared by cleaving a Si bar *in situ* ; the equilibrium 7x7 reconstruction was produced by thermally annealing the cleaved surfaces. SH radiation was produced by normally incident excitation from a Q-switched Nd:YAG laser operating at 1.06 μm. Further discussion of the experimental procedure can be found in Refs. 1-3.

Figs. 1 and 2 display the observed polarization dependences for the two reconstructed surfaces. The strong dissimilarity in the SH response arises from the difference in surface symmetry in the two cases. The results for the 7x7 reconstruction correspond to the theoretical predictions for a surface with 3m symmetry [Eq. (3)], while the data for the 2x1 reconstruction imply a surface with only m symmetry [Eq. (4)]. Relying on the fact that for pump excitation polarized perpendicular to a mirror plane no SH light can be radiated with the same polarization, we have verified directly the existence of the three planes of mirror symmetry for the 7x7 reconstructed surface and the single plane of mirror symmetry for the 2x1 reconstructed surface. The implications of these results for models of the structure of the reconstructed surfaces are considered in Ref. 1

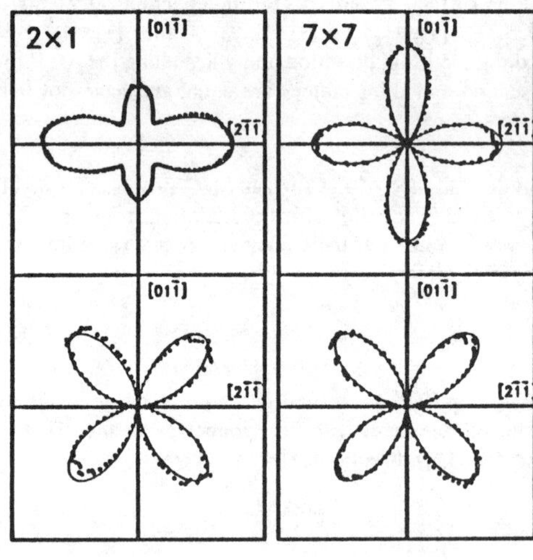

Fig. 1 Fig. 2

Figs. 1-2. SH intensity for Si(111)-2x1 and 7x7 surfaces as a function of pump polarization. The upper and lower panels display the SH signal polarized along the $[2\bar{1}\bar{1}]$ and $[01\bar{1}]$ directions, respectively. The solid lines are fits to theory from the symmetry analysis discussed in the text.

The dependence of the SH radiation on the surface structure has also been utilized in a number of studies of reactions and transformations occurring on Si(111) surfaces. These measurements include monitoring of the 2x1 → 7x7 surface phase transformation,[1] disordering of the surface induced by ion bombardment [2] and room-temperature Si deposition,[3] and oxidation of the Si surface. [2,7] As an optical probe, surface SHG could be applied for *in-situ* investigations under a variety of ambient conditions.

CONCLUSIONS

Studies of the 2x1 and 7x7 reconstructions of clean Si(111) surfaces have demonstrated the sensitivity of the SHG technique to the atomic arrangement of the surface layer. Accurate measurement of the symmetries of the surface have been obtained and a number of surface reactions and transformations have been monitored. The method should be applicable to a range of ordered surfaces and interfaces.

This work was supported in part by the U.S. Office of Naval Research

REFERENCES

1. T.F. Heinz, M.M.T. Loy, and W.A. Thompson, Phys. Rev. Lett. 54, 63 (1985).
2. T.F. Heinz, M.M.T. Loy, and W.A. Thompson, J. Vac. Sci. Technol. B 3, 1467 (1985).
3. T.F. Heinz, M.M.T. Loy, and W.A. Thompson, in *Laser Spectroscopy VII*, edited by T.W. Hansch and Y.R. Shen (Springer, Berlin, 1985), p.311.
4. For a references to this work, see Y.R. Shen, J. Vac. Sci. Technol. B 3, 1464 (1985).
5. See, for example, T.F. Heinz, H.W.K. Tom, and Y.R. Shen, Phys. Rev. A 28, 1883 (1983).
6. For SHG measurements of oxidized Si samples, see H.W.K. Tom, T.F. Heinz, and Y.R. Shen, Phys. Rev. Lett. 51, 1983 (1983).
7. H.W.K. Tom, X.D. Zhu, Y.R. Shen, and G.A. Somorjai, in *Proceedings of the XVII International Confernece on the Physics of Semiconductors* (Springer, Berlin, 1984), p.99.

Enhancement of Surface Optical Second Harmonic Generation by Transient Resonances on an Electrode Surface

C.D. Marshall and G.M. Korenowski
Rensselaer Polytechnic Institute, Troy, NY 12180-3590

ABSTRACT

Molecular size Ag clusters generated on a Ag electrode are found to form chemical complexes with adsorbed AgCl that give a resonance enhancement for optical surface second harmonic generation. Laser induced luminescence also confirms the presence and transient stability of these small Ag clusters.

INTRODUCTION

In the following material, we present studies of the Ag electrode in contact with a KCl electrolyte. Reflected optical second harmonic generation (SHG) and laser induced luminescence (LIL) are used to follow oxidation and reduction at the electrode surface and formation of reactive Ag clusters of molecular dimensions. These Ag clusters and adatoms, formed during electrochemical reduction, are found to enhance SHG from the electrode surface, and persist on the surface for seconds or possibly longer.

EXPERIMENTAL, RESULTS AND DISCUSSION

We present results and interpretations beyond those of previous studies on this system.[1,2] Apparatus description and experimental procedures will be presented in a future paper. The electrochemical system consisted of Ag/0.1M KCl/Ag. The electrode potential was ramped at 5mV per second. A typical oxidation-reduction cycle (ORC), no laser irradiation, is shown in Fig. 1. The scan is from left to right. The fundamental of a Nd:YAG laser provided the incident probe beam and the SHG signal from the Ag electrode was detected at 532 nm. Fig. 2 gives the SHG signal as a function of applied potential for 3.0 mJ/cm^2 per pulse of incident 1064 nm light. Cycle 1 begins with a polished electrode. Superimposed on the SHG signals are curves which model the oxidation product (AgCl) thickness on the electrode surface (L-with laser, NL-no laser). As seen from cycle 1, the SHG signal during oxidation and early stages of reduction is related to AgCl thickness. The effect of an electromagnetic enhancement from the electrochemically formed surface roughness is seen in the early stage of oxidation for cycle 2 as an early

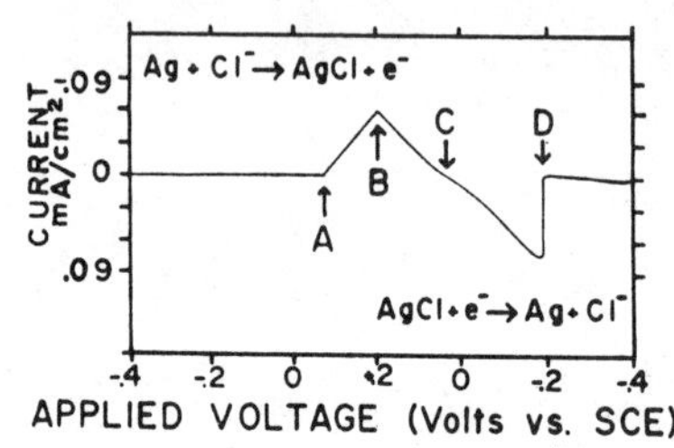

Fig. 1. Voltammogram of Ag/0.1M KCl/Ag system

0094-243X/86/1460716-2$3.00

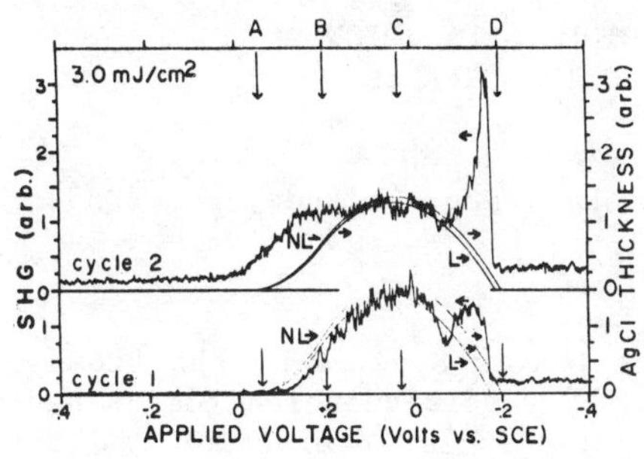

Fig. 2. SHG from Ag electrode during ORC

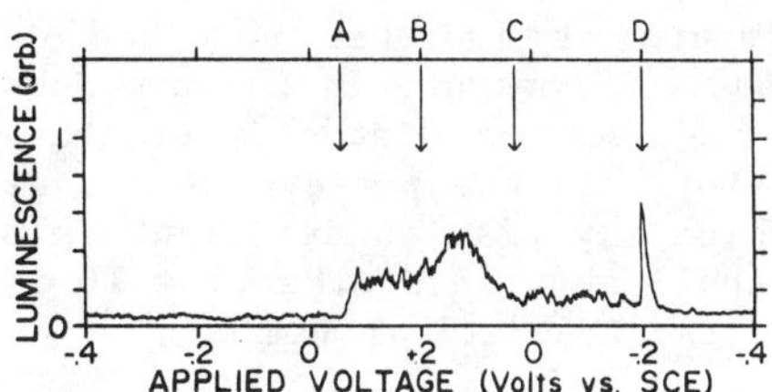

Fig. 3. Ag cluster LIL (360-400nm region) during ORC

rise of the SHG signal. This rise just preceding reduction completion, near arrow D, is a resonance enhanced signal. The resonance, believed to be at 532 nm, results from electronic states of Ag cluster-AgCl complexes formed on the surface. (Ag clusters have also recently been found to give an enhanced surface Raman signal).[3] As the AgCl thickness decreases, more probe laser reaches the metal surface where cluster-AgCl complexes form. This accounts for the increase in SHG signal as reduction nears completion. The roughened electrode, with its classical electromagnetic enhancement, makes the cluster-AgCl resonance enhancement more pronounced in cycle 2. With termination of the reduction current, a simultaneous drop in the SHG signal occurs as the AgCl and cluster-AgCl complexes disappear leaving "clean" small Ag clusters.

Fig. 3 gives the LIL from the Ag clusters during an ORC. Ag clusters up to approximately ten atoms absorb light in the near UV.[4] LIL is induced via multiphoton excitation using 1064 nm light at 13.2 mJ/cm^2 per pulse. Luminescence is observed for surface defects and clusters generated during oxidation. Once reduction begins, the LIL decreases. The Ag clusters formed during reduction are in intimate contact with AgCl and readily form complexes. When excited, the cluster-AgCl complexes undergo nonradiative decay through photoreduction of the AgCl thus decreasing the observed LIL. As the reduction current terminates and the SHG signal falls, the LIL from the "clean" Ag clusters rises abruptly and then falls as the growing clusters take on bulk metal properties.

These studies demonstrate that SHG and LIL can be used to study reactive small metal clusters and surface nucleation and growth at the solid-liquid interface.

REFERENCES

1. C.K. Chen, T.K. Heinz, D. Richard and Y.R. Shen, Phys. Rev. B27. 1965 (1983).
2. D.V. Murphy, K.V. VonRaben, T.T. Chen, J.F. Owen and R.K. Chang. Surf. Sci. 124, 529 (1983).
3. D. Roy and T.E. Furtak, (to be published).
4. W. Schulze, H.V. Becker and H. Abe, Chem. Phys. 157, 208 (1985).

LASER/SURFACE-ENHANCED ISOTOPE SEPARATION OF SPECIES ADSORBED ON SOLID SURFACES

J.T. Lin
Littons Systems, Inc., Laser Systems Division, Orlando, FL 32854

ABSTRACT

Isotope separation of adsorbed species is studied by the steady-state total excitation spectra governed by the frequencies (detunings) of the isotope species, the direct dipole-dipole interaction and the phonon-mediated indirect coupling. Under certain conditions, theoretical prediction of infinite steady-state excitation is presented.

INTRODUCTION AND EQUATION OF MOTION

Infrared laser excitation of mixture of isotopes absorbed on solid surfaces has been studied previously[1]. However, in the previous model, the indirect coupling (IC) or the phonon-mediated interaction among the isotopes has not been included. In the present paper, we shall include the effects of the IC which may play an important role in the isotope selectivity (IS). We shall show later that the IC is indeed the origin of the infinite IS in the theoretical limit.

Following the technique used in Reference 1, the dipole operators of the isotopes A and B, in the harmonic case, are given by a set of coupled equations:

$$i\dot{a}_A = (\Delta_A - i\delta_A)a_A + Da_B + V_A, \tag{1}$$

$$i\dot{a}_B = (\Delta_B - i\delta_B)a_B + Da_A + V_B, \tag{2}$$

where $\Delta_{A,B}$ are the detuning of the laser field w.r.t. the fundamental frequency of the species A and B; the phonon relaxation rates for the isolated (or diluted) adspecies are given by $\delta_{A,B}$; $V_{A,B}$ are the infrared excitation source terms; finally, D is the overall coupling constant consisting of the direct dipole-dipole coupling D_{AB} and the phonon-mediated IC as follows:

$$D = D_{AB} - (d_0 + id_1). \tag{3}$$

RESULTS AND DISCUSSION

The coupled equations of motion may be exactly solved for the harmonic case. When the laser pulse width is much larger than the phonon relaxation times, the steady-state total excitation is also available in an analytical form:

$$x = (x_A + x_B)/x_0, \tag{4}$$

where $x_{A,B}$ are the steady-state excitations for the interacting case normalized by the isolated case at resonance, x_0.

Results based on the analytic expression of Equation (4) are shown in Figure 1, where we have chosen $\Delta_A = -\Delta_B = 4$ and $\delta_A = \delta_B = V_A = V_B = 1$. In Figures A and B, we show the results for systems without the IC, noting that in Figure A the spectrum peaks at the isotope frequency W_A and W_B. These peaks are shifted when the couplings are included. In Figures E and F, the intensities of the peaks are equal and shifted closely caused by the IC terms. Figure G shows the theoretical limit where an infinite total excitation is found when the IC coupling reaches a value of $d_1^* = (\delta_A\delta_B - \Delta_A\Delta_B)^{\frac{1}{2}}$, which may be exactly calculated. These features of "giant excitation" may be realized by the fact that the IC coupling plays the role of frequency-shift, as shown by the imaginary part of Equation (3). At some point where the phonon damping factors, $\delta_{A,B}$, and the detunings, $\Delta_{A,B}$ are simultaneously compensated by the IC, the total excitation is then governed by an "effective resonance" state. We note that in the usually laser excitation of species in the gas-phase, the "giant peak" will not occur even at the resonance, since the system is dissipated by a finite damping factor which will not be "removed" if there is no other "interaction force" included.

In conclusion, we note that the "giant peak" feature governed by the phonon-mediated indirect coupling (IC) may be achieved at the appropriate species coverage, laser detuning and the substrate temperature. This theoretical limit of infinite excitation may not be reached experimentally due to the laser field induced broadening or other surface-induced effects; however, excitations with significant enhancement should be observed under the appropriate condition.

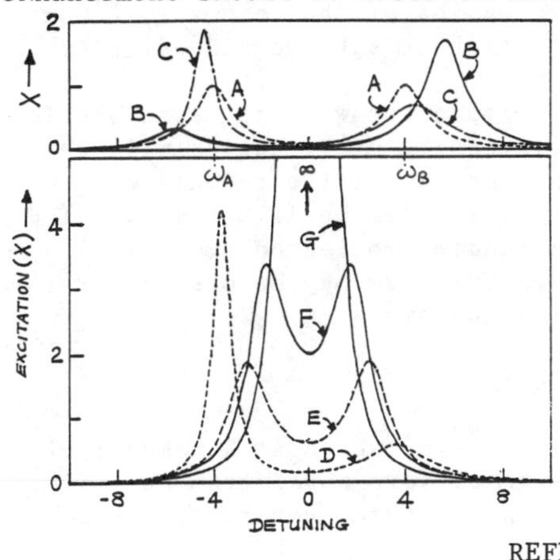

Fig. 1. Steady-state excitation spectra showing the effects due to the direct and indirect couplings for (d_0, d_1) = (A) (0,0), (B) (4,0), (C) (2,1), (D) (1,2), (E) (0,3), (F) (0,3.5) and (G) $(0, d_1^*)$.

REFERENCE

1. J.T. Lin and T.F. George, J. Chem. Phys. 78, 5197 (1983).

THEORY OF MORPHOLOGY DEPENDENT RESONANCES

Peter W. Barber
Clarkson University, Potsdam, NY. 13676

ABSTRACT

Dielectric particles exhibit natural resonant frequencies in the optical spectrum. The wavelength location of these natural resonances depends upon the morphology of each particular particle, i.e. its size, shape, and internal structure. Recent light scattering measurements and calculations will be presented which quantify the morphology dependence of the resonances.

INTRODUCTION

For dielectric spheres and cylinders with radius a and an incident plane wave of wavelength λ, the Lorenz-Mie formulism can be used to calculate the precise size parameters ($x=2\pi a/\lambda$) at which peaks will exist in the elastic-scattering spectrum[1]. Peaks associated with morphology-dependent resonances (MDR's) have been experimentally observed in elastic scattering[2], optical levitation force[3], fluorescence and laser emission[4], and Raman scattering[5].

Two features of the MDR's are particularly important. First, the resonance spectrum depends strongly on the physical features of the scattering particle - its size, shape, and index of refraction. Second, the intensity at resonance within and around a particle can be orders of magnitude greater than the incident intensity. Related to this feature is a spatial redistribution of the intensity. For example, the internal intensity for resonant spheres and cylinders is concentrated at the surface.

The mathematical origin of the resonances will be described for spherical particles. The internal intensity distribution will be shown for a resonant glass fiber and the use of the resonance spectrum to size a glass fiber accurately will be illustrated. The calculated resonance spectrum for a randomly-oriented prolate spheroid indicates that the resonance spectrum may be useful for the characterization of randomly oriented objects.

RESULTS

The scattering efficiency for a sphere, which is a measure of the ability of a sphere of radius a to capture power from an incident plane wave and redirect it as scattered power, is given by

$$Q_s = \frac{2}{x^2} \sum_{n=1}^{\infty} (2n+1)(|a_n|^2 + |b_n|^2), \tag{1}$$

where the size parameter $x = 2\pi a/\lambda$ and λ is the wavelength in the surrounding medium. The expansion coefficients a_n and b_n are given by

$$a_n = \frac{j_n(x)[mxj_n(mx)]' - m^2 j_n(mx)[xj_n(x)]'}{h_n^{(2)}(x)[mxj_n(mx)]' - m^2 j_n(mx)[xh_n^{(2)}(x)]'} \tag{2}$$

and

$$b_n = \frac{j_n(x)[mxj_n(mx)]' - j_n(mx)[xj_n(x)]'}{h_n^{(2)}(x)[mxj_n(mx)]' - j_n(mx)[xh_n^{(2)}(x)]'}, \tag{3}$$

where m is the relative refractive index and j_n and $h_n^{(2)}$ are the spherical Bessel functions and Hankel functions of the second kind, respectively.

The calculated scattering efficiency as a function of size parameter is shown in Fig. 1 for a dielectric sphere with an index of refraction of 1.4.

The detailed behavior of the rapid oscillation can be seen in Fig. 2, which is an expanded view of a portion of Fig. 1. Here we note further fine structure consisting of narrow peaks on the rapid oscillations that were seen in Fig. 1. The peaks in Fig. 2 (both narrow and broad) are a manifestation of the resonance behavior of the a_n and b_n coefficients given by Eqs. (2) and (3). Figure 2 is called the resonance spectrum. Each of the coefficients a_n and b_n can be associated with a structural mode of electromagnetic vibration of the sphere. The peaks in Fig. 2 are labeled according to the coefficient that is responsible for the resonance. The subscripts n and l denote the lth resonance of the nth mode.

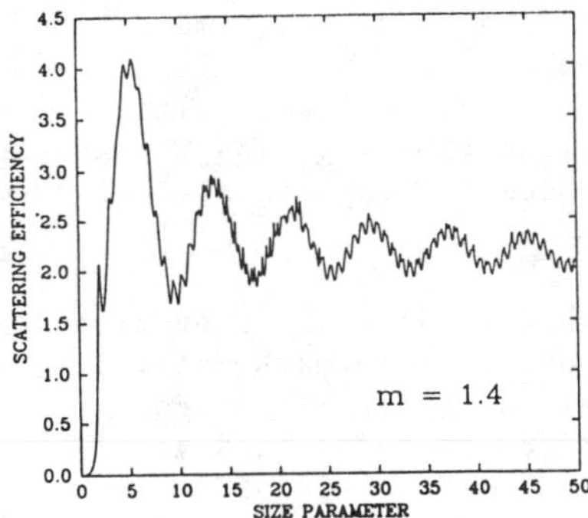

Fig. 1. Scattering efficiency versus size parameter for a dielectric sphere.

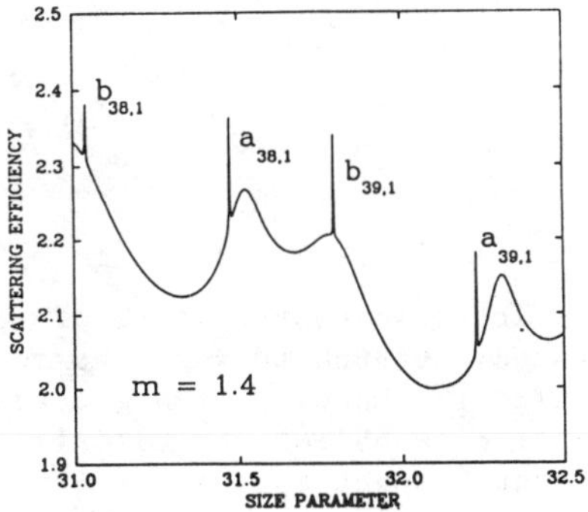

Fig. 2. An expanded view of Fig. 1 over a narrow size-parameter range. The n,l subscripts indicate the mode and the order of each resonance, respectively.

The solution for the circular cylinder parallels that for the sphere. Results of calculations of the total TM electric field intensity inside and around a cylindrical glass fiber for x in resonance with modes n=53, 1=3 and for a nonresonant value of x are presented in Fig. 3. Note the increase in the internal intensity and the peaking near the surface for the resonant case.

The resonance spectrum has been used to characterize both spheres and cylindrical fibers. Experimental measurements can easily be made using conventional spectroscopic techniques. Figure 4 shows a measured spectrum and the best-fit calculated spectrum for a glass fiber. The spectral scattering approach has permitted a very accurate size measurement.

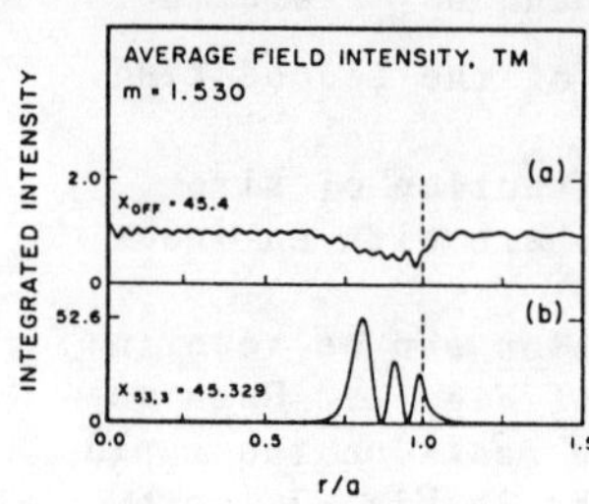

Fig. 3. The average TM field intensity integrated around a circle centered at the cylinder axis $[(2\pi)^{-1}\int E(r,\phi).E^*(r,\phi)d\phi]$ as a function of the radius r of the circle. Results are shown for an off-resonance condition and for x values in resonance with mode n = 53, 1 = 3.

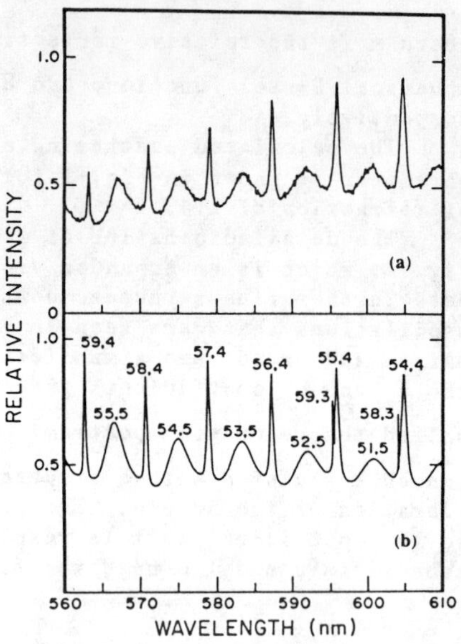

Fig. 4. (a) Experimental and (b) calculated (m = 1.5265, d = 9.51 μm) spectral scattering from a glass fiber, giving TE resonant modes for an infinite cylinder.

The resonance spectra of spheres and cylinders are convenient to study because they can be related to closed-form mathematical solutions. However, the general application of the resonance spectrum to object characterization requires an understanding of the resonance spectra of arbitrary geometries. An example of a resonance spectrum for a more complicated geometry is the calculated result shown in Fig. 5 for a randomly oriented prolate spheroidal particle.

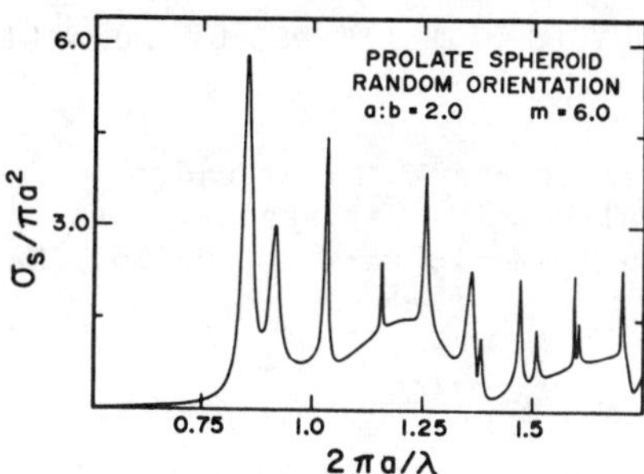

Fig. 5. Calculated spectra for randomly oriented 2:1 prolate spheroid

The existence of sharp resonances indicates that the particle characterization applications and enhanced intensity results, which have already been exploited for spheres and cylinders, are also applicable to a broader class of particle geometries.

REFERENCES

1. M. Kerker, The Scattering of Light and Other Electromagnetic Radiation (Academic, New York, 1969); C.F. Bohren and D.R. Huffman, Absorption and Scattering of Light by Small Particles (Wiley, New York, 1983.)
2. A. Ashkin and J.M. Dziedzic, and R.H. Stolen, Appl. Opt. 20, 1803 (1981); A. Ashkin J.M. Dziedzic, and R.H. Stolen, Appl. Opt. 20, 2299 (1981); J.F. Owen, P.W. Barber, B.J. Messinger, and R.K. Chang, Opt. Lett. 6, 272 (1981).
3. A. Ashkin and J.M. Dziedzic, Phys. Rev. Lett. 38, 1351 (1977); A. Ashkin, Science 210, 1081 (1980); P. Chylek, J.T. Kiehl, J.K.W. Ko, and A. Ashkin, in Light Scattering by Irregularly Shaped Particles, D.W. Schuerman, ed. (Plenum, New York, 1980) p. 153.
4. R.E. Benner, P.W. Barber, J.F. Owen, and R.K. Chang, Phys. Rev. Lett. 44, 475 (1980); J.F. Owen, P.W. Barber, P.B. Dorain, and R.K. Chang, Phys. Rev. Lett. 47, 1075 (1981); S.C. Hill, R.E. Benner, C.K. Rushforth, and P.R. Conwell, Appl. Opt. 23, 1680 (1984);H.-M. Tzeng, K.F. Wall, M.B. Long, and R.K. Chang, Opt. Lett. 9, 499 (1984).
5. J.F. Owen, R.K. Chang, and P.W. Barber, Aerosol Sci. Technol. 1, 293 (1982); R. Thurn and W.Kiefer, Appl. Spectrosc. 38, 78 (1984).

GENTLE SURFACE ANALYSIS WITH INTENSE UV LASER LIGHT

C. H. Becker and M. M. Freund,
Chemical Physics Laboratory,
SRI International, Menlo Park, CA 94025, USA

ABSTRACT

The surface analysis by laser ionization (SALI) method is described which can provide very gentle, effectively nondestructive, surface analysis by virtue of its high sensitivity. When combined with continuous sputtering (ion milling), the method gives material composition information as a function of depth (depth profiling). Examples are presented of depth profiling through the near surface region of three 304 stainless steel samples each with a different chemical treatment for corrosion resistance.

INTRODUCTION

The SALI approach to surface analysis developed at SRI[1-3] uses a variety of types of particles to irradiate the surface, usually a keV energy ion beam or photon (laser) beam at low or modest doses, to remove atoms and molecules from the surface for subsequent mass spectrometry. These desorbed particles (usually neutral in charge) are intersected above the surface by a focused pulsed intense UV laser beam of power typically $\gtrsim 10^9$ W/cm^2 for nonresonant MPI, often saturating the ionization. The photoions are then mass analyzed by reflecting TOF. The high sensitivity observed (demonstrated[1] ppm analysis for all elements simultaneously while removing about 1% of a monolayer over 10^{-1} cm^2, as well as 10^{-17} mole detection[2]) enables minimum irradiation of the surface itself and hence "gentle" surface analysis to be performed (negligible modification of the the top monolayer). Laser ionization of neutrals usually offers much more quantitative surface composition spectra than obtained by measuring the secondary ions directly as in SIMS.[3]

Gentle surface analysis with SALI is particularly easily obtained by keeping the surface irradiating beam dose low by pulsing the ion beam. When composition as a function of depth into the material is desired (depth profile), a straightforward approach is the use of a continuous ion beam to sputter, i.e. ion mill, into the material.

An interesting and important application for SALI presented below is that of compositional analysis of the near surface region of stainless steels. The spectra below show the chemical effects, as a function of depth into the material, of three chemical pretreatments of 304 stainless steel intended to retard degradation reactions that can occur under harsh conditions.

EXPERIMENTAL DETAILS

The experimental arrangement has been described in some detail previously.[1] Briefly, sputtering was performed using a rastered and scanned Ar^+ beam at 3 keV incident at about 70 degrees from normal. The ion beam was differentially pumped and the chamber base pressure was in the 10^{-9} torr regime. The laser ionization was performed with a 193 nm excimer laser beam focused by a 40 cm focal length suprasil lens, to give a power density of about 5 x 10^8 W/cm^2. The photoions were detected by a chevron microchannel plate assembly with the signal attenuated by 20 dB to maintain a linear response by the detection electronics; the data was recorded in analog form with a 100MHz transient digitizer. Secondary ions were discriminated against by using the TOF reflector as an energy bandpass filter.[1]

Sample A is 304 stainless steel that has been electropolished with phosphoric acid with chromium trioxide electrolyte for 10 minutes at 4 ampere/in^2 at room temperature. Sample B has been electropolished with 85% phosphoric acid with 2% dissolved type 304 stainless steel at 2.29A/in^2 for 10 minutes at 70 C; the sample was then oxidized in air for 1 hour at 400 C. Sample C was electropolished at 120 C with the same chemistry as for sample B.

To quantify the data accurately, comparison has been made with NBS standard stainless steel C1154. The semiquantitative raw data can then be scaled quantitatively for the elements in the metallic state with estimated accuracy of 10%; the elemental data plotted below has been so scaled as atomic per cent. The molecular information describing the oxide layer (such as sputtered diatomic oxides) was not quantified using oxide standards. However the raw data which is plotted for the oxide molecules give a qualitative picture of the varying oxide composition, and comparison from one sample to another will give quantitative values of the oxides' relative compositions.

The depth scale for the stainless steel sputtering was not determined directly. However comparison was made with a Ta_2O_5 film standard. The 21 minute sputtering plotted corresponds to about 700Å depth of sputtering of the Ta_2O_5 film. The sputter yields of Ta_2O_5 and stainless steel have been reported to be approximately equal.[4]

RESULTS

The depth profiles of Fe, Ni, Cr, V, Mn, and Co (from the atomic photoions) as well as FeO^+, CrO^+, and NiO^+ signal profiles for samples A, B, and C are shown in Figs. 1-3, respectively. The depth profiles are reported beginning with initial sputtering through a carbonaceous overlayer commonly associated with any sample introduced from laboratory air with no precleaning. That is why the signals for metal atomic and diatomic oxide photoions have zero intensity upon initial sputtering. The comparison to the NBS standard has been applied to the Fe, Ni, Cr, V, Mn, and Co atomic ions at long sputtering times; the form for each depth profile represents the relative signal for that element as a function of sputter time or depth. The diatomic oxide ions are scaled as directly measured with the scaled profile of Cr replotted for reference. The diatomic oxide ions are by far the most prominent molecular ions observed. Other elements were detected and could have been plotted as well, such as Si and Cu.

Careful examination of these figures reveals numerous interesting features in the depth distributions. Comments on some of the prominent features follow.

It is apparent that sample B is Fe surface rich, with sample C being Cr surface rich. The Ni concentration reaches its maximum at greater depths than for Cr, for all samples. For all samples, the Mn profile resembles that of Cr, while V and Co resemble Ni.

Examination of the atomic and especially the diatomic oxide profiles shows that sample B has much more iron oxide than the other two samples, and the iron and chromium oxides are most separated for sample B; this apparently is the result of the heat treatment in air. For samples A and C, though the iron oxide signal is much reduced compared to sample B, there still appears to be a partial separation of iron and chromium oxide regions with iron oxide positioned nearer the surface.

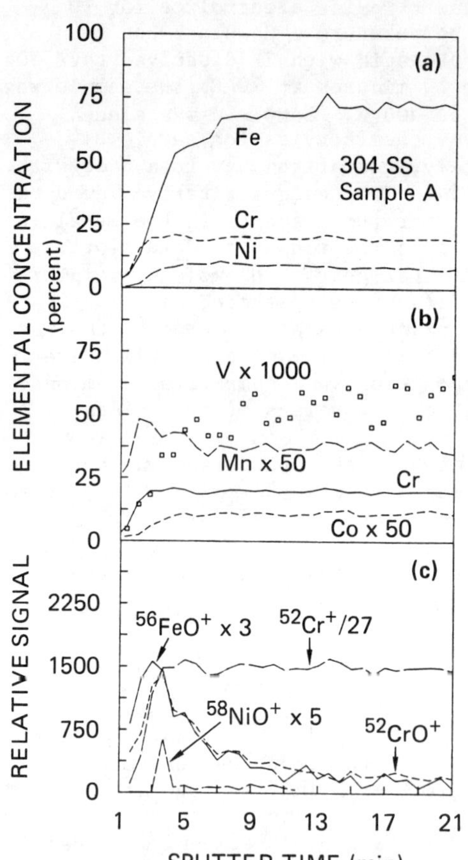

Fig. 1. Intensity as a function of Ar^+ sputtering time (depth) for sample A for (a) the atomic Fe, Cr, and Ni photoions, (b) the atomic V, Mn, Co and Cr photoions, and (c) the relative intensities as measured of the indicated diatomic oxide photoions. Sections (a) and (b) have been calibrated for long sputtering times to give atomic percentages based on an NBS C1154 stainless steel standard. Laser ionization was performed with 193 nm radiation.

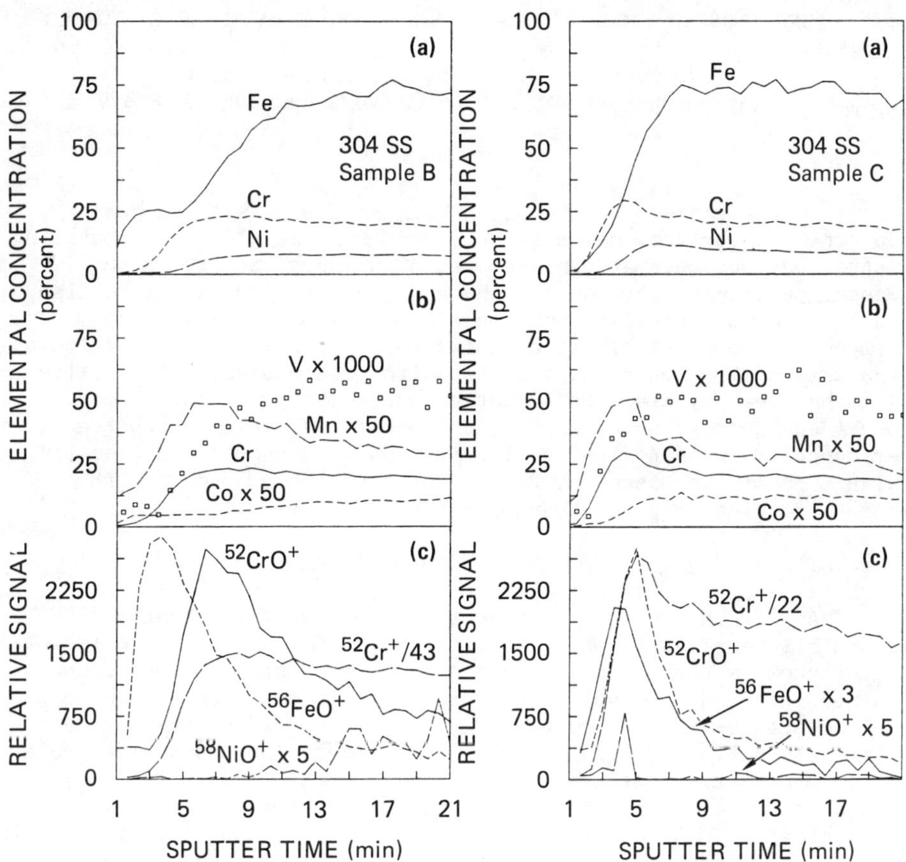

Fig. 2. Same as for Fig. 1 except for sample B.

Fig. 3. Same as for Fig. 1 except for sample C.

Support from Electric Power Research Institute and National Science Foundation, Division of Materials Research, is gratefully acknowledged.

REFERENCES

1. C. H. Becker and K. T. Gillen, Anal. Chem. 56, 1671 (1984).
2. C. H. Becker and K. T. Gillen, J. Opt. Soc. Am. B2, 1438 (1985).
3. C. H. Becker and K. T. Gillen, J. Vac. Sci. Technol. A3, 1347 (1985).
4. H. H. Andersen and H. L. Bay, in Sputtering by Particle Bombardment I, R. Behrisch ed., (Springer-Verlag, N.Y., 1981), Chap. 4, pp. 173, 184; G. Betz and G. K. Wehner, in Sputtering by Particle Bombardment II, R. Behrisch ed., (Springer-Verlag, N.Y., 1983), Chap. 2, p. 58.

LASER-INDUCED VAPORIZATION MASS SPECTROMETRY OF REFRACTORIES

D.W. Bonnell, P.K. Schenck, and J.W. Hastie
National Bureau of Standards, Gaithersburg, MD 20899

ABSTRACT

A Nd/YAG laser, coupled to a High Pressure Sampling Mass Spectrometer system, has been developed for the study of vaporizing refractory surfaces at temperatures high enough to attain ca. 10^5 Pascal pressures. Data on both BN and graphite indicate establishment of thermal equilibrium among the evolving species at a temperature characteristic of known phase transformations. Velocity analysis of the neutrals, together with mass analysis of positive ions produced by electron impact of the neutral molecular beam, provided unambiguous assignment of species, as well as additional evidence of interspecies equilibrium. BN(g) was observed directly, with a partial pressure at 2900 K in excellent agreement with spectroscopically-based thermodynamic data.

INTRODUCTION AND RESULTS

Many of the problems inherent in the study of refractory materials (especially those melting above 2500 K) can, in principle, be overcome with the use of lasers as heat sources. Use of focused, pulsed laser energy avoids the usual containment problem as the sample acts as its own container. In addition, laser heating can be extremely rapid and highly surface localized. The short time scale also permits time-resolved studies and the attainment of high pressures while maintaining a surrounding vacuum. Use of a mass spectrometer (MS) adds the advantage of species-specificity. This is particularly essential for refractory-material vapors, as many complex species can co-exist at very high temperatures.

Figure 1 shows schematically the data collection arrangement with a typical time-resolved MS display for C^+ (C). The MS system is based on modular quadrupole components, with a conventional electron impact ion source and channel-type electron multiplier detector. A specially designed vacuum system was used to maximize gas handling capability/pumping speed with a minimum molecular beam path (see refs 1 and 2 for further details on the MS and vacuum system, together with principles of high pressure sampling and quantitative high temperature MS). The measured net pumping speed in the sample region is ~2000 ℓ/s, at an ambient pressure of 0.003 Pa, and the path length to the first collimating orifice is less than 7 cm. These conditions lead to an insignificant pressure increase in the background pressure outside the gas-dynamic shock boundaries, and mean-free-path calculations indicate that less than 3% of the collisions occur outside the plume boundary. Velocity measurements yield an expansion gas cooling factor of ~14, indicative of a well-developed supersonic expansion with $\Delta P > 10^5$.

The laser is a Nd/YAG pulsed system which delivers up to 200 mJ in the ~ 10 ns pulses used at repetition rates up to 20 Hz. Typical experimental conditions deposited 20-80 J/cm^2 at 532 nm into a spot size of $5x10^{-4}$ cm^2, yielding strong MS signals for species from both BN and graphite.

Experimental data was collected by tuning the MS to a specific mass, and signal averaging (1K points at 5μs/point) up to 1000 laser shots. See inset C^+ trace in Fig. 1.

Derived temperatures in both systems were found to be independent of laser fluence over the order-of-magnitude investigated. This observation was confirmed both by species time-of-flight (TOF) and species ratios. Similar effects have been noted by other workers[3]. Temperature determinations were made by a variety of methods, including relative TOF, species intensity ratios, and gravimetric comparison of material loss with ion intensity, using an independently obtained MS calibration constant. Excellent internal agreement was found. In the graphite system, C_n species, with n up to 9, were observed. From BN, the species B, B_2, BN, BO, N_2, BO_2, and B_2O_3 were observed. $(BN)_2$, O_2, H_2O, N, and were not observed at the 1 Pa detection limit. The B-O species arose from oxygen lattice impurities in the BN fabrication process and totaled less than 5% of the vapor. The $C_1{}^+$ intensity was observed to vary quadratically with laser fluence, unlike higher carbon polymers, suggesting photodissociation of C_n to C_1. This explains our observation of a significant excess of $C_1{}^+$, clearly identified as due to a C_1 precursor. Additional detail on the initial results of these studies may be found elsewhere[4].

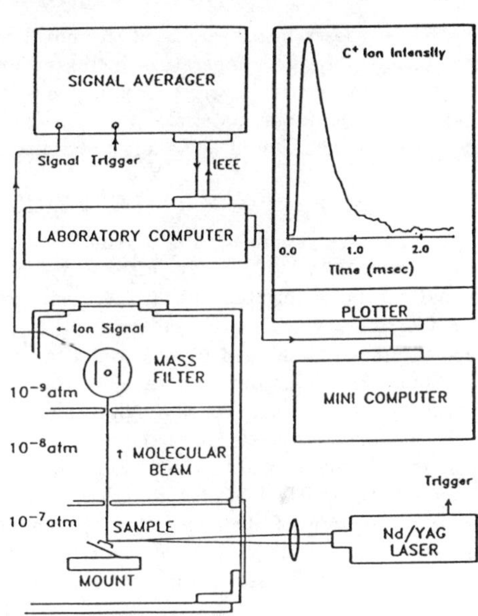

FIGURE 1.

REFERENCES

1. J.W. Hastie and D.W. Bonnell, in Thermochemistry Today and its Role in the Immediate Future, ed. M.A.V. Ribeiro da Silva, (Reidel Publ., Boston, 1983).
2. J.W. Hastie, Pure and Applied Chem., 56,1583 (1984).
3. J.E. Rothenberg, G. Koren, and J.J. Ritso, J. Appl. Phys. 57, 5072 (1985).
4. J.W. Hastie, D.W. Bonnell, and P.K. Schenck, NBSIR 84-2983 (1984).

RAMAN MICROSCOPY OF SOLID SURFACES AFTER LASER IRRADIATION

P.M. Fauchet, I.H. Campbell
Princeton University, Princeton, NJ 08544

F. Adar
Instruments S.A., Metuchen, NJ 08840

ABSTRACT

We demonstrate that Raman scattering with one micron spatial resolution is an ideal tool to characterize the effects of pulsed laser irradiation of solid surfaces. In particular, we observe changes in crystallinity, stress, and uniformity which extend many microns beyond the visible boundary between processed and unprocessed silicon thin films.

INTRODUCTION

Focused laser beams are used to anneal wafers, repair integrated circuits, or deposit metal lines [1]. Large temperature increases and large temperature gradients are induced locally and transiently. Laser processing may be performed within microns of pre-existing or future devices. If the material's properties are affected in some way beyond the processed area, then the operation of devices in proximity may be altered.

EXPERIMENTS

Thin polycrystalline films of silicon on insulator (SOI) were irradiated with short pulses from a Nd:YAG laser [2]. In the experiments discussed here, the wavelength was 1064 nm and the pulse duration 100 ns. We will report elsewhere on similar experiments performed at other wavelengths and with other pulselengths [3]. Transient melting was produced in areas that ranged from 1 to 100 microns. The surface is observed by high resolution Nomarski microscopy and then by Raman microscopy [4]. The Ar^+ probe laser is focused to a one micron spot through a microscope objective. Three parameters characterize the one-phonon Stokes line : the peak frequency, the full-width half-maximum, and the asymmetry factor defined as the ratio of the half-width half-maximum (HWHM) on the low energy side to the HWHM on the high energy side. In Figure 1, we plot these three quantities as a function of distance from the center of an irradiated area. The dashed line represents the boundary between melted and unmelted regions, as observed by Nomarski microscopy. The full lines on the right represent the parameters of the virgin SOI film.

DISCUSSION

The visible boundary is accompanied by an abrupt jump in the three parameters. After melting and recrystallization, the silicon regrows with larger grains and is under a considerable and quite homogeneous tensile stress. We infer this from the much reduced asymmetry, the depressed peak frequency, and the relatively narrow FWHM, respectively [2]. The stress is partly relieved when the surface appears rough (in the center of the melted region). The recovery of the Raman line is not complete immediately after the boundary and continues for over 20 microns. In particular, the stress decreases at an approximate rate of 10^8dyn/cm^2/μm, and the average microcrystal size decreases toward the initial 20 nm size. From the constant FWHM, we also conclude that the film is quite inhomogeneous, probably under the combined action of nonuniform stress and microcrystal size.

These results, which appear to depend on the processing laser, nevertheless indicate that laser irradiation may alter the material's properties beyond the processed area. These alterations will probably exist and be detrimental to device operation only under some conditions which we are investigating.

We acknowledge support from NSF grant ECS-8504202.

REFERENCES

1. see the proceedings of the fall meeting of the Materials Research Society, held annually in Boston.

2. P.M. Fauchet, I.H. Campbell, and F. Adar, Appl. Phys. Lett.**47**, 479 (1985).

3. I.H. Campbell, P.M. Fauchet, and F. Adar, to be published in the proceedings of Symposium C of the 1985 MRS fall meeting, Boston.

4. P.M. Fauchet, IEEE Circuits and Devices, January 1986.

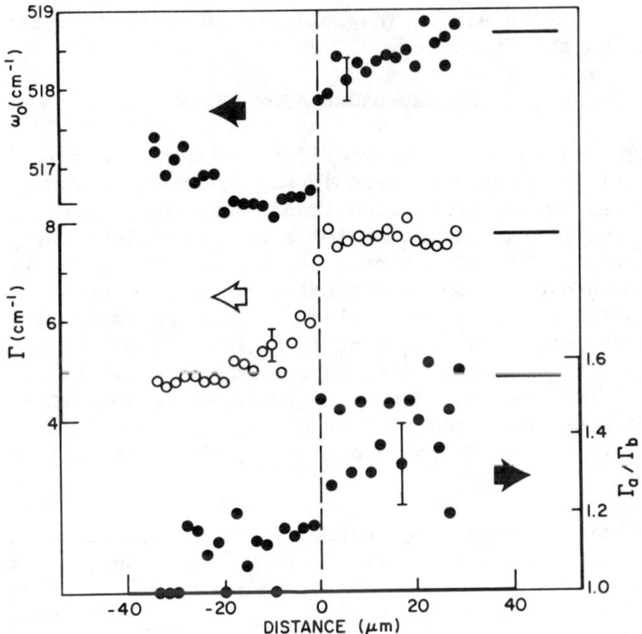

Figure 1

Stokes line parameters inside (x<0) and outside (x>0) an area that has been melted after irradiation by 600 laser shots at 1.06μm. The material's properties become undistinguishable from those of virgin SOI (shown by the full lines) after 20μm. ω_0 = peak frequency; Γ = FWHM; Γ_a/Γ_b = asymmetry factor.

ANALYSIS OF SOLID SURFACES USING A PULSED-LASER TIME-OF-FLIGHT ATOM-PROBE*

T. T. Tsong, Y. Liu and S. B. McLane
Pennsylvania State University, University Park, PA 16802

ABSTRACT

A high resolution pulsed-laser time-of-flight atom-probe field ion microscope has been developed which is a surface analytical tool of single atom detection sensitivity.[1] It is a mass and energy analyzer with a resolution and accuracy of ~5 parts in 10^5, and an ion reaction time measuring device of 20 fs time resolution. This system can be used to analyze surface atoms one by one and suface atomic layers one by one also. The application of this system to study solid surfaces are described.

PULSED-LASER ATOM-PROBE

We have developed a time-of-flight atom-probe using laser pulse stimulated field desorption technique. As shown in Fig. 1, the system uses laser pulses of wavelength 337 nm and pulse width 300 ps to produce pulsed field desorption for a time-of-flight mass and ion energy analysis. The effect of laser pulses is mainly thermal, although photo-excitation effect can also occur especially for semiconductor samples. Ions are collected from the sample surface within an area of one to several atom diameter. Since the system counts ions one at a time, the conditions are adjusted so that on average only one ion is detected every 5 to 20 laser pulses. The flight times of ions are measured with an electronic time-digital-converter of 1 ns resolution which can count flight times of up to 8 ions in one trigger.

In ordinary atom-probes, pulsed field desorption is done by ns high voltage pulses, which induce a very large energy spread of ions. Such systems have very poor mass resolution unless a 163° Porschenrier electrostatic flight time focusing lens is used. Unfortunately with such an artificial scheme, information on ion energy distribution and ion

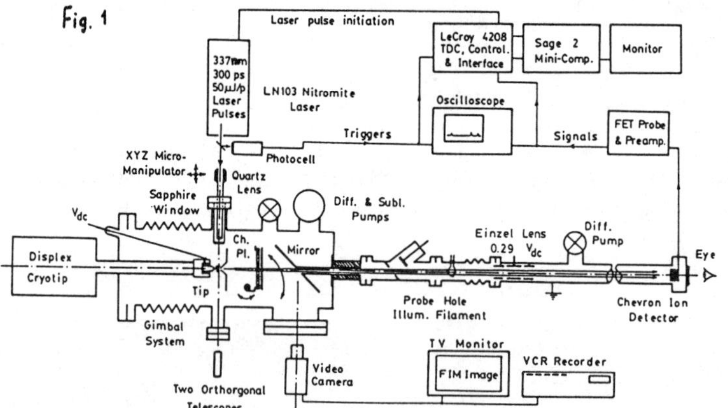

0094-243X/86/1460732-2$3.00

reaction time is completely lost. Also such systems cannot study poor conductivity materials such as a high purity Si. These difficulties are removed in the pulsed-laser atom-probe. Our system, with an overall resolution of 1 ns of the time measuring devices and a careful procedure for determining the flight path constant and the time-delay constant, is capable of achieving an accuracy and resolution of 5 parts in 10^5 in ion energy and mass measurements. By adopting an ion reaction time amplification scheme[2] to the system, a time resolution of 20 fs has been achieved in an ion reaction time measurement.[3]

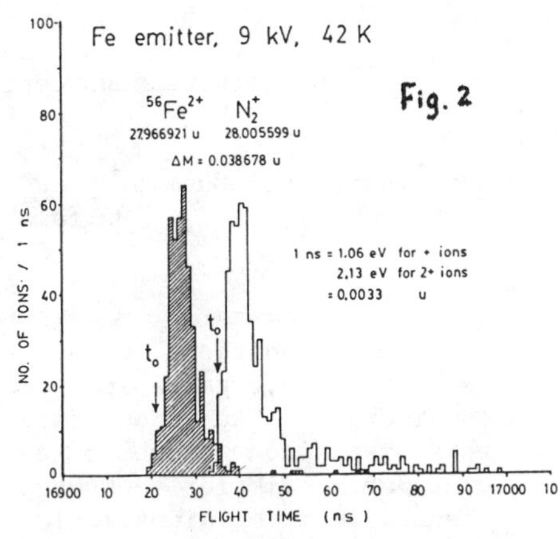

APPLICATIONS

The system can be used for mass analysis of the surfaces of metals and alloys, atom by atom and atomic layer by atomic layer, with excellent mass resolution, e.g. N_2^+ and $^{56}Fe^{2+}$ can be resolved as shown in Fig. 2. In addition it can be used to mass analyze materials of poor electrical conductivity such as a high purity Si, and it has also been successfully used to analyze reaction intermediates in surface catalyzed chemical reactions such as the synthesis of ammonia from N_2 and H_2. These cannot be done with the high-voltage pulse atom-probe. The atom-probe, originally conceived for the purpose of mass analysis alone, is now capable of doing ion energy analysis and recently it has been successfully applied to do ion reaction time analysis also. In particular the dissociation time of $^4HeRh^{2+}$ by atomic tunneling in a field of 4.5 to 3.0 V/Å has been measured to be 790 ± 21 fs, which is the rotation time of 180° of the ion.[3] It is possible that with this system, ion reaction time spectroscopy can emerge, and other applications are also expected.

* This work was supported by DOE and NSF.

REFERENCES

1. T. T. Tsong, S. B. McLane and T. J. Kinkus, Rev. Sci. Instrum. 53, 1442 (1982); T. T. Tsong, Y. Liou and S. B. McLane, Rev. Sci. Instrum. 55, 1246 (1984).
2. T. T. Tsong, J. Appl. Phys. 58, 2404 (1985).
3. T. T. Tsong and Y. Liou, Phys. Rev. Lett. 55. 2180 (1985); T. T. Tsong, Phys. Rev. Lett. (December, 1985).

LIGHT SCATTERING FROM A DEEP METALLIC GRATING

Dan Agassi
Naval Surface Weapons Center, Silver Spring, Maryland 20910

Thomas F. George
Departments of Chemistry and Physics, State University of New York
Buffalo, New York 14260

ABSTRACT

The conditions under which the Rayleigh hypothesis is exact and the convergence properties of the Rayleigh expansion are considered. The identification of the cause of the deficiency of this expansion suggests an alternative "dressed" expansion with presumably simpler convergence properties. This proposition is checked for a sinusoidal grating (SG), for which convergence is found for an arbitrary value of $\beta = 2\pi g/d$, where g and d denote the height and periodicity of the SG, respectively. The dressed expansion is used to analyze the surface plasmon dispersion and local field enhancement distribution pertaining to the SG in the limit as β goes to infinity. The dispersion relation is comprised of two bands. The local field enhancement predicts stronger fields at the bottoms of the troughs than at the peaks of the SG.

INTRODUCTION

Light scattering from a grating is a well-developed subject in classical optics.[1] The underlining, well-known physical process is (elastic) Bragg scattering, i.e., the exchange of surface-parallel momentum between the grating and the incident light by an amount nk_G, where $n = 0,\pm1,\pm2,...$ and the grating momentum is $k_G = 2\pi/d$. All notations are defined in Fig. 1. It is, however, important to

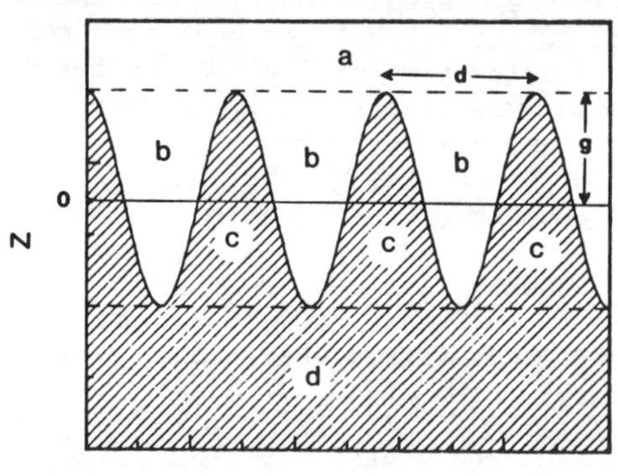

Fig. 1. Generic grating defining the notations. The incident light is directed downward, and the hatched area indicates the metal. The domains "a"-"d" pertain to the discussion of the Rayleigh expansion.

realize that, barring a few recent studies,[2-6] most of the work pertains to the restricted physical domain of shallow gratings where $\beta = 2\pi g/d < 1$. Light scattering from such gratings is qualitatively different from that pertaining to deep gratings, i.e., when $\beta >> 1$. In the former case, the grating can exchange only a limited number of k_G quanta, i.e., $n \sim \mathcal{O}(1)$. Consequently, the scattered light is comprised of a dominant specular and a few Bragg components. On the other hand, for deep gratings, $n \sim \mathcal{O}(\beta) >> 1$ (see below). Hence, the scattered light is comprised of many, equally important and strongly interfering Bragg components, which gives the scattered light a new qualitative character. The theoretical description of the deep grating domain is at the focus of this paper.

RAYLEIGH EXPANSION AND HYPOTHESIS

The Rayleigh expansion[1,8] provides a framework to the shallow grating work and a starting point to our analysis. It results from the underlying symmetry of the grating: Since it is invariant under translations $x \rightarrow x+nd$, where n is any integer, the Floquet-Bloch theorem implies that the electromagnetic field $\vec{\Psi}(x,z)$ has the property $\vec{\Psi}(x+d,z) = \exp[ik_{\parallel}d]\vec{\Psi}(x,z)$ where $k_{\parallel}$ is a surface-parallel (Fig. 1) momentum label.[4] Consequently, the Fourier series expansion of the fields in a "unit cell" $0 \leq x \leq d$ (the z-dependence is determined from the wave equation) suffices to describe the field throughout the xz-plane. When the proper boundary conditions at infinity and the vector character are incorporated, the ensuing p-wave expansion is the Rayleigh expansion:

$$\vec{E}_\alpha(x,z) = \sum_{\ell=-\infty}^{\infty} \{C_\alpha(\ell)\hat{p}_{\alpha,-}(\ell)e^{i[k_\ell x - W_\alpha(\ell)z]} + A_\alpha(\ell)\hat{p}_{\alpha,+}(\ell)e^{i[k_\ell x + W_\alpha(\ell)z]}\} , \qquad (1)$$

where

$$k(\alpha) = \sqrt{\varepsilon_\alpha}k , \qquad k = \omega_0/c , \qquad k_G = 2\pi/d , \qquad \beta = gkG,$$

$$W_\alpha(\ell) = [k^2(\alpha)-k_\ell^2]^{1/2} , \quad k_\ell = k + \ell k_G , \quad \mathrm{Re}(W_d),\mathrm{Im}(W_\sigma) \geq 0,$$

$$\hat{p}_{\alpha,\pm}(\ell) = \frac{1}{k(\alpha)}[k_\ell\hat{z} \mp W_\alpha(\ell)\hat{x}], \quad \hat{s} = \hat{x} \times \hat{z} . \qquad (2)$$

In (2), $\hat{x}$ and $\hat{z}$ are unit vectors in the x- and z- directions (Fig.1), and α denotes a domain in the xz-plane with a constant dielectric constant ε_α. Once a convenient partition of the xz-plane into domains has been chosen, the coefficients C_α, A_α are determined by matching the boundary conditions across the domain boundaries. To avoid mathematical pitfalls in the subsequent discussion, we confine ourselves henceforth to gratings such that the fields are non-singular everywhere.

The Rayleigh hypothesis can now be introduced in terms of expansions (1). For example, the grating in Fig. 1 calls for four expansions (1) while the Rayleigh hypothesis asserts that only two are needed since the expansions in domains "a" and "b" are identical and likewise for domains "c" and "d". By matching the boundary conditions across the (z=g)-plane, we conclude that the hypothesis is exact for gratings whose profile is expressible in terms of a finite Fourier series. For other gratings, the Rayleigh hypothesis may, or may not be exact.

DRESSED RAYLEIGH EXPANSION

Attempts to apply the Rayleigh expansion to deep-grating calculations inidicate its divergence for $\beta > 0.6$ (sinusoidal gratings).[11] This situation has motivated the introduction of alternative schemes[2-6,12] in the quest to enlarge the β-range accessible for calculations. Our approach is to expose the origin of the instability of (1) when $\beta \gg 1$, which then naturally leads to an alternative expansion --[10] the dressed expansion -- with excellent convergence properties.

For the grating in Fig. 1, the expansion in domain "0", which combines domains "a" and "b", is

$$\vec{E}_0(x,z) = \vec{E}_{in} + \sum_{\ell=-\infty}^{\infty} A_0(\ell)\hat{p}_{0,+}(\ell)e^{i[k_\ell x + W_0(\ell)z]} \quad , \tag{3}$$

Disregarding the uninteresting phase $\exp(ik_\ell x)$ and geometrical factor $\hat{p}_{0,+}(\ell)$ in (3), the $|\ell| \to \infty$ behavior is (see (2)):

$$A_0(\ell)e^{iW_0(\ell)z} \xrightarrow[|\ell|\to\infty]{} A_0(\ell)e^{-k_G|\ell|z} \tag{4}$$

Consequently, at the bottom of a trough, where $-g \leq z \leq 0$, the z-dependent factor in (4) diverges exponentially. Therefore, to render the total field $\vec{E}_0(x,z)$ finite, the <u>exact</u> $A_0(\ell)$ must converge at least as $\exp(-\beta|\ell|)$. Hence for $z < 0$,(3) is a sum of energy terms, most of which are products of exponentially large factors times exponentially small numbers. The <u>calculated</u> $A_0(\ell)$, however, always entail an error which, when multiplied by $\exp(\beta|\ell|)$, yields large erros in $\vec{E}_0(x,z)$. Thus, (3) is intrinsically unstable as a scheme for $\beta \gg 1$ (many $|\ell|$'s) calculations.

This deficiency is easily remedied by rewriting (4) as

$$A_0(\ell)e^{iW_0(\ell)z} = \alpha_0(\ell)e^{iW_0(\ell)(z+g)} \quad , \tag{5}$$

where $\alpha_0(\ell) = A_0(\ell)\exp[-iW_0(\ell)z]$. By construction the RHS of (5) always converges for $z+g > 0$, and when $z+g = 0$, the $\alpha_0(\ell)$ converge to render, by assumption, a finite $\vec{E}_0(x,z)$ on the grating's surface. Therefore, by transcribing (1) into the "dressed" expansion, as defined by the transformation (5), a major flaw in (1) is eliminated.

The convergence of the dressed expansion can be explicitly demonstrated for a sinusoidal metallic grating.[10] Lack of space does not allow us to outline the analysis. The results are that the dressed expansion converges for an arbitrary large β, and the number of significant components in the dressed expansion is on the order $N \sim 4\beta$. This analysis also[11] explains why the (bare) Rayleigh expansion diverges for $\beta \gtrsim 0.6$.

SURFACE PLASMON (SP) DISPERSION RELATION IN THE $\beta \to \infty$ LIMIT

As an application of the dressed expansion, we consider the SP dispersion relation[13] in the $\beta \to \infty$ limit for a[13] sinusoidal grating. Lack of space allows us only quote the result:

$$\frac{[3k^2(0) + k^2(1))(k^2(0) + 3k^2(1)]}{3[k^2(0) + k^2(1)]^2}$$

$$= \left\{\frac{\sin[L(\alpha+\gamma)]\sin[\frac{\alpha-\gamma}{2}] - \sin[L(a-\gamma)]\sin[\frac{\alpha+\gamma}{2}]}{\sin[L(\alpha+\gamma)]\sin[\frac{\alpha-\gamma}{2}] + \sin[L(\alpha-\gamma)]\sin[\frac{\alpha+\gamma}{2}]}\right\}^2 , \quad (6)$$

where $k(i) = k\sqrt{\varepsilon_i}$ from Eq. (1), $\gamma = 2\pi k/k_G$,

$$\alpha = \frac{\pi}{k_G}\{-[k^2(0) + k^2(1)]\}^{1/2} , \quad (7)$$

and $L \sim \Theta(\beta) >> 1$, the precise value of which is immaterial.

The analysis of (6) is straightforward: The LHS is a smooth function of $x = k/k_p$, where $k_p = \omega_p/c$ and ω_p is the (volume) plasmon frequency. At $x = 0$ it is unity, decreasing monotonically to $-\infty$ at $x = 1/\sqrt{2}$ and subsequently increasing and becoming positive for $x \geq \sqrt{3}/2$. The RHS of (6), on the other hand, is rapidly oscillating with frequency $\sim 1/\beta$ and amplitude $\sim\beta$. Furthermore, since α (Eq. (7)) is real for $x < 1/\sqrt{2}$ and imaginary for $x > 1/\sqrt{2}$, the RHS of (6) is positive and negative, respectively. Consequently, (6) has a large number of solutions ($\Theta(\beta)$) in the domains $0 \leq x \leq 1/2$ and $\sqrt{2}/2 \leq x \leq \sqrt{3}/2$.[13] The frequency band gap between these two band is hence $\Delta\omega/\omega_p = (\sqrt{2}-1)/2$. This outcome is in keeping with the known behavior for[14] $\beta << 1$.

In summary, we have exposed the reason for the poor convergence of the Rayleigh expansion when $\beta >> 1$, and introduced an alternative expansion -- the dressed expansion -- which has allegedly very good convergence properties. This premise is explicitly demonstrated for a sinusoidal grating. We have also discussed the surface plasmon dispersion relation in the $\beta \to \infty$ limit, where we find that the branches tend to cluster into two bands separated by $\Delta\omega/\omega_p = (\sqrt{2}-1)/2$.

ACKNOWLEDGMENTS

This research was supported by the Office of Naval Research and the Air Force Office of Scientific Research (AFSC), the United States Air Force, under Contract F49620-86-C-009.

REFERENCES

1. See, for example, Electromagnetic Theory of Diffraction Gratings, edited by R. Petit (Springer, New York, 1980).
2. N. E. Glass and A. A. Maradudin, Phys. Rev. B 24, 595 (1981).
3. P. Sheng, R. S. Stepleman and P. N. Sanda, Phys. Rev. B 26, 2907 (1982).
4. M. Weber and D. L. Mills, Phys. Rev. B 27, 2698 (1983).
5. A. Wirgin and T. Lopez-Rios, Opt. Commun. 48, 416 (1984).
6. N. Garcia, G. Diaz, J. J. Saenz and C. Ocal, Surf. Sci. 143, 338 (1985).
7. See the paper by S. R. J. Brueck in this Conference Proceedings for a discussion of experimental work on deep gratings.
8. Lord Rayleigh, Theory of Sound, 2nd Ed. (Dover, New York, 1945), Vol. II, p. 89.
9. J. E. Sipe, Surf. Sci. 105, 489 (1981).
10. D. Agassi and T. F. George, Phys. Rev. B, in press.
11. N. Garcia, V. Celli, N. R. Hill and N. Cabrera, Phys. Rev. B 18, 5184 (1978).
12. F. Toigo, A. Marvin, V. Celli and N. R. Hill, Phys. Rev. B 15, 5618 (1977).
13. D. Agassi and T. F. George, Surf. Sci., in press.
14. B. Laks, D. L. Mills and A. A. Maradudin, Phys. Rev. B 23, 4965 (1981).

PROPERTIES OF GRATINGS WRITTEN WITH A SINGLE LASER BEAM

P.M. Fauchet

Department of Electrical Engineering
Princeton University, Princeton, NJ 08544

ABSTRACT

During the past few years, laser-induced surface ripples have been studied by several groups. A stage has been reached where most properties are understood, although some significant observations remain unexplained. After a brief historical review followed by a summary of our present understanding, we discuss some results that we have obtained very recently using a Raman microprobe.

HISTORICAL REVIEW

If the laser is 25 years old, laser-induced surface ripples are only five years younger [1]. The first observations of this interesting phenomena, which is an example of spontaneous order appearing in a nominally uniform system, remained unexplained until the work of Emmony and coworkers ten years ago [2]. They were the first to propose that interference between the incoming wave and some diffracted wave (perhaps due to a localized defect or scratch) could produce these periodic surface structures.

In the late seventies and early eighties, laser processing became very popular and surface ripples were observed during laser annealing [3], laser-assisted surface deposition [4], surface etching [5], and laser-induced surface damage [6]. Furthermore, they have been produced with picosecond [7] to cw [8] lasers, from the mid-ir [9] to the VUV [10], in metals [11], semiconductors [2], and dielectrics [6]. Several detailed theoretical models have been published which all describe rather successfully this universal phenomenon [10,12,13]. Although they differ in the exact treatment of the electromagnetic waves that interfere with the incident wave, all models are in a sense only rigorous and detailed extensions of the original idea suggested by Emmony.

THEORY

The nature of the diffracted wave which produces the interference pattern with the incident wave has been the subject of some controversy. It is now clear that different types of waves play a role, and that the nature of the dominant wave is dictated by the details of the experiment [14]. For example, if the material is metallic-like, surface plasmon-polaritons are produced by coupling of the incident beam through initial surface roughness. If however the material is non metallic, the wave scattered along the surface is of a different nature and has been christened radiation remnants [13]. Our own analysis is valid when any of these waves dominates, since we have solved Maxwell's equations at the interface between two media having arbitrary dielectric functions [12].

All the models predict successfully the spacing and orientation of the most common ripples. They also give a consistant picture of the growth of the ripples, including the range of intensity and wavelength required for each material. A number of experimental observations remains however unexplained by any of these models. These nonstandard ripples can be divided in three categories:

1. Perpendicular ripples [2,4,7,15]: these are periodic surface structures that are

oriented perpendicular to the standard ripples. Their spacing can be larger, equal to, or smaller than that of the standard ripples.

2. Nonperpendicular ripples [15,16]: these are periodic structures that run at some angle compared to the standard ripples.

3. Ripples with wrong spacing [17,18]: these ripples have the same orientation as the standard ripples but have a different periodicity.

We have previously reviewed these unexplained ripple structures [19]. One of the major reasons why theory fails to explain these observations is that all models use perturbation theory whereas we have shown that the modulation of the incident power can easily reach 100 percent. This is especially true when the aspect ratio of the corrugation (height vs period) exceeds 0.1. Another important reason is that the material's response has not been adequately described. Let us consider two situations. First, at the laser intensity often encountered here, the dielectric function of the solid has a significant nonlinear component that has been neglected. Second, the dielectric function of the surface becomes inhomogeneous during illumination: for example, melting threshold can be exceeded on the hill (or the valley) of the corrugation, and both the surface profile and the dielectric function become periodic. Each effect has been modeled separately but they have not been considered together.

PROBING STRUCTURAL VARIATIONS

We have recently measured the structural variations across periodic ripples. Ultimately, these measurements should give us a clearer picture of ripple's growth and may help us solve some of the problems mentioned above. The probe is a Raman microprobe [20], in which the interrogating beam is focused through a microscope onto a one μm spot at the sample surface and the Raman photons are collected by the same optics and analyzed by a monochromator. A map of the sample structure can thus be obtained with a resolution better than or comparable to the period of most of the ripples.

In this study, we are primarily concerned with variations in composition, crystallinity, and stress. The relation between the Raman Stokes line and these properties are discussed in detail elsewhere [20]. Here, we mention that composition changes lead to new Raman lines, that the amorphous, polycrystalline (grain size less than 20 nm), and single crystal phases have distinctly recognizable Raman lineshapes, and that compressive or tensile stress smaller than 10^9dyn/cm^2 leads to a measurable shift of the Raman peak.

Our results are discussed at greater length elsewhere [21]. We have observed no measurable change in the Raman Stokes line across ripples written on Ge wafers. To check if large inhomogeneities in energy deposition could be "frozen-in" after resolidification, we wrote permanent grating lines of the same spacing (4 microns) by interfering two equal-intensity beams from the same pulsed laser. Again, no measurable change in the Raman line was observed between regions of constructive interference (where melting threshold had been exceeded) and regions of destructive interference (where no energy was deposited directly from the laser beams). We repeated this interference experiment on a thin polycrystalline Si film on fused silica. Here, both regions displayed a Raman line that differed from that of the starting material. The tensile stress of the illuminated region was increased by 5x10^9dyn/cm^2 and the average grain size exceeded the 20 nm detection limit. Our interpretation is that heat transport can only occur in the plane (i.e., across the ripples) and that enough energy is carried from the constructive interference regions to the destructive interference regions to transform the whole layer. The lateral temperature gradient is in excess of 300K/μm, which nevertheless is still small compared to the gradient in the direction perpendicular to the wafer surface. Thus, for the Ge wafers, lateral heat transport does not take place and molten stripes alternate with unmolten stripes. If the initial material is polycrystalline, melting and recrystallization lead to larger grains and a modified Raman line. Note that in the destructive interference regions of the poly-Si films, the observed larger low wavenumber tail indicates the presence of grains of smaller sizes, which is not in disagreement with the indirect and presumably less perfect melting and recrystallization in these regions.

We have also examined the Raman spectrum across ripples formed on films of Si and C produced by laser-assisted chemical vapor deposition. These deposits were grown in Dr. Allen's laboratory at USC. They had been produced in a Si-rich environment and had been examined by Auger sputtering prior to the Raman microprobe experiments. We observed a strong Si line present all across the surface. No SiC line was detected and a weak graphite line at 1580 wavenumbers was observed at the peak of the corrugation only. These preliminary results indicate that some composition variation exists across ripples observed in laser-assisted CVD. Wilson and Houle [22] very recently observed structural variations across ripples obtained in laser-assisted Cu/C film deposition. They used transmission electron microscopy and Auger electron microprobe to detect composition and grain size variations. More systematic experiments of this type are needed if we want to understand better the coupling of intense laser beams to solids, which leads to ripple's formation.

CONCLUSIONS

Although the formation of periodic laser structures, commonly called surface ripples, is much better understood now than it was ten years ago, some problems remain, especially concerning the material's response. We and others have now shown that it is possible to detect variations in the structure of the material *post-mortem,* although more work is needed before we can interpret the results. Ultimately, detection of these changes *in-situ* and in real-time is desirable, although it is a much more formidable task.

I thank Ian Campbell, Fran Adar, Susan Allen, Loren Pfeiffer and Anthony Siegman who contributed to this work, which was supported in part by NSF grant ECS-8504202.

REFERENCES

1. M. Birnbaum, J. Appl. Phys. **36**, 3688 (1965).

2. D.C. Emmony, R.P. Howson, L.J. Lewis, Appl. Phys. Lett. **23**, 598 (1973); L.J. Lewis, D.C. Emmony, Optics and Laser Technol. **7**, 222 (1975).

3. H.J. Leamy, G.A. Rozgonyi, T.T. Sheng, G.K. Celler, Appl. Phys. Lett. **32**, 535 (1978).

4. R.M. Osgood Jr., D.J. Ehrlich, Optics Lett. **7**, 385 (1982).

5. N. Tsukada, S. Sugata, H. Saitoh, Y. Mita, Appl. Phys. Lett. **43**, 189 (1983).

6. P.A. Temple, M.J. Soileau, IEEE J. Quantum Electron. **17**, 2067 (1981).

7. P.M. Fauchet, A.E. Siegman, Appl. Phys. Lett. **40**, 824 (1982).

8. R.J. Nemanich, D.K. Biegelsen, W.G. Hawkins, Phys. Rev. **B 27**, 7817 (1983).

9. F. Keilmann, Y.H. Bai, Appl. Phys. **A 29**, 9 (1983).

10. S.R.J. Brueck, D.J. Ehrlich, Phys. Rev. Lett. **48**, 678 (1982).

11. A.K. Jain, V.N. Kulkarni, D.K. Sood, J.S. Uppal, J. Appl. Phys. **52**, 4882 (1981).

12. Z. Guosheng, P.M. Fauchet, A.E. Siegman, Phys. Rev. **B 26**, 5366 (1982).

13. J.E. Sipe, J.F. Young, J.S. Preston, H.M. van Driel, Phys. Rev. **B 27**, 1141 (1983).

14. J.F. Young, J.E. Sipe, H.M. van Driel, Phys. Rev. **B 30**, 2001 (1984).

15. P.M. Fauchet, A.E. Siegman, Appl. Phys. **A 32**, 135 (1983).

16. D. Haneman, R.J. Nemanich, Solid State Commun. **43**, 203 (1982).

17. G.N. Maracas, G.L. Harris, C.A. Lee, R.A. McFarlane, Appl. Phys. Lett. **33**, 453 (1978).

18. N. Mansour, G. Reali, P. Aiello, M.J. Soileau, to be published in the Proceedings of the 1984 Laser Damage Conference (NBS Boulder).

19. P.M. Fauchet, A.E. Siegman, in *Energy Beam-Solid Interactions and Transient Thermal Processing 1984*, Biegelsen *et al* editors, Materials Research Society, 1985, pp 199-204.

20. P.M. Fauchet, to be published in IEEE Circuits and Devices Magazine, January 1986.

21. P.M. Fauchet, to be published in *Beam-Solid Interactions and Phase Transformations 1985*, Kurz *et al* editors, Materials Research Society, 1986.

22. R.J. Wilson, F.A. Houle, Phys. Rev. Lett **55**, 2184 (1985).

DIFFRACTION OF LIGHT FROM A CLASSICAL GRATING AND A BIGRATING

N. E. Glass
U. S. Naval Postgraduate School, Monterey, CA 93943, U.S.A.

A. A. Maradudin
University of California, Irvine, CA 92717, U.S.A.

A. Wirgin
Universite Paris VI, 75230 Paris Cedex 05, France

ABSTRACT

Two problems in the diffraction of light from periodically corrugated metal surfaces are discussed. The first is the diffraction of s-polarized (TE) light from a metallic lamellar grating. It is shown that cavity resonances in the grooves of the grating can be excited thereby, if the grooves are deep enough. These resonances do not show up as well-defined features ("anomalies") in the reflectivity spectrum. They do, however, manifest themselves by a significant enhancement of the squared modulus of the electric field within the grooves (by as much as a factor of 440) when the wavelength of the incident light is near the resonance wavelength. In the second problem a nonperturbative theory of light diffraction from a metallic bigrating with resonant excitation of surface polaritons is applied to the fitting of recent experimental results of Inagaki et al. The exact calculations described here account for some, but not all, of the discrepancies between experiment and the predictions of a first-order perturbation theory of such diffraction. This indicates that these discrepancies are not entirely due to the underestimation of higher order processes by perturbation theory. Some other causes of these discrepancies are suggested.

INTRODUCTION

Much recent research on surface-enhanced effects connected with Raman scattering(1), second harmonic generation in reflection(2), and photoelectric yield(3), for example, focuses on the role of rough surfaces, and gratings in particular, for coupling radiant energy into the surface electromagnetic waves responsible for electric field amplification near air-metal (or lossy dielectric) interfaces. In this paper we briefly consider this coupling in two cases of experimental interest.

DIFFRACTION OF S-POLARIZED LIGHT BY DEEP LAMELLAR GRATINGS

Until recently it has been thought that all resonant, near-field enhancement processes at bare classical gratings are possible only when the incident radiation is p(TM)-polarized, when the plane of incidence is perpendicular to the grooves of the grating. This is because no surface electromagnetic wave of s(TE)-polarization can exist at a planar metal-air interface,[4] and none had been found at a periodically corrugated metal-air interface.[5]

The first theoretical indication that classical gratings can produce cavity-type resonances when illuminated by s-polarized light whose plane of incidence is perpendicular to the grooves of the grating was provided by Hessel and Oliner[6] on the basis of a periodic surface impedance boundary condition. They predicted that such resonances occur only for gratings whose groove depth h satisfies $m\lambda_g/2 > h > (2m-1)\lambda_g/4$ (m = 1, 2, 3,...), but did not explain the origin of this criterion nor did they define λ_g.

Subsequently Andrewartha _et al._[7] verified, on the example of a perfectly conducting lamellar grating, that gratings can give rise to s-resonances. They also showed that the imaginary part of a wavelength resonant pole is generally so large that the corresponding s-resonance goes unnoticed (i.e. does not appear as an "anomaly" with sharp features as do p-resonances) in a far field observable such as the specular reflectance. How, then, can these s-resonances be observed?

It has recently been shown, both for perfectly conducting lamellar gratings of the kind studied by Andrewartha _et al._[7] and for (finitely conducting) noble metal (e.g. silver) gratings, that s-resonances manifest themselves in the immediate vicinity of the grating, in particular within the grooves where they give rise to large field enhancements when the groove depth satisfies the Hessel-Oliner criterion.[8]-[10] The latter is shown to arise from the dispersion relation of a slow guided wave analogous to the s-polarized guided wave that is supported by a flat, lossless dielectric film overlying a flat infinitely conducting substrate. The enhancement in the squared magnitude of the electric field within the grooves has been found to be as large as by a factor of 440. At the wavelengths of the incident light at which these s-resonances occur the grating acts as a planar mirror as concerns the electromagnetic field above the grooves of the grating.[9]-[10]

In recent work Glass _et al._[11] found that the field enhancement associated with the grating-induced excitation of the plane surface polariton of a silver grating by incident light of p-polarization first increases with increasing groove depth h and then past some critical value of the latter begins to decrease with further increase in h due to radiation damping. In contrast, it is found in Refs. 9 and 10 that the field enhancement at the resonance wavelength of the incident light increases monotonically with increasing h, with some tendency toward saturation noted in the results of Ref. 10, but with no evidence of a decrease for h greater than some critical value. This behavior is consistent with the finding of Andrewartha _et al._[7] that the imaginary part of the complex zero of the dispersion relation that gives the wavelengths at which s-cavity resonances occur tends to zero as $h \to \infty$.

A possible way of observing the s-resonances is to measure the enhancement in the intensity of light Raman scattered from molecules adsorbed within the grooves of a grating.

DIFFRACTION OF LIGHT FROM A BIGRATING

In studies of the coupling of light to surface polaritons through grating surfaces bigratings have acquired interest recently because (1) they give rise to increased coupling efficiency (both p- and s-polarized light can couple to surface polaritons irrespective of the azimuthal angle of the incident light), and (2) because a bigrating allows one to model an array of real protuberances on a planar surface.

Recently Inagaki *et al.*[12] used a photoacoustic method to determine the total optical absorptance of a bigrating fabricated on silver. For a fixed value of the azimuthal angle $\theta_1 = 90^o - \phi$, which measures the angle between the plane of incidence and the x_2-axis, the absorptance $A(\theta)$ was measured as a function of the polar angle of incidence θ at a wavelength of the incident light of 633 nm. The incident light was either purely p-polarized or s-polarized. The peaks in $A(\theta)$ correspond to the excitation of surface polaritons, and occur when the condition $\vec{k}_{\parallel} + \vec{G}_{\parallel} = \vec{k}_{sp}$ is satisfied, where $\vec{k}_{\parallel}$ is the surface projection of the wave vector of the incident light, $\vec{k}_{\parallel} = (\omega/c)\sin\theta(\cos\phi, \sin\phi)$, and $\vec{k}_{sp}$ is the two dimensional wave vector of the surface polariton excited in the scattering process. Such a peak can thus be labeled by the two integers (m_1, m_2) defining the reciprocal lattice vector. The peak value of A for a given (m_1, m_2) was then plotted versus θ_1. Inagaki *et al.* used a first-order (in the surface profile function) perturbation theory of electromagnetic wave absorption due to Elson and Sung[13] to fit their data. The first order theory gave a good fit except at azimuthal angles of incidence corresponding to the simultaneous excitation of two surface polaritons. Inagaki *et al.* suggested that these discrepancies are due to grating-induced surface polariton interactions that are dominated by higher order processes not adequately described by the first order theory. Moreover, the measured angles of incidence at which the surface polariton resonances are observed were 1.0-1.5^o larger than those calculated from the dispersion relation for surface polaritons on a flat surface. This was attributed to the corrugation-induced shift in the surface polariton dispersion curve[14].

The data of Inagaki *et al.*[12] have been analyzed[15] on the basis of an exact, nonperturbative theory of light scattering from a finitely conducting bigrating[16], with the aims of determining the range of validity of the first-order theory of light scattering from a bigrating; of seeing if the discrepancies at the two-surface polariton peaks are due simply to the lack of higher order processes inherent in the perturbation theory; and of determining whether the angular shift of the resonance is due to the bigrating itself.

The equation for the surface of the bigrating was initially chosen to have the form $x_3 = \zeta(x_1, x_2) = h_1(\cos(2\pi x_1/a) + \cos(2\pi x_2/a)) + h_2(\cos(4\pi x_1/a) + \cos(4\pi x_2/a)) + h_{11}\cos(2\pi x_1/a)\cdot\cos(2\pi x_2/a)$. The translation vectors of the reciprocal lattice defined by the grating are therefore $\vec{G}_{\parallel}(m_1, m_2) = (2\pi/a)\,(m_1, m_2)$, where $m_1, m_2 = 0, \pm 1, \pm 2, \ldots$ The value of a, taken from Ref. 12, was 2186 nm.

The Fourier coefficients of $\zeta(x_1,x_2)$ obtained by the use of the first-order theory[12] are $h_1 = 15.4$ nm, $h_2 = 2.66$ nm, and $h_{11} = 5.14$ nm. A direct fit of the exact theory to the measured absorptance gave $h_1 = 13.1$ nm, $h_2 = 1.9$ nm, and $h_{11} = 8.2$ nm. When the results of the exact theory were corrected for the background absorptance, due to the imaginary part of the dielectric constant, and estimated to be 0.0116 for p-polarized and 0.0107 for s-polarized incident radiation, and refitted to the experimental data, the values $h_1 = 13.7$ nm, $h_2 = 2.42$ nm, and $h_{11} = 9.1$ nm were obtained. The results are thus sensitive to the value of the dielectric constant of silver used. Values of ε for silver are tabulated by Christy and Johnson at $\lambda = 617$ and 659 nm. The value used for $\lambda = 633$ nm was obtained from these data by linear interpolation. To gauge the extent to which the uncertainty in ε affects the comparison between experiment, the first-order theory, and the exact theory, the exact calculations were repeated for the value of ε corresponding to $\lambda = 617$ nm. The results yielded $h_1 = 14.3$ nm, $h_2 = 2.6$ nm, and $h_{11} = 9.5$ nm. These changes are small compared with the differences between the results of the first-order and exact theories. However, the change in ε led to changes in the peak position upwards by 0.1-0.2^o , which are much larger than the grating-induced shifts of $\sim 0.01^o$, but which are still smaller than the discrepancies seen by Inagaki *et al.*[12]. The results of Ref. 15 thus indicate that a very precise knowledge of the complex dielectric constant of the grating material at the exact wavelength of the incident light is required for an accurate comparison of the theory and experiment.

The exact theory does only slightly better than the first order perturbation theory in fitting the experimental data. This observation, however, does not validate the perturbation-theoretic results, because of the important differences in the values of the Fourier coefficients h_1, h_2, h_{11} noted above that are obtained by the two approaches.

It is found that higher order processes are important in obtaining h_{11}. They also explain some, but not all, of the discrepancies between the results of first order perturbation theory and the experimental data for the two-surface polariton peaks.

Explanations for the discrepancies remaining between the results of the exact theory and the experimental data might include the failure to include asymmetric terms, e.g. $h_s(\sin(2\pi x_1/a) + \sin(2\pi x_2/a))$, in the expression for the surface profile function. Although Inagaki *et al.* found no asymmetry in the intensities of the diffracted beams at normal incidence, it is possible that a very small grating asymmetry could make a negligible change in scattering at normal incidence, while strongly affecting the two-surface polariton peaks.[11] This was confirmed by a calculation in which a contribution of the form $h_s(\sin(2\pi x_1/a) + \sin(2\pi x_2/a))$ was added to $\zeta(x_1, x_2)$. A decrease in A of 0.01 for the (1, 0) - (0, 1) two-surface polariton peak was produced, while the reflectances of the (1, 0) and (-1, 0) beams and of the (0,1) and (0, -1) beams at normal incidence in p-polarization differed only by 0.0017. While the decrease in A is smaller than needed to explain the experimental data, these results show that asymmetry in the surface profile can have a significant effect on the two-surface polariton peak. A study of the effects of antisymmetric cross terms of the type of $h_{cs}\cos(2\pi x_1/a)\sin(2\pi x_2/a) + h_{sc}\sin(2\pi x_1/a) \times \cos(2\pi x_2/a) + h_{ss}\sin(2\pi x_1/a)\sin(2\pi x_2/a)$ in $\zeta(x_1,x_2)$ would be of particular interest in this context.

Finally, we note that even for very small corrugation strengths (e.g. the value $\zeta_{max}/a \cong 0.02$ used in Ref. 15) the use of first order perturbation theory for treating optical scattering from a silver bigrating does not seem to be reliable to better than 10%, and can be worse for determining the cross terms in the grating profile function.

ACKNOWLEDGEMENT

This research was supported in part by NSF Grant No. DMR 82-14214.

REFERENCES

1. See, for example, H. Metiu, in Surface Enhanced Raman Scattering, R. K. Chang and T. E. Furtak (eds.) (Plenum, New York, 1982), p. 1.
2. C. K. Chen, A. R. B. de Castro, and Y. R. Shen, Phys. Rev. Lett. 46, 145 (1981).
3. J. G. Endriz and W. E. Spicer, Phys. Rev. B, 4, 4159 (1971).
4. See, for example, A. A. Maradudin, In Surface Polaritons, V. M. Agranovich and D. L. Mills (eds.) (North-Holland, 1982) p. 405.
5. An example of an unsuccessful search for waves of this type is provided by W. Zierau, A. A. Maradudin, and C. Falter, J. Phys. (Paris) Colloq. 45, C5-225 (1984).
6. A. Hessel and A. A. Oliner, Appl. Opt. 4, 1275 (1965).
7. J. R. Andrewartha, J. R. Fox, and I. J. Wilson, Optica Acta 26, 69 (1979).
8. A. Wirgin and A. A. Maradudin, Phys. Rev. B31, 5573 (1985).
9. A. A. Maradudin and A. Wirgin, Surf. Sci. 162, 980 (1985).
10. A. Wirgin and A. A. Maradudin, "Resonant Response of a Bare Metallic Grating to S-Polarized Light" (preprint, 1985).
11. N. E. Glass, M. Weber, and D. L. Mills, Phys. Rev. B29, 6548 (1984).
12. T. Inagaki, J. P. Goudonet, J. W. Little, and E. T. Arakawa, J. Opt. Soc. Am. B2, 433 (1985).
13. J. M. Elson and C. C. Sung, Appl. Opt. 21, 1496 (1982).
14. N. E. Glass, A. A. Maradudin, and V. Celli, Phys. Rev. B26, 5367 (1982).
15. N. E. Glass and A. A. Maradudin, Opt. Commun. (to appear).
16. N. E. Glass, A. A. Maradudin, and V. Celli, J. Opt. Soc. Am. 73, 1240 (1983).

ENERGY AND PHASE RELAXATION IN LASER-INDUCED ADMOLECULAR PROCESSES

Xi-Yi Huang
Department of Chemistry, University of Rochester
Rochester, New York 14627

Thomas F. George
Departments of Chemistry and Physics, State University of New York
Buffalo, New York 14260

ABSTRACT

Vibrational excitation and relaxation of a molecule adsorbed on a surface is investigated. The phase relaxation is seen to assist the laser-driving force in overcoming the anharmonic and/or detuning bottleneck. The incoherent thermalization processes in the initial stage of the admolecular multiphoton excitation are evaluated. It is found that phase relaxation (dephasing) due to the surface and admolecular perturbation plays an important role in the thermalization processes.

MODEL

The IR-laser-driven resonant active mode absorbs photons and transfers its energy to other vibrational modes or the surface. Statistical thermodynamical behavior may occur in the resulting desorption/dissociation processes. We confine ourselves to the excitation and relaxation of the active anharmonic vibrational mode which is driven by a nearly resonant coherent field and controlled by both energy (T_1) and phase (T_2) relaxation. We treat the many other inactive modes and surface excitation as a large thermal reservoir, by which dissipation mechanisms are provided.

A generalized master equation which considers the anharmonicity and phase and energy dissipation of the active mode has been derived.[1-3] For the case of (1) low bath temperatures, i.e., where the average excitation of the bath quantum $\bar{m}_B$ is much less than unity, and (2) where the rate of phase relaxation η is much larger than the energy relaxation κ and the Rabi frequency Ω_R, the generalized master equation leads to the following coupled differential equations:

$$\dot{W}_m = 2\kappa(\bar{m}_B+1)[(m+1)\,W_{m+1} - mW_m] + 2\kappa\bar{m}_B[mW_{m-1} - (m+1)W_m]$$

$$+ 2\eta\Omega_R^2(t)\left\{\frac{m(W_m - W_{m-1})}{\eta^2 + [2\varepsilon(m-1)-\Delta]^2} + \frac{(m+1)(W_m - W_{m+1})}{\eta^2 + (2\varepsilon m-\Delta)^2}\right\} \quad . \quad (1)$$

Here W_m is the population of the m-th vibrational level, ε is the anharmonicity, and Δ is the laser detuning. These equations take both the laser coherent excitation and the thermal excitation and relaxation into account.

Phase relaxation (dephasing) is seen to be more important than energy relaxation in the active-mode thermalization process. We have found that when η is large, such as $\eta = 15$ shown in Fig. 1, the population distribution in the active mode is close to a thermal Planck distribution. We have also found that the energy relaxation constant κ has a minor influence on the thermalization of the driven active mode.

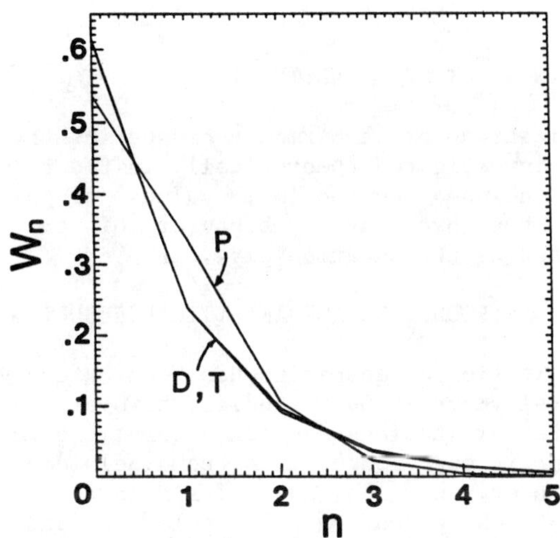

Fig. 1. Level population W_m for : (a) a driven damped anharmonic oscillator (D); (b) Poisson distribution (P); and (c) thermal distribution (T). The same $\bar{m}$ is used for all cases, and $\eta = 15$ is chosen.

ACKNOWLEDGMENTS

This research was supported by the Office of Naval Research and the Air Force Office of Scientific Research (AFSC), the United States Air Force, under Contract F49620-86-C-009.

REFERENCES

1. X. Y. Huang, L. M. Narducci and J. M. Yuan, Phys. Rev. A 23, 3084 (1981).
2. X. Y. Huang, T. F. George, J. M. Yuan and L. M. Narducci, J. Phys. Chem. 88, 5772 (1984).
3. X. Y. Huang, T. F. George and J. M. Yuan, J. Opt. Soc. Am. B 2, 985 (1985).

TRANSIENT EXCITATION OF ANHARMONIC ADSPECIES BY PULSED LASER RADIATION

J. T. Lin
Litton Laser Systems, P. O. Box 7300, Orlando, Florida 32854

Xi-Yi Huang
Department of Chemistry, University of Rochester
Rochester, New York 14627

Thomas F. George
Departments of Chemistry and Physics, State University of New York
Buffalo, New York 14260

ABSTRACT

The excitation of anharmonic adspecies by pulsed laser radiation is investigated theoretically in the transient regimes. New features include the optimum values of pulse duration and detuning and the oscillatory behavior of the time-dependent excitation caused by the anharmonicity.

INTRODUCTION AND ANALYTICAL RESULTS

Laser excitation of adspecies has been extensively studied in the past several years.[1] Such studies, however, have been limited to steady-state excitations and/or adiabatic processes where the laser duration is much longer than the dipole dephasing time. In the present paper, we show an analysis in the transient regimes where new features given by the pulse duration, detuning and anharmonicity are studied both analytically and numerically. The key equations describing the laser excitation of adspecies are given by[1]

$$\frac{da}{dt} = -\left(i\Delta_{eff} + \frac{\gamma_1 + \gamma_2}{2}\right) a - iV(t) \quad , \tag{1}$$

$$\frac{dN}{dt} = iV(t)\,(a-a^\dagger) - \gamma_1(N-N_0) \quad , \tag{2}$$

$$\Delta_{eff} = \Delta - 2\varepsilon^* N \quad . \tag{3}$$

Here a and $a^\dagger$ are the dipole operators of the active mode, where $N \equiv \langle a^\dagger a \rangle$, γ_1 and γ_2 are damping factors defined by the inverse of the energy (T_1) and the phase (T_2) relaxation time, respectively, and V(t) is the excitation source term proportional to the laser electric field. The initial phonon-bath excitation is given by N_0, and the nonlinear effects due to the anharmonicity (ε^*) are given by Eq. (3), where Δ is the laser detuning for the harmonic case. We shall consider a Gaussian laser profile with pulse duration t_p.

0094-243X/86/1460750-3$3.00

Analytical results are available only for the harmonic cases ($\varepsilon^*=0$), where the excitations at three regimes are found:

I. Steady-state ($t_p \gg T_{1,2}$): $N_{s.s.} \propto I(t)$ (laser intensity) .

II. Adiabatic-state ($T_1 > t_p \gg T_2$): $N(t) \propto \phi$ (laser energy) .

III. Transient-state ($t_p < T_{1,2}$): $N(t) \propto t_p\phi$.

NUMERICAL RESULTS

The role of pulse duration is shown in Fig. 1 for $\gamma_1 = \gamma_2 = 1$ and $\Delta = \varepsilon^* = 0$. It is seen that, for a given laser energy, there is an optimum pulse duration and the behavior of the excitations is consistent with the analytical results in the various regimes. The effects of changing the detuning are shown in Fig 2 for $\gamma_1 = \gamma_2 = \varepsilon^* = t_p = 1$, and the excitations at various anharmonicities are shown in Fig. 3 for $\Delta = 20$. We note that for a given anharmonicity (detuning), an optimum detuning (anharmonicity) is found accordingly. These features may be realized by the fact that maximal excitation is achieved when the detuning is compensated by the anharmonic excitation term ($2\varepsilon^*N$), such that the effective detuning (Δ_{eff}) is suppressed. Furthermore, the significant oscillatory behavior, which is absent in Fig. 1 for the harmonic resonance cases, is found in Fig. 3.

As a concluding remark, we note that the effects caused by an increase in the substrate temperature may be included by introducing a temperature-dependent phonon-bath through N_0 in Eq. (2). Numerical results based upon Eqs. (1)-(3) combined with a heat diffusion equation will be shown elsewhere.

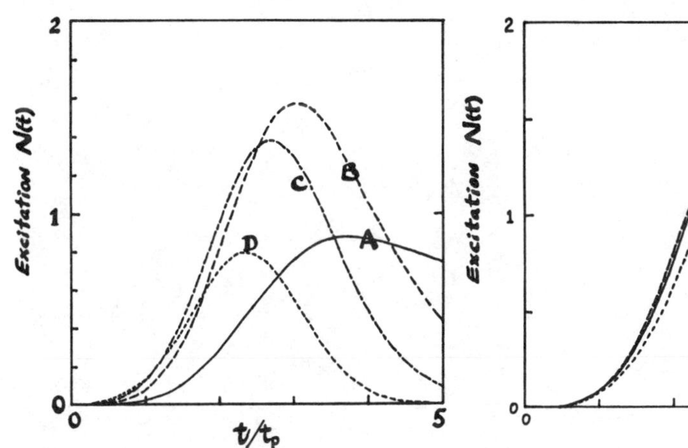

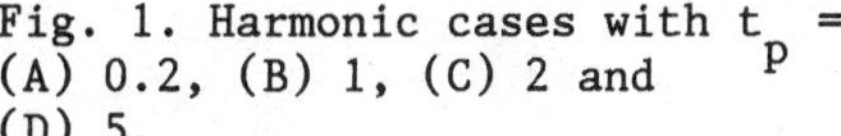

Fig. 1. Harmonic cases with t_p = (A) 0.2, (B) 1, (C) 2 and (D) 5.

Fig. 2. Anharmonic cases with Δ = (A) 0.5, (B) 1.0 and (C) 1.5.

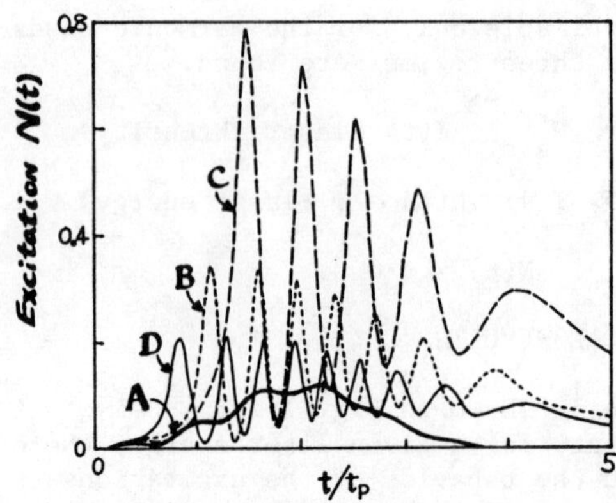

Fig. 3. Oscillatory features for ε* = (A) 0, (B) 20, (C) 50 and (D) 80.

ACKNOWLEDGMENTS

This research was supported by Air Force Office of Scientific Research (AFSC), the United States Air Force, under Contract F49620-86-C-009, and the Office of Naval Research.

REFERENCES

1. For a recent review, see J. T. Lin, M. Hutchinson and T. F. George, in Multi-Photon Processes and Spectroscopy, ed. by S. H. Lin (World Scientific, Singapore, 1984), pp. 105-237.

GAUSSIAN BEAM DIVERGENCE MEASUREMENTS USING SURFACE-ACOUSTIC WAVE MODULATION

T. D. Black, D. A. Larson, and T. Haghighatjou
The University of Texas at Arlington, P.O. Box 19059
Arlington, Texas 76019

ABSTRACT

A surface acoustic wave (SAW) acts as a spatial phase modulator of an incident laser beam. If the laser beam is also directed onto a stationary reference grating (SRG) such that the wave vectors of the SAW and SRG are parallel and have the same magnitude, then the resulting spatial intermodulation produces a direct frequency modulation of the zero and higher order diffracted beams at the frequency of the SAW. Furthermore, if the two "gratings" are in proximity, that is separated, the intensity and phase of the modulation is dependent on the incident angle of the laser beam. Consequently, since a diverging beam can be considered as a superposition of different incident angles or spatial frequencies, it was found that the detected signal was extremely sensitive to the divergence of the incident beam.

INTRODUCTION

Experimental measurements were made in a transmission schematic previously reported[1] using $LiNbO_3$ at frequencies of 17.4 MHz and 87.2 MHz corresponding to SAW wavelengths, Λ, of 200 μm and 40 μm respectively. The essential features of the analysis were confirmed for a HeNe laser for beam divergences from 150 μrad to 20 m rad.

The method of detecting SAW using reference gratings was first developed for an SRG placed directly on the surface of the SAW substrate.[2] The transmission case of the proximity probing technique is essentially insensitive to the probing beam's obliquity, γ . For beams with low divergence the sensitivity of this technique to changes in L, the separation between the SRG and SAW planes, was found to increase dramatically for higher frequency SAW. However, it was found that the effect of the incident beam's divergence induced only a gradual fall off in sensitivity as L increased. A detailed investigation and analysis of these effects has provided the basis for a new dynamic technique for determining laser beam divergence.

ANALYSIS

In order to analyze a divergent laser beam, it is necessary to consider the beam as the superposition of a continuum of oblique plane waves. A general analysis to be presented elsewhere[3] has shown that for a Gaussian beam of low divergence the rationalized spatial average of these plane wave terms is of the form,

$$S(L,\theta,\Lambda) = B \exp\left\{-\frac{1}{2}\left(\frac{\pi\theta L}{\Lambda}\right)^2\right\} \tag{1}$$

where θ is the half apex angle of the asymptotic cone of the Gaussian beam. The value of this integral corresponds to the integrated signal strength and is basically in excellent agreement with experimental results that follow.

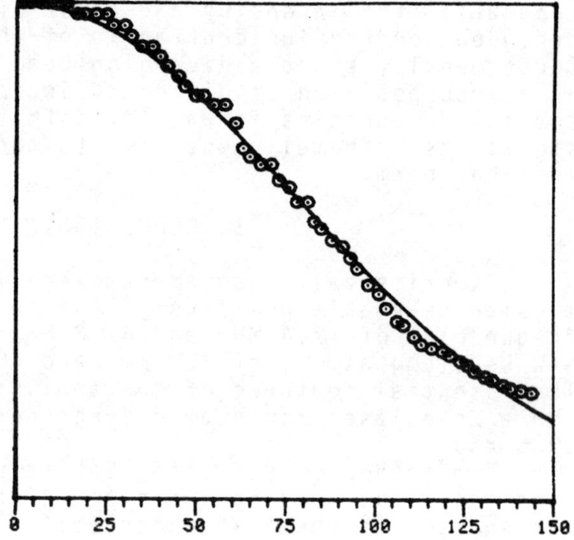

Fig. 1. Dependence on separation, L, between the SAW and SRG in mm;
$\Lambda = 40\ \mu m$
$m = 0$
$\theta = 160\ \mu rad$.

REFERENCES

1. T. D. Black, V. A. Komotskii, and D. A. Larson, Proc. IEEE Ultrasonics Symposium, p. 274 (1984).

2. V. A. Komotskii and T. D. Black, J. Appl. Phys., 52, 129 (1981).

3. T. D. Black, T. Haghighatjou, and D. A. Larson, Proc. IEEE Ultrasonics Symposium, PC-2, (1985).

MODE SELECTIVITY AND MISALIGNMENT SENSITIVITY IN NON-DEGENERATE PHASE-CONJUGATORS

J.T. Lin
Litton Systems, Inc., Laser Systems Division, Orlando, FL 32854-7300

ABSTRACT

Eigenvalue equation for a resonator system with a phase-conjugate mirror at one end is solved for the nth-order transverse mode in the degenerate and the non-degenerate cases. The effects of the finite size soft-apertures and the mirror misalignment on the mode discrimination and the resonator stability are discussed. An optimal condition for a TEM_{00} mode operation utilizing a phase-conjugate mirror is obtained.

INTRODUCTION AND EIGENVALUE EQUATION

We consider a resonator system consisting of a conventional mirror (CM) and a phase-conjugate mirror (PCM) with the CM tilted as shown in Figure 1. The integral equation describing the propagation of the field inside the CM-PCM strip resonator may be expressed as:

$$\gamma u(x) = \int_{-a_2}^{a_2} R_2(y)\, K^{(2)}(x,y)\, U^*(y)\, dy, \tag{1}$$

where $R_2(y)$ is the reflectivity of the CM with aperture size $2a_2$ and $K^{(2)}(x,y)$ is the round-trip kernel function related to the one-way kernel K_j (j=1,2 for the forward and backward propagation) by, in the paraxial ABCD matrix experssion,

$$K^{(2)}(x,y) = \int_{-a_1}^{a_1} K_1^*(x,x')\, K_2(y,x')\, R_1(x')dx' \tag{2}$$

$$K_j(x_1,x_2) = \left(\frac{ik_j}{2\pi B_j}\right)^{\frac{1}{2}} \exp\left[-\left(\frac{ik_j}{2B_j}\right)(A_j\, x_1^2 - 2x_1x_2 + D_jx_2^2) + i\phi_j\right] \tag{3}$$

where $k_{1,2}$ are the wavenumbers for the incident field and the reflected field from the PCM which is separated from the CM by a distance L. The phase-shift $\phi_j=2k_j\theta x_1$ is introduced to account for the phase variation of the kernel caused by the tilt angle θ of the CM. In general, Eq. (4) can only be solved numerically. For analytic results, we shall impose a Gaussian soft aperture for the CM and the PCM and extend a_1 and a_2 to infinity.

Solving the eigenvalue equation by a self-consistent procedure, we obtain the mode loss, caused by the soft apertures and the mis-

0094-243X/86/1460755-3$3.00

alignment, for the nth-order mode in a close form

$$L_n = 1 - [1-(2n+1)\varepsilon f_1]\,|\gamma_n|^2\, e^{-\Delta^2 f_2},\quad \Delta=2\theta L(k_2/k_1-1), \tag{4}$$

where the finite size of the CM is accounted by εf_1, with $\varepsilon=(2/k_1a_1a_2)^2$; the misalignment loss given by the exponential factor is governed by an effective tilt angle Δ; $f_{1,2}$ are appropriate functions related to the PCM parameter $\delta=2g_2L/k_1a_1^2$, with g_2 being the g-factor of the CM; finally, γ_n is the eigenvalue for the unbounded cavity (ε=0), which, for the degenerate operation, is given by

$$\gamma_n = [(1+\delta^2)^{\frac{1}{2}} - |\delta|]^{(\frac{1}{2}+n)} \tag{5}$$

We readily see that the eigenvalue goes to 1, i.e., there is no mode loss, when the PCM is a perfect one and when both the CM and the PCM have infinite aperture.

DISCUSSION

Equation (4) provides an analytic experssion for the nth-order mode loss of a misaligned, bounded CM-PCM resonator for arbitrary operation wavenumbers $K_{1,2}$. Some of the important features of the misalignment effects on the mode loss and the mode characteristics are summarized as follows:

a) For degenerate operation, k_2/k_1=1 and Δ=0. The mode loss is insensitve to misalignment and a full cancellation of the linear phase-shift caused by the tilted CM occurs as expected within the paraxial approximation. In contrast to that of a tilted CM-CM resonator, the fundamental mode of the tilted CM-PCM resonator is not distorted by the misalignment and the Gaussian profile remains symmetric to the optical axis.

b) For the non-degenerate case, $k_2 \neq k_1$. The mode loss is quite sensitive to the tile angle, increasing exponentially with the effective angle Δ. For small tilt angle or near degenerate operation, the loss increases linearly with Δ^2, in a manner similar to that obtained by first-order perturbation in a tilted CM-CM resonator,[2] if we replace Δ by the actual tilt angle. Furthermore, the Gaussian profile is shifted by an amount proportional to Δ.

c) From Equation (5), we see that the larger loss for higher-order modes provides us the possible stable TEM_{oo} mode oeprating in a CM-PCM resonator, which for the degenerate case, is insensitive to the MM and the distortion due to the cavity and/or the optics. The maximum mode discrimination of the nth-order mode is expected when the mode-loss related to the order parameter (n) by

$$\delta = (1-B^2)/2B,\quad B=1/(2n+1)^m,\quad m=1/(2n).$$

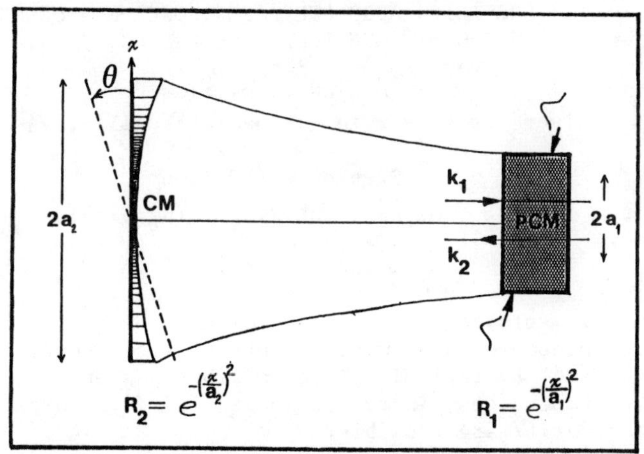

Fig. 1 - Schematic of misaligned CM-PCM resonator with Gaussian soft-reflectivity $R_{1,2}$ as indicated.

We note that due to the intrinsic features of PCM, the eigenvalues in the CM-PCM resonator do not depend on the distortion in the PCM and is only governed by the g-factor of the CM.

For completeness, we shall show the mode loss for resonators of CM-CM and PCM-PCM as follows:

i) for CM-CM case, $L_n = 1 - M^{-2(n+\frac{1}{2})}$, where M is the magnification of the telescope CM-CM resonator.

ii) for PCM-PCM case, $L_n = 1 - |\beta|^{-2(n+\frac{1}{2})}$, where β is a parameter depending only on the loss due to the finite aperture size of the conjugate mirrors and is independent of the wavefront curvature of the field inside the cavity, since the PCMs will make the self-compensation for each double-pass.

REFERENCES

1. A.E. Siegman, P.A. Belanger and A. Hardy, Optical Phase Conjugation, ed. by R.A. Fisher (Academic Press, NY, 1983), Chap. 13.

2. R. Hauck, H.P. Kortz and H. Weber, Appl. Opt. 19, 598 (1980).

HIGH PRECISION ABSOLUTE LASER POWER MEASUREMENTS USING A MAGNETICALLY SUSPENDED ROTOR

George T. Gillies
University of Virginia, Charlottesville, VA 22901

Stephen W. Allison
ORGDP*, Oak Ridge, TN 37831

ABSTRACT

Improvements to a torsion pendulum technique for making radiation pressure based measurements of laser power absolutely are presented. By replacing torsion fiber suspensions with magnetic suspensions, increases in sensitivity and reduced systematic drift are possible.

INTRODUCTION

There has been substantial interest recently in making absolute measurements of a laser's optical output power by means of a torsion balance. Aida and Bouzidi[1] have made one with a very delicate fiber suspension and use it in the angle-of-deflection mode. Roosen and Imbert[2] have taken this a step farther by designing an instrument whose output is independent of the system's mechanical lever arm.

In both cases, and here too, the fundamental principle of operation is elementary: A beam of light produces a force (due to its momentum transfer) on a mirror. The mirror is situated on one end of a rod, the other end of which is counterbalanced. The force produces a torque about the suspension's axis, and the system's angular deflection is, then, proportional to the power produced by the laser. Measurements of the deflection are usually made interferometrically. The suspension's restoring force (which controls the system sensitivity) is determined by measuring the balance beam's moment of inertia and timing its oscillation frequency. In this way, the requirements of a system capable of making absolute laser power measurements are satisfied.

RESULTS AND DISCUSSION

Figure 1 shows the magnetically suspended balance beam used in these studies. The suspension system is

* Operated by Martin Marietta Energy Systems, Inc., under contract with the U. S. Department of Energy.

Figure 1.

of standard design[3]. The position sensor is an optical autocollimator whose output signal is differentiated and fedback to the servo-coil which then stabilizes the suspended body's potision. The balance beam consists of an inverted "T" made of stainless steel tubing. The horizontal arm is 15 cm long. On one end is a front surface mirror, and it is counterbalanced with an aluminum weight on the other end. The system must be operated in vacuum to prevent convection currents from driving it.

Using an annealed steel sphere as the suspension mass, restoring torque constants as small as 7.4×10^{-10} Nm/rad have been obtained, over a factor of 200 times more sensitive than the best fiber measurements to date[3]. We hope to obtain absolute laser power measurement accuracies of $\Delta P/P = 10^{-5}$ with this technique in the near future.

1. Y. Aida and M. Bouzidi, IEEE Catalogue 80CH1497-71M, 269 (1980).
2. G. Roosen and C. Imbert, Opt. Eng. 20, 437 (1981).
3. W. Cheung, et al, Prec. Engng. 2, 183 (1980).

Author Index

A

B

C

D

E

F

G

H

T

U

V

W

X

Y

Z

AIP Conference Proceedings

		L.C. Number	ISBN
No. 1	Feedback and Dynamic Control of Plasmas – 1970	70-141596	0-88318-100-2
No. 2	Particles and Fields – 1971 (Rochester)	71-184662	0-88318-101-0
No. 3	Thermal Expansion – 1971 (Corning)	72-76970	0-88318-102-9
No. 4	Superconductivity in *d*- and *f*-Band Metals (Rochester, 1971)	74-18879	0-88318-103-7
No. 5	Magnetism and Magnetic Materials – 1971 (2 parts) (Chicago)	59-2468	0-88318-104-5
No. 6	Particle Physics (Irvine, 1971)	72-81239	0-88318-105-3
No. 7	Exploring the History of Nuclear Physics – 1972	72-81883	0-88318-106-1
No. 8	Experimental Meson Spectroscopy –1972	72-88226	0-88318-107-X
No. 9	Cyclotrons – 1972 (Vancouver)	72-92798	0-88318-108-8
No. 10	Magnetism and Magnetic Materials – 1972	72-623469	0-88318-109-6
No. 11	Transport Phenomena – 1973 (Brown University Conference)	73-80682	0-88318-110-X
No. 12	Experiments on High Energy Particle Collisions – 1973 (Vanderbilt Conference)	73-81705	0-88318-111-8
No. 13	π-π Scattering – 1973 (Tallahassee Conference)	73-81704	0-88318-112-6
No. 14	Particles and Fields – 1973 (APS/DPF Berkeley)	73-91923	0-88318-113-4
No. 15	High Energy Collisions – 1973 (Stony Brook)	73-92324	0-88318-114-2
No. 16	Causality and Physical Theories (Wayne State University, 1973)	73-93420	0-88318-115-0
No. 17	Thermal Expansion – 1973 (Lake of the Ozarks)	73-94415	0-88318-116-9
No. 18	Magnetism and Magnetic Materials – 1973 (2 parts) (Boston)	59-2468	0-88318-117-7
No. 19	Physics and the Energy Problem – 1974 (APS Chicago)	73-94416	0-88318-118-5
No. 20	Tetrahedrally Bonded Amorphous Semiconductors (Yorktown Heights, 1974)	74-80145	0-88318-119-3
No. 21	Experimental Meson Spectroscopy – 1974 (Boston)	74-82628	0-88318-120-7
No. 22	Neutrinos – 1974 (Philadelphia)	74-82413	0-88318-121-5
No. 23	Particles and Fields – 1974 (APS/DPF Williamsburg)	74-27575	0-88318-122-3
No. 24	Magnetism and Magnetic Materials – 1974 (20th Annual Conference, San Francisco)	75-2647	0-88318-123-1

No. 25	Efficient Use of Energy (The APS Studies on the Technical Aspects of the More Efficient Use of Energy)	75-18227	0-88318-124-X
No. 26	High-Energy Physics and Nuclear Structure – 1975 (Santa Fe and Los Alamos)	75-26411	0-88318-125-8
No. 27	Topics in Statistical Mechanics and Biophysics: A Memorial to Julius L. Jackson (Wayne State University, 1975)	75-36309	0-88318-126-6
No. 28	Physics and Our World: A Symposium in Honor of Victor F. Weisskopf (M.I.T., 1974)	76-7207	0-88318-127-4
No. 29	Magnetism and Magnetic Materials – 1975 (21st Annual Conference, Philadelphia)	76-10931	0-88318-128-2
No. 30	Particle Searches and Discoveries – 1976 (Vanderbilt Conference)	76-19949	0-88318-129-0
No. 31	Structure and Excitations of Amorphous Solids (Williamsburg, VA, 1976)	76-22279	0-88318-130-4
No. 32	Materials Technology – 1976 (APS New York Meeting)	76-27967	0-88318-131-2
No. 33	Meson-Nuclear Physics – 1976 (Carnegie-Mellon Conference)	76-26811	0-88318-132-0
No. 34	Magnetism and Magnetic Materials – 1976 (Joint MMM-Intermag Conference, Pittsburgh)	76-47106	0-88318-133-9
No. 35	High Energy Physics with Polarized Beams and Targets (Argonne, 1976)	76-50181	0-88318-134-7
No. 36	Momentum Wave Functions – 1976 (Indiana University)	77-82145	0-88318-135-5
No. 37	Weak Interaction Physics – 1977 (Indiana University)	77-83344	0-88318-136-3
No. 38	Workshop on New Directions in Mossbauer Spectroscopy (Argonne, 1977)	77-90635	0-88318-137-1
No. 39	Physics Careers, Employment and Education (Penn State, 1977)	77-94053	0-88318-138-X
No. 40	Electrical Transport and Optical Properties of Inhomogeneous Media (Ohio State University, 1977)	78-54319	0-88318-139-8
No. 41	Nucleon-Nucleon Interactions – 1977 (Vancouver)	78-54249	0-88318-140-1
No. 42	Higher Energy Polarized Proton Beams (Ann Arbor, 1977)	78-55682	0-88318-141-X
No. 43	Particles and Fields – 1977 (APS/DPF, Argonne)	78-55683	0-88318-142-8
No. 44	Future Trends in Superconductive Electronics (Charlottesville, 1978)	77-9240	0-88318-143-6
No. 45	New Results in High Energy Physics – 1978 (Vanderbilt Conference)	78-67196	0-88318-144-4

No. 46	Topics in Nonlinear Dynamics (La Jolla Institute)	78-57870	0-88318-145-2
No. 47	Clustering Aspects of Nuclear Structure and Nuclear Reactions (Winnepeg, 1978)	78-64942	0-88318-146-0
No. 48	Current Trends in the Theory of Fields (Tallahassee, 1978)	78-72948	0-88318-147-9
No. 49	Cosmic Rays and Particle Physics – 1978 (Bartol Conference)	79-50489	0-88318-148-7
No. 50	Laser-Solid Interactions and Laser Processing – 1978 (Boston)	79-51564	0-88318-149-5
No. 51	High Energy Physics with Polarized Beams and Polarized Targets (Argonne, 1978)	79-64565	0-88318-150-9
No. 52	Long-Distance Neutrino Detection – 1978 (C.L. Cowan Memorial Symposium)	79-52078	0-88318-151-7
No. 53	Modulated Structures – 1979 (Kailua Kona, Hawaii)	79-53846	0-88318-152-5
No. 54	Meson-Nuclear Physics – 1979 (Houston)	79-53978	0-88318-153-3
No. 55	Quantum Chromodynamics (La Jolla, 1978)	79-54969	0-88318-154-1
No. 56	Particle Acceleration Mechanisms in Astrophysics (La Jolla, 1979)	79-55844	0-88318-155-X
No. 57	Nonlinear Dynamics and the Beam-Beam Interaction (Brookhaven, 1979)	79-57341	0-88318-156-8
No. 58	Inhomogeneous Superconductors – 1979 (Berkeley Springs, W.V.)	79-57620	0-88318-157-6
No. 59	Particles and Fields – 1979 (APS/DPF Montreal)	80-66631	0-88318-158-4
No. 60	History of the ZGS (Argonne, 1979)	80-67694	0-88318-159-2
No. 61	Aspects of the Kinetics and Dynamics of Surface Reactions (La Jolla Institute, 1979)	80-68004	0-88318-160-6
No. 62	High Energy e^+e^- Interactions (Vanderbilt, 1980)	80-53377	0-88318-161-4
No. 63	Supernovae Spectra (La Jolla, 1980)	80-70019	0-88318-162-2
No. 64	Laboratory EXAFS Facilities – 1980 (Univ. of Washington)	80-70579	0-88318-163-0
No. 65	Optics in Four Dimensions – 1980 (ICO, Ensenada)	80-70771	0-88318-164-9
No. 66	Physics in the Automotive Industry – 1980 (APS/AAPT Topical Conference)	80-70987	0-88318-165-7
No. 67	Experimental Meson Spectroscopy – 1980 (Sixth International Conference, Brookhaven)	80-71123	0-88318-166-5
No. 68	High Energy Physics – 1980 (XX International Conference, Madison)	81-65032	0-88318-167-3
No. 69	Polarization Phenomena in Nuclear Physics – 1980 (Fifth International Symposium, Santa Fe)	81-65107	0-88318-168-1

No.	Title	L.C. No.	ISBN
No. 70	Chemistry and Physics of Coal Utilization – 1980 (APS, Morgantown)	81-65106	0-88318-169-X
No. 71	Group Theory and its Applications in Physics – 1980 (Latin American School of Physics, Mexico City)	81-66132	0-88318-170-3
No. 72	Weak Interactions as a Probe of Unification (Virginia Polytechnic Institute – 1980)	81-67184	0-88318-171-1
No. 73	Tetrahedrally Bonded Amorphous Semiconductors (Carefree, Arizona, 1981)	81-67419	0-88318-172-X
No. 74	Perturbative Quantum Chromodynamics (Tallahassee, 1981)	81-70372	0-88318-173-8
No. 75	Low Energy X-Ray Diagnostics – 1981 (Monterey)	81-69841	0-88318-174-6
No. 76	Nonlinear Properties of Internal Waves (La Jolla Institute, 1981)	81-71062	0-88318-175-4
No. 77	Gamma Ray Transients and Related Astrophysical Phenomena (La Jolla Institute, 1981)	81-71543	0-88318-176-2
No. 78	Shock Waves in Condensed Mater – 1981 (Menlo Park)	82-70014	0-88318-177-0
No. 79	Pion Production and Absorption in Nuclei – 1981 (Indiana University Cyclotron Facility)	82-70678	0-88318-178-9
No. 80	Polarized Proton Ion Sources (Ann Arbor, 1981)	82-71025	0-88318-179-7
No. 81	Particles and Fields –1981: Testing the Standard Model (APS/DPF, Santa Cruz)	82-71156	0-88318-180-0
No. 82	Interpretation of Climate and Photochemical Models, Ozone and Temperature Measurements (La Jolla Institute, 1981)	82-71345	0-88318-181-9
No. 83	The Galactic Center (Cal. Inst. of Tech., 1982)	82-71635	0-88318-182-7
No. 84	Physics in the Steel Industry (APS/AISI, Lehigh University, 1981)	82-72033	0-88318-183-5
No. 85	Proton-Antiproton Collider Physics –1981 (Madison, Wisconsin)	82-72141	0-88318-184-3
No. 86	Momentum Wave Functions – 1982 (Adelaide, Australia)	82-72375	0-88318-185-1
No. 87	Physics of High Energy Particle Accelerators (Fermilab Summer School, 1981)	82-72421	0-88318-186-X
No. 88	Mathematical Methods in Hydrodynamics and Integrability in Dynamical Systems (La Jolla Institute, 1981)	82-72462	0-88318-187-8
No. 89	Neutron Scattering – 1981 (Argonne National Laboratory)	82-73094	0-88318-188-6
No. 90	Laser Techniques for Extreme Ultraviolt Spectroscopy (Boulder, 1982)	82-73205	0-88318-189-4

No. 91	Laser Acceleration of Particles (Los Alamos, 1982)	82-73361	0-88318-190-8
No. 92	The State of Particle Accelerators and High Energy Physics (Fermilab, 1981)	82-73861	0-88318-191-6
No. 93	Novel Results in Particle Physics (Vanderbilt, 1982)	82-73954	0-88318-192-4
No. 94	X-Ray and Atomic Inner-Shell Physics – 1982 (International Conference, U. of Oregon)	82-74075	0-88318-193-2
No. 95	High Energy Spin Physics – 1982 (Brookhaven National Laboratory)	83-70154	0-88318-194-0
No. 96	Science Underground (Los Alamos, 1982)	83-70377	0-88318-195-9
No. 97	The Interaction Between Medium Energy Nucleons in Nuclei – 1982 (Indiana University)	83-70649	0-88318-196-7
No. 98	Particles and Fields – 1982 (APS/DPF University of Maryland)	83-70807	0-88318-197-5
No. 99	Neutrino Mass and Gauge Structure of Weak Interactions (Telemark, 1982)	83-71072	0-88318-198-3
No. 100	Excimer Lasers – 1983 (OSA, Lake Tahoe, Nevada)	83-71437	0-88318-199-1
No. 101	Positron-Electron Pairs in Astrophysics (Goddard Space Flight Center, 1983)	83-71926	0-88318-200-9
No. 102	Intense Medium Energy Sources of Strangeness (UC-Sant Cruz, 1983)	83-72261	0-88318-201-7
No. 103	Quantum Fluids and Solids – 1983 (Sanibel Island, Florida)	83-72440	0-88318-202-5
No. 104	Physics, Technology and the Nuclear Arms Race (APS Baltimore –1983)	83-72533	0-88318-203-3
No. 105	Physics of High Energy Particle Accelerators (SLAC Summer School, 1982)	83-72986	0-88318-304-8
No. 106	Predictability of Fluid Motions (La Jolla Institute, 1983)	83-73641	0-88318-305-6
No. 107	Physics and Chemistry of Porous Media (Schlumberger-Doll Research, 1983)	83-73640	0-88318-306-4
No. 108	The Time Projection Chamber (TRIUMF, Vancouver, 1983)	83-83445	0-88318-307-2
No. 109	Random Walks and Their Applications in the Physical and Biological Sciences (NBS/La Jolla Institute, 1982)	84-70208	0-88318-308-0
No. 110	Hadron Substructure in Nuclear Physics (Indiana University, 1983)	84-70165	0-88318-309-9
No. 111	Production and Neutralization of Negative Ions and Beams (3rd Int'l Symposium, Brookhaven, 1983)	84-70379	0-88318-310-2

No. 112	Particles and Fields – 1983 (APS/DPF, Blacksburg, VA)	84-70378	0-88318-311-0
No. 113	Experimental Meson Spectroscopy – 1983 (Seventh International Conference, Brookhaven)	84-70910	0-88318-312-9
No. 114	Low Energy Tests of Conservation Laws in Particle Physics (Blacksburg, VA, 1983)	84-71157	0-88318-313-7
No. 115	High Energy Transients in Astrophysics (Santa Cruz, CA, 1983)	84-71205	0-88318-314-5
No. 116	Problems in Unification and Supergravity (La Jolla Institute, 1983)	84-71246	0-88318-315-3
No. 117	Polarized Proton Ion Sources (TRIUMF, Vancouver, 1983)	84-71235	0-88318-316-1
No. 118	Free Electron Generation of Extreme Ultraviolet Coherent Radiation (Brookhaven/OSA, 1983)	84-71539	0-88318-317-X
No. 119	Laser Techniques in the Extreme Ultraviolet (OSA, Boulder, Colorado, 1984)	84-72128	0-88318-318-8
No. 120	Optical Effects in Amorphous Semiconductors (Snowbird, Utah, 1984)	84-72419	0-88318-319-6
No. 121	High Energy e^+e^- Interactions (Vanderbilt, 1984)	84-72632	0-88318-320-X
No. 122	The Physics of VLSI (Xerox, Palo Alto, 1984)	84-72729	0-88318-321-8
No. 123	Intersections Between Particle and Nuclear Physics (Steamboat Springs, 1984)	84-72790	0-88318-322-6
No. 124	Neutron-Nucleus Collisions – A Probe of Nuclear Structure (Burr Oak State Park - 1984)	84-73216	0-88318-323-4
No. 125	Capture Gamma-Ray Spectroscopy and Related Topics – 1984 (Internat. Symposium, Knoxville)	84-73303	0-88318-324-2
No. 126	Solar Neutrinos and Neutrino Astronomy (Homestake, 1984)	84-63143	0-88318-325-0
No. 127	Physics of High Energy Particle Accelerators (BNL/SUNY Summer School, 1983)	85-70057	0-88318-326-9
No. 128	Nuclear Physics with Stored, Cooled Beams (McCormick's Creek State Park, Indiana, 1984)	85-71167	0-88318-327-7
No. 129	Radiofrequency Plasma Heating (Sixth Topical Conference, Callaway Gardens, GA, 1985)	85-48027	0-88318-328-5
No. 130	Laser Acceleration of Particles (Malibu, California, 1985)	85-48028	0-88318-329-3
No. 131	Workshop on Polarized ^{3}He Beams and Targets (Princeton, New Jersey, 1984)	85-48026	0-88318-330-7
No. 132	Hadron Spectroscopy–1985 (International Conference, Univ. of Maryland)	85-72537	0-88318-331-5

No. 133	Hadronic Probes and Nuclear Interactions (Arizona State University, 1985)	85-72638	0-88318-332-3
No. 134	The State of High Energy Physics (BNL/SUNY Summer School, 1983)	85-73170	0-88318-333-1
No. 135	Energy Sources: Conservation and Renewables (APS, Washington, DC, 1985)	85-73019	0-88318-334-X
No. 136	Atomic Theory Workshop on Relativistic and QED Effects in Heavy Atoms	85-73790	0-88318-335-8
No. 137	Polymer-Flow Interaction (La Jolla Institute, 1985)	85-73915	0-88318-336-6
No. 138	Frontiers in Electronic Materials and Processing (Houston, TX, 1985)	86-70108	0-88318-337-4
No. 139	High-Current, High-Brightness, and High-Duty Factor Ion Injectors (La Jolla Institute, 1985)	86-70245	0-88318-338-2
No. 140	Boron-Rich Solids (Albuquerque, NM, 1985)	86-70246	0-88318-339-0
No. 141	Gamma-Ray Bursts (Stanford, CA, 1984)	86-70761	0-88318-340-4
No. 142	Nuclear Structure at High Spin, Excitation, and Momentum Transfer (Indiana University, 1985)	86-70837	0-88318-341-2
No. 143	Mexican School of Particles and Fields (Oaxtepec, México, 1984)	86-81187	0-88318-342-0
No. 144	Magnetospheric Phenomena in Astrophysics (Los Alamos, 1984)	86-71149	0-88318-343-9
No. 145	Polarized Beams at SSC & Polarized Antiprotons (Ann Arbor, MI & Bodega Bay, CA, 1985)	86-71343	0-88318-344-7